LOGIC, EPISTEMOLOGY, AND THE UNITY OF SCIENCE

LOGIC, EPISTEMOLOGY, AND THE UNITY OF SCIENCE

VOLUME 1

Logic, Epistemology, and the Unity of Science aims to reconsider the question of the unity of science in light of recent developments in logic. At present, no single logical, semantical or methodological framework dominates the philosophy of science. However, the editors of this series believe that formal techniques like, for example, independence friendly logic, dialogical logics, multimodal logics, game theoretic semantics and linear logics, have the potential to cast new light on basic issues in the discussion of the unity of science.
This series provides a venue where philosophers and logicians can apply specific technical insights to fundamental philosophical problems. While the series is open to a wide variety of perspectives, including the study and analysis of argumentation and the critical discussion of the relationship between logic and the philosophy of science, the aim is to provide an integrated picture of the scientific enterprise in all its diversity.

Logic, Epistemology, and the Unity of Science

Edited by

Shahid Rahman
Université Lille 3,
France

John Symons
University of Texas,
El Paso, U.S.A.

Dov M. Gabbay
King's College London,
U.K.

and

Jean Paul van Bendegem
Vrije Universiteit Brussel,
Belgium

KLUWER ACADEMIC PUBLISHERS
DORDRECHT / BOSTON / LONDON

A C.I.P. Catalogue record for this book is available from the Library of Congress.

ISBN 1-4020-2807-5 (HB)
ISBN 1-4020-2808-3 (e-book)

Published by Kluwer Academic Publishers,
P.O. Box 17, 3300 AA Dordrecht, The Netherlands.

Sold and distributed in North, Central and South America
by Kluwer Academic Publishers,
101 Philip Drive, Norwell, MA 02061, U.S.A.

In all other countries, sold and distributed
by Kluwer Academic Publishers,
P.O. Box 322, 3300 AH Dordrecht, The Netherlands.

Printed on acid-free paper

Printed in the Netherlands.

For Marie and Charo

TABLE OF CONTENTS

PART I
SOME PROGRAMMATIC COMMENTS

LOGIC, EPISTEMOLOGY AND THE UNITY OF SCIENCE: AN ENCYCLOPEDIC PROJECT IN THE SPIRIT OF NEURATH AND DIDEROT

SHAHID RAHMAN[1] and JOHN SYMONS[2]
[1]*Université Charles-de-Gaulle, Lille 3, France, E-mail:* rahman@univ-lille3.fr;
[2]*University of Texas, El Paso, USA, E-mail:* jsymons@utep.edu

> [...] *on devra cependant se garder de dissimuler l'ambiguïté de certains énoncés, et de vouloir esquisser un système unitaire* [...] *Pour nous au contraire, nous voudrions déclarer d'emblée que la forme de l'encyclopédie est la plus parfaite que nous puissions jamais atteindre pour exposer l'ensemble de la science, opposant ainsi expressément au pseudo-rationalisme de toutes les philosophies "centralistes", notre travail scientifique concret; Qui se garde soigneusement d'anticiper la systématisation générale de la science.* (Otto Neurath ([1935], II.3)

> *It is against the principle of encyclopedism to imagine one "could" eliminate all such difficulties. To believe this is to entertain a variation of Laplace's famous demon who was supposed to have a complete knowledge of present facts sufficient for making complete predictions of the future. Such is the idea of "the system" in contrast to the idea of "an encyclopaedia"; the anticipated completeness of the system is opposed to the stressed incompleteness of an encyclopedia.* (Otto Neurath ([1938], 20–21)

> *The Encyclopedia presents a contemporary version of the ancient encyclopedic ideal of Aristotle; the Scholastics; Leibniz, The Encyclopedists and Comte. It wishes to give satisfaction to the pervasive human interest in intellectual unity, but its common point of view permits divergences and differences in emphasis and does not blur the fact that an inseparable feature of the institution of science is constant growth. It aims to provide a basis for co-operative activity and not a panacea.* (Charles W. Morris [1938], 75)

1. Unity: An Unfashionable Notion

The idea that the unity of science can be achieved by means of logical analysis, an idea widely associated with the Vienna Circle, has fallen into disrepute. Today, logic plays a relatively minimal role in mainstream philosophy of science and no single approach to logic or semantics can claim to dominate the field. Logic, according to popular wisdom, has done more harm than good, abstracting us from the important subtleties of scientific investigation and mistakenly forcing a frozen universal structure on the dynamic process of knowledge-seeking. While many philosophers of science have shied away from logic there have, in the meantime, been many important new developments in logic, some of which may have extremely significant applications to questions in epistemology and general philosophy of science. These developments have gone virtually unnoticed in the broader

S. Rahman et al. (eds.), Logic, Epistemology, and the Unity of Science, pp. 3–15.

philosophical community. One of the purposes of this volume is to encourage philosophers to recognize the potential riches to be found in recent work in logic, for instance in the plethora of non-classical logics, including, prominently, game theoretical semantics and independence-friendly logic. The rhetoric of systematicity and formalism continues to characterize analytic philosophy of science and it is likely that the gradual withering of logic as a significant part of the philosophy of science is due more to neglect than to any serious argument.

By contrast, there has been no shortage of arguments against the notion of unity. It has been subject to ceaseless attack in recent history and philosophy of science. Evidence of the disunity of the sciences has been easy for critics to muster and arguments against specific attempts to achieve the unity of science are often quite devastating. However, while some of these criticisms are sound and will give pause to any prospective unifier, it is a mistake to ignore the preponderance of unified theories in science.[1] Much of the development of physics in the second half of the twentieth century was motivated by the desire to unify quantum mechanics and general relativity. It would be difficult to understand the conceptual framework or the historical development of quantum field theory or string theory without recognizing the important role of unification in science. One can find many other important examples that show the abiding interest in, and heuristic function of, unification in scientific practice.

It is likely that many arguments for the disunity of science are directed towards a strawman. In particular, we would suggest that there is a widespread misreading of all talk of unity in science as an insistence on the unity of scientific method. On the contrary, as we show below, one can find prominent examples of unification projects that emphasize instead the danger of forcing scientific practice into conformity with favored a priori conceptions of rationality. This aspect of Neurath's vision of the unity of science, is clearly articulated in the short article translated below. In this introduction we will show a less familiar face of the unity of science movement by highlighting the encyclopedic spirit of philosophers like Neurath and Morris, along with those earlier great Encyclopedists Diderot and D'Alembert. This is the encyclopedic spirit that, we hope, informs the present volume. On our view, to consider the knowledge-seeking enterprise as a single human endeavor does not involve rigidly adhering or enforcing a simplified model of science.

To understand the kind of methodological and even ontological diversity that a particular unification project might encompass, consider, for instance, the relationship between genetics and evolutionary theory. It is likely that any attempted unification of these fields would preserve their methodologically distinct features while allowing communication and cooperation between them. The great evolutionary synthesis in biology in the 1930s and 1940s can, perhaps, be understood along these lines.

Similarly, unification is not necessarily a matter of subsuming distinct fields under a single set of laws. Rather, such unification may involve the discovery of a third set of phenomena or regularities or even the development of a third discipline. By way of illustration, consider for instance the emergence in the 1980's and 1990's of Cognitive Neuroscience. Such unifying developments will often leave the original sciences essentially intact while permitting an exchange of contributions between them.

This rough discussion of two possible responses to anti-unity arguments is meant merely as a sketch of possible lines of argument. At this stage it would be a mistake to prescribe the possible outcome of unification projects in detail. To do so would be tantamount to prematurely adopting the dogmatism of what Neurath called 'centralist philosophies'. As we shall show below, allowing for the co-operation across scientific specializations was one of the original goals of the Vienna Circle's Encyclopedia project, at least as articulated by Neurath.

The current emphasis on disunity in science studies and the philosophy of science is also due to the belief that an attempt to unify automatically entails an absence of historical or practical understanding of the sciences. This is simply false. One of the goals of the unifiers, exemplified by some of our best historians and philosophers of science, is to satisfy an abiding human need to understand connections and similarities in contexts of great diversity and apparent disunity. Again, it is not necessary to prejudge in detail what unity will look like or what functions it will serve in order to make the case for its intellectual significance. While unification projects will vary from science to science, there is a set of important and interesting general lines of philosophical inquiry that are opened up once one takes the possibility of unity seriously. Consider two recent and widely acclaimed works in the history of science: Jean Gayon's *Darwinism's Struggle for Survival* and TianYu Cao's *The Conceptual Development of 20th Century Field Theories*. Both are exquisitely researched works by acknowledged masters of the history of their respective fields of study. Both are also intent on providing readers with an understanding of the continuity that characterizes the history of those fields. Specifically, Cao's approach explicitly involves looking to the details of scientific change in order to find continuity in the preserved mathematical structures employed by field theorists. Similarly, Gayon traces the central conceptual core of Darwinism through a variety of historical mutations, from Darwin to Kimura. So, while the devil may be in the details, those same details may also be the best place to look for unity.

2. The Collapse of the Second Wave

As discussed above, in epistemology and general philosophy of science, many philosophers clearly seem to regard the present situation as a reaction to, and rejection of logical positivism. Thus, it might seem a very inauspicious time to launch a project dedicated to the relationship between logical investigations and the unity of science. Indeed, our interest in the application of insights from logic to problems in philosophy of science will undoubtedly strike some critics as oddly old-fashioned. However, against the apparent consensus that has emerged over the past thirty years, we believe that the widespread abandonment of logic in the philosophy of science has been a serious mistake. This is especially true, given the important, though often neglected developments that have taken place in logic since the 1960's. As we hope this volume demonstrates, recent progress supports our view that investigations at the logical or conceptual level can play a vital role in epistemology and the philosophy of science. The power of recent insights, encourages us to believe that the time is ripe for philosophers to take up anew the challenge of considering the scientific enterprise in its entirety. In fact, we would suggest that investigations into the unity of science are more intriguing and potentially fruitful now than they could have been in the 20th century.

In place of the familiar diagnosis of the contemporary situation, we suggest that we are not witnessing the burial of logical positivism but rather the last days of the movements that claim to have surpassed the Vienna Circle. These movements might be called for brevity the "second wave" of analytic philosophy. They include prominently the "New Philosophy of Science" that followed in the footsteps of Kuhn, Lakatos and others, various forms of skepticism or nihilism, along with sundry eclectic tendencies in contemporary epistemology. As an intentional provocation to readers, we are tempted to issue a challenge and declare the second wave of analytic philosophy to be dying, or at least exhausted. While there may be valuable lessons to be drawn from these self-styled post-positivist movements, our view is that, in general, the rejection of logical positivism has been accompanied by a shift from a logic-oriented methodology to a mélange of psychological and sociological reductionism and anti-systematic rhetoric. This has led to a highly damaging impoverishment of philosophical argumentation. Even if contributing to the progress of science was not their goal, we suggest that a greater awareness of the logical and methodological issues they have spurned could even have helped those who reject the heritage of the Vienna Circle in their own historical and interpretive work.

3. Is the Unity of Science Project a Form of Scientism?

Our efforts should not be misinterpreted as an uncritical return to some kind of dogmatic scientism. Instead, one of the central tenets of our enterprise is that philosophical investigations be conducted in a spirit of openness that recognizes the essential pluralism of knowledge-seeking practices. Scientism, one could say, is obviously an unscientific view. At the same time, we reject the excessive modesty which is often exhibited by contemporary philosophers of science in relation to the natural sciences and instead hope to provide a venue for philosophers who genuinely engage with natural science in the sense of being bolder and more actively involved in scientific disputes. One could argue that, by contrast with the robust philosophical engagement made possible via the application of logical insights to science, much of the recent case-study style work in science studies and the philosophy of science looks passively scientistic.[2]

Philosophers of science should not see themselves as underlaborers for scientific geniuses or as critics who claim to stand apart from the scientific process. Instead, let's take that great critic (and student) of the Vienna Circle, W. V. Quine seriously when he argues for the continuity of philosophical inquiry and the natural sciences. Quine's criticism of the analytic-synthetic distinction is widely thought to deny philosophy a privileged domain of purely conceptual investigation. But one could also read Quine's argument as an acknowledgment that philosophy is neither below nor above the wider scientific enterprise. If one agrees with Quine, then one could see his criticism as licensing the development of a truly scientific philosophy akin to that of the great seventeenth and eighteenth century pre-Kantian thinkers. In this spirit, we hope that this series 'Logic Epistemology and the Unity of Science' provides a venue for the development of (or perhaps the return to) a self-critical natural philosophy.

Distancing ourselves from scientism does not mean separating ourselves from the project of the Vienna Circle. Quite the opposite. We reject the easy caricatures that too often pass for history with respect to these philosophers and insist instead on emphasizing the progressive, collaborative and optimistic spirit of their *Encyclopedia* project. As we emphasize below, we see our work, inspired by the Vienna Circle's promotion of the scientific attitude in philosophy, as part of a very traditional, though radical, Enlightenment project. Like Diderot and D'Alembert before them, Neurath's vision of the Encyclopedia is a cooperative and ambitious enterprise. We would like to see our series continue this tradition. This is why we take Otto Neurath as the inspiration for this series and look to his initial statement of the Encyclopedia of the Unified Sciences as our model.

While we believe that the history of recent philosophy of science is due for revision, we should not be seen as rejecting everything that has happened over the past forty years. Rather, it is our view that many of what are now seen as the important lessons of recent philosophy of science can be retained within the general perspective we are advocating. In fact, many of these insights were already an integral part of the original work of the Vienna Circle. As we return to the projects of Neurath and Diderot, we certainly do not lose the kind of self-critic al capacity with respect to the natural sciences that modern philosophers and scholars of science studies sometimes seem to believe was invented in the 1960's. Thomas Kuhn's work, for instance, has played a central role in the second wave of analytic philosophy. (Perhaps, after reading Steve Fuller (2000), we should say that Kuhn's *Structure of Scientific Revolutions* played a variety of roles in this history.) No matter how one understands the meaning and consequences of Kuhn's work, it is common knowledge that Structure was published in the *International Encyclopedia for the Unified Sciences* and was seen by Carnap as a natural component of that project. It would be an interesting project to see how many of the allegedly revolutionary ideas of the second wave, were already anticipated by members of the first.

Perhaps there is no better example of this debt to the Vienna Circle than the fact that the central anti-foundationalist metaphor in contemporary post-positivism is drawn from Neurath, namely that of the community of inquirers afloat on their rickety communal boat. Neurath's boat, like the scientific project itself, is an improvised assembly of components, adjusted on the fly and always subject to reconstruction. Nevertheless, as long as the sailors survive, they find their community engaged in a single unified project.

In addition to Neurath's own apparently anti-fundamentalist position when one considers the origins of the *Encyclopaedia of the Unified Sciences* more generally one finds a marked emphasis on interdisciplinarity and the role of scientific practice in reflections on the philosophy of science. This strongly pragmatic dimension, due at least in part, in its early stages, to the influence of Charles Morris, does not jibe with the widely accepted caricature of the *Encyclopaedia*. Morris recognized that logical and epistemological reflections on science should be informed by reflection on the dynamical nature of science and scientific co-operation. While this pragmatic dimension of the project is often overlooked, if one actually reads the early volumes it is hard to ignore.

We are eager to provide a venue for works that include this pragmatic dimension. It is our view that the logic of *knowing that* in philosophy of science must be connected with the logic of *knowing how* within the natural sciences. The real practice of scientific inquiry as embodied in a living scientific community, diachronically and synchronically studied; should furnish the starting point for any logic of science. And yet, this starting point is not sacrosanct. Scientists are as fallible as the rest of us and just as liable to conceptual or logical confusion. Therefore, the influence should work in both directions, without any presumed fundamentality of one over the other.

In presenting the *Encyclopaedia* project, Neurath and Morris frequently referred to the facilitation of scientific co-operation as one of their goals. Cross-fertilization is one of the advantages of seeing commonalities across the sciences. Given the increasing level of specialization in the natural sciences, interdisciplinarity is increasingly recognized as an urgent necessity and a quality to be fostered in scientific work. Here we would like to note that interdisciplinarity, understood as capacity for systematic interaction and communication across the disciplinary boundaries of scientific investigation does not necessarily presuppose that all collaboration is equivalent to or presupposes some form of reductionism. Rather, the possibility of interdisciplinary co-operation or communication of some sort can be understood as a necessary condition for inclusion as a genuine natural science. If a discipline entirely lacked the ability to communicate and cooperate it would surely not be considered a science by the wider scientific community.

4. The Idea of the Encyclopedia

Co-operation and interdisciplinarity are two important virtues which were central to the Vienna Circle's *Encyclopedia* project. But of course this virtue is not unique to the Vienna Circle and, in fact, has been an abiding quality of natural science since the Enlightenment. Another important influence informing our project stems from our view that the Vienna Circle's *Encyclopaedia* project shares deep affinities with its eighteenth century French predecessor. When one reads the early papers in the Vienna Circle's Encyclopedia project it is hard to ignore the connection.

One symbolic, though perhaps trivial point is the French origin of the Vienna Circle's Encyclopedia project. The idea for an *Encyclopaedia of the Unified Sciences* was officially presented and approved at the Sorbonne in 1935 during the *Congrès International de Philosophie Scientifique*, the first international congress of the "*Science unitaire*".[3] The proposal was presented by Morris and defended in French by Neurath. That important paper was published in turn in *the Actes du congrès international de philosophie scientifique* (Paris: Hermann & Cie, 1936, II.1–II.6). A translation of that lecture is included in this volume. The ideas presented in Neurath's paper were further developed three years later by Neurath, Morris and Carnap in the first volume of the *Encyclopedia and Unified Science* (Chicago, Chicago University Press 1938).

As we understand matters, one of the most significant points from this 1935 paper comes when Neurath argues that "*in order to present science as a whole the Encyclopedic form is the best we can expect*". Clearly, it is important to note that for Neurath, the concept of the encyclopaedia should be differentiated from the rationalist task of building a general system of science. The remainder of this introduction is intended to give readers some idea of what we understand by this encyclopedic spirit. While disagreeing on a number of points, the present authors share a deep admiration for Neurath's vision and can certainly agree that his project serves as a useful model for us as we work to bring developments in logic to bear on scientific problems.

That congress in Paris was the first official meeting of the logical positivists ("scientific empiricists" was the name preferred by Morris) and Bertrand Russell. Russell (who calls the group "scientific philosophers") announces the birth of a new child who will inherit the virtues of both empiricist and rationalist philosophical traditions. With Russell as the witness, France can seen as the official birthplace or at least the site of the Baptism, of the new unity of science project. And while the logical positivists acknowledged their relation to the *characteristica universalis* of Leibniz and the positivism of Comte, they were keen to stress that they considered themselves offspring of the Encyclopaedia of Diderot and D'Alembert as energized by Comte and Darwin.

In his talk at the Paris Conference of 1935 Russell points to the core of the Vienna Circle's identity by pointing out where they diverge from the Leibnizian project: "Scientific philosophy" should combine empiricism with mathematical method in the same way that the impressive progress of physics resulted from the interaction of empiricism and rationalism. This same point is stressed in Neurath's 1938 paper. This is also the paper in which Neurath clearly stresses the connections between his project and the *Encyclopaedia* of D'Alembert and Diderot.[4]

From our perspective, it is important to emphasize the pluralistic quality of their unification project. Morris, for instance, stresses the point that the project "permits divergences and differences" and "aims to provide a basis for co-operative activity and not a panacea". In the same spirit, Neurath mentions the example of D'Alembert who "in his introduction opposed Rousseau's attack against science and yet expressed his pleasure that Rousseau had become a collaborator in the work: Neurath writes: *As a modern scientific man, D'Alembert stressed the degree to which all scientific activities depend upon social institutions* ([1938], 2–3).[5] In Neurath's view, while the French Encyclopaedists had an organic idea of unity and social co-operation; the rigidity of their method of classification proved an impediment to the kind of dynamical interaction and progress implicit in their ideas of social co-operation. The notion of an encyclopaedia permitting what Shahid Rahman has called "dynamical unity in diversity" was made possible by means of the introduction of a variety of historically and biologically oriented ideas from nineteenth century thought. These include ideas from Comte, Peirce, elements of Spencerianism and the influence of Darwinism.

This dynamical understanding of scientific co-operation was understood to rest on the ability of investigators in various sciences to interact and communicate with one another. Additionally, co-operation is understood as a contribution to a larger social project. There are important socio-political commitments and implications involved in such a project which could be explored at great length in both the 18th and 20th century cases. However, for our purposes, an important point to note with respect to this collective project was that it did not necessarily involve subordinating a plurality of sub-sciences to a master science. Rather, one of the key notions at work here is the idea of a contribution from one scientific enterprise to another. In more familiar philosophical terms, this emphasis on contribution entails the notion that one science can contribute to another without necessarily reducing to the other and without losing its own distinctive methodological identity – in fact Morris explicitly argued that "unity does not exclude differentiation" (1938, 73). A science that fully reduces to a more general or fundamental science can hardly be said to contribute to that science. Indeed, the interaction between sciences and the contribution of one science to another can potentially lead to the development of an entirely new discipline. Hence rather than homogenizing the practice of scientific inquiry, the early Vienna Circle model can be understood as encouraging a proliferation of approaches and interests.

Perhaps the most basic feature of their conception of the unity of science project is the important role played by communication, in particular the important role played by the interaction between scientific specializations. We would like to suggest that this was one of the first and maybe even one of the best formulations of the demarcation problem. Perhaps we can say that, for Neurath and Morris at least, it is the interactions within the scientific community and between the sciences which serves to determine the significance of the scientific concepts which are at stake. Seen in this light, the role of the Encyclopedia becomes extremely important to the progress of science. The concept of an interdisciplinary Encyclopaedia as envisaged by this project offered coherence to the development of scientific inquiry. Especially when scientific inquiry is understood in this dynamical and communicative form.

In characterizing the process of scientific inquiry, Neurath uses the image of a mosaic, where each science contributes to the shape of the whole pattern. Neurath underlines the fact that the image will perhaps never be finished and even that it could constantly change: *the generations of the mosaicists are not only inlaying the stones but varying the whole pattern*. As Neurath himself contends, the very idea of an Encyclopedia in the sense of the French tradition stands in direct opposition to the idea of a unification achieved by means of a monolithic, philosophical system from which all scientific and metaphysical principles could be derived. (cf. the quotation at the beginning of this introduction). The obvious contrast to this Encyclopedic spirit would be the work of philosophers like Hegel. By contrast with, for example, an Hegelian approach, logical analysis and an emphasis on language can be understood as part of the process of facilitating co-operation in a pluralistic context. This is not the only role logic can play, but it is certainly a central one for Neurath and colleagues.

5. Logic and the Practice of Scientific Inquiry

> *[T]he International Encyclopaedia of the Unified Sciences hopes to avoid becoming a mausoleum or an herbarium, and to remain a living intellectual force growing out of a living need of men; and so in turn serving humanity* (Neurath 1938, 26)

Briefly surveying the contributions of Neurath, Carnap and Morris as presented in the 1938 volume we find some significant differences between the three. While Neurath stresses the sociological and historical dynamism of the encyclopaedic project, Carnap, by contrast, emphasizes the difference between what Reichenbach later called the *context of justification* and the *context of discovery*. Morris differs with Carnap on this distinction. Instead, to use Ryle's formulation, he seeks a logic of science where *knowing that* is integrated with *knowing how*. Additionally, his pragmatist bent leads him to argue that *knowing that* should be put in the service of improving or facilitating our "know-how". An emphasis on the importance of scientific practice has been a familiar theme in philosophy of science since Kuhn. However, once again, it is important to recognize that practice was already given great weight in the early days of the Vienna Circle.

The application of logic to science and epistemology is unavoidably entangled with the process and practice of scientific investigation. Inevitably, a certain level of acquaintance with the basic skills and techniques of knowledge-seeking has to be assumed prior to treating science as an object of philosophical inquiry. Following this line of reasoning, Neurath emphasizes that one can only do justice to science if an historical dimension is included in one's analysis. This places Neurath's views squarely in the same tradition as Pierre Duhem and Henry Poincaré, both of whom are explicitly mentioned by Neurath in the 1938 paper. Indeed, in that same paper Neurath writes that "one can hope for great success if scientists analyzing various sciences co-operate with men concerned with the history and logic of science" (1938, 14).

Science is not only a bloodless body of knowledge but a process by means of which knowledge is won. Admittedly, the logical analysis of this process – which in traditional philosophy was seen as the job of "gnoseology" and was badly lacking in the times of logical positivism – can be seen as one of the main tasks for philosophers . Moreover; this task, which can be seen as providing the link between sciences and philosophy – can nowaday be accomplished for the first time thanks to the pioneering work of Jaakko Hintikka on epistemic logic and game-theoretic semantics.

Gerhard Heinzmann, who in his Habilitationsvortrag stressed the importance of the scientific community to the French encyclopedists, suggests there that the role of the philosopher is to function as a mediator within the broader scientific community. On his view the philosophers role is to make trans- and interdisciplinary co-operation possible. It looks as though Neurath and Morris thought along similar lines about the role of the philosopher. However, both Neurath and Morris would add that this role should be performed with help of a logical analysis of the language of the sciences. In Neurath's papers of 1935 and 1938 the so-called linguistic turn – or more precisely the use of a logical language – is thought of more as an instrument of unification, a kind of *lingua franca*, or even lingua universalis à la Lullus or Leibniz, which, as already mentioned, would provide the philosopher an instrument to perform his role as a mediator and motivator of co-operation between sciences.

In addition to this rather modest role of mediator, there is another important role to be played by the philosopher. Conceptual work of various kinds is involved not only in the interaction between the sciences, but also with respect to specific problems in the sciences themselves. Thus, in addition to bridging the sciences, the pragmatic or Quinean leveling of the sciences places philosophy in the position to actively contribute to solving problems in the sciences themselves. So, in addition to mediating between various scientific disciplines, there seems to be no principled reason to exclude philosophers from actively intervening in scientific investigation.

The pragmatic approach of Neurath and Morris echo the main characteristics of the tradition of various encyclopaedic or "*orbis doctrinae*" projects not only that of Diderot and D'Alembert's famous *Encyclopédie ou Dictionnaire Raisonné des Sciences; des Arts et des Métiers* but also from Varro's *Rerum Divinorum et Humanorum Antiquitates* (116–24 BC.), through, among others, St. Isidorus' (560–

636 AD) *Etimologiae*, Yung-Loh Ta Tien's Encyclopaedia of 11,995 volumes of the 15th century, Bacon's *Instauratio Magna*, the Leibnizian project, Tu Shu Chi Ch'engs Encyclopaedia of 5,020 volumes edited in Shanghai some years before to the Encyclopaedia of the Vienna Circle.[6] Indeed, in these works the Encyclopaedia is thought of as displaying an essential openness, both in terms of subject matter and contents. As such, these Encyclopedia projects can be understood as permitting a variety of ways to fix the standards of use or practice that, at least in part, help to determine the significance of the concepts in question.

Such projects have a prospective or perhaps progressive role: By establishing cross-connections they have a positive heuristic value, not only by allowing local systematizations and reorganizations, but also by showing gaps in our present state of knowledge: D'Alembert formulates this beautifully at the end of his preliminaries: *Voila le peu que vous avez appris, voici ce qui vous reste à chercher*: Here is the little you have learned and here too what remains to be done.

The foregoing presents a view of the philosophy of science quite different from the one ordinarily associated with the logical positivists. We are not historians and do not claim to have told the whole story with respect to the Encyclopedia project of the Vienna Circle. Admittedly, for example, we have emphasized the pragmatist tendancies as represented by Neurath and Morris, over the more logicist elements of the Circle. Furthermore, there is clearly some tension between these great early papers of Neurath, Carnap and Morris. There is the obvious tension between Neurath's emphasis on the dynamic sociological aspect of the project and Carnap's antipsychologism and his focus on logic. With Morris' contribution, we see an attempt to overcome this tension via pragmatism. However, an analogous tension can be found between D'Alembert's Cartesian and mathematical approach and Diderot's more biological and dynamic one. Perhaps one can even say that this tension was part of the richness of the French Encyclopaedia. In any event, our point is a philosophical one and not entirely a historical one. We aim, with this introduction to draw some parallels between our goals with this series and what we see as the encyclopedic spirit which informs the early Vienna Circle and the French Encyclopedists.

6. A New Series in the Enyclopedic Spirit

The volume which launches this series gathers like-minded philosophers from across the philosophical community. While our contributors and even the editors of this book disagree on many things, we believe they share the goal of overcoming the arbitrary barriers that sometimes result from professional specialization. While the work gathered here is written in the spirit of integration and synthesis, it is not simply an exercise in interdisciplinarity for its own sake. The chapters, which constitute the backbone of our volume, should suggest the first steps towards an integration of the concept of the encyclopaedia as we have discussed it above and the various recent developments in logic represented by our contributors.

The task of investigating the logic of the knowledge-seeking enterprise in its entirety is being facilitated by the application of a range of new logical insights to questions in philosophy of science. Unlike the days of the Vienna Circle a range of different logics – suitable for a variety of different scientific contexts, are available to us. These developments are the subject of many of the papers in the present volume.

Acknowledgements

This volume has been two years in the making and we would like to thank everyone involved for their patience and effort. Academic life is often a solitary pursuit and perhaps this is why one of the most enjoyable aspects of this project has been its thoroughly collaborative nature. Each of the papers included here has been through a rigorous referee process and most have undergone at least one revision. While this process has certainly improved the book, it has also come at the cost of a great deal of time and hard work for all concerned. We would especially like to express our gratitude to the members of the editorial board for their encouragement and their active participation as referees and advisors.

Thanks also to Emmanuel Genot (Lille3), Laurent Keiff (Lille3), Alain Lecomte (Grenoble), Kuno Lorenz (Saarbrücken), Holger Sturm (Konstanz) Daniel Schoch (Saarbrücken), Charles Wolfe (Toulouse and Boston) and Juan Ferret (El Paso) for their help in the refereeing (in addition to efforts undertaken by contributors to the volume and members of the editorial board). Thanks also to Floor Oosting and Ingrid Krabbenbos at Kluwer, for their efficient editorial assistance and patient encouragement, to Laurent Keiff (Lille3), who lent his artistic skills to the design of the cover and Emmanuel Pruvost (Lille3) who organized the indexes.

Shahid Rahman is grateful for enriching discussions concerning the reconstruction of the non-official story of the Vienna Circle to Narahari Rao (Saarbrücken), for his lessons on pragmatism; Daniel Vanderveken (Montreal) Jean Caelen (Grenoble) and Denis Vernant (Grenoble) for their support, to his colleagues at Lille3, François De Gandt for his advice concerning the French Encyclopaedia and André Laks director of the Maison des Sciences de l'Homme du Nord-Pas de Calais (MSH) – this volume is part of the research projects *Preuve and La science et ses contexts*, attached to the MSH-Nord-Pas de Calais. Rahman also would like to thank to his students at the university: of Lille3: Emmanuel Genot, Emmanuel Pruvost; Laurent Keiff; Hassan Tahiri and Alexandre Thiercelin

In addition to those listed above, John Symons would like to thank Tian Yu Cao, Bob Cohen, Juliet Floyd, Dermot Moran and Fred Tauber for their wise advice and assistance. Thanks also to the students and faculty members of the department of philosophy at University of Texas at El Paso, for fostering a friendly and collegial environment and to Ann Lee for invaluable secretarial support. Cliff Hill and Jorge Tarin also provided excellent assistance. Most importantly, Jaakko Hintikka and the late Burt Dreben have been cherished teachers under whose guidance this project was born. While few who knew Dreben are likely to find much of his influence in this volume or in this series, others will recognize his spirit in our obstinate effort to avoid drawing a hard and fast line between sense and nonsense! Hintikka puts us all to shame with his brilliance, depth and energy. Not only is he an awe-inspiring source of new ideas and insight, he is also one of the most generous and decent members of the philosophical community.

Special thanks to Brian Hamilton, without his expert typesetting skills and help with the index, this book would never have been finished.

Finally, Symons gratefully acknowledges financial support from the University Research Institute at the University of Texas at El Paso.

Notes

[1] Margaret Morrison (2000) provides an excellent recent discussion of unification in scientific theories. Her views are more or less consonant with our own.

[2] A case along these lines has been made repeatedly by Steve Fuller (cf. Fuller 2000).

[3] Neurath stresses this fact in the first volume of the encyclopaedia edited later, where he remarks that in the Prague Congress of 1934 only the preliminaries were stated (Neurath [1938], 26).

[4] Neurath and Morris (1938) note the biological origin of the classification system employed by the French Encyclopaedists. Strikingly, they stress that they would like to emphasize the non-reductive character of the idea of "organized co-operation" and "scientific tolerance' of the French tradition.

[5] The influence of Bacon's conception of *idola*, acknowledged in the encyclopedia article "Préjugé" (written by Chevalier de Jaucourt) on later discussions of the demarcation problem by the logical positivists is worthy of further historical study.

[6] These features have been worked out by Olga Pombo in her paper *Leibniz and the Encyclopaedic Project* (Valencia: Editorial de la Universidad Politécnica de Valencia, 2002, 267–278). Rahman would like to thank her for making available her paper and for enriching discussions concerning Encyclopedism.

References

Bohr, Niels, Rudolf Carnap, John Dewey, Charles W. Morris and Otto Neurath: 1938, *Encyclopedia and Unified Science*, Chicago, Chicago University Press.

Cao, TianYu: 1998, *The Conceptual Development of 20th Century Field Theories*, Cambridge, Cambridge University Press.

Fuller, Steven: 2000, *Thomas Kuhn: A Philosophical Life For Our Times*, Chicago, Chicago University Press.

Gayon, Jean: 1998, *Darwinism's Struggle for Survival*, Cambridge, Cambridge University Press.

Neurath, Otto: 1936, *Une Encyclopédie internationale de la science unitaire, Actes due congrès international de philosophie scientifique – Sorbonne 1935*, Paris, Hermann & Cie, pp. 1–6.

Morrison, Margaret: 2000, *Unifying Scientific Theories: Physical Concepts and Mathematical Structures*, Cambridge, Cambridge University Press.

Pombo, Olga: 2002, *Leibniz and the Encyclopaedic Project*, Valencia, Editorial de la Universidad Politécnica de Valencia.

[illegible] HEM [illegible] provided [illegible] assistance [illegible] the [illegible] Blanc [illegible] [illegible] [illegible] [illegible] [illegible] [illegible] [illegible] [illegible] [illegible] of new ideas and [illegible] and [illegible] the [illegible] [illegible]

Special thanks to Brian [illegible] without [illegible] and help [illegible] the index this book would never have been [illegible]

Finally [illegible] acknowledge [illegible] the University Research [illegible] the University of [illegible]

Notes

[illegible]

References

[illegible]

AN INTERNATIONAL ENCYCLOPEDIA OF THE UNIFIED SCIENCES[1, *]

OTTO NEURATH
The Hague

From the point of view of scientific empiricism, one can say that the notion of "encyclopedia" rather than "system" offers us the correct model of science taken as a whole. In the spirit of scientific empiricism, the MUNDANEUM Institute of The Hague is preparing an *International Encyclopedia of the Unified Sciences* which is intended to serve as a compliment to existing encyclopedias.

The best encyclopedias of our time present each of the branches of knowledge in a vast tableau wherein recognized specialists show what has been achieved and what one should make of these results. In proportion to their development, different disciplines have elaborated distinct scientific languages making it difficult today to find the points of contact between them. Certain eminent thinkers continue to accentuate and underline these differences. However, from the point of view of scientific empiricism it must be emphasized that it is possible to alleviate this plurality of languages and that bridges can be built between the sciences. Today, either insufficient attention is paid to such bridges or, just as likely, they do not yet exist. It is precisely this matter that the International Congress for the Unity of Science met to address when it was held for the first time in Paris in 1935 with scientific philosophy as its main theme.

Thus, the creation of the encyclopedia comes in response to a contemporary need. It is destined to complete its predecessors and to show the extent to which one can unify contemporary science and make its internal structure apparent. To begin with, it can be argued by means of concrete examples that it is already possible to unify scientific language while avoiding metaphysical formulations.

It is not within the scope of this work to describe the achievements of particular disciplines. Instead, to the extent possible, it will present the many branches of science as a whole. In particular, it must be seen to what extent logico-scientific analysis can be put in the service of the unification of science.

While pointing out the essential unity of auxiliary scientific procedures, it is also necessary to emphasize that even within domains where axiomatization and other forms of systematic deduction exist, we have made only partial headway. Quite obviously there remain lacunae, since serious scientific studies undertaken success-

* The original text of Neurath's lecture was published in *Actes du congrès international de philosophie scientifique* (Paris, Hermann & Cie, 1936, II.1–II.6).

S. Rahman et al. (eds.), Logic, Epistemology, and the Unity of Science, pp. 17–21.

fully within different branches of science can still contradict one another. While other encyclopedias provide a retrospective synthesis, so to speak, this new work must begin by showing what new directions are opening up, where the problems lie, and where, from the point of view of a unified science, unforeseen possibilities can be detected. While general encyclopedias that are intended as presentations of the totality of our knowledge have, up to now, been shaped to the needs of a specific country or set of countries, subordinating most of their assertions to this point of view, the vast international encyclopedia that we are elaborating must apply itself primarily to showing the profound unity of the general idea of science, leaving its diversity for later.[2]

The program being formulated implies considerable cooperation between specialists of different disciplines. But this is precisely where partisans of scientific empiricism can hope for success since scientific empiricism is well-suited to the joint efforts of experts. Thus, provided that we succeed in unifying scientific terminology and symbolism, it seems legitimate and timely to undertake this encyclopedia. While striving to use the results of contemporary logic for the entirety of science, we must nevertheless be careful not to conceal the ambiguity of certain terms in pursuit of a unitary system, while having, in fact, only certain initial elements of such a system in place. Such elements may themselves be highly perfected, without being amenable to coordination.

Prior encyclopedias have often been regarded as partial, eclectic samplings whose imperfection one had to accept with resignation while acknowledging that the true ideal had been precisely a "system". However, we want to declare, above all, that the form of encyclopedia is the most perfect that we could ever attain in presenting the whole of science. In clear opposition to the pseudo-rationalism of all centralist philosophies, our concrete scientific labor is careful not to attempt to forecast the overall systematization of science.

It will not be enough to bring out the unity of auxiliary scientific procedures in each discipline; they will each have to be treated separately in a systematic study. The new encyclopedia must show in detail, for example, how calculations of probabilities or how certain methods of coordination, are applicable to all possible domains, in a way in which one should be able to establish a kind of *instrumentarium* for science in general and where one would at the same time show their effective uses.

Besides this general tableau of scientific instruments, one can strive to increase the uniformity of scientific language generally or in specialized areas, and it must to be seen which of the many possible approaches to unification is most suited to the concrete goals of the Encyclopedia. It is thus not only a matter of proposing the principle of unity as the title of this program; it must also be proven through actions. By common agreement, it will be obligatory to avoid particular terms and formulas within the encyclopedia. One must renounce terms suited only to specialized sciences, if one is to accommodate oneself to a general terminology that is convenient to all the sciences.

These are fundamental guidelines that can hardly be contested. But the concrete execution requires serious efforts of organization. While until now encyclopedias limit themselves to recommending to their collaborators that each subject be treated with care and discernment, this new encyclopedia has to bring its collaborators to a common understanding in order to push contributions as much as possible towards formal unity. Clearly such uniformity has its limits. But a great deal would already be accomplished for communication if fundamental terms or "model-words" to be adopted within different areas could be settled upon. Thus the Encyclopedia is not to be arranged alphabetically, but by subject matter, and its general index will be the expression of a strictly scientific attitude with respect to the ideal of unity.

This non-alphabetized Encyclopedia will publish a few volumes per year, each composed of 3 to 6 short monographs. As such, it will be several years before the project is completed. But as each monograph will form a whole, and as the most recent accomplishments of science can always be presented in the form of supplements, the reader will always have at hand an Encyclopedia which will indeed be partial, but each part of which will be complete unto itself, and incorporated into an ordered whole. The plan of the encyclopedia is to have an initial series of volumes as a basic layer, that will provide the general perspective. It will be possible to add layers later and nothing prevents us from continuing in this manner, if one is so inclined, until sufficiently specific issues are published. These later works will find a well-determined place within the general plan although it will not be necessary for every discipline to include these particular studies. In many branches one can already lean on existing work, while in others it will be necessary to create works in accordance with the specific intention of this encyclopedia in order to highlight the internal links, the unity, and the vertical and horizontal relations. Such works must be able to find place within the framework of the encyclopedia.

Each of the volumes will appear in the three languages used by the International Congress for the Unity of Science: German, French and English. It will be an opportunity to create a little trilingual lexicon of the most relevant terms. This lexicon will not consist of terms that are already familiar, but only those that have been chosen in a common agreement to be used in the encyclopedia. If this agreement is actually continued and realized, it will be a precious contribution to international understanding on the terrain of logical empiricism.

Clearly, this encyclopedia will focus not only on the unification of scientific language, but also in the unification of graphical representation. Curves and other figures are also tools of scientific expression and all the images in the encyclopedia will be produced using standardized elements, whether it be to represent technical, biological or sociological objects. These standardized elements can be catalogued within a sort of "sign lexicon" and one can combine them with each other according to a sign grammar. The *Mundaneum Institute* of the Hague will adapt the figurative language called ISOTYPE (**I**nternational **S**ystem **O**f **Ty**pographic **P**ictorial **E**ducation), which it is already using elsewhere, to the particular needs of the encyclopedia. With this, the encyclopedia, which addresses a very large public will gain in intelligibility.[3]

Since this encyclopedia does not want to present each discipline as part of a finished tableau, but rather to show precisely the lacunae and deficiencies of current knowledge, it will emphasize the "contingent" aspects of research and the fact that all science depends on historical conditions. It will also be equally necessary to note the direct connections between practical life and science. Yet these efforts to point out that scientific thought is close to ordinary life should not be converted into "imperatives". The encyclopedia must take as a principle the effort to avoid any affirmation stained with emotionalism, blame or praise. It is clear that such an enterprise is conditioned by emotional elements, as is any historical human behavior, to the extent that the simple choice of the matters in question are not themselves scientifically justifiable. Yet this does not change the essential difference between a mode of expression that uses emotional means and one which is careful to avoid them. Naturally, emotional elements will figure in the historical exposition of the encyclopedia, where the evolution of all modes of expression that are not part of scientific empiricism will be treated. A historical survey may also be required in order to determine whether primitive languages already contain metaphysical sentences or whether such sentences appeared only until later and if so, under what conditions.

Because of its fundamental logical attitude, this encyclopedia goes back to a certain extent to Leibniz, who in his projects had dreamed of figurative representations. But the general tendency to favor the cultivation of intuitive processes of learning goes back to Comenius' *Orbis Pictus* and to Paul Otlet's efforts to present all of contemporary knowledge by modern intuitive methods (*The Global Village*). In a sense, this encyclopedia of the unified sciences also continues the work of Auguste Comte and that of Herbert Spencer, who wanted to give an image of the whole of science in a purely empiricist spirit. However, this is a job that could not have been carried out methodically before now because we only now have at our disposal the resources of the new logic and modern media for figurative representations (visualization).

The committee of the Encyclopedia at the Mundaneum institute of the Hague, comprised of Carnap, Frank, Jorgenson, Morris, Neurath and Rougier, will not be required to find representatives from every discipline and to convince them of a new doctrine. Rather, it will seek collaborators that strive, via cooperative work, to show what can be achieved even today with the aid of the logical empiricism, and how the results of science can be incorporated within this new framework. This is a new path that opens up for young people. In order to facilitate access to the Encyclopedia to the largest possible number of people, especially the young one must take account of certain pedagogical requirements while also respecting scientific rigor. The goal is less to refine disciplines that are already established, than to engage with those branches that up to now have been somewhat marginal, like psychology, biology and sociology. This will be one of the important tasks of this Encyclopedia, to show to what extent these disciplines can share a common language with physics and how nevertheless the laws of the different sciences present distinctive particularities.

Within the context of the search for unity, we will emphasize all the problems faced by those who are sensitive to what is new in logical empiricism. They are precisely the ones who will appreciate this way of presenting the unity of knowledge: "to he who has arrived, no satisfaction can be given, whereas he who is 'in progress' will always be grateful". Indeed, this encyclopedia of the unified sciences is conceived precisely as an Encyclopedia in progress.

Notes

[1] At the suggestion of Professor Charles W. Morris of Chicago, the Congress has given its approval to the *International Encyclopedia of the Unified Sciences* project of the MUNDANEUM Institute of the Hague.

[2] This new encyclopedia, unlike other projects with the same title, does not seek to provide an overview of the totality of knowledge but rather to show the structure of our science. It will therefore not be as large as ordinary encyclopedias and will possess its own distinctive character.

[3] Cf. Otto Neurath, *International Picture Language. The First Rules of ISOTYPE*, London, Kegan Paul, 1936.

Translated by
John Symons and Ramon Alvarado

PART II
GAME THEORY AND INDEPENDENCE FRIENDLY LOGIC AS A UNIFYING FRAMEWORK

TOWARDS A UNITY OF THE HUMAN BEHAVIORAL SCIENCES

HERBERT GINTIS[1]
Santa Fe Institute and University of Massachusetts, USA

Abstract. Despite their distinct objects of study, the human behavioral sciences all include models of individual human behavior. Unity in the behavioral sciences requires that there be a common underlying model of individual human behavior, specialized and enriched to meet the particular needs of each discipline. Such unity does not exist, and cannot be easily attained, since the various disciplines have incompatible models and disparate research methodologies. Yet recent theoretical and empirical developments have created the conditions for unity in the behavioral sciences, incorporating core principles from all fields, and based upon theoretical tools (game theory and the rational actor model) and data gathering techniques (experimental games in laboratory and field) that transcend disciplinary boundaries. This paper sketches a set of principles aimed at fostering such a unity. They include: (a) evolutionary and behavioral game theory provides a transdisciplinary lexicon for communication and model-building; (b) the rational actor model, rooted in biology but developed in economic theory, applies to all the human behavioral disciplines. This model treats actions as instrumental towards satisfying preferences. However, the content of preferences must be empirically determined. Moreover, the rational actor model is based on a notion of preference consistency that is not universally satisfied, so its range of applicability must also be empirically determined; (c) controlled experiments have been underutilized in most behavioral disciplines. Game theory and the rational actor model can be used as the basis for formulating, deploying, and analyzing data generated from controlled experiments with human subjects.

1. Introduction

The human behavioral sciences include economics, human biology, anthropology, sociology, behavioral psychology, and political science.[2] We may consider a set of disciplines as *unified* if they are (a) *consistent*, so that in cases where two disciplines deal with the same social phenomena, their models are equivalent, and *synergic*, each discipline being substantively enriched by the scientific content of the others. The natural sciences achieved unity with the development of quantum mechanics, elementary particle and solid state physics, and the big bang model of the universe. Such unity is lacking in the human behavioral sciences. Each behavioral discipline models *individual human behavior*, and construct models of aggregate social behavior compatible with, and often derived from, a model of individual behavior. Unity in the human behavioral sciences requires a common underlying model of individual behavior, which each discipline specializes and enriches for its particular purposes. No current model enjoys such transdisciplinary status.

S. Rahman et al. (eds.), Logic, Epistemology, and the Unity of Science, pp. 25–39.

Yet, recent developments reveal links across the behavioral sciences sufficiently deep to establish the preconditions for unity. Both sociology (Hechter and Kanazawa 1997) and political science (Monroe 1991), following the pioneering contributions of Coleman (1990), Downs (1957), Olson (1965), Buchanan and Tollison (1984) and others, have begun to adopt the rational actor model, previously espoused virtually exclusively in economics. Game theory, a central element of economic theory, was introduced to biology by Lewontin (1961), Hamilton (1967) and Maynard Smith and Price (1973), subsequently maturing into an invaluable behavioral tool (Alcock 1993; Dugatkin and Reeve 1998; Gintis et al. 2001; Gintis 2003a). In anthropology, the application of experimental game theory to understanding cultural variation is rather new, but quite promising (Henrich et al. 2001, Henrich et al. 2004). Conversely, increasing numbers of economists develop behavioral models of social interaction, and draw upon e vidence from experimental game theory in modeling behavior. This development is evidenced by the Nobel prize in economics for the year 2002, awarded jointly to two experimentalists: psychologist Daniel Kahneman and economist Vernon Smith.

In this paper I will sketch a set of principles that express my current conception of unity. I will argue the following points. First, game theory provides a transdisciplanary behavioral lexicon for communication and model-building. For many years it was widely thought that game theory presupposes methodological individualism and a high level of cognitive functioning on the part of subjects. Were this the case, game theory would be inapplicable to settings where emotion, traditional, and heuristic behaviors are prominent, and where group-level processes and dynamic interactions are common. Contemporary evolutionary and behavioral game theory, however, extends classical game theory to cover such settings.

Second, evolutionary biology underlies all behavioral disciplines because *Homo sapiens* is an evolved species whose major characteristics are the product of its particular evolutionary history.

Third, evolutionary and behavioral game theory provide the substantive framework for the biology of human behavior.

Fourth, the rational actor model, developed in economic theory, is a flexible tool that applies to all the human behavioral disciplines. This model treats actions as instrumental towards satisfying preferences. However, the content of preferences must be empirically determined, and agents may have preferences over actions as well as their outcomes. Moreover, the rational actor model is based on a notion of preference consistency that is not universally satisfied, so its range of applicability must also be empirically determined.

Fifth, progress in modeling human behavior has been hampered by the *underutilization of controlled experiments*, which are common only in behavioral psychology. Game theory and the rational actor model can be used as the basis for formulating, deploying, and analyzing data generated from controlled experiments in social interaction. Such controlled experiments are replicable across laboratories and foster cumulative knowledge relevant to all behavioral disciplines.

Sixth, progress in modeling human behavior has been hampered by the *artificially restricted range of social situations studied by behavioral scientists*. Only anthropology has systematically studied the effects of cultural differences *across societies* on human behavior, only sociology has systematically studied the effects of cultural differences *within societies* on human behavior, and only behavioral psychology has systematically studied the effects of *personality differences* on social interaction. A unified model of human behavior is fostered by taking controlled experiments to the field, and deploying such experiments in a variety of cross-cultural settings across and within societies.

Seventh, The demographic success of *Homo sapiens* is due to the ability to of humans to sustain a high level of cooperation among non-kin. Whereas biology and economics explains this ability in terms of exchange among self-interested agents, the facts are in line with basic sociology and behavioral psychology: humans often display altruistically prosocial behavior, especially in a form that I call *strong reciprocity* – a predisposition to cooperate and to punish non-cooperators at personal cost (Gintis et al. 2004).

Finally, prosocial behavior in humans can be modeled biologically using the tools of gene-culture coevolution, but the social mechanisms involved must include using the sociological notions of *socialization* and the *internalization of norms*.

2. The Language of Game Theory

Communication across disciplines presupposes a common language. *Game theory* is a universal behavioral lexicon that offers such a common language. In the language of game theory, *players* (or *agents*) are endowed with a set of available *strategies*, and have a range of *information* concerning the rules of the game, the nature of the other players and their available strategies, as well as the structure of payoffs. Finally, for each combination of strategy choices by the players, the game specifies a distribution of *individual payoffs* to the players. If the game is accurately specified, we can predict the behavior of the players by assuming they attempt to maximize some preference function involving their personal payoffs, their chosen strategies, the personal payoffs to other agents, and the actions of the other agents (Gintis 2000). Self-regarding agents maximize their personal payoffs, while other types of agents may care about fairness, the intentions of other agents, the sum of all payoffs, their relative personal payoff, and other aspects of the array of payoffs.

Developments within game theory in recent years have considerably enhanced its value to behavioral disciplines that have traditionally found little use for this analytical tool. First, it is now widely recognized that in many social interactions, agents are not self-regarding, but rather care about the payoffs to and intentions of other players (Rabin 1993; Bergstrom and Stark 1993; Andreoni and Miller 2002; Fehr and Gächter 2002; Wood 2003). Second, human actors care not only about material payoffs, but power, self-esteem, and behaving morally (Gintis 2003b;

Bowles and Gintis 2003; Wood 2003), goals that are recognized as central to many behavioral disciplines. Third, evolutionary and behavioral game theory do not require the extensive cognitive and information processing capacities of classical game theory, so disciplines in which it is recognized that cognition is a scarce and costly good can make use of game-theoretic models (Young 1998; Gintis 2000; Gigerenzer and Selten 2001). Thus, agents may consider only a restricted subset of strategies (Winter 1971; Simon 1972), and they may use by rule-of-thumb heuristics rather than maximization techniques (Gigerenzer and Selten 2001). Game theory is thus a generalized schema that permits the precise framing of meaningful empirical assertions, but imposes no particular structure on the predicted behavior.

3. The Rational Actor Model

The rational actor model assumes that agents have *preferences* reflecting their wants and the tradeoffs among these wants, and that agents maximize their *utility* by choosing from an *action set* that is limited by available information, material resources and time, cognitive capacity, and the agent's physical capacities. Choice is also contingent upon *beliefs* concerning the probabilities of various states of nature, the frequency distribution of types of agents with whom they interact, and the relative effectiveness of different actions. The rational actor model is most highly developed in economics, but it applies to all the disciplines dealing with human behavior.

The rational actor model appears *prima facie* to apply only when extremely stringent conditions are satisfied. However, the model can be shown to apply over any domain in which the agent has *transitive preferences*, in the sense that if he prefers A to B and he prefers B to C, then he prefers A to C, and the agent can *make tradeoffs among outcomes* in the sense that for any finite set of outcomes $A_1, \ldots, A_n$, if A_1 is the least preferred and A_n the most preferred outcome, than for any A_i, $1 \leq i \leq n$ there is a probability p_i, $0 \leq p_i \leq 1$ such that the agent is indifferent between A_i and a lottery that pays A_1 with probability p_i and pays A_n with probability $1 - p_i$ (Kreps 1990). Clearly, these assumptions are often extremely plausible. When applicable, the rational actor model's transitivity assumption strongly enhances explanatory power, even in areas that have traditionally abjured the model (Coleman 1990; Kollock 1997; Hechter and Kanazawa 1997).

The rational actor model has been underutilized in some behavioral disciplines through several prominent misunderstandings. First, the rational actor model does not require that agents be self-interested. There is no connection between the notion of the transitivity of preferences and the notion that preferences are purely self-regarding. Indeed, one can apply standard choice theory, including the derivation of demand curves, plotting concave indifference curves, and finding price elasticities, for such preferences as charitable giving and punitive retribution (Andreoni and

Miller 2002). Second, because the rational actor model treats action as instrumental towards achieving rewards, it is often inferred that action itself cannot have reward value. This is an unwarranted inference. For instance, the rational actor model can be used to explain the expressive motivation in rational action, including collective action, that is precluded by the assumption that agents act instrumentally towards satisfying their material needs (Olson 1965), since agents may place positive value on the process of acquisition (for instance, "fighting for one's rights"), and can value punishing those who refuse to join in the collective action (Moore Jr. 1978; Wood 2003). Third, the areas over which the transitivity postulate holds must be *empirically determined.*

Broadening the rational actor model beyond its traditional form in neoclassical economics run the risk of developing unverifiable and *post hoc* theories, as our ability to theorize outpaces our ability to test theories. To avoid this, and following the lead of behavioral psychology, we must expand the use of controlled experiments, as suggested above. Often we find that the appropriate experimental design can generate new data to distinguish among models that are equally powerful in explaining the existing data (Tversky and Kahneman 1981; Kiyonari et al. 2000).

4. Game Theory and Biology

The analysis of living systems includes one analytical element that does not occur in the non-living world, and is not analytically represented in the natural sciences. This is the notion of a *replicator* (Schrödinger (1958) called this an "aperiodic crystal"), which is a physical system capable of drawing energy from its environment to make relatively accurate copies of itself. The dynamics of replicators are described by the evolutionary notions of replication, mutation, selection, and adaptation (Lewontin 1974).

The most natural setting for replicator dynamics is game theoretic. Replicators endow copies of themselves with a repertoire of strategic responses to environmental conditions, including information concerning the conditions under which each is to be deployed in response to character and density of competing replicators. Mutations included replacement of strategies by modified strategies, and the "survival of the fittest" dynamic (formally called a *replicator dynamic*) ensures that replicators with more successful strategies replace those with less successful (Taylor and Jonker 1978).

5. Gene-Culture Coevolution

Genetic replicators transmit information encoded in DNA sequences, through a germ line that is unaffected by environmental conditions. Genetic adaptation to new environments then takes the form of shifts in allele frequencies, and promotion of mutations that better exploit the new environment. In the context of rapidly changing environments, there is a fitness benefit to the transmission of *epigenetic* information concerning the current state of the environment. Such epigenetic information is quite common (Jablonka and Lamb 1995), but achieves its highest and most flexible form in *cultural transmission* in humans and to a considerably lesser extent in other primates (Bonner 1984; Richerson and Boyd 1998).

There are several basic categories of culture: *conventions* (e.g., language use), *techniques and practices* (e.g., how to prepare food, how to make and use tools, how to treat illnesses), *ethical values* (e.g., norms of fairness, reciprocity, justice) and *transcendental beliefs* (e.g., sickness is caused by angering the gods, good deeds are rewarded in the afterlife). A transcendental belief is the assertion of a causal relationship or a state of affairs that has a truth value, but whose truth holders either cannot or choose not to test. There are of course other types of beliefs, but these appear to be subsumable under other cultural categories. For instance, one may believe that a certain convention exists, a certain technique is effective, or a certain ethical value is justifiable. To avoid confusion, we treat such beliefs as part of the conventions, techniques and practices, and values that they affirm.

Conforming to *conventions* is adaptive because it is payoff-maximizing to conform when all others are doing so. When an agent must determine the most effective of several alternative *techniques* or *practices*, and if experimentation is costly, it may be payoff-maximizing to copy others rather than incur the costs of experimenting (Boyd and Richerson 1985; Conlisk 1988). If everyone else experiments to find the superior technique, it will generally pay simply to follow the majority. By contrast, if everyone else conforms to a single technique in a situation where different techniques are best suited to different environments, then when the environment changes an individual who experiments may do better than the conformists. Thus, in general there will be a cultural equilibrium with a positive fraction of both conformists and experimenters. In this sense, the genetic machinery for a predisposition to conform to conventions and to imitate techniques is biologically adaptive.

It is plausible to extend this explanation to transcendental beliefs as well. such beliefs affirm techniques where the cost of experimentation is extremely high or infinite, and the cost of making errors is high as well. This is, in effect, Blaise Pascal's argument for the belief in God and the resolve to follow His precepts. It is supported by Boyer (2001), who models religion as a set of cognitive beliefs that coexist and interact with our other more mundane and testable beliefs. In this view, one conforms to transcendental beliefs because their t ruth value has been ascertained by others (relatives, ancestors, prophets), and are as worthy of affirmation as the techniques and practices (such as norms of personal hygiene, that one accepts on faith, without personal verification.

Conventions, techniques, and beliefs are *instrumental* in the sense that they specify how best to achieve certain end or goals. The remaining cultural category, *ethical norms and values*, is *final* in the sense of specifying what ends or goals to embrace. I discuss the place of cultural values in human behavioral theory below.

6. The Puzzle of Prosocial Behavior in Humans

The success of *Homo sapiens*, as measured by its broad geographical distribution and its considerable share of the Earth's biomass, is based on its unique capacity to use cultural forms to transmit technical knowledge accurately across generations, and its unique ability to sustain cooperation through space and across time among large numbers of unrelated individuals (Richerson and Boyd 1998). How do we explain this cooperation?

Biologists maintain that cooperation can be sustained based by *inclusive fitness*, or cooperation among kin (Hamilton 1964), by and individual self-interest in the form of *reciprocal altruism* (Trivers 1971). Reciprocal altruism occurs when an agent helps another agent, at a fitness cost to itself, with the expectation that the beneficiary will return the favor in a future period. The explanatory power of inclusive fitness theory and reciprocal altruism convinced a generation of biologists that what appears to be altruism – personal sacrifice on behalf of others – is really just long-run self-interest. Economics has developed a similar model of cooperation, based on the notion of long-term, enlightened self-interest (Arrow and Debreu 1954; Axelrod and Hamilton 1981; Fudenberg and Maskin 1986), an idea that goes back to Bernard Mandeville's concept of "private vices, public virtues" (1729) and Adam Smith's notion of the "invisible hand" (2000 [1759]).

Sociology, by contrast, has used the *socialization* to explain cooperation among non-kin. According to Durkheim (1951), the division of labor in society involves assigning individuals to specific *roles*. Individuals are inculcated with *values* and *norms* that induce them to conform to the duties and obligations of the role-positions they occupy. This is altruism.

A key tenet of socialization theory is that a society's values are passed from generation to generation through the *internalization of norms* (Durkheim 1951; Benedict 1934; Mead 1963; Parsons 1967; Grusec and Kuczynski 1997). In the language of optimization theory, internalized norms are accepted not as instruments towards and constraints upon achieving other ends, but rather as *arguments in the preference function that the individual maximizes*. Internalized norms are thus what we termed *ethical values* in our lexicon of cultural forms. In true gene-culture coevolutionary form, a variety of uniquely human prosocial emotions come into play, including prominently *shame*, *guilt*, and *empathy*, directly reinforcing internalized norms.

The programmability of the preference function appears in the form of the human *capacity to internalize norms*, which consists in an older generation instilling the values and objectives of a younger generation through an extended series of personal interactions, relying on a complex interplay of affect and authority. Agents conform to an internalized norm because so doing is an end to itself, and not merely because of the material rewards that follow from norm compliance or punishments that follow from norm violation. For instance, an individual who has internalized the value of "speaking truthfully" will do so even in some cases where the net payoff to speaking truthfully would otherwise be negative. It follows that where people internalize a norm, the frequency of its occurrence in the population will be higher than if people follow the norm only instrumentally; i.e., when they perceive it to be in their narrow material interest to do so. The capacity to internalize is based on a distinctively human psychological predisposition, unrecognized in biology and economics.

7. The Internalization of Norms

An "altruistic norm", when acted upon, reduces the bearer's individual fitness or material well-being, but increases the fitness or well-being of other, unrelated, group members. The internalization of altruistic norms appears to be an evolutionary *curiosum* because agents who internalize such norms should be at a fitness disadvantage in comparison with self-interested actors. A closer look at the cultural transmission process, however, offers a resolution to this problem (Gintis 2003a).

Suppose there is an altruistic behavior A that imposes fitness cost s on those who embrace it. Suppose also that only a fraction of youth have the genetic capacity to accept ethical norms, and this fraction increases or decreases over time according to the biological fitness of its bearers. Suppose further that altruistic behavior A is transmitted to offspring with this genetic capacity by their parents in an unbiased manner (i.e., if both or neither parents embraces A, all of their genetically enabled offspring do the same, and if only one parent embraces A, half of such offspring embrace A). In addition, suppose there is *extraparental transmission* of A, in the form of social pressure (rumor, shunning, and ostracism), rituals (dancing, prayer, marriage, birth, and death), and in modern societies, formalized institutions (schools, churches, sacred texts). Such extraparental transmission is itself altruistic, since it will generally be individually costly while the benefits, in the form of a higher frequency of altruism in the group, accrue to unrelated others. We handle this, plausibly, we believe, by assuming that the altruistic norm is both to embrace A and to encourage others to embrace it as well, and we include the cost of extraparental transmission in s, the cost of altruism. We measure the strength of extraparental transmission by a parameter γ, such that if the fraction of altruists in the older generation is p_A, then $\gamma\ p_A$ is the probability that a given non-altruistic child with the genetic capacity to acquire the altruistic norm, will in fact be induced to embrace the altruistic norm.

Suppose, further, that an altruist who meets a nonaltruist, which we assume occurs with a probability proportional to the fraction of altruists, switches to the nonaltruist's behavior with probability α. Gintis (2003a) then shows that if α satisfies the inequality

$$\alpha < \frac{\gamma - s}{1 - \gamma}$$

then the altruistic cultural equilibrium, in which all agents have the genetic capacity to embrace ethical norms, and all actually embrace A, is evolutionarily stable. Note that (a) the larger the fitness cost s of altruism, and (b) the smaller the rate γ of oblique transmission, the lower the maximal rate of "moral defection" α to the nonaltruistic that is compatible with an altruistic cultural equilibrium. Note also that if γ is sufficiently large (specifically, if $\gamma > (1+s)/2$) then no rate of defection can undermine the altruistic equilibrium, because agents rarely meet nonaltruists with whom they can compare their fitness.

The substantive questions, then, are (a) why γ might be positive and large, and (b) why the rate α of moral defection might be low. To address (a), note that the rate of extraparental transmission depends not only on the willingness of individuals to sacrifice on behalf of the group by engaging in extraparental socialization and by rewarding others who do the same, but also on the structure of social institutions that routinize cultural transmission. There is thus no guarantee that γ will be high, but societies that do effectively organize cultural transmission, and stress ethical norms that are heavily prosocial, will tend to grow and otherwise outcompete societies that do not (this process is referred to above as *weak* group-level selection).

To address (b), we must explain why agents might not defect at a very high rate to fitness-maximizing behavior. The following argument suggests that the psychological constitution of *Homo sapiens* is conducive a high rate of adherence to moral norms, and hence to the satisfaction of Equation (1). While nearly everyone behaves amorally on some occasions, and some behave amorally much of the time, there is normally a sufficient reserve of moral behavior, including the motivation to punish the moral transgressions of others, to maintain a high level of conformity with group morality.

As we have noted, humans do not maximize fitness, but rather a preference function that is but a rough proxy for fitness under constant environmental conditions. The rapid pace of environmental change and cultural innovation over the past 100,000 years has produced a situation in which the set of needs, desires, drives, pleasures, and pains associated with the human preference function is out of line with the dictates of fitness maximization (Richerson and Boyd 1998). Even a random deviation of the human preference function from fitness maximization towards other goals, such as power, esteem, wealth, and pleasure, might be conducive to a relatively slow rate of rejection of moral norms, of which altruistic norms might figure prominently.

However, there is evidence of a more systematic force intervening between biological fitness and human preferences: as we have seen, the human preference function is, to some considerable extent *programmable*, in the sense that human goals can be altered by socialization. The notion of a programmable preference function is sufficiently unusual that such a mechanism must have arisen as an adaptation, and hence the content of socialization, the actual internalized norms themselves, must be, at least on balance, fitness enhancing. Yet standard sociological theory has not supplied an argument as to why it might be adaptive, and indeed have generally ignored evolutionary arguments altogether. We can, however, supply such an argument.

A programmable preference function is the most complex instrument facilitating epigenetic information flows, all of which represent means of transferring information across generations in a manner complementary to, and often more flexible than, genetic transmission (Bonner 1984; Boyd and Richerson 1985; Jablonka and Lamb 1995, 1998). The form that this epigenetic transmission process takes in the case of the internalization of norms is a protracted series of interactions, controlled by parents and influential elders, undertaken at considerable cost, and reinforced by a complex web of informal sanctions. While cultural learning occurs in many species, programmability of goals is virtually limited to humans because the capacity to be socialized presupposes a high level of cognitive capacity (Tomasello 1999), as well as specialized mental circuitry for valuing interpersonal relationships and making informed social judgments (Damasio 1994), and specialized emotional capacities that enhance the individual's capacity to attain internalized goals, such as pride, shame, empathy, and remorse (Bowles and Gintis 2003). The genetic basis for prosocial emotions is clear from the fact that the inability to experience prosocial emotions, associated with sociopathic personality types, is partially heritable (Mealey 1995), and is deficient in individuals with damage to specific regions of the brain's frontal lobes (Damasio 1994).

The capacity to program changes in the preference function *culturally* indeed has great adaptive value. By redirecting human goals, and thereby curbing, repressing, and channeling an agent's basic impulses, the agent will have higher fitness than another agent who lacks this capacity. Included among the norms that are commonly internalized are thus norms of personal hygiene, concern for the approval of others, control of temptation, cultivation of a work ethic, and maintaining a long time horizon in decision-making. Such norms are upheld and transmitted in virtually all societies (Brown 1991), though a breakdown of cultural transmission in this area occurs in some poorly functioning societies (Edgerton 1992).

For an example of the fitness-enhancing capacity of internalization, note that a sophisticated weapon, such as a sharp knife, may aid an individual in taking revenge upon a transgressor, but the spontaneous impulse to attack an enemy may be fitness-reducing when such weapons are widely available. However, parents can instill in their offspring the norms of "love thy neighbor" and "be slow to anger". Individuals who have acquired this genetic predisposition to internalize norms will pass both this capacity and its content – the conflict-limiting norm itself – to their offspring. For this reason, the internalization of norms may be fitness-enhancing.

For a second example, suppose someone invents an aerodynamic spear that is extremely effective in the hunt, but requires daily practice to hone the throwing skills needed to use the spear effectively (Calvin 1983). Since agents primordially prefer less expenditure of energy to more and have inappropriately short time horizons, they will skimp on daily practice. The hunter who internalizes the norm "good hunters like to practice" will have an adaptive advantage.

One might object that a non-internalizer could always mimic the behavior of internalizers when it suits his purposes, and do better by violating the norm strategically when it is in his interest to do so. In fact, the noninternalizer *could*, but will not *want* to emulate internalizers, in the sense that emulating their behavior simply does not maximize his preference function. To pursue the first example above, curbing one's violent tendencies may improve fitness, but the primordial preference function is not geared towards maximizing fitness, but rather a set of "fitness proxies" that entail being violent under earlier evolutionary but not contemporary circumstances. The noninternalizer will, of course curb his violence for prudential reasons, but not, because he in addition values peace, his neighbor's well-being, or even his biological fitness. In the second example, the noninternalizer will prefer the larger portion of meat, and the greater prestige that follows from a rigorous practice routine, but nevertheless, not enough to engage in such a routine.

Once genes for norm internalization are in place, there is nothing preventing altruistic norms from being culturally transmitted, internalized, and acted on in the same manner as personally fitness-enhancing norms. Altruism thus 'hitchhikes' on the personal fitness-enhancing capacity of norm internalization, and hence is an *exaption*, in the sense of Gould and Vrba (1981). It is for this reason that the rate of defection from altruistic norms might be sufficiently low that Equation (1) might hold, as long as fitness costs are not too high and there is some positive level of oblique transmission.[3]

It might be suggested that in a cultural equilibrium with internalized altruistic norms, a mutant family that teaches it children to internalize the personally fitness-enhancing norms but not the altruistic ones would out-compete families that transmit both personally fitness-enhancing and altruistic norms. However, if part of the ethic of altruism is to punish selfish types, even selfish types will act altruistically, so under plausible conditions, the mutant may have no adaptive advantage. Moreover, we show in Gintis (2003a) that, using the above notation, if

$$\alpha < \frac{\gamma - s}{1 + \gamma - s}$$

selfish internalizers are positively disadvantaged with respect to altruistic internalizers.

8. Conclusion

Each of the behavioral disciplines contributes strongly to human behavioral science. Taken separately and at face value, however, they offer partial, conflicting, and incompatible models of human behavior. From a scientific point of view, it is scandalous that this situation was tolerated throughout most of the Twentieth Century. Fortunately, there is currently a strong current of unification based on both mathematical models and common methodological principles for gathering empirical data on human behavior and human nature.

The true power of each discipline's contribution to knowledge will only appear when suitably qualified and deepened by the contribution of the others. For instance, the economist's model of rational choice behavior must be qualified by a biological appreciation that preference consistency is the result of strong evolutionary forces, and where such forces are absent, consistency will be imperfect and behavior must be augmented by empirical evidence. Moreover, *aprioristic* notions that preferences are self-regarding must be abandoned. These are the key tenets of behavioral economics. Second, the sociologist's notion of internalization of norms is generally rejected by the other behavioral disciplines because the ease with which diverse values can be internalized depends on human nature (Cosmides and Tooby 1992; Pinker 2002), and the rate at which values are acquired and abandoned depends on their contribution to fitness and well-being (Gintis 2003b; Gintis 2003a). Finally, there are often swift society-wide value changes that cannot be accounted for by socialization theory (Wrong 1961; Gintis 1975). When properly qualified, however, and appropriately related to the general theory of cultural evolution and strategic learning, the socialization theory is considerably strengthened.

Disciplinary boundaries in the behavioral sciences have been determined historically, rather than conforming to some consistent scientific logic. Perhaps for the first time, we are in a position to rectify this situation.

Notes

[1] I would like to thank Samuel Bowles and Suresh Naidu for helpful comments and the John D. and Catherine T. MacArthur Foundation for financial support.

[2] By 'human biology' I mean the application of biological techniques to modeling human behavior. I use the term 'behavioral psychology' to mean social psychology and psychological decision theory.

[3] By the same token, even antisocial norms can hitchhike on the internalization capacity, and it not infrequently does just that (Edgerton 1992). We show in Gintis (2003a) that the tendency of higher-fitness groups to out-compete lower fitness groups provides a strong tendency towards the circumscription of anti-social norms.

References

Alcock, John: 1993, *Animal Behavior: An Evolutionary Approach*, Sunderland, MA, Sinauer.

Andreoni, James and John H. Miller: 2002, 'Giving According to GARP: An Experimental Test of the Consistency of Preferences for Altruism', *Econometrica* **70**(2), 737–753.

Arrow, Kenneth J. and Gerard Debreu: 1954, 'Existence of an Equilibrium for a Competitive Economy', *Econometrica* 265–290.

Axelrod, Robert and William D. Hamilton: 1981, 'The Evolution of Cooperation', *Science* **211**, 1390–1396.

Benedict, Ruth: 1934, *Patterns of Culture*, Boston, Houghton Mifflin.

Bergstrom, Theodore C. and Oded Stark: 1993, 'How Altruism can Prevail in an Evolutionary Environment', *American Economic Review* **83**(2), 149–155.

Bonner, John Tyler: 1984, *The Evolution of Culture in Animals*, Princeton, NJ, Princeton University Press.

Bowles, Samuel and Herbert Gintis: 2003, 'Prosocial Emotions', in Lawrence Blume and Steven Durlauf (eds.), *Complex Nonlinear Systems III* (under submission).

Boyd, Robert and Peter J. Richerson: 1985, *Culture and the Evolutionary Process*, Chicago, University of Chicago Press.

Boyer, Pascal: 2001, *Religion Explained: The Human Instincts That Fashion Gods, Spirits and Ancestors*, London, William Heinemann.

Brown, Donald E.: 1991, *Human Universals*, New York, McGraw-Hill.

Buchanan, James M. and R. D. Tollison: 1984, *The Theory of Public Choice*, Ann Arbor, MI, University of Michigan Press.

Calvin, William H.: 1983, 'A Stone's Throw and its Launch Window: Timing Precision and its Implications for Language and Hominid Brains', *Journal of Theoretical Biology* **104**, 121–135.

Coleman, James S.: 1990, *Foundations of Social Theory*, Cambridge, MA, Belknap.

Conlisk, John: 1988, 'Optimization Cost', *Journal of Economic Behavior and Organization* **9**, 213–228.

Cosmides, Leda and John Tooby: 1992, 'The Psychological Foundations of Culture', in Jerome H. Barkow, Leda Cosmides, and John Tooby (eds.), *The Adapted Mind: Evolutionary Psychology and the Generation of Culture*, New York, Oxford University Press, pp. 19–136.

Damasio, Antonio R.: 1994, *Descartes' Error: Emotion, Reason, and the Human Brain*, New York, Avon Books.

Downs, Anthony: 1957, *An Economic Theory of Democracy*, Boston, Harper & Row.

Dugatkin, Lee Alan and Hudson Kern Reeve: 1998, *Game Theory and Animal Behavior*, Oxford, Oxford University Press.

Durkheim, Emile: 1951, *Suicide, a Study in Sociology*, New York, Free Press.

Edgerton, Robert B.: 1992, *Sick Societies: Challenging the Myth of Primitive Harmony*, New York, The Free Press.

Fehr, Ernst and Simon Gächter: 2002, 'Altruistic Punishment in Humans', *Nature* **415**, 137–140.

Fudenberg, Drew and Eric Maskin: 1986, 'The Folk Theorem in Repeated Games with Discounting or with Incomplete Information', *Econometrica* **54**(3), 533–554.

Gigerenzer, Gerd and Reinhard Selten: 2001, *Bounded Rationality*, Cambridge, MA, MIT Press.

Gintis, Herbert: 1975, 'Welfare Economics and Individual Development: A Reply to Talcott Parsons', *Quarterly Journal of Economics* **89**(2), 291–302.

Gintis, Herbert: 2000, *Game Theory Evolving*, Princeton, NJ, Princeton University Press.

Gintis, Herbert: 2003, 'The Hitchhiker's Guide to Altruism: Genes, Culture, and the Internalization of Norms', *Journal of Theoretical Biology* **220**(4), 407–418.

Gintis, Herbert: 2003, 'Solving the Puzzle of Human Prosociality', *Rationality and Society* **15**(2).

Gintis, Herbert, Eric Alden Smith, and Samuel Bowles: 2001, 'Costly Signaling and Cooperation', *Journal of Theoretical Biology* **213**, 103–119.

Gintis, Herbert, Samuel Bowles, Richard Boyd and Ernst Fehr: 2004, *The Moral Sentiments: Modeling the Roots of Cooperative Exchange*, The MIT Press.

Gould, Stephen J. and Elizabeth Vrba: 1981, 'Exaption: A Missing Term in the Science of Form', *Paleobiology* **8**, 4–15.

Grusec, Joan E. and Leon Kuczynski: 1997, *Parenting and Children's Internalization of Values: A Handbook of Contemporary Theory*, New York, John Wily & Sons.

Hamilton, W. D.: 1964, 'The Genetical Evolution of Social Behavior', *Journal of Theoretical Biology* **37**, 1–16, 17–52.

Hamilton, William D.: 1967, 'Extraordinary Sex Ratios', *Science* **156**, 477–488.

Hechter, Michael and Satoshi Kanazawa: 1997, 'Sociological Rational Choice', *Annual Review of Sociology* **23**, 199–214.

Henrich, Joe, Robert Boyd, Samuel Bowles, Colin Camerer, Ernst Fehr and Herbert Gintis: 2004, *Foundations of Human Sociality*, Oxford, Oxford University Press.

Henrich, Joseph, Robert Boyd, Samuel Bowles, Colin Camerer, Ernst Fehr, Herbert Gintis, and Richard McElreath: 2001, 'Cooperation, Reciprocity and Punishment in Fifteen Small-scale Societies', *American Economic Review* **91**, 73–78.

Jablonka, Eva and Marion J. Lamb: 1995, *Epigenetic Inheritance and Evolution: The Lamarckian Case*, Oxford, Oxford University Press.

Jablonka, Eva and Marion J. Lamb: 1998, 'Epigenetic Interitance in Evolution', *Journal of Evolutionary Biology* **11**, 159–183.

Kiyonari, Toko, Shigehito Tanida, and Toshio Yamagishi: 2000, 'Social Exchange and Reciprocity: Confusion or a Heuristic?', *Evolution and Human Behavior* **21**, 411–427.

Kollock, Peter: 1997, 'Transforming Social Dilemmas: Group Identity and Cooperation', in Peter Danielson (ed.), *Modeling Rational and Moral Agents*, Oxford, Oxford University Press.

Kreps, David M.: 1990, *A Course in Microeconomic Theory*, Princeton, NJ, Princeton University Press.

Lewontin, Richard C.: 1961, 'Evolution and the Theory of Games', *Journal of Theoretical Biology* **1**, 382–403.

Lewontin, Richard C.: 1974, *The Genetic Basis of Evolutionary Change*, New York, Columbia University Press.

Mandeville, Bernard: 1729, *The Fable of the Bees: Private Vices, Publick Benefits*.

Maynard Smith, John and G. R. Price: 1973, 'The Logic of Animal Conflict', *Nature* **246**, 15–18.

Mead, Margaret: 1963, *Sex and Temperament in Three Primitive Societies*, New York, Morrow.

Mealey, Linda: 1995, 'The Sociobiology of Sociopathy', *Behavioral and Brain Sciences* **18**, 523–541.

Monroe, Kristen Renwick: 1991, *The Economic Approach to Politics*, Reading, MA: Addison Wesley.

Moore, Jr., Barrington: 1978, *Injustice: The Social Bases of Obedience and Revolt*, White Plains, M. E. Sharpe.

Olson, Mancur: 1965, *The Logic of Collective Action: Public Goods and the Theory of Groups*, Cambridge, MA, Harvard University Press.

Parsons, Talcott: 1967, *Sociological Theory and Modern Society*, New York, Free Press.

Pinker, Steven: 2002, *The Blank Slate: The Modern Denial of Human Nature*, New York, Viking.

Rabin, Matthew: 1993, 'Incorporating Fairness into Game Theory and Economics', *American Economic Review* **83**(5), 1281–1302.

Richerson, Peter J. and Robert Boyd: 1998, 'The Evolution of Ultrasociality', in I. Eibl-Eibesfeldt and F. K. Salter (eds.), *Indoctrinability, Idology and Warfare*, New York, Berghahn Books, pp. 71–96.

Schrödinger, Edwin: 1958, *What is Life?: The Physical Aspect of the Living Cell*, Cambridge, Cambridge University Press.

Simon, Herbert: 1972, 'Theories of Bounded Rationality', in C. B. McGuire and Roy Radner (eds.), *Decision and Organization*, New York, American Elsevier, pp. 161–176.
Smith, Adam: 2000[1759], *The Theory of Moral Sentiments*, New York, Prometheus.
Taylor, P. and L. Jonker: 1978, 'Evolutionarily Stable Strategies and Game Dynamics', *Mathematical Biosciences* **40**, 145–156.
Tomasello, Michael: 1999, *The Cultural Origins of Human Cognition*, Cambridge, MA, Harvard University Press.
Trivers, R. L.: 1971, 'The Evolution of Reciprocal Altruism', *Quarterly Review of Biology* **46**, 35–57.
Tversky, Amos and Daniel Kahneman: 1981, 'Loss Aversion in Riskless Choice: A Reference-Dependent Model', *Quarterly Journal of Economics* **106**(4), 1039–1061.
Winter, Sidney G.: 1971, 'Satisficing, Selection and the Innovating Remnant', *Quarterly Journal of Economics* **85**, 237–261.
Wood, Elisabeth Jean: 2003, *Insurgent Collective Action and Civil War in El Salvador*, Cambridge, Cambridge University Press.
Wrong, Dennis H.: 1961, 'The Oversocialized Conception of Man in Modern Sociology', *American Sociological Review* **26**, 183–193.
Young, H. Peyton: 1998, *Individual Strategy and Social Structure: An Evolutionary Theory of Institutions*, Princeton, NJ, Princeton University Press.

SOME COLOURED REMARKS ON THE FOUNDATIONS OF MATHEMATICS IN THE 20TH CENTURY

GERHARD HEINZMANN
Department of Philosophy, University of Nancy 2, Laboratoire de Philosophie et d'Histoire des Sciences – Archives Henri Poincaré, UMR 7117 du CNRS, 23, Bd. Albert-Ier, F-54015 Nancy Cedex, France, E-mail: Gerhard.Heinzmann@univ-nancy2.fr

Abstract. According to the mainstream in the 20th century, the foundations of mathematics were identified with logic and set theory. Indeed, results concerning philosophically most interesting questions are often negative: the first order axiomatic set-theoretical universe is deductively incomplete, inevitably non-standard, and we have no clear idea of what the intended models of set theory are (part I). So, the foundational view of mathematics itself might be suspect. But in the spirit of Poincaré, one should look for an other solution. He remarks that the varieties of classical first order theories is unable to deal with the most common modes of mathematical reasoning such as complete induction and model building. For such a purpose, Hintikka's IF-Logic seems to be an adequate way-out.

1. Towards a Foundational Echec

1.1.

The best known studies on the foundations of mathematics in the end of the 19th century were less written by philosophers as by philosophically minded mathematicians: the titles I have in mind concern the foundations of arithmetic and geometry, and were written by Frege (Grundlagen der Arithmetik 1884), Dedekind (Was sind und was sollen die Zahlen? 1888) and Peano (*Arithmetices principia*, 1888; *I principii di geometria*, 1889), by Poincaré (On the Foundations of Geometry, 1898) and by Hilbert (Grundlagen der Geometrie, 1899). The genesis of the questions underlying these studies is well known. The studies themselves are already partially concerned with what I call the principal tools of foundational studies in the 20th century, i.e., unification and presentation concerns of the whole field (such as generality, minimum number of primitive terms and axioms, clear language), epistemological concerns (consistency, existence, relation between mathematical and physical objects, what is a proof?) and historical aspects (How does mathematical knowledge grow? How are methods of proofs modified?). Historical aspects are present in Frege's booklet, but absent in the leading movement of foundational studies in the first half of the century: I allude naturally to Logical Empiricism. Now, if the concept underlying the historical interest is not reduced to a strictly expository accumulation of the details of past science (cf. Shea 1983, 3),

S. Rahman et al. (eds.), Logic, Epistemology, and the Unity of Science, pp. 41–50.

and if, nevertheless, the history is written from some point of view and with some unifying theme without being reduced to a simple confirmation theory (cf. Grattan-Guinness 1996), then historical studies fall under the domain of foundations. I think we all can agree with the observation that the equilibrium of historical and philosophical aspects and the addition of a sociological component separates more recent views about the foundations of mathematics[1] from the studies in the first half of this century (cf. Echeverria et al. 1992, XI).

More generally, the mentioned foundational concerns relate – like mathematical sciences themselves – to questions of content or of method. When we examine content we find *natural numbers*, *special sets*, *relations*, *operations*; in the case of method, basic elements are *definition*, *proof* and *construction* (cf. Henkin 1967, 116). In each case, the analysis might proceed either along unificational, historical or epistemological concerns. The first one emphasises more mathematical, the last one more philosophical interests and the historical concerns is of a mixed form. The line is, of course, difficult to draw on all sides, but the attempt may be reasonable in order to keep a clear idea.

It is not surprising, but perhaps not worthless to mention, that the birth of symbolic logic and set theory together with the new elucidation role of the axiomatic method is accompanied by an unificational aspect which is already manifest by the choice of titles: *Foundations of Mathematics* replaces since Peano's *Formulaire de mathématique* (1895ff) the *Foundations of Geometry and Arithmetic* and remains current during the whole century. Beginning with Russell's *The Principles of Mathematics* (1903) and ending with Hintikka's *The Principles of Mathematics Revisited* (1996) one finds a great number of prominent essays entitled *Foundations, Philosophy, Principles or Elements of Mathematics*.[2]

Let's not exaggerate, but the large majority of these studies by philosophers and mathematicians concern Set theory and Logic. Non-historical books on the foundations including other fields than set theory or Logic and perhaps elementary arithmetic or the birth of non-Euclidean geometry are rather seldom[3] and have often the form of readings.[4] Nevertheless, in the last thirty years we find a crowing number of conference-proceedings often structured around a special topic, concerned with a larger part of mathematics and mixed in character, i.e., systematic and historic.[5] Naturally, there are a lot of textbooks in special mathematical domains containing a chapter on foundations and there are a large number of historically or systematically oriented studies and readings in special mathematical domains.

1.2.

But let us return to Set Theory and Logic. There are at least three different approaches in these fields. One can lead the research

(a) with a foundation or justification interest
(b) with a unification and clarification interest
(c) in a technical view.

From a classical point of view, the two first approaches were pursued by both, philosophers and mathematicians, until about 1930 to respect the question of completeness, until about 1933–1935 to respect the vision of a single universal language, until about 1948 to respect the question of existence, until 1963 to respect the question of axiomatization. Concerning consistency, there are two dates, one, 1931, for philosophers, the other, 1940, for mathematicians. Concerning the works of Church and Curry, the original goal was first to encompass with their calculi all of mathematics but, after the discovery of the Kleene-Rosser paradoxes, the research on the lambda calculus was restricted on the notion of computability that gives the technical foundation of computer science and concerns mostly interest (c).

Apart from the questions of semantic completeness, universal language and computability, all other questions have their roots in the nineteenth-century discussion about monsters and the formalisation-tendency, itself motivated by the critical assessment of classical rationalism pretending the possibility of a clear and distinct apperception of concepts by rational evidence (cf. Vuillemin 1979).

Studies on consistency are directly motivated by Hilbert's Program to eliminate intuition from mathematics and the logical antinomies discovered by Zermelo and Russell (the "paradoxes" of Cantor and Burali-Forti were originally perhaps only interpreted as a step in a *reductio ad absurdum* procedure[6]). Now, the consistency of a theory is made philosophically credible only to the extent that the principles used in the consistency proof are trustworthy. So, after Gödel's 1931 proof, everybody, except philosophers and constructivists, have deserted the epistemological *reductive proof theory* in favour of what Prawitz calls *general proof theory* analysing our understanding of proof-complexity, derivability in a given calculus and relative consistency (cf. Prawitz 1974). Most prominent results are here Gentzen's 1936 consistency-proof of elementary arithmetic and, above all, Gödel's 1940 proof of the consistency of the axiom of choice and of the continuum hypothesis with the axioms of set theory. Herewith we have reached the technical view. In this approach, modern formal logic has turned into a mathematical field and has essentially cut the bounds to its original task of serving as a means to mathematical inference and also lost its relevance for investigations into the foundations of mathematics. – I will later return to the relation of logic and mathematics.

Concerning the completeness of the first-order calculus, proven in 1930 by Gödel, it compensates, so to speak, the undecidability of the syntactical deduction concept, proven only in 1936 by Church. Now all logical truth is deductible. But is this a result providing more clarification? I think, Dummett is right in emphasising (cf. Dummett 1973, 204), that Gödel's result which presupposes set theoretical means in its proof, should not be called "semantic" but "algebraic" completeness in order to indicate its defect of foundational force but its technical usefulness. Contrary to its first order analogue, the pure *second* order calculus is "algebraically" incomplete and cannot be made complete by the addition of any finite number of axioms (cf. Henkin 1967, 124).

The Logical Empiricists' vision of a single universal language was negatively resolved by Tarski's theorem (1933–1935) of the indefinability of truth (given a language rich enough, truth for this language is inexpressible in it), or more generally, by the ineffability of semantic concepts such as truth, designation, satisfaction etc.

Concerning the axiomatisation of set theory, there were proposed various manner to axiomatize the original idea of Cantor: the systems of ZF, of von Neumann, Bernays and Gödel (NBC) or of Quine (NF) are not equivalent but all seem adequate for providing a basis for the traditional mathematical theories of analysis, algebra and geometry (cf. Henkin, 127–128). Nevertheless, it is well known that the axiomatisation of set theory in a first order language is not only characterised by the not intended existence of non-standard models (Gödel 1931) and by the general fact that deductive first order theories cannot provide an adequate description of mathematical structure (cf. Beth 1965, 643) – a consequence of the Löwenheim-Skolem-Tarski theorem (1915–1920) – but the axiomatisation of set theory has even another defect: its originates from Cohen's 1963 independence proof of the axiom of choice (ZF+¬AC is consistent). Clearly, NF seems not be concerned because AC is not compatible with it. Nevertheless, chapter 10, entitled *Mathematics without choice*, of Thomas Jech's well known book *The axiom of Choice* (Jech 1973) illustrates very well the consequences for, say, an algebra without choice: there exists a vector space which has no basis, a field which has no algebraic closure, a free group whose subgroup is not free etc. On the other hand, the addition of AC to ZF leads to paradoxical consequences, too: one is, for example, Hausdorff's discovery that a half of a surface is congruent to a third of it (1914), another is the Banach-Tarski paradox that a sphere with a given radius may be decomposed in such a way that, after rotations and translations, one can obtain two spheres of the same radius (1924). So, Cohen's work created a to-day undecided question: should one accept the existence of two incompatible set theories, each one being useful concerning certain interests (although this view leaves the success unexplained), should one find new axioms, which eliminate one of the incompatible theories (cf. Mostowski 1966, 149), should one search after another framework for mathematics (e.g., category theory) or, finally, should one argue that the difficulties are inherent in the very nature of mathematics (cf. Cohen 1966, 1)?

Concerning existence, Quine's celebrated ontological criterion *To be is to be the value of a variable* (Quine 1953, 15), installing classical first order logic as ontological measure,[7] specifies technically the philosophical open-end discussion about Platonism (Logicism), Intuitionism and Formalism, standardised 1931 by Carnap's, Heyting's and von Neumann's articles published in *Erkenntnis*: naturally, Quine's criterion doesn't resolve the philosophical question what there is but gives "to know what a given remark or doctrine [...] *says* there is" (ibid.). "Platonism condones the use of bound variables to refer to abstract entities [...] specifiable and unspecifiable, indiscriminately. [...] Intuitionism [...] countenances the use of bound variables to refer to abstract entities only when those entities are capable of being cooked up individually from ingredients specified in advance". Nominalism "object to admitting abstract entities at all. [... According to this view], an adequate basis for agreement among mathematicians can be found simply in the rules which govern the manipulation of the notations" (ibid. 14). Significant are not the notations themselves, but only the syntactical rules. The philosophical literature in this domain is immense.

Whereas the axiomatic method was in the both discussed perspectives (a) and (b) used for the purpose of elucidating the foundations on which mathematicians build (Hilbert's position), it has become, according to the technical view (c), a tool for concrete mathematical research; while formerly Axiomatics was concerned with axioms which determine the structure of the system, axiomatic systems are now the common basis for the investigation of individual entities arising by specified constructions (cf. Weyl 1985, 13), such as the study of definable sets of real numbers (descriptive set theory), and differentiation or by the variety of models of a given system. The first orientation stand close to the inheritance of Cantor's and Hausdorff's studies on ordinal and cardinal numbers on the basis of set theoretical axioms supplemented by various extensions, the second line is the birth of model theory. Both domains are themselves considered as part of mathematics, and must also have their proper foundations although they may have consequences for the foundations: the theorem of Löwenheim-Skolem is such an example.

In the 19th century, the relation of mathematics to general formal logic was impressed on by "Boole's fundamental idea that the language of mathematics is the most perfect form of the universal language of thought, and that general logic, therefore, is mathematics with all conceptions of quantity struck out" (Bryant 1902). In the 20th century, Quine's idea is pre-eminent: a constant is called logic iff it is invariant in all uses. The language of logic contains only brackets, logical constants and schema-letters for propositions and relations, the language of mathematics in addition at least one special binary predicate, the $\in$-relation. Hence, if one would qualify today logic (*in a strong sense*) as mathematics, it should not have the same significance as in Boole. It would signify that mathematics is an auxiliary science in order to formulate or to prove – with the induction principle or structures like group theory – logical facts. One sees that in this interpretation logic could not have a foundational function in an epistemological sense.

It is also not surprising, that there exist works, many works, about the foundations of Logic and Set Theory, written either in a systematic or historical perspective. Here we are sent back to parts (a) and (b). Nevertheless, we have already seen that technically motivated investigations can have interesting consequences for foundational questions.

On the other hand, it is right that mathematical Logic (in a large sense) as mathematical discipline whose branches set theory, recursion theory, proof theory and model theory are described in Barwise's *Handbook of Mathematical Logic* (1977), has no properly intrinsic interest for the foundations of mathematics.

Mathematical Logic is a very marginal mathematical discipline – Dieudonné used to prove this fact by brandishing the *Zentralblatt as* witness for the small percentage of logical studies in respect to the totality of mathematical publications. Indeed, although the historical question at issue here is not the foundational one, the history of Mathematical Logic is probably the best documented of 20th century's mathematics. From Lewis' *A Survey of Symbolic Logic* (1918) to now there is a great number of competent surveys and histories.[8]

In recapitulating now the mainstream of Set Theory and Logic, one can see that there is a tendency to distinguish most important *positive* results which are overall interesting from a technical point of view from most important *negative* results which are even of first importance from a philosophical point of view. More precisely, the situation amounts to the following table:

Some classical results which are most important	
From a technical point of view	Even from a philosophical point of view
– 1930: completeness (Gödel)	– 1915–1920: Löwenheim-Skolem theorem
– 1936: consistency of elementary arithmetic (Gentzen)	– 1931: consistency (negative) (Gödel)
– 1936: Church's Thesis	– 1933: undefinability of truth (Tarski)
– 1940: consistency of ZF+AC+CH (Gödel)	– 1936: undecidability (Church)
	– 1939: existence-question (Quine)
	– 1963: independence of the axiom of choice (Cohen)

Surely, technical results, as for example the consistency proof for elementary arithmetic, can by their of certain technical tools again generate most interesting philosophical questions, in the given example the differentiation of the infinite domain. A corresponding transformation can even observed for philosophical results: first negatively considered, the existence of non-standard models can even work as an aspect for technical problems, in the given example the development of the theory of category.[9] But these are, so to speak, not intended "second" order results.

So, in reading this table, which is of course only selective, one can perhaps not suppress the following question: is the translatability in the language of set theory and logic really the exclusive form of justification and rigour in mathematics? In fact, the first order axiomatic set-theoretical universe is deductively incomplete, inevitably non-standard, and that we have no unproblematically clear idea of what the intended models of set theory are (cf. Hintikka 1996, 166ff). In other terms, we cannot have confidence in a formal theory without having confidence in our intuition of a conceptual content.

One possible answer could be: "Mathematics without foundations", and it could be evidenced by the fact that the existence of formally undecidable propositions (within a given arithmetical system) or of problems unsettled by standard axioms (within set theory) does not obstruct the development of a viable and, in fact, powerful science. Accordingly, the foundational view of mathematics itself might then be suspect. Mathematics can to be understood from mathematical praxis alone. Indeed, there is even another possibility.

2. The Poïetic Revision

Since Poincaré there have always been some outsiders who rejected the standard view about the foundations of mathematics. Formulated in modern terms, Poincaré held that the varieties of formal logical theories – which he thought to be considerably attached to set operations – don't express the structure which is essential for a genuine understanding of mathematical proof (cf. Poincaré 1908, 149). Surely, his critic cannot mean that we should fully give up formal languages as the corresponding well justified critic of rational evidence did not mean to give up all evidence in favour of logical reasoning. What happens is that the new formal categories of deductive mathematical reasoning are no longer specific enough for the idea of a comprehensive formalised language. For this reason, one should perhaps make a step from a first "intuitive" epistemological level to a "theoretical" epistemological level. On this latter, one should proceed by a revision of the criterion of deductive rigour. This revision affords a way of pragmatically circumscribing the correlation of the concepts involved. Such consequences seem to be drawn from the core of Poincaré's critic on logic. Voices in this sense were more and more audible at the end of the century. Poincaré was fighting with the logicists in a large sense, Gonseth with the logical empiricists, Beth with Church, Steiner argued in its *Mathematical Knowledge* (1975) for the importance of non deductive arguments in mathematical justification (cf. Aspray and Kitcher 1988, 18) and more generally, philosophers of mathematics argued against mathematical logicians (cf. Detlefsen's collection *Proof, Logic and Formalization*, 1992).

But nevertheless, one will reduce Arithmetic and Analysis to set theory! Why? According to Hintikka, one relevant reason may be the important limitation theorem of Tarski (1933): the theoretical truth predicate of a first-order language containing elementary arithmetic is not definable in that language itself. Now, we would like to formalise truth: all model theory depends on truth definitions. As long as these definitions can only be given on second order level or in set theory, then model theory depends on second order logic or set theory (Hintikka 1996, VIII).

It is even very well known that first order logic without individual variables including higher-order entities cannot deal with the most characteristic modes of concept and inference in mathematics such us complete induction, well-ordering or power-set formation. So, mathematical induction is not generated but only represented by indefinite repetition of different levels (cf. Heinzmann 1987, 72), that means that the separation between object and symbol is not yet accomplished: it postulates a survey of a potentially reiterated stroke-concatenation or something analogous and a survey of a potentially reiterated *modus ponens*. An analogues principle consist in the fact that mathematical symbols can be read with respect to different contents without the possibility to check this ambiguity on the level of the given notation. Herewith we have a typical case of partial information. Poincaré suggests that this difficult situation including semantic ambiguity should be overcome by the introduction of aesthetic feeling in mathematics. The mastering of simultaneous reasoning about different contents, provoked by the lack of perfect information in one field, requires the acquisition of a practice and can be made only explicit by model theoretical means. What are requested are new deductive methods corresponding to factual mathematical reasoning. Is there any formal way-out?

In one sense, yes: if the "correspondence" is interpreted large enough. The solution is suggested in the last fight of the XXth century: Hintikka IF-Logic against Frege's first order logic. By formulating the logic of Henkin or branching quantifiers in a first order notation,[10] Hintikka proved that the game theoretical truth predicate of a so called Independence-friendly first-order language containing elementary arithmetic is definable in that language itself in the sense "that there is a complex predicate (of the Gödel numbers of the sentences of the given first-order language) which applies to a number if and only if it is the Gödel number of a true sentence" (cf. Hintikka 1996, 118f).

We just have to be carefully with respect to the informational independence of quantifiers in their role of lying numbers as numbers and numbers as Gödel codifications. IF-Logic is a conservative extension of ordinary first-order logic, but it does not admit of a complete axiomatization (cf. Hintikka 1996, 65, 88): "There is no finite (or recursive) set of axioms from which all valid sentences of this logic can be derived as theorems by means of completely formal (recursive) rules of inference". So, perhaps the efforts made during all the century along the axiomatisation-line was from the justificational point of view doomed to failure as a foundational enterprise.

Notes

[1] Livingston's *Ethnomethodological Foundations of Mathematics* (Livingston 1986) goes in this sense to extremes.

[2] I recall the works of Brouwer (*Over de Grondslagen der Wiskunde*, 1907), Winter* (*La méthode dans la philosophie des mathématiques*, 1911), Hölder (*Die mathematische Methode*, 1924), Ramsey (*The Foundations of Mathematics*, 1925), Gonseth* (*Les Fondements des mathematiques. De la géometrie d'Euclide à la rélativité générale et à l'intuitionnisme*, 1926), Weyl* (*Philosophie der Mathematik und Naturwissenschaft*, 1928), Dubislav (*Die Philosophie der Matheamtik in der Gegenwart*, 1932), Hilbert and Bernays (*Grundlagen der Mathematik*, 1934), Heyting (*Mathematische Grundlagenforschung, Intuitionismus, Beweistheorie*, 1934), Wittgenstein (*Remarks on the Foundations of Mathematics*, 1937–1944), Gentzen (*Die gegenwärtige Lage der mathematischen Grundlagenforschung*, 1938), Carnap (*Foundations of Logic and Mathematics*, 1939), Curry (*Outline of a Formalist Philosophy of Mathematics*, 1951), Becker (*Grundlagen der Mathematik*, 1954), Körner (*The Philosophy of Mathematics*, 1960), Cavaillès (*Philosophie Mathématique*, 1962 (1938)), Beth (*The Foundations of Mathematics*, 1965), Bar-Hillel et al. (eds) (*Essays on the Foundations of Mathematics*, 1962), Kneebone (*Mathematical Logic and the Foundations of Mathematics*, 1963), Goodstein (*Essays in the Philosophy of Mathematics*, 1965), Putnam (*Mathematics without Foundations*, 1967), Bernays (*Abhandlungen zur Philosophie der Mathematik*, 1976), Dummett (*Elements of Intuitionism*, 1977), Field* (*Science without numbers*, 1980), Kitcher* (The Nature of Mathematical Knowledge, 1982), Parsons (*Mathematics in Philosophy*, 1983), Tymoczko* (ed., *New Directions*

in the Philosophy of Mathematics, 1985) and Hersh* (ed., *New Directions in the Philosophy of Mathematics*, Synthese, 1991).

[3] In note 2 above, they are indicated by *.

[4] So, in French we have, for example, Le Lionnais (*Les Grands courants de la pensée mathématique*, 1948) and collections of writings from Borel and Fréchet (Les mathématiques et le concret, 1955), in German we can mention *Mathematiker über die Mathematik* (edited in 1974 by Otte).

[5] I mention the collection of readings edited 1988 by Aspray and Kitcher (*History and Philosophy of Modern Mathematics*), *The Space of Mathematics*, edited in 1992 by Echeverria and the *Revolutions in Mathematics*, edited in the same year by Gillies. I should also add here Merten's monography *Moderne Sprache Mathematik* (1990).

[6] This is the thesis of (Garcia Diego 1992).

[7] For a discussion cf. (Heinzmann 2002).

[8] For example Cavaillès' history of set theory, Kneebone's (period 1899–1931), Mostowski's (period 1930–1964), Hermes' and Wang's surveys of symbolic logic ((Cavaillès 1938), (Kneebone 1963), Mostowski (1966), (Hermes 1956), (Wang 1964)). There are also the Ω-*Bibliography*, edited by Gert-Heinz Müller, and the bibliometrical study of (Wagner and Döbler 1993).

[9] This will be analysed in the doctoral thesis of Ralf Krömer (Nancy, May 2004).

[10] I want to thank John Burgess for a comment on this point.

References

Aspray, W. and P. Kitcher (eds.): 1988, *History and Philosophy of Modern Mathematics*, Minneapolis, Minnesota Press.

Beth, Evert W.: 1965, *The Foundations of Mathematics ([1] 1959)*, Amsterdam, North-Holland.

Bryant, Sophie: 1902, 'The Relation of Mathematics to General Formal Logic', *Proceedings of the Aristotelian Society* **2**, 105–134, cf. *The Journal of Symbolic Logic* **1**(4), 139.

Cavaillès, Jean: 1938, *Remarques sur la formation de la théorie abstraite des ensembles. Etude historique et critique*, Paris, Hermann.

Cohen, Paul: 1966, *Set Theory and the Continuum Hypothesis*, New York, Amsterdam, Benjamin.

Dummett, Michael A. E.: 1973, 'The Justification of Deduction', *Proceedings of the British Academy* **59**, 201–232.

Echeverria, Javier et al. (ed.): 1992, *The Space of Mathematics. Philosophical Epistemological and Historical Explorations*, Berlin, New York, De Gruyter.

Garcia Diego, Alejandro R.: 1992, *Bertrand Russell and the Origins of Set-theoretic 'Paradoxes'*, Basel, Boston, Berlin, Birkhäuser.

Grattan-Guinness, Ivor: 1996, 'Normal Mathematics and its Histo(iograph)y: The Tenacity of Algebraic Styles', in E. Ausejo and M. Hormigon (eds.), *Paradigms and Mathematics*, Madrid, Siglo XXT de Espana Editores, pp. 203–213.

Heinzmann, Gerhard: 1987, 'Philosophical Pragmatism in Poincaré', in J. Srzednicki (ed.), *Reason and Argument, Initiatives in Logic*, Dordrecht, Boston, Lancaster, Nijhoff, pp. 70–80.

Heinzmann, Gerhard: 2002, 'Les dogmes rationaliste et empiriste face à leur révision poiétique en philosophie des mathématiques', in E. Schwartz (ed.), *Actes du Colloque Jules Vuillemin*, Hildesheim, Olms (forthcoming).

Henkin, Leon: 1967, 'The Foundations of Mathematics', in R. Klibansky (ed.), *Philosophy in the Mid Century*, Firenze, La Nuova Italia Editrice, pp. 116–129.

Hermes, Hans: 1956, 'Über die gegenwärtige Lage der mathematischen Logik und Grundlagenforschung', *Jahresbericht der Deutschen Mathematiker Vereinigung* **59**, 49–69.

Hintikka, Jaakko: 1996, *The Principle of Mathematics Revisited*, Cambridge, Cambridge University Press.

Jech, Thomas: 1973, *The Axiom of Choice*, Amsterdam, London, North-Holland.

Kneebone, G. T.: 1963, *Mathematical Logic and the Foundations of Mathematics. An Introductory Survey*, London, Van Nostrand.

Livingston, Eric: 1986, *The Ethnomethodological Foundations of Mathematics*, London, Boston, Henly, Routledge.

Mostowski, Andrzej: 1966, *Thirty Years of Foundational Studies*, Oxford, Basil Blackwell.

Müller, Gert-Heinz and Lenski, Wolfgang: 1987, *Ω-Bibliography of Mathematical Logic*, Vols I–VI, Berlin, Heidelberg, Springer.

Poincaré, Henri: 1908, *Science et méthode*, Paris, Flammarion.

Prawitz, Dag: 1974, 'On the Idea of a General Proof Theory', *Synthese* **27**, 63–77.

Quine, Willard Van Orman: 1953, 'On What There Is', in Quine (ed.), *From a Logical Point of View*, Cambridge MA, London, Harvard University Press, pp. 1–19.

Shea, William, R.: 1983, 'Do Historians and Philosophers of Science Share the Same Heritage?', in W. Shea (ed.), *Nature Mathematized*, Dordrecht, London, Reidel.

Vuillemin, Jules: 1979, 'La raison au regard de l'instauration et du développement scientifiques', in Th. Geraets (ed.), *La rationalité aujourd'hui*, Editions de l'Université, Ottawa, pp. 67–84.

Wagner, Roland and Döbler, Jan Berg: 1993, *Mathematische Logik von 1847 bis zur Gegenwart*, Berlin, New York, De Gruyter.

Wang, Hao: 1964, *A Survey of Mathematical Logic*, Peking, Amsterdam, Science Press, North-Holland.

Weyl, Hermann: 1985, 'Axiomatic versus Constructive Procedures in Mathematics', *The Mathematical Intelligenzer* **7**, 10–17, 38.

LOGICAL VS. NONLOGICAL CONCEPTS: AN UNTENABLE DUALISM?

JAAKKO HINTIKKA
Department of Philosophy, Boston University

One of the greatest disasters that befell twentieth-century analytic philosophy was Quine's (1953) rejection of the distinction between analytic and synthetic truths as an "untenable dualism", to use Morton White's (1950) phrase. Or rather, the disaster was the widespread acceptance of this untenable rejection. It deprived philosophers of the means of mastering the defining concept of our era, the notion of information. It made it apparently pointless for them to use axiomatization or other kind of logical systematization as a tool of serious philosophical analysis, and hence encouraged the currently popular no-brainer appeals to "intuitions". (cf. Hintikka 1999) These unfortunate consequences follow because on Quine's view a deduction of a theorem from axioms (and more generally a deduction of a consequence from premises) can introduce what for us is new factual information, for according to Quine such information cannot be separated from what for us is purely linguistic information. Hence the cognitive content (factual information) of a theory cannot be summed up in its axioms, for new assumptions can be introduced by the logical and mathematical methods used in the derivation of the theorems. This was precisely the kind of conundrum that the great David Hilbert sought to eliminate by means of his axiomatic approach, as shown by the sixth problem in his famous list of open problems. (See here e.g., Cory 1997, 1998; Majer 2001; and Yandell 2002)

In this paper, I will discuss one of the assumptions on which Quine's argument for rejecting the analytic-synthetic distinction rests. As I have pointed out before (Hintikka forthcoming (c)) the terms "analytic" and "synthetic" are most unfortunate from a historical point of view. What is meant is in fact a distinction between conceptual and factual information. Quine is right in effect pointing out that one cannot tell from a person's behavior whether the information he is relying on is factual or conceptual. But this interwovenness of factual and conceptual information can be much better explained by distinguishing from each other two different kinds of logical truths and accordingly two different kinds of information – much better than by abolishing the borderline between logical (conceptual) truths and factual truths, as Quine wants to do. (See here Hintikka, forthcoming (c)) Accordingly, Quine's accurate insight into (in effect) the behavioral indistinguishability of the two kinds of information does not mean that one cannot define the distinction by some other means. Indeed, it can be shown beyond any reasonable doubt that the

S. Rahman et al. (eds.), Logic, Epistemology, and the Unity of Science, pp. 51–56.

usual logical truths of first-order logic are uninformative aka "tautological" in one basic sense of information.

This sense means exhausting all the possibilities that are expressible in the language in question. Quine (1970, pp. 98–99) denies this because he thinks that the only relevant possibilities are all the different states of the physical universe. He overlooks the incontrovertible fact that distinctions between the relevant possibilities are built into the very language of first-order theories. (See here Hintikka, forthcoming (c)) These possibilities are expressed by the consistent constituents of a first-order language. Then a logical truth is one which admits all the relevant possibilities, which means that it is uninformative in a striking sense. The only further qualification needed here is that the elimination of inconsistent constituents yields information in a different sense. However, this kind of information is in a perfectly obvious sense conceptual.

Quine seems to be on a firmer ground in discussing prima facie conceptual ("analytic") truths which are not logical. In order to maintain a strict distinction, one has to maintain that such truths are also factually uninformative and this seems to be a much taller order than to argue that logical truths are factually uninformative.

What is involved here? Logical truths depend only on the way in which logical constants occur in them. (Philosophers frequently try to reify this "way in which" into the notion of logical form.) The allegedly analytical but nonlogical truths depend also on the way nonlogical constants occur in them. Quine's way of thinking, and that of many other contemporary philosophers, is hence predicated on a distinction between logical and nonlogical constants. This distinction plays an important role in many arguments of philosophers, for instance in John Etchemendy's (1990) criticism of the notion of logical consequence. This distinction is the subject of the present paper.

The first observation that can be made here is that nonlogical analytical truths sometimes turn out to be logical ones when their structure is analyzed properly. In his *Tractatus*, Wittgenstein apparently assumed that this can be done for all conceptual truths. Within his truth-functional logic this amounted to the independence of atomic (elementary) propositions of each other. He ran into difficulties, however, in connection with color concepts, and was ultimately led to change his entire philosophy because of such difficulties. Wittgenstein despaired too soon, however, for at least in his paradigm case, the conceptual incompatibility of color terms can be turned into a logical truth simply by conceptualizing the concept of *color* as a function mapping points in a visual space into color space. (See Hintikka and Hintikka 1986, pp. 123–132.) This is an instructive example of how nonlogical but "analytic" truths can be interpreted as logical ones. Then they are uninformative ("tautological") by the same token and in the same sense as logical truths.

But can this be done in all cases? This question depends on the semantical framework used. For instance, in Wittgenstein's case he was prevented from availing himself to the analysis just mentioned because he did not accept functions as basic nonlogical constants. In this respect, the distinction between logical and nonlogical constants is put to a new light by the approach known as game-theoretical semantics (GTS). (See here and in the following Hintikka and Sandu 1996) In the game-theoretical semantics and in its outgrowth, independence-friendly (IF) first-order logic, the truth of a sentence S is defined as the existence of a winning strategy in a correlated two-person game G(S) for one of the players which is variously called the verifier, the inquirer, myself, Eloise or simply E. The other player is correspondingly called the falsifier, nature Abelard or A. Of these the least misleading terms seem to me to be *the inquirer* and *nature*. For in the intended applications a play of G(S) can be thought of as a probe on the inquirer's part as to whether S is true or not. This probe is codified in the inquirer's strategy, and nature's strategies can be thought of as the different courses that the probe can take depending on nature's behavior. If some particular probe always gives a positive result, no matter how nature behaves, the sentence S is true. This is not an unrealistic model of what happens in actual epistemological (for instance scientific) inquiry. (cf. Hintikka, forthcoming (a))

The meaning of each logical constant is defined by the game rule that applies to a logical constant when it occurs in a given sentence. In the simplest rules, this constant is the dominating one, but this is not an indispensable restriction.

A winning strategy in a game with a first-order sentence is expressed by an array of what are known as its Skolem functions. These functions serve to provide as their values the "witness individuals" that show the truth of the sentence in question. Skolem functions are functions because such witness individuals depend often on other witness individuals. A winning array of Skolem functions can be called a Skolem operator. It becomes an operator in the usual mathematical sense if sequences of variables are considered as vectors in a logical space. Hence technically the truth of S can be equivalently defined as the existence of a Skolem operator for it. A second-order sentence Sk(S) expressing this existence is called the Skolemization of S. If the underlying logic includes IF first-order logic, Sk(S) is equivalent to a first-order sentence, and its contradictory negation is equivalent to a sentence in terms of what is known as extended IF first-order logic. It is obtained from the plain IF first-order logic by admitting to it a sentence-initial contradictory negation.

The consistency (satisfiability) of S in a given domain can be expressed by a first-order sentence S* obtained by replacing all nonlogical constants by variables of the same type and then binding these variables to sentence-initial existential quantifiers. This sentence S* can again be translated back into the corresponding IF first-order language. Indeed, it can be shown that satisfiability in a given domain can be defined for such a language in the same language. As might be expected, all that the resulting sentences do is to express the cardinality of the given domain (as fully as it can be expressed in the language). By the same token, the validity of a given IF first-order sentence can be expressed by a sentence of an extended

IF language. Again, all that this resulting sentence does is to express conditions on the cardinality of the domain.

Thus not only the truth of a sentence whose logic is IF first-order logic can be defined in the same language. Its validity (analyticity) can be defined in an ever so slightly extended language. In such languages, validity (analyticity) can thus be mastered eminently well. Such languages are beginning to resemble in some ways the universal language of science which Carnap envisaged but could not construct. (Cf. here Carnap 1937 and Köhler 1991)

But how can such a treatment be extended by taking into account the conceptual behavior of a nonlogical constant? This would extend the notions of truth, satisfaction and validity (analyticity) to apply to sentences containing the new constant. An answer is obvious: by formulating a game rule that characterizes the meaning of the new constant. For the definition of truth as the existence of a winning strategy is independent of the particular game rules used and therefore independent of the selection of logical or nonlogical constants which trigger moves in accordance of the rules. In other words, from the vantage point of GTS there is no difference between logical and nonlogical constants as long as their semantics can be captured game-theoretically.

Can this be done generally? Obviously we have to restrict our attention to constants which serve the purpose of fact-stating discourse. For the semantical games must be capable of being thought of as truth-testing activities. But what can be said with this restriction in mind? I do not have any transcendental arguments to offer to such a universal possibility. However, it should not be possible to persuade a Wittgensteinean philosopher or at least make the universal applicability of GTS to fact-stating discourse plausible. Such a discourse relies on the notion of truth. Now the notion of truth, as all semantical notions, concerns the relations holding between language and the world. According to Wittgenstein's basic vision, all such relations are mediated by certain rule-governed human activities that he called language-games. (This is the sense of the frequently misunderstood slogan of meaning as use.) In the case of the notion of truth the comparison activities depend, of course, on nature's reactions. But it is not unnatural to assume that those comparison activities are such that if (and only if) they tend to the desired result independently of what nature does, the sentence whose status is at issue is true.

This is not a strict proof. However, it gains enhanced credence by noting that the GTS approach has already been extended successfully to a number of nonlogical concepts. They include epistemic and temporal concepts. These generalizations presuppose extensions of what counts as the "world" to which a language is applied. But, apart from such changes, the extension of GTS to such concepts takes place precisely along the lines sketched above.

The naturalness of this game-theoretical approach is also attested to by the fact that it gives rise to a clear-cut and interesting type of epistemology, which I have tentatively explored in an earlier paper. (Hintikka forthcoming (a)).

Along different lines, IF logic has led to the assimilation of certain interesting concepts to logical ones even though they were previously assumed to be clearly nonlogical ones. This development starts from the observation that the negation which narrowly ensues from the game rules of GTS is a strong (dual) negation that does not obey the law of excluded middle. If we then add a sentence-initial contradictory negation (strictly speaking, a contradictory negation which does not occur within the scope of quantifiers) we obtain an algebraic structure which is a Boolean algebra with an operator in Tarski's sense. (See here Hintikka, forthcoming (b)) Hence it is by Tarski's results isomorphic to a set algebra. This associates the basic logical operators to the traditional set-theoretical notions of complementation, set union and set intersection. But what about the set-theoretical or geometric counterpart of the strong negation? I have shown that it can be considered in certain special cases as a generalization of the geometrical notion of orthocomplementation. (See Hintikka 2002) This motivates considering orthogonality in the usual sense as a special case of IF negation, which is never but a generalization of this familiar notion. Thus orthogonality, which used to be considered as the geometrical rather than logical notion, can be extended so as to become a logical one, which coincides with the old geometrical one when the logical space happens to have the structure of a geometrical space. By means of the notion of orthogonality we can then define for instance the notion of dimensionality (number of dimensions) of a logical space, thus bringing still further concepts to the ambit of logical treatment. (See Hintikka forthcoming (b)) These developments open up the allegedly strict distinction between logical and nonlogical notions still further.

References

Carnap, Rudolf: 1937 (original 1934), *The Logical Syntax of Language*, London, Routledge & Kegan Paul.

Cory, Leo: 1998, 'Hilbert on Kinetic Theory and Radiation Theory (1912–1914)', *The Mathematical Intelligencer* **20**, 52–58.

Cory, Leo: 1997, 'David Hilbert and the Axiomatization of Physics (1894–1905)', *Archive of the History of Exact Sciences* **51**, 83–198.

Etchemendy, John: 1990, *The Concept of Logical Consequence*, Cambridge, Harvard University Press.

Hintikka, Jaakko: forthcoming a, 'An Epistemology for Game-theoretical Semantics'.

Hintikka, Jaakko: forthcoming b, 'What is the True Algebra of Logic?'.

Hintikka, Jaakko: forthcoming c, 'A Distinction Too Many or Too Few?.

Hintikka, Jaakko: 2002, 'Quantum Logic as a Fragment of Independence-friendly Logic', *Journal of Philosophical Logic* **31**, 197–209.

Hintikka, Jaakko: 1999, 'The Emperor's New Intuitions', *Journal of Philosophy* **96**, 127–147.

Hintikka, Jaakko: 1996, *Ludwig Wittgenstein: Half-Truths and One-and-a Half Truths*, Dordrecht, Kluwer Academic.

Hintikka, Jaakko, and Gabriel Sandu: 1996, 'Game-theoretical Semantics', in Johan van Benthem and Alice ter Meulen (eds.), *Handbook of Logic and Language*, Amsterdam, Elsevier, pp. 361–410.

Hintikka, Merrill B. and Jaakko Hintikka: 1986, *Investigating Wittgenstein*, Oxford, Basil Blackwell.
Köhler, Eckehart: 1991, 'Gödel und der Wiener Kreis', in Paul Kruntorad (ed.), *Jour fixe der Vernunft: Der Wiener Kreis und die Folgen*, Vienna, Hölder-Pichler-Tempsky, pp. 127–158.
Majer, Ulrich: 2001, 'The Axiomatic Method and the Foundations of Science', in Miklós Redei and Michael Stöltzner (eds.), *John von Neumann and the Foundations of Quantum Physics*, Dordrecht, Kluwer Academic, pp. 11–33.
Quine, W. V.: 1953 (original 1951), 'Two Dogmas of Empiricism', in *From a Logical Point of View*, Cambridge, Harvard University Press, pp. 20–46.
Quine, W. V.: 1970, *Philosophy of Logic*, Englewood Cliffs, NJ, Prentice-Hall.
White, Morton, N. J.: 1950, 'The Analytic and the Synthetic: An Untenable Dualism', in Sidney Hook (ed.), *John Dewey: Philosopher of Science and Freedom*, New York, Dial Press, pp. 316–330.
Wittgenstein, Ludwig: 192?, *Tractatus Logico-Philosophicus*, London, Routledge and Kegan Paul.
Wittgenstein, Ludwig: 1953, *Philosophical Investigations*, Oxfored, Basil Blackwell.
Yandell, Benjamin H., 2002, *The Honors Class: Hilbert's Problems and Their Solvers*, Natck, MA, A.K. Peters.

SEMANTIC GAMES IN LOGIC AND EPISTEMOLOGY

AHTI-VEIKKO PIETARINEN
Department of Philosophy, University of Helsinki, P.O. Box 9, FIN-00014, Finland, E-mail:
pietarin@cc.helsinki.fi

Abstract. The purpose of this paper is to introduce the reader to game-theoretic semantics (GTS), and to chart some of its current directions, with a focus on epistemological issues. GTS was originally developed by Jaakko Hintikka in the 1960s and became one of the main approaches in logical and linguistic semantics. The theory has been researched in numerous publications. I will put games in a wider historical and systematic perspective within the overall development of logic, and explore some of the recent advances.

1. Introduction

1.1. FOUR QUESTIONS

Four major questions are addressed here: (i) What kinds of tools and doctrines semantic games provide for the scientific study of logic and language? (ii) What is the structure of such games? (iii) What is the relation between logic, language and games? (iv) What is the relevance of semantic games to epistemology?

The following responses are proposed.

(i) GTS makes available a formal apparatus that can be put to use in logic in new ways, unifying different semantic outlooks on natural language. Its philosophical component is to be found in the analysis of lexical and logical meaning in terms of enriched game-theoretic content.

(ii) Semantic games may be viewed as a special class of extensive forms of games that show the flow of semantic information and the distribution of the strategic actions of the players during the actual playing of a game. Variations in the information structure of the players give rise to different kinds of logics, including the IF (independence-friendly) logics introduced in Hintikka and Sandu (1989) and studied further in Hintikka (1996) and Hintikka and Sandu (1997), for example. Briefly, IF logic is capable of expressing various informational independencies, and its formulas are correlated with games of imperfect information.

(iii) Various logical semantics may be distilled from different classes of games, which also proves to be useful for the study of language. When games are varied, different logics, other than the classical propositional, first-order or modal, are seen to emerge. This again allows us to perceive much more in the structure and semantics of natural language than is currently believed to exist.

S. Rahman et al. (eds.), Logic, Epistemology, and the Unity of Science, pp. 57–103.

(iv) IF logic and the associated semantic games bring in new logical perspectives to epistemology. This can be attained within the context of epistemic logic. By relaxing the assumption of perfect information in epistemic logic we admit that the knowing agents may not be able to establish the truths of every construction of their knowledge. Although being thus forced to make some concessions to the Skeptic, the Inquirer's process of trying to find out the truth of agent's knowledge statements remains one of the defining characteristics of semantic games: analogously to games for extensional logic (Hintikka 1973), in epistemic logic they serve as enriched mediators between different kinds of knowledge and the world by seeking and finding possible worlds.

1.2. Some Recent Literature

In order to set this paper within the context of current research, I mention here some of the more specific results that have been obtained. (a) Extensive semantic games have a subclass of extensive games of imperfect information satisfying non-repetition, consistency, the von Neumann-Morgenstern condition, and imperfect recall (Pietarinen and Sandu 1999; Sandu and Pietarinen 2001). (b) Hodges' uniformity problem arises from violations of game-theoretic consistency in the propositional IF fragment (Sandu and Pietarinen 2001). (c) A new four-place connective ('transjunction') of propositional logic of imperfect information gives rise to a functionally complete set of connectives for all partial functions together with the usual Boolean ones. In addition, compositional semantics can be given to a propositional IF fragment (Sandu and Pietarinen 2003). (e) Epistemic (multi-agent, first-order) language of informational independence captures the phenomenon of intentional identity, dispensing with pragmatic concerns (Pietarinen 2001a). Implications to knowledge in multi-agent systems are evident (Pietarinen 2002b, 2003d). (f) GTS may be defined for both monadic and polyadic generalised quantifiers, and for many other cross-categorial linguistic items (such as negative polarity items, adverbs of quantification, the morphemes *even* and *not even*, and eventualities), with consequences for linguistic theorising (Pietarinen 2003f). Furthermore, generalised quantifiers and eventualities are affected by the phenomenon of informational independence (Pietarinen 2001b).

1.3. Wider Perspectives

In order to make the broader scientific picture easier to discern, it is necessary to outline also some of the wider goals and prospects.

Since the early 1980s, theories of discourse representation, dynamic semantics (or the dynamic theory of meaning), and relational generalised quantifier theory have been the linguistically-driven approaches to semantics that have dominated the main research fields in logic and the semantics and pragmatics of natural language. These approaches have been complemented more recently by theories referring to the concept of choice functions. While all of these theories have led to many interesting insights in terms of how logic and language may work, their supremacy is unfair. GTS is probably the first dynamic system that was successfully applied to the study of logic and language. Choice functions, in turn, are simply special cases of the game-theoretic concept of a strategy. This explains what the linguistic role of such functions in the theory ought to be.[1] However, GTS did not take off to the same extent as the other semantic frameworks, despite the fact that there is a vast array of natural-language expressions in the purview of semantic games. This holds even if the expressions let in a modicum of strategic meaning. In addition, a somewhat less-known but widespread phenomenon in language is the cross-categorial notion of informational independence, the treatment of which is typically successful only via the game-theoretic apparatus, and which may be put into a unified perspective by such a game-theoretic analysis.

An instance of informational independence is the branched organisation of quantifier phrases. In addition, there are several other fields in which games may turn out to be at least as useful and versatile as discourse-representation theory, dynamic logic and dynamic semantics, or the theory of generalised quantifiers. These include issues to do with anaphora and functional dependencies, tense and aspect, and a logical representation of eventualities. This wider story remains largely untold.

One current effort in the study of language involves locating the semantics/pragmatics interface (Turner 1999) and charting the phenomena within it. Such crossing points are where games have always naturally operated. Any cast-iron division here may, in any case, turn out to be quite artificial and uninspiring.

The third answer could be supplemented with the remark that, perhaps in its most general sense, the notion of a game could be thought of as a regimentation of the idea that whenever two forms contact one another, the befalling mutual action gives rise to content. The forms in question can be a language and its users or a single communicator, patterns of logic, or a computational system and its environment.

One of my core underlying theses is that games provide a first-rate insight into the different aspects of information flow in logical semantics. Within the present context, these streams and their fluctuation will be harnessed for the most part by the theory of extensive games, intermingled with imperfect information and other phenomena that increase their applicability.

1.4. The Double Role of Games

What is it that makes games helpful in the study of logic and language? This question will be addressed here, with an emphasis on games that are not just the best known and studied two-player perfect-information ones, but also those involving teams of players and imperfect information. (For a more comprehensive exposition and survey of formal theories that resort to game-theoretic concepts from logical, mathematical and computational perspectives, see Pietarinen (2003e).) Accordingly, the following sections will mostly concern such i mperfect-information team games, the logic they are associated with, and applications to epistemology.

Games in general are widely used across a broad intellectual territory. Game-related ideas are found in philosophy, logic, mathematics, cognitive science, artificial intelligence, computation, linguistics, and of course economics. It is something of a paradox that games are one of the oldest paradigms in the study of human cognition, behaviour and reasoning, going back to the art of argumentation in Aristotle's *Topica* and Socratic *elenchus*. To date, however, this paradigm has not been fully understood. One reason for this might be the alleged dispensability of games: such terminology may sometimes be brushed aside in favour of better-understood notions. The other reason is that games bring together a loose category of formal techniques. They may have remarkably dissimilar characteristics and only some minimal set of common elements, such as a universe, positions, well-defined move rules, winning and losing conventions, and strategies. Therefore games run the risk of becoming 'abstract nonsense' (to borrow a description of category theory) with a frail theoretical status and only a minor importance in their own right.

This fear is an illusion. An example is provided by logic, in which the notion of game has found a home in a number of areas. Game-theoretic concepts have frequently been resorted to when traditional methods have not easily applied.

The other reason for availing ourselves of games is methodological: they guide us towards a deeper understanding of the concepts and activities involved in cognitive reasoning processes, by providing accounts for the existence of such processes in terms of the meaning of logical constants, logical rules, and natural-language expressions.

2. A Brief History of Game Theories in Logic

2.1. The Legacy of Aristotle

The analytical and formal use of games is certainly not a twentieth-century invention. In *Topica VIII*, Aristotle discusses dialectical situations and duties that participants in such situations must respect. Aristotle's set up involves the Answerer, who must defend his or her thesis or *positum*, and Questioner, who tries to make the Answerer change his positum, that is, to grant the opposite of the thesis.[2]

An anonymous text, *Abbreviatio Montana*, written in the middle of the twelfth century, describes the art of dialectics as follows (Kretzmann and Stump 1988, 40):

> In order to discern the purpose of this art, you have to know that there are two practitioners of the art. Who are they? There is one who acts on the basis of the art, who disputes in accordance with the rules and precepts of the art, and he is called a dialectician – i.e., a disputant. The one who acts in a way that concerns the art is the one who teaches the art and expounds its rules and precepts, and he is named either a master or an expositor

(*demonstrator*). And so we ascribe different purposes in association with the different practitioners.

The text goes on to describe the (i) purpose, (ii) function, (iii) subject matter, and (iv) termination of both participants' activities. For the Disputant, they are (i) to prove on the basis of readily believable arguments a thesis that has been proposed, (ii) to dispute properly in keeping with the rules and precepts of the art, (iii) the proof of a thesis, which is the central issue of the dialectical disputation, (iv) to induce belief in the proposed question. For the Expositor, they are (i) to teach the art, (ii) to expound the rules and precepts of the art and to add new ones, (iii) to put forth utterances and to discover things signified by the utterances, (iv) to discover the judgements of reasons for the induced beliefs.

The writings on *ars obligatoria* put across typical features of a game. The Opponent attacks a thesis defended by the Respondent. The Respondent then has at least two duties: first, he must grant the thesis in the sense that whatever seems to be true of it must be defended. Second, whatever seems to follow from what he has already granted must also be defended. It might thus happen that the Respondent has to defend a false *positum*. In this case, he would have the new task of trying to keep his answers consistent. This is an interesting feature of *ars obligatoria*, for in such *position falsa* the Respondent may still survive by keeping the set of answers free of contradictions.

Later on, a connection between mathematical reasoning and game-theoretic thinking was discovered by Gottfried Wilhelm Leibniz, who invented the 'epsilon–delta definition' of continuity and explicated it as a game between two players: one player uses a function value $f(x)$ to bring home the value of his epsilon-move, while the other offsets it with delta about x.

Interestingly, Leibniz was also one of the key contributors to the early dawn of game theory, urging his colleagues to develop "a new kind of logic, concerned with degrees of probability, [. . .] to pursue the investigation of games of chance" (Leibniz 1981, 467). His wider perspective was that the art of invention (or discovery, *inventer*) would be improved, since the human mind "is more thoroughly displayed in games" ["paraissant mieux dans les jeux"] "than in the most serious pursuits" (*ibid.*, p. 467).

2.2. Peirce's Game-theoretic Ideas

Besides the above names, a logician who contributed significantly to the development of logic was Charles S. Peirce (1839–1914). Now game theory, as we recognise it today, was not yet developed during Peirce's lifetime. However, Peirce did conceive his logical and semeiotic ideas in ways that allow faithful translation into game-theoretic terminology. Hilpinen (1982) has shown that in Peirce's logical system – the system that through its later developments came to be known as first-order logic – existential and universal quantifiers, as indeed connectives and negation, are understood as integral parts of a dialogue between two functionaries, the Utterer (the Assertor, the Defender) and the Interpreter (the Critic, the Attacker).[3]

This idea can be found in his published writings as well:

> Begin by saying: "Take any things you please, namely," and name the letters representing bonds not encircled; [...] each hecceity [proper name, – *A.-V.P.*] corresponding to a letter encircled odd times is to be suitably chosen according to the intent of the assertor of the medad proposition, while each hecceity corresponding to a bond encircled even times is to be taken as the interpreter or the opponent of the proposition pleases. (CP 3.479, c.1896)

To the same effect, consider also: "In the sentence "Every man dies," "Every man" implies that the interpreter is at liberty to pick out a man and consider the proposition as applying to him" (CP 5.542, c.1902).

Such an affinity between games and logic is also found in Peirce's diagrammatic approach to logic in his influential theory of existential graphs. These facets are explored more fully in Pietarinen (2003a, c, 2004a).

Yet, Peirce's logic was not able to come down on many of the key features of modern game theory. In particular, the concept of winning strategy, crucial in defining truth and falsity in GTS, is conspicuously absent. However, an early anticipation of the notion of strategy can be found in Peirce's concept of a habit.

Some preliminary evidence for this is to be found in places in which he describes habits of interpretation: "The interpreter will have formed the habit of acting in a given way whenever he may desire a given kind of result" (CP 5.491, 1907). This statement is interesting, because here he addresses one participant of the game of language, the interpreter, and emphasises his or her decisions based on the concept of desire. In addition, he wrote earlier: "A habit arises, when, having had the sensation of performing a certain act, m, on several occasions a, b, c, we come to do it upon every occurrence of the general event, l, of which a, b and c are special cases" (CP 5.297, 1868). One possible interpretation of this is to refer to the character of a strategy as an abstraction of a rule that looks away from any single position.

There is further evidence in Peirce's writings to support the view that habit contains at least implicit aspects of strategic behaviour and action: "Action cannot be a logical interpretant, because it lacks generality. [...] But how otherwise can a habit be described than by a description of the kind of action to which it gives rise, with the specification of the conditions and of the motive?" (CP 5.491). In the terminology of semantic games, the motives Peirce refers to are the purposes of the two players, the verifier and the falsifier, the former aiming to verify a sentence or an expression and the latter aiming to falsify it. Further, in CP 2.665 [1910] we find (emphasis in the original): "It would be necessary, in order to define a man's habit, to describe how it would lead him to behave and upon what sort of occasion – albeit this statement would by no means imply that the habit *consists* in that action". If we take it that a logical interpretant of a sign is its meaning, what Peirce is in effect saying is that no single action or sequence of actions, that is, no choice or sequence of choices, can spell out the meaning of the signs in question, because it does not put in the picture how one arrives at such choices. In order to do that one would need to effectively use a strategy that leads to those actions.

However, Peirce had some fragmentary indications on the connections between elementary game-like ideas and the notion of truth: "The duality of the *ego* and *non-ego* is the chief constituent of the idea of the Truth" (MS 515: 24, n.p., n.d.). This duality and the ensuing dialectic subject of thought have much wider significance in Peirce's general theory of logic and semeiotics. For instance, an experience of an event is a duality between consciousness and the object of consciousness: the new excitement appears as *non-ego*, opposing the old *ego* and instantly passing into it.

An important character of the game-theoretic interpretation of logic is that it evaluates formulas by starting with the outermost component and then proceeding from the outside in, ending when an atomic formula is reached. This idea can be traced back to Peirce's theory of existential graphs. He coined the method "endoporeutic", (*endon* 'within'; *poros* 'passage, pore', see CP 4.561, 4.568, MS 293: 51,53, MS 514: 16), and took it to be at work in the evaluation of proper names, for instance.

In many places in which the term of endoporeutic is not explicitly mentioned, it is still clearly assumed as the reason behind the expected direction of the flow of information. For example: "The rule of interpretation which necessarily follows from the diagrammatization is that the interpretation is "endoporeutic" (or proceeds inwardly)" (MS 514: 16, 1909).

Had the endoporeutic method become more popular, we might have witnessed the game-theoretic development of logic in full, instead of the more prevalent Tarski semantics.[4] It is of some interest that it was only much later that the usefulness of game-theoretic methods was demonstrated in corners of logic in which the more prevalent methods failed. In retrospect, such developments have vindicated Peirce in that one of the most prominent methods in logical semantics in the early part of the last century only merits an isolated chapter in the study of logic in general, and a fortiori was only a special case in Peirce's general semeiotic and endoporeutic programme of logic.

2.3. The Rise of Modern Game Theory

Among the early *ludents* was Ernst Zermelo, who showed that for a two-player strictly competitive game with finitely many possible positions, a player can avoid losing for only finitely many moves (if his opponent plays correctly), if and only if the opponent is able to force a win (Zermelo 1913).

The received version of the theorem states that every finite, strictly competitive perfect-information two-player game is determined: either player 1 or player 2 has a winning strategy.

Game theory truly kicked off with Émile Borel (1921) and John von Neumann (1928), supported by contributions from László Kalmár (1928-1929) and Dénes

König (1927). One of the driving motivations in König's and Kalmar's papers was to improve upon Zermelo's earlier work. As to the von Neumann and Morgenstern's contribution, the game-theoretic concepts put forward in von Neumann (1928) were, according to the author himself, discovered independently of Borel's earlier discovery of pure and mixed strategies: "I developed my ideas on the subject before I read [Borel's] papers" (von Neumann 1953, 124). According to Ulam (1958), however, "Early in his work, a paper by Borel on the minimax property lead [sic] [von Neumann] to develop ... ideas which culminated later in one of his most original creations, the theory of games". (Kalmár acknowledges von Neumann's work in his 1928 paper, though.) All the same, games were doubtless developed into a fully-fledged theory in von Neumann and Morgenstern (1944).

After Zermelo, Thoralf Skolem introduced what is known as the Skolem normal form for first-order logic. Although aware of Zermelo's work, Skolem did not explore possible connections between logic and games. The development of the Skolem normal form is nonetheless interesting, and its exact history still needs to be documented. According to Skolem (1920, 254), "Löwenheim proves his theorem by means of Schröder's "development" ["Ausführung"] of products and sums, a procedure that takes a Π sign across and to the left of a Σ sign, or vice versa". Schröder used an awkward (sub)subscripting notation adopted by Löwenheim. Interpreted as (existentially quantified) functions, they become what are known as Skolem functions, previously also known as the 'fleeing subscripts'. The works of Schröder and his contemporary Peirce are related, but their mutual influence is still somewhat unclear.[5]

The first explicit connection between the Skolem functions and games appeared in Henkin (1961). According to him, the Skolem normal forms, and infinite quantifier strings in particular, could be conceived as games. As is well-known, David Hilbert used game-inspired ideas in his approach to the foundations of mathematics, and, to a degree, so did Gerhard Gentzen.[6]

The modern era of games and logic started with Henkin (1961), Hintikka (1973) and Scott (1993). Dana Scott presented the earliest game-theoretic elucidation of logic, based on an interpretation of Kurt Gödel's Dialectica (functional) translation of first-order logic and arithmetic into a higher-order language. The Dialectica interpretation has resurfaced since in various guises, such as in category theory, delivering abstract notions of games as Chu spaces or Dialectica categories that are used to model linear logic, and in consistency proofs for constructive theories. The connection between the truth-values and the existence of winning strategies was noted in Hintikka (1973).

Wittgenstein's far-reaching notion of a language game offers a concept that could be compared with semantic games. In addition to his remarks that at least some language games are ones of verification and falsification, the purposes of players in semantic games can be best accounted for in terms of the activities of "showing or telling what one sees". What the players try to achieve is to bring to the fore what they see to be the case in the context of an assertion. They have been prompted to do this by a specific expression, and they aim to show or say what is the case by instantiations of suitable elements: "“Surely if he knows anything he must know that he sees!” – It is true that the game of “showing or telling what one sees” is one of the most fundamental language games, which means that what we in ordinary life call using language mostly presupposes this game" (Wittgenstein 2000, item 149, 1). Contrary to what is claimed in Hodges (1997), the attributes of winning and losing were made applicable to language games by way of Wittgenstein's own remarks. Wittgenstein's *Nachlass* reveals further that game theory was not completely alien territory to him, for he remarked that the theory of the game is not arbitrary, although a game itself is (Wittgenstein 2000, item 161, 15r). He did not show particularly keen interest in such theorising, however.

2.4. Dialogues and Logic

Since the 1950s, dialogues and dialogical processes have earned a place in logic, as well as in applications involving formal procedures for reasoning and argumentation. The key players have been Paul Lorenzen and Kuno Lorenz (Lorenzen 1955; Lorenzen and Lorenz 1978; Lorenz 2001; Mann 1988).

There are two participants in dialogical logic, the Proponent and the Opponent (misleadingly sometimes called the Defender and the Attacker). The former proposes a claim while the latter challenges it. The moves are made according to logical and procedural rules. Informally, the logical rules consist of rules for (i) conjunction, prompting a challenge by the Opponent, the chosen conjunct becoming available to be defended by the Proponent; (ii) disjunction, according to which the Proponent chooses one of the disjuncts for the defence, and (iii) negation, which, as in GTS (see below) is a signal to change roles. In other words, negated statements are challenged by defending the statement governed by the negation. An existential statement is a request for a witness produced by the Proponent, instantiated as the value of the quantified variable to serve as a claim to be defended in the future. Likewise, a challenge on universal quantification asks for an individual produced by the Opponent, and the result of the instantiation will be the next challenge.

The Proponent is taken to have lost if the claim can no longer be defended, and the Opponent is taken to have lost if the claim can no longer be challenged. As in semantic games, the key concept here is the existence of winning strategies, which prescribes when the formulas will be valid. An analogous result to that of GTS is that a first-order sentence S can be deduced from the set of first-order sentences Γ ($\Gamma \vdash S$) if and only if $\Gamma \vdash S$ is valid in intuitionist logic.

Procedural conventions place some restrictions on how games are played. For example, it is often stipulated that a challenging claim may be answered at most once (or vice versa), or that responses by the Opponent are restricted to the latest challenge not yet defended. There are significant choices to be made between these conventions, as shown by the fact that classical logic can be reproduced by a suitable combination of these rules (Lorenz 1961; Rahman and Rückert 2001).

Interestingly, Peirce had ideas that came close to applying dialogue to logic, remarking, "Thinking always proceeds in the form of a dialogue, – a dialogue between different phases of the *ego*, – so that, being dialogical, it is essentially composed of signs, as its Matter, in the sense in which a game of chess has the chessmen for its matter" (MS 298: 6; CP 4.6, c.1906). Numerous related passages in Peirce's published and unpublished papers suggest that Peirce took logic (semeiotics) and thinking in general to be closely related to dialogue between the Utterer and the Interpreter, and in many instances he viewed these actors as the actual users of language. However, as I argued in the first section, Peirce presented several semantic ideas concerning what subsequently has became known as the semantic game approach to logic. Indeed, in the previous quotation he intended the Ego and the Non-Ego to transpire within a single mind or a quasi-mind. Thus, Peirce's dialogues were not always games for actual language users. In any case, what Peirce seems to have anticipated was not only the semantic but also the dialogical application of games to logical matters.

In his comprehensive survey on dialogues in logic, Felscher (2002, 125) notes,

> Lorenz (1961) observed that a change of dialogue rules would give rise to a type of dialogue the strategies for which would prove precisely the *classically* provable formulas. For this situation made it perfectly clear that the mathematical arbitrariness of the Theory of Games, being a tool to describe formally such different ways of reasoning as are classical logic and intuitionist logic, could not possibly produce a philosophical foundation for either of them.

Contrary to this view, however, one aspect in which dialogical games differ from the theory of semantic games is simply that actual game-theoretic concepts have proved instructive in the latter. As will be shown here, such concepts include extensive-form representations of games, uncertainty, information sets, payoffs vs. winning, competitiveness, and aspects of team theory. Consequently, his claim concerning semantic games is, in the end, obsolete: "Hintikka restricts his attention to the single argumentation forms and nowhere cares to formulate game rules proper (such that the implied reference to mathematical games remains but an incantation)" (Felscher 2002, 126). Another counterexample to this is provided by game semantics and ludics in computation and the overarching programme of 'geometry of interaction'.

3. Game-theoretic Semantics: Some Main Ideas

3.1. Semantic Games

What is it that makes games a powerful tool in logic? The basic idea is somewhat simple. You and I confront one another, observing a set of rules telling us which moves are legal, and with the same purpose. We both try to win the game by winning any play of it, and if one of us finds a systematic way of doing so, he or she has a winning strategy. The set of game rules is fixed by the logically active components in language, which in the case of first-order languages comprise the two quantifiers ∃ and ∀ and sentential connectives.

3.1.1. *Rules*

Let us assume that the structure $\mathfrak{A}$ is a τ-structure with a signature τ of a nonempty domain $|\mathfrak{A}|$ on which the game is being played. A valuation g is a mapping from terms of a language L to the domain of the model, restricted to the free variables of every $\varphi \in L$. In the game, the formulas are evaluated according to the rules prompted by the logical ingredients encountered in them, starting with the outermost one. Game $\mathcal{G}$ involves player V (the V∃rifier, H∃loïsé, Myself) and player F (the F∀lsifier, ∀bélard, Nature).

The aim of F is to falsify the formula (i.e. to show that it is false in $\mathfrak{A}$), and the aim of V is to verify it (i.e. to show that it is true in $\mathfrak{A}$). For the sake of simplicity, it is assumed, without loss of generality, that the first-order language $\mathcal{L}_{\omega\omega}$ does not contain $\rightarrow$ or $\leftrightarrow$. The symbols $\forall$ and $\wedge$ prompt a move by F, and $\exists$ and $\vee$ prompt a move by V. When players come across negation, they change roles, and winning conventions will also change. Each move reduces the c omplexity of the formula, and hence an atomic formula is finally reached. The truth-value of an atomic formula, as established by a given interpretation, determines which player wins the play of a game.

Let $\mathcal{L}_{\omega\omega}$ be a standard first-order language with $\{\vee, \wedge, \neg, \exists, \forall\}$. A strictly competitive non-cooperative game $\mathcal{G}(\varphi, g, \mathfrak{A})$ is defined by induction on the complexity of each $\mathcal{L}_{\omega\omega}$-formula φ between V and F:

(G.¬): If $\varphi = \neg\psi$, V and F change roles, and the next choice is in $\mathcal{G}(\psi, g, \mathfrak{A})$.

(G.∨): If $\varphi = \theta \vee \psi$, V chooses either Left or Right, and the next choice is in $\mathcal{G}(\theta, g, \mathfrak{A})$ if Left, and in $\mathcal{G}(\psi, g, \mathfrak{A})$ if Right.

(G.∧): If $\varphi = \theta \wedge \psi$, F chooses either Left or Right, and the next choice is in $\mathcal{G}(\theta, g, \mathfrak{A})$ if Left, and in $\mathcal{G}(\psi, g, \mathfrak{A})$ if Right.

(G.∃): If $\varphi = \exists x\, \psi$, V chooses an individual of the domain of the structure $\mathfrak{A}$, and the next choice is in $\mathcal{G}(\psi, g \cup \{(x, a)\}, \mathfrak{A})$.

(G.∀): If $\varphi = \forall x\, \psi$, F chooses an individual of the domain of the structure $\mathfrak{A}$, and the next choice is in $\mathcal{G}(\psi, g \cup \{(x, a)\}, \mathfrak{A})$.

(G.atom): If φ is atomic, the game ends, and V wins if φ is true, and F wins if φ is false.

Strict competitiveness means that if V loses then F wins, and if F wins, then V loses. Non cooperation roughly means that players decide the action they take alone. According to the rules for connectives, rather than choosing subformulas, players choose elements from one domain split into two.

3.1.2. *Strategies*

The strategy for each player in game $\mathcal{G}(\varphi, g, \mathfrak{A})$ is a complete rule indicating at every contingency in which the player is required to move what his or her choice is. A winning strategy is a sequence of strategies by which a player may make operational choices such that every play of the game results in a win for him or her, no matter how the opponent chooses.

Let $\mathcal{G}(\varphi, g, \mathfrak{A})$ be a game for $\mathcal{L}_{\omega\omega}$-sentences φ, and f a strategy.

- $(\mathfrak{A}, g) \models \varphi$ if and only if a strategy f exists which is winning for V in $\mathcal{G}(\varphi, g, \mathfrak{A})$;
- $(\mathfrak{A}, g) \not\models \varphi$ if and only if a strategy f exists which is winning for F in $\mathcal{G}(\varphi, g, \mathfrak{A})$.

The game-theoretic notion of truth invokes the key notion of strategies, which may be viewed as Skolem functions. Moreover, an existential quantifier that is within the scope of a universal quantifier (in the sense of scope expressing the logical priority order of components) is functionally dependent on the universal quantifier. For example, if Pxy is atomic, then $(\mathfrak{A}, g) \models \forall x \exists y\, Pxy$, if and only if there exists a one-place function f such that for any individual chosen by F (say, a), $Paf(a)$ is true in $\mathfrak{A}$.

The distinction in the next two sections between 'games that are not played' and 'games that are played' is analogous to the distinction between normal forms and extensive forms of games.

3.2. Games That 'Are Not Played'

According to the Skolem normal-form theorem, every $\mathcal{L}_{\omega\omega}$-formula φ is equisatisfiable (satisfiable in the same models) with the existential second-order Σ^1_1-formula of the form:

$$\exists f_1 \ldots \exists f_m \forall x_1 \ldots \forall x_n\, \psi, \tag{1}$$

where $f_1 \ldots f_m, m \in \omega$ are new function symbols and ψ is a quantifier-free formula. Such normal forms are effectively to be found for every first-order sentence. The resulting Σ^1_1-formula states the existence of a winning strategy for V. Assuming the axiom of choice, by the Skolem normal form theorem, it follows that

$$\exists f \forall x\, Pxf(x) \equiv \forall x \exists y\, Pxy. \tag{2}$$

A special type of Skolem normal-form theorem may be used in skolemising connectives. The only difference is that it is possible to conjoin to each disjunct a Skolem function f which has its value in a set of two elements, say {Left, Right}.

For example, let

$$\forall x \exists y \forall z (P_1 xyz \vee P_2 xy), \tag{3}$$

where P_1xyz and P_2xy are atomic. This can then be skolemised to

$$\begin{aligned} \exists f \exists g_1 \exists g_2 \forall x \forall z ((P_1 x f(x) z \wedge g_1(x, z) &= \text{Left}) \vee \\ \vee (P_2 x f(x) \wedge g_2(x, z) &= \text{Right})). \end{aligned} \tag{4}$$

However, one might find the use of Skolem functions as winning strategies somewhat restrictive and not able to capture the true strategic nature of interactive moves. Indeed, these functions can express only functional dependencies, namely the existential quantifiers or disjunctions that are within the scope of universal quantifiers.

Further, if there exists a winning strategy for one of the players, what interest can the other player have in playing the game off against such an invincible opponent? Since all games for first-order logic are determined, that is, there always exists a winning strategy for one of the players (and thus the other, given that the games are strictly competitive, loses), the idea of a game as a set of dynamically evolving plays with truly interacting players tends to recede.

3.3. From Incantation to Cantata: Games That 'Are Played'

These qualms are allayed as soon as semantic games are viewed as extensive-form games in the sense of the classical theory of games. In such a framework, one could think of logical games entirely in terms of how information flows in a formula from one component to another, and study various ways in which this flow can be controlled and regulated. This perspective is not confined to functional dependencies, and thus one is able to say something more about the game-theoretic interpretation of logic than would be possible by merely using the existence of winning strategies. This adds credence to the true strategic content of semantic games without suppressing their dynamics.

Extensive games go beyond the normal (strategic) form in the sense that, whereas normal forms conveniently show at a glance, so to speak, which strategies are the winning ones for which player, strategies in extensive games are generated as the game moves on.[7]

In general, extensive games capture the sequential structure of players' strategic decision problems. They may be represented as (finite) trees with decision nodes (histories) and actions labelling the edges departing from them. The game starts at the root of the tree and ends at the terminal nodes. At each non-terminal node or decision point, the player has to make a decision as to what to choose. The outcome of this decision in a particular play is a choice, while the set of all choices from a node determines a move.

Extensive games were first formulated (set-theoretically) in von Neumann (1928), although the (graph-theoretic) presentation in Kuhn (1953) has become commonplace. von Neumann and Morgenstern (1944) set out the essentials of the graphical conception. Applied to logic, the key definitions are as follows.

3.4. Games in Extensive Forms

3.4.1. *Perfect Information*

Let us suppose a family of actions A, in which the finite sequence $\langle a^i \rangle_{i=1}^n, n \in \omega$ represents the consecutive actions of the players in N (no chance moves), $a^i \in A$. An extensive game $\mathcal{G}$ with perfect information is a five-tuple

$$\mathcal{G}_A = \langle H, Z, P, N, (u_i)_{i \in N} \rangle,$$

such that

- H is a set of finite sequences of actions $h = \langle a^i \rangle_{i=1}^n$ from A, called histories of the game. It is required that:
  - the empty sequence $\langle \rangle$ is in H;
  - if $h \in H$, then any initial segment of h is in H too, that is, if $h = \langle a^i \rangle_{i=1}^n \in H$ then $pr(h) = \langle a^i \rangle_{i=1}^{n-1} \in H$ for all n, where $pr(h)$ is the immediate predecessor of h ($= \emptyset$ for $h = \emptyset$).
- Z is a set of m aximal histories (complete plays) of the game. If a history $h = \langle a^i \rangle_{i=1}^n \in H$ can continue as $h' = \langle a^i \rangle_{i=1}^{n+1} \in H$, h is a non-terminal history and $a^n \in A$ is a non-terminal element. Otherwise they are terminal. Any $h \in Z$ is terminal.
- $P: H \setminus Z \to N$ is the player function which assigns to every non-terminal history a player in N whose turn it is to move.
- each $u_i, i \in N$ is the payoff function, that is, a function which specifies for each maximal history the payoff for player i.

For any non-terminal history $h \in H$,

$$A(h) = \{x \in A \mid h \frown x \in H\}.$$

A (pure) strategy for a player i is any function

$$f_i: P^{-1}(\{i\}) \to A$$

such that $f_i(h) \in A(h)$, where $P^{-1}(\{i\})$ is the set of all histories in which player i is to move. A strategy also specifies an action for histories that may never be reached.

In a strictly competitive game, $N = \{V, F\}$ and in addition:

- $u_V(h) = -u_F(h)$;
- either $u_V(h) = 1$ or $u_V(h) = -1$ (that is, V either wins or loses);

for all terminal histories $h \in Z$.

3.4.2. *Imperfect Information*

Let $\mathcal{G}_A$ be a perfect-information game. To represent imperfect information, let us extend $\mathcal{G}_A$ to a six-tuple

$$\mathcal{G}_A^* = \langle H, Z, P, N, (u_i)_{i \in N}, (\mathfrak{I}_i)_{i \in N} \rangle,$$

in which $\mathfrak{I}_i$ is an information partition of $P^{-1}(\{i\})$ (the set of histories in which i moves), such that for all $h, h' \in S^i_j$, $h \frown x \in H$ if and only if $h' \frown x \in H$, $x \in A$, $j = 1 \ldots m$, $i = 1 \ldots k$, $m \leq k$. S^i_j is called an information set.

The games are exactly as before, except that now the players may not have all the information about the past features. This is brought out by an information partition $\mathfrak{I}_i$ of histories into information sets (equivalence classes). The histories that belong to the same information set are indistinguishable to the players, and thus a player takes no notice of what the histories are that have been played.

In imperfect-information games, the strategy function is required to be uniform on indistinguishable histories:

$$\text{If } h, h' \in S^i_j \text{ then } f_i(h) = f_i(h'), \text{ for all } i \in N.$$

The notion of uniformity is customarily disposed of in game theory, because strategies are defined on information sets rather than on individual histories.

3.5. Semantic Games in Extensive Forms

3.5.1. *Perfect Information*

Let Sub(φ) denote a set of subformulas of φ. An extensive-form semantic game $\mathcal{G}(\varphi, g, \mathfrak{A})$ associated with an $\mathcal{L}_{\omega\omega}$-formula φ is exactly like the game $\mathcal{G}_A$ defined above, except that it has one extra element: a labelling function $L\colon H \to \text{Sub}(\varphi)$ such that

- $L(\langle\rangle) = \varphi$ (the root);
- for every terminal history $h \in Z$, $L(h)$ is an atomic formula or its negation.

In addition, the components H, L, P, u_V and u_F jointly satisfy the following:

- if $L(h) = \neg\varphi$ and $P(h) = V$, then $h \frown \varphi \in H$, $L(h \frown \varphi) = \varphi$, $P(h \frown \varphi) = F$;
- if $L(h) = \neg\varphi$ and $P(h) = F$, then $h \frown \varphi \in H$, $L(h \frown \varphi) = \varphi$, $P(h \frown \varphi) = V$;
- if $L(h) = \psi \vee \theta$ or $L(h) = \psi \wedge \theta$, then $h \frown$ Left $\in H$, $h \frown$ Right $\in H$, $L(h \frown$ Left$) = \psi$, and $L(h \frown$ Right$) = \theta$;
- if $L(h) = \psi \vee \theta$, then $P(h) = V$;
- if $L(h) = \psi \wedge \theta$, then $P(h) = F$;
- if $L(h) = \exists x\varphi$ or $L(h) = \forall x\varphi$, then $h \frown a \in H$ for every $a \in |\mathfrak{A}|$;
- if $L(h) = \exists x\varphi$, then $P(h) = V$;

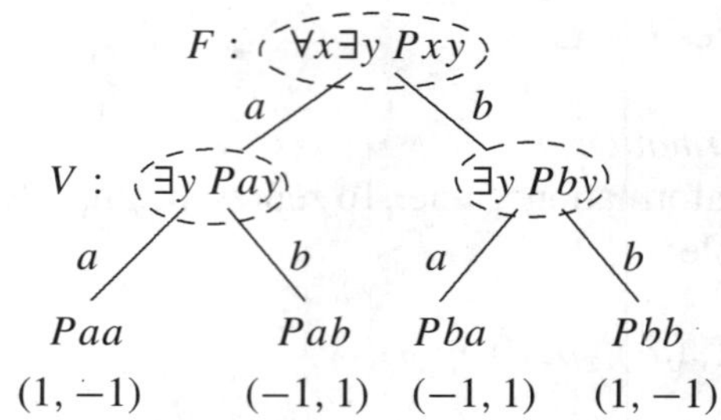

Figure 1. A perfect-information semantic game $\mathcal{G}(\phi, g, \mathfrak{A})$.

- if $L(h) = \forall x\varphi$, then $P(h) = F$;
- for every terminal history $h \in Z$:
  - if $L(h) = Pt_1 \dots t_m$ and $(\mathfrak{A}, g) \models Pt_1 \dots t_m$, then $u_V(h) = 1$ and $u_F(h) = -1$;
  - if $L(h) = Pt_1 \dots t_m$ and $(\mathfrak{A}, g) \not\models Pt_1 \dots t_m$, then $u_V(h) = -1$ and $u_F(h) = 1$.

The notion of strategy is defined in the same way as before. A winning strategy for $i \in \{V, F\}$ is a set of strategies f_i that leads i to $u_i(h) = 1$ no matter how the player $-i$ (the player other than i) decides to act.

3.5.2. *An Example*

An example of an extensive perfect-information semantic game for an $\mathcal{L}_{\omega\omega}$-formula $\forall x\exists y\, Pxy$, on a two-element domain $|\mathfrak{A}| = \{a, b\}$, is depicted in Figure 1. Since this is a game of perfect information, each non-terminal history forms its own singleton information set. (Hereafter, singleton information sets will be omitted in general.) The choices are marked on the edges of the game tree, and they correspond to the choices made by the player acting at the histories from which these edges depart. The atomic formulas label the terminal histories. Depending on the truth or falsity of atomic formulas, either F or V can win particular plays as seen from the payoffs. In this case, V wins the plays Paa and Pbb and F loses them, and F wins Pab and Pba while V loses them. There exists a winning strategy for V in this game, choosing a when F has chosen a, and b when F has chosen b, while there does not exist a winning strategy for F.

3.5.3. *Imperfect Information*

If there is imperfect information, the players may not be able to distinguish between some of the game histories. This is indicated by the information partition $(\mathfrak{I}_i)_{i\in N}$, where the information sets S^i_j spell out the information available to the players when making their moves. When there are only singleton information sets, that is, no two histories belong to the same set, the game is one of perfect information, otherwise it is one of imperfect information. Semantic games of imperfect information are denoted by $\mathcal{G}^*(\varphi, g, \mathfrak{A})$.

What happens in these semantic games is that the partition may have different properties depending on the language in question and on what syntactic restrictions there might be. I will return to these issues towards the end of the next section.

4. Logics of Imperfect Information

4.1. Independence-Friendly Logics

There are languages in which perfect information fails. One example is the extension of ordinary first-order language with the Henkin (finite, partially-ordered, branching, parallel) quantifiers (see Section 4.2 below).

Imperfect-information games also provide semantics for independence-friendly (IF) logics (Hintikka 1996; Hintikka and Sandu 1989). IF logics use a forward-slash notation that linearises Henkin quantifiers, but makes the information regulations more liberal.

Let $Qx\psi$, $Q \in \{\forall, \exists\}$ and $\phi \between \psi$, $\between \in \{\wedge, \vee\}$ be $\mathcal{L}_{\omega\omega}$-formulas in the scope of $Q_1x_1 \dots Q_nx_n$, where $A = \{x_1 \dots x_n\}$. Then the first-order language $\mathcal{L}^*_{\omega\omega}$ with informational independence is formed as follows:

- if $B \subseteq A$, then $(Qx/B)\,\psi$ and $\phi\,(\between/B)\,\psi$ are wffs of $\mathcal{L}^*_{\omega\omega}$.

Let us call '/' an outscoping device and customarily write $\{x_1 \dots x_n\}$ as $x_1 \dots x_n$.

For example, the following are wffs of $\mathcal{L}^*_{\omega\omega}$:

$$\forall x(\exists y/x)\,Pxy.$$
$$\exists x\,(P_1x\,(\vee/x)\,P_2x).$$
$$\neg\forall x_1 \dots \forall x_n(\exists y/x_1 \dots x_n)\,Px_1 \dots x_n y.$$

The semantics of an $\mathcal{L}^*_{\omega\omega}$-formula φ is given by the game $\mathcal{G}^*(\varphi, g, \mathfrak{A})$. As before, let us define an $\mathcal{L}_{\omega\omega}$-formula φ as true (resp. false) if and only if there exists a strategy in $\mathcal{G}^*(\varphi, g, \mathfrak{A})$ that is a winning one for V (resp. F) in $\mathcal{G}^*(\varphi, g, \mathfrak{A})$.

The same outscoping notation can be applied to propositional and modal logics, for instance. As with propositional fragments, the application of the slash gives rise to formulas in which $\vee$ and $\wedge$ may be replaced by $(\vee/\wedge)$ and $(\wedge/\vee)$. As with quantifiers, in encountering $(\vee/\wedge)$, V is not informed about the choice of conjunction, and in encountering $(\wedge/\vee)$, F is not informed about the choice of disjunction. For more complex expressions, we need to distinguish the connective tokens, and the best way of doing that is to think of disjunctions (resp. conjunctions) as restricted existential (resp. universal) quantifiers over a domain with a designated individual. I will not go into detail about these extensions here.[8] As will be seen anon, modal extensions have special significance in terms of the semantics of quantified notions in epistemic logic, the problem of intentional identity, and general epistemological questions.

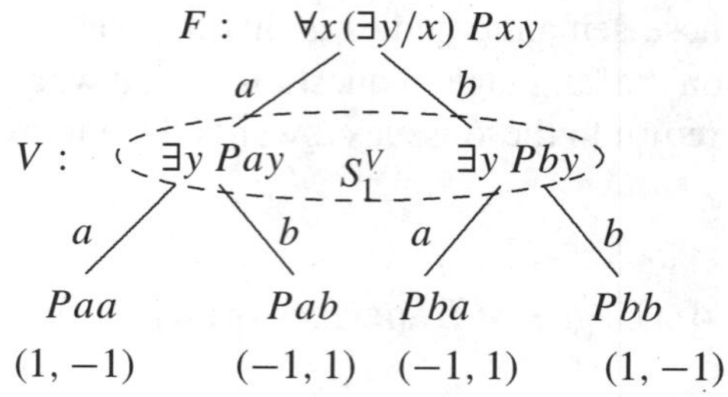

Figure 2. Imperfect-information semantic game $\mathcal{G}^*(\phi, g, \mathfrak{A})$ with one non-trivial information set S_1^V annotated for V.

The technique of information hiding may, in principle, be applied to all logics that allow a coherent game-theoretic interpretation, including the theory of generalised quantifiers (Pietarinen 2001b) and non-monotonic logics interpreted via the modal 'only knowing' of inaccessible worlds (Pietarinen 2002a).

Of particular interest in IF logics is the behaviour of negation. As such, negation $\neg$ denotes strong, game-theoretic negation, prompting a role switch between the two players. If we introduce weak contradictory negation $\neg_w$, then $\neg_w\varphi$ is true if and only if φ is not true. The linguistic 'not' is also a contradictory negation.[9] All common laws involving negation, including de Morgan laws and the law of double negation, remain valid, but the law of excluded middle fails. This is because semantic games for IF logic are not determined, in that if there is no winning strategy for one of the players it does not follow that there is a winning strategy for his, her or its antagonist. An example of such an IF formula in which the law of excluded middle fails is $\forall x(\exists y/x)\, x = y$, interpreted over a two-element domain.

Pietarinen and Sandu (1999) explore some implications of IF logic, and aim to set straight some of the misunderstandings that have occurred in the literature concerning it and its relation to GTS, most notably the misunderstandings and *fauxity* in Tennant (1998). The topics addressed include intuitionism, constructivism, compositionality, truth definitions, mathematical prose, negation in IF logic, and the status of set theory. Janssen (2002) discusses the game-theoretic interpretation of IF first-order logic, and proposes that there is a difference between informational independence and imperfect information, thus putting forward semantics based on the idea of subgames.

An example of an imperfect-information semantic game for an IF sentence $\phi = \forall x(\exists y/x)\, Pxy$ is given in Figure 2. There is now one non-trivial information set that includes within it all the histories in which V is to move. Given the same truth conditions for atomic formulas as in the previous example (Figure 1), it is clear that neither F nor V has a winning strategy in this game.

4.2. Henkin Quantifiers

Leon Henkin considered the possibility of extending first-order logic with partially ordered quantifiers in (Henkin 1961):

$$\begin{matrix} \forall x \exists y \\ \forall z \exists w \end{matrix} Pxyzw. \tag{5}$$

The meaning of (5) can be given by the Skolem normal form

$$\exists f \exists g \forall x \forall z\, Pxf(x)zg(z). \tag{6}$$

Formula (5) is true if for every x there exists y such that for every z there exists w whose choice depends only on z and not on x and y, such that $Pxyzw$. The crucial point is that, whereas in ordinary predicate logic the number of arguments in Skolem functions replacing existential quantifiers corresponds to the number of universal quantifiers within the scope of which the existential quantifiers occur, partially-ordered quantifiers have a reduced number of such arguments.

Henkin had the idea of interpreting quantifiers, and infinitely alternating and parallelly-ordered Henkin quantifiers in particular, through a game on a structure (Henkin 1961, 179):

> Imagine, for instance, a "game" in which a First Player and a Second Player alternate in choosing an element from a set I; the infinite sequence generated by this alternation of choices then determines the winner. If we let π to denote the class of all those sequences for which the First Player is the winner, then the formula $[\exists v_1 \forall v_2 \exists v_3 \forall v_4 \ldots (\pi v_1 v_2 v_3 \ldots)]$ simply expresses the fact that the First Player has a winning strategy.

Henkin quantifiers have been extensively studied, but unlike IF logic, they remain partially ordered and hence do not admit of, say, non-transitive, non-Euclidean, or cyclic quantifier orderings. These further structures may nonetheless be scrutinised in IF logics.

4.3. Constraints on Information

IF logics promote an informational outlook on logic and games. The notion of information may be studied from both logical and game-theoretic perspectives. In particular, I distinguish the following three interrelated notions:

(i) *Uniformity* is a property of strategy functions in a semantic game of imperfect information. This means that the outcome of an action has to be the same across the indistinguishable histories in such a game.

(ii) *The assumption of common actions* is a property of imperfect-information games. This means that the set of available actions has to be the same across the indistinguishable histories.

(iii) *The principle of observed actions* concerns players' knowledge about the game. It means that, whenever a player has to make a decision, he or she can observe and identify the totality of available options.

These notions delineate different levels of representing information. For (i) pertains to the player's strategies, (ii) concerns the ways in which the structure of extensive games of imperfect information is defined, and (iii) concerns the player's perception of epistemic features related to the game.

As noted above, the notion of 'choosing independently' in IF logic has sometimes been explicated in terms of the uniformity of the strategy functions (i). The idea is that nothing in the strategy may signal to the player his or her actual location within an information set. Uniformity turns out to be superfluous, because in game theory no separate property is needed for the obvious reason that strategies are defined on information sets, not on individual histories.

Assumption (ii) of common actions, in turn, means that for all $h, h' \in H$: if $h, h' \in S_j^i$ then $A(h) = A(h')$. The idea here is that if a player cannot distinguish between two histories h and h', then the choices available to him or her after h must be the same as those available after h'. For, if $A(h) \neq A(h')$, then by the assumption of the observability of the available options the player could recover the difference between h and h'.

On the other hand, according to (iii), a player can (and must) observe his or her available options when planning a move. It is of interest to observe that this principle is, in fact, not needed in perfect-information games in which all information sets are singletons. It is thus perfectly legitimate to ask why we suddenly need it in imperfect-information games, which are supposed to solely concern the players' information concerning past actions and not upcoming actions.

By posing such questions one is re-kindling some time-honoured controversies that arose in economics in the pre-games era. In the late 19th century Léon Walras and Wilfredo Pareto were struggling with questions to do with what an agent can foretell in decision-making situations. After them, perfect foresight was long thought to be a precondition for equilibrium. In economics, such an assumption is all the more dubious the more parameters there are for a *homo oeconomicus* to consider, including allocated time, prices, production, income, propositional attitudes involving higher-order beliefs and expectations, and so on. According to Morgenstern (1976), who was unhappy with this general situation, such a gents were tantamount to "demi-gods", and the term is indeed quite apt in describing hyper-rational agents.

Given Morgenstern's axiomatic leanings, he soon sought to fix models of such situations by imposing strict limits to the phenomena, or system, that is tried to be theoretically captured. The method of fixing the boundaries was at that time heavily influenced by the Frege–Russell conception of logic, which Morgenstern was eager to promote as a conceptual breakthrough in economics. The conception of logic was no longer the Peirce–Peano one, which was dominant until after 1910, when Russell rashly decided to boast about the then-quite-chimerical awareness of

Frege's work. However, my issue is not the complexity of the situation under attack, but the somewhat inconsistent way of determining whether perfect foresight is assumed in a like manner in perfect-information and imperfect-information games that has transpired in the literature.

Furthermore, in the light of (iii), it is not invariably clear that (ii) should hold. The usual argument for (ii) is that otherwise a player could carry out an infeasible action at some $k \in S^i_j$. If such choices were excluded from the scope of strategies, infeasible yet unattainable actions would ensue, in which case we would have to assume that (iii) is thus invalid, too.

Yet, am I able to choose an action if I do not know what it is? Is the identity of actions all the players need to know when planning a move? Do they not need to know the consequences of that action too, enabling them to assess the value or the practical bearings of the observable outcomes, making decisions pragmatically feasible, and thus enabling them to make finer distinctions and inferences concerning the actual locations in the game?

Such foresight concerning not only the identification of (immediately) available action but also the practical effects of those actions, should not be seen as tantamount to the principle of the rationality of players. A player may be rational even if he or she has a limited possibility of building a model of the future due to limited information. Foresight is not a precondition for the existence of winning strategies (or more generally equilibrium points), either, because it is something that is inbuilt into the very notion of strategy, be it a function or a non-deterministic set of relations between the decision point and the available actions.

It is worth observing how close this problem of supposed epistemic states of players concerning available actions is to the problem of cross-identification in the semantics of modal notions in terms of possible-worlds semantics for quantified epistemic logic. Assuming $A(h) = A(h')$ for $h, h' \in S^i_j$ is in modal terms analogous to assuming constant domains whenever questions concerning the identification of objects in A may arise (or more precisely, whenever $|h| = |h'|$, that is, whenever the 'modal depths' in the game histories coincide). Such a domain restriction amounts to a new type of quantified modal logic, the models of which have 'stratified' layers of individuals.

More generally, game theory is in need of a theory of knowledge that could address these issues. The programme of 'interactive epistemology' has so far brought only inadequately expressive propositional logics to bear on related problems. There has been no attempt to analyse the questions such as different notions of foresight in terms of much stronger quantified epistemic logics. It is interesting to observe that such a need was already expressed in the writings of Morgenstern and others in the 1930s. The emergence of logical notions of knowledge in the late 1950s and early 1960s, in tandem with the possible-worlds breakthrough of semantics for modal logics, did not draw its motivation from these prevailing problems in economics, however.

I will refrain from further comment on these questions here (see Pietarinen and Sandu (2004) for some philosophical correlates), and merely observe that the condition met by any IF formula, rather than the limiting the foresight, constrains the past of indistinguishable histories. To capture the fact that all indistinguishable histories should be composed of indistinguishable pasts, one states that if $h, h' \in S_j^i$ then $|h| = |h'|$, for all $i \in N$. This condition, sometimes called the von Neumann and Morgenstern condition, also excludes cases of absentmindedness, which are not excluded by (i) or (iii): Let $\preccurlyeq$ be a partial order on the tree structure of extensive games $\mathcal{G}$ and $\mathcal{G}^*$, and let the game satisfy the non-absentmindedness condition: $h, h' \in S_j^i$, if $h \preccurlyeq h'$ then $h = h'$. Let depth $d(Q)$ of logical component Q in an IF formula φ be defined inductively in a standard way. Then $\mathcal{G}^*(\varphi, g, \mathfrak{A})$ satisfies non-absentmindedness, because all of the components in Q have a unique depth $d(Q)$, and so every subformula of φ has a unique position in $\mathcal{G}^*$ given by $L(h)$. Thus, for any two subformulas of φ at $h, h' \in H$ within S_j^i, it holds that $h \not\preccurlyeq h'$ and $h' \not\preccurlyeq h$.

5. Directions in Game-theoretic Semantics

The above description merely places semantic games at the starting point from which to investigate, implement and modify the available game-theoretic apparatus. This variability in the notion of a game has several implications for logic.

5.1. Perfect or Imperfect Information?

The first possibility is to drop the assumption that players have perfect information.

5.1.1. *The Modus Operandi*

- Semantic games are of *perfect information* whenever the flow of information is not constrained. Otherwise they are of *imperfect information.*

In game-theoretic terminology, perfect information means singleton information sets, whereas imperfect information also permits non-singleton information sets.

Perfect-information games are customarily associated with formulas of ordinary first-order logic, while imperfect-information games are associated with IF formulas and Henkin quantifiers.

5.1.2. *Informational Independence*

The assumption of unregulated information flow in logic is one aspect of the informational-dependence assumption, and hence the move to IF languages marks a step from informational dependence to informational independence (Hintikka 1996; Sandu 1993). It is not the rules of the game that one needs to modify for IF logic, but the strategic component pertaining to the information available to the players.

For instance, the impact of informational independence is clear from its role in the resolution of the problem of intentional identity (Pietarinen 2001a). It also gives rise to partiality in logic.

5.2. Partiality and Games

The field of partial logics has arisen as an independent object of study in logic and linguistics in recent years.

5.2.1. *The Modus Operandi*

- Logic is *partial* whenever it has a third truth-value, Undefined, or has truth-value gaps.

(See Langholm (1996) for arguments that truth-value Undefined and the notion of a truth-value gap do not coincide.)

5.2.2. *Remarks*

Partial logics are customarily taken to have multiple values. In addition to the two truth-values True and False, there are truth-value gaps, or a third truth-value Undefined. However, partiality should be studied independently of the question of whether logic has two values or more, because what is termed partiality in the literature emerges from the game-theoretic interpretation, as soon as the transmission of information between participants is not perfect, that is, the players are not perfectly informed of the past features of the game. Partiality is thus a consequence of entirely classical premises concerning the interpretation of language, without any additional postulation of truth-value gaps or third or fourth truth-values. The lack of an existing winning strategy for one of the players does not presume a winning strategy for the adversary.

Two notions of logical consequence are thus distinguished in partial logics. The notion $\models^+$ means a positive logical consequence (a formula being true in a model), and the notion $\models^-$ means a negative logical consequence (a formula being false in a model). Logical equivalence splits in two, weak equivalence (true in the same models) and strong equivalence (true in the same models and false in the same models).

5.2.3. *Partiality in Propositional Logic*

The following example clarifies the relation between partiality and games. In a sentence of propositional logic that also applies the slash notation to propositional connectives, for instance, in the sentence

$$(\varphi\ (\wedge/\vee)\ \psi) \vee (\theta\ (\wedge/\vee)\ \chi), \tag{7}$$

we may think of disjunction as prompting a choice between the two disjuncts (Left or Right), and similarly for conjunction. However, the latter choice is independent of the earlier choice, and hence the second player does not 'know' the earlier choice. We may view (7) as a new four-place connective $W(\varphi, \psi, \theta, \chi)$, a 'transjunction' (Sandu and Pietarinen 2001), and create undefined values by adding to a propositional logic with complete models. It can be shown that, together with the usual Boolean connectives, this set of connectives is functionally complete for all partial functions.

Formula (7) gives rise to a connective that is motivated from a game-theoretic perspective. It emerges from a two-stage extensive-form semantic game of imperfect information between two players. Partiality is generated as a property of the non-determinacy of games.

With regard to the particular example above, one may ask how the second player is supposed to know that it is his turn to move, without knowing the previous choice. For instance, if the second player is supposed to know that he has to choose between θ and χ, then he can infer that the first player has chosen Right, and if the choice is between φ and ψ, then the inference is that the previous choice was Left. The idea of informational independence brings in some complications involving the notion of the information the players have, their knowledge of the game, and so on. Sandu and Pietarinen (2001) tackle these issues by allowing players to choose elements from a two-sorted domain rather than operating on subformulas. In the light of the informational partition of histories, there is no way a player could recover such information concerning the earlier choices.

In general, particular forms of extensive imperfect-information games give rise to partial propositional logic through which various forms of informational dependencies and independencies of connectives are studied (Sandu and Pietarinen 2001, 2003; Pietarinen 2001c). Moreover, it is possible to apply the analysis to partialised logic in which the law of excluded middle also fails for atomic formulas. In this case, the payoffs for both players are negative, namely $u_V(h) = -1$ and $u_F(h) = -1, h \in H$. According to IF logic, when the law of excluded middle fails for complex formulas, no such payoff arises. But it is perfectly possible to combine the two, one outcome being that the two notions of negation do not coincide even if the logic contains no slashed expressions, provided that some terminal histories are labelled with payoffs of $(-1, -1)$.

5.2.4. *Partially-interpreted Logics and Games*

What makes failure of the law of excluded middle possible in atomic formulas? One answer is that it is not only logically active expressions, but also non-logical constants that may enjoy independence. The question that then arises is how we would interpret independence in, say, the formula $\forall x S[(c/x), x]$. What does it mean that constant c is independent of the quantified variable x? This is a derivative of the question of what the game rules for non-logical constants ought to be. Such rules pertain to the interpretation of language. When the atomic formula S of, say, an IF first-order formula φ is reached in the game $G(\varphi, M)$, we play a low-level atomic game $G^{\text{Atom}}(S, M)$ so that whenever a constant is encountered in S, a value is assigned to it by one of the players.

As soon as we have implemented such rules, further modifications are needed. The result of playing low-level atomic games is that, instead of the usual winning conventions, according to which

- if S is true then V wins the play of the game $G(S, M)$,
- if S is false then F wins the play of the game $G(S, M)$,

the converse holds:

- if V wins the play of $G^{\text{Atom}}(S, M)$, then S receives the payoff 1 in the parental game $G(\varphi, M)$,
- if F wins the play of $G^{\text{Atom}}(S, M)$, then S receives the payoff -1 in the parental game $G(\varphi, M)$.

The notion of 'winning' according to these converse rules means that the assignment to non-logical constants will produce values that are right in the sense that they are in accordance with what is given by the 'natural history' of those formulas and its use as an assertive proposition by the players in which the constants appear. In Peirce's semeiotics, the notion of winning refers to the process taking place between the interpreter and the utterer (Pietarinen 2003c).

By doing this, we are in effect making game-theoretically meaningful distinctions between partially and totally interpreted languages, in which the law of excluded middle fails either at the level of atomic formulas (partial interpretation) or at the level of complex formulas (total interpretation). In the former case, M is differentiated in two parts, M^+ and M^-, where M^+ is the model in which the formulas of the language are true and M^- is the model in which they are false. If the intersection of M^+ and M^- is non-empty, it gives the set of atomic formulas that lack interpretation. These receive the payoff $(1, -1)$ in the correlated extensive game.

5.2.5. *Non-standard Partiality*

There are non-standard ways to partiality both at the levels of winning conventions and truth conditions. In the latter case, they emerge via the existence of Skolem functions (winning strategies). For instance:

- If V wins $G^{\text{Atom}}(S, M)$ then S is not false (alternatively: true). If F wins $G^{\text{Atom}}(S, M)$ then S is false (alternatively: not true).
- If V wins $G^{\text{Atom}}(S, M)$ then S is true (alternatively: not false). If F wins $G^{\text{Atom}}(S, M)$ then S is not true (alternatively: not false).
- φ is not false in M if and only if there exists a winning strategy for the player who started the game by playing the role of V.
- φ is not true in M if and only if there exists a winning strategy for the player who started the game by playing the role of F.

The players V and F are perhaps best seen as the Non-Falsifier and the Non-Verifier in the clauses defining truth conditions, respectively. These non-standard clauses – which resemble the so-called 'no-counterexample' interpretations – can then be applied in both IF ('hyperclassical') and non-IF ('classical') logics.

5.3. Perfect or Imperfect Recall?

We have seen how semantic games of imperfect information relate to logics. However, the difference between perfect and imperfect information is seen to be subject to further qualification. The class of imperfect-recall games often has to be taken into account too.

5.3.1. *The Modus Operandi*

- Games are of *perfect recall* whenever the players do not forget their previous information or their previous choices. Otherwise they are of *imperfect recall.*

An illustrative real-life example of the distinction between perfect and imperfect information is the game of chess versus the game of poker. The game of bridge, on the other hand, could be considered a game of both imperfect information and imperfect recall, in that two teams play off against each other.

Imperfect recall is a recurring theme in IF languages and games of imperfect information. The distinction between the player's information and choices can be further characterised in a formal precision (Pietarinen 2001c), but I refrain from doing so here. Imperfect recall is also relevant to the following question.

5.4. Two or More Players?

If the players in semantic games have non-persistent information and hence imperfect recall, we need an effective way of modelling such a phenomenon. A natural way of doing so in game theory is to divide the two principal players into multiple-selves or members of teams.

5.4.1. *The Modus Operandi*

- A *team* is a (finite) set of non-coordinating players who have identical payoffs but who act individually.

The teams V ('Us') and F ('Them') consist of a finite number of individual members $V_l \in V$ and $F_k \in F$, for the finite positive integers l, k.

The members of a team are not allowed to communicate because this destroys the team's ability, when viewed as one player, to forget something about which it has had information. The members of the same team all receive the payoff $u_i(h)$ when the outcome of a play is solved.[10]

5.4.2. *Remarks*

The team approach provides a way to prevent players signalling their choices to other players. The information for the individual team members remains persistent, although the teams, viewed as single players, do not forget it. Hence all the moves made by the individual members are assumed to be member-specific, which means that information sets are assigned to them. However, when each team is viewed as a single player, one could think in terms of an implicit map from the 'information set' ($\mathfrak{I}_{i=V}$) or ($\mathfrak{I}_{i=F}$) of all the information sets of the respective player to the information sets of its members. Thus V and F could coordinate the individual players, and determine who makes the next move or who is to be introduced next, although the decision for the actual choices is made by the individual agents.

The idea of the team (or multi-person or multi-self) approach to games of imperfect recall goes back to the early works of von Neumann and Morgenstern (1944), Strotz (1956) and Isbell (1957). According to von Neumann and Morgenstern 1944, 53), "It is worth noting how the necessary "forgetting" of [move$_2$] between [move$_1$] and [move$_3$] was achieved by "splitting the personality" of [V] into [V_1] and [V_2]".

It should be emphasised that the team approach is not, strictly speaking, technically necessary, but it is rather an implementation device to enable us understand the transmission of information in games of imperfect recall. Viewing imperfect recall as a team-theoretic game aims at explaining what happens when information is dispelled from the player's memory. We want information to be persistent for real decision makers, and the team approach provides a way of understanding semantic games for imperfect-information logics. What one eventually may arrive at is an agent normal form for extensive games, whenever each information set is associated with a separate player in a team.

There is a concrete twist in semantic games of imperfect recall, however. Consider the game correlated with the IF formula $\exists x(\exists y/x)\, x = y$. Is this sentence true over the structure of natural numbers? The answer is affirmative, given the basic notion of semantic games with two players with persistent information, but negative, if the players are considered as teams. In the latter case, team V cannot pool the information (here, two Skolem constants) to yield a winning strategy, in other words it does not have persistent information and thus, not unlike typical approaches to automata, exhibits non-persistent memory storages.

5.5. Complete or Incomplete Information?

Despite their generality, imperfect-information games, together with their various refinements, provide just one way of looking at the streams of information between players, or the logical independence and dependence relations thereof. Given any logic that admits of coherent semantic rules, and hence a game-theoretic interpretation, another form of independence, or informational regulation, is seen to emerge. Namely, there may be *lack of information about the mathematical structure of the game* itself – defined, for example, by its extensive form. This paucity may take many forms. The players may not be fully informed about the other players' payoff functions, about the strategies available to them, about the knowledge other players have about the game, and so on.

5.5.1. *The Modus Operandi*

- The game is one of *incomplete information*, if there is a chance move by NATURE that is unobserved by at least one player.[11]

The connection between incomplete-information games and logic is to be found in the behaviour of negation. For negation, like quantifiers and connectives, may be hidden in suitable extensions of IF logic. What such hiding boils down to is that the information about the role switch may not be available to players at the later stages. This means that the players are uncertain about which role they should assume, that is, they are not informed about whether they are to act as the verifiers or as the falsifiers of a given formula.[12]

5.5.2. *Remarks*

These games were distinguished by von Neumann and Morgenstern (1944, 30) from those that do not lack such information, by calling the former games of incomplete information and the latter of complete information. The received explorations in GTS and IF logic have by contrast been confined to games of imperfect information.

There is an evident reason why incomplete-information games have not been studied in logic. The slash-notation '/', as it is defined in IF logic, does not extend to representations concerning the lack of information about the structure of the game. This notation is intended to express which quantifiers, connectives and constants are hidden from which other operators, the cases in point being the subformulas $(\forall x/B)\,\psi$ and $(\exists x/B)\,\psi$ of an IF first-order formula φ, $B \subset \mathsf{BoundVar}(\varphi)$, $x \notin B$ (the strong negation is defined as a role exchange). However, if the slash notation is extended to cover negative operations, namely by (i) hiding negations by extending first-order logic with expressions such as $(\forall x/\neg_i)\,\psi$, $(\exists x/\neg_i)\,\psi$), (ii) letting something off from the scope of negations by $(\neg_i/\forall x)\,\psi$, $(\neg_i/\exists x)\,\psi$ etc., or (iii) combining (i) and (ii) by expressions of the kind $(\neg_i/\neg_j)\,\psi, i \neq j$, one needs to reorganise the game structure itself. What this boils down to is that the notion of scope of negation – different occurrences of negation signs being distinguished from each other by a suitable indexing – needs to be revised, as such scopes may become partially overlapping, thus no longer being transitively nested.

One consequence is that the ordinary law of double negation becomes a limiting case in which $\neg_1(\neg_2/\emptyset)\,\varphi \;\rightarrow\; \varphi$. If there are negations on the right-hand side of the slash, then, for instance, $\neg_1(\neg_2/\neg_1)\,\varphi$ is randomly equal to $\neg_1\neg_2\varphi$ and φ. Consequently, the usual approaches to logical equivalence break down.

Recall that Harsanyi (1967, 167–168) has shown that all forms of incomplete information can be reduced to the case in which players are not fully informed of each other's payoff functions. What thus ensues from the above considerations is a logic of *payoff independence* (Pietarinen 2004b). Formulas of such a logic may be constructed from the expressions described in the previous paragraph, or from some suitable subset of them. To provide semantics, the extensive-form representation of semantic games will be invaluable. Given $\mathcal{G}^*$, it is extended with chance moves by Nature who uses common priors for probabilities and chooses the types of players from a two-element set {V, F} by random. Depending on this choice, the players receive allocations of their payoff functions $u_i(h)$ on terminal histories $h \in H$. Chance moves may also determine the amount of information that the player i obtains about $-i$'s payoff functions. Payoffs are then defined in accordance with the chosen types.

Like perfect information, complete information is thus a mammoth romanticization in logical semantics. The physical outcome of strategy combinations, attitudes to risk, and strategies available to other players are all notions that can be subsumed under the auspices of incomplete information. In logic, what this essentially means is that the hidden chance moves by Nature may alter the way truth (and falsity) of formulas are defined. In contrast to the received game-theoretic definition of truth, we would rather say that the formula is true in M, if and only if in the subgame G' of $G(\varphi, M)$ (which is not led to by the choice of the type F), there exists a pure optimal strategy for the player whose type Nature chose to be V (i.e. the player would act as the verifying player). Likewise, the formula is false in M, if and only if in the subgame G' of $G(\varphi, M)$ (which is not led to by the choice of the type V), there exists a pure optimal strategy for the player whose type Nature chose to be F (i.e. the player would act as the falsifrer).[13]

It is even conceivable that Nature has some preference toward one type on the expense of the other. This amounts to a somewhat recondite but interesting *weighed* notion of negation.

This kind of reversed approach to what is called 'interactive epistemology' in economics is related to modal logic via sample space of subjective probabilities and measurable subsets of events chosen by Nature. Furthermore, in this, but not only in this, sense extensive games can interestingly be viewed as ancestors of the possible-worlds semantics for modal logic.[14]

In the light of the received slash-notation for IF logic and formulas extended with expressions such as $(\forall x/\neg_i)\,\psi$, $(\exists x/\neg_i)\,\psi$, $(\neg_i/\forall x)\,\psi$, $(\neg_i/\exists x)\,\psi$ and $(\neg_i/\neg_j)\,\psi$ and in modal contexts with $(K_j/\neg_i)\,\psi$, $(\neg_i/K_j)\,\psi$, payoff independence explains what happens whenever negation is subject to informational regulations similar to other IF formulas.

Games of incomplete information and the properties of independent negation deserve more attention than can be provided in this paper. For one thing, game-negation is minimal in the sense of being a presupposition-preserving operator. In general, payoff independence is related to NEG-rising in natural language, and is the gist in explaining away the confusions commonly called the Kripkean puzzles of belief.

5.6. Strictly or Non-strictly Competitive Games?

Whereas the classes of semantic games mentioned above deal with various notions of information that the players may possess, a different variant may be created that alters the objectual attitudes they have towards each other – concerning competitiveness, for example. Indeed, strictly competitive games, as the above-mentioned semantic games are, are rarely considered in game theory and non-strictly competitive games such as variable-sum and mixed-motive games are much more common.

One important feature of IF logics is that negation denotes a strong game-theoretic negation. It is possible to introduce a weak contradictory negation $\neg_w$, but this cannot be captured by any game rules (Hintikka 1996, 131–162). The behaviour of classical negation is instead captured by:[15]

(i) $(\mathfrak{A}, g) \models^+ \neg_w\varphi$ if and only if not $(\mathfrak{A}, g) \models^+ \varphi$
(ii) $(\mathfrak{A}, g) \models^- \neg_w\varphi$ if and only if not $(\mathfrak{A}, g) \models^- \varphi$.

In clause (i), the sentence $\neg_w\varphi$ being a truth-consequence of $\mathfrak{A}$ says that φ cannot be verified, and in the latter, $\neg_w\varphi$, being a falsity-consequence of $\mathfrak{A}$, asserts that φ cannot be falsified. Therefore sentences prefixed with weak negation become assertions about games, indicating when a winning verifying or falsifying strategy does not exist.

Consequently, the occurrence of weak negation introduces the fourth truth-value Over-defined. Such an introduction is somewhat limited (albeit not without interest in relation to applications such as some truth theories). Hence, an alternative approach is to take games to be non-strictly competitive.

5.6.1. *The Modus Operandi*

- For any $\mathcal{L}_{\omega\omega}$ or $\mathcal{L}^*_{\omega\omega}$-formula φ, the game $\mathcal{G}(\varphi, g, \mathfrak{A})$ or $\mathcal{G}^*(\varphi, g, \mathfrak{A})$ is *strictly competitive*,
  - if strategy f exists which is winning for V then strategy g does not exist which is winning for F, and
  - if strategy g exists which is winning for F then strategy f does not exist which is winning for V.

In non-strictly competitive games, it may happen that both players have a winning strategy. For instance, there are some terminal nodes that are winning for both V and F. Consequently, atomic formulas ψ are interpreted as having the truth-values True and False, that is, they also have the value of Over-defined.

5.6.2. *Remarks*

Non-strictly competitive games are useful in distinguishing between various notions of consistency: although a version of *ex falso sequitur quodlibet* could be tolerable as $\varphi \land \neg\varphi$, it is never the case that $\varphi \land \neg_w\varphi$, for it does not make sense to assert that 'there exists a winning strategy for V in φ, but there does not exist a winning strategy for V in φ', which would denote strong inconsistency.

Given the zero-sum property of strictly competitive games, the partial truth value Undefined means that the attempts of both V and F can be frustrated. Furthermore, in strictly competitive games, players' preferences could become inverses, but if the preferences are not assumed to be strictly oppositional, the presence of a definite truth value in a sentence does not necessarily mean serious deprivation in terms of the purposes and motivation of the other player. For example, the strategy functions in non-strictly competitive games could at times be (partially) revealed to the opponent.

Non-coherence arises as soon as the assumption that the games are strictly competitive is dropped. How viable is this assumption? A number of real-life situations relate to games that are not strictly competitive, such as the prisoner's dilemma, differential games, bargaining and negotiation games or argumentation. These suggest the emergence of yet another 'non-classic' logic.[16]

Interestingly enough, Aristotle observed the game-like character of competitiveness in relation to certain characteristics of an argument:

> If it [what a question says] is partly true and partly false, he [the answerer] must add a remark that it has several meanings and that in one meaning it is false, in the other true. (*Topica VIII*, sect. VII)
>
> He who hinders the common task is a bad partner, and the same is true in argument; for here, too, there is a common purpose, unless the parties are merely competing against one another; for then they cannot both reach the same goal; since more than one cannot be victorious. (*ibid.*, sect. XI)

The point Aristotle makes here is that if argumentative situations are seen as competitions, then only one player can come out as the winner. There are reasons why such a situation is not preferred over mutually beneficial ones, in which participants may concede points made by an opponent. This is commonly viewed as a disputational rule, a rule for dialectics rather than a rule for logic. Nevertheless, Aristotle devoted considerable energy to it in discussing possible exceptions or qualifications to the law of contradiction in the context of logical investigation.

6. Semantic Games in Epistemology

6.1. Epistemic Logic Generalised

Modern epistemology has wrestled with concepts such as reliability of knowledge, scepticism (in all its colours form Plato to Davidson), processes of justification in scientific inquiry, and more recently with those deploying evolutional models and metaphors. These notions, with a possibly exception of evolutionary models, have been thoroughly investigated using formal tools, and at least sporadically if not systematically, within the context of epistemic logic (logic of knowledge and belief). It may thus appear that almost everything has already been said on the subject of knowledge or, for that matter, on justified, true belief in logic and formal epistemology.

This is far from being true. Epistemic logic is kicking in its most recent developments employing tools and methods from game theory. Indeed, game theory itself has sought its own answers to epistemological problems under the auspices of interactive epistemology, a programme the key idea of which is to apply the toolkit of epistemic logic to the analysis of game-theoretical problems (Bacharach et al. 1997). Here I propose to do the converse and apply game theory to epistemic logic. What this amounts to is that the entire framework of possible-worlds semantics is given an added but far-reaching twist. If we take possible worlds and accessibility relations to refer to agents' range of knowledge as an elimination of uncertainty, the game-theoretic evaluation of that structure is an additional superstructure also involving knowledge, this time integral to the epistemic concepts of those who are playing such a semantic game.[17]

This puts traditional epistemological issues under a new light. Examples of one kind of convergence between logic, epistemology and games are found in Hintikka (1996), within the context of quantified epistemic logic concerning mathematical knowledge. These examples aim to show that by means of the IF notation applied to predicate epistemic logic one is able to distinguish between the notions of 'knowledge of mathematical objects or things' (such as functions), and 'knowledge of mathematical truths, propositions or facts'. The former comes close to intuitionist mathematics, while the latter is what suffices for classical mathematics. Knowledge of a function can be represented by the formula $K_1\forall x(\exists y/K_1)\,(f(x) = y)$, which in (M, w_0) is equisatisfiable to $\exists f\forall g\forall x\,(g \in [w_0]_{\rho_1} \wedge (f(x) = g(x)))$.[18]

The basic model of epistemology in semantic games proceeds as follows. Given a sentence of epistemic logic the truth of which has to be checked, whenever a knowledge operator is encountered, the Skeptic (Malin Genie, the Lightning/the Swampman) purports to find a possible world in which the remaining sentence would come out as false. His adversary, the Inquirer, is allowed to choose worlds for the dual L_i of knowledge, defined by $L_i\psi := \neg K_i\neg\psi$ (following Hintikka (1962), this is read as 'it is possible, for all that i knows, that ψ'). As usual, conjunction and universal quantifier prompt a move by the Skeptic, and disjunction and existential quantifier prompt a move by the Inquirer. The winning rule is given by an interpretation of atomic formulas determined by Nature. The truth and winning strategies are as before: the knowledge sentence is true (false) if and only if there exists a winning strategy for the player who initiated the game as playing the role of the Skeptic (the Inquirer).

For the sake of conceptual clarity, at this point it should be noted that, although there is a theory of epistemology known as the interrogative model of inquiry Hintikka et al. (2002) – which also seeks to articulate scientific enterprise by a game conceptualisation, but by those between Inquirer and the Oracle playing a questioning game where the Inquirer gets new information by the series of questions put to the Oracle and by logical inferences – the interrogative games are different from semantic games. In interrogation, actual epistemological practices of the players are at work, and they can be said to aim at knowledge of truths. How does this differ from semantic games, for is it not the case that we are here defining precisely such games for epistemic concepts? The answer is that it is not the games of actual verification or falsification for sentences involving epistemic terms that are suggested. I do not promote an Aristotelian approach to reasoning and epistemology in the sense of interrogation here. The games for epistemic concepts that aim at settling the truth of sentences of epistemic logic remain purely semantic. They are games of seeking and finding of possible worlds, to pull out the scope of the paradigm argued for in Hintikka (1973) a bit.

At no point am I assuming that the knowing agents mingle with players in this quest for knowledge. It is agents' knowledge that the game checks and shows whether it is true or false. What is not involved is knowledge of something being true or false. Yet, the entrepreneurs of this game come interspersed with their own epistemology, as in the ongoing search for 'interactive' modes of epistemology and game-theoretic solution concepts. The Skeptic and the Inquirer may well have formed their own beliefs and expectations concerning each other's beliefs and expectations.

Furthermore, once one of them makes a choice, the other may or may not have been shown or communicated this choice. If not, the characteristics of the game will radically change, and so will the process of how the solution concepts are formed. On the other hand, if the choices of possible worlds are revealed to the adversary, he or she can be said to *learn* something about the opponent's aims, hence affecting the formation of solution concepts. This fact of there being an element of learning within possible worlds is itself a many-faceted and important issue in game-theoretically interpreted epistemic logics.[19]

One may nonetheless raise a question that, if I do not know what world You has produced, how can I continue playing a game like this? The answer is that the notions of a player knowing or not knowing something and player's position in a game say no same thing. If some information is not visible, the players are choosing actions within any one world, but they may be restricted to actions that have to be legitimate also at certain other worlds, because they could have been positioned to those other states with an equal probability.

As soon as the players' range of actions is limited, important consequences will ensue, since such restrictions impose conditions on admissible models. For instance, the set of available actions will have to coincide for worlds that are equally legitimate under uniform choices. In quantified epistemic logic, this – in addition to the limitation of actions to uniform accessible worlds – will mean uniform (common) domains for those subsets of the set of possible worlds that lie at the endpoints of the histories within the same equivalence class in which the player is making his or her moves. This domain restriction will imply a novel type of *stratified domains assumption*, which means that the domains will be have to coincide for certain modal depths given by the length of the histories in the extensive game for formulas labelled on the histories within a given equivalence class.

6.2. Applications and Consequences

As soon as epistemic logic has multiple agents and iterated modalities referring, in addition to knowledge, to other propositional attitudes such as belief, obligation or temporal concepts, we need to be open to the ideas of taking knowledge and other attitudes of different knowers into account, and to how parts of knowledge come into life in any given flow of time. Even more generally, this mixture goes beyond the sentence level to account for anaphoric relations in discourse involving propositional attitudes.

6.2.1. *Intentional Identity*

An illustrative example of the kind of step needed is provided by the problem of intentional identity of formalising anaphora in trans-sentential multi-agent contexts, originally discovered in Geach (1967). The problem asserts that there may be an anaphoric link between an indefinite term and a pronoun across a sentential boundary and across propositional attitude contexts, so that actual existence of an individual for the indefinite term it not presupposed. An example of this is the following two-agent, two-attitude construction of 'modal coreference' in discourse:

> Einstein thinks that there is a noncovariant solution to the gravitational field equations of general relativity theory. Hilbert thinks that it (the same solution) works for the equations, too. (8)

The resolution to the elusive puzzle of finding a semantic explication for the coreference phenomenon in sentences like this is based on a quantified epistemic logic of imperfect information. I have presented details and various generalisations in Pietarinen (2001a). The idea is that (8) is symbolised by the following two-dimensional operator-quantifier structure:

$$\begin{matrix} K_{\text{Einstein}} \exists x \\ K_{\text{Hilbert}} \exists y, \end{matrix} \tag{9}$$

followed by the matrix 'x is the noncovariant solution to the gravitational field equations of general relativity theory, y is the noncovariant solution to the gravitational field equations of general relativity theory, and x and y are the same'. Overall, the symbolisation resorts to the phenomenon of 'quantifying-out' in modal contexts. When many agents have thoughts on the same individual or entity, these attitudes cease to be non-specific in the sense of constructing a shared individual 'solution-concept' for the two agents. Furthermore, anaphoric pronouns may occur in sentences of different parts of discourse almost arbitrarily far from those items that prompt their intended value.

One consequence of the game-theoretic approach to modal coreference is that various pragmatic factors typically drawn in in tackling these kinds of identities become of secondary concern: it is possible to symbolise intentional identities and resolve the puzzles associated with them by logico-semantic methods, albeit somewhat radically renewed ones.

6.2.2. *Focused Attitudes in Anaphoric Environments*

The other implication is that the notion of *focused attitudes* arises as a stimulating object of study. As soon as there are at least two agents that share objects so that their attitudes are meant to be about the same individual, albeit possibly in different ways, such focused notions of attitudes will ensue. This is old lore if we take all agents to have specific (*de re*) attitudes towards an individual, or if we take all agents to have non-specific (*de dicto*) attitudes (as indeed in (8)). Real novelties arise in multi-agent contexts that mix these two attitudes. It is perfectly conceivable that some agents focus specifically and some non-specifically on a shared individual. This happens in anaphoric sentences such as:

> Einstein knows that a function solves the equations and Hilbert knows what it is. (10)

Some formalisations of these kinds of sentences are given in Pietarinen (2001a, 2002a). In brief, informational independence is needed in order to undo the excess nesting of attitudes of which there is no trace in (10).

The overall motto in relation to predicate epistemic logic with informational independence is that agents' independent thought-spheres may after all be focused towards shared individuals, and this can be done without dispatching pragmatic cargo into the vessel of meaning and ontology.

6.3. A Simplified Propositional Epistemic Language and its Applications

6.3.1. *The Main Idea*

Apart from these quantificational novelties, informational independence in epistemic logic opens up new perspectives to iterated attitudes on the propositional level, too. Let the superscripts in $K_1^a\,\varphi$ be syntactic labels $a, b, \ldots$ to distinguish between different occurrences of epistemic operators. (They can be disposed of if the confusion about which operators are meant by the right hand side of the slash does not arise.) The subscripts may denote the knowing agents or perhaps methods of scientific inquiry. When there is imperfect information, it is no longer the case that nested attitudes such as $K_1^a K_2^b\,\varphi$ and the informationally independent $K_1^a(K_2^b/K_1^a)\,\varphi$ (that is, the branching version $\begin{smallmatrix}K_1^a\\K_2^b\end{smallmatrix}\varphi$) coincide, for the latter has to be evaluated in models with several actual worlds (e.g., in $(M, (w_0, w_0'))$). What this amounts to is that models for informationally independent formulas may consist of independent, detached submodels with multiple designated worlds.

6.3.2. *The KK-thesis Revisited*

Far from being a technical gimmick for creating recondite logics, informational independence has epistemological consequences for the KK-thesis, for example. This thesis states that if an agent knows φ, then he or she knows that he or she knows φ. The status of this principle has been disputed since the inauguration of epistemic logic in Hintikka (1962). However, we need to keep apart the iterated reading of KK-thesis and its branched or informationally independent reading. Recalling the remarks I made on imperfect recall in sect. 5, this difference arises because of the attitudes of the same agent need to be evaluated with respect to a sequence of designated worlds (since there is nothing to distinguish multiple selves within a single agent from those of the selves of many agents). Yet, $K_1^a(K_1^b/K_1^a)\,\varphi$ does not reduce to $K_1^a\,\varphi \wedge K_1^b\,\varphi$ and hence to $K_1\varphi$, because each independent attitude that does *not* depend on any mediating attitudes has to start off from its own designated world.[20] In case the model consists of a single designated state w_0 in which $K_1\varphi$ is true, $K_1^a(K_1^b/K_1^a)\,\varphi$ would be false in it.[21] Hence, even though $K_1\varphi \rightarrow K_1^a K_1^b\,\varphi$ would be a valid axiom of some epistemic system, $K_1\varphi \rightarrow K_1^a(K_1^b/K_1^a)\,\varphi$ is not.

An alternative interpretation of the non-iterated reading of the KK-thesis, in other words, of the sentence $K_1^a(K_1^b/K_1^a)\,\varphi$ is that the Skeptic does not have perfect memory and at K_1^b forgets the possible world that was just chosen for K_1^a. This is in line with semantic games exhibiting imperfect recall. Again, this kind of imperfect recall should not be confused with one that may obtain on the level of knowing agents. The latter need to be captured by epistemic rules such as $K_1\varphi \rightarrow P_1\neg K_1\varphi$, in which P_1 is a Priorean tense operator denoting a point in the future: ‘if 1 knows φ then at some time in the future he will not know φ’.

Similar non-iterated readings are available for other axiom systems of epistemic logic, such as negative introspection. Negative introspection asserts that not knowing a hypothesis implies knowledge of not knowing it: $\neg K_1^a\,\varphi \rightarrow K_1^a\neg K_1^b\,\varphi$. Unlike what goes on in the informationally independent relaxation of the KK-thesis, here the Skeptic does not need to recall the world produced for K_1^a in the consequent, since when scanning a suitable world for K_1^b, this activity is overturned to the Inquirer who seeks to falsify φ. In other words, we would have $K_1^a\neg(K_1^b/K_1^a)\,\varphi$ in the consequent of the negative introspection axiom.[22]

Agent’s iterated knowledge is thus restricted in the sense that the legitimate set of possible worlds that may be chosen for the operator next to the atomic predicate, and in which the predicate thus has a valuation, is the same set which is accessible for all worlds chosen for certain outer operators. The idea thus is to look for such a reading of the KK-thesis that preserves the transitivity of frames while making the thesis less vulnerable to traditional skeptical objections.

6.3.3. *The Varieties of Successful Inquiry*

Instead of knowing agents we may speak of a method $i \in \{1 \ldots n\}$ solving the inductive problem φ. Expressing this as before by $K_i\varphi$, we now say that given a model M and a world w, i is *reliable in solving* φ in w just in case the method succeeds in all worlds in a conceivable epistemic relation to w (Kelly 1996). In this case it is the inductive problem that specifies the range of accessible worlds that are in epistemic relationship labelled by the method (or by several methods). The notion of correctness of output generated by the problem is given by the Environment who gets to decide where the atomic sentences are deemed true. The method that refers to the success of inquiry is constrained by the set of game-theoretic counter-strategic decisions made by the Skeptic whose aim is to try to prevent any such success taking place.

Within this model, the players may entertain different strategic aims within the overall notion of success: we may for instance require them to try to *decide* φ, *verify* it, *falsify* it, or *refute* it. Let us suppose the game on M has reached w. The varying notions of success and failure necessitate a redefinition of winning conventions for atomic ψ:

(G.atom°): If ψ° is atomic, the game ends. The Inquirer wins if ψ° is true in w, and the Skeptic wins if ψ° is not true in w.

(G.atom⋆): If $\psi^\star$ is atomic, the game ends. The Inquirer wins if $\psi^\star$ is not false in w, and the Skeptic wins if $\psi^\star$ is false in w.

The formula $K_i\psi^\circ$ now captures the two facts that (i) the method i *verifies* ψ° in w_0, if and only if ψ° is true in all epistemically i-accessible world of w_0, and (ii) i *refutes* ψ° in w_0, if and only if ψ° is not true in some i-accessible world. On the other hand, the rule (G.atom⋆) has the effect on the truth-conditions so that $K_i\psi^\star$ expresses the cases in which (i) the method i *decides* $\psi^\star$ in w_0, if and only if $\psi^\star$ is not false in every epistemically i-accessible world, and (ii) i *falsifies* $\psi^\star$ in w_0, if and only if $\psi^\star$ is false in some i-accessible world.

As the terminology here suggests, the standard interpretation subsumes the truth-conditions for $K_i\psi^\circ$ and the falsity-conditions for $K_i\psi^\star$. Hence, by changing the game rules we weaken the rules for winning. This illustrates yet another dimension in the pursuit of reliable interactive epistemology, as soon as the game-theoretic parameters for winning are allowed to reflect different, dynamic aspects of success.

Let us finally make a brief remark concerning the complexity of settling the truths of inquiry. Kelly and Glymour (1990) argue that hypotheses have quantificational structures that fit certain set-theoretic and topologic patterns. What ensues from Kant's hypotheses? He argued that there are hypotheses that are not verified by experience, namely antinomies. Kelly (1996) notes that they are of the form 'for each instant, there is an earlier instant', or 'for all entities, there is another entity on which it is contingent'. At first blush, these may seem to exhibit a '∀∃' pattern of quantification. In other words, we may hope to capture antinomies by the general schema 'For all A, there exists B', simply by replacing A and B with the categories in question. Yet, this pattern does not capture one particular sense of antinomies, namely that *there are no more A's than B's* (for monadic A, B) (Boolos 1981).[23] But if so, what we are dealing with are generalised quantifiers expressing facts like 'There are no more Bs than As' or 'At least as many As as Bs'. In particular, the quantifier 'At least as many As as Bs', expressing the comparison between the two cardinalities of A and B, is captured by the Henkin quantifier $\begin{matrix} \forall x & \exists y \\ \forall z & \exists u \end{matrix}((x = z \leftrightarrow y = u) \wedge (Ax \rightarrow By))$, which is equivalent to the IF formula $\forall x \exists y (\forall z/xy)(\exists u/xy)((x = z \leftrightarrow y = u) \wedge (Ax \rightarrow By))$, but it is not reducible to any linear first-order formula. Yet, the complexity of IF first-order logic is enormous: Väänänen (2001) shows that the validity problem for IF logic is not in Σ^n_m for any $n, m \in \omega$.

7. Conclusions and Further Developments

The starting point of this paper was the identification of a class of games that could serve as a semantic framework for all kinds of IF logics. Consequently, some propositional, first-order, extensional and intensional variations were proposed. GTS was applied to a number of natural-language expressions, and a suggestion was made to look at semantic games in logic and language through the lens of extensive games, with their way of representing notions such as information and memory, partiality, winning and losing, and strategies. Some epistemological topics were brought to the light by applying the game-theoretic apparatus to epistemic logic.

7.1. Semantic Games in Logic

Once IF formulas are associated with extensive games of imperfect information, the question arises of what the logics are whose formulas give rise to semantic games that satisfy the required consistency, non-repetition and von Neumann–Morgenstern conditions, but which do not correspond to any known IF formulas. For instance, the sheer existence of imperfect information may be conditionalised for later stages of the game by a single action, which restrict the players' strategy set and their information in novel ways. This shows that there is much more to the IF phenomena than is currently believed.

There are also several implications for game theory. For one thing, how should the assumption of observed options (item (iii) in Section 4.3) be understood? In traditional game theory, 'uniformity' means that the number of available actions has to coincide for histories within an information set.[24] But how can we distinguish between different situations just by counting the number of immediately available actions? It seems that the identities of these actions have to be taken into account as well. Is the inspection of alternatives something that can invariably be accomplished? Precisely what are the identity criteria for the alternatives within an information set?

While these are important questions, their status has to be weighed against other presuppositions of semantic games, and need to be examined in another study. Among such presuppositions is the idea that the domain of the game as not a completed totality, but a figment of reality that the players examine and become only gradually aware.

Moreover, information sets are more dynamic objects than usually assumed. There just does not seem to be any compelling reason to assimilate a player's decision nodes, that is nodes where he or she has to act, into those where his or her information sets are drawn (Pietarinen 2003b). One general implication for game theory is the possibility of representing the phenomenon of a player being absentminded about another player's moves, and not only about his or her own moves. It is possible to accomplish this by means of dynamic information sets that include multiple nodes along the same history within one set.

7.2. Semantic Games in Epistemology

What is the relation between GTS and the epistemic notions it evaluates? On the one hand, there are agents with propositional attitudes, and on the other hand, there are players interpreting these attitudes, with their own epistemology. This distinction was made in relation to the resolution of intentional identity, for instance. To what extent does the external description of attitudes by GTS reflect the internal epistemics, that is, the agent's attitudes and his or her identification capabilities? Does the external evaluation receive some fresh cognitive significance? In preliminary terms, one could look at the dynamics of payoffs (the interpretations of atomic formulas), so that they can be made to depend on the particular sequence of worlds that has been traversed in order to arrive at the terminal formulas.

Furthermore, what is the epistemological impact of the possibility of changing the characteristics of the game in order to have different classes of games at our disposal? One answer is that, once we admit the existence of the Skeptic, he will have some power to control the amount of disturbance to the Inquirer in the possible-worlds structure, and hence make the knower liable to err. For instance, computational (formal) learning theory has some suggestive ideas as to how elements of noise can be introduced to the learning environment. Far from being confined to the idea of the opposing roles of the wicked Skeptic and angelic Inquirer, the disturbance in the process of finding out the truth may, instead, come in the form of regulated or obstructed information between the opposing players. This may amount to undetermined games, wherefore the truth of sentences may remain unknown even if the atomic formulas would be completely interpreted. It is in this way that partiality arises in epistemic logic. Unlike its previous treatment (Doherty 1996), I have produced it for complex formulas, using the class of games of imperfect information as the descriptive foundation.

There are many unexplored combinations including an epistemic logic of in*complete* information, and non-strictly competitive games for modalities. These two possibilities remind us of some curiosities as to how Nature may behave. Does she deceive us? Does she cooperate?

7.3. Further Developments

In order to make imperfect information more viable, game theory has sought further solution concepts to complement the traditional winning and losing positions. One refined solution concept is sequential equilibrium (Kreps and Wilson 1982), according to which players need to form expectations concerning the behaviour and beliefs of other players. Since not all previous moves are known in imperfect-information games, players cannot be certain about opponent's intentions and plans, yet there need to be strategies defined on all decision points, even on out-of-equilibrium ones. Attempts could then be made to capture such twists of uncertain expectations by applying the notion of sequential equilibrium.

It is of some interest that formulas with informational independence may give rise to extensive games that are not structurally consistent, which means that there exists a belief system that does not incorporate the fact that there are strategy profiles according to which some information sets are reached with a positive probability. In general, these further solutions concepts may then be studied in relation to the notion of truth in logic.

Yet another appealing topic is the proximity of evolutionary game theory to logic (Maynard Smith 1982). While most research so far has concentrated on cooperative evolutionary games, there is also a paradigm of non-cooperative games, even within the framework of extensive games. It is worthwhile investigating how concepts such as evolutionary stable strategies relate to logic. The foundational value of such an enterprise is in the concept of evolutionary language games, which aims to establish how humans actually acquired their language. These evolutionary language-games need to accomplish much more than just the "naming games" referred to in Steels (1998). One is reminded of Wittgenstein, who noted that in merely naming something, we have not yet made a genuine move in a language game.

Looking ahead, how far can the idea of controlled information flow be pushed in logic? If we take games themselves as objects of study, could we entertain a notion of independent formulas (or sets of formulas) as well, with concatenated games preserving imperfect information? What if quantifier-free matrices are informationally independent, too, amounting to formulas such as $Qx_1 \ldots Qx_n \genfrac{}{}{0pt}{}{P_1x_1\ldots x_k}{P_2x_{k+1}\ldots x_l} P_3x_{l+1\ldots n}$ $(k < l < n,\ k, l, n \geq 1)$? There is nothing sacred about independent atomic formulas as such, and within extensive games it makes perfect sense to have single instantiations independent of some previous choices (such as $P_2a_1 \ldots (a_l/x_1)$), even if logical constants elsewhere were linearly ordered.

However, as soon as there is informational independence in logic, why not go all the way? Semantic flows in IF logic are regulated by a special notation, usually confined – at least as far as the truth is concerned – to the relaxation of the dependence of existential quantifiers on universal ones. In its most general form, however, independence means that the formulas themselves show all kinds of dependence and non-dependence relations between quantified variables and connectives (and even further, between non-logical constants and subformulas).

To attain the most general form of independence, I suggest to represent formulas themselves by pairs $\langle G, \varphi \rangle$, in which G is a directed graph and φ is the formula with no presuppositions about the relations between its constituents. The relation between two nodes in G would then mean that the information concerning the value of the variable that is instantiated to the variable of a starting node of the relation is transmitted to the ending node of that relation. The graph closed under equivalence relation represents the case in which all of the variables and connectives depend on themselves and on all the others, and the disjoint graph represents the case in which no variable and no connective depends on anything else, not even on itself. The associated semantic games need to be adjusted to reflect these generalisations, by playing concurrent games for components not in any relation. In disjoint graphs, there are no reflexive relations and the game would thus even dispense with singleton information sets.

Acknowledgements

The work on this essay was made possible by a three-year scholarship from the Osk. Huttunen Foundation, and by the Academy of Finland (Project no. 1178561). I am grateful to Jaakko Hintikka, Tapio Janasik, Mika Oksanen, Panu Raatikainen, Shahid Rahman, Veikko Rantala, Gabriel Sandu, Tero Tulenheimo, and the anonymous referee who all helped to improve earlier versions of these ideas.

Notes

1 It has sometimes even been claimed that choice functions are somehow 'deictic', viz. prone to pragmatic considerations (Kratzer 1998). This is not a meaningful claim in GTS, although strategies may of course get deictic information as input. Overall, this concerns the strategic meaning of utterances.

2 Aristotle, *Topica*, edited and translated by E. S. Forster, London, Harvard University Press, 1960.

3 See MS 290 for a connective interpretation in terms of a dialogue, and CP 3.480–482 for a dialogical conception of negation. The references CP are to Peirce (1931–1935) by volume and paragraph number, and the references MS are to Peirce (1967) by manuscript and page number. MS 290 is published, only in part, as CP 5.402n.

4 Following Peirce, we may hence dub Tarski semantics 'ectoporeutic'.

[5] Early sketches of the Skolem normal form are found, for instance in CP 3.505 [1896], where Peirce urges his readers "to place Σ's as far to the left and Π's as far to the right as possible".
[6] See Gentzen (1969), cf. Jervell (1985).
[7] But see Stalnaker (1999), who put forward the view that, even from a strategic viewpoint, the distinction between these two forms is somewhat immaterial.
[8] See Pietarinen (2001a, 2002a) for IF epistemic logic, and Sandu and Pietarinen (2001, 2003) for IF propositional logic.
[9] This kind of clause is not a definition of negation, because it constitutes negation. Or, as Wittgenstein wrote, "We would like to say: "Negation has the property that when it is doubled it yields an affirmation". But the rule doesn't give a further description of negation, it constitutes negation" (Wittgenstein 1978, 7).
[10] Contrary to what was suggested in van Benthem (2003), coalition games typically assume coordination and hence do not provide proper models for understanding informationally independent logics with imperfect recall. Accordingly, they have not been considered in relation to imperfect recall in game-theoretic literature.
[11] Note the introduction of the third player. This new player should not be confused with the player playing the role of the falsifier, also called Nature in previous literature on GTS.
[12] That the lack of information about the mathematical structure of the game itself connects with the ignorance of the roles of the players is of course not obvious. Nonetheless, the notion earned its inventor John C. Harsanyi a Nobel Prize.
[13] A curious question is that, what does the formula show, that is, what is its 'truth-value', had Nature chosen differently?
[14] See Copeland (2002) for a detailed hunt-down of the history of possible-worlds semantics. This investigation would still be needed to be complemented with a systematic study of related and complementary ideas that led to the birth of possible-worlds semantics and accessibilities between states in relation to various notions of modalities, such as the development of the theory of probability and statistics, the early history of game theory in the 1920–1960, and the invention of the theory of personal construct psychology. Surprisingly many of these ideas date back to Peirce's investigations.
[15] It is also possible to study just one direction of these definitions:

- if not $(\mathfrak{A}, g) \models^{+} \varphi$ then $(\mathfrak{A}, g) \models^{+} \neg_{w}\varphi$
- if not $(\mathfrak{A}, g) \models^{-} \varphi$ then $(\mathfrak{A}, g) \models^{-} \neg_{w}\varphi$.

This kind of unidirectional non-truth-functional definition of classical negation does not seem to have been studied in the context of partial and IF logics before.
[16] In Pietarinen (2002c) it is argued that the kind of non-coherence that may result from non-strict winning strategies that transmit potential contradictions (nonzero-sum payoffs) to the level of complex formulas, may be eliminated by evoking 'negotiation games' of alternating offers. These games, metaphorically speaking, aim at bridging corrupt links between language and reality. Cf. Pietarinen (2000).
[17] This means that in any non-epistemic extensional logic, the notion of players' knowledge is material. Furthermore, I will ignore the problem of logical omniscience here. If needed, one may resort to 'impossible possible worlds' (Hintikka 1975; Rantala 1982), or to assume that all attitudes are implicit and that for explicit attitudes, there exists a special 'awareness filter'. Yet again, logical omniscience on the level of agents is to be distinguished from the possibility of having it on the level of players.
[18] As usual, ρ is the accessibility relation of agent i, and $g \in [w_0]_{\rho_i}$ means that g is i-accessible from w_0. Further criteria concerning this type of knowledge are given by different ways of identifying the instances of the known object across possible worlds.
[19] Learning itself can be construed as a game. In particular the notion of PAC ('probably approximately correct') learning from computational learning theory can be viewed as a two-person game

where the environment selects an example e that splits the concept class C that is to be learned into two sets: the set C_0 of concepts that label e negative, and the set C_1 of concepts that label e positive. Then the learner chooses one of these slices which becomes the new C, and throws away the other. The payoff is the length of the play of the game – the shorter the length the better the value for the learner. This is not exactly a semantic game, but it is instructive to recognise the impact of game conceptualisations in relation to learning processes that also raise their heads in epistemic logic (Pietarinen 2003e).

[20] Attitudes that depend on some mediating attitudes but are independent of attitudes that precede those mediating ones – such as $K_1^a K_1^b (K_1^c / K_1^a)\, \varphi$ – are interpreted so that the strategies will get as input the whole information sets.

[21] According to Suppe (1989), skeptical arguments in epistemology are based on the KK-thesis. Thus, if the thesis is not true, and if it is true that a skeptic hinges on iterated knowledge of some sort, skepticism does not need to detain the epistemologists. One understanding of the KK-thesis involving informational independence is a 'confused Skeptic' playing off against the Inquirer in the game in which the Knower entertains multiple realisations of the hypothesis φ. In a sense, we can allow him to 'know again', but not to simple-mindedly 'know that he knows'.

[22] See Pietarinen and Sandu (1999) for some indications as to how formulas of propositional epistemic logic with informational independence may be read in natural language.

[23] Similar argument was observed also by Peirce in CP 4.470: "There is a relation in which every man stands to some woman to whom no other man stands in the same relation; that is, there is a woman corresponding to every man or, in other words, there are at least as many women as men". Peirce was referring to the "kind of graphs which may go under the general head of second intentional graphs" (CP 4.469). Peirce's term of "second intentional", adopted from medieval writers, came subsequently to mean 'second order'.

[24] Luce and Raiffa (1957, 43) express this as follows:

> Each of the moves must have exactly the same number of alternatives. For if one move has r alternatives and another s, where $r \neq s$, then the player would need only count the number of alternatives he actually has in order to eliminate the possibility of being at one move or at another.

Even though they later go on to speak of identification, we are not told what kind of presuppositions it bears, or what the identification criteria is assumed to be.

References

Bacharach, M. O. L., Gérard-Varet, L.-A., Mongin, P., and Shin, H. S., (eds.): 1997, *Epistemic Logic and the Theory of Games and Decisions*, Dordrecht, Kluwer.

van Benthem, Johan: 2003, 'Hintikka Self-applied', to appear in R. E. Auxier and L. E. Hahn (eds.), *Library of Living Philosophers: Jaakko Hintikka*. Available electronically at `http://turing.wins.uva.nl/~johan/H-H.ps`.

Boolos, George: 1981, 'For All A there is a B', *Linguistic Inquiry* **12**, 465–467.

Borel, Emil: 1921, 'La théorie du jeu et les equations intégrales à noyau symétrique', *Comptes Rendus Hebdomadaires des Séances de l'Académie des Sciences* **173**, 1304–1308. (Translation by L. J. Savage: 1953. 'The Theory of Play and Integral Equations with Skew Symmetric Kernels', *Econometrica* **21**, 97–100.)

Copeland, B. Jack: 2002, 'The Genesis of Possible Worlds Semantics', *Journal of Philosophical Logic* **31**, 99–137.

Doherty, Patrik (ed.): 1996, *Partiality, Modality, and Nonmonotonicity*, Stanford, CSLI Publications.

Felscher, Walter: 2002, 'Dialogues as a Foundation for Intuitionistic Logic', in Gabbay, D. and Guenthner, F. (eds.), *Handbook of Philosophical Logic 5*, (2nd edn), Dordrecht, Kluwer, pp. 115–146.

Geach, Peter: 1967, 'Intentional Identity', *Journal of Philosophy* **64**, 627–632.

Gentzen, Gerhard: 1969, 'The Consistency of Elementary Number Theory', in M. E. Szabo (ed.), *The Collected Works of Gerhard Gentzen*, Amsterdam, North-Holland, pp. 132–213.

Harsanyi, John C.: 1967, 'Games with Incomplete Information Played by 'Bayesian' Players. Part I: The Basic Model', *Management Science* **14**, 159–182.

Henkin, Leon: 1961, 'Some Remarks on Infinitely Long Formulas', in (no editor given) *Infinistic Methods. Proceedings of the Symposium on Foundations of Mathematics, Warsaw, Panstwowe* (2–9 September 1959), Naukowe: Wydawnictwo, New York, Pergamon Press, pp. 167–183.

Hilpinen, Risto: 1982, 'On C. S. Peirce's Theory of the Proposition: Peirce as a Precursor of Game-theoretical Semantics', *The Monist* **62**, 182–189.

Hintikka, Jaakko: 1962, *Knowledge and Belief: An Introduction to the Logic of the Two Notions*, Ithaca, Cornell University Press.

Hintikka, Jaakko: 1973, *Logic, Language-Games and Information*, Oxford, Oxford University Press.

Hintikka, Jaakko: 1975, 'Impossible Possible Worlds Vindicated', *Journal of Philosophical Logic* **4**, 475–484.

Hintikka, Jaakko: 1996, *The Principles of Mathematics Revisited*, New York, Cambridge University Press.

Hintikka, Jaakko and Gabriel Sandu: 1989, 'Informational Independence as a Semantical Phenomenon', in J. E. Fenstad, I. T. Frolov and R. Hilpinen (eds.), *Logic, Methodology and Philosophy of Science*, Vol. 8, Amsterdam, North-Holland, pp. 571–589.

Hintikka, Jaakko and Gabriel Sandu: 1997, 'Game-theoretical Semantics', in J. van Benthem, and A. ter Meulen (eds.), *Handbook of Logic and Language*, Amsterdam, Elsevier, pp. 361–410.

Hintikka, Jaakko, Ilpo Halonen and Arto Mutanen: 2002, 'Interrogative Logic', in D. M. Gabbay, R. H. Johnson, H. J. Ohlbach and J. Woods (eds.), *Handbook of the Logic of Argument and Inference. The Turn Towards the Practical*, Dordrecht, Kluwer, pp. 295–337.

Hodges, Wilfrid: 1997, 'Games in Logic', in P. Dekker, M. Stokhof and Y. Venema (eds.), *Proceedings of the 11th Amsterdam Colloquium*, Amsterdam, University of Amsterdam, pp. 13–18.

Isbell, John: 1957, 'Finitary Games', in D. Dresher, A. W. Tucker and P. Wolfe (eds.), *Contributions to the Theory of Games*, Vol. 3, Princeton, Princeton University Press, pp. 79–96.

Janssen, Theo M. V.: 2002. 'On the Interpretation of IF Logic', *Journal of Logic, Language and Information* **11**, 367–387.

Jervell, Hermann R.: 1985, 'Gentzen Games', *Zeitschrift für Mathematische Logic und Grundlagen der Mathematik* **31**, 431–439.

Kalmár, László: 1928–1929, 'Zur Theorie der abstrakten Spiele', *Acta Scientiarum Mathematicarum (Szeged)*, **4**, 65–85. (Translation: 'On the Theory of Abstract Games', in M. A. Dimand, and R. W. Dimand (eds.), *The Foundations of Game Theory*, Vol. 1, Cheltenham, Edward Elgar, 1997, pp. 247–262.)

Kelly, Kevin: 1996, *The Logic of Reliable Inquiry*, New York, Oxford University Press.

Kelly, Kevin and Clark Glymour: 1990, 'Theory Discovery form Data with Mixed Quantifiers', *Journal of Philosophical Logic* **19**, 1–33.

König, Dénes: 1927, 'Über eine Schlußweise aus dem Endlichen ins Unendliche' ('On a Consequence of Passing from the Finite to the Infinite'), *Acta Scientiarum Mathematicarum (Szeged)* **3**, 121–130.

Kratzer, Angelica: 1998, 'Scope or Pseudoscope? Are there Wide-scope Indefinites?', in S. Rothstein (ed.), *Events in Grammar*, Dordrecht, Kluwer, pp. 163–196.

Kreps, David M. and Robert B. Wilson: 1982, 'Sequential Equilibria', *Econometrica* **50**, 863–894.

Kretzmann, Norman and Eleonore Stump: 1988, *The Cambridge Translations of Medieval Philosophical Texts*, Vol. 1, Melbourne, Cambridge University Press.

Kuhn, Harold W.: 1953, 'Extensive Games and the Problem of Information', in H. W. Kuhn, and A. W. Tucker (eds.), *Contributions to the Theory of Games*, Vol. 2, Princeton, Princeton University Press, pp. 193–216.

Langholm, Tore: 1996, 'How Different is Partial Logic?', in P. Doherty (ed.), *Partiality, Modality, and Nonmonotonicity*, Stanford, CSLI, pp. 3–43.

Leibniz, Gottfried W.: 1981, *New Essays on Human Understanding* (Translated and edited by P. Remnant, P. and J. Bennett), Cambridge, Cambridge University Press.

Lorenz, Kuno: 1961, *Arithmetic und Logic als Spiele*, dissertation, Universität Kiel. (Partially reprinted in Lorenzen and Lorenz 1978).

Lorenz, Kuno: 2001, 'Basic Objectives of Dialogue Logic in Historical Perspective', *Synthese* **127**, 255–263.

Lorenzen, Paul: 1955, 'Einführung in die operative Logik und Mathematik', *Die Grundlehren der mathematischen wissenschaften* **78**, Berlin, Springer.

Lorenzen, Paul and Lorenz, Kuno: 1978, *Dialogische Logic*, Darmstadt, Wissenschaftliche Buchgesellschaft.

Luce, R. Duncan and Howard Raiffa: 1957, *Games and Decisions*, New York, John Wiley.

Mann, William C.: 1988, 'Dialogue Games: Conventions of Human Interaction', *Argumentation* **2**, 511–532.

Maynard Smith, John: 1982, *Evolution and the Theory of Games*, Cambridge, Cambridge University Press.

Morgenstern, Oskar: 1976, *Selected Economic Writings of Oskar Morgenstern*, Andrew Schotter (ed.), New York, New York University Press.

von Neumann, John: 1928, 'Zur Theorie der Gesellschaftsspiele', *Mathematische Annalen* **100**, 295–320. (Translation by S. Bargmann: 'On the Theory of Games of Strategy', in A. W. Tucker, and R. D. Luce (eds.), *Contributions to the Theory of Games*, Vol. 4, Princeton, Princeton University Press, 1959, pp. 13–42.)

von Neumann, John: 1953, 'Communication on the Borel Notes', *Econometrica* **21**, 124–125.

von Neumann, John and Oskar Morgenstern: 1944, *Theory of Games and Economic Behavior*, New York, John Wiley.

Peirce, Charles S.: 1931–1935, in C. Hartshorne and P. Weiss (eds.), *Collected Papers*, Vols. 1–6, Cambridge, MA: Harvard University Press.

Peirce, Charles S.: 1967, Manuscripts in the Houghton Library of Harvard University, as identified by Richard Robin, *Annotated Catalogue of the Papers of Charles S. Peirce* (Amherst, University of Massachusettes Press, 1967), and in 'The Peirce Papers: A Supplementary Catalogue', *Transactions of the C. S. Peirce Society* **7** (1971), 37–57.

Pietarinen, Ahti-Veikko: 2000, 'Logic and Coherence in the Light of Competitive Games', *Logique et Analyse* **171–172**, 371–391.

Pietarinen, Ahti-Veikko: 2001a, 'Intentional Identity Revisited', *Nordic Journal of Philosophical Logic* **6**, 144–188.

Pietarinen, Ahti-Veikko: 2001b, 'Most Even Budged Yet: Some Cases for Game-theoretic Semantics in Natural Language', *Theoretical Linguistics* **27**, 20–54.

Pietarinen, Ahti-Veikko: 2001c, 'Propositional Logic of Imperfect Information: Foundations and Applications', *Notre Dame Journal of Formal Logic* **42**, 193–210.

Pietarinen, Ahti-Veikko: 2002a, 'Knowledge Constructions for Artificial Intelligence', in M.-S. Hacid, Z. W. Ras, D. A. Zighed and Y. Kodratoff (eds.), *Foundations of Intelligent Systems, Lecture Notes in Artificial Intelligence* 2366, Springer, pp. 303–311.

Pietarinen, Ahti-Veikko: 2002b, 'Games and Logics of Knowledge for Multi-agent Systems', in C. A. Coello Coello, A. de Albornoz, L. E. Sucar and O. C. Battistutti (eds.), *Advances in Artificial Intelligence, Lecture Notes in Artificial Intelligence* 2313, Springer, pp. 214–223.

Pietarinen, Ahti-Veikko: 2002c, 'Negotiation Games and Conflict Resolution in Logical Semantics', in Paolo Boquet (ed.), *Meaning Negotiation: Papers from the AAAI Workshop (MeaN-02)*, Technical Report WS-02-09, AAAI Press, pp. 25–31.

Pietarinen, Ahti-Veikko: 2003a, 'Peirce's Game-theoretic Ideas in Logic', *Semiotica* **144**, 33–47.

Pietarinen, Ahti-Veikko: 2003b, 'A Note on the Structural Notion of Information in Extensive Games', *Quality & Quantity* **37**, 91–98.

Pietarinen, Ahti-Veikko: 2003c, 'Peirce's Theory of Communication and its Contemporary Relevance', in Kristóf Nyíri (ed.), *Mobile Learning: Essays on Philosophy, Psychology and Education*, Vienna, Passagen Verlag, pp. 81–98.

Pietarinen, Ahti-Veikko: 2003d, 'What do Epistemic Logic and Cognitive Science have to Do with Each Other?', *Cognitive Systems Research* **4**, 169–190.

Pietarinen, Ahti-Veikko: 2003e, 'Games and Formal Tools Versus Games as Explanations in Logic and Science', *Foundations of Science* **8**, 317–364.

Pietarinen, Ahti-Veikko: 2003f, 'What is a Negative Polarity Item?', *Linguistic Analysis* **31**, 165–200.

Pietarinen, Ahti-Veikko: 2004a, 'Diagrammatic Logic and Game-playing', to appear in Grant Malcolm (ed.), *Multidisciplinary Approaches to Visual Representations and Interpretations*, Elsevier.

Pietarinen, Ahti-Veikko: 2004b, 'IF Logic and Incomplete Information', to appear in J. van Benthem et al. (eds.), *The Age of Alternative Logics: Assessing Philosophy of Logic and Mathematics Today*, Kluwer.

Pietarinen, Ahti-Veikko and Gabriel Sandu: 1999, 'Games in Philosophical Logic', *Nordic Journal of Philosophical Logic* **4**, 143–173.

Pietarinen, Ahti-Veikko and Gabriel Sandu: 2004, 'IF Logic, Game-theoretical Semantics and the Philosophy of Science', this volume.

Rahman, Shahid and Helge Rückert: 2001, 'Dialogical Connexive Logic', *Synthese* **127**, 105–139.

Rantala, Veikko: 1982, 'Impossible Worlds and Logical Omniscience', *Acta Philosophica Fennica* **35**, 106–115.

Sandu, Gabriel: 1993, 'On the Logic of Informational Independence and its Applications', *Journal of Philosophical Logic* **22**, 29–60.

Sandu, Gabriel and Ahti-Veikko Pietarinen: 2001, 'Partiality and Games: Propositional Logic', *Logic Journal of the IGPL* **9**, 107–127.

Sandu, Gabriel and Ahti-Veikko Pietarinen: 2003, 'Informationally Independent Connectives', in G. Mints and R. Muskens (eds.), *Games, Logic, and Constructive Sets*, Stanford, CSLI Publications, pp. 23–41.

Scott, Dana: 1993, 'A Game-theoretical Interpretation of Logical Formulae' (manuscript original 1968), *Jahrbuch 1991 der Kurt-Goedel-Gesellschaft*, Wien, Kurt-Goedel-Gesellschaft, pp. 47–48.

Skolem, Thoralf: 1920, 'Logico-combinatorial Investigations in the Satisfiability or Provability of Mathematical Propositions: A Simplified Proof of a Theorem by L. Löwenheim and Generalizations of the Theorem', in J. van Heijenoort (ed.), 1967, *From Frege to Gödel: A Source Book in Mathematical Logic, 1879–1931*, Cambridge, MA, Harvard University Press, pp. 254–263. (Original: 'Logisch-kombinatorische Untersuchungen über die Erfüllbarkeit und Beweisbarkeit mathematischen Sätze nebst einem Theoreme über dichte Mengen', *Skrifter utgit av Videnskabsselskapet i Kristiania*, Vol. 1, Matematisk-naturvidenskabelig klasse **4**, 1–36. Reprinted in Skolem, T.: 1970, in J. E. Fenstad (ed.), *Selected Works in Logic*, Oslo, Universitetsforlaget, pp. 103–136.)

Stalnaker, Robert: 1999, 'Extensive and Strategic Forms: Games and Models for Games', *Research in Economics* **53**, 293–319.

Steels, Luc: 1998, 'Synthesizing the Origins of Language and Meaning using Coevolution, Self-organization and Level Formation', in J. R. Hurford, M. Studdert-Kennedy and C. Knight (eds.),

Evolution of Language: Social and Cognitive Bases, Cambridge, Cambridge University Press, pp. 384–404.

Strotz, Robert H.: 1956, 'Myopia and Inconsistency in Dynamic Utility Maximization', *Review of Economic Studies* **23**, 165–180.

Suppe, Frederick: 1989, *The Semantic Conception of Theories and Scientific Realism*, University of Illinois Press.

Tennant, Neil: 1998, 'Games Some People Would Have All of Us Play', Critical Study/Book Review of Hintikka 1996', *Philosophia Mathematica* **6**, 90–115.

Turner, Ken: 1999, *The Semantics/Pragmatics Interface from Different Points of View*, Current Research in the Semantics/Pragmatics Interface Vol. 1, Oxford, Elsevier.

Ulam, Stanislaw M.: 1958, 'John von Neumann, 1903–1957', *Bulletin of the American Mathematical Society* **64**, 1–49.

Väänänen, Jouko: 2001, 'Second-order Logic and the Foundations of Mathematics', *Bulletin of Symbolic Logic* **7**, 504–520.

Wittgenstein, Ludwig: 1978, *Philosophical Grammar*, Columbia, University of California Press.

Wittgenstein, Ludwig: 2000, *Wittgenstein's Nachlass, The Bergen Electronic Edition*, The Wittgenstein Trustees, The University of Bergen, Oxford University Press. (The transcription used is the diplomatic transcription.)

Zermelo, Ernst: 1913, 'Über eine Anwendung der Mengenlehre auf die Theorie des Schachspiels', in E. W. Hobson, and A. E. H. Love (eds.), *Proceedings of the Fifth International Congress of Mathematicians*, Vol. 2, Cambridge, Cambridge University Press, pp. 501–504. (Translation by Schwalbe, Ulrich and Paul Walker: 'On an Application of Set Theory to the Theory of the Game of Chess', in Schwalbe, Ulrich and Paul Walker: 2001, 'Zermelo and the Early History of Game Theory', *Games and Economic Behaviour* **34**, 123–137.)

IF LOGIC, GAME-THEORETICAL SEMANTICS, AND THE PHILOSOPHY OF SCIENCE

AHTI-VEIKKO PIETARINEN and GABRIEL SANDU
Department of Philosophy, University of Helsinki, P.O. Box 9, FIN-00014 University of Helsinki, Finland, E-mails: pietarin@cc.helsinki.fi; Sandu@elo.helsinki.fi

Abstract. IF (independence-friendly) logic is about informational independence that may take place between any components that admit of an interpretation in terms of game-theoretical semantics. These two approaches are seen to provide integrative tools and methods across individual sciences, including strategic meaning in linguistics, concurrency in computation, knowledge in multi-agent systems, and quantum information. An overarching theme is to get less-than-hyper-rational, decentralised decision makers to agree on the truth of statements codifying central structural features of these individual sciences. One upshot is that semantic games call for a re-examination of some basic assumptions in game theory.

1. New Prospects for the Philosophy of Science?

How can we come to propose new prospects for such an aged authority as the philosophy of science, given such a recent, even juvenile theory as IF (independence-friendly) logic? The reason is that IF logic exceeds classical logic not unlike the way in which non-commutative probability theory exceeds classical Kolmogovarian probability, or the way in which quantum mechanics exceeds classical Newtonian mechanics.

What is IF logic? We will review some of its basic concepts below. Briefly, it is a conservative extension of traditional first-order logic that liberates first-order logic from the confines of linearity. By linearity, it is meant the reflexive, asymmetric and transitive dependence relations between logically active components of a formula, the chief components being the archetypal universal and existential quantifiers.

IF logic has to go together with a semantic theory from which its expressions derive their meaning. We will use game-theoretical semantics (GTS), an overall approach to logical and linguistic semantics developed in Hintikka (1973) and in many subsequent publications. It is this combination of IF logic plus GTS that, we believe, will stimulate new questions in the philosophy of science and in the philosophy of particular sciences. These contributions include, but are not limited to, issues in the foundations of logic, mathematics, linguistics, logical approaches to quantum theory and the philosophy of physics, and in the structural design of parallel forms of processing in computer science. Some of these issues are investigated here.

S. Rahman et al. (eds.), Logic, Epistemology, and the Unity of Science, pp. 105–138.

What are the overall invariants that one could hope to persist across the multiplicity of these disparate fields of inquiry? One such topic took shape in the 1950s when economics was observed to be that particularly captivating arena in which to formalise notions of rationality, decision making, or anything from operations research to cybernetics, including general equilibrium theory, non-linear programming, control and measure theory, and optimal allocation. From economics it escalated to other sciences, ranging from AI to logic and logical epistemology, from game theory to physics, and from cognitive science to evolutionary biology and genetics. It is the notion of limited or bounded rationality, typically ascribed to individual decision makers.

By this notion, we nonetheless do not mean what Herb Simon long ago put forward as a response to rational analysis roaming the post-war Princeton campus, a view opposed to the idea that an organism or artificial system should be capable of optimising its behaviour, especially when it comes to problem solving. In Simon's view, agents are at best local optimisers, with a limited supply of resources unveiling their inherent impediments in problem solving.

In reality, one has to be prepared to face aspects of this notion whatever the field of inquiry related to games is going to be. It does not respect the distinction between exact and non-exact sciences, and is as likely to arise in exact scientific topics such as logic or physics as in societal and psychological phenomena, including cognitive science.

Yet, precisely how does the notion of bounded rationality arise in the topic of this paper, IF logic and GTS? For isn't what we call the meaning in logic and language, from the game-theoretic perspective, timeless and abstract, to the extent that there is hardly anything 'bounded' in the resources of a real agent, acting in the real world, that could be adduced in corners of logic in the first place?

This is indeed where the departure that modern game theory took from Simon's sayings proves instructive. In the writings of game theorists, most notably perhaps by Robert Aumann,[1] the concept of bounded rationality was introduced to bolster a Nash-type analysis, reinforced by the injection of memory and information into game-theoretic argumentation. This broader perspective has subsequently been vindicated in Rubinstein (1998) and elsewhere.

Even further, while the idea of limited agenthood is rife in economics, especially in the field of 'interactive epistemology', its logical investigation has been limited to the neighbouring fields of bounded reasoning capabilities of agents in modal logics of knowledge and belief (Bacharach et al. 1997). According to this view, epistemic logic is thought to be the main arena in which hyper-rational reasoning and logical omniscience has to be confronted head-on. But this confines different modes of rationality inside the realm of logical reasoning. Following the game-theoretic conduct described in the previous paragraph, in logic, too, rational agenthood may be looked into from a broader perspective, taking the notion of information and agents' access to it as one of the prime motifs.

Contributions of IF logic and GTS to the philosophy of science are, in their most part, found within individual sciences. In the following pages we present a few samples of these contributions. Precisely how these support arguments pro and con the unity, or pro and con the disunity, of science remains aloof, even though the overarching methods and tools will be similar across individual sciences. The problem with the possibility of there being some grand unisonance of scientific theories is that science itself may not provide any clear and distinct signal in case some sense of a unity was reached. But then, one ought not to be led to reckon that science would, at some point, signal some miserable disunity and an increasing fragmentation as the only option, either.

This agnostic stance is hardly novel. It can be characterised as a passage from the individual truth of how some thinkers, C. S. Peirce the pragmaticist leading the way, understood the concept of the scientific method and its limitations, to the generic truth concerning the unity of scientific theories.

Yet how can anyone claim to know *this*? (For is it not the case that knowledge of the fact that science is not showing off definite signs concerning the truth or falsity of our theories is also beyond the reach of our understanding?) We don't, but the philosophy of science, as indeed scientific methods, has to begin in the beginning. Since the preference is that the beginning is in the current (actual, institutionalised) state the science is in, scientific philosophising may be applied to science as any scientific method may be applied to inquiry, namely by looking at and studying what is done in the individual disciplines, adding logical analysis as the need arises. If this preference, via the promotion of increased communication and cooperation between individual fields, leads to some convergence, then the states that were reached need assessment. Meanwhile, many of the perspectives herein are only just evolving.

2. IF Logic and Semantic Games

Independence-friendly (IF) logic alias hyperclassical logic (Hintikka 1996, 2002; Hintikka and Sandu 1997) is an extension of first-order logic with Henkin quantifier prefixes (finite partially-ordered quantifiers) of the following form:

$$\begin{matrix} \forall x_1 \ldots x_n \, \exists y \\ \forall z_1 \ldots z_m \, \exists w \end{matrix} \quad (\text{for some } n, m \in \omega). \tag{1}$$

The Henkin quantifier prefix is here in what we call the Krynicki normal form for Henkin quantifiers.[2]

IF logic is a generalisation of Henkin quantifiers in three main respects: it allows (i) non-transitive quantifier/connective ordering, (ii) cyclic or mutual dependencies between quantifiers/connectives, and (iii) imperfect information extending to individual non-logical constants. IF logic is also known as a logic of informational independence, which underscores the fact that it is the semantic information flow within formulas, or dependencies and independencies between quantified variables or connectives, that are liberated from classical linearity and perfect information. Furthermore, the idea of informational independence may be extended to apply to formulas of modal and epistemic logics (Pietarinen 2001a, 2002b).

In the syntax of IF first-order logic, expressions of the following kind may replace their 'slash-free' counterparts: $(\forall x/W)$, $(\exists x/W)$, $(\vee/W)$, $(\wedge/W)$, where W is a subset of bound variables of a formula φ containing at least one of these expressions. A propositional IF fragment is derived by restricting quantifiers to selecting from a set of two elements (e.g., a designated plus any other element of the domain), and the interpreting $\vee$ and $\wedge$ as restricted quantifiers of this sort. The language may then be closed under strong negation.

For example, the formula

$$\forall x_1 \ldots x_n \exists y (\forall z_1 \ldots z_m / x_1 \ldots x_n y)(\exists w / x_1 \ldots x_n y)\, Sx_1 \ldots x_n y z_1 \ldots z_m w$$

is an equivalent IF version of

$$\mathbf{H}^*\, Sx_1 \ldots x_n y z_1 \ldots z_m w,$$

where $\mathbf{H}^*$ is the Krynicki normal form quantifier prefix as in (1).

The semantics for IF logic is given by means of games. We prefer the extensive games approach (an alternative is to stick to Skolem functions throughout). A finite sequence $\langle a^i \rangle_{i=1}^n$, $n \in \omega$ represents the consecutive actions of players in N (no chance moves), $a^i \in A$. An extensive game $\mathcal{G}$ of perfect information is a five-tuple $\mathcal{G}_A = \langle H, Z, P, N, (u_i)_{i \in N} \rangle$ such that H is a set of finite sequences of actions $h = \langle a^i \rangle_{i=1}^n$ from A, called histories of the game, so that the empty sequence $\langle\rangle$ is in H, and if $h \in H$, then any initial segment of h is in H too, that is, if $h = \langle a^i \rangle_{i=1}^n \in H$ then $pr(h) = \langle a^i \rangle_{i=1}^{n-1} \in H$ for all n, where $pr(h)$ is the immediate predecessor of h ($= \emptyset$ for $h = \emptyset$); Z is a set of maximal histories (complete plays) of the game. If a history $h = \langle a^i \rangle_{i=1}^n \in H$ can continue as $h' = \langle a^i \rangle_{i=1}^{n+1} \in H$, h is a non-terminal history and $a^n \in A$ is a non-terminal element. Otherwise they are terminal. Any $h \in Z$ is terminal; $P \colon H \setminus Z \to N$ is the player function that assigns to every non-terminal history a player in N whose turn is to move; each u_i, $i \in N$ is the payoff function that specifies for each maximal history the payoff for player i.

For any non-terminal history $h \in H$ define $A(h) = \{x \in A \mid h \frown x \in H\}$. A strategy for a player i is any function $f_i \colon P^{-1}(\{i\}) \to A$ such that $f_i(h) \in A(h)$, where $P^{-1}(\{i\})$ is the set of all histories where player i is to move. A strategy specifies an action also for histories that may never be reached.

In a strictly competitive game, $N = \{V, F\}$ and in addition, $u_V(h) = -u_F(h)$, and either $u_V(h) = 1$ or $u_V(h) = -1$ (that is, V either wins or loses) for all terminal histories $h \in Z$.

Given a perfect information game $\mathcal{G}_A$, we represent imperfect information by extending $\mathcal{G}_A$ to $\mathcal{G}_A^* = \langle H, Z, P, N, (u_i)_{i \in N}, (\mathfrak{I}_i)_{i \in N} \rangle$, where $\mathfrak{I}_i$ is an information partition of $P^{-1}(\{i\})$ such that for all $h, h' \in S^i_j$, $h \frown x \in H$ if and only if $h' \frown x \in H, x \in A, j = 1 \ldots m, i = 1 \ldots k, m \leq k$. S^i_j is called an information set. In imperfect-information games, the strategy functions are defined on the information sets of the partition. A winning strategy for $i \in \{V, F\}$ is a set of strategies f_i that leads i to $u_i(h) = 1$ no matter how the player $-i$ decides to act.

Let Sub(φ) denote a set of subformulas of a formula φ. An extensive semantic game $\mathcal{G}(\varphi, g, \mathfrak{A})$, associated with an $\mathcal{L}_{\omega\omega}$-formula φ, is exactly like $\mathcal{G}_A$ except that it has a labelling function $L: H \rightarrow \text{Sub}(\varphi)$ such that $L(\langle\rangle) = \varphi$; for every terminal history $h \in Z$, $L(h)$ is an atomic formula or its negation. In addition, the components H, L, P, u_V and u_F jointly satisfy that: if $L(h) = \neg\varphi$ and $P(h) = V$, then $h \frown \varphi \in H, L(h \frown \varphi) = \varphi, P(h \frown \varphi) = F$; if $L(h) = \neg\varphi$ and $P(h) = F$, then $h \frown \varphi \in H, L(h \frown \varphi) = \varphi, P(h \frown \varphi) = V$; if $L(h) = \psi \vee \theta$ or $L(h) = \psi \wedge \theta$, then $h \frown \mathsf{Left} \in H$, $h \frown \mathsf{Right} \in H$, $L(h \frown \mathsf{Left}) = \psi$, and $L(h \frown \mathsf{Right}) = \theta$; if $L(h) = \psi \vee \theta$, then $P(h) = V$; if $L(h) = \psi \wedge \theta$, then $P(h) = F$; if $L(h) = \exists x\varphi$ or $L(h) = \forall x\varphi$, then $h \frown a \in H$ for every $a \in |\mathfrak{A}|$; if $L(h) = \exists x\varphi$, then $P(h) = V$; if $L(h) = \forall x\varphi$, then $P(h) = F$. Furthermore, for every terminal history $h \in Z$, if $L(h) = Pt_1 \ldots t_m$ and $(\mathfrak{A}, g) \models Pt_1 \ldots t_m$, then $u_V(h) = 1$ and $u_F(h) = -1$, and if $L(h) = Pt_1 \ldots t_m$ and $(\mathfrak{A}, g) \not\models Pt_1 \ldots t_m$, then $u_V(h) = -1$ and $u_F(h) = 1$.

3. Facets of Bounded Rationality

Agents who need to operate within pre-defined limits in their representational and cognitive scenery, make ideal rationality as assumed by the traditional game theory look plainly false. How should we cope with any less-than-ideal rationality in semantic games? If the players – in Aumann's words, "do not scan the choice set and consciously pick a maximal element from it" (Aumann 1992, 108) – what is there to reflect this in GTS? IF logic already provides a preliminary answer: it is semantic information that is suppressed from a decision-maker, which, among other things, may turn the logic partial.

However, the concept of bounded rationality may refer to all sorts of limitations subordinate to a variety of interpretations. The restriction on information is certainly illustrative in devising partial logics, but it throws light only on one particular aspect of it. Even though partiality is one of the main characteristics of IF logic, there is more to the notion of boundedness than meets the eye in the earlier literature, such as other kinds of informational losses and increases, including imperfect and bounded recall of actions, and knowledge of and information about other players' actions (Pietarinen 2001c; Pietarinen and Sandu 1999).

Other, low-level characteristics of bounded rationality include the following:

- Decision makers recognise the environment within which they operate, and make inferences on the basis of that recognition.
- Negative knowledge may lead to falsehood of propositions.

- Complicated strategies and equilibria refinements may be replaced by tractable protocols, rules of thumb, habits, customs, institutions, and so on.

The first item implies that the surroundings at which the player is located deliver only partial or, at best, uncertain information. In other words, what we are dealing with is what Savage (1954) coined the doctrine of "small worlds" of a decision maker. What does this mean? Savage considers decision problems that take place within such "small worlds" (*ibid.*, pp. 8–10, 15–17, 82–91). By this he means that sequence of events that contains sets of states are temporally and spatially limited and do not encompass all the information about the actual or "grand world". As far as the individual decision maker is concerned, attention needs to be restricted to "relatively simple situations" (*ibid.*, p. 82), or at least such situations have to be isolated from larger contexts. To describe a particular state of a small world is to say which possibilities are included in it. To describe an act in it is to state which particular possibility is performed within a world.

Game theorists have largely approved such foundations in at least non-cooperative decision problems.[3] The Savagian approach is essentially also a Bayesian one; laid bare it means that the uncertainty faced by the players concerns the strategy choices of their opponents. In particular, in Bayesian types of reasoning, each player forms a prior expectation over the strategy profile of the opponent, each player has some uncertainty over this prior, and each player has some uncertainty over the other players' priors. Therefore, players' beliefs and expectations form an infinite hierarchy with inherent uncertainty measures. The problem is, where do these priors come from? Several answers, such as maximum entropy measures, have been suggested, but with little conceptual consensus.

The first item in our list thus concerns agents' limited ability to analyse the status of their environment. This is indeed reflected in the partitional information structure of games of imperfect information for IF logics. However, bounded rationality concerning agents' environment also concerns the sets of strategies to which a player has an access. It affects the equilibrium selection and hence the winning strategies. For example, a player may not know his or her current equilibria. Consequently, game-theoretic notions of truth rely on assumptions concerning players' rationality.[4]

As far as the second item is concerned, if it is assumed that negative knowledge of a proposition does not lead to the falsehood of that proposition, negative knowledge may lead at least to indifference concerning the truth-value of the propositions. That is, it may lead to some propositions being neither true nor false. This is the characteristic feature of partial and imperfect information logics, including IF logic. However, this feature ought to be contrasted with another outcome of negative knowledge – a distant relative to the small worlds doctrine – namely the closed world assumption (CWA). Its *sine qua non* is that anything one does not know to be true is false.

CWA is pretty much the official doctrine in areas of AI such as logic programming and nonmonotonic logics. This is not surprising, given that in these fields the meaning of the negation is usually not the classical, contradictory operation. In order to ensure that there is limited rationality in transforming negative knowledge to the falsehood of propositions, the CWA maintains that propositions that do not exist, or are not found, or are not provable (verifiable), in a given system, data- or knowledge base, will be interpreted as false.

Yet, in logic programming, even though negation is not classical in the sense that whenever it is encountered in the body of a clause, it is interpreted as a failure to unificate goals. Nonetheless, it is interpreted as a contradiction-forming operator under the CWA. Hence, 'negation as failure' is rendered equivalent to classical negation whenever attached to propositions that are asserted not to exist within the confines of a given system.

In contrast, what contradictory negation says in IF logic is that there does not exist a winning strategy for the verifier of the proposition to which the negation is prefixed. This weak concept of negation is related to the idea of negation as a failure to unificate, although it is more general. The purpose may refer to a proof, verification, resolution, or any argumentative or informal warrant for moving from premisses to the conclusion. A weak concept of negation is inevitable in order to sustain the CWA, even though it is optional logically. For example, in IF logic contradictory negation is added to the language that already contains the strong, game-theoretically produced concept of negation.

This shows that the CWA does not itself yield to outright partiality, although it may yield to nonmonotonicity, viz. a logic in which conclusions may be defeated and revised on the basis of additional inferences, by providing a default in the absence of a better hypothesis. Yet, the CWA and partiality are not unrelated. They are both motivated by considerations of bounded rationality in the spirit of the canon of small worlds. This is not to say that CWA is unproblematic from the philosophical or linguistic point of view. For one thing, in the distinction between disbelieving a statement and believing its negation is often marred in nonmonotonic modal reasoning – but an all-important difference in doxastic logics all the same. For instance, this distinction has to be kept in mind both in assessing the adequacy of the so-called puzzles of belief, as well as in coping with the related syntactic phenomenon of NEG-rising in natural language.

IF logic is thus a candidate for knowledge-based systems in its capacity of representing inexact concepts. As far as complex formulas are at issue, inexactness can be interpreted as undefinedness in the sense of game-theoretic non-determinacy.

Furthermore, rational agents ought to notice and draw inferences from the facts that nothing happens, examples of which range from climate change and ethical paradoxes to the curious case of the Silver Blaze.

As to the third item, taking bounded rationality seriously will ultimately question the need for the rationality principles in game theory and a fortiori in GTS. This is not as radical as it may sound, since for instance in evolutionary game theory, actors can arrive at equilibria even if no ordinary sense of rationality is involved. We will return to this point in the last section.[5]

Further along the road to limited information processing, one needs to cope with non-partitional information, which has repercussions to logics extending the above-mentioned IF first-order logic. These logics are realistic for what the 'practical turn' in logic calls, since players would have an imperfect understanding of their own information processing.[6]

4. Partiality, Coherence and IF logic

What else transpires in non-ideal situations that may crop up in a semantic game? As remarked above, not all semantic games are determined, and hence the law of excluded middle fails and the logics become partial. In partial logics, formulas may either receive a truth-value of Undefined or lack a truth value altogether.[7]

Yet, if this much is the case, why is not the law of non-contradiction invalidated? The reason is that semantic games are strictly competitive, in other words both players cannot come out as winners. In no semantic game winning strategies exist for both players. However, this holds only if the class of games is limited to strict competition. But such class is to general game theory what W. C. Handy was to Satchmo. It is perfectly possible to relax this assumption. In that case, the following no longer holds: if there exists a winning strategy f_V then there does not exist a winning strategy g_F, and if there exists a winning strategy g_F then there does not exist a winning strategy f_V.

If the game is not strictly competitive, call it non-strictly competitive. To implement this, one needs to stipulate the existence of terminal histories in Z that are winning for both players, namely the values of the utility $u_i(h)$ may be $(1, 1)$ for some $h \in Z$. Consequently, given a literal, it may be interpreted so that it has both the truth-value True and the truth-value False, and hence has a truth-value of Over-defined.

One consequence is that, with respect to determinacy, the presence of nonzero-sum payoffs may cancel the effect of imperfect information, which otherwise would have turned a strictly competitive game into a non-determined one. To see this, let the strategies that force a non-determined extensive game into a determined one be winning strategies for determinacy. Assume that $u_V(h) = u_F(h) = 1$ for some $h \in Z$, and that the rest of the payoffs at Z are strict. Then all $h' \in Z$ reached from an immediate predecessor $pr(h) \in H$ of h have to get $u_V(h') = -1$ or $u_F(h') = -1$, because otherwise a player would have a winning move at $pr(h)$. Suppose that $pr(h)$ is contained in a non-singleton information set S^i_j. Then every action from $pr(h)$ has to have a corresponding action at k, such that $pr(h), k \in S^i_j, i \in \{V, F\}$. But any action $a \in A$ corresponding to the action that leads to h can lead only to $k' = k \frown a$ that has either $u_V(k') = 1$ or $u_F(k') = 1$, which hence constitutes a winning step for either V or F.

It follows that what is needed in order to restore non-determinacy in non-strict games are partially interpreted models. That is, one needs partial logics in which atomic formulas are partially interpreted.

5. Conflict Resolution by Negotiation

It is precisely here that the main thrust of the previous section lies. For if there are nonzero-sum payoffs suggesting a 'division of surplus', and if the terminal histories of such payoffs are reached with a positive probability, there will inevitably be a potential conflict. In such a case the commodity needs to be redistributed, which in logical terms is a consequence of the fact that the truth – and likewise the falsity – of the sentences of the underlying logic are unevenly agreed upon. Nonetheless, this is merely a potential conflict or potential non-coherence, because it is not yet stated that there actually exist duplicate winning strategies. However, any potential conflict always runs a risk of becoming actual.

The difference between two notions of conflict is that in the case of the potential, the only formulas giving rise to non-coherence are literals. If in the non-strictly competitive game both nonzero-sum payoffs and winning strategies exist, the potential contradictions are transmitted to complex sentences and thus become actualised.

Suppose that actual non-coherence falls out from formulas of the form $S \wedge \neg S$, in which S are non-atomic and '$\neg$' is strong negation. This is surely not a welcomed feature. However, what may be done here is to call on a 'negotiation process' in order to try to resolve the conflict.

This prompts some fundamental questions concerning the resolution of contradictory statements in logic. What, in fact, are the kinds of negotiations that are to be carried out concerning the meaning of a complex contradictory formula? Shouldn't we make our life easier and forestall negotiations by sticking to the class of strictly competitive games in the first place? Who plays the negotiation games? What are their characteristics, and what, if anything, do they have to do with the theory of semantic games?

To outline some partial answers, the commodity is over truth-values of complex sentences. Hence conflicts may be resolved even if we were not to dispense with nonzero-sum payoffs of atomic predicates, provided that we dispense with non-strict winning strategies. It is nevertheless not clear whether it is possible to dispense with something that pertains to the 'existence' of such strategies, because existence is an objective property of the part of the reality or the model in question, not any epistemological quiz of coming to know what such strategies are.

It would be tempting to conclude that logical conflicts arise because of some cognitive or epistemological restraints, such as players' imperfect knowledge of the model or noisy communication between the partners. However, as far as semantic games are concerned, these informational limitations give rise to partiality of the underlying logic rather than non-coherence. There may be partiality even if the language was completely interpreted. Admittedly, it is technically perfectly possible to map 'over-defined truth values' to 'truth-value gaps' (Langholm 1988), but this is merely a technical roundabout tactics for substituting non-coherence in favour of partiality.

There is an alternative in resolving actual logical conflicts, however, which resorts to a negotiation game modelled by an alternating sequence of actions consisting of players' acceptances and rejections.[8] In this game, V and F make alternating offers according to some schedule of integers. The first move in the schedule takes place whenever the first player in either the team of the Verifiers or the team of the Falsifiers makes an offer, and the first player in the adversary team chooses either to accept or reject the offer.[9] If the choice is to accept, the game ends, and if it is to reject, then the schedule moves to the next stage according to a common clock. The negotiation then repeats. There is a possibility of the negotiation breaking down, and if there is no acceptance there will be no agreement.

We need not incorporate any specific notion of 'offers' into this model. In principle, they refer to choices that have led to non-zero sum payoffs. It suffices that the parties either stick to or throw away those winning strategies that have actually led to conflicting situations. These negotiations differ from semantic games in running through the choices that have been made in an alternating fashion.

The Nash solution and other solution concepts are known only if the negotiation game is one of perfect information. It is thus an open question how such solution concepts that try to account for the players' beliefs and information under uncertainty (i.e., sequential equilibria) could be incorporated into the model. This does not affect the possibility of there being semantic games for a logic that are of imperfect information, because unresolved questions about solution concepts concern the negotiation phase that takes place after the semantic game has been completed.

Therefore, negotiation games are parasitic on semantic games with nonzero-sum payoffs. Moreover, the terminal histories in which such payoffs are found have to be reachable by winning strategies in a non-strictly game. The impact is that contradictory constituents may be passable, provided that not both players' winning strategies lead to them. If that were the case, then some such winning strategies will have to be voted off.

This negotiation phase is not a version of any dialogue game between actual utterers and interpreters of language (contrary to suggestions to a similar effect in Hulstijn 2000), but a language game of conflict resolution, in which non-coherence results not from any contradictory game rules, but from the existence of certain sets of strategies. If a slogan is needed, negotiation aims at bridging corrupt links between language and reality.

Aside from semantic games, in classical game theory the idea of negotiation has been taken to imply some social connotations not only in the sense of negotiations taking place among groups of actual participants, but also in the sense that it is typically assumed to be rational for actual players to resort to posturing, information concealing, exaggeration, threat or deception. On the face of it, this seems far removed from the goals of meaning of expressions of one's language in its truth-conditional sense. For surely there are no such things in the medium that links language to the reality it aims to describe, one might – naïvely perhaps – argue. Yet, how do we know that Nature does not apply these? And if she does, why should not I engage in similar activities?[10]

Non-coherence in the game-theoretic sense encodes features of environment compatible with Peirce's understanding of vagueness of signs in logic and semeiotics:

> A proposition is vague when there are possible states of things concerning which it is intrinsically uncertain whether, had they been contemplated by the speaker, he would have regarded them as excluded or allowed by the proposition. By intrinsically uncertain we mean not uncertain in consequence of any ignorance of the interpreter, but because the speaker's habits of language were indeterminate; so that one day he would regard the proposition as excluding, another as admitting, those states of things. (Peirce 1902, 748)

In our context, this definition may be deciphered by taking speaker's habits of language to reflect strategies that allow for semeiotic 'latitude' (Peirce's term) in affecting the achievement of players' aspirations.

6. Bringing Wittgenstein In

Together with Peirce, also Wittgenstein would have been content with the kind of strategic outlook on contradictions outlined above. For the previous observations may be complemented with yet another, and as far as we know, not previously noted, character of Wittgenstein's language games, namely competitiveness.[11] The place in which this is emphasised refers to Wittgenstein's remarks on the "civil" nature of strategies in language games (Wittgenstein 1953, 125):

> We lay down rules, a technique, for a game, and that then when we follow the rules, things do not turn out as we assumed. That we are therefore as it were entangled in our rules. [...] It throws light on our concept of *meaning* something. For in those cases things turn out otherwise than we had meant, foreseen. That is just what we say when, for example, a contradiction appears: "I didn't mean it like that." The civil status of a contradiction, or its status in civil life: there is the philosophical problem.

As Wittgenstein was right in noting, there does not have to be anything inconsistent in the *rules* of the language game in order for us to end up with non-coherent formulas in which both participants may claim success for their own purposes. Yet, a great deal of recent discussion concerning his views on contradictories as a result of his way of setting up games presupposes that contradictories should somehow be the end-products of contradictory game rules (see e.g., Goldstein 1989). Such a presupposition is not warranted, as shown by the possibility of having semantic games with characteristics that are different from those of ordinary games, which result in inconsistencies simply by altering the class of games in question. Moreover, a steadfast refutation of the assumption comes from Wittgenstein himself: "Why may not the rules contradict each other? Because otherwise they wouldn't be rules" (Wittgenstein 1978, 305).[12]

7. Towards Decentralised Processing in Logic

While there is likeness in Wittgenstein and Peirce's views on logic and language, the topic of this section can perhaps be best delineated by a quotation from Peirce. In CP 4.240 [c. 1902] he remarked "Formal logic, however, is by no means the whole of logic, or even its principal part. It is hardly to be reckoned as a part of logic proper. Logic has to define its aim; and in doing so is even more dependent upon ethics, or the philosophy of aims, by far, than it is, in the methodeutic branch, upon mathematics".[13] This quote is instructive in pointing out the generality of the science of logic beyond the purview of its purely formal or mathematical use.

To illustrate just one example revolving around normativity of logic we note that, as soon as we allow unrestricted notation in representing various ways of expressing variable dependencies and semantic information flow within formulas, IF logic becomes equipped with a way of capturing the phenomenon of forgetting information – or imperfect recall, as game theorists prefer saying. Without going into the details of this notion and its consequences here (Pietarinen 2001c; Pietarinen and Sandu 1999), imperfect recall follows not only from semantic games of imperfect information in which independencies exist between existentially or between universally quantified variables (as in IF formulas $\exists x(\exists y/x)\, Sxy$ or $\forall x(\forall y/x)\, Sxy$), but also from games for IF formulas such as $\forall x \exists y(\exists z/x)\, Sxyz$. What the former mean is that players forget some *actions* they have made before, while in the latter, V forgets *information* she held at $\exists y$ while choosing a value for z. These can be accounted for by viewing players as teams of players, or multiple-selves of a single player, in which members of a team are responsible for individual decisions. The team approach is by far the most common and natural way of capturing the game-theoretic notion of forgetting, and is spontaneously resorted to in a number of game and decision-theoretic problems (Piccione and Rubinstein 1997; Rubinstein 1998).

There is a tradition in the game-theoretic literature known as team theory (Bacharach 2001, Ho and Chu 1972; Kim and Roush 1987; Marschak and Radner 1972; Witsenhausen 1968). A team is a finite set of non-coordinating players $i = \{1 \ldots n\}$ who have identical payoffs $u_i(h)$ but who act individually. Thus the teams V and F consist of a finite number of individual members. They are groups of individuals with a common goal but individual information, knowledge and actions. The central result of team theory says that solutions of two-person zero-sum games hold for games played by teams (Ho and Sun 1974).[14]

Semantic games for IF logic may thus be set broadly in line with team theory, which sees teams as groups of agents with identical interests but individual actions and individual information. Furthermore, strategies are still based on previous information within a game, but not on the information other members of the team might have had. If we take these games to be strictly competitive, it follows that the basic solution concept, the existence of winning strategies, is formed in games played by teams precisely as if there were just two players.[15]

In IF logic the members of a team are not allowed to communicate with one another because this would destroy the team's ability, when viewed as one player, to genuinely forget something. Hence the semantics is one of decentralised processing. The members of the same team all receive the payoff $u_i(h)$ when the outcome of a play is resolved. In addition, the information for individual team members remains persistent although the teams, when viewed as single players, do not forget information. Hence, whenever a move associated with the team V or the team F is regarded as independent of the move made by the member of the same team, we capture that by introducing a new member who makes the new move in question. Alternatively, some communication within teams may be permitted.

Some consequences and concrete examples of team actions in IF logic are presented in Pietarinen (2001c). Just to mention a few, team games do not presuppose that every logical component is assigned a distinct member. Only in the case of failures to recall, a new member will be produced to account for such loss. The game still contains just two players who, upon reassessing their plans and actions when moving from one information set to another, are able to control their behaviour at future information sets. Therefore, semantic games for IF logic rarely form what are known as agent normal forms, that is, extensive games in which each information set is assigned a distinct player.[16]

According to team games, the semantic information is persistent and the players do not forget information on the level of individual players. On the level of principle players, they exhibit imperfect recall. One can think of an implicit map from the 'information set' containing all the information sets of the respective player to the information sets of the members of a team; in this way coordination takes place. From a slightly different perspective, one can think of players as playing the roles of all of the members, one at the time. When a subformula has the first component associated with a member of either of the teams, the player in question assumes the role of a single member. As it happens, she or he is seen to forget information, since the players are not, during a particular turn, allowed to use the information available to the other members of the team. Viewing teams as single players usually gives away the coordination aspect and introduces some excess strategies.

Some further evidence for the team perspective has been provided in Koller and Megiddo (1992) and von Stenger and Koller (1997), albeit indirectly. They show that games of imperfect recall should use strategies more appropriate than just the traditional mixed ones, for instance team-maxmin strategy profiles. This need arises in a game of one team playing off against a single player, which in IF logic corresponds to the semantic game for weak equivalence.[17]

Logical representation of teams and actions has scores of potential applications in system and organisation theory as well as in distributed computing and routing problems in communicating networks, constantly in the need of teams or groups of agents in their decentralised modelling tasks. For instance, in heterogeneous agent societies (Subrahmanian et al. 2000), despite concerning groups of agents, individualistic rather than collective strategies are commonplace. Unlike coalitional games, such societies would fall quite naturally within the realm of team theories with collective strategies.

8. Two Cases of Strategic Meaning: Aspect and Anaphora

What strategies cannot accomplish, there is little the players can do. Yet, there is an important distinction between using a strategy that is 'up for grabs' and knowing some vital things about it. This is roughly what the distinction between abstract meaning and strategic meaning of an expression tries to capture. Bifurcations to these two senses of meaning are abundant in and around the semantics/pragmatics interface.[18]

8.1. Aspect

What does it mean that in the temporal system of our language ('primary aspect') the ways of coding and expressing time is language-internal, as it often has been said to be?[19] The answer is that language-internality means an interpretation of primary aspect that takes place within the actual world. It does not need possible worlds in the same sense as other, exogenous temporal expressions such as ones that the Prior-type temporal logic or one of its extensions tries to cover. The interpreter needs to look at the actual state of affairs in order to see what it is that makes a proposition with an aspectual verb phrase to hold. What does this 'not in the same sense' mean? Surely primary aspect, referring to temporal constructions, also has to resort to at least some sense of possible worlds?

This much may be true, but it would be far from providing a conclusive answer. An interpretation of primary aspect does not survive just within a structure of time given by the possible-worlds construction. This is best seen from natural language expressions involving aspectual particles. In interpreting such particles, among which are *still*, *already*, *yet* and *anymore*, a player picks a time point from a preferred time structure, say, from a homogeneous reference interval I. In addition to this choice, however, another, contrastive assertion is made in which no reference to the aspectual particle is made, as it is replaced by a certain assertion on the part of the interpreter with reference to that time point chosen earlier. We may even hypothesise that it is this contrastive sentence that is the primary component in interpreting aspectual particles.

An example is the meaning of *already*, which denotes properties on inceptive time scales. The game rule is the following.

(G.already): If the game has reached a sentence of the form

X – already Y – Z,

V chooses a time t_1, whereupon F chooses a time t_2 from a reference interval I, $t_1 < t_2$, and the game continues with respect to the sentence

X – Y – Z at t_1, and X – was expected to Y – Z at t_2.

Here $t_1 < t_2$ means that the time point t_1 occurs earlier than t_2. X, Y and Z are arbitrary linguistic contexts. An example of an application of this rule renders 'John already did the job' as 'John did the job on Monday, and John was expected to do the job on Friday'.

As far as continuative scales are concerned, the following rule may be formulated for *still*:

(G.still): If the game has reached a sentence of the form

X – still Y – Z,

V chooses a time t_1, whereupon F chooses t_2 from a reference interval I, where $t_1 > t_2$ or $t_1 = t_2$, and the game continues with respect to the sentence

At t_1, X – Y – Z, and X – was expected to neg(Y' – Z) at t_2.

Here Y′ is otherwise like Y but the main verb is not progressive. (See Pietarinen (2001b) for further rules and explanations.)

The notion of expectation on which these rules fall back is intrinsic to player's grasp of strategic meaning. When, and why, do the speakers or interpreters of the given sentence or discourse expect that some contrastive assertion holds or does not hold? Any answer to these depends on what there is in their strategies that guide their courses of action.

The fact that the main component in understanding primary and also to some extent secondary aspect does not really refer to the notion of possibility, albeit in an indirect, derived sense of 'recycling' the time point that has been selected earlier, is shown among other things by the difficulties and controversies that the proposed model-theoretic treatments have provoked in the literature, especially when aspect is found in event reports. How the tensed propositions containing aspect need to be understood turn on the strategic meaning. This is the principle reason why both model-theoretic semantics based on intervals (Humberstone 1979) and axiomatic (Galton 1984) theories attempting to define new aspectual operators, fall short of providing comprehensive methods of treating aspectual time.

The notion of expectation in the game rules may further be analysed by means of suitable nonmonotonic systems based on preferential models, in which models are ordered by partial inclusion and logical consequence defined only in minimal models given by such an order (Lin and Shoham 1989). Thus expectation boils down to the fact that normal or typical situations are preferred over atypical ones.

8.2. ANAPHORA

GTS has proved resourceful also in explaining anaphora. The basic mechanism may be illustrated by the analysis of a simple conditional $S_1 \rightarrow S_2$ (Hintikka and Kulas 1985). The game $\mathcal{G}(S_1)$ on the antecedent is played first with the players' roles reversed. If S_1 turns out true, the players move on to play the game $\mathcal{G}(S_2)$ on the consequent. The strategy used in $\mathcal{G}(S_1)$ by player i for verifying S_1 is then available for, or 'remembered' by, player $-i$ in $\mathcal{G}(S_2)$ who in turn sets out to verify S_2.[20] For instance, in GTS (2) is symbolised by something like (3).

If a man owns a donkey, he beats it. (2)

$$\exists F \exists G \forall f_i \forall g_i (S_1[Mf_i \wedge Dg_i \wedge O[F(f_i, g_i)]] \rightarrow S_2[B(G(g_i), f_i)] \quad (3)$$

Yet, the previous expositions have left the notion of the player $-i$ 'remembering' the verification strategies f_i, g_i in $\mathcal{G}(S_1)$ informal. The notion can nonetheless be captured in the extensive-form representation of a semantic game. There are, in fact, two stages toward a comprehensive theory of anaphora. The first is related to singular anaphora, and its aim is to derive anaphoric information from game histories. This means that subgames and operations on them are defined so that the remembering of a strategy amounts to the inheritance of assignments from the top node downwards. The second is related to functional anaphora, in which anaphoric information is derived from players' knowledge and information. In brief, this means that a strategy is remembered if the player's 'local state' contains information about that strategy, and the relevant information sets for $-i$ in the subgame corresponding to $\mathcal{G}(S_2)$ are singletons.

As to the first step, operations on subgames are defined so that a consequent subgame is augmented with the terminal histories of the antecedent subgame. Then the consequent is played with the assignment inherited from the antecedent. The information about anaphoric relations is thus captured in terms of the histories of the game. In addition, given an input assignment at the start of the game, what the play in effect produces is an output assignment that captures the anaphoric information (Janasik and Sandu 2003). As far as the mechanism of anaphora is concerned, this method sets GTS on a par with theories of dynamic semantics. Moreover, the notion of a choice set will become redundant.

As to the second step, a general definition of what we call hyper-extensive games is needed to represent complex anaphora involving functional dependencies. Alongside with individuals, such games allow us to refer to strategies in the course of the game (Janasik et al. 2003). For example, in (4) the values for *a gun* and *it* have to be interpreted as given by a function from men to guns, producing for each man a particular gun.

Every man carried a gun. Most of them used it. (4)

Such a definition extends the usual game-theoretic definition of extensive games in the sense that at any non-terminal position, in addition to individuals, players' information state has to contain strategies applied earlier. Consequently, not only assignments but also strategies may be remembered and forgotten as the game goes on. (Individuals may moreover be deictically introduced or just picked from the domain.)

In general, it is useful to think of games as systems in the sense closely related to their computational interpretation (Fagin et al. 1995). Given an extensive game $\mathcal{G}$, we define a hyper-extensive game $\mathfrak{G}$ consisting of the following three components. (i) A set of local states $\{l_1 \dots l_n\}$ for players $j, j \in \{V, F\}$. A local state l_j is a description of the information a player j has at any history, not given by the history in the sense of traditional extensive games alone. The set l_j is built up from a set of actions $B \subseteq A$, a set of strategies $S \subseteq F$, and a set of deictic individuals $E \subset \mathcal{E}$ given by the linguistic context or environment. The set E can also be taken to contain players world knowledge, scripts, schemes or episodic memory, symbolised, if need be, in a suitable knowledge representation language such as epistemic logic. (ii) Ordered tuples $\langle l_V, l_F \rangle$ of local states, one for each player, called global states. A global state is thus a tuple of local states. A global state captures the state of the game as viewed from outside (modeller's perspective). A global state says what the information any player possesses is at any point of the game. (iii) Functions $f\colon H \rightarrow G$ associating to any history $h \in H$ a global state g, or 'information flows'. When h' is the root, the global state $g(h')$ is likely to contain only local states that are made up of the sets E. When $k \in Z$, the local state also contains the payoffs u_j associated to that terminal history k.

A local state l_j is thus a set $\{B, S, E, u_j\}$ of actions, strategies, environmental elements and, for terminal nodes, payoffs that the player with l_j is aware of (or has an access to). Since there are just two teams, each global state at any $h \in H$ consists of tuples of local states. A game is essentially just the set of information flows. The notion of a strategy is likewise generalised in the sense that it gets as input the local states whenever a player is planning his or her decisions. Thus a strategic decision may involve an assessment of those other strategies to which a player according to a local state has an access. A strategy $s_j \in F$ is a functional from a local state l_j to the set of actions in A.

Let us confine ourselves to hyper-extensive games of perfect information.[21] Even so, we need to capture the notion of players 'remembering' the strategies in the game, and one way of doing this is to use imperfect information in the sense that given $P(h) = P(h') = j$, j remembers a strategy $s_j \in F$ at $h \in H$, if $g(h) \sim_j g(h')$ then $h = h'$. That is, the player remembers the history h because there is nothing to distinguish it from h', in other words the equivalence relation $\sim_j$ for j does not do any work.

This is not the only, and probably not even the most common, case of remembering strategies in anaphoric discourse. Sometimes the strategy is relegated to the local state associated with the history that emanates from the different part of the split discourse, as is the case in (4). The reason for the split is just the same as in simple anaphora, namely that the choice for *every* prompts a move by F. The hyper-extensive games capture this by including the relevant strategies that arise from the functional dependency in the former clause to the specification of the player's local states at the history in which the latter clause is evaluated. For instance, in (4) *most* prompts a move by V from the set of men carrying a gun, and *it* is interpreted by applying the same strategy that V used in the subgame at the history in which she had chosen for the indefinite *a gun*.

These games for functional anaphora have vast expressive resources. Among them is plural anaphora. A general way of dealing with plural anaphora is obtained as soon as we assume strategies to be set-valued functions from sets of individuals to sets of individuals (a collection is logically an individual). Since also generalised quantifiers may be given suitable game rules, plural anaphora may be treated alongside with the singular.

Moreover, the 'proportion problem' (Heim 1982) need not detain us. This is because in the game rules for generalised quantifiers such as for the quantifier *most* (Pietarinen 2001b), the players will choose sequences of individuals, and quantifiers do not quantify over pairs. Likewise, the requirement of antecedents for anaphora (the 'familiarity principle', Heim 1982) turns out to be inadequate in plural contexts.[22] For example, in

> Of course there is live music in our night-club. Unfortunately, tonight they have a night off, (5)

the generic *live music* in the antecedent cannot function as the intended head that possesses the value for *they*. However, the semantic game rule for *they* mandates a uniqueness condition similar to that of definite descriptions, in terms of a two-stage game between the players. This is shown by the close proximity of (5) to a sentence with a definite description, in other words to its paraphrase in which, inter alia, *they* is replaced by *the band*. The proposal is thus also relevant for what is called 'bridging cross-reference'.

It may nonetheless be asked what the actual linguistic mechanisms are that 'account for' this variability in remembering (or any kind of transmission of) the strategies in different parts of the game. The answer is that often the precise nature of such linguistic mechanisms is in the strategic meaning rather than in the abstract meaning, the latter of which merely showing when the piece of discourse involving anaphora is true and when it is false. As soon as claims are made concerning the knowledge of what the strategies that a player uses are, we are dealing with strategic meaning. This knowledge may concern his or her own as well as the adversary's strategies. There is thus little hope to account for functional anaphora by means of strict rules spelling out how functions may be transmitted in discourse.

Precisely how difficult it is to make do with abstract meaning alone is shown by anaphora that appears to exhibit functional dependency, but in which what is expressed by the posterior clause is not a consequence but an antecedent of the fact given in the former clause:

Yesterday, every student failed an examination. The brains just did not work. (6)

Furthermore, it is not inconceivable to even have functional cataphora:

Most students did not get high grades. But everyone passed a math examination last week. (7)

It is possible to read (7) so that the functional dependency is of a reversed sort: Whatever *most students* denotes has to be chosen among those individuals who passed a math exam last week. How this is done in hyper-extensive games is such that discourse splits in two even if the universal clause would exists in the anterior clause. It then gets evaluated, and the function induced in the anterior is included to the local state of *V* choosing for *most* in the antecedent.

Because a number of strategies from which linguistic meaning is derived are not just abstract, global options up for grabs but refer to subjective and epistemic elements, we can never be absolutely precise about the processes and the linguistic mechanisms that are responsible for the transmission of certain strategies from some parts of discourse to other, anaphoric ones. The transmission may, among other things, be constrained by things like agent's range of attention and awareness, short-term memory concerning text processing, or any other capacity in retrieving strategies linked with other parts of the game. Thus, what it means that certain strategy is 'remembered', actually subsumes a range of phenomena. Variables to be instantiated in the game are rather like memory registers with pointers. By not assuming too much on the relation between the registers and pointers, we leave ample space for further consideration on strategic aspects of anaphora and the theory of strategic meaning.[23]

9. IF Logic and Computation: A View from Concurrency

In recent years, there has been an increasing want for logical approaches to concurrency and parallel architectures of computerised systems. Yet, logical languages have not developed in a comparable pace in order to be able to reflect the needs of concurrency theory and multi-processing systems (Cleaveland and Smolka 1996). To date, it has remained by and large uncertain what the true logic of concurrency is. Even further, if we want to rigorously model knowledge and information in multi-agent (multi-processor) message-passing systems, the received knowledge representation languages do not seem to provide enough expressivity to capture all the interesting configurations that may arise in such systems. But now, we have logics that facilitate informational independence. Apart from the logical need to model parallel and distributed systems, novel uses for IF logics can be found in three-valued logic design and parallel logic programming.

9.1. The Logic of Concurrency

In Hintikka and Sandu (1995), the logical representation of the configuration of parallel processors is argued to be better captured by the logic with Henkin quantifiers (or IF logic) than by traditional linear formulas, because the latter forces the configuration to be serial. For example, the Henkin quantifier formula $\begin{matrix}\forall x\ \exists y\\ \forall z\ \exists u\end{matrix}\ Sxyzu$ is associated with the following simple system: two processes (say 1 and 2) are running in parallel, and the inputs for 1 and 2 are x and z, respectively, and the outputs of 1 and 2 are y and u, respectively. Since 1 and 2 are in a parallel configuration, y may not depend on z, and u may not depend on x. Hence also the logical description of the system needs to entertain some sense of concurrency.

There are two problems with this argument, however. First, IF formulas are represented by processes that can compute suitable (recursive) functions; in this case the previous formula is skolemised to $\exists f_1 \exists f_2 \forall x \forall z S x f_1(x) z f_2(z)$. However, one should not associate functions with processes, since they may have multiple outputs, or not been designed to halt and give an output.

Second, if there is communication in the system, this may destroy independence and reduce the formulas into ordinary first-order ones. Often, there is communication between parallel processors, or at least they are synchronous in the sense that separate units are no longer entirely independent from each other. For instance, in concurrency, one needs to be able to observe when a computation terminates, the specification of these termination rules depending on the states of other processors in the system.

By way of an example, then, let us see how communication creates various dependence relations between units: let the predicates R_1xu and R_2zy express requests to send values of u and y in variables x and z, respectively, and let the predicates S_1u and S_2y say that the values of u and y have been sent to the processes 1 and 2. Further, let the predicates E_1xu and E_2zy describe executions of these processes (u (resp. y) is stored in 1's (resp. 2's) input x (resp. z)). In order to execute something, the process 1, for example, needs to submit a request to the other process and to receive a message from it: $(R_1xu \wedge S_1u) \rightarrow E_1xu$. In first-order logic, the formula associated with the system is now

$$\forall x \forall z \exists y \exists u (((R_1xu \wedge S_1u) \rightarrow E_1xu) \wedge ((R_2zy \wedge S_2y) \rightarrow E_2zy)), \tag{8}$$

which skolemises to

$$\begin{aligned}\exists f_1 \exists f_2 \forall x \forall z (((R_1 x f_1(x,z)\ \wedge\ S_1 f_1(x,z)) \rightarrow E_1 x f_1(x,z)) \wedge \\ \wedge((R_2 z f_2(z,x)\ \wedge\ S_2 f_2(z,x)) \rightarrow E_2 z f_2(z,x))).\end{aligned} \tag{9}$$

Hence IF logic suits for parallel systems in which processes do not communicate and in which processes, like programs, may be viewed as functions.

Despite these initial considerations, the proposal in Hintikka and Sandu (1995) is, in principle, on the right track. There is irreducible independence in the logical representation of concurrent systems, but it should be looked for in the *control* of concurrent, synchronous processes rather than in the configuration of the processes.

That one needs to resort to independent quantifiers in control problems has been shown in de Alfaro et al. (2000). The point is that, to capture the notion of dependency between system outputs and controller inputs, and likewise between controller inputs and system inputs, a type system may be used, which formalises the dependencies of the composite system. In case the type of the controller is known, the states that can be controlled by fixed types are characterised by

$$M \models \begin{matrix} \forall o_1 \exists i_1 \\ \forall o_2 \exists i_2 \end{matrix} So_1 i_1 o_2 i_2 \quad \text{iff} \quad \exists f \exists g \forall o_1 \forall o_2 \, So_1 f(o_1) o_2 g(o_2), \tag{10}$$

where controller output i_1 depends only on system output o_1, and controller output i_2 depends only on system output o_2.

The synchronous control systems in the previous example are non-blocking. Usually, one wants systems to be non-blocking, in other words to be free from deadlocks in the sense that the controller does not control merely by blocking the unwanted behaviour of the system.

What kind of logic, then, could correspond to control modules that are blocking? The following IF formula fits the bill:

$$\forall x_1 \exists y_1 \forall x_2 (\exists y_2 / x_1 y_1)((x_2 = y_1 \wedge x_1 = y_2) \to Px_1 y_1 x_2 y_2). \tag{11}$$

Here x_1 and x_2 contain the outputs of the control module and y_1 and y_2 contain the inputs. If the control modules are of a non-static type, they would block if either of the universally quatified variables or either of the existentially quantified variables receives the value 1.

The logic in the previous examples was first-order, but blocking may be represented at the propositional level, too, by means of restricted quantifiers:

$$\exists f_1 \exists f_2 \forall i_1 \forall i_2 ((i_1 = f_1(i_2) \wedge i_2 = f_1(i_1)) \to p_{i_1 f_1(i_1) i_2 f_2(i_2)}). \tag{12}$$

The propositional case is significant, since in the design of logical circuits, for example, the novelty is that since the logic is partial, it gives rise to non-determined (three-valued) formulas. The design of three (or in general multiple) valued digital circuits is topical (Epstein 1993). IF logic nonetheless differs from other three-valued logics in that atomic formulas do not have to be partially interpreted in order to attain general partiality. The undetermined truth-values are due to the non-determined nature of the correlated semantic games. Hence the inputs that are given to logical circuits do not have to be undetermined in order to arrive at indeterminate outputs of a complex circuit. In other words, the input voltage may be just either 1 or 0 and yet, one might receive the undetermined value Undetermined.

A further consequence is that, by means of IF logic, it is possible to extend the design of logical circuits to architectures that involve cyclic dependencies between inputs and outputs. Such circuits are found in many typed control systems for parallel processes.

9.2. Parallel Logic Programming

Yet another illustration of how imperfect information and independent logics may be employed in parallellism is found in parallel logic programming. As it is known, the (independent) AND-parallelism is usually more complex than OR-parallelism, since in the former, conjuncts are constrained in a special way (Chassin de Kergommeaux and Codognet 1994). In brief, in parallel logic programming one tries to unificate goals in a body of a rule or a query concurrently. The restriction says that goals may not share variables because they are bound run-time, whereas no information may be transmitting between parallelly processed clauses that potentially lead to inconsistent states. A possible solution to this is to use identities:

$$\forall x \forall y \exists z \exists u ((Cyz \wedge Sxu \wedge z = u) \rightarrow Uxy). \tag{13}$$

To resolve this requires a lot of bookkeeping in AND-parallel processing. Therefore, we skolemise the formula by disregarding unnecessary information:

$$\exists f_1 \exists f_2 \forall x \forall y ((Cyf_1(y) \wedge Sxf_2(x) \wedge f_1(y) = f_2(x)) \rightarrow Uxy). \tag{14}$$

This simplifies bookkeeping, since for unification purposes, f_1 and f_2 no longer have to depend on all input variables.

9.3. Knowledge in Multi-agent Systems

Like classical first-order and propositional logics, the received modal and epistemic logics have been of perfect information: each evaluation step is revealed to the next level. The assumption of perfect information is inadequate for multi-agent systems, in which information is often uncertain and hidden from other parties.

In the field of knowledge representation, communicating multi-agent systems will profit from imperfect information. To see this, suppose that a process U_2 sends a message x to U_1. We ought to report this by 'U_2 knows what x is' and 'U_1 knows that it (the same message) has been sent'. U_1 might knows this, say, because the communication channel is open. This is already a rich situation involving knowledge that cannot be captured in ordinary (first-order) epistemic logic.

What is involved are the clauses 'U_2 knows what has been sent' and 'U_1 knows that something has been sent', but not 'U_1 knows that U_2 knows', nor 'U_2 knows that U_1 knows'. The question is, how do we combine these clauses? It is easy to see that $\exists x K_{U_2}\mathsf{Mess}(x) \wedge K_{U_1}\exists y\mathsf{Mess}(y)$, $K_{U_1}\exists x(\mathsf{Mess}(x) \wedge K_{U_2}\mathsf{Mess}(x))$ and $\exists x(K_{U_2}\mathsf{Mess}(x) \wedge K_{U_1}\exists y\mathsf{Mess}(y) \wedge x = y)$ fail. So does the attempt to use two variables to distinguish between a message the content of which is known ($\mathsf{Content}(x)$), and a message that has been sent ($\mathsf{Sent}(y)$): $\exists x\exists y((K_{U_1}\mathsf{Content}(x) = y) \wedge K_{U_2}\mathsf{Sent}(x))$. This does not work because now U_2 comes to know what has been sent, which is too strong.

What is needed is information hiding concerning choices for possible worlds and individuals:

$$\begin{matrix} \exists x & K_{U_2} \\ K_{U_1}\exists y & \end{matrix} (\mathsf{Mess}(x) \wedge x = y). \tag{15}$$

This formula is equivalent to the IF version

$$K_{U_1}\exists y(\exists x/K_{U_1}y)(K_{U_2}/K_{U_1}y)(\mathsf{Mess}(x) \wedge x = y), \tag{16}$$

which hides information concerning the choices for K_{U_1} and y at K_{U_2}, x.[24]

Informational independence in quantified epistemic logics gives rise to a novel type of focussed knowledge whenever there are two or more agents involved.[25]

9.4. Concurrency in Epistemic Logic

Every sentence of modal logic defines a game $\mathcal{G}(\varphi, w, M)$ on a model M at a possible worlds $w \in \mathcal{W}$ between two players. For classical (epistemic) modalities, the following rule is needed:

(G.K_i): If $\varphi = K_i\psi$, and the game has reached w, F chooses $w_1 \in [w]_{\rho_i}$, and the next choice is in $\mathcal{G}(\psi, w_1, M)$.

In quantified epistemic logic with imperfect information, whenever K_i is in the priority scope of $\exists x$ and the game has reached w, the individual picked for x by V has to be defined and exist in all worlds accessible from the current one. This assumption is motivated by the fact that the course of the play reached at a certain point in the game is unbeknownst to F choosing for K_i. This leads to specific knowledge (*de re*) of individuals, correlated with games of imperfect information in their extensive form. On the other hand, whenever $\exists x$ lies in the priority scope of K_i, the individual picked for x has to be defined and exist in the world chosen for K_i. This, in turn, will lead to the notion of non-specific (*de dicto*) type of knowledge.

Let B be an ordered set of modal operators and variables occurred in the game when an expression of the form (Q/B) is encountered. The rule for the hidden information is:

(G.Q/B): If $\varphi = (Q/B)\psi$, $Q \in \{\forall x, K_i\}$, and the game has reached w, then if $Q = \forall x$, F chooses an individual from the domain D_{w_1} of individuals, in which

w_1 is the world from which the world chosen for the first modal operator in B departed. The next choice is in $\mathcal{G}(\psi, w, M)$.

If $Q = K_1$, then F chooses a world $w_1 \in \mathcal{W}$ in the model M 'independently' of the choices made for the elements in B, and the next choice is in $\mathcal{G}(\psi, w_1, M)$. Likewise for V.

This can be deciphered by writing the game out in its extensive form. In such a form, we will automatically have a bookkeeping system of derivational histories of the plays of the game that keeps track of the previously chosen worlds as well as the values for the quantifiers and connectives. Thus the notion of 'choosing independently' with reference to the choices of worlds may, but does not have to, mean the existence of non-singleton information sets on which strategies are defined. It may also mean that such worlds are picked that are accessible from the worlds found by backtracking to the history from which the first operator in the sequence B departs (which is unique). The task of choosing between these differing interpretations is left for the modeller to decide.[26]

Yet, in a sense imperfect-information games are not sequential, since players move simultaneously. For if I am ignorant of the earlier move, I can as well make the choice before that move, or concurrently with it. The structure of these extensive games is just an artificial way of depicting the actions of players in a superficially sequential format.[27]

10. Quantum Phenomena in IF Perspective

The purpose of this section is to bring forth quantum logic and quantum theory as a field of inquiry in which IF logic and the correlated game-theoretic tools turn out to be beneficial. The main points concern non-locality, EPR-type phenomena, quantum logic, and quantum interference.

10.1. Non-locality and the EPR-phenomena

Non-locality is a property of entangled quantum systems. Coined by Schrödinger, entanglement ('Verschrankung') is the characteristic trait of quantum mechanics. It refers to the system that consists of two or more particles forming a singlet quantum system. Whenever two space-like separated subatomic particles are in a 'pre-established harmony' sharing the same history of the source, and one particle is manipulated – if the particle is a photon – with a polarisation filter that changes the polarisation of the particle, there is a 100% anti-correlation to that polarisation in the other particle. This is because the polarised photon has to maintain its original correlations with the other photon in the entangled quantum system.

Non-locality can be translated to the game-theoretic language of simultaneous action. Let the propositions be: φ – the measurement outcome of a photon x being left-polarised; ψ – the measurement outcome of a photon x not polarised; θ

– the measurement outcome of a photon y anti-correlated (right-polarised); χ – the measurement outcome of a photon y not anti-correlated (not polarised). The propositional IF formula reflecting non-locality is: $(\varphi\ (\vee/\wedge)\ \psi) \wedge (\theta\ (\vee/\wedge)\ \chi)$.

Logically, then, non-locality means that being space-like separated but correlated refers to the fact that no physical information passes between subsystems, and in this sense the two particles are separated. However, entanglement states that the outcome of the measurement on one of the particles is not independent of how the measurement is performed on the other, separated particle. Game-theoretically, when simultaneous action is represented, the outcome of one of the actions determines the winner – actions on which the winning strategies and hence the t ruth-values of the propositions are based, which in turn depends on other actions in the game – quite independently of whether the actions are taken to be hidden or not. Such barriers to information trespassing in the evaluation of the above formula can be brought out by saying that the information regarding V's choice of the disjunction may not be used when F plans a decision between the conjuncts (and vice versa).

Information encapsulation is not a sufficient reason to account for non-locality. In entangled systems, some further effect such as a quantum field is needed to correlate the separated systems. In logical terms, however, non-locality means that in order to make $(\varphi\ (\vee/\wedge)\ \psi) \wedge (\theta\ (\vee/\wedge)\ \chi)$ true, one has to make at least one atomic formula in both conjuncts true, and this in turn means that, despite the hidden information regarding the conjunct that has been chosen at the other history, both conjuncts representing the states of two separated systems are needed.

Nothing in this argument – purported, in the end, to show that logic of quantum mechanics goes beyond not only the purviews of classical but to some extend also received quantum logics – hinges on this traditional EPR formulation of non-locality concerning two separated but correlated particle systems. Similar remarks carry over to other descriptions of non-locality. An example is the Greenberger–Horne–Zeilinger (GHZ) experiment (Pietarinen 2002a).

10.2. Rules of Quantum Logic and Propositional IF Logic

One classically valid propositional rule is commutativity $\varphi \wedge \psi = \psi \wedge \varphi$, not valid in quantum logic. There is a game-theoretic reason for its failure. In extensive games of imperfect information, commutativity is constrained by the existence of non-singleton information sets. Connectives influenced by imperfect information affect the outcome of actions and hence do not permit commutation, because in the correlated histories, the strategy functions are defined on whole information sets, and hence not only the identity of actions but also their order has to coincide for all the histories within the same information set.

A game-theoretic basis lurks behind the failure of distributivity, that is, the law $p \wedge (q \vee r) = (p \wedge q) \vee (p \wedge r)$, too. In quantum logical terms, distributivity holds just in case the propositions are not members of a common sublattice, that is, they denote incompatible subspaces (observables). The left and the right-hand side sentences thus say that the subspaces in question are different, and hence the identity should fail. How is this explained in the game-theoretic jargon? In short, the explanation is that distributivity changes the order in which players move. The left and right-hand sides do not exemplify the same logical situation: Incompatible subspaces cannot be conjoined, since from $p \wedge (q \vee r)$ one may not infer $(p \wedge q) \vee (p \wedge r)$, because the pairs $\{p, q\}$ and $\{p, r\}$ are mutually incompatible. Indeed, these laws do not hold in quantum theoretic algebra that is non-Boolean.

What about the law of modularity $p\wedge(q\vee r) = (p\wedge q)\vee r$ (assuming $p \leq r$, i.e., element p of a lattice is a subspace of element r), which is weaker than distributivity? Again, modularity illustrates an imperfect information phenomenon. It boils down to $p\wedge(q\ (\vee/\wedge)\ r) = (p\wedge q)\vee r$. For if you choose disjunction independently of conjunction, you can as well go ahead and choose it before conjunction. Again, this may be spelled out in semantic games of imperfect information.

However, in quantum logic one is typically interested in orthomodular structures, which have the following order between elements in a lattice:[28] If $p \leq q$, then $q = p \vee (q \wedge p^{\perp})$. (The operation '$\perp$' marks singular orthocomplementation, corresponding to game-theoretic negation.) Logically, orthomodularity replaces distributivity, since it does not form conjunctions of mutually incompatible proposition, that is, propositions that are not members of a common Boolean sublattice. The conjugation is legitimate only if any two propositions in $p \wedge (q \vee r)$ are complements of each other, in which case distributivity is retained. Like modularity, orthomodularity thus illustrates a relative independence phenomenon, albeit in a weaker sense than full modularity.

The main architect of both the mathematical foundations for quantum logic and the theory of games was János von Neumann. On the face of it, the two theories do not seem to have much in common. However, the connections are more than skin deep. Simultaneous games interpret quantum phenomena, and imperfect information plays a central role in quantum logic. The theory of games and the foundations of quantum mechanics share many other common elements and inspirations. Among others, what are the insights in the quantum theoretic notion of mixed states in relation to the game-theoretic uncertainty by mixed strategies (probability distributions over the set of pure strategies)?

10.3. Quantum Interference: The Square Root of Not

Yet another suggestive logical perspective is related to quantum computation, and to the behaviour of quantum logical gates in particular. One may think of quantum switches as randomising devices mapping $\{0, 1\}^n$ into $\{0, 1\}^m$. That is, each of the four possibilities for a particle (say, a photon) has an identical probability of 0.5. However, when two such identical machines are concatenated, the net effect of the combined system is logical complementation instead of randomisation. This prima facie surprising phenomenon contradicts the usual additivity of the received probability calculus, because the probability of the combined event is not the sum of two mutually exclusive constituent events. Indeed, this connective has received a special symbol "$\sqrt{\neg}$" in the literature (Deutsch et al. 2000).

What is going on? In Deutsch et al. (2000, 269) it is claimed that there exists no corresponding operator in logic, or a priori mathematical construction, that could capture the nature of such a randomising device. Yet, physicists have directly observed exactly this type of single-particle interference behaviour.

Contrary to these sentiments, IF logic throws light on this issue. First of all, the simultaneous nature of single-photon trajectories in quantum interference devices takes place inside quantum gates; there is no interaction between gates and the environment. If one interprets these simultaneous actions as uncertainty links, the third choice of action in the concatenated system, although simultaneous with respect to the second choice, is not simultaneous with respect to the choice of action at the first interference gate of the concatenated system. Thus the strategy by which the third action is executed carries information concerning the first action, thereby complementing the input signal.

The corresponding game may be constructed as a three-stage game in which there is imperfect information between the second and the third move, but not between the first and the third move. This game captures the behaviour of the connective $\sqrt{\neg}$ of quantum interference in concatenated gates. In IF notation, two concatenated $\sqrt{\neg}$'s are symbolised by $\forall i_1 \exists i_2 (\exists i_3 / i_2)\, \psi(i_1, i_2, i_3)$, and interpreted by a three-layer extensive game over two-element models.

Accordingly, the information sets may be viewed as superpositions produced by the first gate of the system.

By way of concluding this section, what we have is a logical perspective to imperfect information (uncertainty) in quantum mechanics. This should not be interpreted in any unmotivated epistemic sense like the lack of knowledge or information about some physical phenomenon. The uncertainty goes deeper. It refers to the information transmission between the players playing the semantic games on quantum logical proposition describing physical phenomena. As soon as the transmission is imperfect, uncertainty affects the laws of quantum logic and the associated propositional representation of the subspaces. This is not unlike what happens in experimental games of inquiry between Nature and the Experimenter (Frieden and Soffer 1995), but certain differences have to be recognised. Quantum logic virtually arises from imperfect-information games – hence the parallels between physical information flow and the game-like structures in extracting information measures from Nature. Unlike what happens in experimentation games, one may perhaps say that the physical reality itself may be held responsible for the failure of complete information in quantum logics.

11. Language Games and Logical Semantics in a Game-theoretic Perspective

Let us conclude with a couple of remarks that aim to put the concept of the semantic game into a wider philosophical and historical perspective. Wittgenstein's concept of a language game shares some significant parallels with Peirce's ideas on dialogical semeiotics. For both Peirce and Wittgenstein, the concept of interaction, dialogue, or game, regardless of who or what are participating, was fundamental to the understanding of the concept of meaning in logic or in the language of our natural discourse. Thus, these philosophers offered some fundamental insights into the relation between such activities and logic, and it is these insights that are needed in order to understand different positions that may be taken up in assessing the game-theoretic import for logic.

The idea of a logic game or a language game of Peirce–Wittgenstein origin should first of all be contrasted with an important distinction between two broad kinds of such games. Hintikka and Hintikka (1986) argue that Wittgenstein's language games fall broadly within two categories, the primary and the secondary. Primary games operate by means of spontaneous responses. They do not involve propositional, let alone epistemic attitudes, and they do not seem to have room for any traditional concept of a strategy. Secondary games bank on rationality in the sense of making use of player's knowledge of his or her own strategies. Since secondary language games do not operate independently of identity criteria for actions, many of the epistemic concepts of our discourse derive their meaning from these games.

In view of this, it is the secondary notion of games that we might attempt to relate to the received notion of games as conceived in game theory. Does this render the theory of games non-viable in the study of logic and language, especially since, in order to make sense of the theoretical notion of a game, surely some rationality postulates ought to be presupposed?

It quickly becomes evident, however, that there is plenty of room in modern game theory for the concept of a strategy that does not presuppose rationality on the part of the players. The assumption that the strategic evolution of thought is not an exclusive province of the human brain has often proved useful, a case in point being evolutionary game theory (Maynard Smith and Price 1973). This theory does not advocate winning strategies, but requires strategies to be stable, which means that they should resist any attempt at invasion by adversary strategies. Stable strategies are associated with non-human actors such as populations, computers, systems and agents. Hence the usage of the term 'game' is not, strictly speaking, a necessity, either. To be sure, the term does not surface in Peirce's writings on logic, although it is rife in his ample writings on recreational matters. For Wittgenstein, the term 'game' sprang to his mind, according to the anecdote in Malcolm (1958, 65) – reporting what Wittgenstein once told Freeman Dyson –, when he was passing a pitch on which a football game was in progress. In fact, Wittgenstein was aware of the economics-focussed atmosphere in 1930s Vienna, was well connected though ambivalent with the philosophical ideas of the Vienna Circle, and so may have taken the game idea from his associations with that environment. So his comment to Dyson may have been a hoodwink. What is nonetheless essential in Wittgenstein is the idea of language as a rule-governed system or process with variable meaning relations. What is essential in Peirce is the idea of thought as a dialogue between different phases of a mind, or, concerning any agent, entity or role in general, between the quasi-utterers and the quasi-interpreters of a quasi-mind.

The possibility of applying the idea of a strategy to situations in which non-hyper-rational players take part in the process of interpretation took root in Peirce's evolutionary philosophy of signs, habits and dialogues, and recurred in Wittgenstein's language games as primitive, instinctive behaviour. Pietarinen (2003a) argues that Peirce's concept of a habit was in no way restricted to rational human agents. This anticipated evolutionary games in biology, players acting not for their own good but for the good of a population, the *summum bonum*, which is congenial to Peirce's evolutionary agapism. Furthermore, in evolutionary games, agents no longer have similar perfect foresight as classical rational players do. Since evolutionary games are played repeatedly, the processes for arriving at stable equilibria (or focal points) are in a sense mechanical, namely not based on calculations concerning unlimited access to strategies.[29]

Apart from the differences in the concept of strategy, the division of games into two main categories is strongly reflected in assumptions concerning the structure of the games themselves. This comes to light as soon as we think of semantic games in their extensive form. Primary language games are those in which the players do not identify the actions available to them across the non-terminal histories in which they move. Secondary language games build identification of actions into the game in the sense that strategies cease to be operational if not presented with a range of options. A related distinction is proposed in Pietarinen (2004a) to reflect the different notions of information that players may have regarding past moves and also regarding the question of what the legitimate future actions are, given their knowledge about them. As far as identity criteria are concerned, in games of imperfect information, for instance, some actions have to be identified across multiple histories within an information set. It is worth observing that games in the customary account of extensive games are, in this sense, secondary, as it is assumed that the set of legitimate actions is available to the players so that they are able to choose their optimal actions from the set of alternatives.

The upshot is that semantic games call for a re-examination of some of the basic assumptions in game theory. They are not secondary *simpliciter*, but make public some fundamental hidden assumptions concerning the received notion of a game in the theory of games in general.

Acknowledgements

Partial support has been received from the Academy of Finland (Project no. 1178561) and from the Ella and Georg Ehrnrooth Foundation (Ahti-Veikko Pietarinen).

Notes

1 See e.g., Aumann (1987, 1992).

2 In Krynicki (1993) it is shown that each complex Henkin quantifier prefix can be defined by a Krynicki normal form.

3 With the possible exception of Aumann, who has postulated all-inclusive, full descriptions of the states of the worlds (Aumann 1987), including the fact that information sets, defined as functions on histories of the game (i.e., the states of the world) are known to all players. Logically, this means that players would known all the Skolem functions, which turns the game from the semantic into the epistemological.

4 The issue of agent's limited ability to analyse the environment has further repercussions. It is customary to assume that players are not only informed about the totality of their available actions, but are also able to identify the actions in the sense of recognising what counts as the same action across different decision nodes within the same information set. Imperfect foresight dispenses with such uniformity.

5 Even if one is to retain 'full' rationality, strategies themselves may be subject to several regimentations. One may, for instance, require strategies in a semantic game to be recursive (Hintikka 1996). Within the loose-fitting limits of computability, various notions of learning may be entertained.

6 See Lipman (1995) on a game-theoretic exploration of non-partitional information structures. One of its several logical counterparts is the fact that individual actions may dictate whether there is going to be imperfect information (i.e., slashed expressions) later on in the formula.

7 See Sandu and Pietarinen (2001) on partiality and its relations to semantic games with respect to sentential logic, and Pietarinen (2002b) on partiality and modality.

8 See Rubinstein (1998) for a basic negotiation model in terms of alternating offers.

9 The terminology of teams is just a generalisation of semantic games to accommodate imperfect information. If the game is one of perfect information, the same member of the team get to choose repeatedly.

10 See Pietarinen (2000) for further discussion on games and the notion of non-coherence in logic.

11 Further remarks concerning games that resemble linguistic patterns characterised by their winning, losing, or competitiveness, are found in Wittgenstein (2000, item 226, 48).

12 See Pietarinen (2003b) on Wittgensteinian piquancy in recent computational theories.

13 The reference CP is to Peirce (1931–1966) by volume and paragraph number.

14 Team theory is a fairly heterogeneous field that aims to bring together decision and systems theory, operations research, dynamic games, search and coordination, and parallel processing. Such generosity has its advantages as shown by the present-day popularity of multi-agent systems (Pietarinen 2004b).

15 As far as we know, it is an open question whether these results may be applied also to non-strictly competitive games.

16 This suggests that the decision maker at one information set is able to control his behaviour at some future information sets (cf. Rubinstein 1998, 78). An example of an imperfect-recall game in

which two consecutive moves are made by the same player of the team is provided by formulas with quantifier segments of three existential quantifiers and two slashes: $\exists x(\exists y/x)(\exists z/x)$.

[17] That is, for equivalence with respect to the truth of the formulas in a model, or with respect to the falsity of the formulas in a model, but not both.

[18] Examples of strategic meaning have been studied in Hintikka (1987), Hintikka and Kulas (1985), for example.

[19] By primary aspect, we mean reference to time in pure verbs or verb phrases, in contrast to secondary, morphological and morphosyntactic aspect such as progressive. This distinction is sometimes drawn in terms of actionality (Aktionsart) and aspect. A caveat is that these distinctions are, to some extent, language-dependent.

[20] Falsification strategies are not taken to carry over.

[21] If there is imperfect information, global states become 'stretched' sequences reflecting the fact that *V* and *F* are actually teams of players.

[22] In Hintikka and Kulas (1985), a similar phenomenon was discussed in the context of singular anaphora.

[23] Among the phenomena that could be analysed from the game-theoretic perspective of strategic meaning include 'salience' of indefinites, choice functions, and the topic/focus contrast. Choice functions in particular are 'mutilated' strategy functions incapable of reproducing the dependence structure of variables.

[24] The meaning of the identity is given by 'world lines', i.e., functions from worlds to extensions that coincide with one another, see Pietarinen (2001a).

[25] The notion of focus is further studied in Pietarinen (2001a, 2002b). Bradfield (2001) shades propositional modal logics involving concurrency, that is, the Henkin quantifier or IF type of modalities.

[26] Since (G.Q/B) refers to locution 'the world from which the world chosen for the first modal operator in *B* departed from', we come close to hybrid modal logics, in which terms in the object language refer to individual worlds.

[27] It has been claimed that simultaneous moves in an extensive game relate to concurrency, provided that every history crosses all information sets (Bonanno 1992). Richer notions of concurrent games have been developed for computational purposes in Abramsky and Jagadeesan (1994) and de Alfaro and Henzinger (2000).

[28] But see Pavicić and Megill (1999), in which a family of orthomodularity laws are devised and a quantum logic formulated that dispenses with orthomodularity.

[29] Such processes are chief constituents of the notion of linguistic meaning, for example, by virtue of reinforcing certain meanings among populations of language users against mutants.

References

Abramsky, Samson and Radha Jagadeesan: 1994, 'Games and Full Completeness for Multiplicative Linear Logic', *Journal of Symbolic Logic* **59**, 543–574.

de Alfaro, Luca and Tom A. Henzinger: 2000, 'Concurrent Omega-regular Games', in *Proceedings of the 15th Annual IEEE Symposium on Logic in Computer Science*, IEEE Computer Society Press, pp. 141–154.

de Alfaro, Luca, Tom A. Henzinger and F. Y. C. Mang: 2000, 'The Control of Synchronous Systems', *Proceedings of the 11th International Conference on Concurrency Theory, Lecture Notes in Computer Science 1877*, Berlin, Springer-Verlag, pp. 458–473.

Aumann, Robert: 1987, 'Correlated Equilibrium as an Expression of Bayesian Rationality', *Econometrica* **55**, 1–18.

Aumann, Robert: 1992, 'Perspectives on Bounded Rationality', in Yoram Moses, (ed.), *Proceedings of the 4th Conference on Theoretical Aspects of Reasoning about Knowledge*, Monterey, CA, Morgan Kaufmann, pp. 108–117.

Bacharach, Michael: 2001, 'Superagency: Beyond an Individualistic Theory of Games', in J. van Benthem (ed.), *Theoretical Aspects of Rationality and Knowledge*, San Francisco, Morgan Kaufmann.

Bacharach, Michael, L.-A. Gérard-Varet, P. Mongin and H. S. Shin (eds.): 1997, *Epistemic Logic and the Theory of Games and Decisions*, Dordrecht, Kluwer.

Bonanno, Giacomo: 1992, 'Rational Belief in Extensive Games', *Theory and Decision* **33**, 153–176.

Bradfield, Julian: 2001, 'Independence: Logic and Concurrency', in P. G. Clote and H. Schwichtenberg (eds.), *Proceedings of the 14th International Workshop on Computer Science Logic, Lecture Notes in Computer Science 1862*, Berlin, Springer-Verlag.

Chassin de Kergommeaux, J. and P. Codognet: 1994, 'Parallel Logic Programming Systems', *ACM Computing Surveys* **26**, 295–336.

Cleaveland, R. and S. A. Smolka: 1996, 'Strategic Directions in Concurrency Research', in *ACM Computing Surveys* **28**, 607–625.

Deutsch, David, A. Ekert and R. Lupacchini: 2000, 'Machines, Logic and Quantum Physics', *Bulletin of Symbolic Logic* **6**, 265–283.

Epstein, George: 1993, *Multiple-Valued Logic Design*, Bristol, Institute of Physics Publishing.

Fagin, R., J. Y. Halpern, Y. Moses and M. Y. Vardi: 1995, *Reasoning about Knowledge*, Cambridge, MA, MIT Press.

Frieden, B. Roy and B. H. Soffer: 1995, 'Lagrangians of Physics and the Game of Fisher-information Transfer', *Physical Review* **E 52**, 2274–2286.

Galton, Anthony: 1984, *The Logic of Aspect*, Oxford, Oxford University Press.

Goldstein, Lawrence: 1989, 'Wittgenstein and Paraconsistency', in G. Priest, F. R. Routley and J. Norman (eds.), *Paraconsistent Logic. Essays on the Inconsistent*, Munich, Philosophia Verlag, pp. 540–562.

Heim, Irene: 1982, *The Semantics of Definite and Indefinite Noun Phrases*, Dissertation, University of Massachusetts at Amherst.

Hintikka, Jaakko: 1973, *Logic, Language Games and Information*, Oxford, Oxford University Press.

Hintikka, Jaakko: 1987, 'Language Understanding and Strategic Meaning', *Synthese* **73**, 497–529.

Hintikka, Jaakko: 1996, *The Principles of Mathematics Revisited*, New York, Cambridge University Press.

Hintikka, Jaakko: 2002, 'Hyperclassical Logic (aka IF Logic) and its Implications for Logical Theory', *Bulletin of Symbolic Logic* **8**, 404–423.

Hintikka, Jaakko and Merril B. Hintikka: 1986, *Investigating Wittgenstein*, Oxford, Basil Blackwell.

Hintikka, Jaakko and Jack Kulas: 1985, *Anaphora and Definite Descriptions*, Dordrecht, D. Reidel.

Hintikka, Jaakko and Gabriel Sandu: 1995, 'What is the Logic of Parallel Processing?', *International Journal of the Foundations of Computer Science* **6**, 27–49.

Hintikka, Jaakko and Gabriel Sandu: 1997, 'Game-theoretical Semantics', in J. van Benthem, and A. ter Meulen (eds.), *Handbook of Logic and Language*, Amsterdam, Elsevier, pp. 361–410.

Ho, Y. C. and K. C. Chu: 1972, 'Team Decision Theory and Information Structures in Optimal Control Problems I', *IEEE Transactions on Automatic Control* **17**, 15–22.

Ho, Y. C. and F. K. Sun: 1974, 'Value of Information in Two-team Zero-sum Problems', *Journal of Optimization Theory and Applications* **14**, 557–571.

Hulstijn, Joris: 2000, *Dialogue Models for Inquiry and Transaction*, Dissertation, University of Twente.

Humberstone, Lloyd: 1979, 'Interval Semantics for Tense Logic: Some Remarks', *Journal of Philosophical Logic* **8**, 171–196.

Janasik, Tapio and Gabriel Sandu: 2003, 'Dynamic Game Semantics', in J. Peregrin (ed.), *Meaning: The Dynamic Turn*, Dordrecht, Kluwer, pp. 215–240.

Janasik, Tapio, Ahti-Veikko Pietarinen and Gabriel Sandu: 2003, 'Anaphora and Extensive Games', in M. Andronis et al. (eds), *Papers from the 38th Meeting of the Chicago Linguistic Society*, Chicago, Chicago Linguistic Society.

Kim, K. H. and F. W. Roush: 1987, *Team Theory*, New York, Ellis Horwood.

Koller, Daphne and N. Megiddo: 1992, 'The Complexity of Two-person Zero-sum Games in Extensive Form', *Games and Economic Behavior* **4**, 528–552.

Krynicki, Michail: 1993, 'Hierarchies of Partially Ordered Connectives and Quantifiers', *Mathematical Logic Quarterly* **39**, 287–294.

Langholm, Tore: 1988, *Partiality, Truth and Persistence*, Stanford, CSLI Publications.

Lin, F. and Y. Shoham: 1989, 'Argument Systems – A Uniform Basis for Non-monotonic Reasoning', in *Proceedings of the 1st International Conference on Principles of Knowledge Representation and Reasoning*, pp. 245–255.

Lipman, Barton L.: 1995, 'Information Processing and Bounded Rationality: A Survey', *Canadian Journal of Economics, Revue canadienne d'Economique* **28**, 42–67.

Malcolm, Norman: 1958, *Ludwig Wittgenstein: A Memoir*, London, Oxford University Press.

Marschak, J., and R. Radner: 1972, *Economic Theory of Teams*, Yale University Press, New Haven.

Maynard Smith, John and G. Price: 1973, 'The Logic of Animal Conflict', *Nature* **246**, 15–18.

Pavicić, Mladen and Norman D. Megill: 1999, 'Non-orthomodular Models for Both Standard Quantum Logic and Standard Classical Logic: Repercussions for Quantum Computers', *Helvetical Physica Acta* **72**, 189–210.

Peirce, Charles S.: 1902, 'Vague', in J. M. Baldwin (ed.), *Dictionary of Philosophy and Psychology*, New York, MacMillan, p. 748.

Peirce, Charles S.: 1931–1966, in Charles Hartshorne, Paul Weiss, and A. W. Burks (eds.), *Collected Papers of Charles Sanders Peirce*, 8 Vols., Cambridge, MA, Harvard University Press.

Piccione, Michael and Ariel Rubinstein: 1997, 'On the Interpretation of Decision Problems with Imperfect Recall', *Games and Economic Behavior* **20**, 3–24.

Pietarinen, Ahti-Veikko: 2000, 'Logic and Coherence in the Light of Competitive Games', *Logique et Analyse* **171–172**, 371–391.

Pietarinen, Ahti-Veikko: 2001a, 'Intentional Identity Revisited', *Nordic Journal of Philosophical Logic* **6**, 144–188.

Pietarinen, Ahti-Veikko: 2001b, 'Most Even Budged Yet: Some Cases for Game-theoretic Semantics in Natural Language', *Theoretical Linguistics* **27**, 20–54.

Pietarinen, Ahti-Veikko: 2001c, 'Varieties of IFing', in M. Pauly and G. Sandu (eds.), *Proceedings of the ESSLLI 2001 Workshop on Logic and Games*, University of Helsinki.

Pietarinen, Ahti-Veikko: 2002a, 'Quantum Logic and Quantum Theory in a Game-theoretic Perspective', *Open Systems & Information Dynamics* **9**, 273–290.

Pietarinen, Ahti-Veikko: 2002b, 'Knowledge Constructions for Artificial Intelligence', in M.-S. Hacid, Z. W. Ras, D. A. Zighed and Y. Kodratoff (eds.), *Foundations of Intelligent Systems, Lecture Notes in Artificial Intelligence*, 2366, Springer, pp. 303–311.

Pietarinen, Ahti-Veikko: 2003a, 'Peirce's Game-theoretic Ideas in Logic', *Semiotica* **144**, 33–47.

Pietarinen, Ahti-Veikko: 2003b, 'Logic, Language Games and Ludics', to appear in *Acta Informatica* **18**.

Pietarinen, Ahti-Veikko: 2004a, 'Semantic Games in Logic and Epistemology', this volume.

Pietarinen, Ahti-Veikko: 2004b, 'Multi-agent systems and Game Theory – A Peircean Manifesto', *International Journal of General Systems*.

Pietarinen, Ahti-Veikko and Gabriel Sandu: 1999, 'Games in Philosophical Logic', *Nordic Journal of Philosophical Logic* **4**, 143–173.

Rubinstein, Ariel: 1998, *Modeling Bounded Rationality*, Cambridge, MA, MIT Press.

Sandu, Gabriel and Ahti-Veikko Pietarinen: 2001, 'Partiality and Games: Propositional Logic', *Logic Journal of the IGPL* **9**, 107–127.

Savage, Leonard J.: 1954, *The Foundations of Statistics*, New York, Dover.

von Stenger, B. and Daphne Koller: 1997, 'Team-maxmin Equilibria', *Games and Economic Behavior* **21**, 309–321.

Subrahmanian, V. S. et al.: 2000, *Heterogeneous Agent Systems*, Cambridge, MA, MIT Press.

Witsenhausen, H. S.: 1968, 'A Counterexample in Stochastic Optimum Control', *Siam Journal of Control* **6**, 131–147.

Wittgenstein, Ludwig: 1953, *Philosophical Investigations*, (third edition 1967), Oxford, Blackwell.

Wittgenstein, Ludwig: 1978, *Philosophical Grammar*, Columbia, University of California Press.

Wittgenstein, Ludwig: 2000, *Wittgenstein's Nachlass, The Bergen Electronic Edition*, The Wittgenstein Trustees, The University of Bergen, Oxford University Press. (The transcription used is the diplomatic transcription.)

PART III
UNITY AND PLURALITY IN SCIENCE AND IN LOGIC

CONCEPTS STRUCTURED THROUGH REDUCTION: A STRUCTURALIST RESOURCE ILLUMINATES THE CONSOLIDATION-LONG-TERM POTENTIATION (LTP) LINK

JOHN BICKLE
Department of Philosophy and Neuroscience Graduate Program, University of Cincinnati, P.O. Box 210374, Cincinnati, OH 45221-0374, USA, E-mail: bicklejw@email.uc.edu

Abstract. The structuralist program has developed a useful metascientific resource: *ontological reductive links* (ORLs) between the constituents of the potential models of reduced and reducing theories. This resource was developed initially to overcome an objection to structuralist "global" accounts of the intertheoretic reduction relation. But it also illuminates the way that concepts at a higher level of scientific investigation (e.g., cognitive psychology) become "structured through reduction" to lower-level investigations (e.g., cellular/molecular neuroscience). After (briefly) explaining this structuralist background, I demonstrate how this resource illuminates an actual, emerging scientific example: the link between the psychological concept of a "consolidation switch" from short-term to long-term memory and the cellular/molecular mechanisms of the transition from early- to late-phases of long-term potentiation (LTP) (an important type of synaptic plasticity in mammalian hippocampus and cortex).

1. Structuralist Background

Historically, structuralists have viewed intertheoretic reduction as a "global" relation. They eschew the standard Anglo-American account of theory structure as a set of sentences, propositions, or other linguistic items. But they retain the standard idea that reduction is a relation over entire classes of entities that constitute theories (i.e., models). Although structuralists place far stronger restrictions on the reduction relation than Suppes's (1956, chapters 8 and 9) "isomorphism" condition – see, e.g., Balzer et al. (1987, chapter 6), Mormann (1988), and Mayr (1976) – this "global" feature still leaves the accounts open to Schaffner's (1967) devastating "too weak to be adequate" challenge (Bickle 1998, chapter 3). There exists, or can be contrived, obvious cases of nonreduction that meet all the formal (set-theoretic) conditions structuralists lay down on the relation.

Although he didn't mention Schaffner by name, C. U. Moulines (1984) first developed a structuralist resource that can be tailored to address Schaffner's challenge (Bickle 1998, chapter 3). More interesting for my purposes in this essay, this same resource also illuminates another puzzling metascientific notion. Sometimes in the process of being related to scientific developments in other disciplines (especially "lower-level" ones), concepts from one theory become "structured through

S. Rahman et al. (eds.), Logic, Epistemology, and the Unity of Science, pp. 141–150.

reduction". In the laws or generalizations of the original theory, the concept is characterized *functionally*: as an *entity or process* fully defined by its *causes and effects*, with little or no concern for the underlying mechanisms that yield this functional profile. 'Gene expression' in Mendelian through transmission genetics might be a good recent example. This process, characterized in terms of phenotypic ratios and theoretical posits tied directly to these (e.g., 'dominance', 'allele'), gets redefined within current molecular genetics in terms of elaborate sequences of molecular and (increasingly) biochemical transcriptional, translational, and recombinant processes (Lewin 1999). However, philosophers of science have not carried an analysis of "concepts structured through reduction" much beyond the vague hints and example just mentioned.

The structuralist concept of theory is an ordered set of classes of models. More precisely, a theory **T** is an ordered triple $\langle \mathbf{M_p}, \mathbf{M}, \mathbf{I} \rangle$ where $\mathbf{M_p}$ is a set of *potential models*, **M** is a set of *actual models*, and **I** is a set of *intended empirical applications*. Intuitively, potential models are entities with the appropriate "candidate structure" (defined set-theoretically) to be investigated as actual models of **T** but which may not meet the conditions specified by **T**'s laws or generalizations. Models are potential models that also meet the conditions specified by the laws or generalizations. Intended empirical applications are "real-world" systems that are expected to be revealed by investigation to be actual models of **T**. $\mathbf{M} \subseteq \mathbf{M_p}$ and **T**'s *empirical claim* is that $\mathbf{I} \subset \mathbf{M}$, although at any given time for any actual theory **T**, at best $\mathbf{I} \cup \mathbf{M} \neq \varnothing$.[1]

To reconstruct actual cases, it is convenient to specify the appropriate classes by defining a set-theoretic predicate. Consider a simple example: the theory of classical collision mechanics (**CCM**). We define 'x is a (model of) **CCM**' iff there exist P, T, v, m such that

(1) $x = \langle P, T, \mathfrak{R}, v, m \rangle$
(2) P is a finite, nonempty set
(3) T is an ordered pair set $\langle t_1, t_2 \rangle$
(4) $v : P \times T \rightarrow \mathfrak{R} \times \mathfrak{R} \times \mathfrak{R}$
(5) $m : P \rightarrow \mathfrak{R}^+$
(6) $\sum_{p \in P} m(p) \times v(p, t_1) = \sum_{p \in P} m(p) \times v(p, t_2)$

On the standard interpretation P is a set of particles, T a set of time instances (t_1 before, t_2 after the collision), v is the velocity function, m is the mass function, and (6) specifies the law of conservation of momentum before and after the collision. The class $\mathbf{M_p}$(**CCM**) contains all those structures ("real world" and "purely mathematical") that meet conditions (1)–(5). To use Stegmüller's (1976) helpful phrase, these are the structures "about which it makes sense to ask" whether they are actual models of **CCM**. **M**(**CCM**) contains all those members of $\mathbf{M_p}$(**CCM**) that also meet lawful condition (6). **I**(**CCM**) contains all the "real world" systems we expect to be confirmed empirically to meet conditions (1)–(6).

Intertheoretic reduction can then be specified as a relation ρ whose domain is $\mathbf{M_p}(\mathbf{T_B})$ (where **TB** is the "reducing" or "base" theory) and whose range is $\mathbf{M_p}(\mathbf{T_R})$ (where $\mathbf{T_R}$ is the "reduced" theory). Structuralists and their sympathizers have proposed a variety of conditions restricting ρ (see especially Balzer et al. 1987, chapter 6; Mayr 1976; and Bickle 1998, chapter 3). However, despite the increasingly sophisticated set-theoretic conditions and applications to historical cases, Moulines pointed out the essential flaw with such "global" accounts:

> For a complete picture of a reductive relationship between two theories, one has to take into account some sort of relation between the respective domains. Otherwise, when confronted with a particular example of a reductive pair, we would feel that all we have is an *ad hoc* mathematical relationship between two sets of structures, perhaps by chance having the mathematical properties we require of reduction but not really telling us something about "the world". ... The possibility that we find a formally appropriate ρ just by chance or by constructing it in an *ad hoc* way cannot be ruled out in general. ... I think we would feel that such a reduction is not "serious". (1984, 55)[2]

According to Moulines, this problem arises because ρ is defined "too globally", i.e., as a relation over entire sets of potential models comprising theories. But we can take our structuralist analysis of theories "down a level" (so to speak), to the constituents of individual potential models. ρ can then be construed as constructed, at least in part, out of cross-theory *links* between (some of) these "ontological" constituents.

Any $x \in \mathbf{M_p}(\mathbf{T})$ will have the following general form:

$$x = \langle D1, \ldots, D_n, A_1, \ldots, A_m, r_1, \ldots, r_p \rangle$$

where the D_is are the "real" base sets, the A_is are auxiliary base sets (formal or mathematical spaces), and the r_is are relations or functions typified by the base sets (i.e., constructed out of the base sets using possibly repeated operations of power set and Cartesian product) (Balzer et al. 1987, chapter 1). In the **CCM** example, P and T are "real" base sets, $\mathfrak{R}$ is an auxiliary base set, and v and m are relations typified by P, T, and $\mathfrak{R}$.

Consider now a $\rho \subseteq \mathbf{M_p}(\mathbf{TB}) \times \mathbf{M_p}(\mathbf{TR})$. Moulines (1984) defines ñ as an *ontological reduction link* (ORL) just in case ρ meets all the conditions on the reduction relation and is partly composed of relations between the D_is constituting the potential models of $\mathbf{T_R}$ and at least some of the D'_js constituting the potential models of $\mathbf{T_B}$. There are two types of ORLs. *Homogeneous* ORLs are total or partial identity relations between the base sets: total when the D_i is identical (in the extensional, set-theoretic sense) with some D'_j, partial when the D_i is identical with some proper subset of some D'_j. *Heterogeneous* ORLs link at least one real base set of $\mathbf{T_R}$ to one or more of $\mathbf{T_B}$ in a way that does not imply identity of elements. A global reduction link ρ can be composed entirely of homogeneous ORLs, entirely of heterogeneous ORLs, or of some combination of both types. Moulines calls the last type mixed (*ontological*) *reduction* and insists that "in real science ... it is likely that mixed reduction is the most frequent case" (1984, 60).

He cites the rigid body mechanics-to-Newtonian particle mechanics and Newtonian particle mechanics-to-special relativity theory reductions as accomplished mixed cases. In the first, although the base sets of space points and time points are linked homogeneously across the two theories, elements of the set of rigid bodies do not belong to any base set in the reducing theory. The base set of rigid bodies is linked heterogeneously with the base set of Newtonian particles. And in the second case, the set of particles is linked homogeneously across related potential models of the two theories, but elements of the separate Newtonian base sets of space points and time points don't belong to any base sets of special relativity. The former get linked heterogeneously to the base set of Minkowskian spacetime points. Moulines also points out a number of reductions "in progress" that appear to be either mixed or completely heterogeneous, including simple thermodynamics of gases-to-kinetic gas theory (see Bickle 1998, chapters 2 and 3 for further work on this example), wave optics-to-classical electrodynamics, and Mendelian genetics-to-molecular biology (1984, 60–62).

Heterogeneous reductions come in many varieties. The simplest relates a D_i of $\mathbf{T_R}$ to a single D'_j of $\mathbf{T_B}$. More complex examples include those that relate a single D_i to a sequence of elements from some D'_j, to a sequence of elements from several base sets $D'_i, \ldots, D'_j$, or even to a sequence of elements from several base sets and relations $r'_k, \ldots, r'_m$ of $\mathbf{T_B}$. Moulines points out that these complexities arose in initial structuralist attempts to reconstruct the Mendelian genetics-to-molecular biology reduction: "If D_i is the set of genes of an organism and D'_j a certain set of organic molecules, then to each gene a sequence of organic molecules is supposed to correspond biunivoquely" (1984, 64). Elements of the base set of genes, however, don't belong to any base set of organic molecules. And if we attempt to reconstruct the reduction of Mendelian/transmission theory of gene expression (a process) to molecular/biochemical mechanisms, the base sets of the former will be heterogeneously linked to both base sets and relations from molecular biology.

Structuralists typically think of a theory's relations and functions as typifications, constructions out of the real and auxiliary base sets using only repeated applications of power set and Cartesian product. This makes it unnecessary to specify ORLs for the reduced theory's relations and functions, and implies that heterogeneous ORLs that link a base set of $\mathbf{T_R}$ to a combination of base sets and relations/functions of $\mathbf{T_B}$ is "reducible" in principle to one that links the former only to some sequence or combination of the base sets of $\mathbf{T_B}$ (Moulines 1984, 66–67). But we need not assume so conservative a view of theory relations and functions. Some are clearly typifications (especially in mathematical physics, the "natural home" of the structuralist approach). But for "process-focused" sciences like genetics, molecular biology, psychology, and physiology, we might need to treat some theoretical functions more on a par with base sets. They can then be constituents in genuine ("unanalyzable") ORLs. The formal definitions that Moulines (1984) provides of homogeneous and heterogeneous ORLs could be extended easily to accommodate this view.

ORLs add a condition on structuralist reduction concepts that enable them to overcome Schaffner's "too weak to be adequate" challenge. Both actual and contrived cases that meet the mathematical conditions on the reduction relation ρ but which aren't genuine reductions will not be genuine ORLs: such links will not obtain across the base sets in the intended empirical applications of the two theories (Bickle 1998, chapter 3). But in heterogeneous cases where some base set of $\mathbf{T_R}$ gets linked to a sequence of elements from base sets and relations from $\mathbf{T_B}$, we also get an account of what it is for "the amorphous basic entities of the reduced theory [to] become structured through reduction" (Moulines 1984, 67–68). An entity (or process, in light of the previous paragraph), characterized entirely by way of the relations and laws/generalizations of $\mathbf{T_R}$, comes to be related *in a domain eliminating way* to sequences of entities and processes characterized by the relations and laws/generalizations of $\mathbf{T_B}$. There are no rigid bodies, separate space points and time points, or Mendelian genes, at least not in the way that there remain particles in special relativity theory and planets in Newton's celestial mechanics. The former aren't part of the way that the $\mathbf{T_B}$ "carves up the world" – although the roles that these base sets play in the relations and laws/generalizations of $\mathbf{T_R}$ might bear interesting structural similarities to the roles played by the base sets (and possibly theoretical relations and functions) in $\mathbf{T_B}$ to which they are linked by heterogeneous ORLs.

2. The Consolidation-LTP Link[3]

Armed with this structurtalist resource, I next turn to a recent development across psychology and neuroscience. My hope is that the structuralist resource developed above can illuminate an emerging intertheoretic link. Since the seminal work of Ebbinghaus, Müller, and Pilzecker in the 1880s, and elaborated by James in his classic *Principles of Psychology* (1890), psychologists have distinguished *short-term* from *long-term memory*. The former is transient, lasting anywhere from the immediate present to several minutes ("working memory") up to an hour or more with rehearsal. The latter is stable, lasting for weeks, months, years, sometimes even decades, and typically requires stimulus repetition to induce this stability. Furthermore, as Müller and Pilzecker demonstrated experimentally more than one century ago, the conversion from short-term to long-term memory can be disrupted by *retrograde interference*, distractions introduced after the initial items had been stored in short-term memory. They coined the phrase *consolidation period* to refer to the time needed for the short-term "memory trace" to achieve stable long-term form.

Other than careful exploration of the time course of consolidation for different memory items, the nature and timing of effective retrograde interference, and the amount of repetition required to convert short-term to long-term memory, psychologists have been unable to explain satisfactorily the *consolidation pro-*

cess or *switch*. Recent neuroscience has made greater progress. Pharmacological manipulations dating back nearly forty years have produced animals (including mammals) with intact learning and short-term memory capacities but profoundly deficient long-term memories. Over the past decade, in keeping with biotechnology's expansion, these manipulations are now carried out using genetic knockout and transgenic rats and mice.

The current state of theory about learning and memory in mainstream neuroscience follows a lead first developed by Donald Hebb in his classic book, *The Organization of Behavior* (1949). (By mainstream neuroscience I mean the "Society for Neuroscience" crowd, to be distinguished for the most part from self-described "cognitive neuroscientists".). Hebb recommended that we think of learning and memory in terms of *synaptic strength* and *plasticity*, the changeable effect that a given neuron has on inducing a change in membrane potential in neurons with which it shares an active synapse. Mammalian *long-term potentiation* (LTP), a type of synaptic plasticity now documented in hippocampus, cerebellum, and cortex, is a promising candidate for the cellular mechanism of certain types of long-term memory. It is rapidly induced, specific only to activated synapses ("associative"), enhanced by repetition, lasts for as long as can be measured, is selectively blocked by treatments that block behavioral learning, and is induced by physiological inputs that also give rise to learning in behaving animals.

Recent work by Eric Kandel and his colleagues has examined LTP in the Schaffer collateral pathway of the rat hippocampus. The hippocampus is a bilateral structure in the subcortical medial temporal lobe. It is known to play a crucial role in long-term memory storage and access. Hippocampal ablation in experimental animals produces little deficiency in initial learning and short-term recall, but profound deficits on certain types of long-term recall tasks. The human neuropsychological syndrome of global amnesia results from bilateral damage to hippocampus (and some surrounding tissue in the medial temporal lobe). Medial temporal lobe amnesics, like their experimental animal counterparts, have intact short-term memory but profound long-term memory deficits for "declarative" items (Squire 1987). The Schaffer collateral pathway is a bundle of axons from cells in the hippocampal CA3 region that project excitatory synapses to the hippocampal CA1 region. This has been a common site for studying LTP for nearly thirty years.

Kandel and his colleagues found evidence for two distinct phases of LTP. The early phase (E-LTP) begins immediately after a single high-frequency electric pulse train to the pre-synaptic axon and lasts from one to three hours. Increased glutamate release (an excitatory neurotransmitter) by the stimulated presynaptic neuron binds to postsynaptic AMPA receptors (α-amino-3-hydroxy-5-methyl-4-isoxazole proprionic acid) that directly open Na+ (sodium ion) gates. This produces enhanced depolarization of membrane potential in the vicinity of the bound receptors, releasing voltage gated Mg^{++} (magnesium) ions that under normal membrane potentials block NMDA (N-methyl-D-aspartate) postsynaptic receptors. Unblocked NMDA receptors bound by glutamate open gates for Ca^{++} (calcium ion) influx into the postsynaptic cell. Increased intracellular Ca^{++} concentration activates a Ca^{++}-calmodulin cascade that in turn activates a set of protein kinases (enzymes) that (1) phosphorylate AMPA receptors, making them even more efficient gates for Na^{+} influx, and (2) drive the production of *retrograde transmitters* (from post- back to presynaptic cell). One of these, nitric oxide (NO), is a gas that readily diffuses across cell membranes but has a very limited diffusion area and is only active in presynaptic terminals releasing glutamate. Although its mechanisms of action are not yet entirely clear, NO enhances glutamate release from presynaptic terminals. The result of these activity-driven cellular mechanisms is a sharp increase in transmission capacity at affected synapses, measurable from roughly one up to three hours.

The late phase of LTP (L-LTP) requires a series of electric pulse trains to the presynaptic axon (the laboratory analog of "repetition" known from psychological studies to be important for consolidation to long-term memory!). This increases further the rate of postsynaptic Ca^{++} influx through the open voltage-gated NMDA receptors and in turn the amount of Ca^{++} – calmodulin. The latter, in conjunction with a second messenger receptor activated by input from modulatory interneurons, activate G proteins in the postsynaptic cell that convert ATP molecules (adenosine triphosphate) into cAMP (cyclic adenosine monophosphate) . cAMP binds to the regulatory subunits of PKA (protein kinase A) molecules, freeing the catalytic subunits. In sufficient numbers, these freed PKA catalytic subunits translocate into the postsynaptic cell nucleus, where they have two principal effects. First, they phosphorylate CREB-1 (cAMP-response element binding protein-1), enabling this molecule when bound to CRE (cAMP-response element), a subregion in the regulatory region of two important classes of *immediate early genes*, to initiate transcription of both regulatory proteins that maintain the PKA in a persistently active state and proteins that lead to the growth of new postsynaptic sites. Second, by interacting with MAP kinase (mitogen-activated protein kinase) in the cell's nucleus, the catalytic PKA subunits inhibit CREB-2 (cAMP-response element binding protein-2). CREB-2 is an inhibitory transcription regulator (a "repressor"). It is thought to inhibit the facilitating action of CREB-1 at the two classes of immediate early genes by binding to both the CREB-1 molecule and the CRE regulatory subregion. The PKA-MAP kinase interaction blocks CREB-2's repressive effects.

In a nutshell, the "consolidation switch" of psychology yields to a three-part sequence of processes described within contemporary cellular and molecular neuroscience: the activity-induced enhancement of PKA leading to (1) increased binding of CREB-1 to CRE regulatory subregions on a class of immediate early genes that transcribe regulatory proteins for persistently active PKA and another class that transcribes proteins needed for the growth of new postsynaptic sites; (2) the inhibition of CREB-2, a transcription repressor for these immediate early genes; and (3) the production of protein products transcribed by these immediate early genes. L-LTP requires new gene transcription and protein synthesis. The biochemistry of affected neurons changes permanently when L-LTP is induced, enhancing the probability of successful neural transmission for long periods. What psychologists call "retrograde interference" turns out to be any process that interferes with any of these steps after initial (repeated) stimulus presentation.

But is L-LTP really the mechanism for long-term memory in behaving animals? Work with transgenic mice and an ingenious behavioral paradigm shows convincingly that it is. Kandel's group generated mutant mice partially expressing a gene that blocked the action of the catalytic subunit of PKA. Rusiko Bourtchouladze and Alcino Silva studied mice in which the gene expressing CREB-1 was partially knocked out. In both groups the transgene or knockout was specific to hippocampus. Both groups along with controls were subjected to a novel environment for two minutes, followed by a sound (CS) for thirty seconds, followed by a foot shock (US) for 2 seconds. When placed back in the same box a few minutes later, normal mice display a defensive reaction (freezing); memory for environmental cues requires an intact hippocampus. Similarly, normal mice will freeze to the tone when it is presented a few minutes later in any context; this type of CS-US fear conditioning requires an intact amygdala (another bilateral structure in the subcortical medial temporal lobe). Both types of genetically altered mice learned both tasks as easily as normal mice and still showed normal freezing to the environmental cues and the CS when tested one hour after initial training. But 24 hours later, unlike normals, both groups of genetic mutants showed significantly reduced freezing to the environment. They were deficit in a long-term memory task that requires the hippocampus. But they displayed normal frezzing 24 hours later to the CS. They were intact on a long-term memory task that requires the amygdala (a region where the transgene was not expressed). On the other hand, normal mice given a protein synthesis inhibitor after initial training that acts on both hippocampus and amygdala are deficient at both long-term memory tasks. Manipulating steps in the process that yields L-LTP not only blocks that cell-physiological/molecular process. It also produces selective deficits in long-term memory.

From the structuralist background presented above, what can we say about the LTP-consolidation link? Cognitive psychology, through its base sets, fundamental and derived theory relations and functions, and generalizations in terms of these, characterizes an entity/process, the consolidation switch. But it does so only in terms of the time course and amount of repetition needed to convert a given type of memory item from short-term to long-term memory (with the latter concepts also characterized primarily in terms of duration of recall after initial presentation) and the behavioral efficacy of different types of retrograde interference. In other words, psychology characterizes this entity/process in purely functional fashion, with little regard for the *causal mechanisms* generating this functional profile. The link between this base set or process and those containing the cellular/molecular sequences signaling the transition from E-LTP to L-LTP and the maintenance of the latter is a heterogeneous ORL. The elements of the former are not even elements of partial subsets of the latter. Cellular and molecular neuroscience "carves up the world" in a fundamentally different way than does cognitive psychology, even though the intended empirical applications of the two theories overlap significantly; the two theories are intended to apply to roughly the same set of "real world" systems.

3. Conclusion

From the structuralist perspective articulated here, is there any such thing as "psychology's consolidation switch"? No, in the sense that neither that concept nor its affiliated ontology within cognitive psychology are part of the base sets and fundamental theory relations and functions of contemporary cellular and molecular neuroscience. But the emerging cellular/molecular story nevertheless still puts its constituents together into a structure abstractly similar (at a coarse-grained level) to psychology's functional concept. The sequence of cellular/molecular mechanisms even explains the sort of behavioral data that psychologists use to study the duration of the consolidation process and methods of retrograde interference. In an important sense, cognitive psychology's notion of a consolidation switch is an important functional *approximation* of the cellular/molecular mechanisms that signal the switch from E-LTP to L-LTP and maintain the latter.

So do we here have cross-theoretic identification or elimination? Based on the application of the structuralist resource presented above, I am inclined to answer the latter. But the answer we give to that question is less important than the project of clarifying intertheoretic links, both "global" and "local," in interesting scientific cases. The question now strikes me as metaphysical in the perjorative logical-positivist sense. On the other hand, structuralist philosophy of science provides fruitful resources for tackling the significant metascientific project.

Notes

[1] This is a simplification of the structuralist concept of *theory-element*, the simplest concept that corresponds to one of the meanings of "theory". It is common in structuralist writings on intertheoretic relations to work with this simplification. (Moulines 1984 uses it in the original paper where he develops the resource I am about to explain.) For the full structuralist account of theory and the relation between theory-element and theory, see Moulines (1996) and Balzer et al. (1987, chapter 1).

[2] Compare Moulines's worry with Schaffner's well-known critique of Suppes's Reduction Paradigm, which like structuralist accounts treated reduction as a relation across "global" theories characterized set-theoretically: "Different and nonreducible (at least to one another) physical theories can have the same formal structure – e.g., the theory of heat and hydrodynamics – and yet one would not wish to claim that any reduction could be constructed here" (1967, 143). In Bickle (1998) I call this the "too weak to be adequate challenge" to this approach to reduction.

[3] The scientific details in this section draw heavily on Larry Squire's and Eric Kandel's recent book (1999). It is the best sustained introductory treatment of neuropsychological and neurobiological work on memory that is available today. Squire is one of the world's leading neuropsychologists and Kandel recently shared the 2000 Nobel Prize for Physiology and Medicine for his work on the

cellular and molecular basis of learning and memory. Together they cover the full range of levels at which memory is researched. All the scientific details I present below are at least mentioned in that book, primarily in chapters 6 and 7. However, the book has only a short list of further readings for each chapter. For those seeking a deeper level of scientific detail or a more extensive reference list, consult the chapters in the last part of Kandel et al. 2000. Bickle (2003), chapters 2 and 3, provides an account of these experiments and results drawn from the primary scientific literature.

References

Balzer, W., C. U. Moulines and J. D. Sneed: 1987, *An Architectonic for Science*, Dordrecht, D. Reidel.
Bickle, J.: 1998, *Psychoneural Reduction*, Cambridge, MA, MIT Press.
Bickle, J.: 2003, *Philosophy and Neuroscience: A Ruthlessly Reductive Account*, Dordrecht, Kluwer.
Kandel, E., J. Schwartz and T. Jessell (eds.): 2000, *Principles of Neural Science*, 4th edn. New York, McGraw-Hill.
Lewin, B.: 1999, *Genes VII*, Oxford, Oxford University Press.
Mayr, D.: 1976, 'Investigations of the Concept of Reduction I', *Erkenntnis* **10**, 275–294.
Mormann, T.: 1988, 'Structuralist Reduction Concepts as Structure-Preserving Maps', *Synthese* **77**, 215–250.
Moulines, C. U.: 1984, 'Ontological Reduction in the Natural Sciences', in W. Balzer, D. Pearce and H. J. Schmidt (eds.), *Reduction in Science*, Dordrecht, D. Reidel, pp. 51–70.
Moulines, C. U.: 1996, 'Structuralism: The Basic Ideas', in W Balzer and C. U. Moulines (eds.), *Structuralist Theory of Science*, Berlin, Walter de Gruyter, pp. 1–13.
Schaffner, K.: 1967, 'Approaches to Reduction', *Philosophy of Science* **34**, 137–147.
Squire, L.: 1987, *Memory and Brain*, Oxford, Oxford University Press.
Squire, L. and E. Kandel: 1999, *Memory: From Mind to Molecules*, New York, Scientific American Library.
Stegmüller, W.: 1976, *The Structure and Dynamics of Theories*, Berlin, Springer-Verlag.
Suppes, P.: 1956, *Introduction to Logic*, Princeton, NJ, van Nostrand.

THE UNITY OF SCIENCE AND THE UNITY OF BEING: A SKETCH OF A FORMAL APPROACH[1]

C. ULISES MOULINES

Seminar für Philosophie, Logik und Wissenschaftstheorie, Universität München

Abstract. It is argued that (philosophical) ontology is supervenient on the ontological commitments of empirical science, and that therefore the idea of the unity of being depends on the unity of science. What prospects are there for the latter? The aim of this paper is not to provide an ultimate answer to this question but rather to sketch the conceptual framework with in which the question can be discussed in precise terms. Four notions are decisive for such a framework: ontological commitment of an empirical theory, reduction, fundamental theory, and compatibility of theories. By using the idea that every theory is uniquely associated with a class of models (in the sense of formal semantics), an attempt is made to explicate formally these four notions; this, in turn, provides the adequate base to decide the question of the unity of science, and therefore of being. In particular, it is shown that, even though it might be unavoidable to have several, not mutually reducible, fundamental theories, there could still be a sense in which we might speak of a unified ontological system, when the fundamental theories are not mutually incompatible (in the sense specified in this article).

Ontology is the philosophical discipline dealing, in very general terms, with the very general question of what there is – Quine *dixit*.[2] Perhaps, it would be more precise to say that it deals with the question of what **kinds** of things there are.[3] This characterization keeps the spirit, though not the letter, of Quine's *dictum*; the philosopher, as a philosopher, is not interested in knowing, say, whether there is life in Mars but rather in finding out whether the class of all living beings should be considered as fundamentally different from the class of non-living beings; nor is he/she interested in discovering a particular mental state accompanying sleep-walking but rather in the question whether mental states as such exist on their own. The first sort of questions belong to particular empirical sciences – the second sort to the empirical sciences **as well as** to philosophical ontology.

This doesn't mean, however, that we are entitled to engage in a discussion of ontological matters without taking science into consideration. Quite the contrary. A philosopher who would propose an ontological system incompatible with the results of science would not be taken seriously by the vast majority of more or less cultivated persons, not even by the vast majority of philosophers. Suppose, for example, a philosopher would claim that the fundamental ontology of real beings is such that empirically accessible reality consists of at least two kinds of beings: let's call them "*A*-type beings" and "*B*-type beings", where both *A* and *B* are supposed to be properties that are (more or less directly or indirectly) accessible

S. Rahman et al. (eds.), Logic, Epistemology, and the Unity of Science, pp. 151–161.

to human experience. Suppose, further, that the philosopher claims that these two kinds of beings are ontologically incompatible, i.e., anything that is an *A* is not a *B*. (The variables *A* and *B* may be instantiated by such properties as *mental* and *physical*, or *living organism* and *inert matter*, or *cultural* and *natural*, or any other of the dichotomies with empirical meaning philosophers have been fond of in the course of history.) On the other hand, suppose all those scientific theories which are well-established at the time the philosopher makes his/her claim and which deal, at least in part, with things, that are *A* and/or *B* allow us to draw the conclusion that everything that is an *A* is also a *B*. Faced with this situation, the philosophical ontologist has three options: (These options are not only conceptual possibilities but are also found in the actual history of ideas.)

(1) To dismiss the results of empirical science and keep his/her categorization of being as the true one.[4]

(2) To "reinterpret" properties *A* and *B* as not corresponding to the same kind of properties empirical science deals with, but rather as some sort of "transcendental" properties, only metaphysics has access to.

(3) To abandon the own philosophical doctrine and revise his/her ontological system in the light of the generally accepted scientific results (which doesn't necessarily presuppose that the latter are definitive).

I don't think it is necessary in the present context to argue at length why options (1) and (2) above are not to be taken seriously. I think they are manifestly untenable for any friend of rational philosophising. Position (1) reflects an arrogant attitude that, after the immense successes of empirical science in the last four centuries, and the total failure of a supposedly independent *Naturphilosophie*, can only be regarded as ridiculous. Position (2) represents a pretty example of (philosophical) *mauvaise foi*: it amounts to arbitrarily changing the rules of the game when one is losing. It seems clear to me that, in the situation just depicted, the only intellectually honest reaction of the ontologist would be to take path (3). This may be dubbed a form of "scientism", or even of "positivism", but we should not be afraid of more or less ill-sounding labels here. As terminology already suggests, physics comes first, metaphysics afterwards – and not the other way around. At least in matters of being, *philosophia est ancilla scientiae*.

This doesn't mean that a scientifically-conscious ontology comes out almost automatically from science itself, so to speak, as a corollary. Here, as everywhere else, the philosophical task is one of patient conceptual analysis, explication, and reconstruction of what is already given as a cultural item – in this case: science. But, at any rate, science should be the starting point of our ontological endeavour. The first thing to be taken into account are the "ontological commitments" of scientific theories.

What I envisage to do (in a very rough way) in this paper is to lay out the formal structure our task as science-based ontologists may take and what general methodological problems may be lurking there. Before that, however, let me make two remarks on the scope of the task envisaged. The first remark is that it does not cover (at least not at the beginning) a particular realm of being – if it is "real being" at all: the realm of mathematical entities. That is, I won't say a word about the problems posed by taking seriously the ontological commitments of the theories of pure mathematics, if we take them side by side with the ontological commitments of theories of empirical science. My reason for leaving this problem aside (besides obvious constraints of space) is not that I think the question of the ontology of mathematics is trivial, or meaningless, or whatever; it is just that it seems to me that it is a very difficult problem which is only a part of a more general, and still more difficult, problem: the nature of purely mathematical thought. And, for the time being, I see no prospects of a fruitful connection between this problem and the ontological issues around empirical science. At any rate, it would be already a considerable progress to get clear ideas about the formal structure and the methodological problems of the ontological commitments of empirical science before we let mathematics come into the scene. The second remark is that the task envisaged is conceived as belonging to substantial (or "material') ontology – not to what is nowadays usually described as "formal (analytic) ontology". This means that no thesis is defended here as to the particular frame of formal categories to be chosen to deal with ontological matters. To make our ideas as precise as possible I use here the language of (naive) set theory, also because it is the most divulged. But, as far as I can see, the main points laid out in the pages to follow could be easily recast in terms of mereology, category theory, tropes, or perhaps other formal categories of ontology one might prefer for some reason or other. What I want to discuss here is how the task of analyzing the ontology of empirical science looks like in general (but substantial) terms.

The ontologist's task would be greatly facilitated if science were a single, homogeneous, neatly constructed system – something like a *Bauhaus* building. Then, we could concentrate on the foundations of this building to find out what the stuff of reality is and how it can be categorized. The task would still not be quite trivial but we would have solid ground on which to work. Suppose, for example, we would agree that all of science is Newtonian mechanics. Then, the most plausible ontology would presumably be one that says that being manifests itself in three different categories: a discrete category of matter and two continuous categories: space and time. (Numbers would probably be seen as purely heuristic artifacts – not to be taken seriously from an ontological point of new.) Philosophers could then still discuss further about different ways of explicating, in a more fundamental way, what matter, space and time **really** are. But at least we would have a quite firm starting point, a neat picture of the world in the sense of the kinds of things there are and how they can be categorized. This is the reason why the idea of the **unity of science** is so attractive **also** for ontological matters. The unity of science would still not imply the unity of being but it would make plausible a single system of what there is, a unique ontology. If science were a *Bauhaus* building, then we would have a guarantee that the question of Being would have a definite answer – whatever Heidegger's ghost might still grumble...

Unfortunately, we all know that science is **not** a *Bauhaus* building. There is not just Newtonian mechanics in the world of (scientific) culture. There are hundreds of different theories – each one of them with its own ontological commitment. Now these commitments need not be all mutually incompatible, many of them may prove to be actually identical, or at least tightly related; nevertheless, one has to analyze them carefully. In order to do ontology we are forced to do a bit of empirical metascience. This might cause bad feelings to the traditional ontologist but this is his/her problem.

Now, the presence of many different theories in really existing science is not, by itself, a deadly blow to the unitarian ontologist, it just makes his/her task somewhat harder. His/her endeavour may still be saved by means of a magical remedy: **reduction**. If all those hundreds of theories, even those having apparently different ontological commitments were step-by-step reducible to a single theory, then the unity of science would be restored and with it the underlying unique system of Being. But even if there is no such single theory reducing all the rest, the unitarian ontologist should not immediately give up his/her hopes: there might be several fundamental theories reducing the rest but if they are not mutually incompatible, then we might still construct a unique ontological system of reality where the ontological categories are extracted from the ontological commitments of the several fundamental theories. Reality would be then definable as the disjunction of the domains of several fundamental theories. All of this admittedly sounds rather metaphorical and fuzzy, and it is so. Let's try to be more precise.

There are at least four terms of the problem needing urgent clarification: "ontological commitment of a theory", "reduction", "fundamental theory", and "compatibility of theories". To this end, a bit of formal semantics might help.

A theory is something uniquely associated with a class of models.[5] Some might even say that a theory **is** just a class of models but we need not go so far in the present context. Let's just admit that there is always a definite class M of models associated with a theory T. If the theory purports to say something about what there is, then at least some of its models have to be regarded as more or less good representations of a given domain of our experience. This means that some of the experiences you have when seeing with the naked eye or through a telescope, a microscope, or any other device of the sort, can be subsumed under one of the models of the theory in such a way that some predictions you make really fit, at least approximately.

The process of subsumption of a bit of experience under a theory's model is a quite involved and ill-understood matter that we cannot discuss in detail here. However, let me sketch very briefly the essential aspects of this relation of subsumption of experience under theory according to the semantic or structuralist view of science. Suppose you have a theory T determined by a class of models M, each one of them having a structure of the type

$$m = \langle D_1, \ldots, D_n, (A_1, \ldots, A_p), R_1, \ldots, R_q \rangle,$$

where the D_i are variables for each model's base sets (the theory's "universe of discourse"), the A_j are variables for auxiliary base sets (mostly sets of numbers or similar "purely mathematical" entities), which in some theories may be missing, and the R_k are relations or functions defined over some of the (auxiliary) base sets – they represent the theory's specific empirical relations or magnitudes. What is common to all elements of M is that they have the same type of structure as given by m and satisfy the same set of axioms (ideally expressed in a convenient formal language), where some or all of the variables in m essentially occur.

Now, suppose that, for whatever reasons, you think that a particular piece of experience, call it E, you (or your colleagues in the "scientific community") have – that is, something you can see, or touch, or hear, either with your "naked" sensory organs or with the aid of telescopes, microscopes, computers, etc., or that you can manipulate in the laboratory, or in your kitchen, or wherever – can be fruitfully *subsumed* under a structure of type m, satisfying the appropriate axioms; an alternative, but equivalent, way to express this is to say that you expect T to be *applicable* to E. However, E will normally not be described in terms of T (i.e., as a structure of type m), but rather in the more or less fuzzy terminology of everyday language or (more often in the case of advanced present-day science) of some already well-established "underlying" theory T' different from T. Consequently, in order to have some chance of success in your endeavour of subsuming E under T, the whole process of subsumption must consist of at least three different partial operations:

(01) You "re-conceptualize" E in (some of the) terms of T, which means that you re-interpret E as a *substructure* d of a particular structure of type m. You may say that you have reconstructed E now as a "model of data" for T.[6] This process of "re-conceptualizing", or "re-interpreting", or "reconstructing" experience in terms of the substructures of a particular theory you want to apply, is one of the least well-understood matters in the philosophical foundations of empirical science, and it is not at all clear what the admissible criteria for this operation are. Certainly, the "suboperations" of *idealization* and/or *approximation* play an important role here, but very likely they are not the only ones. For the rest of the present discussion, let's just assume that we already know how and when the operation (01) successfully works.

(02) By adequately choosing particular values of the variables $D_1, \ldots, R_q$ (which don't necessarily have a direct empirical counterpart in E), you *expand* the substructure d into a full structure m. Now you have structured your experience E as a kind of structure that *could be* a real model of T, that is, an element of M. To check this, you have to proceed to the third operation,

(03) which consists in finding out whether the chosen m really satisfies T"s axioms, i.e., whether the chosen values of the variables $D_1, \ldots, R - q$ really cohere together so as to constitute a model of T, that is, something satisfying T's axioms.

Now, suppose the operations (01), (02), (03) have already been performed successfully. What are the ontological implications of this? The answer is quite simple:

you are entitled to say now that your experience confirms that the "real things behind" your experience are the way the theory says they are. And the things that really exist, are those and only those included in the model(s) representing your experiences. They are the theory's ontological commitment – and, insofar as you take the theory's claims literally, they are also **your** ontological commitment.

The last point is important in order to make clear that the present discussion of a theory's ontological commitments is independent of the current realism/instrumentalism debate – at least in the way this issue is usually put. A theory's ontological commitments are one thing, the ontological commitments of the theory's **users** quite another. For example, a thermodynamicist applying both phenomenological thermodynamics and statistical mechanics might be a "realist" with respect to the ontological commitments of the first theory because he/she takes them literally (i.e., he/she really believes there are gases with different states) while he/she may come out as an "instrumentalist" with respect to the second theory (i.e., he/she doesn't think there "really" are molecules, a space continuum or a time continuum). The reasons why the theory's user might take such a stance, however, are of no concern to us here. It may be because of his/her favorite metaphysics (for example, because he/she thinks that only the middle-sized objects he/she can manipulate in the laboratory do really exist while microphysical entities and continua are only *façons de parler*); but this has nothing to do with the ontological commitments of the used theories themselves. It's only the latter which interest us here. Indeed, the sterility of the current "realism vs. instrumentalism" issue for a cogent analysis of the ontology of science comes from the fact that the debate is usually stated outside the particular theories making up present-day science. Of course, someone may claim to be a realist with respect to **most** actually used scientific theories and someone else may advocate, on the contrary, an instrumentalist position with respect to **most** theories used, but, again, such a position should have to be defended or dismissed for reasons independent of the ontological analysis of really existing science. Note that it wouldn't be reasonable to be a realist with respect to **all** parts of **all** theories used, nor to be an instrumentalist with respect to **all** of them. Even a tough realist will sometimes use some parts of a theory he/she doesn't take literally (e.g., the theory saying that the sun moves westwards everyday). And even a radical instrumentalist will take some parts of some theories literally (e.g., the theory stating that all human beings are mortal). Thus, being a realist or an instrumentalist is just a matter of degree. In particular, any reasonable instrumentalist will have to take at least one theory literally (perhaps a theory about the existence and relationships of ordinary middle-sized objects, or a theory about sense-data, for that matter). And the interesting question for us is what ontological consequences does the literal acceptance of at least one theory have.

Let's clarify this general point by means of a kind of "metatheoretical *Gedankenexperiment*". Suppose there were just one theory – call it "the theory of dots", T_d – under which **all** of your experiences could be subsumed. That is, whether you get wet when it's raining, or you lose your money playing at the stock-market, or you go through a sleepless night after quarreling with your neighbour – all of these experiences, and any other you may have, can be successfully represented by picking out particular models of T_d and by subsuming your experiences under them. Suppose, moreover, T_d has a simple structure: its models have the form

$$\langle D; \rho_1, \ldots, \rho_n \rangle$$

where D is a non-empty, finite set of entities (you call them "dots" but, of course, you could give them any other name), and ρ_i are relations defined over D satisfying some axioms $\alpha_1(\rho_1, \ldots, \rho_n), \ldots, \alpha_m(\rho_1, \ldots, \rho_n)$. In such a case, you would be entitled to say that the world ("your world") just consists of dots. To be is to be a dot. That is, to be is to be the value of a bound variable running over the elements of D. Presumably, a traditional ontologist would still show up and ask strange questions like: "But what is the hidden being (*verborgenes Sein*) behind a dot?", or "Isn't it the case that dots are essentially constituted by nothingness?", or similar ones. However, as an obstinate analytical, pre-postmodern ontologist you have to stick to your scientifically-oriented ontological system and make clear that the only justified ontological commitment is the one with dots. All you can say about Being is that it is to be a dot. And the unity of science, revealed by the uniqueness of T_d, would imply the unity of Being in a clear-cut fashion.

Let us refer to the possible ontological situation just depicted by this schematic example as "Situation #1". Another situation, however, is also imaginable. Not all of your experiences are such that you can always find a model of T_d to successfully represent them. Some stubbornly resist being subsumed under T_d. You conclude then that not all beings are dots, that to be is not always to be a dot. The unity of science, and with it the unity of the world, appears to be destroyed. You start looking for a different theory to determine the non-dottical part of Being.

Suppose, however, that while you are worrying about these things, someone else comes along and offers you a new theory T_{bs} – call it "the theory of beams and slices". T_{bs}'s models have the form

$$\langle B, S; \sigma_1, \ldots, \sigma_p \rangle,$$

the σ_i satisfying certain axioms $\beta_1, \ldots, \beta_q$.

Suppose T_{bs} has the following properties with respect to T_d:

(1) D can be defined as a so-called **echelon set** over B and S, i.e., as a set constructed out of the successive application of the set-theoretical operations of power-set construction and Cartesian product:

$$D \in \wp^n(B^k \times S^{k'})$$

(2) For any axiom α_i of T_d there are some particular conditions $\gamma_1, \ldots, \gamma_h$ which are actually accepted and used in scientific practice and which are such that, when added to some of the axioms β_j of T_{bs}, you obtain

$$\ldots \wedge \beta_j \wedge \ldots \wedge \gamma_1 \ldots \wedge \gamma_h \mid\!\cong \alpha_1,$$

where the symbol "$|\cong$" means that the conjunction to the left at least approximately implies α_i, i.e., it implies a statement which stands in a particular topological relationship with α_i we call "approximation".

If these two conditions are satisfied then we may say that T_d is ontologically **reducible** to T_{bs}. T_d is not a fundamental theory anymore; the **fundamental** theory is T_{bs}.[7]

Suppose, further, T_{bs} would not only reduce T_d but would also be such that it has models successfully representing those experiences not covered by T_d. Then, we would have the unity of science and of the world restored. But we would have to say that there is not just one fundamental ontological category ("dots") but rather two: beams and slices. Of course, from a purely formal point of view, one could also restore the unicity of ontology by defining a set A, call it of "beam-slices", as the union $B \cup S$. But if we assume that some of the relations σ_i and axioms β_j only apply to elements of B and some others only to elements of S, then the alleged ontological unity is spurious: from a material, ontological point of view, we should still differentiate the elements of A satisfying the first set of relations and axioms from those satisfying the second set. Typically, there would be a third set of relations and axioms applying simultaneously to elements of B and S but this still would not restore the unity of ontology. One would be still constrained to register that there are two fundamentally different kinds of things: beams and slices. To be is not to be a dot but to be rather is either to be a beam or to be a slice. Nevertheless, though we have here two fundamental ontological categories we still have a unitary ontological system, since both categories pertain to just one fundamental theory. Let's call this kind of situation, where you have several ontological categories but only one fundamental theory reducing all other ontological categories and theories "Situation #2".

Suppose, however, that you don't find a theory of beams and slices reducing your initial theory of dots but you find one or several other theories $T_1, \ldots, T_h$, each one of them having some model that subsumes some of the experiences left out by T_d and none of them being reducible to the others. Here we have a set of several fundamental theories. Can we still speak of a unitary ontological system? The answer is: maybe. It depends on whether some of the theories in this set are **ontologically incompatible** or not.

Now I have to explain what I mean by "ontologically incompatible". Let me first say what I don't understand by it. That T and T' are ontologically incompatible does **not** mean that T contradicts T'; since both T and T' are fundamental and are determined by different axioms, the entities of the respective domains of the models are different **things** and therefore what you say about one sort of things in one theory cannot contradict what you say about another sort of things in the other. What ontological incompatibility means is that the same sort of experience can be subsumed both under models of T **and** models of T' – T and T' being thereby assumed to be fundamental, i.e., neither is T reducible to T' nor is T' reducible to T, and of course they are not equivalent. Suppose, for example, that

T only speaks about dots whereas T' only speaks about beams. And that dots and beams are really different things (and not just different names for the same thing) because different relations are established between them satisfying different axioms. Then, if the same experience E can be represented by both a model of T and a model of T', this implies that E justifies, or makes plausible, or supports both the idea that being is just being a dot **and** the idea that being is just being a beam. And this is a sort of contradiction – not a formal contradiction, for sure, but an ontological (or "onto-epistemological") one. On the other hand, T and T' would be ontologically **compatible** whenever this doesn't happen, that is, whenever they speak of different things **and** represent different ranges of experiences; or, still in other words, whenever each one minds its own business.

Let's call a situation where you have a series of different but mutually compatible fundamental theories "Situation #3"; and a situation where you have several fundamental theories some of them being incompatible with others "Situation #4".

Now it seems to me that in Situation #3 we would still speak of a unitary ontological system, and therefore of one world, and therefore of being. Being would be determined in this case by different categories. The categories would come sometimes from one and the same theory, some other times from different theories. But there would not be an essential ontological problem in this. Experience would sometimes support the claim that something that is, is a dot, sometimes that something that is, is a beam, sometimes that it is a slice, and so on. But as long as the experiences are essentially different we can live with that. According to the experiences we have, we categorize being in different ways. Surely, the resulting ontological system is less well-unified than in Situation #1 or #2, since we lack a unifying theory. However, it is still **one** big ontological system. There is just one Being of beings.

The picture changes drastically when we go over to Situation #4. It doesn't make sense here to speak of one ontological system. Experience would tell that we live simultaneously in fundamentally different, incompatible worlds. The notion of one and the same world itself would cease to make sense and with it the notion of Being. Ontology in other than a Pickwickian sense would become impossible.

The interesting question, of course, is what is really the case. Are we in Situation #1, 2, 3 or 4? What does actually existing science suggest that is the truth of the matter? Now it seems pretty clear that we don't live in Situation #1. This is certainly not a logical truth, it is an empirical fact; but it is so obvious that it almost has the force of a logical truth. What about Situation #2? Since Newton's times, the most brilliant physicists and natural philosophers have invested tremendous efforts to bring it about that Situation #2 becomes realized. *GUT* is just the most recent example. And along with physicists, also chemists, biochemists, sociobiologists, etc. have tried to give good reasons to endorse the same world picture: reductionism. At certain times they seem to be on the verge of fulfilling the utopia of ONE THEORY – ONE WORLD. But, then, once again some trouble-makers appear on the scene with other theories representing well other sorts of experiences and definitely **not** reducible to the pretended one big theory. The history of science of the last 300 years tells us that we should be rather cautious in expecting something close to Situation #2.

At the present stage of our scientific worldview, the real issue seems to be posed between Situations #3 and #4. Whatever the fans of the "theory of everything" might contend, it seems obvious to me that we are confronted with **several** fundamental theories in the sense explained here. Two fundamental theories certainly are general relativity and the electroweak theory. Actually, one should add a third branch of fundamental physical theories: thermodynamics. As Lawrence Sklar has pointed out,[8] it is far from clear that equilibrium thermodynamics is really reducible to **any** mechanical theory – whatever popular scientific textbooks might contend. And even if equilibrium thermodynamics really were reducible to statistical mechanics, the latter is **not** reducible to quantum mechanics, and *a fortiori* not reducible to electroweak theory. If we go over to the thermodynamics of irreversible processes, the prospects are even worse for the reductionist. And if we take into consideration the really existing theories of biology, psychology, linguistics, etc., the reductionistic unity of science talk appears as a bad joke. Thus, we have no reason to expect seriously that the theory of everything is around the corner. So, the real question is whether these numerous fundamental theories are incompatible or not – in the proposed sense of "incompatibility".

Now, many people contend nowadays that at least two of the existing fundamental theories in physics, viz. general relativity and electroweak theory are incompatible. Some people might also add thermodynamics, especially irreversible thermodynamics (and, with it, important theories of chemistry, biochemistry, etc.), to the list of mutually incompatible theories. If this were the case we would have Situation #4 and, therefore, the idea of a general ontology would appear as quite dubious. However, it is not clear that people asserting the incompatibility of, say, general relativity and quantum electrodynamics understand by it the same intertheoretical relation I have described as incompatibility here. Rather, it seems that what they mean is that neither the principles of general relativity are derivable from quantum electrodynamics nor the other way around, and that these theories are based on very different conceptual frameworks. But this still is no sufficient reason for asserting incompatibility in the present sense. If the experiments and observations subsumed under the models of quantum electrodynamics (or electroweak theory) are clearly distinct from the portion of experience subsumed under general relativity, then we have Situation #3 rather than #4. The same goes for the experiments and observations for which irreversible thermodynamics, etc. have been designed. It seems to me that this is actually the more realistic interpretation of what is going on in present-day science. If this were admitted, then we still would be able to build up a single ontological system, though of course at the price of admitting that Being sometimes is said to be in the sense of general relativity, sometimes in the sense of particle physics, sometimes in the sense of irreversible thermodynamics, and perhaps in still other senses. There would still be one world but this world would be rather like a de-centralized federal system and not like a centralized State. Whether this is a plausible interpretation of present-day science, however, ultimately depends on the results of a careful ontological analysis of the existing fundamental theories and their interrelationships. This still has to be done.

Notes

[1] The present article is a substantially revised and expanded version of the paper "Ontology, Reduction, and the Unity of Science", delivered at the 20th World Congress of Philosophy (Boston 1998) see Moulines (2001). The author would like to thank an anonymous referee for some remarks which led to an improvement of the original text.

[2] See Quine (1971, 1).

[3] See Moulines (1998, 319 f).

[4] This seems to have been Hegel's actual position: As he was claiming that, for "deep philosophical reasons", there could only be seven planets, when someone told him that a nasty astronomer had just discovered an eighth one, he is supposed to have answered that, if reality didn't agree with his theory, so much the worse for reality.

[5] The vast majority of semantic or structuralist approaches in present-day philosophy of science takes the class of models of a theory to be an essential element of its identity. For a detailed exposition of this see, e.g., Balzer et al. (1987, Ch. I).

[6] As far as I know, the first author to have introduced this way of speaking was the precursor of the semantic and/or structuralist view of science, Patrick Suppes – see his article Suppes (1962).

[7] There is a rich literature on the formal or semi-formal explication of the reduction of theories, starting with the classical Nagel (1961). The two conditions laid out here may be seen as the "core idea" of ontological reduction and are filtered out of previous explications; see, particularly, Moulines (1984) and Balzer et al. (1987).

[8] See Sklar (1992, 137).

References

Balzer, W., C. U. Moulines and J. D. Sneed: 1987, *An Architectonic for Science*, Dordrecht, Reidel.

Moulines, C. U.: 1998, 'What Classes of Things are There?', in: C. Martínez, U. Rivas and L.Villegas-Forero (eds.), *Truth in Perspective*, Hants, Ashgate.

Moulines, C. U.: 'Ontological Reduction in the Natural Sciences', in: W. Balzer, D. Pearce and H.-J. Schmidt (eds.), *Reduction in Science*, Dordrecht, Reidel.

Moulines, C. U.: 2001, 'Ontology, Reduction and the Unity of Science', in Tian Yu Cao (ed.), *The Proceedings of the Twentieth World Congress of Philosophy*, Vol. 10, *Philosophy of Science*, Bowling, Green, Philosophy Documentation Center.

Nagel, E.: 1961, *The Structure of Science*, New York et al., Harcourt, Brace & World.

Quine, W. V. O.: 1971 (1953), 'On What There Is', in *From a Logical Point of View*, Cambridge, MA, Harvard University Press.

Sklar, L.: 1992, *Philosophy of Physics*, Oxford, Oxford University Press.

Suppes, P.: 1962, 'Models of Data', in E. Nagel, P. Suppes and A. Tarski (eds.), *Logic, Methodology and Philosophy of Science*, Stanford, Stanford University Press.

LOGICAL PLURALISM AND THE PRESERVATION OF WARRANT

GREG RESTALL
Philosophy Department, The University of Melbourne, Australia, E-mail:
restall@unimelb.edu.au[1]

Abstract. I defend logical pluralism against the charge that ONE TRUE LOGIC is motivated by considerations of warrant transfer. On the way I attempt to clarify just a little the connections between deductive validity and epistemology.

1. Pluralism

I am a logical pluralist. With JC Beall, I have argued that there is no single relation of logical consequence, but rather, there are different consequence relations, equally adequately formalising deductive validity (Ball and Restall 2000, 2001; Restall 1999; 2002). There is no *one true logic* but rather, there are *many*. Logic is a matter of "truth preservation in all cases" in the sense that

▷ An argument is (deductively) *valid* if and only if in any *case* in which all its premises are true, its conclusion is true too.

The variety of relations of logical consequence spring from the variety of appropriate ways to make the pretheoretic notion of a "case" precise. Classical predicate logic stems from taking cases to be the consistent complete worldlike entities of Tarski's model theory of classical logic. Constructive logic is given when you take cases to be possibly incomplete bodies of information or warrants or constructions that you might find in Kripke models for intuitionistic logic. Relevant logic is given when you take cases to be possibly incomplete or inconsistent (or both) ways the world might or might not be.

I provide this very swift story to set the scene. I will not be *arguing for* logical pluralism here, except in the attenuated sense of defending it against some very particular objections. I will focus here on just one issue for logical pluralism: The connection between logic and warrant or justification. On the way, I will hopefully clarify this connection, and show that, instead of leading one away from pluralism, a proper understanding of the epistemology of inference actually supports pluralism with respect to logical consequence.

S. Rahman et al. (eds.), Logic, Epistemology, and the Unity of Science, pp. 163–173.

2. The Dog and Disjunctive Syllogism

Once there was a Dog, pursuing a Man down a bush track. The Dog follows the scent until she gets to a fork in the track. The Dog believes that either the Man went down the left fork (call this proposition A) or the right fork (call this proposition B). So the Dog believes $A \vee B$. The Dog sniffs for a while down the left fork, and detects no scent. The Dog thus also believes $\sim A$: the Man did not go down the left fork. Without hesitating to check for the scent down the right fork (and thus, without trying to find any independent *evidence* or *warrant* for B), the Dog continues her pursuit down the right fork in the track, inferring B, the claim that the Man has gone that way (Anderson and Belnap 1975; Anderson et al. 1992; Belnap and Dunn 1981). On this account of the Dog's behaviour, the Dog is reasoning, and indeed, she seems to be reasoning well. She has used the inference *disjunctive syllogism*, which has the following form:

$$\frac{A \vee B \quad \sim A}{B}\,[\textsc{ds}]$$

Proponents of *relevant* logics reject disjunctive syllogism, because when you combine it with certain unproblematic inferences (here, conjunction elimination and disjunction introduction) you can make the following implausible inference:

$$\dfrac{\dfrac{\dfrac{A \wedge \sim A}{A}\,[\wedge E]}{A \vee B}\,[\vee I] \qquad \dfrac{A \wedge \sim A}{\sim A}\,[\wedge E]}{B}\,[\textsc{ds}]$$

That is, disjunctive syllogism (together with conjunction elimination and disjunction introduction) leads to the odd conclusion that *anything whatsoever* (here B) follows from a contradiction (here $A \wedge \sim A$). That doesn't seem right (at least to relevant logicians) and so, disjunctive syllogism is rejected.

I have no doubt that there is something *right* about the relevantist's rejection of disjunctive syllogism, especially in the context of the inference of B from $A \wedge \sim A$. When we assume $A \wedge \sim A$, it seems correct to deduce A (and hence $A \vee B$) and $\sim A$, but to deduce from $A \vee B$ and $\sim A$ that *it is* B that is true seems out of place. How could the reasoning go? $A \vee B$ is true, and A isn't true, so it must be B that makes $A \vee B$ true. But that's not right in this context. It's A that makes $A \vee B$ true (Dunn 1986). Of course, we are inconsistent about the truth of A, but that is simply a reflection of the starting point of our inference: we were reasoning in the context of an inconsistent assumption.

My analysis, as a pluralist, of the appeal of this rejection of disjunctive syllogism is simple. We admit an inconsistent situation (or circumstance, or whatever you like to call these things) in which the contradiction $A \wedge \sim A$ is true, but in which B is not. In this situation, the inference of disjunctive syllogism is blocked in the straightforward sense that the premises are true (for in this situation $A \vee B$ is true, given $A \wedge \sim A$, and $\sim A$ is true, also given $A \wedge \sim A$) but the conclusion is not true. It need not follow from this that I think that it is in any way *possible* for $A \wedge \sim A$ to be true. A pluralist must simply think that there are cases (in the sense of used to define logical consequence) in which contradictions are true. These cases or situations need not be *possible* any more than possible worlds need be *actual*.

An aside may help allay any fears about accepting the kind of *impossibilia* countenanced here: This acceptance of *impossibilia* is much less radical than it might seem, for proponents of classical predicate logic are also committed to cases in which impossibilities are true. For the inference from "x is red" to "x is coloured" is classically invalid, as there are models of classical predicate logic which allow objects into the extension of "red" which are not in the extension of "coloured." Models for predicate logic may involve impossibilities in much the same way as models for relevant logics. It is of course true that the behaviour of negation in an impossibility differs somewhat from its behaviour in what we might call "the actual world". The same is true for the behaviour of "is red" and "is coloured" in models for predicate logic. (A defence must be mounted to the effect that this variant behaviour is acceptable – in particular that it does not "change the subject" but is indeed providing an account of the behaviour of *negation* and not something else (Quine 1970). Beall and I have mounted such a defence elsewhere (Beall and Restall 2000; Restall 1999).)

As much as I find impossible situations important in the semantics of relevant logics, they are not the only way to understand the failure of disjunctive syllogism. The *Scottish Plan* of Stephen Read (Read 1988) takes the inference to fail because the conditional "if $A \vee B$ and $\sim A$ then B" just fails to be true (on grounds of relevance) without there being any possible counterexample in which the premise is true and the conclusion false. The difference between Read's Scottish Plan and my pluralism is important, but not relevant to the task at hand. However you understand the failure of disjunctive syllogism, you need to give a plausible explanation of what is going on in the case of the Dog's reasoning.

The problem this poses for the pluralist (and the relevantist too) is straightforward one. Arguments about relevance (or constructivity) seem like mere quibbles when confronted with the Dog's reasoning. The classical acceptance of disjunctive syllogism seems to be completely correct. The Dog is not making a mistake by inferring B from $A \vee B$ and $\sim A$. Given that her beliefs that $A \vee B$ and that $\sim A$ were warranted, then so is her belief that B. Conversely, if it turns out that she was somehow mistaken in her belief that B, this will be because one of her prior (but perhaps still justified) beliefs that $A \vee B$ and $\sim A$ was also mistaken. Her inference clearly seems a good one. The move from premises to conclusion seems to preserve any warrant the Dog has for the premises into warrant for the conclusion too.

What can the proponent of relevant logics say about such a case? The *relevantist* like Read, for whom the One True Logic is a relevant one seem to be forced to say that the Dog is indeed making a mistake in inferring B from her premises. The inference is invalid, and you make a mistake in stepping from premises to the conclusion. This does not mean that it is possible that the premises be true while the conclusion is false. Not at all. The relevantist may agree that it is impossible that the premises be true and the conclusion untrue – but the relevantist will go on to say that the impossibility of true premises and false conclusion is not a correct analysis of validity. More is needed for the validity of the argument, in particular the relevance of the premises to the conclusion.

Read's own analysis is that while the argument as we presented it is invalid, a closely related one is valid. This is the *intensional* version of disjunctive syllogism, which replaces the extensional disjunction in $A \vee B$ with its intensional cousin.

$$\frac{A + B \quad {\sim}A}{B}\,[\textsc{ids}]$$

Here, $A + B$ is the intensional disjunction of A and B, which is true if and only if ${\sim}A \to B$ is true (and equivalently, if and only if ${\sim}B \to A$ is true, at least according to the standard analysis of the relevant conditional $\to$). Intensional disjunction differs from the standard extensional disjunction in a number of respects. First, the inference IDS is relevantly valid, where DS is not. IDS is a form of *modus ponens*: from ${\sim}A \to B$ and ${\sim}A$ to infer B. Second, intensional disjunction is properly *stronger* than extensional disjunction. While the truth of A or of B is sufficient for the truth of $A \vee B$, neither need be sufficient for the truth of $A + B$. To infer $A + B$ from A would be to infer ${\sim}A \to B$ from A. To infer $A + B$ from B would be to infer ${\sim}A \to B$ from B. Both of these inferences fail for a relevant conditional.

We can agree with Read that IDS is relevantly valid, but this fact does not help in the case of the Dog's inference. For this to help, we would need to have reason to take analyse the Dog's inference as being an instance of IDS. It is by no means clear that this is the case.

A direct argument might be thought to show that the Dog must be inferring using IDS: The Dog correctly infers B from ${\sim}A$ and something else. This inference wouldn't be correct (so the relevantist says) were the other premise $A \vee B$. What premise *would* suffice to render the argument valid? The weakest such premise is ${\sim}A \to B$. In other words, the intensional disjunction $A + B$.

This argument clearly explains why the relevantist must appeal to IDS or something like it. However, in our dialectical context, it fails to give an independent reason to take the original argument, DS to be somehow inappropriate. Furthermore, it is by no means clear that IDS is the appropriate analysis of the Dog's reasoning. Consider the following elaboration of the scenario.[2] The Man went down the right fork in the track. He was always planning to go down the right fork. He would never even consider going down the left fork. If his way down the right fork was barred, he would have gone back home, in the other direction. Now consider the Dog's epistemic state as she sniffs at the fork in the road. She considers that the Man has gone down one or other fork. She verifies (to her satisfaction at least) that it was not the left fork (that $\sim A$ is true) she infers that it was the right (that B is true). Was she right that the Man went down one or other fork? Clearly the answer is yes – she was correct, for the Man went down the right fork. But if the claim that the Man went down one or other fork is the intensional disjunction $A + B$ it is by no means clear that this is the case! For $A + B$ is equivalent to $\sim B \rightarrow A$, the claim that *if the Man did not go down the right fork, he went down the left*. But it is no longer clear that this is, in fact, true. At the very least, it cannot be confirmed as true on the basis that B is true, for B does not suffice for the truth of $A + B$. In fact, knowing what we know about the Man and his intentions, it seems as if $\sim B \rightarrow A$ is plainly false. It's not true that if the Man didn't go down the right track he went down the left, for under *no* circumstances was he planning to go there. If he were not togo down the right track, he'd go back home. So, if the Dog were to infer using IDS her inference may be relevantly valid, but it trades in this virtue for the vice of unsoundness. The premise $A + B$ is no longer true.

The relevantist has more to worry about. Read agrees that it is impossible for the premises to be true and the conclusion false. The Dog will *never* (and indeed, *can* never) step from truth to falsity in making the inference DS. Why does this not count as validity? A technique of reasoning which is assured to never step from truth to falsity would at the very least seem to be useful for extending our body of knowledge, and to be the kind of thing logicians have attempted to develop and understand. Why this is not called *validity* seems unclear. Perhaps some account of relevance could be developed according to which inferences such as disjunctive syllogism make mistakes of relevance, but it is something else to say that there is *no* sense in which the Dog reasons well in using disjunctive syllogism. That is an altogether more difficult position to hold. I must leave it to relevantists such as Read to defend this position.

The pluralist's position seems at least at the onset more simple. A pluralist can make the quite simple case that while the Dog's reasoning is not *relevantly* valid (as there are *impossible* situations in which the premises are true and the conclusion not true, and this explains the failure of relevance of the argument) it is valid in some other sense. One example is that it is classically valid, in the simple sense that it is impossible that the premises be true and the conclusion false. This too is a sense of validity worth the name, and the Dog is warranted in drawing the conclusion of B from its premises. The warrant of the premises "flows through" to the conclusion.

But now the pluralist has a problem not shared by the relevantist. We have a classical inference which preserves warrant in this sense. Will this not single out one logic as the logic which preserves warrant in just this sense in all reasoning situations? Will this not deserve to be called the One True Logic, since it seems best placed to guide reasoning? If warrant is preserved in cases such as the Dog's reasoning, then this seems to point to a logic stronger than relevant logics as appropriate for guiding and analysing reasoning. The initial appeal of pluralism might wear thin.[3]

In the rest of this paper I will defend pluralism against this objection, chiefly by showing that the issue of warrant preservation does not point the reasoner to a particular logic. Many logics can be used to guide and analyse reasoning, and many logics will play their part the transfer of warrant from premises to conclusions. There is no One True Logic of warrant transfer.

3. Problem 1: Logical Omniscience

Suppose, for the sake of the argument, that there *is* a logic of warrant transfer, and, for the sake of the argument, that it is something like classical predicate logic or one of its extensions. If this is the case, then for any argument from premises X to conclusion A, if all of the premises in X are warranted (for some believer) so is the conclusion. Now, if just *one* premise (say B) is warranted then since all arguments from B to a tautology are valid, it will follow that all tautologies are warranted. But this cannot be correct. Not everyone has warrant for believing *every* tautology of classical predicate logic. Some tautologies are so complex that you only ought believe them when you have taken the time to read and understand what they might mean. Here is an example. Just *when* do you have a warrant for believing this?[4]

$$(\exists x){\sim}Rxx \supset (\exists x){\sim}\,((\forall y)(\forall z)(Rxy \supset (Ryx \wedge ({\sim}Ryz \vee Rxz)) \wedge (\exists y)Rxy)$$

Or in a simpler case, this?

$$(\exists x)\,(Fx \supset (\forall y)Fy)$$

Of course we need not deny that any tautology in classical predicate logic is *potentially* warranted, for if it is a tautology, it is provable, and any proof can at least in theory be surveyed and give reason for its conclusion. However, it does not follow that for any tautology we *actually possess* a warrant for believing it. But this is what follows, at least if a logic of warrant transfer is something like classical logic. For any argument to a tautology is valid, and so, if warrant is transferred from premises to conclusion, simply take an argument with a warranted premises and a tautology as a conclusion. Since warrant is transferred to the conclusion, it follows that the tautology is warranted. This is, of course, an all too swift way of demonstrating the warrant attaching to tautologies.

It may be thought that we are biting off more than we need to chew in applying *all* of the validities of classical logic to the transfer of warrant. Perhaps it is simply the *small*, *basic* argument forms which preserve warrant: argument forms such as conjunction elimination, disjunction introduction and disjunctive syllogism. In these cases we do not take wild jumps from premises to irrelevant (but tautologous) conclusions. We take much more measured baby steps, from $A \wedge B$ to A, or from $A \vee B$ and ${\sim}A$ to B. Here it seems much more likely that warrant is preserved from premises to the conclusion.

If this is to be used as a defence of a *logic* such as classical logic as the logic of the preservation of warrant, then the baby steps used must add up to the whole logic. The collection of basic (warrant preserving) inferences must be a complete selection of rules so that any argument which is valid in the logic can be made up of a chain (or a tree) of these basic inferences. One candidate set of basic rules might be that of a natural deduction presentation of the logic in question, with introduction and elimination rules which present the simple sense of each logical connective (Prawitz 1965). Plausibly, to know the meaning of the connective is to know how to use it in these basic inferences, so if anything preserves warrant *these* rules must.

This is an appealing picture, but it too must be flawed. If each of the baby steps in a proof preserves warrant, in the sense that if the premises have warrant so does the conclusion, then *any* argument, no matter how complex or convoluted, no matter how unsurveyably large, preserves warrant. The warrant provided for the premises filters down uninterrupted to the conclusion, passing through each rule along the way. If the warrant stops, it must stop at some inference, and this inference will fail to preserve warrant.[5]

Let me make this example more vivid. I take it that we have warrant for enough basic claims of arithmetic to do simple calculations involving addition and multiplication. From these basic claims, using logic alone we can prove all sorts of things. Andrew Wiles has shown that Fermat's Last Theorem can be proved (Wiles 195; Wiles and Taylor 1995). It seems that we are all warranted in believing that Fermat's Last Theorem is indeed true. However, we weren't so warranted *before* the proof was produced. Mere entailment is not enough to guarantee the preservation of warrant. If there is a logic of warrant preservation, it is nothing like classical predicate logic or any of its neighbours.

4. Problem 2: Inconsistency and Warrants in Tension

The reasons we have seen so far lead us away from the view that warrant preservation is tracked by classical predicate logic or anything like it. Any logic in which tautologies are entailed by anything whatsoever will give us warrants for all tautologies. Perhaps a *relevant* logic, in which tautologies need not follow from anything and everything, might do the trick.

This, however, seems also to not be the case. Epistemic agents like you and me not only fall below a logical ideal by not being logically omniscient, we also fail by having inconsistent beliefs, and even by having *warranted* inconsistent beliefs. I may have reason to believe *A* and also, reason to believe a claim *B* inconsistent with *A*. David Lewis provides a clear example

> I used to think that Nassau Street ran roughly east-west; that the railroad nearby ran roughly north-south; and that the two were roughly parallel. (By "roughly" I mean "to within 20°".)
> (Lewis 1982, 436)

Here, Lewis believes three mutually inconsistent propositions. I take it that the example can be extended into one in which he has *warrant* for these three propositions.[6] Each of the three can be warranted beliefs. If any false belief is ever warranted, and if any simple observational beliefs like these are warranted, then I take it that these three beliefs may be warranted, and may each be warranted *together*. If a belief must be *consistent* with the rest of an agent's beliefs in order to be warranted, then it seems to follow that *nothing* is ever warranted, as consistency of a belief set is so difficult to attain.

So, in this example, Lewis has three warranted beliefs, A, B and C, such that the three are jointly inconsistent. Nothing forces us to conclude that the *conjunction* $A \wedge B \wedge C$ is one of Lewis's warranted beliefs. In fact, it seems implausible to suppose that the conjunction need be one of his beliefs, *or* that it has any warrant. If you were looking for a warrant for the belief that the world was inconsistent in just that way, you would need a lot more than simply the three warrants Lewis might have had for the beliefs A, B and C. (Despite this hesitancy about the conjunction, there is still a sense that Lewis was *committed* to the inconsistent conjunction of his three beliefs. It is when Lewis realises that he is committed to such an inconsistency that he revises away one of the beliefs, or at least attempts to do so.[7]) This means that even simple logical inferences such as conjunction introduction

$$\frac{A \qquad B}{A \wedge B}\,[\wedge I]$$

do not preserve warrantedness. We may have warrant for each of the premises of an inferences such as this, yet it may not be preserved to the conclusion. If there is anything like a logic of preservation of warrant, it will look *nothing* like traditional classical, constructive or relevant logics.

It is easy to see how this failure of simple rules comes about. What is desired is a logic of preservation of *warrantedness*. We wish to know that if we have warrants for the premises of an argument, we have a warrant for the conclusion too. If an argument has more than one premise, then the warrant for each premise might be in *tension*, and not combine to provide a warrant for the conclusion. The warrant Lewis might have for taking Nassau Street to run east-west was one thing (some experience of the street?), the warrant for taking the railroad to run north-south was another (some experience of the railroad?), and the warrant for taking them to be parallel a third (seeing them run in the same direction?). Taking these warrants together does not give you a warrant for the conjunction – it gives you instead a reason to reject at least one of the conjuncts.

The argument shows that despite having a reasons to believe A, B and C, Lewis has no reason to believe A, B and C together. Each of the premises is warranted, but the conclusion is not. The logic of warrant preservation, if you wish to call it a "logic," is not classical, and it is not like any non-classical logic I defend. Considerations of warrant preservation do not lead a logical pluralist to favour one system over another. They point in a very different direction again.

5. The Dog Returns

Now let us return to the case of the Dog. She infers B from $A \vee B$ and $\sim A$. In this case, her warrant for the premises is preserved as warrant for the conclusion. We may agree that this step was a good one. But it does not follow that inferences of the form of disjunctive syllogism *always* preserve warrant. We may have a situation similar to Lewis's inconsistent triad of beliefs.

Perhaps one kind of warrant leads me to believe $A \vee B$ on the basis of A. Perhaps I believe that Socrates was either an Athenian or a Spartan, on the basis that I have heard that he was an Athenian. Perhaps I also come to believe $\sim A$ (and to have warrant for such a belief) by some other route. Someone gives me reason to believe that Socrates was not an Athenian. Now *if* in this circumstance I still have warrant for my belief that he was Athenian (so I have warrant for A and warrant for $\sim A$) then the inference fails to preserve warrant. There is no way that this difficult epistemic situation gives me *any* reason to believe that Socrates was a Spartan. Warrant here for $A \vee B$ and for $\sim A$ does not transfer to warrant for B.[8] Disjunctive syllogism, like other truth preserving inferences, may fail to preserve warrant if the epistemic situation is right.

We have gone some way in showing what considerations of preservation of warrant do *not* tell us. They do not force us away from a pluralist position into One True Logic, for considerations of preservation of warrant lead us away from logic altogether. What can we say positively about the connection between logic and the preservation of warrant? For this we need a coherent story of the nature of warrant and what it is for claims to be warranted. Different positions will give different conclusions at this point. Externalisms about warrant will tell us that a belief is warranted if it is acquired in the right kind of way. Given this account of warrant it will be clear that inference will play an important role, but validity will not track warrantedness in anything other than special cases. For an externalist story will tell us only what it is for a *belief* to be warranted, and it does not tell much about warrant for propositions not yet encountered or considered.

Internalisms about warrant might do a little more. Internalist accounts of warrant utilise logical relationships among beliefs, and proof (or *recognition* of proof) is an important part of internal justification and acquisition. Perhaps here we will find more to say about the relationship of validity and warrant preservation. The considerations we have seen so far, however, give us good cause to think that pluralism is in no way at threat, given a proper understanding of the role of logic in evaluating the preservation of warrant.

Notes

[1] Thanks to JC Beall, Pragati Jain, Gary Kemp, Stephen Read, an audience at the Australian National University and an anonymous referee for help in clarifying the issues discussed here. This research has been supported by the Australian Research Council, through Large Grant A00000348.

[2] I will continue to talk of what the Dog believes, without any hesitation. If you think that the Dog is not the right kind of creature to have this repertoire of beliefs, retell the story in more comfortable terms.

[3] I thank Gary Kemp and Stephen Read for both putting this objection in discussion in February 2000.

[4] I am not sure that I am warranted in believing it. I got it by transforming the tautology $((\forall x)(\forall y)(Rxy \supset Ryx) \wedge (\forall x)(\exists y)Rxy \wedge (\forall x)(\forall y)((Rxy \wedge Ryz) \supset Rxz)) \supset (\forall x)Rxx$ (to the effect that if R is symmetric, without dead ends, and transitive, then it is also reflexive) to a non-obvious one by simple predicate logic transformations. However, I am not at all confident that I have not made a mistake along the way.

[5] You may notice the similarity with sorites paradoxes. Perhaps some case may be made to the effect that warrantedness is vague, and that there is no individual step where we pass from warranted premises to an unwarranted conclusion.

[6] I say "extended", because Lewis's discussion does not introduce the notion of warrant. He seeks merely to explore the notion of what might be "true according to his beliefs."

[7] I have defended a simple paraconsistent logic for the purpose of analysing truth according to a body of beliefs, or more simply, a believer's commitments (Restall 1997).

[8] If I replace the disjunction $A \vee B$ with $A + B$ defined using robust *conditional* (particularly with a conditional possessing some counterfactual force) then warrant may well be preserved, since it is less likely that the warrant we have for $\sim A$ will undercut the warrant for $\sim A \rightarrow B$. Yet inference too might fail to be warrant preserving given unfortunate epistemic conditions.

References

Anderson, Alan Ross and Nuel D. Belnap: 1975, *Entailment: The Logic of Relevance and Necessity*, Vol. 1, Princeton, Princeton University Press.

Anderson, Alan Ross, Nuel D. Belnap and J. Michael Dunn: 1992, *Entailment: The Logic of Relevance and Necessity*, Vol 2, Princeton, Princeton University Press.

Beall, JC and Greg Restall: 2000, 'Logical Pluralism', *Australasian Journal of Philosophy* **78**, 475–493.

Beall, JC and Greg Restall: 2001, 'Defending Logical Pluralism', in Bryson Brown and John Woods (eds.), *Logical Consequence: Rival Approaches. Proceedings of the 1999 Conference of the Society of Exact Philosophy*, Stanmore, Hermes, pp. 1–22.

Belnap, Nuel D. and J. Michael Dunn: 1981, 'Entailment and the Disjunctive Syllogism', in F. Fløistad and G. H. von Wright (eds.), *Philosophy of Language/Philosophical Logic*, The Hague, Martinus Nijhoff, pp. 337–366. Reprinted as Section 80 in *Entailment* Volume 2, Anderson et al. (1992).

Dunn, J. Michael: 1986, 'Relevance Logic and Entailment', in Dov M. Gabbay and Franz Günthner (eds.), *Handbook of Philosophical Logic*, Vol. 3, Reidel, Dordrecht, pp. 117–229

Lewic, David: 1982, 'Logic for Equivocators', *Noûs* **16**, 431–441.

Prawitz, Dag: 1965, *Natural Deduction: A Proof Theoretical Study*, stockholm, Almqvist and Wiksell.

van Orman Quine, Willard: 1970, *Philosophy of Logic*, Englewood Cliffs, NJ, Prentice-Hall.

Read, Stephen: 1988, *Relevant Logic*, Oxford, Basil Blackwell.

Restall, Greg: 1999, 'Ways Things Can't Be', *Notre Dame Journal of Formal Logic* **39**, 583–596.

Restall, Greg: 1999, 'Negation in Relevant Logics: How I Stopped Worrying and Learned to Love the Routley Star', in Dov Gabbay and Heinrich Wansing (eds.), *What is Negation?*, Vol. 13 of *Applied Logic Series*, Dordrecht, Kluwer Academic Publishers, pp. 53–76.

Restal, Greg: 2002, 'Carnap's Tolerance, Meaning and Logical Pluralism', *Journal of Philosophy* **99**, 426–443.

Wiles, Andrew: 1995, 'Modular elliptic curves and Fermat's Last Theorem', *Ann. Math.* **141**, 443–551.

Wiles, Andrew and Richard Taylor: 1995, 'Ring-theoretic Properties of Certain Hecke Algebras', *Ann. Math.* **141**, 553–572.

IN DEFENCE OF THE DOG: RESPONSE TO RESTALL

STEPHEN READ
Department of Logic and Metaphysics, University of St Andrews, Fife KY16 8RA, Scotland, U.K.
E-mail: `slr@st-and.ac.uk`

Abstract. Greg Restall challenges the relevantist to explain the Dog's reasoning in pursuing the Man down the right fork, having only verified that the Man did not take the left fork. I do so, showing thereby not only what the relevantist must mean by validity, but why Restall's pluralism is an incoherent and untenable position.

1. The Dog

Greg Restall challenges the relevantist (and me in particular) "to give a plausible explanation of what is going on in the case of the Dog's reasoning" (Restall 2004, 165). I have discussed the relevant issues in various places before (see, e.g., Read (1981a, b, 1988)), but never directly applied to the famous case of the Dog. So it may be helpful to repeat those points in analysis of the Dog's reasoning, especially since many of the views Restall attributes to me are not ones I recognise myself as ever having held.

The Dog, chasing the Man down a path, comes to a fork. Restall writes: "The Dog believes that either the Man went down the left fork (call this proposition A) or the right fork (call this proposition B). So the Dog believes $A \vee B$" (Restall 2004, 164). Only later does Restall confirm that he is using '$\vee$' in its nowadays familiar truth-functional sense. What warrants him in using $\vee$ to represent the Dog's belief? That the Dog believes A or B is not because it (Restall's Dog is a politically correct bitch, but mine will be neutered) believes A or believes B. So far, before it sniffs the left fork, it appears to believe neither. Faced simply with a fork in the track, it believes the Man went down one or other track – if the Man has not turned round and retraced his steps, which he clearly has not or the Dog would have met up with him already. So the evidence proclaims loudly that if the man has not gone down one track, he has gone down the other. The Dog's belief is properly represented by $A + B$, on my analysis, as Restall later recognises (Restall 2004, 166). '$+$' here is intensional disjunction, explained in, e.g., Read (1988, 53) or Restall (2004, 166). Moreover, from $A + B$, given its subsequent verification of $\sim A$, the Dog validly infers B.

S. Rahman et al. (eds.), Logic, Epistemology, and the Unity of Science, pp. 175–180.

2. Validity

What morals may we draw from this analysis?

The Dog is not mistaken in its inference. Nonetheless, the inference-pattern, 'A or B. Not-A. So B' is equivocal, since the disjunction may be intensional or extensional. As shown in Section 1, the Dog's beliefs are properly represented as '$A + B$' and '$\sim A$', where '$+$' is an intensional disjunction. The inference-pattern, '$A + B$. $\sim A$. So B' is valid, for it preserves truth. Whenever the premises are true, so is the conclusion.

The inference-pattern DS, '$A \vee B$. $\sim A$. So B' is not valid. It is not the case that whenever the premises are true, so is the conclusion. But the reason it is not valid is not the lack of relevance of the premises to the conclusion. That is, rather, a consequence of its invalidity. It is invalid because it does not preserve truth. I have never said that "more is needed for the validity of the argument" (as Restall attributes to me), "in particular the relevance of the premises to the conclusion" (Restall 2004, 166). Relevance is not a sieve on truth-preservation, to separate out the valid arguments from the truth-preserving ones. Valid arguments are those which preserve truth. This becomes clear once one has obtained a proper analysis of the concept of truth-preservation.

What I characterised as "Geach's challenge" (Read 1988, Section 7.1) was the mistaken demand he made of the relevantist to provide a counterexample to DS where $A \vee B$ and $\sim A$ were true and B false. Let '$p \rightarrow q$' represent 'if p then q'. The classicist and the relevantist disagree over the correct logic of '$\rightarrow$'. For the classical logician, '$p \rightarrow q$' is false only if '$p \wedge \sim q$' is true. Hence, if someone denies that q is true whenever p is, that is, that '$p \rightarrow q$' is not always (or necessarily) true, the classical logician in the person of Geach expects to be given a counterexample in which '$p \wedge \sim q$' is true. But the relevantist claims that there are false conditionals with true consequent, some with true and some with false antecedent. Just as, for example, q does not follow from '$p \wedge \sim p$', so too 'if $p \wedge \sim p$ then q' is false. Thus Geach's challenge is unreasonable, as Geach meant it.

What the relevantist does believe is that '$p \rightarrow q$' is false only if '$p \circ \sim q$' is true, where '$\circ$' is an intensional conjunction corresponding (by a De Morgan equivalence) to '$+$' in the same way that '$\wedge$' corresponds to '$\vee$'. Thus I rose to Geach's challenge by providing a counterexample where '$((A \vee B) \wedge \sim A) \circ \sim B$' was true. This was the heart of the Scottish plan (Read 1988, Section 7.8). Classical logic represents logical theory wrongly in a way not dissimilar to the way Restall misrepresents the Dog's beliefs. The argument from p to q is valid if and only if q is true whenever p is, that is, iff '$p \rightarrow q$' is necessarily true, i.e., iff '$\sim(p \circ \sim q)$' is necessarily true, which is the case just when '$p \circ \sim q$' is impossible, i.e., iff it is impossible that p be true and q not. 'And' in the familiar formula, "the premises cannot be true and the conclusion false", is properly represented by '$\circ$', not by '$\wedge$', just as the Dog's belief that the Man went down either the left-hand fork or the right is properly represented by '+', not by '∨' (see Read (1981a)).

Restall elaborates the Dog's tale in an attempt to fend off this account of the Dog's beliefs. Suppose the Man would "never even consider going down the left fork", he suggests (Restall 2004, 166). Thus '$A + B$' is false. But that does not show that the Dog does not believe it. I am not only willing to credit the Dog with beliefs, I am happy to attribute to it false beliefs. What is implausible is that the Dog somehow intuit the Man's unexplained aversion to the left fork. How could the Dog know that he was always planning to go down the right fork? Restall seems to think that (my) ascribing a valid inference from false beliefs is somehow a vice. But what in the original story told us that the Dog's beliefs were true? The challenge was to justify the Dog's reasoning, not to determine whether its beliefs were true or not.

Hence, contrary to Restall's puzzlement (Restall 2004, 167), I do call truth-preservation *validity*. It would be mad not to. The suggestion that "more is needed" is an error Restall himself once made, one which I myself accused him (and others) of making (Read 1998). Restall wrote: "it is not sufficient for B to be a consequence of A that, necessarily, given that A is true, so is B (or more crudely, that B is true in all worlds in which A is true)" (Restall 1997, 158). He call this, simple truth-preservation in all worlds, *classical validity* (Restall 2004, 163 *et passim*); preservation of truth in all cases, including incomplete or inconsistent ones, he dubs *relevant validity*. Logical pluralism allows him to embrace both concepts of validity. But as I observed (Read 1998), this threatens to be incoherent. For suppose an argument with true premises is classically but not relevantly valid. Does it follow that the conclusion is true? According to Restall, it will follow classically, but not relevantly, that it is true. But that is not really an answer to our question. We do not want to know what a classicist or a relevantist thinks. We want to know the truth, or at least, we want to know what Restall thinks. We have described the situation: true premises and classically valid. Query: is the conclusion true?

3. Pluralism

To repeat: suppose an argument has true premises and is classically valid. Is its conclusion true? One might be tempted to respond that the question is ill-defined. Classical validity is truth-preservation over possible worlds – "consistent complete worldlike entities". Thus to say the premises are true is equivocal: are they classically true, or merely true in some incomplete or inconsistent case? If the former, then the answer is clear: the conclusion is classically true too. If the latter, the answer is not determined, since the argument *ex hypothesi* does not preserve incomplete or inconsistent truth. Pluralism about validity seems to lead to relativism about truth. There is no longer a notion of absolute truth, or truth *simpliciter*. Rather, there is only truth relative to a case, matching the different kinds of validity as "truth-preservation in all cases" (Restall 2004, 163). There is no single property of truth, but many, springing from the many "ways of making the pretheoretic notion of 'case' precise" (*loc.cit.*).

But this is not Restall's position, it seems. For he defends what he calls non-dialethic paraconsistency (Restall 1997, 159), which claims that contradictions cannot be true (while denying that they entail triviality). Truth is consistent truth. Moreover, he says, "truth-value gaps are hard to come by" (Restall 2001, 474). What are incomplete are warrants. Hence there is, he concedes, an absolute notion of truth, truth in the cases considered in classical validity. Nonetheless, he denies that classical validity is real validity: "there is no further fact of the matter as to whether [an] argument is *really* valid" (Restall 2002, 426). All valid arguments – classically valid, relevantly valid, or intuitionistically valid – guarantee never to take one from truth to untruth. But some, such as relevantly valid arguments, do more. They preserve "truth-in-a-situation".

In the second half of his paper (Restall 2004), Restall turns to preservation of warrants, considering whether there is only one logic of warrant transfer. His conclusion is that there are many. What his arguments show, however, is that there is none. For first he shows that there are no tautologies of warrant. They clearly could not be arbitrarily complex, for no one has a warrant to believe sufficiently complex tautologies. But neither can they be simple, for warrants to believe simple tautologies build up in simple steps to warrant complex ones.

But not all our beliefs are tautologies, or even consistent. Relevant validity, by Restall's definition, preserves truth-in-a-situation, including inconsistent situations. Does it then preserve warrant? No, for one can, he claims, not even be warranted in believing a conjunction for each of whose conjuncts one has a warrant. He concludes: "if there is anything like a logic of preservation of warrant, it will look *nothing* like traditional classical, constructive or relevant logics" (Restall 2004, 170). I conclude: if this is what he is searching for, there is no logic of warrant transfer.

So what is the purpose of relevant validity (as Restall defines it)? It apparently preserves more than just truth; but it does not preserve warrant. It preserves "truth-in-a-situation". But what is the point of that, since Restall believes we are never (just) in a situation? Either we are in a complete consistent situation (Restall and I agree that dialethism is false); or we have a warrant to believe we are in one. By Restall's lights, relevant validity is suited to neither case.

The fact is that the phrase 'logic of warrant' is an oxymoron. Logic preserves truth, not warrant. Consider again our puzzle: an argument whose premises cannot be true and conclusion false, and whose premises are true. Is its conclusion true? Neither classical validity, nor relevant validity (as Restall defines it) are of any assistance here. If the argument from, say, p to q, is classically valid, that tells us only that '$p \wedge \sim q$' is impossible, which says nothing about the truth of '$p \rightarrow q$', whose truth we need to know in order to infer q from p. If relevant validity, on the other hand, required more than the impossibility of the truth of p and falsity of q, then we could find ourselves in the absurd position of conceding that q must be true (for p is true, and it is impossible for p to be true and q false), but refusing to infer q from that fact (on the grounds that the argument is not relevantly valid – it lacks that vital relevant ingredient).

There is one true logic, the correct logic of truth-preservation. It is given by a correct analysis of the connectives '$\rightarrow$' and '$\circ$', one which does not conflate them with material implication and extensional conjunction, respectively. This analysis reveals the true meaning, and extension, of the familiar phrase "impossible for the premises to be true and conclusion false" (cf. Read (1981a)).

4. Conclusion

Restall's pluralism rests on a confusion, confusing logic as preservation of truth with logic as preservation of warrant. What appears to connect the two is the notion of "truth-in-a-situation". Restall misunderstood Meyer's Sermon to the Gentiles (cf. Read (1988, 144)). To give a classical semantics (with classical validity) in which, e.g., extensional Disjunctive Syllogism (DS) is invalid, Meyer and others introduced incomplete and inconsistent situations (not unlike the strange worlds of Kripke's semantics for non-normal modal logics and for intuitionistic logic), answering Geach's challenge directly by providing a model in which, in this case, $A \vee B$ was true but A and B were false (in the sense that '$\sim A$' and '$\sim B$' were both true). But validity is not preservation of "truth-in-a-situation" any more than it is preservation of warrant. Validity is preservation of truth, and tells us what we have a warrant to infer given what we believe is true.

The classical logician believes he has a warrant to believe anything whatever if he has a warrant to believe a contradiction. He is wrong. Chrysippus believed the Dog had a warrant to believe the Man went down the right fork given its belief that he went left or right and not left. He was right. He was right because what the Dog had warrant to believe was that if the Man did not go left, he went right. It is impossible that he went left or right and did not go left, or right.

Spare a thought for Restall's Dog, however. She comes to the fork, believing $A \vee B$, but $\sim A$. She looks to her master. 'B follows classically, but not relevantly', she is told. 'But does it really follow?', she asks. 'There is no further fact of the matter', comes the reply. She lies down with a soft moan.

References

Read, S.: 1988, *Relevant Logic*, Oxford, Blackwell.

Read, S.: 1998, 'The Irrelevance of the Concept of Relevance to the Concept of Relevant Consequence', *Conference on Knowledge, Logic, Information*, Darmstadt.

Read, S.: 1981a, 'Validity and the Intensional Sense of 'and' ', *Australasian Journal of Philosophy* **59**, 301–307.

Read, S.: 1981b, 'What is Wrong with Disjunctive Syllogism?', *Analysis* **41**, 66–70.
Restall, G.: 2002, 'Carnap's Tolerance, Meaning and Logical Pluralism', *Journal of Philosophy* **99**, 426–443.
Restall, G.: 2001, 'Constructive Logic, Truth and Warranted Assertability', *The Philosophical Quarterly* **51**, 474–483.
Restall, G.: 2004, 'Logical Pluralism and the Preservation of Warrant', this volume, pp. 163–173.
Restall, G.: 1997, 'Paraconsistent Logics!', *Bulletin of the Section of Logic of the Polish Academy of Sciences* **26**, 156–163.

NORMIC LAWS, NON-MONOTONIC REASONING, AND THE UNITY OF SCIENCE

GERHARD SCHURZ
Chair of Theoretical Philosophy, Department of Philosophy, University of Dusseldorf, Universitaetsstrasse 1, Geb.23.21, D-40225 Duesseldorf, Germany,
E-mail: gerhard.schurz@phil-fax.uni-dusseldorf.de

Abstract. *Normic* laws have the form "if A, then normally B". This paper attempts to show that if a philosophical analysis of normic laws (Sections 1, 5) is combined with certain developments in *non-monotonic logic* (Sections 2, 4), then both the unity and the diversity of scientific disciplines can be seen in a new perspective (Section 8.9). In particular, this perspective may shed new light on various received questions such as the importance of individual case understanding in the humanities (Section 2), theory-protection through addition of auxiliary hypotheses (Section 3), the fundamental role of *evolution* in the explanation of normic laws and their relation to *statistical normality* (Section 5), the different nature of *ceteris paribus laws* (Section 6) and of the underlying system *laws* (Sections 7, 8) in physical versus non-physical sciences. The resulting picture is one of unity on the background of diversity on the background of unity (Section 9).

1. Normic Laws – Introduction and History

Most law hypotheses in everyday life and (applied) sciences are not *strictly* universal but admit of exceptions. Their linguistic form is not "All As are Bs", formally, $\forall x(Ax \rightarrow Bx)$, but "As are normally Bs". Following Scriven (1959) I call these loose laws *normic* laws and represent them as $Ax \Rightarrow Bx$ (for "As are normally Bs"; '$\Rightarrow$' is a variable-binding conditional).

Not much attention has been paid to normic laws in the history of philosophy. Neither in Descartes or Leibniz, nor in Locke or Hume, normic laws played a significant role, and the same is true for Kant. The thoughts of these philosophers were occupied with the *epistemological* distinction between necessary (a priori) versus contingent (a posteriori) laws, rather than with the *ontological* distinction between strict (exceptionless) versus normic (exception-admitting) laws. All the more remarkable is the fact that *Aristotle* did introduce the ontological distinction between strict and normic laws in the 6th book of his *Metaphysics*. He also observes that the realm of being which is governed by normic laws covers the entire 'sublunar' (earthly) region and, hence, is more comprehensive than the realm of being which is governed by strict laws – namely mathematics and astronomy. However, also for Aristotle the role of normic laws was minor: he mentions them only as the reason why, besides the principle of essentiality, a mechanism of accidentality has to be assumed as operating in nature.

S. Rahman et al. (eds.), Logic, Epistemology, and the Unity of Science, pp. 181–211.

In the 20th century, normic laws were (re)discovered in the 1950's, when philosophers of history (and the humanities) discussed the Popper-Hempel model of *deductive-nomological* explanation. *Deductive* explanation requires strict laws: Ba, because Aa, and As are *always* Bs. Gardiner and Dray pointed out that there are no strict laws in the historical sciences. Yet historians do explain. And when they do, they refer to loose laws such as the following normic rationality principle (cf. Gardiner 1952, 124f; Dray 1957, 132, 137):

(1) People's actions are normally goal-oriented, in the sense that if person x wants A and believes B to be an optimal means for achieving A, then x will normally attempt to do B.

Note that (1) governs various special normic laws such as "People who want water normally try to get water" (Fodor 1991, 28).

Unfortunately, the dominant attitude at that time was to regard normic laws as *pseudo-laws*, void of empirical content, because they are *not falsifiable*: by proclaiming counterexamples as exceptions we can always protect a normic law from falsification (Dray 1957, 132; Scriven 1959, 466). The non-falsifiability argument rested on Popper's *demarcation* criterion, which equates empirical content with falsifiability. Popperians argued that the only way to turn normic laws into 'genuinely' scientific laws would be their *strict completion*, by strengthening the antecedent of the law so that exceptions are excluded (Albert 1957, 132ff). Since then, however, it was pointed out by several philosophers that, typically, the class of possible exceptions is unknown and potentially infinite, and hence not definable by a finite list (cf., e.g., Hempel 1988; Rescher 1994). So the strategy of strict completion must fail. This is not only obvious for laws of 'folk' psychology such as (5), but also for biological laws such as (4) or technical laws such as (3):

(2) Birds normally can fly.
(3) If a match is struck it normally will light.

There are two kinds of reasons for the impossibility of strict completion: (i) reasons of practical *efficiency* and (ii) reasons of *principle*. Reasons of efficiency are based on the consideration that if we had to verify the absence of all possible exceptions before coming up with a prediction, our practical reasoning would be hopelessly inefficient and could not contribute to survival in a moderately complex environment. For the program of *non-monotonic logic* as being part of the *cognitivistic paradigm* (cf., e.g., Goldman 1986), reasons of efficiency constitute the main motivation of the use of normic laws or 'rules': assume the most normal case *by default*, and don't waste your time with reasoning about possible exceptions *except* they are *forced* on you by evidence! However, there are also strong reasons of principle based on the non-determinism of nature which will be discussed in Section 6.

Let us clarify our notion of "exception", to avoid possible misunderstandings. With a *strict* exception to a law one usually means a true singular statement which falsifies the law, i.e., logically entails its negation. If we speak of normic laws as laws which admit exceptions, then we do of course not mean strict exceptions, but *loose* exceptions in the following sense: a loose exception to a normic law of the form $Ax \Rightarrow Bx$ is a true singular (basic) statement S which logically entails an *abnormal L-instance*, which is a statement of the form $Aa \wedge \neg Ba$ for a an individual constant. Only strict laws may have strict exceptions. Normic laws never have strict exceptions, because they are not falsifiable. But they have loose exceptions.

Today there is agreement that Popper's identification of empirical content with falsifiability is too narrow. It is well-known that also all numerical-statistical generalizations over a possibly infinite domain are not falsifiable by any finite observation sample. And yet they *do* have empirical content, because they may get *gradually disconfirmed* by the observation of sample frequencies which significantly deviate from the probability value predicted by the law. The same gradual (dis)confirmation argument could be applied to normic laws – *provided a connection* between *normic* laws and *statistical* generalizations can be established.

First suggestions along this line have been made by Hempel (1965, ch. 12.3). His proposal was simply to *replace* normic laws by numerical-statistical laws of the form "the probability of Bx, given Ax, is r, more formally $p(Bx/Ax) = r$, where 'p' is an objective-statistical probability operator and 'r' a real number between 0 and 1, e.g. 93%. However, Hempel's replacement strategy was obstructed by three severe problems.

First of all, many authors[1] have pointed out that normic laws express a kind of *prototypical* (or 'ideal') normality which is different from statistical normality. For example, the ability to fly is a biologically prototypical property of birds, and this remains true *even if*, by some major disaster, the majority of birds would lose their flying ability. I do not fully agree with this objection. I agree as far as prototypical normality cannot be *reduced* to statistical normality – it is *more* than that: as we shall see, it makes normic generalizations *non-accidental* (cf. Section 5). However, the conclusion drawn by these authors (cf. fn. 1) that prototypical normality has nothing to do with statistical normality is rash. Prototypical normality may nevertheless *imply* statistical normality – more precisely, the fact that '$Ax \Rightarrow Bx$' is a true normic law may generally imply that the conditional statistical probability $p(Bx/Ax)$ is *high*, without that the reverse implication generally holds. This thesis would be sufficient to establish the empirical content and, hence, the scientific status of normic laws. For if the number of (weak) exceptions increases in relation to the normal cases, our belief in the normic law will become increasingly weaker, until finally, we give it up. I call this thesis the *statistical consequence thesis*, and I will defend it in this paper.

The statistical consequences thesis can be defended in two different ways: (i) *methodologically* and (ii) *ontologically*. The methodological justification argues that the statistical consequence thesis is a necessary condition for both the *reliability* and the *efficiency* of reasoning from normic laws (cf. Schurz 1997a). This kind of justification has been put forward by Pearl (1988, 477–480) against McCarthy

(1986, 91) and Reiter (1987, 149f, 180f) who have supported the view that normic laws such as "birds normally can fly" do not express statistical facts, but *linguistic conventions* of the following sort: "when receiving information about birds, you may assume by default that the bird is a flyer; exceptions will be stated explicitly". Pearl defeats this view as follows. First of all, he argues, for the purpose of *hunting* birds, as opposed to *talking* about birds, certainly not linguistic conventions but statistical facts are relevant. But even for the purpose of *talking* about birds, he adds, default conventions are *only* reasonable if the default cases are in the statistical majority. Linguistic conventions like "humans are by default assumed to be older than 100 years" would force us into permanent exception statements which would make communication hopelessly inefficient.

This methodological justification is certainly very clever. However, it tells us only why the statistical consequence thesis *should* hold, but not whether it *in fact* holds, i.e., whether it is satisfied by the majority of the normic laws on which practical and scientific reasoning actually relies. Although there exists certain evidence in favour of this assumption (cf. Pelletier and Elio 1997, 180), it is well-known that severe doubts have been raised about the probabilistic reliability of intuitive reasoning (cf. Kahneman et al. 1982). So, what we need is a *factual* justification of the statistical consequence thesis which tells us *why* normic laws and their statistical consequences are an objective feature of reality. This kind of justification will be given in Section 5.

If the statistical consequence thesis is accepted and Hempel's original suggestion is replaced by the weaker requirement that normic laws should imply statistical generalizations, then there is still a *second* problem waiting for the adherents of probability laws. Usually we do not know any *numerical* values of the conditional probabilities corresponding to normic laws, like "$x\%$ of all birds can fly' or "$y\%$ of all matches light when being struck" (when?, where?). All we know or assume is that these probabilities are *high* – but exactly *how* high they are is strongly dependent on domain-specific circumstances. This explains why random samples of 'arbitrary birds' or 'arbitrary matches' do not make much sense (cf. Millikan 1989, 281; Hempel 1988, 25). One may reply that we know at least some *lower bound* of the unknown probabilities: the *minimal acceptability condition* for a normic law $Ax \Rightarrow Bx$ is the so-called 'Leibniz-condition': $p(Bx/Ax) > p(\neg Bx/Ax)$, or equivalently, $p(Bx/Ax) > 0.5$. Of course, this is true – but the Leibniz-condition is much too uninformative. Usually, the normal case probabilities are much higher than 0.5, close to 1. They have to be that high in order to ensure practical success. But exactly how high they are, or have to be, depends on the *domain of application*: compare the breaking of a match with the crashing of a plane. On this reason it has been suggested in the area of *expert systems* that the fixation of minimal probability thresholds should be left to the *user* instead of being predetermined by a *system-inbuilt* threshold-value (cf. Schurz 1997a).

The *third* and hardest problem was that for a long time, a *logic* of normic reasoning was missing. Toulmin (1958, 166) had argued that normic reasoning is irreducibly 'substantial', being incapable of formalization. His own attempts at formalizing normic reasoning have been proven to be inconsistent by Hempel (1965, ch. 2). This situation, however, has changed. Beginning in the late 70's, various *logical systems* for normic reasoning have been developed, which have been called *non-monotonic* or *default* logics. Although the idea of non-monotonic reasoning had its roots in philosophy, the main development of non-monotonic logics (NMLs) has taken place in AI (artificial intelligence) – which does not exclude that also several philosophers have been engaged in it. I do not claim that *all* NML systems which have been developed are suitable candidates for a philosophical logic of normic laws.[2] Many of the earlier systems (e.g., some type (1), (2) or (5) approaches of fn. 2) suffer from too much syntactic ad-hocery or are too tightly connected to purely computational purposes. In the meanwhile, however, there is a considerable consensus in the NML community about the so-called system **P** and its neighbours (type (3) approaches of fn. 2). These systems have a philosophically transparent semantics and will be discussed in Section 4. What I want to demonstrate in this paper is that this family of NML systems, if combined with a thorough philosophical analysis, may shed new light on the significance of normic laws for science and its philosophy.

In order to avoid misunderstandings associated with the very *name* "non-monotonic logic", let me add two important points. *First*, NML in the sense of Section 4 is not supposed to replace classical monotonic logic – it is not a *non-classical* logic but an *extension* of classical logic, which introduces rules for 'normic' conditionals and inference operators. In the core of NML, there still is a monotonic *conditional* logic, whence NML can be seen as a research program in conditional logic. *Second*, NML is not supposed to *compete* with classical probability theory (like, e.g., fuzzy logic); in the contrast, it is complete for the semantics of *high probability preservation* (Section 4). So NML can also be seen as a research program in probability logic (cf. Pearl 1988, ch. 10, 1990; Bacchus 1990, ch. 5.7). *Nevertheless* there is a clear sense in which NML goes *beyond* both traditional conditional logic and probability theory (vide Section 4).

Let me summarize the two main characteristics of normic laws pointed out so far: (i) Normic laws claim a *prototypical* if-then relation, which admits of an indefinite range of exceptions, whence their strict completion is (usually) impossible; and (ii) normic laws imply high conditional statistical probabilities, but their numerical values are (usually) unknown; tight lower probability bounds depend on specific application domains. The prototypicality-claim in (i) explains why normic laws are *lawlike* (non-accidental), and the statistical consequences-thesis in (ii) explains why normic laws have *empirical content*. Before we proceed to a deeper justification of (i) and (ii) we analyse how non-monotonic reasoning from normic laws really works.

2. Non-monotonic Reasoning and the Understanding of the Individual Case

The semantic idea which underlies deductive logic is *strict truth preservation*. This idea does not apply to normic reasoning: even if $Ax \Rightarrow Bx$ and Aa are true, Ba may be false. So a different semantic idea is needed. Although a variety of semantics for NML has been developed, they all are (more or less) reducible to two semantic ideas. In *normal-world* (or preferential model) *semantics*, a non-monotonic inference from a set of normic laws $\boldsymbol{L}$ and factual knowledge $\boldsymbol{F}$ is regarded as correct iff it preserves truth (not in all, but) in all $\boldsymbol{F}$-worlds that are most normal. In *probability semantics*, such an inference is regarded as correct iff it preserves truth (not in all, but) in most of the $\boldsymbol{F}$-instantiating situations. Precise definitions will be given in Section 4.2.

This semantic difference has a drastic consequence for the non-monotonic inference relation $|\!\sim\!\sim$. Classical *deductive* inference $\vdash$ is *monotonic*: if a deductive inference is correct, then it remains correct whatever new knowledge is added to its premises. Formally, if Prem $\subseteq$ Prem*, then Prem $\vdash$ Con implies Prem* $\vdash$ Con, i.e., $Cn_{\vdash}$ (Prem) $\subseteq$ $Cn_{\vdash}$ (Prem*). In normic inferences, monotonicity is violated. As an illustration, consider the example of a knowledge base consisting of the factual knowledge $A1$: my pet can fly, and the three normic laws $L1$: flying animals are normally day-active, $L2$: bats are normally not day-active, and $L3$: bats normally can fly. As long as nothing else is known about my pet than that it is a flyer, it is correct to infer by default that my pet is a normal flyer and hence is day-active (D): $L1, L2, L3, A1 \;|\!\sim\!\sim D$. Now suppose you acquire the additional evidence $A2$: my pet is a bat. Then it is no longer correct to infer D: $L1, L2, L3, A1, A2 \;|\!\sim\!/\!\sim D$. Exactly this is non-monotonicity. Now the normal-case law $L1$ gets *blocked* by the exception-case law $L2$ which has become instantiated by $A2$ and *fires*: $L1$, $L2$, $L3$, $A1, A2 \;|\!\sim\!\sim \neg D$.

The "normality" of a normic law $Ax \Rightarrow Bx$ is *relative* to both the antecedent predicate Ax and the consequent predicate Bx. For example, "birds normally can fly" speaks about what is normal *for birds* and not about what is normal for arbitrary animals; e.g., it is normal for fishes to be able to swim but not to fly. Moreover, "birds normally can fly" tells us what it means for a bird to be normal w.r.t. (with respect to) its way of locomotion, but not necessarily w.r.t. other property-families; for example, a bird which can fly but is infertile is normal w.r.t. its way of locomotion but abnormal w.r.t. its reproduction ability. This demonstrates that "$\Rightarrow$" is a *genuine conditional operator* which *cannot* be adequately understood either as a special 'predicate' "$Ax \wedge \text{Norm}(x) \rightarrow Bx$", or as an unary 'normality-operator' attached to an ordinary material implication "$\text{Norm}(Ax \rightarrow Bx)$". For both of these latter reconstructions would imply a monotonic behaviour of the resulting conditional, because if "$Ax \rightarrow Bx$" is true in every most normal world, then also "$Ax \wedge Cx \rightarrow Bx$" is true in every most normal world. In contrast, "$Ax \Rightarrow Bx$" is non-monotonic, because the truth of Bx in all most normal Ax-worlds does not imply the truth of Bx in all most normal $(Ax \wedge Cx)$-worlds. The probabilistic counterpart is the fact that a high *unconditional* probability of the material implication $p(Ax \rightarrow Bx)$ differs in the same drastic way from a high *conditional* probability $p(Bx/Ax)$.[3]

Because of their antecedent-relativization, normic laws may also describe the ('normal') behaviour of exceptional cases. What counts as normal and what as exceptional cannot be read off from a normic law in isolation, but only from the entire *system* of *normic laws* over a domain A which forms a normic *theory*. This system is organized by a *hierarchy* of *levels* (or degrees) of exceptionality: *Level 0* contains normal-case laws in the proper sense: $Ax \Rightarrow Bx$; *level 1* contains exception-laws of degree 1: $Ax \wedge Xx \Rightarrow \neg Bx \wedge Cx$, *level 2* exception-laws of degree 2: $Ax \wedge Xx \wedge Yx \Rightarrow \neg Cx \wedge Dx$, etc. In the above example, exceptions of degree 2 would be bats which are day-active, e.g. by training. Instead of $Ax \wedge Xx$ we may also have A^*x – the only important thing is that A^*x is known to be *more specific* than Ax. This means that $A^*x \Rightarrow Ax$ but not $Ax \Rightarrow A^*x$ is known to hold. $A^*x \Rightarrow Ax$ (in our example $L3$) may also be strictly or even logically true. The levels of exceptionality are coordinated by the *rule of specificity*: in cases of *conflict* (as in our example with $L1$, $A1$, $L2$, $A2$), the law with the more specific antecedent blocks that with the less specific one. Of course, it also happens that normic laws with contradicting consequents *cannot* be prioritized by the relation of specificity. Most of the more recent accounts opt for *scepticism* in such cases: if $\{Ax \Rightarrow Bx, Cx \Rightarrow \neg Bx, Aa, Ca\}$ but nothing else is known, then neither Ba nor $\neg Ba$ is non-monotonically inferable.

A deep consequence of non-monotonic as opposed to deductive inference concerns the different role that our *understanding* of the *individual case* plays in reasoning. Let a represent the individual case about which we are to infer a conclusion. In deductive reasoning, we may apply every strict law L to any factual knowledge Fa which *matches* with L (by some rule of deduction). As long as the local premises are granted as true, we may *detach* the conclusion and *forget* about the rest of our knowledge about a. The context-dependency of deductive reasoning is merely *local*: we may split off the 'part' from the 'whole'. In non-monotonic reasoning we cannot simply apply every normic law to any factual knowledge which matches with it. Before we do this, we have to find out which normic laws *fire* and which of them get *blocked*. We must do this in the light of our *total knowledge* about a. Thus, non-monotonic reasoning is *globally context-dependent*. Every new piece of incoming evidence may overthrow our previous inferences, even if it *does not touch* the truth of previous premises. The 'part' is inseparable from the 'whole', at least in principle. The more we *understand* about a, the more reliable our inferences will be. For example, before we predict or explain a person's behavior by the normic law (1), we must ask whether emotional influences or cognitive mistakes cause exceptions (of degree 1), which in turn may be overruled if the person trusts good advisors (exceptions of degree 2), etc. In this way, non-monotonic reasoning provides a reconstruction of a *certain aspect* of the 'antipositivistic' method of *understanding* – the *significance* of understanding the *individual case* – without deviation from the logical and scientific spirit.

The phenomenon of non-monotonicity has been known in philosophy of science for a long time, but under different labels. Carnap (1950, 211) has discovered it in the area of inductive logic as the "principle of total evidence", and Hempel (1965, ch.12.3.2.2) in the area of inductive explanation as the "principle of maximal specificity". Since non-monotonic inferences are not strictly truth-preserving, some philosophers would presumably prefer not to speak of NML-systems as of logics in the narrow *deductive* sense, but rather, of a species of *inductive logics* (cf. Tan 1997) – and I would agree, but these are more terminological than substantial questions. Certainly, many insights of the philosophical debate can be directly transferred to non-monotonic logic: for example, that the *predicates* of normic laws have to be *nomological* – otherwise the rule of specificity may quickly be reduced to absurdity (Hempel 1968, 124, 127; Schurz 1995b, 452). The *main* point in which the non-monotonic research program goes far beyond these earlier achievements is that while Hempel's inductive-statistical explanation model and its relatives remained in the context of 'quasi'-modus ponens, the NML-approaches provide a complete set of *rules of reasoning* from normic laws. Before we turn to them (in Section 4), we mention a further area of science where non-monotonic reasoning plays an important role.

3. Theory Protection by Introduction of Auxiliary Hypotheses

Popper's falsificationist model has been critized by Kuhn, Quine, and in particular Lakatos (1970), in a well-known way. Recall Lakatos' argument. Assume Newtonian physicists predict the orbit of a newly discovered planet a from their background theory T and antecedent Aa, but observations yield significant deviations from the prediction Pa. Do the physicists now consider T as having been falsified, as they should according to the falsificationist model? By no means. What physicists usually do in such a situation is to introduce an auxiliary hypothesis which protects the theory from falsification – for example, by postulating a small and hitherto unobserved planet which perturbs a's orbit (Lakatos 1970, 16f). Lakatos' argument and his examples from history of physics have become 'folklore' among philosophers of science.

However, in the deductive setting, avoidance of theory falsification by the mere introduction, i.e. *addition*, of auxiliary hypotheses is impossible. This has been pointed out by Coffa (1968). For deductive reasoning is *monotonic*: if T and Aa deductively imply a false prediction Pa, then they imply Pa also when auxiliary hypotheses *Aux* of whatever sort are added to the premises. Hence, the popular view of theory protection through *addition* of auxiliary hypotheses *contradicts* the deductivistic model.

As Lakatos' concludes in his appendix (1970, 184ff), in order to protect a theory from falsification we must always *change some* part of our premises. This is an unavoidable consequence of the *deductivistic* reconstruction. But, as I wish to point out here, this is *not* a necessary consequence of the *non-monotonic* reconstruction. To explain the general point, let us distinguish between the (knowledge) *base* B (the set of nonderived statements) and the (consequential) *hull* H (the set of derived consequences of B; $B \subseteq H$). In the deductive case ($H = Cn_{\vdash}(B)$), the *dynamic connection* between B and H is monotonic: *if B expands, then H must expand.* As an example, assume $B = \{Aa, T\}$ where $T \vdash \forall x(Ax \rightarrow Px)$. Then $Pa \in H$. If we expand B to $B' = B \cup \{Aux\}$, for *Aux* some auxiliary hypothesis, then clearly $H \subseteq H' = Cn_{\vdash}(B')$; hence also $Pa \in H'$. Vice versa, in order to eliminate the unwelcome element Pa from H, we must *contract* or revise the basis B. The so-called *Duhem-holism* of deductive theory-falsification consists in the fact that one may contract B in several different ways in order to remove Pa from H: we may either remove Aa from B, or T, or at least some part of T such that $\forall x(Ax \rightarrow Px)$ does not follow any longer.

In contrast, in the normic situation the dynamic connection between base and hull is non-monotonic ($H = Cn_{|\sim\sim}(B)$). Here it is indeed possible to *eliminate* an unwelcome element Pa from the hull by a mere *expansion* of the basis, for example the addition of an auxiliary hypothesis *Aux*. For example, assume again $B = \{Aa, T\}$, where T now is or at least entails the normic law or generalization $Ax \Rightarrow Px$. Then again, $Pa \in H$. But if we expand B to $B' = B \cup \{Ea, Ax \wedge Ex \Rightarrow \neg Px\}$ (where "Ex" denotes some *exceptional* property), and let the new hull be $H' = Cn_{|\sim\sim}(B')$, then Pa will no longer be an element of H' – in the contrary, $P \notin H'$, hence $H \subseteq H'$ will *not* hold. For, the more specific exception law $Ax \wedge Ex \Rightarrow \neg Px$ is instantiated in B' by Aa and Ea and hence *blocks* the inference of Pa, as outlined in Section 2. Let me add that I have sometimes heard deductivists arguing that also in the non-monotonic case something 'basic' has been given up – namely the default assumption that a is a *normal* A. But this is a misunderstanding: the default assumption of 'normality' is *not* an explicit premise but an implicit and derived assumption which is *generated* during the inference process (cf. Section 4) – in other words, non-monotonic logic is *indeed* non-monotonic.

Which is the better reconstruction of how scientists really proceed? *It depends.* In the planet example, if we let Aa be the theoretical antecedent condition $A_t a$ ("x approximately obeys two-body-mechanics"), then – as it will emerge from Section 7 – the relation between $A_t x$ and the theoretical prediction of the planet's trajectory is indeed mathematically-deductive. So in this case, Lakatos' reconstruction is correct and contains the following important insight: by adding the assumption of a disturbing planet, we have also *changed* one the of original theoretical premises in $A_t a$ concerning the total force which acts upon the planet.

On the other hand, if we assume that Aa is an empirical antecedent description $A_e x$ (saying just that a is a planet of our solar system), and that the uncertain or normic *application claim* $A_e x \Rightarrow A_t x$ (saying that planets of our solar system are normally only influenced by the sun) counts as a *peripherical part* of our theory T, then it is possible by the mere *addition* of auxiliary information *to defeat* the inference from $A_e a$ to $A_t a$ and hence to *block* the inference to the false prediction Pa. Such defeating auxiliary information may be, e.g., the statement Da ("a is disturbed by another planet") plus the more specific application claim "$A_e x \wedge Dx \Rightarrow A_t^*$" with $A_t^* x$ for "x obeys three-body mechanics". I leave it open whether this reconstruction is indeed desirable in the planet example. One might alternatively replace the original application claim ($A_e x \Rightarrow A_t x$) by the more specific but *strict* application claim "All planets except Uranus and Mercury are approximately described by two-body mechanics". However, in other examples where no control over exceptions is available, scientists *do* proceed non-monotonically. As we shall see, in non-physical disciplines this non-monotonic procedure is the usual case. For example, if biologist have accumulated certain normic core laws characterizing a certain biological class such as Mammalia, then they may discover new normic laws characterizing exceptional subclasses (such as Monotremata or Tachyglossidae) without any need of changing the normic core laws about Mammalia. But cases of this sort even appear in chemistry. For example, when chemists discovered in 1962 the possibility of a chemical bonding between the inert gases Xenon and Fluor, they just added this new kind of molecule into their list of exceptions to Lewis' theory of molecular bond formation, without changing *anything* in this theory which was always understood as a normic and *defeasible* generalization (cf. also Schurz 1995a).

We finally mention that the non-monotonic case of belief-revision has also caused troubles for the so-called AGM-approach to belief revision (Alchourrón et al. 1985; Gärdenfors 1988). According to this approach, K^*A (the revision of belief system K by proposition A) contains $K \cup \{A\}$ as a subset whenever K is consistent with A. This implies a monotonic expansion relation between base and hull, which can only hold if the consequence relation over the elements of a belief system is monotonic. Since the so-called Ramsey-conditionals depend non-monotonically on what is contained in a belief system, their addition to belief systems cannot satisfy the monotonic expansion principle – which is formally reflected in Gärdenfors' famous 'impossibility theorem' (1986). Since then, various attempts have been made to incorporate non-monotonic belief revisions into the AGM research program (cf., e.g., Fuhrmann and Rott 1996).

4. Three Modules of Non-monotonic Reasoning

As the formal language *Lan* of this section we choose a *propositional fragment* of monadic 1st order logic, with just *one* individual variable x and one individual constant a, and extended by the normic conditional operator $\Rightarrow$ which *binds* the variable x (hence, $Fx \Rightarrow Gx$ is a closed formula). The reason of this choice is that (i) statistically interpreted normic laws apply to open formulas, and (ii) for reasons of simplicity we restrict our attention to the essentially propositional case.[4] $\boldsymbol{L}$ ranges over finite sets of normic conditionals ('laws') $L_i \in \boldsymbol{L}$; Aa, Ba, ... range over closed *factual* (= nonconditional) formulas, $\boldsymbol{F}(a)$ ranges over finite *conjunctions* of factual formulas, and A, B, ... range over *open* factual formulas, so the variable 'x' is omitted. (The restriction to *finite* premise sets is only needed for the non-infinitesimal probability semantics).

It is useful to distinguish between a formal *inference relation*, i.e., a subset of Pow(*Lan*) × *Lan*, and a corresponding *pragmatic application* principle for this inference relation with respect to a *knowledge system* $\boldsymbol{K}$. The application principle for monotonic (deductive) inference relations $\vdash$ is so simple that it is usually not explicitly stated: given $Prem \vdash Aa$, then you may infer Aa iff $Prem$ is *included* in your knowledge: $Prem \subseteq \boldsymbol{K}$. In contrast, the application principle for non-monotonic inference relations $|\!\sim\!\sim$ is much more complicated – it is the *principle of total knowledge*: given $Prem \;|\!\sim\!\sim\; Aa$, then you may *infer* Aa from $Prem$ if and only if $Prem$ comprises *all* of your (relevant) knowledge $\boldsymbol{K}$. Without this principle, applications of $|\!\sim\!\sim$ may quickly produce inconsistencies, for there may be subsets $Prem'$, $Prem'' \subseteq \boldsymbol{K}$ such that $Prem' \;|\!\sim\!\sim\; Aa$ and $Prem'' \;|\!\sim\!\sim\; \neg Aa$. To make this principle practically *operable*, non-monotonic reasoning is structured into three modules:

4.1. Module 1

Module 1 is the principle of *total evidence conditionalization*.[5] It consists of two parts. *First*, it requires that the factual premises $\boldsymbol{F}(a)$ of a normic inference must comprise all known facts about the individual case a. *Second*, this principle *reduces* the inference relation between a factual conclusion Ca, factual premises $\boldsymbol{F}(a)$ and normic premises $\boldsymbol{L}$ to an inference relation solely defined over normic conditionals, in the following way:

$$\boldsymbol{L}, \boldsymbol{F}(a) \;|\!\sim\!\sim\; Ca \quad \text{iff} \quad \boldsymbol{L} \;|\!\sim\!\sim\; \boldsymbol{F} \Rightarrow C.$$

We call $\boldsymbol{F} \Rightarrow C$ the *total evidence law* (w.r.t. Ca).

4.2. Module 2

Module 2 is a monotonic *core* logic for inference among normic conditionals, the so-called system **P** (for preferential entailment), which goes back to Adams (1975). Today, there is broad agreement about **P** (at least in approaches (3) of fn. 2). The rules of **P** are stated below. They are *weaker* than the corresponding rules for material implication – only 'cautious' versions of them hold. Note that although $\vdash_P$ is monotonic, $\Rightarrow$ is non-monotonic (i.e., $A \Rightarrow B \nvdash_P A \wedge C \Rightarrow B$). By module 1, this makes $|\!\sim\!\sim$ non-monotonic with respect to factual knowledge. The *extended* system $\mathbf{P}^+$ admits of truth-functional compounds of conditionals (Adams 1984; Schurz 1998). (CM) implies the mentioned rule of specificity:

$\{A \Rightarrow \neg B, C \Rightarrow B, C \Rightarrow A\} \vdash_P A \wedge C \Rightarrow B$, which enables to infer Ba from $\{Aa \wedge Ca\}$ by module 1. A stronger version of it, where $C \Rightarrow A$ is replaced by $\neg(C \Rightarrow \neg A)$, is provided by RM.

System P – Basic Rules:	
(Cautious Cut CC):	$A \Rightarrow B, A \wedge B \Rightarrow C \vdash_P A \Rightarrow C$
(Cautious Monotonicity CM):	$A \Rightarrow B, A \Rightarrow C \vdash_P A \wedge B \Rightarrow C$
(Cautious Disjunction CD):	$A \Rightarrow C, B \Rightarrow C \vdash_P A \vee B \Rightarrow C$
(Supraclassicality SC):	If $\vdash A \rightarrow B$, then $\vdash_P A \Rightarrow B$
Some Derived Rules:	
(Conjunction C):	$A \Rightarrow B, A \Rightarrow C \vdash_P A \Rightarrow B \wedge C$
(Left Logical Equivalence LLE):	If $\vdash A \leftrightarrow B$, then $A \Rightarrow C \vdash_P B \Rightarrow C$
(Right Weakening RW):	If $\vdash B \rightarrow C$, then $A \Rightarrow B \vdash_P A \Rightarrow C$
(Cautious Conditional Proof CP):	$A \wedge C \Rightarrow B \vdash_P A \Rightarrow (B \rightarrow C)$
Additional rule of $\mathbf{P}^+$:	
(Rational Monotonicity RM):	$A \Rightarrow B, \neg(A \Rightarrow \neg C) \vdash_{P+} A \wedge C \Rightarrow B$

The system **P** is correct and complete for three kinds of semantics: *normal world* semantics, *infinitesimal* and *noninfinitesimal probability* semantics. This remarkable fact is stated in the theorem below. An analogous theorem holds for $\mathbf{P}^+$ (cf. Schurz 1998). Some terminology: A *normal world model* $\langle W, r\rangle$ is a set of possible worlds W together with a ranking function r: $W \rightarrow \{0, 1, 2, \ldots\}$ attaching to each world its degree of exceptionality. (Worlds of degree i verify the antecedents of normic laws of degree i, cf. Section 2). A normic law $A \Rightarrow B$ holds in $\langle W, r\rangle$ iff Ba is true in all Aa-worlds of W with lowest rank.

Central P-Theorem. *The following four conditions are equivalent:*

(1) (*Calculus:*) L is derivable from $\mathbf{L}$ in the calculus **P**.

(2) (*Normal world semantics:*) For all normal world models M: if all $L' \in \mathbf{L}$ hold in M, then L holds in M.

(3) (*Infinitesimal probability semantics:*) For all probability models P: if, for all $L' \in \mathbf{L}$, $u[L'/P]$ converges to zero, then $u[L/P]$ converges to zero.

(4) (*Noninfinitesimal probability semantics:*) For all probability models P : $u[L/P] \leq$ the sum of all $u[L'/P]$, for $L' \in \mathbf{L}$.

A *probability model* P attaches to each open formula A a statistical probability $p(A)$ and is constructed in the usual way over the finite set of all possible states of x definable in the language. For $L = A \Rightarrow B$, $p[L/P] = p(B/A)$ is the conditional probability *associated* with the normic law L in the probability model P, and $u[L/P] = 1 - p(B/A)$ is the conditional uncertainty associated with it.

Note that ranked models semantics (2) is equivalent with partially ordered model semantics (Kraus et al. 1990). The infinitesimal condition (3) has two major interpretations: Adams' limes-condition ($\forall\delta\exists\varepsilon\forall P$: if $\forall L' \in \mathbf{L}$, $u[L'/P] \leq \varepsilon$, then $u[L/P] \leq \delta$) and (ii) Lehmann and Magidor's nonstandard infinitesimal semantics (1992). Noninfinitesimal probability semantics (4) adds to non-monotonic reasoning a *probabilistic reliability component*. Given lower bounds of the conditional probabilities associated with the premise laws in $\mathbf{L}$, then 1 minus the sum of these bounds is an *approximately tight* lower bound of the conditional probability associated with the total evidence law $\boldsymbol{F} \Rightarrow C$ (cf. Schurz 1997a). This bound is *transferred* to the singular conclusion Ca as a lower bound of its *degree of belief* relative to the knowledge base $\langle \mathbf{L}, \boldsymbol{F}(a)\rangle$; so that module 1 takes the following form: $\mathbf{L}, \boldsymbol{F}(a) \mid\!\sim\sim [Ca, 1-\varepsilon]$ iff $\mathbf{L} \mid\!\sim\sim \boldsymbol{F} \Rightarrow_{1-\varepsilon} C$. Note also that by dropping the rule (CD) one obtains the weaker system **C** (for "cumulativity") which has been suggested by Gabbay (1984) as a basis for plausible reasoning independent from any semantical treatment; a complete semantics for **C** (and some neighbours) in terms of labelled normal world models has been developed by Kraus et al. (1990). A new semantics for the systems **C** and **P** (and their relatives) in terms of *qualitative neuronal networks* has been developed by Leitgeb (2001).

4.3. Module 3

Module 3 is a mechanism of default assumptions of normality. Mainly, they concern *assumptions of irrelevance*. Suppose we wish to derive Ba from the normic law $A \Rightarrow B$ and the fact Aa in the presence of further facts $Ca(\ldots)$. For this purpose, we assume by default that these further facts Ca are irrelevant, as long as our knowledge base contains no information to the contrary. A very elegant realization of module 3 is Pearl's system **Z** (1990; Goldszmidt and Pearl 1996) and the equivalent *Rational Closure* of Lehmann and Magidor (1992). Here one puts a *global semantical* constraint on the admitted normal world models: the sum of the exceptionality degrees of their worlds has to be minimal. Due to its computational simplicity, the system **Z** has also various drawbacks (e.g., failure of property-inheritance to exceptional subclasses). An alternative are *local syntactical* irrelevance assumptions as proposed in Schurz (1994, 1997a). Here, in order to infer Ba from $A \Rightarrow B$, Aa in the presence of further facts Ca, the irrelevance assumption $Irr(C : A \Rightarrow B)$ is generated by default; it justifies the intermediate step from $A \Rightarrow B$ to $A \wedge C \Rightarrow B$ which (together with Aa, Ca) yields Ba by module 1. Schurz (1994) shows how default-reasoning in the style of Reiter (1980) can be made 'probabilistically safe' by translating it into **P**-reasoning extended by irrelevance-assumptions. Schurz (1997a) demonstrates the same for default reasoning the style of Poole (1988); here one needs additional relevance assumptions (expressed by negated normic laws in system $\mathbf{P}^{+}$) in order to validate contraposition steps. Irrelevance and relevance assumptions make it possible to 'blow up' the normic P-consequences of $\boldsymbol{L}$ into a reasonable total evidence law. Both are globally context-dependent and make normic reasoning non-monotonic with respect to new normic knowledge. For example, if we add $C \Rightarrow \neg B$ or $\neg(A \wedge C \Rightarrow B)$ to our knowledge base, the default generation of the irrelevance assumption $Irr(C : A \Rightarrow B)$ is no longer allowed.

As emerging from the previous description, NML shares features with both *deductive logic* and numerical *probability theory*. With the former it shares the idea of *rule-guided* reasoning. With the latter it shares uncertainty and non-monotonicity. But what is more important, NML fills a gap which is left open not only by deductive logic (which is obvious), but also by standard probability theory. From the viewpoint of probability theory, normic laws correspond to conditional probability inequalities of the form $p(Gx/Fx) \geq 1-\varepsilon$ (for ε a small number) which *constrain* the space of *possible* probability distributions. Standard probabilistic accounts, such as Bayes nets, do not tell us how to infer from this *partial* probabilistic information non-trivial probability assertions conditional on the total factual evidence, because they assume *complete* information about the probability function. As an example, assume we wish to infer Ha from the knowledge base $\langle \boldsymbol{L}, \boldsymbol{F}(a)\rangle = \langle\{F \Rightarrow_{1-\varepsilon_1} G, G \Rightarrow_{1-\varepsilon_2} H\}, \{Fa\}\rangle$. In order to draw such an inference with help of a Bayes net, we would need to know the numerical probability values of all $2^3 = 8$ cells of the partition $p(\pm F \wedge \pm G \pm H)$. In probability logic (module $1 + 2$), it would be sufficient to know in addition that $F \wedge G \Rightarrow_{1-\varepsilon_3} H$ holds; from this we can infer Ha with a lower bound $1 - \varepsilon_1 - \varepsilon_3$ by (CC) and the central **P**-theorem. In NML (modules $1 + 2 + 3$), we add the irrelevance assumption $Irr(F : G \Rightarrow H)$ by default and immediately obtain Ha with lower bound $1 - \varepsilon_1 - \varepsilon_2$.

To avoid misunderstanding, we finally mention some relations to other formal presentations of NML. For example, Gärdenfors and Makinson (1994) and other authors use the non-monotonic entailment operator "$|\!\sim\!\sim$" instead of our conditional operator $\Rightarrow$; e.g., they write Bird(a) $|\!\sim\!\sim$ Canfly(a). But their "$|\!\sim\!\sim$" is a 'material' entailment, not a logical one, because a minimal condition for *logical* entailment relations is that they are *formal* in the sense of being closed under syntactically isomorphic substitutions (cf. Schurz 2001c). It seems to be semantically more transparent to represent this relation as a conditional. On the other hand, the '2nd degree entailments' of these authors, such as "If $p \;|\!\sim\!\sim q$, then $p \;|\!\sim\!\sim q \vee r$" etc. correspond to what we call non-monotonic entailment relations ($p \Rightarrow q \;|\!\sim\!\sim p \Rightarrow q \vee r$). In many of the earlier systems, normic laws or rules look superficially as if they were 'deductivistic'; for example the counterpart of $Ax \Rightarrow Bx$ in McDermott and Doyle (1980) is the modal statement $Ax \wedge \Diamond Bx \rightarrow Bx$, and in Reiter (1980) it is the rule $Ax, \Diamond Bx/Bx$ (a so-called 'normal' default). However, $\Diamond Bx$ does not figure here as an antecedent conjunct with ordinary semantic interpretation, and $\Diamond Ba$ does not appear as a singular premise in a "quasi-Modus Ponens" with conclusion Ba; rather, "$\Diamond Ba$" means roughly that "$\neg Ba$ is not derivable" and $\Diamond Ba$ is implicitly generated in the process of forming a so called non-monotonic extension of the knowledge base. This semantic interpretation turns $Ax \wedge \Diamond Bx \rightarrow Bx$ into a non-monotonic conditional assertion.

The described logical system applies to genuine normic laws such as "birds normally can fly" and to accidental normic generalizations such as "most people don't live in Europe" in just the same way. This difference does not concern the logic but the philosophical interpretation of normic laws. To this we turn now.

5. Evolution-theoretic Foundation of Normic Laws

Normic laws are virtually *omnipresent* in everyday life and in all 'special' sciences. Is their omnipresence the result of an *objective* feature of reality which could justify the statistical consequence thesis? Or is it merely the product of our *subjective framing* of a world which is in fact too complex to be understood? In other words, are normic laws *genuine* laws? In the case of strict laws it was suggested to distinguish genuine laws from merely *accidental* regularities by the fact that the former have a *unified explanation* by general *theories* (Earman 1986, 87, speaks here of the 'Mill-Ramsey-Lewis account'). Is such an objective theoretical foundation also possible for normic laws? In Schurz (2001b) such a foundation is developed. For the following we need a brief sketch of it.

All 'higher' sciences, from biology 'upwards', are concerned with living systems in a general sense (biology, psychology, social sciences, history. etc.), or with their cultural and technical products (humanities and arts; geography, technology, etc.). What these systems have in common is the characteristic capacity of *self-regulation* under the permanent 'pressure' of their environment. Our start is the thesis that *normic laws* are the phenomenological laws of self-regulatory systems. According to the framework of *cybernetics* (cf. Ashby 1961), the identity of self-regulatory systems is governed by certain prototypical *norm states*, which these systems constantly try to approximate by means of their *real states*. They manage this with the help of certain *subsystems* (organs) performing the necessary *regulatory mechanisms* (functions) which *compensate* for *disturbing influences* of the environment by producing counteracting processes.

But what explains the omnipresence of self-regulatory systems in our world? The answer is *evolution* in the generalized 'Darwinian' sense. Self-regulatory systems which have evolved through a recursive process of reproduction, variation and natural or cultural selection are called *evolutionary systems*. Evolution theory explains why evolutionary systems obey normic laws which imply high conditional statistical probabilities. The prototypical (norm) states and self-regulatory mechanisms of evolutionary systems have been gradually selected according to their contribution to reproductive success. Due to the *limited* compensatory power of evolutionary systems, dysfunctions may occur, hence their normic behaviour may have various *exceptions*. Yet it must be the case that these systems are in their prototypical norm states in the high *statistical majority* of cases and times.

For otherwise, they could not *survive* in evolution and would die out. Birds, for instance, can normally fly. Of course, it is possible that due to an environmental catastrophe all birds suddenly loose their ability to fly. But then (with high probability) the species of birds will become extinct after a short period of evolution. For similar reasons, electric installations normally work, for they are constructed in that way, and if this were not so, they could not survive in the economic market. Put into a nutshell: *prototypical and statistical normality are connected by the law of evolutionary selection.*

Of course, this is a rather simplified presentation of the evolution-theoretic account to normic laws. This account has been elaborated in Schurz (2001b). To be applicable to normic laws of all higher sciences, the account is based on the *generalized* theory of evolution (cf. Dawkins (1989, ch. 11; Boyd and Richerson 1984; Blackmore 2000). In contrast to *socio-biology*, generalized evolution theory does not intend to explain cultural (including technical) evolution by the evolution of *genes*, but assumes it as an independent level based on the evolution of so-called *memes*. Moreover, in Schurz (2001b) the statistical consequence thesis is *proved* to be a consequence of the evolution-theoretic definition of prototypical normality.

Not all prototypical characteristics of individuals have (or had) a direct selective advantage in evolution. They may only be causal *side-effects* of characteristics which have such a direct selective advantage. For example, the prototypical sound of the heart beat is a *side-effect* of the heart's selectively relevant function of pumping the blood. To cover this difference, I distinguish between *fundamental* and *derived* normic laws (derivable from fundamental normic laws and boundary conditions). While the former express prototypical characteristics which confer a direct selective advantage, the latter express prototypical characteristics which are mere causal side-effects of the former. What is characteristic for *both* kinds of normic laws is that the normality which they express is not accidental but has a – direct or more indirect – evolution-theoretic explanation.

Let me finally mention that besides prototypical and statistical normality there exists a *third* notion of normality in the *normative-ethical* sense. In spite of Neander's emphasis of the *normative* character of proper (i.e., prototypical) functions (1991, 180f) I think that there exists no *direct* relation between prototypical and ethical normality (cf. Wachbroit 1994, 580, who makes the same point.). Possession of a prototypical function is a *descriptive* property, because it is defined in terms of the *factual* evolution of an organism. So, an inference from prototypical to ethical normality would be an *is-ought-inference*, which is a logical *fallacy* (cf. Schurz 1997b). In order to infer ethical normality from evolution-theoretic normality one would have to presuppose the analytical validity of the following is-ought bridge-principle: "whatever increases the fitness (i.e., the number of offspring) of a species is ethically good". However, this principle can hardly be 'analytically valid' – it is even not be contingently true, because in a situation of limited resources, the population increase of one species goes on the cost of other competing species.

6. On the Relation between Normic Laws and Ceteris Paribus Laws

6.1. Deductivistic Reconstruction of Ceteris Paribus Laws

Ceteris paribus laws, in short CP-laws, have constantly provoked discussions. Churchland (1981, 71) and Fodor (1989, 75, 1991) have argued that laws of (folk) psychology are always furnished with 'invisible' CP-clauses. Canfield and Lehrer (1961) and Lakatos (1970, 17f) have made the same assertion for fundamental laws of physics, and so did Cartwright (1983, 52) and Joseph (1980, 777), with the only difference that these authors draw rather sceptical conclusions. Hempel (1988), finally, has generalized this sceptical diagnosis to strict laws of any sort.

"Ceteris paribus" is itself an *unsharp* notion (Hempel 1988, 29). In (Schurz 1995a, 2002) it is suggested that one should distinguish between two very different conceptions of CP-law: comparative versus exclusive. The comparative sense derives from the literal meaning of "ceteris paribus' as "the others being equal". A *comparative CP-law* claims that if all other (unknown) parameters of an underlying system x are held constant, then the increase (or decrease) of one (quantitative) parameter leads to an increase (or decrease) of another parameter. Thus, a comparative CP-law does not exclude the presence of other 'disturbing' factors, but merely requires to *keep them constant*. Therefore, comparative CP-laws are testable by the methods of statistical experiment (Schurz 2002).

In the philosophical debate, however, CP-laws have usually been understood in the different exclusive sense. An *exclusive CP-law* asserts that a certain state or event-type expressed by a (possibly complex) predicate Ax leads to another state or event-type Cx *provided* disturbing influences are *absent*. I call Ax the *antecedent* and Cx the *consequent* predicate. Hence, an exclusive CP-law does not merely require to keep all other causally interfering factors constant; it rather *excludes* the presence of causally interfering factors. In this sense, Cartwright had remarked that "the literal translation is 'other things being equal'; but it would be more apt to read 'ceteris paribus' as 'other things being right' " (1983, 45). Joseph (1980, 777) has spoken of "ceteris absentibus" clauses, and Hempel (1988, 29) calls exclusive CP-clauses "Provisos" ("... *provided* disturbing factors are absent"). Consider the following two examples of exclusive CP-laws – (4) from physics and (5) from psychology:

(4) Ceteris paribus, planets have elliptical orbits (Lakatos op. cit.).
(5) Ceteris paribus, people act goal-oriented (in the sense example (1); Fodor, op. cit.)

In (4), it is assumed that other forces on the planet except that of the sun are – not merely constant but – *absent*. Likewise (5) requires that all factors causing irrational behaviour are absent, whether they be of physical or psychological sort. Note that (5) is Fodor's "ceteris paribus" version of our normic law (1) (Section 1). Likewise, our examples (2) and (3) from biology and technology are often stated as CP-laws. From now on we understand CP-laws always in the exclusive sense, and we formulate them prima facie as $CP(Ax \rightarrow Cx)$, meaning that "if other interfering ('disturbing') factors are absent, Ax leads to Cx" (where the variable "x" is bound by the CP-quantifier).

An important distinction is that between *definite* and *indefinite* CP-clauses and CP-laws. A definite CP-*clause* can be replaced by a (finite) list of all disturbing factors which are excluded in the antecedent of the CP-law (where it is assumed, of course, that the enriched antecedent does not analytically imply the consequent predicate Cx). Definite CP-laws are harmless insofar they admit a straightforward deductivistic reconstruction as strict (universal) implications of the form $\forall x(Ax \wedge \neg \text{DIST} \rightarrow Cx)$, where DIST is the conjunction of finitely many possible disturbing factors of the form $D_i x$. Such a transformation is called a strict completion of a CP-law. In most examples of CP-laws, however, a strict completion is impossible. This is especially clear for non-physical CP-laws such as (5) above. Whether a strict completion is also impossible for examples of physics such as (4) is a delicate point which is delayed to the next section. Nonetheless, most authors on CP-laws agree that their real philosophical significance lies in situations where a strict completion is impossible (cf. Rescher 1994, 14; Pietroski and Rey 1995, 84, 102; Horgan and Tienson 1996, 119f).

Taken literally, an indefinite CP-law makes a strict assertion *within the CP-scope: if* disturbing factors are excluded, then Ax will *always* imply Cx. So it seems that an indefinite CP-law can be reconstructed *deductivistically* (i.e., within classical deductivistic logic), namely as a universally quantified strict implication of the form "For all x: If 'CP', then if Ax, then Cx", or equivalently "For all x: If Ax, then Cx or else not 'CP' ". Reconstruction (i) has been suggested by Lakatos in (1970) (cf. p. 18, 26 – although Lakatos rejects this suggestion in his appendix, p. 98, fn. 3); it is furthermore supported by Gadenne (1988) and Horgan and Tienson (1996, 138). Pietroski and Rey's definition (1995, 92, cond. ii) corresponds to the equivalent form (ii) (for details cf. Schurz 2001a). The problem, of course, is the precise explication of the phrase 'CP'. Various philosophers have pointed out that indefinite CP-laws are vacuous tautologies, because if we define a "disturbing factor" to be "anything what produces exceptions from the law", then an indefinite CP-laws becomes completely *immune* to empirical criticism (Scriven 1959, 466; Canfield and Lehrer 1961; Albert 1971, 411; Schiffer 1991). In return, several authors have tried to develop non-vacuous deductivistic reconstructions of CP-laws. What their accounts have in common is that an indefinite CP-clause is understood as a *second order quantification* which ranges over arbitrary properties or events.

According to the joint intuition which underlies the accounts of Lakatos (1970) and Pietroski and Rey (1995), $CP(Ax \rightarrow Cx)$ is true iff $\forall x(Ax \wedge CP_{A\rightarrow C}(x) \rightarrow Cx)$ is true, where "$CP_{A\rightarrow C}(x)$" means "no nomological property ψ is present in x which would cause $\neg Cx$" (formally, $\neg\exists\psi(N\psi \wedge \psi x \wedge \forall y\text{Causes}(\psi y, Cy))$; assuming $\forall x(\varphi x \wedge \text{Causes}(\varphi x, \psi x) \rightarrow \psi y))$. Let us first consider the *mono-*

tonic behaviour of these deductivistically reconstructed CP-laws. Assume the premise set $P1 = \{\forall x(Ax \wedge CP_{A\to C}(x) \to Cx), Aa, CP_{A\to C}(a)\}$; it holds that $P1 \vdash Ca$. What happens if we add the 'conflicting' premise set $P2 = \{\forall x(Bx \wedge CP_{B\to\neg C}(x) \to \neg Cx), Ba, CP_{A\to\neg C}(a)\}$? Answer: the conclusion Ca is still monotonically entailed, but now also the conclusion $\neg Ca$ is monotonically entailed, because the resulting premise set $P1 \cup P2$ is logically inconsistent (which is seen when the CP-clauses are replaced by their 2nd order definitions). In other words, the deductivistic reconstruction behaves *indeed* monotonic. Of course, one may suggest to interpret the CP-clause semantically in an implicitly non-monotonic way; but then the reconstruction would no longer be deductivistic.

In Schurz (2001a) the deductivistic reconstruction of CP-laws is proved to suffer from two major defects which are informally explained as follows. *First*, according to this characterization, the truth of a CP-law $CP(Ax \to Cx)$ implies that conditional on Ax, all $\pm Cx$-events have *deterministic causes* (where "$\pm$" stands for "unnegated or negated"). For, if Aa holds, then either Ca is strictly determined by the absence of any disturbing condition, or $\neg Ca$ is strictly determined by the presence of such a disturbing condition. Hence, this understanding of CP-laws is incompatible with the assumption that non-deterministic random processes have an influence on $\pm Cx$. *Second* and even worse, the above characterization makes CP-laws almost empty: a CP-law does not only imply determinism w.r.t. the consequence predicate, it is also implied by it and, hence, is not stronger than it. For, whenever Aa is true (for arbitrary a), then either Ca and hence $CP(Aa \to Ca)$ is true, or $\neg Ca$ is true, in which case $\neg Ca$ must have had a deterministic cause ψa, thus the CP-clause is false and so, $CP(Aa \to Ca)$ is again true. To illustrate the counterintuitive consequences of these results, assume that my and your actions are deterministically caused. Then every CP-law of the form "ceteris paribus, whenever I do X you do Y" is true, e.g., "CP whenever I play tennis you play chess", "CP whenever I play tennis, you don't play chess", etcetera. The same triviality result have been proved for Pietroski and Rey's improved definition (cf. Schurz 2001a) and for Fodor's definition of CP-laws (cf. Schurz 2002).

6.2. Normic Reconstruction of Indefinite CP-Laws in Non-Physical Sciences

The shortcomings of deductivistic reconstructions of indefinite CP-laws can be summarized as follows:

(1.) Deductivistic reconstructions of indefinite CP-laws are *too strong*, because they presuppose that the consequent predicate has deterministic causes. In other words, deductivistic reconstructions exclude *random exceptions*. However, contemporary science makes it plausible that a certain portion of non-determinism does not only occur in quantum mechanics – it reigns in all domains of sufficiently complex systems (cf., e.g., Earman 1986).

(2.) At the same time, deductivistic reconstructions are *too weak* – for as we have seen, they are almost empty. In particular, they imply nothing about the *statistical probability* with which undisturbed antecedent-cases will produce the consequent (cf. Pietroski and Rey 1995, 84; Schiffer 1991, 8). But this is counter-intuitive. At least in the non-physical sciences, CP-laws are usually asserted *only* if the situation without (non-neglectible) disturbing factors is also the *statistically normal* situation. Examples of deductivistically correct CP-laws which are unintuitive because they violate this normality condition can be given in their thousands – e.g., CP no tire blows, CP there are no clouds in the sky, CP every human is naked (because that's how s/he was born), etc.

The above objections point to an alternative suggestion proposed by Schurz (1995a, 2002) and, in different form, by Silverberg (1996): indefinite CP-laws $CP(Ax \rightarrow Cx)$ should be reconstructed as *normic* laws $Ax \Rightarrow Cx$. In this reconstruction, the indefinite CP-clause is understood as a normality clause which does not make up a separate antecedent-conjunct or consequent-disjunct, but is implicitly contained in the normic conditional operator. The normic reconstruction solves the above-mentioned objections: normic laws do not imply determinism w.r.t. the consequent predicate, but they imply a statistical normality condition and hence have empirical content. Moreover, the normic reconstruction is a perfect fit with our non-physical examples of CP-laws, since all these examples are about evolutionary systems. In conclusion, the adequate reconstruction of indefinite CP-laws of *non-physical sciences* is their reconstruction as normic laws.

7. Theoretically Definite *CP*-Laws in Physics – Hempel's and Cartwright's Challenge

Can we escape indefinite ceteris paribus clauses, or normic weakenings, at least in *theoretical physics*, and achieve *genuinely strict* laws? This view is shared by many philosophers of physics.[6] Hempel (1988) opposed this view. We illustrate his argument at hand of our example (4): the prediction of the orbit of a particular planet by Newtonian physics T and antecedent conditions Aa. As we know from Section 6, the application of Newton's total force law requires an exclusive CP-assumption: *all* forces (acting upon the planet in the considered time interval) have been recorded, other 'disturbing' forces are absent. Hempel (1988, 30) has argued that the concept of total force would *transcend* the resources of every physical theory. It would not only require to consider all gravitational, electromagnetical or frictional forces acting upon the planet, but even all kinds of telekinetical or supernatural forces. So, for Hempel, "total force" is a theoretically *indefinite* concept, whence all CP-assumptions w.r.t. force must be indefinite.

Here I disagree. Like Coffa (1968) I think that total force has to be understood as a theoretically definite concept. Otherwise it could not be quantitatively related to mass and acceleration. I agree with Hempel that the concept of total force is not definable within a particular theory such as mechanics. But it is theoretically characterized by the total body of physical knowledge. Physical theories provide a theoretical classification of *all* kinds of elementary forces which is intended to be *complete*. At present, these are the four fundamental interaction forces (gravitational, electromagnetical, strong and weak nuclear force), and the inertial forces arising from collision. The concept of total force is backed up by this theoretical classification and, hence, excludes 'supernatural' forces of any sort.

For this reason, the CP-clause according to which no further forces are present is not an indefinite but a *theoretically definite* CP-clause. My suggestion is to consider all idealized system laws of theoretical physics – if formulated with care – as *theoretically definite CP-laws* (cf. Schurz 2002). If this reconstruction is correct, then – in distinction to the non-physical sciences – the theoretical part of physics is governed by *strict laws* and, modulo mathematics, by *strictly deductive* inference procedures.

However, the *non-strict* part of physics has not been dissolved by these considerations. It reappears when we ask about the relation between the *idealized system* as described in the *theoretical antecedent description* (abbreviated as $A_t x$) and the *real system* as characterized by our empirical evidence – the *empirical antecedent description* (abbreviated as $A_e x$). For planets obeying Kepler's laws, $A_t x$ mainly asserts that the only (non-negligible) force on x is that of the sun, while $A_e x$ contains astronomical data supporting $A_t x$. As Hempel emphasizes in his remarks about "theoretical ascent" (1988, 21f), the inference from $A_e x$ to $A_t x$ must remain *uncertain*, for in the empirical or pre-theoretical language, a complete description of all possible disturbing factors is impossible. We call the implication from $A_e x$ to $A_t x$ an *application claim. How uncertain* are these application claims? Do they at least hold with high conditional probability?

According to Cartwright (1983, 47), the answer is *no*. For, the undisturbed ideal case described by $A_t x$ is a statistical rarity, and often even a physical impossibility (cf. Joseph 1980). However, there exist well-known methods of *approximation*, by which this challenge can be defeated (cf., e.g., Laymon 1989, Hüttemann 1998, including Cartwright 1989). Without going into details, we will just mention three of them: (1.) In *empirical approximation* one describes a *restricted* empirical situation $A_e x$ in which it is known that the disturbing factors are negligible. For example, Newton knew very well that Kepler's laws will be approximately true only if the planets do not come too close to each other (cf. Lakatos 1970, 50). (2.) In *theoretical approximation, correction* terms are added to the original *theoretical* system laws which *improve* the prediction. For example, Newton spent years in calculating the correction terms describing the effects of planet-planet-interactions (Lakatos ibid.). Finally (3.), in *statistical approximation* a layer of statistically distributed measurement errors is added to the *theoretical prediction* $P_t x$. The resulting empirical prediction $P_e x$ asserts that with high probability, the measured value will be found within an error interval around the theoretically predicted value (cf. Kyburg 1988, 87f).

It is important for approximation procedures that the idealized CP-law is *theoretically definite*, i.e., that the possible disturbing factors are theoretically classified and partially controlled by the available background theory. Under this condition, approximation procedures make it possible to turn ideal theories into empirical predictions of $P_e x$ given $A_e x$ which hold at least with high conditional probability.

We have seen that the empirical part of physics is full of non-strict, merely probabilistic generalizations. But does it also contain *normic laws*? While I have defended this claim in (1995a), I present a more differentiated view in the next section.

8. Differences between Physical and Non-Physical Sciences

8.1. Laws of Nature versus System Laws

The difference between CP-laws in physical and non-physical sciences has its roots in the difference between closed or isolated systems of physics and chemistry, and open systems of non-physical sciences. Our starting point is the distinction between *laws of nature* and *system laws* (cf. Josef Schurz 1990). Laws of nature are those fundamental laws of physics which hold everywhere within the universe, or in other words, which are not restricted to special entities. There are only a few of them. In classical physics, the *total force* law $F(x, t) = m(x).d^2s(x, t)/dt^2$ is a law of nature. It is a differential equation in which $F(x, t)$ figures as a *variable* function denoting the sum of *all* forces acting at time t on particle x without saying what these forces are. Another kind of law of nature are *special force laws*, e.g., the classical laws for gravitational force or electric force – provided they are understood as laws about *abstract component forces* or 'capacities' in the sense of Cartwright (1989, 183ff). Laws of nature are *strictly* true, *without* any ceteris paribus clause – but at the cost of not per se being *applicable* to *real* systems, because they do not specify *which* forces are active.

System laws, in contrast, do not apply to the whole universe, but refer to particular systems of a certain kind in a certain time interval Δt. They contain or rely on a *specification* of all forces which act within or upon the system x in the considered time interval Δt – the so-called *boundary conditions*. Examples of system laws in classical physics are Kepler's laws of elliptic planetary orbits, Galileo's law of falling bodies, the classical wave equations, etc. – almost *all* laws in physics textbooks are system laws. Within system laws, we distinguish between *theoretical* and *phenomenological* system laws. Theoretical system laws state the *special* differential equations for the system under consideration, with a specification of boundary conditions (forces). Phenomenological system laws describe the temporal behavior of the system in an empirical or pre-theoretical vocabulary dependent on given initial conditions – in the planet example, these are Kepler's elliptic trajectory laws. For *simple* systems of physics, the phenomenological system laws are literally derivable from the theoretical system laws (as solutions of differential equations), which in turn are literally derivable from laws of nature plus boundary conditions. But as Cartwright has convincingly demonstrated (e.g. 1983, 104f, 113f), for systems of moderate complexity these derivations are not possible without additional assumptions. For non-physical system laws such as that birds can normally fly, derivation attempts of this sort are usually completely hopeless. Phenomenological system laws of this kind are usually obtained by purely empirical-inductive means.

The difference between laws of nature and system laws is neither *logical* nor *epistemological* but *ontological*. It does *not* coincide with the logical distinction between strictly universal versus spatio-temporally restricted laws (cf. Earman 1986, 91), because (theoretical) system laws may also be expressed in a spatio-temporally unrestricted manner. It does also not coincide with Hempel's epistemological distinction between fundamental and derived laws (1965, 272, 292), because many system laws are not derivable and are thus epistemologically fundamental. One may object against our distinction that, from a practical viewpoint, the difference between laws of nature and system laws is merely a gradual one, because also laws of nature are only applicable to certain well-defined subsystems of reality. However, this objection misses our distinction, because before a law of nature can be applied, it first has to be turned into a system law (by replacing the variable force parameters by specific force functions). In other words, the proposed distinction between laws of nature and system laws is not expressed in terms of their applications, but in terms of their *semantic content* – in terms of what they assert. Let us recapitulate two crucial differences between laws of nature and system laws: *(1.)* Laws of nature do not refer to any specific system description in their if-part, not even to the entire universe conceived as a system. For example, that our universe is composed of matter (instead of anti-matter) is not a law of nature but a system law – a universe consisting of anti-matter is allowed by laws of nature. In other words, laws of nature are not only intended to speak about our universe, but also, to speak about other possible universes. *(2.)* Following from (1.), the truth of laws of nature does not depend on a CP-clause. For example, Newton's total force law refers to "all" forces acting within or upon a considered system; this law remains true (if it is true) whatever disturbing factors are present.

It is exactly this second point which makes the distinction important for our topic – while laws of nature go without CP-clauses, system laws always need CP-clauses. This fact may shed new light on some controversial matters. For example, philosophers like Schiffer (1991) who doubt that non-physical sciences contain genuine laws of their own are right if they mean laws of nature. It is indeed true that non-physical sciences do not have laws of nature of their own – but they *do* have system laws of their own. The distinction may also shed some light on a point was made by Joseph (1980, 789) who – after finding that most laws of physics are CP-laws – recognizes that there are also laws which are literally true, such as the conservation laws. This has the following explanation: the conservation laws for energy and momentum are *directly* obtained from laws of nature by integrating the total force law over space or time, respectively, *without* any insertion of special boundary conditions – so these laws are derived laws of nature.

8.2. Closed (Isolated) versus Open (Self-regulatory) Systems

Now we can introduce the *system-theoretic* distinction between *closed* or *isolated* versus *open systems*. In closed systems, there is no exchange between system and environment; in isolated systems, there is exchange of heat-energy, but no exchange of matter (e.g., a gas under isothermic conditions). Only in open systems is there a continuous exchange of both matter and energy between system and environment (cf. Bertalanffy 1979, 39f, 141ff; Rapaport 1986, 177f). The systems studied by *physics* or chemistry are, at least traditionally, closed or isolated systems. Of course, physical system laws are never strictly but at most *approximately* true, because no real system is completely closed. This explains our observations of Section 7, namely, that physical system laws are always *idealizations* which need approximation procedures for their empirical applications. In contrast, all 'higher-level' sciences are concerned with *open* systems, and more specifically, with self-regulatory systems which have come about through evolution.

Very generally, *systems* are physical ensembles composed of parts which preserve a relatively strict *identity in time*, by which they delimit themselves from their (significantly larger) environment (Rapaport 1986, 29ff). For closed systems this preservation of identity follows from their isolation which, in turn, is a matter of *postulate*: that our planetary system is stable is a *frozen accident* of cosmic evolution; should it be once devastated by a gigantic swarm of meteorites, then it stays so forever and will *not* regenerate. But how can we explain the relatively strict identity of *open* systems, which are permanently subject to possibly destructive influences from the environment? We have given the answer in Section 5: the stability of open 'living' systems follows from their self-regulatory capacities, which have arisen through evolution.

This difference between closed (or isolated) and open self-regulatory systems explains the different nature of their CP-laws. For closed (or isolated) systems, a detailed specification of *all* forces ('and nothing else') is needed. For open self-regulatory systems, such a specification is neither possible nor necessary. It suffices to assume that the disturbing influences, whatever they may be, are within the 'manageable range' of the system's self-regulatory compensation power. It is also usually impossible to give an exact theoretical prediction of this 'manageable range'. But evolution-theoretic considerations of Section 5 tell us that *normally* the external influences will be within this manageable range. This explains the *normic* character of the CP-laws of open self-regulatory systems.

A related difference is the following. In physics one traditionally thinks of *complexity* as a source of *disorder* – regularities are obtained by *abstracting away* from complexities. In evolutionary systems, however, complexity is usually a *source of order* – complexity which has been selected by evolution to stabilize normic behaviour. Ideal planets are theoretical abstractions: mass points under the influence of a centripetal force and 'nothing else'. They do not literally exist (cf. Wachbroit 1994, 587f). In contrast, normal birds really *do* exist because they are what has been selected through evolution. When speaking of a normal bird, we do not abstract from its admirable complexity, but we *rely* on it as the cause of its normal behavior. The idealization procedures needed for planets would not make good sense for birds: there are no disturbing parameters which, when going to zero, turn a real bird into an ideal bird which necessarily can fly and which is approximated by the real bird.

To avoid misunderstandings: it is clear that, in principle, one may always try to describe open systems as parts of larger closed physical systems (ultimately the universe) – but in most real examples this would be a theoretically hopeless enterprise. However, there are indeed systems which can be fruitfully described both as closed systems of physics and parts of open evolutionary systems – namely *technical systems*. Consider the systems of electricity which surround us every day. We may consider this automatic dish washer together with its electric circuit as an ideally closed physical system. From this perspective, there are thousands of possible disturbing factors which may prevent our dishes from being cleaned and, amazed, we may ask ourselves how all these electrical systems can be so cheap and yet work so well. Alternatively, we may consider them as part of an evolutionary system – the economic system of production and distribution of electric products. This perspective does not give us detailed knowledge of the physical mechanisms underlying dish washers, but it gives us an evolution-theoretic explanation of their amazing optimization of cheapness and functionality.

9. Consequences for Unity and Diversity in Science

The picture of science which flows from our analysis of normic laws and non-monotonic reasoning embraces unity as well as diversity. More accurately, it is a picture of *unity* on the background of *diversity* on the background of *unity*:

(1.) Ontological unity and diversity. In the picture which I have drawn, physics still has a special status since it is the only science which is concerned with laws of nature. However, all sciences have their own system laws. Due to the differences between closed or isolated physical systems and open evolutionary systems, the system laws of physical and non-physical sciences are different in nature: theoretically definite CP-laws on the one hand, normic laws on the other. Moreover, it is usually impossible to derive system laws from laws of nature and boundary conditions. Nevertheless, phenomenological hypotheses about system laws must not get in conflict with laws of nature; in other words, they must cohere with principles such as conservation of energy, equivalence of matter and energy, or maximal velocity of light, etc. So there is ontological diversity on the background of ontological unity. On the other hand, all non-physical sciences exhibit a striking

similarity because they all are concerned with evolutionary systems in the generalized sense. So we have a third layer of ontological unity on the background of ontological diversity.

(2.) Logical unity and diversity. Classical monotonic logic (or an intuitionistic weakening of it) has still a special status because it is the common core of all extended logics, including the non-monotonic logics. However, due to the different status of strict versus normic laws, theoretical reasoning in physical sciences is typically mathematically-deductive, while reasoning from normic laws in non-physical sciences is non-monotonic and exhibits peculiar features which do not occur in deductive reasoning, such as global background-dependency and the role of individual case understanding (Section 2), or the possibility of removing consequences by the mere expansion of the base set (Section 3). So we have logical diversity on the background of logical unity. On the other hand, all non-physical sciences have these peculiarities of non-monotonic reasoning in common; so again we have a third layer of logical unity on the background of logical diversity. This unity in the third layer is substantiated by the fact that the probability semantics of non-monotonic reasoning establishes a 'grand bridge' between qualitative non-monotonic reasoning and quantitative probabilistic reasoning which in turn plays a fundamental logical role in the methodology of many higher-level sciences.

(3.) Methodological unity and diversity.: Laws and theories of science have to have empirical content, by which their are testable. This is not only true for laws of physical sciences, but also for normic laws of non-physical sciences, since they imply statistical majority claims. Therefore, predictions and explanations based on general lawlike principles are common to all sciences. So we have unity in the first place. But normic laws and non-monotonic prediction and explanation patterns involve various methodological peculiarities which do not arise in their deductive counterparts. So there is methodological diversity on the background of methodological unity. Again, the fundamental similarities between all sciences dealing with evolutionary systems establish a third layer of methodological unity.

In this picture, *microreduction* no longer plays that important role which it had played in earlier programs to the unity of science such as Oppenheim and Putnam (1958) or Causey (1977). In these earlier accounts, unity of science consisted mainly in the possibility of microreduction, i.e., the possibility of reducing the entities and the laws of higher-level sciences to the level of microphysical entities and their laws. I do not deny that microreduction has been successful, e.g., in the area of statistical thermodynamics and physical chemistry, but these disciplines are physical disciplines in my terminology. Microreduction is hopeless in the description of evolutionary systems such as normal birds, human persons or social systems. Nevertheless, there exist also mathematical theories in these areas. They are not based on microreduction but on the assumptions of 'abstract forces'. In our terminology, these are the *theoretical* system laws of the respective disciplines. For example, the theoretical system laws of population dynamics assume abstract 'force' parameters such as growth rates, competition coefficients, or mutation rates (cf. Bertalanffy 1979, ch. 3; Rapaport 1986, ch. 2; Ridley 1993, 108). These 'force' parameters are not microphysically derived but are abstractly assumed. Ontologically, they are based on certain stability properties of the underlying evolutionary systems. Logically, they allow the 'derivation' of the trajectory space which yields a qualitative explanation of the phenomenological behaviour of the underlying system; if this system has normic stability properties then most trajectories turn out to be approximately stable (as opposed to, say, chaotic systems). The unity of science which is reflected in meta-disciplines such as *systems theory* or *synergetics* (Haken 1983) is mainly based on mathematical, i.e., structural, similarities among theoretical system laws rather than on physical micro-reduction. This structural similarity among theoretical system laws constitutes a fourth kind of unity in the 'first layer'.

Acknowledgements

Work on this paper was supported by the research grant F012 of the Austrian Science Funds (FWF).

Notes

[1] Scriven (1959), Dray (1957, 132–137); in Artificial Intelligence, e.g., McCarthy (1986), Reiter (1997); in Philosophy of Biology, e.g., Millikan (1984, 1989), Neander (1991), Laurier (1996).

[2] The presently developed NML-systems can be classified as follows: (1) default reasoning (Reiter 1980; Poole 1988, 1994); (2) autoepistemic (non-monotonic) logic (McDermott and Doyle 1980; Moore 1985); (3) conditional entailment – system **P** (and its relatives), which covers three converging approaches (cf. also Rott 1997): (3.1) conditional logic and probabilistic entailment (Adams 1975, 1986; Pearl 1990; Goldszmidt and Pearl 1996; Delgrande 1988; Schurz 1997a; 1998), (3.2) preferential entailment (Shoham 1988; Kraus et al. 1990; Lehmann and Magidor 1992; Makinson 1994), (3.3) expectation-orderings (Gärdenfors and Makinson 1994; Rott 1997); (4) defeasible reasoning (Pollock 1994; Nute 1994), (5) circumscription (McCarthy 1986) and (6) possibility logic (Dubois et al. 1994). For an overview cf. Brewka (1991) and Gabbay et al. (eds., 1994). Important philosophical forefathers are Adams (e.g., 1975) for type (3.1) approaches, Pollock (e.g., 1974, ch. 3.4) for type (4) approaches, and Rescher (1976) for type (3.2) and type (6) approaches.

[3] In the NML-formalism, "Norm($Ax \to Bx$) is expressible as (i): "$T \Rightarrow (Ax \to Bx)$" ("$T$" for "Verum"); (i) is implied by, but does not imply (ii): "$Ax \Rightarrow Bx$". A similar antecedent-relativity is well-known from the logic of counterfactuals (cf. Lewis 1973). The probabilistic counterpart is the fact that a high $p(Ax \to Bx)$ (which corresponds to (i)) is implied by, but does not imply a high $p(Bx/Ax)$ (which corresponds to (ii)). For example, $p(\text{Tiger}(x) \to \text{CanFly}(x))$ is high, because there exist almost no tigers, although $p(\text{CanFly}(x)/\text{Tiger}(x))$ is zero. This fact was especially emphasized by Adams (1975), and the failure to recognize it was the central deficiency of Nilsson's "Probabilistic Logic" (1993).

[4] Cf. Schurz and Adams (2004) on unrestricted 1st order statistical probability logic.

[5] Under different labels, this principle is stated in many of the approaches in mentioned fn. 2 and is implicitly contained in all of them. A justification of this principle in terms of maximizing utility is found in Good (1983, 177–179).

[6] Also quantum mechanics is governed by strict laws (cf. Earman 1986, 200). Only the relation to measurements is statistical. Note that strict laws do not necessarily imply deterministic trajectory spaces (ibid., ch. III).

References

Adams, E. W. : 1975, *The Logic of Conditionals*, Dordrecht, Reidel.

Adams, E.W.: 1986, 'On the Logic of High Probability', Journal of Philosophical Logic **15**, 255–279.

Alchourón, C. E., P. Gärdenfors and D. Makinson: 1985, 'On the Logic of Theory Change', *Journal of Symbolic Logic* **50**, 510–530.

Ashby, W. R.: 1961, *An Introduction to Cybernetics*, London, Chapman & Hall.

Bertalanffy, L. v.: 1979, *General System Theory*, 6th edn, New York.

Blackmore, S.: 2000, *The Meme Machine*, Oxford, Oxford Paperbacks.

Boyd, R. and P. J. Richerson: 1985, *Culture and the Evolutionary Process*, Chicago, University of Chicago Press.

Brewka, G.: 1991, *Non-monotonic Reasoning. Logical Foundations of Commonsense*, Cambridge University Press.

Canfied, J. and K. Lehrer: 1961, 'A Note on Prediction and Deduction', *Philosophy of Science* **28**, 204–208.

Carnap, R.: 1950, *Logical Foundations of Probability*, Chicago, University of Chicago Press.

Cartwrigt, N.: 1983, *How the Laws of Physics Lie*, Oxford, Clarendon Press.

Cartwright, N.: 1989, *Nature's Capacities and their Measurement*, Oxford, Clarendon Press.

Causey, R.: 1977, *The Unity of Science*, Dordrecht, Reidel.

Churchland, P.: 1981, 'Eliminative Materialism and Propositional Attitudes', *Journal of Philosophy* **78**, 67–90.

Coffa, J. A.: 1968, 'Deductive Predictions', *Philosophy of Science* **35**, 279–283.

Dawkins, Richard: 1989, *The Selfish Gene*, 2nd edn, Oxford, Oxford University Press.

Delgrande, J. P.: 1988, 'An Approach to Default Reasoning Based on a First-Order Conditional Logic: Revised Report', *Artificial Intelligence* **36**, 63–90.

Dray, William: 1957, *Laws and Explanation in History*, Oxford, Oxford University Press.

Dubois, D. et al.: 1994, 'Possibilistic Logic', in (1994, ed.), 439–513.

Earman, J.: 1986, *A Primer on Determinism*, Dordrecht, Reidel.

Fodor, J.: 1989, 'Making Mind Matter More', *Philosophical Topics* **17**, 59–79.

Fodor, J.: 1991, 'You Can Fool Some of the People All of the Time', *Mind* **100**, 19–34.

Fuhrmann, A. and H. Rott: 1996, *Logic, Action and Information*, Berlin, de Gruyter.

Gabbay, D.: 1984, 'Theoretical Foundations for Non-Monotonic Reasoning in Expert Systems', in K. R. Apt (ed.), *Logics and Models for Concurrent Systems*, Berlin, Springer, pp. 439–458.

Gabbay, D. M. et al.: 1994, (eds.), *Handbook of Logic in Artificial Intelligence*, Vol. 3, *Non-monotonic Reasoning and Uncertain Reasoning*, Oxford, Clarendon Press.

Gadenne, V.: 1998, 'Grundprobleme der Prüfung von Theorien', in M. Albert and W. Meyer (eds.), *Theorie, Modell und Erfahrung*, Tübingen, Mohr.

Gardiner, P.: 1952, *The Nature of Historical Explanation*, Oxford, Oxford University Press.

Gärdenfors, P.: 1988, *Knowledge in Flux*, Cambridge, MA, MIT Press.

Gärdenfors, P.: 1986, 'Belief Revisions and the Ramsey Test for Conditionals', *Philosophical Review* **95**, 81–93.

Gärdenfors, P. and D. Makinson: 1994, 'Non-monotonic Inference based on Expectation Orderings', *Artificial Intelligence* **65**, 197–245.

Goldman, A. I.: 1986, *Epistemology and Cognition*, Cambridge, MA, Harvard University Press.

Goldszmidt, M. and J. Pearl: 1996, 'Qualitative Probabilities for Default Reasoning, Belief Revision and Causal Modeling', *Artificial Intelligence* **84**, 57–112.

Good, I. J.: 1983, *Good Thinking. The Foundations of Probability and Its Applications*, Minneapolis, University of Minnesota Press.

Grünbaum, A. and W. Salmon (eds.): 1988, *The Limitations of Deductivism*, Berkeley, University of California Press.

Haken, H.: 1983, *Synergetics*, 3rd edn., Berlin, Springer.

Hempel, C. G.: 1965, *Aspects of Scientific Explanation and Other Essays*, New York, Free Press.

Hempel, C. G.: 1968, 'Maximal Specifity and Lawlikeness in Probabilistic Explanation', *Philosophy of Science* **35**, 116–133.

Hempel, C. G.: 1988, 'Provisos: A Problem Concerning the Inferential Function of Scientific Theories', in Grünbaum and Salmon (eds.), pp. 19–36.

Holzkamp, K.: 1968, *Wissenschaft als Handlung*, Berlin, de Gruyter.

Horgan, T. and J. Tienson: 1996, *Connectionism and the Philosophy of Psychology*, Cambridge, MA, MIT Press.

Hüttemann, A.: 1998, 'Laws and Dispositions', *Philosophy of Science* **65**, 121–135.

Joseph, G.: 1980, 'The Many Sciences and the One World', *Journal of Philosophy* **77**, 773–790.

Kahneman, D., P. Slovic and A. Tversky: 1982, *Judgement Under Uncertainty: Heuristics and Biases*, Cambridge, Cambridge University Press.

Kraus, S., D. Lehmann and M. Magidor: 1990, 'Non-monotonic Reasoning, Preferential Models and Cumulative Logics', *Artificial Intelligence* **44**, 167–207.

Kyburg, H. E. Jr.: 1988, 'The Justification of Deduction in Science', in Grünbaum and Salmon (eds.), pp. 61–94.

Lakatos, I.: 1970, 'Falsification and the Methodology of Scientific Research Programmes', reprinted in Lakatos, I.: 1978, *Philosophical Papers*, Vol 1, Cambridge, Cambridge University Press.

Laurier, D.: 1996, 'Function, Normality, and Temporality', in M. Marion and R. S. Cohen (eds.), *Québec Studies in the Philosophy of Science*, Dordrecht, Kluwer, pp. 25–52.

Laymon, R.: 1989, 'Cartwright and the Lying Laws of Physics', *Journal of Philosophy* **89**, 353–372.

Lehmann, D. and M. Magidor: 1992, 'What does a Conditional Knowledge Base Entail?', *Artificial Intelligence* **55**, 1–60.

Leitgeb, H.: 2001, 'Non-monotonic Reasoning by Inhibitions Nets', *Artificial Intelligence* **128**, 161–201.

Lewis, D.: 1973, *Counterfactuals*, Oxford, Basil Blackwell.

Makinson, D.: 1994, 'General Patterns in Non-monotonic Reasoning', in Gabbay (ed.), pp. 35–110.

McCarthy, J.: 1986, 'Application of Circumscription to Formalizing Common-Sense Knowledge', *Artificial Intelligence* **13**, 89–116.

McDermott, D. and J. Doyle: 1980, 'Non-Monotonic Logic I', *Artificial Intelligence* **25**, 41–72.

Millikan, R. G.: 1984, *Language, Thought, and Other Biological Categories*, Cambridge, MA, MIT Press.

Millikan, R. G.: 1989, 'Biosemantics', *Journal of Philosophy* **86**, 281–297.

Moore, R. C.: 1985, 'Semantical Considerations on Non-monotonic Logic', *Artificial Intelligence* **25**, 75–94.

Neander, K.: 1991, 'Functions as Selected Effects: The Conceptual Analyst's Defense', *Philosophy of Science* **58**, 168–184.

Nilsson, N. J.: 1993, 'Probabilistic Logic Revisited', *Artificial Intelligence* **59**, 39–42.

Nute. D.: 1994, 'Defeasible Logic', in Gabbay (ed.), pp. 353–395.

Oppenheim, P. and H. Putnam: 1958, 'Unity of Science as a Working Hypothesis', in H. Feigl et al. (eds.), *Minnesota Studies in the Philosophy of Science*, Vol. II, Mineapolis, University of Minnesota Press, pp. 3–36.

Pearl, J.: 1988, *Probabilistic Reasoning in Intelligent Systems*, Santa Mateo, CA, Morgan Kaufmann.

Pearl, J.: 1990, 'System Z', *Proceedings of Theoretical Aspects of Reasoning about Knowledge*, Santa Mateo, CA, pp. 21–135.
Pelletier, F. J. and R. Elio: 1997, 'What Should Default Reasoning Be, By Default?', *Computational Intelligence* **13**(2), 165–187.
Pietroski, P. and G. Rey: 1995, 'When Other Things Aren't Equal: Saving Ceteris Paribus Laws from Vacuity', *British Journal for the Philosophy of Science* **46**, 81–110.
Pollock, J.: 1974, *Knowledge and Justification*, Princeton, Princeton University Press,
Poole, D.: 1988, 'A Logical Framework for Default Reasoning', *Artificial Intelligence* **36**, 27–47.
Poole, D.: 1994, 'Default Logic', in Gabbay (ed.), pp. 189–215.
Rapaport, A.: 1986, *General System Theory*, Cambridge, MA, Abacus Press.
Reiter, R.: 1980, 'A Logic for Default Reasoning', *Artificial Intelligence* **13**, 81–132.
Reiter, R.: 1987, 'Non-monotonic Reasoning', *Annual Review of Computer Science*, Vol. 2, Palo Alto, California, pp. 147–186.
Rescher, N.: 1976, *Plausible Reasoning*, Amsterdam, Van Gorcum.
Rescher, N.: 1994, *Philosophical Standardism*, University of Pittsburgh Press.
Ridley, M.: 1993, *Evolution*, Oxford, Blackwell Scientific Publications.
Rott, H.: 1997, 'Drawing Inferences from Conditionals', in E. Ejerhed and S. Lindström (eds.), *Logic, Action and Cognition*, Dordrecht, Kluwer, pp. 149–179.
Schiffer, S.: 1991, 'Ceteris Paribus Laws', *Mind* **100**, 1–17.
Schurz, G.: 1994, 'Probabilistic Justification of Default Reasoning', in B. Nebel and L. Dreschler-Fischer (eds.), *KI-94: Advances of Artificial Intelligence*, Berlin, Springer, pp. 248–259.
Schurz, G.: 1995a, 'Theories and their Applications – A Case of Non-monotonic Reasoning', in W. Herfel et al. (eds.), *Theories and Models in Scientific Processes*, Amsterdam, Rodopi, pp. 69–293.
Schurz, G.: 1995b, 'Scientific Explanation: A Critical Survey', *Foundation of Science* **I/3**, 29–465.
Schurz, G.: 1997a, 'Probabilistic Default Reasoning Based on Relevance- and Irrelevance Assumptions', in D. Gabbay et al. (eds.), *Qualitative and Quantitative Practical Reasoning* (LNAI 1244), Berlin, Springer, pp. 536–553.
Schurz, G.: 1997b, *The Is-Ought Problem. An Investigation in Philosophical Logic*, (Studia Logica Library Vol. 1), Dordrecht, Kluwer.
Schurz, G.: 1998, 'Probabilistic Semantics for Delgrande's Conditional Logic and a Counter-example to his Default Logic', *Artificial Intelligence* **102**, 81–95.
Schurz, G.: 2001a, 'Pietroski and Rey on Ceteris Paribus Laws', *British Journal for the Philosophy of Science* **52**, 359–370.
Schurz, G.: 2001b, 'What is 'Normal'? An Evolution-Theoretic Foundation of Normic Laws and their Relation to Statistical Normality', to appear in *Philosophy of Science*.
Schurz, G.: 2001c, 'Carnap's Modal Logic', in W. Stelzner and M. Stöckler (eds.), *Nichtklassische logische Ansätze im übergang von traditioneller zu moderner Logik*, Paderborn, Mentis Verlag.
Schurz, G.: 2002, 'Ceteris Paribus Laws: Classification and Deconstruction', in J. Earman et al. (eds.), *Ceteris Paribus Laws*, (special volume) *Erkenntnis* **57**(3), 351–372
Schurz, G. and E. Adams: 2004, 'Measure-Entailment and Support in the Logic of Approximate Generalizations', to appear in E. Adams (ed.), *Approximate Generalizations*, Stanford, CSLI Press.
Schurz, Josef (1990), 'Prometheus or Expert-Idiot? Changes in Our Understanding Sciences', *Polymer News* **15**, 232–237.
Shoham, Y.: 1988, *Reasoning about Chance*, Cambridge, MIT Press.
Scriven, M.: 1959, 'Truisms as Grounds for Historical Explanations', in P. Gardiner (ed.), *Theories of History*, New York, The Free Press.
Silverberg, A.: 1996, 'Psychological Laws and Non-monotonic Reasoning', *Erkenntnis* **44**, 199–224.
Tan, Y.-H.: 1997, 'Is Default Logic a Reinvention of Inductive-Statistical Reasoning', *Synthese* **110/3**, 357–379.

Toulmin, S.: 1958, *The Uses of Argument*, Cambridge, Cambridge University Press.
Wachbroit, R.: 1994, 'Normality as a Biological Concept', *Philosophy of Science* **61**, 579–591.

THE PUZZLING ROLE OF PHILOSOPHY IN LIFE SCIENCES

Bases for a Joint Program for Philosophy and History of Science[1]

JUAN MANUEL TORRES

Universidad Nacional del Sur, Centro de Logica y Filosofía de la Ciencia, (8000) Bahia Blanca, Argentina, E-mail: `jmtorres@criba.edu.ar`

Abstract. We identify and analyze the main causes of the great methodological activity observed in life sciences, a fact that is evident from the appearance of continue philosophical contributions in leader biological journals. Though this activity refutes in some way the postkuhnean dictum "philosophy of science methodologically oriented is old fashion", it does not constitute by itself a vein to extract uncontaminated methodological rules in order to build an unbiased philosophy of science. However, the peculiar situation that we describe can serve as example of fruit collaboration between science and philosophy and give a new perspective with regard to the up to now traumatic relationships between history and philosophy of science.

Aims and Motivations

Within the wide world of life sciences, it is easy to detect two empirical facts that are very significant for philosophy of science: (a) In articles published in leading scientific journals, and other less important journals, frequently appear the names of classical philosophers of science and epistemological texts as references. This is even more evident in those journals on evolutionary biology, systematics, and ecology; (b) The same journals often include contributions whose authors are not professional scientists, but philosophers. Both facts create an outstanding situation, especially if we pay attention to the importance that philosophy has in other journals of natural sciences, such as physics. All this can be measured in the contributions of professional philosophers or the explicit use of epistemological bibliography. As we know, this importance is nearly imperceptible.

These two facts evidence a great methodological activity in life sciences, in many cases carried out by the scientists themselves. Scientists assume the role of philosophers of science, whereas the latter, paradoxically, work as scientists. In this article, we will identify and analyze the main causes for this symbiotic phenomenon between life sciences and philosophy of science or, put in other words, scientists and methodologists. Among these causes we shall point out: (i) the status *in fieri* of an evolutionary theory able to gain general consensus; (ii) the crisis of Modern Synthesis and (iii) the urgency to reach final conclusions on those issues that strongly influence public opinion and social organizations. As examples, we can mention the following controversies: evolutionism vs. neo-creationism,

S. Rahman et al. (eds.), Logic, Epistemology, and the Unity of Science, pp. 213–227.

genes *visà vis* IQ, races or deviate behavior, and the precise determination of the beginning and end of a human person.

Certainly, the importance of methodological activity detected in life sciences can be used against some affirmations that became very common after Kuhn, such as the following: "philosophy of science normatively oriented is old fashioned" and so on. Nevertheless, its presence does not constitute a justification in itself but, instead, it is an academic phenomenon. In addition, the fact that life scientists usually deal with methodological tools deprive some classical but contemporary strategies for founding the philosophy of science of its efficiency. Specifically, this supposes that we can get objective criteria for science evaluation from characteristic rules or *conventions* used and generated by the scientific community. Concerning life sciences, the task of philosophers becomes harder as expected: they will not be able to discover pure and original rules or conventions in the scientific community, but rather their own constructions introduced by scientists themselves.

Certainly, this can be rather disappointing for *hunters* of pure and uncontaminated methodological rules in the construction of an unbiased philosophy of science. Nevertheless, this is the only negative face of the matter. Sometimes, the notions and methodological rules introduced into biological sciences can be further refined and enriched by scientists themselves.

Other highly positive result arises from all this situation, a result which paves the way for a promissory research program, at least with regard to the field of philosophical studies in life sciences. This program, in which philosophy and history of science work jointly, aims at determining the extent to which philosophy of science directly influences the architecture of theories, hypotheses, and doctrines of life sciences and, therefore, helps to open or close spaces for both scientific thought and research.

I.

From an abstract or naïve perspective, one might think that philosophy of science and history of science are not opposite, but complementary disciplines. In spite of this, since the 1960s, the philosophical studies of science have been characterized by the sharp dichotomy *prescription versus description*, where description has mostly referred to historical description. Certainly, the existence of other descriptive studies of science, such as those originated in psychological or sociological views, cannot be neglected. However, it is undeniable that history of science is the branch that congregates the greatest number of specialists in the field of descriptive studies of science. Naturally, there are always specialists who try to conciliate evaluative and historical approaches. So did Imre Lakatos. But, I believe, that his attempt should be considered rather exceptional and in no way an example of a general attitude.

If we take a look at what happened in the 20th century in philosophical studies of science regarding the relationships between philosophy and history of science, a very simple scheme immediately appears before us. Naturally, this scheme is, as all schemes, a generalization that leaves aside certain views and developments. But this is the price we have to pay in order to describe such a rich and complex process. The scheme is well known by everybody, but I will include it here for expositive reasons:

(1) Until the first half of the 1960s, philosophical studies of science were dominated by the firm conviction that philosophers should elucidate rules for evaluating scientific knowledge. Paradigmatic examples of this line of thought were Wiener Kreiss' philosophers or Popper and his followers. According to this view, which we can call "the classical view", it was the philosopher the one who should finally decide what was good science and what was not. Certainly, there were those who opposed this conviction. They were important but isolated voices, not listened to in their time, though they were afterwards.
(2) This period, in which philosophical studies of science were oriented towards producing rules for evaluating scientific knowledge, met its abrupt end with the rapid expansion of the ideas and the style contained in *The Structure of Scientific Revolutions.* Certainly, other important factors influenced the dawn of the classical view, in addition to this extraordinary book whose 40th anniversary we celebrate this year. Nevertheless, as I have already pointed out, in this article we should move with a certain degree of generality. Kuhn's work demonstrated to many that history of science had enough substance to exist in itself and, in no way, it should be condemned to be an appendix of methodology books. As a consequence of the changes introduced by Kuhn and others, the methodological or classic view was banned to the defendand's seat, where it was compelled to answer questions such as, From where do we obtain the rules for producing good science? Which is the historical support for those rules? or Why must science and scientists follow a certain methodology?

The turn from the predominance of the classical to the historical approach within philosophical studies of science constitutes the period of dispute of Popper – Kuhn – Lakatos, which lasted more than fifteen years. An important feature of this dispute is that the examples used by the actors came, mostly, from mathematical physics, chemistry or astronomy. Theories belonging to life sciences, among them, those focusing on the organic evolution, genetics or biological taxonomy, were absent and were not used for illustrating the views in conflict. The same can be said of those theories involved in medical sciences, such as the theory of health or that of disease. All these were also ignored in that famous dispute.

It could be objected that biology – evolutionary biology to be more accurate – was present in that dispute, e.g., in Kuhn, providing at least an epistemological pattern for understanding paradigm or theory change. But this objection does not recognize that one thing is the evolutionary theory as an object for capturing the main features of scientific theories and a very different one is the use of it as an analogy for representing knowledge evolution.

Now, let me make some comments on the reasons for the absence of life sciences in the period previous to the dispute and during the dispute itself (Hershers 1988).

II.

It is well-known that Popper's methodology and that of the logical empiricism were developed in a philosophical atmosphere influenced by the scientific revolutions that had taken place in the field of m athematical physics by the late 19th century and the early 20th century. This is the reason why Popper's methodology has frequently been considered an attempt to understand those transcendental scientific achievements. Metaphorically speaking, those methodologies were born in the shadow of great changes in the world of mathematical physics. Therefore, it is not a surprise that in these methodologies examples coming from physics, chemistry or astronomy were totally dominant.

On their side, Kuhn, Lakatos, and many others, who took part in that famous dispute, created their own views taking into account the perspectives of classical methodologies and, in an explicit or implicit dialog with them. Therefore they centered their critical analyses on the same or very similar examples. That is, the examples that the classical view had selected from the history of science. Naturally, this constitutes one of the most important reasons why their analyses also remained tied to the world of mathematical physics.

There is another factor that helps to understand the predominance of physics in philosophy of science. At least until the 1970s there was no theory of biological evolution that could be used as a subject for epistemological reflection and discussion. Perhaps, someone might say that the evolutionary theory is only a part of life sciences and that there were other areas of biology that, at that time, had received the attention of philosophers, such as the works of J. H. Woodger on blood circulation and classic genetics (Woodger 1952). Nevertheless, it is necessary to realize that these and other similar contributions were scarce and did not illustrate the main questions included in the Popper – Kuhn – Lakatos' dispute. Those questions were, for example, the theoretical load, the incommensurability or the relationships between history and philosophy of science.

It could also be objected that the synthetic theory of evolution was already complete, in its basic aspects, by the 1940s. That is, by the time the classical view was consolidated. This could be so because by that time the work of the so-called *Peres Foundateurs* had already been published. I mean the books of Dobzhansky (1937), Huxley (1942), Simpson (1944) and Mayr (1942), which – according to the common opinion- contain the first version of the synthetic theory of evolution. However, this opinion is partially incorrect because it confuses what was an important movement in the unification of biological disciplines – such as systematics, population genetics or paleontology- around some general tenets, with the effective proclamation of the theory. In this connection, we should take into account that many pieces were still necessary in order to complete the neoDarwinian prototheory. Among them, we could mention the identification and nature of the substance of heredity, the relationship between this substance and the phenotypical characteristics or the genetic code. We should remember that the Central Dogma, which established the direction of the flow of biological information, was introduced by Crick (1958) by the end of 1958 and the questions concerning genetic regulation began to be understood from Jacob and Monod (1961)'s research by the early 1960s. I am not saying that the Central Dogma or the hypothesis on genetic regulation are the property of the neoDarwinian theory of evolution. Simply, that these discoveries were necessary for its formulation as a complete theory, although they were unknown by the 1940s, a period in which many mistakenly date the appearance of the synthetic theory of evolution in the academic world.

Regarding this paper, the following issue is of great importance for us. The late and slow appearance of the evolutionary theory on the academic scene helps to explain, in part, the absence of life sciences in the philosophical studies of sciences by the time classical methodologies dominated this field. Logically, this late appearance also explains why philosophy of biology is a recent discipline. Fundamental works on this new branch, such as those of Ruse (1973) and Hull (1974). By that time, contributions in the field of philosophy of biology were rather scarce. However, ever since we have witnessed an extraordinary increase in this type of contributions in philosophical and scientific journals. In the period previous to the publication of those essential books some important contributions were made. In that sense, finalism or teleological explanation are subjects that have always fueled philosophical analysis and discussion. Nevertheless, the dominant subjects in classical methodologies, and also in the dispute Popper – Kuhn – Lakatos, were associated with the development and change of theories. For this reason, they could not be well illustrated with the evolutionary theory because there was no evolutionary theory at all, despite of the fact that scientists spoke of "synthetic theory" or "neoDarwinian theory".

When philosophy of biology suddenly invaded the academic world, following the rise of the synthetic theory, its massive irruption was characterized by a strong methodological modus operandi. That is, a modus operandi mostly characterized by analyses and evaluations which rested on logical empiricism, hypothetical deductivism or other variants of these methodologies. This should be seen as a surprising fact, if we take into account that it occurred just when the ideas of Kuhn and of other opponents to the classical view were expanding; when there was a generalized suspicion about logical treatments, and when some metatheories, as the Received View, had fallen into great discredit. Michel Ruse's cited book and Mary Williams' contributions (Williams 1970) on the structure of evolutionary theory, are paradigmatic in this respect. These authors try to reconstruct evolutionary theory according to the postulates of the *covering law model*, that, by that time, could be considered abandoned. In other words, these contributions of methodological nature appeared just when the poisoned question "are you prescribing or describing?" embarrassed any who worked from pre-kuhnean perspectives.

Before going on, it is necessary to explain what we mean by the expression "methodological modus operandi" as applied to philosophy of biology. With this expression, we want to indicate that the purpose with which philosophers approached biology by the 1970s, was to reconstruct its theories – evolutionary theory and Mendelian genetics – according to classical methodologies. But they did something more than working in an old fashioned way, trying to reconstruct biological theories. Many of them were involved in the construction of the evolutionary theory itself. This is a historical fact that will play a fundamental role in our conclusions. In order to be absolutely clear about what I want to say with the expression "philosophers making the evolutionary theory", I will mention the following example. In an article published in *Science*, Professor Michel Scriven said: "... we have to abandon Darwin's belief that 'every detail of structure in every living creature' has either current or ancestral utility ..." (Scriven 1959). Here we have the testimony of a philosopher telling biologists what they should suppress in their central theory, in the same way Popper's did with the axiom "the survival of the fittest", as well as many other philosophers.

Here we should emphasize the distinction between reconstruction and construction theory. We know well what a reconstruction theory is because books are full of such philosophical achievements. Briefly speaking, a reconstruction is to work – according to a specific methodology – for polishing the laws of a stated theory or for clarifying the relationships of its central tenets with less specific laws. The crucial point is that any reconstruction supposes that the theory has been in some way effectively enunciated by the scientific community and, therefore, can be considered virtually complete. Nevertheless, a reconstruction work was not the kind of task that philosophers of biology could have done with the synthetic theory because, even for the 1970s, this was rather a prototheory, something *in fieri*, that is, it was under construction.

Let's briefly consider three reasons that support our assertion that by the 1960s the evolutionary theory, known as synthetic theory, was still *in fieri*: (a) in first place, by that time there was a wide discussion on the meaning and role of the axiom "the survival of the fittest". For some, this was no more than a tautology and, for others, it was a crucial principle; (b) there was a wide discussion on the role of hazard. For some, it was limited to the lack of relationship between genetic mutations and favorable organic effects and, for others, it was the clue for speciation processes; and (c) there was also a wide discussion about the level at which natural selection works and about the unity of selection of evolutionary process. Is it the individual? the species? or the genome? In this connection, I remember an article published by professor Hull in the *Annual Review of Ecology and Systematics* (Hull 1980), in which he tried to clarify and solve this last problem.

I do not think that we will be able to find in physical or chemical journals so many and frequent philosophical interventions for amending or clarifying theories. Going back to examples (a), (b), and (c), what these, as well as many others, try to prove is that there were important questions not defined in the so-called "synthetic theory" and, therefore, that there was not a complete theory.

More evidence on the situation of "theory *in fieri*" coming across in the Modern Synthesis by the 1960s, is the great disagreement and confusion brought into the biological community by the discoveries of King and Jukes (1969), and after them by Kimura (1979) on the lack of connection that exists between molecular and organic evolution; or those discoveries by Gould and Eldredge (1977) on the discontinuity of the fossil record. For many biologists and philosophers, these empirical findings were totally consistent with the evolutionary theory, whereas for many others, they were inconsistent with it. This critical situation in the scientific community was very different from that depicted by Kuhn in his celebrated book. It is so because there was no evidence threatening a theory, rather there was evidence that divided the community in halves. If we assume that the theory was, in fact, a prototheory or an incomplete theory, then the crisis in the biological community is totally understandable.

Summarizing, in a way different from the one used by the philosophers who approached mathematical physics by the 1930s, those who approached evolutionary theory by the 1960s found a doctrine with many central questions which were unsolved and surrounded by strong discussions. However, this fact, far from being an adverse circumstance, constituted a great opportunity for them. They had the opportunity to take part in the construction of the theory itself and, therefore, to take part in the process of living science.

There was another outstanding situation added to those incursions of philosophers in the biological field. In their methodological efforts for structuring or amending the evolutionary theory, philosophers were not alone. Biologists – certainly more pressed than philosophers for finally establishing the synthetic theory – worked as classical philosophers of science as well. We should remember that at the beginning of the 80s Stebbins and Ayala (1980) on the one hand and Gould (1982), on the other, published two essential articles in *Science* about the nature and architecture of the synthetic theory. Whereas the first two represented the Darwinian orthodoxy, the second represented a critical position. The point for us is that they reached contradictory conclusions on the nature and structure of Modern Synthesis. It is natural for scientists to discuss a theory, but it is not the case when the discussion runs on philosophical ways or involves methodological topics. In the same vein that Stebbins, Ayala and Gould, many other biologists have contributed to the discussion on the evolutionary theory working as classical philosophers of science. These incursions of biologists in their own field, but with epistemological tools, have not been limited to the evolutionary theory. They have also tried to clarify, for example, the kind of discipline that biology is, the nature of its teleological explanations or its relationship with chemistry and physics. That is, the great debate about the character, whether historical or functional, of biology. This was an epistemological debate which received substantial contributions from G. G. Simpson, E. Mayr or Eldredge (1993).

The last case I want to mention in order to show the epistemological way of working of many biologists, can be found in the recent discussions on Cairn's experimental achievements (Cairn 1988). Following his discoveries, it would be seem that are bacteria which function according to Lamarckian mechanisms. The point for us is that this discussion among biologists, on whether there are such mechanisms or not, took place in a methodological fashion and, mostly, in philosophical terms (McPhee 1993). In other words, questions concerning auxiliary hypotheses, *ad-hoc* hypotheses, *ceteris paribus* clauses, competing hypotheses or inference to the best explanations, were the guests at this *biological* debate.

III.

I think, that a first conclusion that we should reach from all that we said up to now, is the following: when a theory is being constructed or in a period of crisis, it is expected that philosophers assume a methodological attitude. The same kind of situations are suitable for an epistemological approach by scientists themselves. As philosophers, these scientists will work from a logical and analytical perspective because they are primarily interested, not in historical issues, but in determining whether their theories, hypotheses or so on, are false, true, probable, etc. Then, Kuhn's famous saying: "when scientists begin to speak philosophically, there is something wrong with the theory", comes to my mind.

It should be noticed that this natural preference of scientists for epistemological work reinforces, in turn, the methodological analyses of philosophers concerned with the same problems. This is so because the methodological perspective is a necessary condition for the dialog between scientists and philosophers, a dialog that philosophers have always whished. In this way, a virtuous circle is generated, a circle in which methodological work in life sciences are required more and more. Under these circumstances, philosophers of science have two very peculiar opportunities: (a) to collaborate in the construction of the science itself, instead of being limited to its reconstruction, and (b) to establishing a dialog with the living science. Making a historical comparison, we can say that this participation of philosophers and scientists alike in epistemological questions of life sciences is similar to the role they played in the 17th century. At that time, Leibniz's time, philosophers and physicists alike discussed the nature and laws of forces and movements.

When we reach this point a natural q uestion arises: do we find this situation of philosophers and scientists playing the same game in other natural disciplines? Are theories under construction or in crisis exclusive of life sciences? Certainly, not. I think that there are additional reasons that justify the usual methodological approach that we observe in philosophical studies of life sciences. An additional reason, and a very important one, is the following: there are strong social demands for clarifying or solving biological controversies that are in close relation to education, politics, ethics or ideology. In some cases, these social demands for scientific answers exert a true pressure on the academic world. They are scientific disputes as well, but different from the ones that take place in other natural sciences, since the academic answers to these questions have a direct influence on public opinion and society. Let's see now some cases which demonstrate the social relevance of some discussions in the realm of life sciences:

(a) We shall begin by remembering the importance that the controversy between evolutionists and creationists has in many countries and regions. I mean, creationism taken in all its variants, specially the so-called "scientific creationism". The social importance of this biological debate is easy to appreciate. To a great extent, the conclusions reached in these debates determine the presence or absence of evolutionary contents in basic education, that is, in education for children and adolescents. We should see what happens to the synthetic theory of evolution in this respect. Whereas many scientists consider that it is a confirmed or highly confirmed theory, others think that it is partially or totally false. Society is well acquainted with this situation of theory in crisis –a fact of which creationists have taken advantage because much criticism came from eminent researchers. Among them, Devillers (1985), Grassé (1978), Lovtrup (1987) or Kaufmann (1993). It is frequent to see comments about these scientific disputes in well known newspapers and journals. In many cases, their conclusions are used in social discussions about whether we should introduce the evolutionary theory in biology, history or anthropology in basic education.

(b) The sociobiology (Dawkins 1976), a sub-theory which partially depends on the synthetic theory of evolution, constitutes our second example about the strong influence that many biological discussions have on society. As you already know, the main thesis of sociobiology is that organisms are designed for optimizing the presence of their own genes on earth. In turn, this hypothesis would apparently imply that the behavior of superior organisms is naturally selfish. In fact, it is not the case that Dawkins himself considers that selfish behavior is something good or desirable. Far from vindicating selfish behaviour, he warns us that "because we are naturally selfish, we must be educated in altruism". Going back to our subject, it is easy to see that if Dawkin's hypothesis is accepted, then some generalized attitudes or ideologies, such as racism, could receive significant support. Therefore, in this case society also looks anxiously at philosophers and scientists alike asking for their verdict about the veritative state of Dawkins' theory. Is it a well-confirmed hypothesis or just wild fantasy?

(c) When we move from biology to medical sciences and anthropology, it is easier to see the social importance of many issues. This fact encourages a classical methodological approach because society and public opinion ask for decisions on competing hypotheses and theories. Let's think for a moment in the pair *genes and education* in relation to intellectual quotient or deviate behavior.

Any of these topics and other related ones, fuel the media weekly and get a great social impact. Here, again, society's glance is directed to epistemologists and scientists in search of an answer for the actions to be taken. In the same way as the question of selfish behavior, the importance of determining the influence of genes on human behavior and capabilities for the different activities is a point of crucial importance. Books like *The Bell Curve* (Herrnstein and Murray 1993) excuse me from insisting on the importance of epistemological work in the realm of life sciences. To those who still think that all these are questions for scientists exclusively, I would advise them to take a look at the article "Race, Gène et QI" published by *La Recherche* a few years ago. Naturally, the author is a philosopher (Block 1997).

(d) Medical sciences constitute my last example. In this case, we are faced with the most conspicuous example of all, because the key notions of medical sciences are not only permanently discussed by philosophers, but are also, in some way, the notions developed by themselves. In this sense, the family of concepts related to medical sciences, such as health, disease, illness, disability, handicap, unhealth, and others, are the products of philosophical activity, as demonstrated by the pioneer contribution of Boorse (1977), an epistemologist who worked in the 1970s under the principles of logical empiricism. However, the work of philosophers of science in medicine, far from being limited to the analysis of those notions, is related to decisive issues of the highest importance for humankind that are related to our daily life. Among these, I could mention the definition of human life and the precise determination of its beginning and end. Both issues are linked to well known problems such as abortion, experimentation with human embryos, organ transplantation, etc. Ethical questions aside, there are controversial opinions regarding these issues which share the same empirical data but reach opposite conclusions. Therefore, there are methodological issues involved.

I would like to finish this part of this paper with the following consideration. We know well that social demands for determining whether a certain hypothesis is finally true or false, or whether a certain definition of human life is adequate or not, are frequently naïve or badly structured. Many times, society has urgency for answers that might never be satisfied. However, our point is different. This state of affairs influences philosophers and scientists alike to work with methodological classical criteria. One of the outstanding results of all this is to see philosophers doing science and life scientists doing philosophy. In my next and last part, we will be able to appreciate the importance of this fact with regard to the relations between philosophy of science and history of science.

IV. The Puzzle

The great methodological activity that can be found in the field of life sciences is a palpable reality that can be easily corroborated. For this, it is enough to take a look at the bibliography and authors of many articles published in monthly leading life science journals, such as *Evolution, Systematic Zoology* or *The Annual Review of Ecology and Systematics*. There, it is common to see names such as Popper, Nagel, Hempel, Kuhn, Koyré, and others recurrently mentioned. Many times, the authors are also philosophers. This is an outstanding academic fact in natural science publications. I think that we will hardly be able to see so much philosophy and philosophers in journals of physics or chemistry. Certainly, this activity could be used against those who still think that philosophy of science normatively oriented is old fashioned or even obsolete. In other words, classical methodology is alive in the fields of life sciences.

Now, it is time to pose a great question, a crucial and eternal question for the philosophy of science. Where do the rules according to which such activity would be possible come from? Where can they be obtained? In other words, where do the foundations of science methodology lie? It is obvious that the fact that biologists themselves and philosophers of biology currently apply methodologies such as the hypothetical deductive methodology, the covering law model, Lakatosian research programs, the Sneed-Stegmüller structuralist view or others does not constitute in itself a justification for any of these schools. I mean, this fact is not proof that this or that methodology is the true one; however, it makes its presence evident in living science only.

Of course, we are not going to answer this paramount question about the foundations of methodology of science. Instead, I will make a critical comment to an answer frequently heard today in some epistemological circles.

As we said at the beginning of this article, after the spreading of Kunhn's ideas and style, a sharp dichotomy was introduced at the bottom of philosophical studies of sciences or, if you wish, into the philosophical community. This dichotomy is well known by all of us: "prescription versus description", being the latter historical description mostly. With his paraphrasing of Kant's famous dictum "Philosophy of science without history of science is empty; history of science without philosophy of science is blind", Lakatos tried to solve this dichotomy and, at the same time, conciliate both positions. But I think that, despite his intentions, what he did, in fact, was to make evident that we are in the presence of a vicious circle from which we cannot easily get out. Shortly speaking, this circle could be described in the following way: the quarry of philosophy of science is history of science; but, in turn, history of science can only succeed with methodological criteria at hand, which indicate what we should send to internal and what to external history.

I would like to analyze a recent attempt (Diez and Moulines 1999) to solve the problem posed by the vicious circle. Briefly speaking, this attempt could be organized and summarized in the following way:

(1) The dichotomy *prescription versus description* has many possible interpretations. It is so, because there are many ways of interpreting its terms. That is, what do "prescription" or "description" mean.
(2) The term "prescription" can be understood at least in two different ways: in the sense of organizing or ruling something from the outside. It is in this way that we can create, for example, card games or structure a set for certain purposes. Naturally, if epistemological criteria for science evaluation come from outside disciplines, then the classical objection will meet its target because what comes from the outside can always be accused of arbitrariness.
(3) However, exterior sources are not necessarily the only place from where rules for prescriptions might come from. The American philosopher David Lewis gives us the clue for solving the question (Lewis 1969). According to the proposal that we are examining, using Lewis' notion of "convention" it would be possible to understand how prescriptions can be compatible with the evaluative tasks of philosophy of science and with the autonomy of science, at the same time. By the expression "the autonomy of science" I mean that science should follow only its own rules and criteria.
(4) Lewis says that in human communities there are generalized practices which prescribe ways of action that are generated in the same community. He calls them "conventions". A community might well represent and explicitly express such practices, and even apply them correctly, but it is not necessary. Ordinary language is a typical example of convention. Any human community has a language that works according to rules that not necessarily someone has explicitly established or encoded in a grammar.
(5) Another example, one that it is specially interesting for us, would be the logic and methods genuinely used by the scientific community for supporting or rejecting hypotheses, deciding between competing theories, accepting or interpreting facts, etc. Therefore, a major mission of philosophy of science would be to abstract and polish such conventions from the scientific community and, after that, to express them in a precise and explicit way in order to use them for the evaluation of scientific products.

I realize, as well as you do, that against this argument and other analogous ones, many serious objections can be quickly raised. In first place, scientific communities may have adopted and, in fact they did, different methodological patterns during the different historical periods and in the different regions of the world. The title of Ampère's book "*Mathematical Theory of Electrodynamic Phenomena Unequivocally Deduced from Experiment*" shows the acceptance of a methodology that many of us would reject. Second, if we distinguish between the methodology that is actually used and the one that is only recited, then, it would always be necessary to carry out a hard hermeneutic work in order to determine the true conventions of the scientific community, if such true conventions exist. We could continue accumulating more and more objections against the strategy based on Lewis' notion of convention. However, the interesting thing for us is to see how this strategy looks like under the light of the facts that we have detected in the realm of life sciences. Under this light, the result is also negative for the stratagem based on Lewis' notion of convention. It is easy to see the reason why. If life sciences, specially, evolutionary theory, systematics and ecology, are strongly contaminated by epistemological elements, as I have extensively demonstrated, then the task to get from these sciences the rules that create the foundations of philosophy of science is damned to fail from the very beginning. Why? Because behind the argument based on Lewis' notion and other similar strategies is the image of the methodologist, a man who goes to the river of science seeking for pure, uncontaminated water. But, following with the analogy, what the methodologist should know is that somewhere, river up there can be secret connections injecting impure liquids into the river. Therefore, what the man finally collects is just the kind of water he wants to leave aside. Leaving the analogy and directly speaking, there are so many epistemological elements in life sciences – introduced by scientists themselves year after year – that with the strategy of abstraction, we will obtain our well known models of methodology, at least in many cases.

V. A Virtuous Circle

However, the impossibility or, better said, the unlikelihood of finding uncontaminated rules or *conventions* in life sciences for building an unbiased methodology is far from being the only result. This is just the negative side of the question. It is possible that the notions and rules brought into the sciences from philosophy – and from philosophy of science in particular – shall be made more accurate, refined, and enriched with new shades by the use of scientists themselves. The results are absolutely positive from any point of view. Next, we present an example to illustrate our thesis.

Finalism is a philosophical concept that was always present in biological thought, at least from the time of Aristotle. This crucial notion was introduced into the discourse on organisms due to the adaptive nature of their peculiar processes and structures. It is evident that the human hand was designed *for* grasping, his eye *for* seeing, and his kidneys *for* regulating blood composition. However, how can the final states influence or determine structures and processes? For centuries, the answer came from theology: God designed living things for succeeding and surviving in the environment and the circumstances in which they live.

We know well that during modern times mechanicist and anti-metaphysicist thought expelled from science not only God, but the *final causes* as well. However, if the latter meant a significant progress for Physics, for Biology it created a new problem: if living structures and processes reveal an evident finalism, then there must be some kind of final causes. Therefore, in one way or another, they should be reintroduced into life sciences.

A great merit of neodarwinian scientists Mayr (1965) and Ayala (1970), among others, has been to demonstrate that finalism or teleology can be consistent with the mechanicist character of Darwin's theory. For this they had to distinguish between internal and external (or natural and artificial) teleology. According to them, the teleological nature of structures and processes of living things can be explained by showing their contribution to reproductive fitness. Even more, it is possible to distinguish between finalism, *qua* heurist principle, something completely acceptable for biologists and *qua* operative cause.

Epistemologists, who want to see how life sciences use "finalism" – a notion that biologists borrowed once again from Philosophy – will find that this concept has received a special refinement in the hands of biologists. In this way, a virtuous circle is closed: life sciences can give back to methodology enhanced versions of its own rules and notions.

With regard to the specific aims of this article, I think that we got another highly positive result. The facts and arguments that I have exposed and analyzed clearly indicate the possibility and also the need of designing a research program with the aim of determining the extent to which philosophy of sciences directly influenced and still influences the architecture of theories, hypotheses, and doctrines of life sciences. As it is evident, such a program would require the joint work of philosophy of science and history of science.

A classical topic among epistemologists for many years has been the determination of the extent to which 20th century philosophy of science became a philosophy of physics. Or, even better, became a philosophy for physics and physicists, specially under the influence of Carnap, Popper, Poincaré or Duhem. Now and regarding life sciences, we can inquire to what extent life science theories and doctrines owed their peculiar architectures, limits, and horizons to the philosophical methodologies that their creators borrowed.

Note

[1] A first version of this paper was presented at the meeting of the *Union International d'Histoire et Philosophie des Sciences*, Paris, 2002. This final version was accomplished during my stay at the *Université de Lille*. I would like to thank Professors A. Fagot-Largeault and D. Andler and Professors A. Laks and B. Joly for their assistance, and specially Professor Shahid Rahman for his useful criticism and suggestions.

References

Ayala, F.: 1970, 'Teleological Explanations in Evolutionary Biology', *Philosophy of Science*, March.
Boorse, Chr.: 1977, 'Health as a Theoretical Concept', *Philosophy of Science*, **44**, 542–573.
Block, Ned.: 1997, 'Race, Gènes et IQ', *La Recherche* **294**, 50–58.
Cairn et al.: 1988, 'The Origin of Mutants', *Nature* **335**, 142–145.
Crick, F.: 1958, 'On Protein Synthesis', *Symp. Soc. Exp. Biol.* **12**, 138–161.

Dawkins, R.: 1976, *The Selfish Gene*, Oxford University Press.
Devillers, C.: 1985, 'Quelques remises en cause de la théoerie synthétique de l'Evolution', *Ann. Biol.* **24**, 153–177.
Diez, J. and U. C. Moulines: 1999, *Fundamentos de Filosofia de la Ciencia*, ed., Barcelona, Ariel, pp. 20–25.
Dozhansky, T.: 1937, *Genetics and the Origins of Species*, Columbia University Press.
Eldredge, N.: 1993, 'History, Function, and Evolutionary Biology', *Evolutionary Biology* **27**, 33–50.
Gould S. and N. Eldredge: 1977, 'Punctuated Equilibria: The Tempo and Mode of Evolution Reconsidered', *Paleobiology* **3**, 115–151.
Gould, S. G.: 1982, 'Darwinism and the Expansion of Evolutionary Theory', *Science* **216**, 380–387.
Grassé, P. P.: 1978, *Biologie moléculaire, mutagènese et evolution*, Paris, Masson.
Herrnstein, R. J. and Ch. Murray: 1993, *The Bell Curve*, New York, The Free Press.
Hersher, L.: 1988, 'On the Absence of Revolutions in Biology', *Perspectives in Biology and Medicine* **31**(3), 318–323.
Hull, D.: 1974, *Philosophy of Biological Science*, Prentice Hall.
Hull, D.: 1980, 'Individuality and Selection', *Annual Reviw of Ecology and Systematics* **11**, 311–332.
Huxley, J.: 1942, *Evolution, the Modern Synthesis*, Allen & Unwin.
Jacob, F. and J. Monod: 1961, 'Genetic Regulatory Mechanisms in the Synthesis of Proteins', *Journal of Molecular Biology* **3**, 318–359.
Kaufmann, S.: 1993, *The Origins of the Order: Self-organization and Selection in Evolution*, Oxford University Press.
Kimura, M.: 1979, 'The Neutral Theory of Molecular Evolution', *Scientific American* **241**, 94–104.
King, J. L. and Jukes, T. H.: 1969, 'Non Darwinian Evolution', *Science* **164**, 788–798.
Lewis, D.: 1969, *Conventions*, Ch. 5, Harvard University Press.
Lovtrup, S.: 1987, *Darwinism: the Refutation of a Myth*, Croom Helm.
Mayr, E.: 1942, *Systematics and the Origins of Species*, Columbia University Press.
Mayr, E.: 1965, 'Cause and Effect in Biology', in *Cause and Effect*, New York, Free Press, pp. 33–50.
McPhee, D.: 1993, 'Directed Evolution Reconsidered', *American Scientist* **81**, 554–561.
Ruse, M.: 1973, *The Philosophy of Biology*, Hutchinson.
Scriven, M.: 1959, 'Explanation and Prediction in Evolutionary Theory', *Science*, **130**, 477–482.
Simpson, G. G.: 1944, *Tempo and Mode in Evolution*, Columbia University Press.
Stebbins L. and F. Ayala: 1980, 'Is a New Evolutionary Synthesis Necessary?', *Science* **213**, 967–971.
Williams, M.: 1970, 'Deducing the Consequences of Evolution: A Mathematical Model', *Journal of Theoretical Biology* **29**, 343–385.
Woodger, J. H.: 1952, *Biology and Language*, Cambridge University Press.

THE CREATIVE GROWTH OF MATHEMATICS

JEAN PAUL VAN BENDEGEM
Center for Logic and Philosophy of Science, Vrije Universiteit Brussel, Belgium,
E-mail: `jpvbende@vub.ac.be`

Abstract. It is a trivial remark that to discuss the philosophy of any topic, one must have at least a good understanding of the topic itself in order to raise philosophical problems about it. However, if the topic happens to be mathematics, this does not seem to be the case. Philosophers are not particularly interested in mathematical practice itself. Often they prefer the reduced and all too simplified picture of mathematicians as "theorem proving machines". In this paper I present a rough sketch on the macro-, meso- and microlevel of a theory of mathematical practice, that does more justice to the amazing and unexpected complexities of the mathematicians' daily universe.

There is quite literally a world of difference between discovery in the (natural) sciences and discovery in mathematics. The former expression suggests a realist interpretation though certainly not a full-scale realism. Even the most moderate realist possible is willing to talk about discovery in some sense. However, mathematics is a different kettle of fish. If one dares to use the term 'discovery', then, necessarily, in one sense or another, one must be a mathematical realist. But this sort of realism is of a rather peculiar kind, as the world wherein the mathematical objects and/or entities are discovered, happens not to be this world. Therefore, there is a strong ontological claim intrinsically tied up with the word 'discovery'. Unfortunately, replacing the word 'discovery' by a more neutral expression such as 'creative growth' (and not by 'construction', for obvious reasons) does not really solve (or avoid) the fundamental problem, that is the following. It would be nice if one could argue that a particular philosophical position concerning the foundations of mathematics will not affect a description of what mathematicians do and why they do what they do, but such is not the case.

Nevertheless, I do believe it is possible to start from a minimal position. I will therefore in this paper assume, as a philosophical framework, a form of 'mild' constructivism, i.e., the position that mathematical objects, entities, including proofs, are human products and should, in first order, be analyzed as such. I consider this view to be minimal because it does not exclude forms of platonism or some other strong ontological claims about mathematical objects.

The structure of the paper is quite simple. I begin at the most general level – the mathematical community as a whole – and I go slowly down to the level of the working mathematician who is (among other things) trying to find a specific proof

S. Rahman et al. (eds.), Logic, Epistemology, and the Unity of Science, pp. 229–255.

for a particular theorem. The 'afterthought' returns briefly to the philosophical question raised in this introduction.

1. Revolutions in Mathematics?

There is a quite intriguing problem with the attempts to describe the development of mathematics at the large-scale level. In clear contrast with what happened in philosophy of science, one cannot help but to notice that in philosophy of mathematics, there is hardly any agreement. Often, Michael Crowe's *Ten 'laws' concerning patterns of change in the history of mathematics* is considered to be the starting point of the discussion about revolutions in mathematics. Rather surprisingly, the tenth and last 'law' states:

'Revolutions never occur in mathematics'. His basic argument is that 'a necessary characteristic of a revolution is that some previously existing entity (be it king, constitution, or theory) must be overthrown and irrevocably discarded.' (Crowe, 1992, 19).

At the same time, however, Joseph Dauben is a firm defender of the occurrence of revolutions in mathematics (see, e.g., Dauben 1992: Chapters 4 and 5):

'Discovery of incommensurable magnitudes and the eventual creation of irrational numbers, the imaginary numbers, the calculus, non-Euclidean geometry, transfinite numbers, the paradoxes of set theory, even Gödel's incompleteness proof, are all revolutionary – they have all changed the content of mathematics and the ways in which mathematics is regarded. They have each done more than simply add to mathematics – they have each transformed it. In each case the old mathematics is no longer what it seemed to be, perhaps no longer even of much interest when compared with the new and revolutionary ideas that supplant it' (Dauben 1992, 64).

Philosophers of science themselves – such as Thomas Kuhn, to quote the most obvious one – if they do talk about mathematics, their main purpose is to make clear that the natural sciences and mathematics should not be confused (see, e.g., Kuhn 1977). Hence they often end up defending the non-revolutionary nature of mathematics.

To further complicate matters, as I said in the beginning, ontological issues are unavoidable. To give but one example: for Crowe, some sort of 'revolution' remains possible, viz. 'revolutions may occur in mathematical nomenclature, symbolism, metamathematics (e.g. the metaphysics of mathematics), methodology (e.g., standards of rigour), and perhaps even in the historiography of mathematics.' (Crowe 1992, 19). In other words, content-wise, there are no revolutions, but anything-else-but-content-wise, you can have as many as you want. I will not explore this fascinating theme any further in this paper, but rather turn to the common elements that all these authors seem to share, viz. the fact that mathematics does possess a large-scale structure.

2. The Large-Scale Structure of Mathematics (If Any)

The last sentence of the preceding paragraph is close to being tautological. For what would have to be the meaning of the statement that mathematics has no large-scale structure? The crucial feature that interests me – and I assume that the authors mentioned above, whatever their views, share this interest – is that this large-scale structure 'affects' the daily practice of mathematicians by imposing constraints on the kind of (more specific) mathematics that is being done. Examples of such constraints are:

(a) What are the relevant mathematical research themes, and research problems to look at?
(b) How are the results already obtained to be systematized?
(c) What are the global aims of a particular area of mathematics?
(d) What is to count as a mathematical proof? What are the standards of rigour, say, for mathematical proof?
(e) How is the history of (a part of) mathematics to be told?

Seen from this perspective, it becomes possible to identify major periods in the development of mathematics. A nice example of such an attempt to identify such periods is to be found in the work of Teun Koetsier. Let me briefly summarize his approach. Although he distinguishes three levels – the micro-level where the mathematician mainly spends his or her time proving theorems, the intermediate level where research projects are formulated, and the macro-level that identifies a particular period – I will focus only on the macro-level. Koetsier, inspired by, though certainly not a blind follower of Imre Lakatos, speaks about *research traditions*: 'A mathematical research tradition is a group research activity, historically identifiable (in a certain period), characterized by common general assumptions (in the form of e.g., definitions and axioms) about the entities that are being studied in a particular fundamental mathematical domain, and it involves assumptions about the appropriate methods to prove properties of those entities' (Koetsier 1991, 151).

An example may help to clarify his approach. Within Greek mathematics, Koetsier distinguishes two traditions which he calls the *Demonstrative Tradition* (DT) and the *Euclidian Tradition* (ET) in chronological order. A major point of difference between DT and ET is the fact that ET introduces the notion of proof as standard method for establishing mathematical truths. Koetsier claims that the method of proof of DT is non-deductive. It is based on a form of 'Anschauung'. The best example to illustrate this is the 'proof' of $(n + 1)^2 = n^2 + 2.n + 1$, in Pythagorean fashion. Thus, to show that $4^2 = (3+1)^2 = 3^2+2.3+1$, it is sufficient to look at these two drawings:

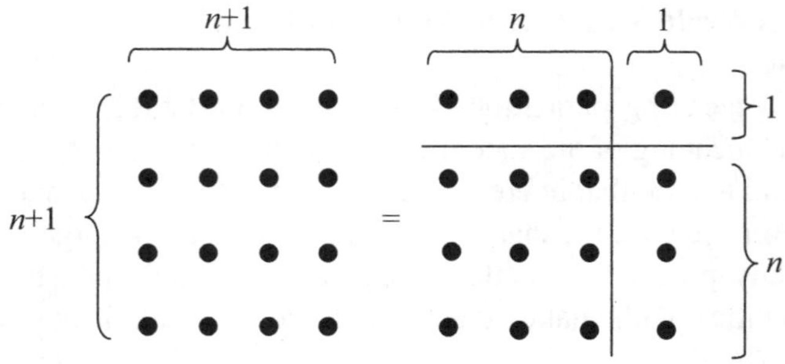

Of course, if this is to count as a convincing method, we must assume that a particular case can be 'seen' as an arbitrary case. That is, I am supposed not only to grasp this figure (or rather its meaning) but also all other cases similar to it. Granted that sense can be made of 'proof by looking',[1] then it is obvious that the transition from DT to ET is a major one indeed. It should also be obvious that it is a progressive move.

As one might expect, not every philosopher and/or historian of mathematics agrees with the picture put forward by Teun Koetsier. I refer the reader to the work of Eduard Glas (especially, 1991a and 1991b) for a critique of Koetsier's approach. To be sure, neither Koetsier nor Glas are the last word on the subject. An approach along the lines of Philip Kitcher (see his 1983) is different from their views and, in addition, it is not straightforward to situate Kitcher's ideas within the broader field of evolutionary and/or naturalist epistemology, to quote but one of the many approaches in the field of epistemology (compare, e.g., with Rav, 1993).

Or perhaps, all of these approaches are fundamentally mistaken as they are looking in the wrong direction. If one is talking about structures, should one not therefore take the idea of structure seriously: in other words, a structuralist approach is what is needed for such a description. To be a bit more concrete, the Bourbaki programme could then be viewed as such a proposal, whether idealist or realist being a matter of further discussion (see, e.g., Corry 1992). Or, for that matter, an approach such as the one promoted by Roman Duda (see Duda 1997), namely, in terms of tensions and polarities: realism-idealism, finite-infinite, discrete-continuous, approximate-exact, certitude-probability, simplicity-complexity, unity-multiplicity.

To complete and complicate matters, I have not said anything – and will not within the framework of this paper – about the multiple relations between, generally speaking, mathematics and society, and between, more specifically and as an example, mathematics and the gender issue. All that has been said up to now, treats mathematics as an autonomous part of society 'obeying' only its internal 'laws', if such exist. But this can only be part of the story, which, once again, I am not going to complete (see Restivo 1983 and 1992, for more details). Nevertheless, from now on, I will assume that a (kind of) large-scale structure has been 'established' and that a mathematician operates within this framework.

3. The Transition from Large-scale to Micro-practice

Given the large-scale structure of the mathematical enterprise, how does it translate into everyday mathematical practice? It cannot be the case that every mathematician has a full-blown view of the whole of mathematics. It is generally agreed that with Henri Poincaré and David Hilbert the last of the generalists have left us. Thus there has to be an intermediate level. One possible way of viewing this level is sketched by Teun Koetsier. According to his model, on this level, we have *research projects*:

> A research project consists of a number of research goals together with a set of hints as to how one can reach the goals. The project includes a paradigmatic solution of a problem that shows the kind of goals and the effectiveness of the hints with respect to the goals. Large projects may very well encompass subprojects. (Koetsier, 1991, 154)

Within a research project operates, what he calls, the MMRT, the *methodology of mathematical research traditions*. Basically, it involves two elements:

(a) 'A mathematical research project or research tradition progresses heuristically if it produces conjectures (theorem candidates) of weight' (Koetsier 1991, 159), and

(b) 'The preference of a rational mathematical community for a research project or a research tradition is proportional to its expected progress' (ibidem).

It is perhaps unnecessary to repeat a comment made before, but Koetsier's model is just one way of looking at things. No doubt different models are possible, but one way or another, they must incorporate some notion, similar to Koetsier's research project.[2] After all, this is the level where the brilliant and promising mathematics student's supervisor decides what topic is worth the effort. This implies in a very precise way, the possibility to evaluate the possible outcomes and the impact of the research to be undertaken on the mathematical community.

3.1. Some Examples of Research Projects

The proof of the pudding, however, remains in the eating. Are there such examples of research projects to be found in 'real' mathematical life? Fortunately, the answer is without any discussion: yes. Here are some:

1. *The Erlanger Program*

No doubt, one of the most famous examples is the *Erlanger Program*, set up by Felix Klein. Saunders MacLane gives a short to the point description of this program:

> 'In geometry, Felix Klein proposed that the many varieties of space provided by non-Euclidean and other geometries could be classified and hence organized in terms of their groups of symmetries – the full linear group, the orthogonal group, the projective group, and others.' (MacLane 1986, 407)

Following the *Erlanger Program*, in more recent times, is the so-called *Langlands Program*. Basically, the idea is to use infinite dimensional representations of Lie groups as a tool to solve problems in number theory. Stephen Gelbart in an excellent expository paper writes the following:

> '... Langlands' program is a synthesis of several important themes in classical number theory. It is also – and more significantly – a program for future research. This program emerged around 1967 in the form of a series of conjectures, and it has subsequently influenced recent research in number theory in much the same way the conjectures of A. Weil shapes the course of algebraic geometry since 1948.' (Gelbart 1984, 178)

In the same paper, the author emphasizes that '... more than one half of this survey will be devoted to material which is quite well known, though perhaps never before presented purely as a vehicle for introducing Langlands' program.' (Gelbart 1984, 179).

2. *Hilbert's Program*

Equally well known in the mathematical community is the general research project outlined by David Hilbert in his famous Paris speech in 1900, 'Mathematische Probleme' at the occasion of the International Congress of Mathematicians. Hilbert discusses twenty-three problems that effectively did determine to a large extent the mathematical activity in the first half of this century. Some of the most famous problems are:

- *Problem 1.* Cantor's continuum hypothesis, i.e., the question whether or not there are cardinalities between the countable and the cardinality of the reals.
- *Problem 2.* The consistency of arithmetic. No comment needed.
- *Problem 8.* The Riemann hypothesis, i.e., given the function $Z(s) = \Sigma 1/n^s$, where s is a complex number and n goes from 1 to infinity, one has to prove that the non-trivial solutions of $Z(s) = 0$ all have 1/2 as the real part.
- *Problem 10.* The Diophantine problem, i.e., to find a method to decide whether a set of equations such that all coefficients are integers (rationals), has integer (rational) solutions. For more details, see Alexandrov (1971) and Browder (1976).

3. *Finite Simple Groups*

As this example shows, it is not necessary for a research project to start with conjectures. A project can be set up around a problem that has been solved. I may add here that few philosophers of mathematics take into account such cases, which I consider to be extremely relevant. The case I am referring to, is the Classification Theorem for Finite Simple Groups, also labelled the *Enormous Theorem*. The existing proof, some 15.000 pages long, consists of a series of papers, most published, though not all, written by a diverse group of mathematicians over a period of thirty years, writing in different styles, using different kinds of proof methods. Such a 'proof' can hardly be called a proof, as says Ronald Solomon:

> 'The state of the original proof is such that if everyone who worked on it should vanish, it would be very hard for future generations of mathematicians to reconstruct the proof out of the literature.' (Cipra 1996, 89)

Part of the explanation is that the simple groups come in four categories: cyclic (of prime order), alternating, Lie-type (to be split up in sixteen families) and sporadic. The first three bring together an infinite number of simple groups, each with their own problems, proof methods and proof techniques, but, in addition, the sporadic simple groups are quite strange. There are precisely 26 of them, and the largest one has no less than some 10^{53} elements, the so-called *Monster*. It is fair to say that in some cases proof methods were designed to deal with this or that specific case. It then becomes clear what the aims of this research project are, started by Daniel Gorenstein (died in 1992), Richard Lyons and Ronald Solomon:

(i) To make uniform the different proof methods that have been used over the thirty year period. The expectation is that this will generate new proof ideas: 'By straightening out the strands of the original proof, Lyons and Solomon have already been able to stretch them further, proving some of the component theorems in considerably greater generality. They and others working on the second-generation proof have also found new applications of the original proof's techniques (Cipra 1996, 89).

(ii) To reduce the size of the proof to something like 5.000 pages, perhaps even shorter and to publish the proof as a *single* proof. This is also a rather surprising aim: apparently, proofs are not perceived as proofs, but are to be presented as such.

(iii) To eliminate all errors present. No comment needed.

For more details see Gorenstein (1986).

4. *Probability Theory Old and New*

Probability theory in the 'old' style was formulated in terms of functions P, usually from a set of sentences S, defined in a particular language L, to the real interval $[0, 1]$, satisfying certain axioms, such as

$$P(A \text{ or not-}A) = 1,$$
$$P(A \text{ and not-}A) = 0,$$
$$P(A \text{ or } B) = P(A) + P(B) - P(A \text{ and } B), \text{ and so on.}$$

This type of approach worked well in discrete cases, but in the continuous case, there were many problems (unless some geometric or other finitely expressible interpretation was available) to determine probabilities.[3]

A. N. Kolmogorov was the first in 1933 to see the connection with measure theory and the theory of integrals. This led to a reformulation of probability theory in such a way that all the results of measure theory could be translated into probabilities. Thus, any handbook today will start with the definition of a probability space PS, being a triple $\langle S, F, P\rangle$, where:

(i) S is a set (actually nothing more is needed, but occasionally this set is referred to as the *sample space*),

(ii) F is a set of subsets of S, satisfying the conditions: (i) $F \neq \varnothing$, (ii) if $A \in F$, then $S\backslash A \in F$, (iii) if $A_i \in F$, for $i = 1, 2, \ldots, n, \ldots$ then $\cup_i A_i \in F$. In other words, F is a σ-algebra, although in probability terms this is called the *event space*,

(iii) P is a *probability measure* on F, such that: (*)$P(A) \geq 0$, for all $A \in F$, (**)$P(S) = 1$, and (***) if $A_i \in F$, for $i = 1, 2, \ldots, n, \ldots$ and $A_i \cap A_j = \varnothing$, for $i \neq j$, then $P(\cup_i A_i) = \Sigma_i P(A_i)$.

Among other things, this new approach makes it possible to talk about singular distribution functions,[4] apart from the already classically known discrete and continuous distributions.

5. *Category Theory*

A recent research project is the project centered around *category theory*. To a certain extent, this may be viewed as the *Erlangen program* for set theory, as is clearly expressed in the words of Saunders MacLane:

> 'The situation bears some resemblance to that in geometry after the discovery of consistency proofs for non-Euclidean geometry showed that there was not one geometry, but many. This meant that geometries could be formulated with many different systems of axioms, some of which were relevant to higher analysis and some to physics. ... Similarly, the initial idea of a collection leads to substantially different versions of set theory, some of which ... have relevance to other parts of Mathematics, though not yet (?) to Physics.' (MacLane 1986, 385–386).

A category C consists of objects $A, B, C, \ldots$ and arrows $f, g, h, \ldots$ from objects to objects, satisfying the conditions:

(i) for each pair of arrows: if $f : A \to B$ and $g : B \to C$, then $g^\circ f : A \to C$ exists, called the composition of f and g,

(ii) for every object A, there is a function $1_A : A \to A$, the identity arrow,

(iii) composition is associative: for all arrows f, g, and h, if the composition is defined, then $(f^\circ g)^\circ h = f^\circ(g^\circ h)$,

(iv) for every arrow f: $A \to B$, it is the case that $f^\circ 1_A = f = 1_B{}^\circ f$.

The power of category theory is truly impressive. Whatever theorem one manages to prove about categories, is applicable to at least the following cases (see MacLane (1986, 387)):

(i) The category where the objects are sets and the arrows functions from sets to sets,

(ii) The category where the objects are groups and the arrows homomorphisms between groups,

(iii) The category where the objects are vector spaces and the arrows linear transformations,

(iv) The category where the objects are topological spaces and the arrows continuous maps.

It is worth mentioning that together with the concepts of category theory, a new way of proving statements was introduced, sometimes referred to as *diagram chasing*. It is absolutely typical for a handbook on category theory to be overloaded with diagrams such as:

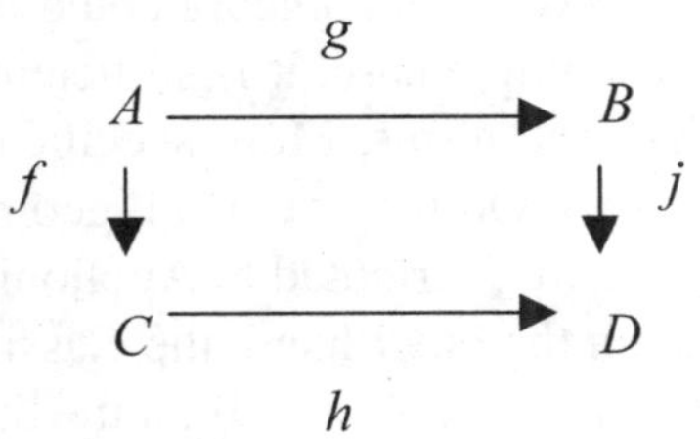

Just by looking at the diagram, one can see that to go from the top left corner to the bottom right corner, can be done in two ways, therefore $h^\circ f = j^\circ g$.

Whether, in terms of Koetsier's model, one is entitled to talk about a research tradition rather than about a research project, is a difficult matter, both for mathematicians and philosophers to decide. For a discussion, see Bell (1994).

This short survey of research projects has no pretence whatsoever of completeness. It is sufficient to consult, e.g., Dieudonné (1987), especially Chapter V ('Nouveaux objets et nouvelles méthodes'), for a wealth of examples.

3.2. The Importance of Proof Methods

All of the above cases have been discussed mainly, though not exclusively, in terms of the problems that had to be solved. But one could equally well look at these examples from the point of view of (novel) proof methods. Very often, the focus of a research project is on the proof methods in the first place and on the problems or conjectures in the second place. Note, additionally, that the proof methods are novel for the domain under discussion. Very often, these methods are already existing in another domain, but the translation was lacking (as is the case, e.g., in the probability research project). Historically speaking, there is a multitude of cases, apart from the one discussed above, to be found:

1. *Some Historical Examples of Proof Methods*

(i) No doubt, the most famous case is the introduction of the proof by *reductio* in Greek mathematics. One might argue about the philosophical significance and the ontological-epistemological implications of this method, but everyone agrees that it marked a new way of looking at and working in mathematics.

(ii) Equally famous is the method of infinite descent, promoted by Pierre de Fermat for proving the non-existence of solutions of Diophantine equations. The basic idea is to prove that if a (integer) solution exists, then there must be another (integer) solution that is strictly smaller, obviously leading to a contradiction.

In terms of translations from one domain to another and thereby importing proof methods from the former to the latter, the two most famous cases are:

(iii) The *reformulation of geometry in algebraic terms* led to an entirely different view of geometry. Whether or not this development is to be situated with Descartes, it is definitely the case that proof methods from algebra could now be used for solving geometrical problems. As a simple case, it is sufficient to think about the classification of curves in algebraic terms. More specifically, think of the classification of curves of degree two. On the one hand, geometrically speaking, there is the well-known cone figure (attributed to Apollonios) intersected by a plane at different angles, and, on the other hand, the classification of curves of the form $ax^2 + bxy + cy^2 + dx + ey + f = 0$ on the basis of determinants and the like.

(iv) The *rigorization of* mathematics in the 19th century, especially of *differential and integral calculus*, made it possible to reformulate questions concerning derivatives and integrals in more abstract terms, thereby making room for proof methods that went beyond the geometrical. Actually, this case is quite similar to the probability example given above.

For (iii) and (iv), see Grattan-Guinness (1997) for more detail and further references.

2. *Some Present-day Examples of Proof Methods*

One might perhaps be tempted to say or to claim that today at least we finally have a single set of proof standards, but that is definitely not the case. Within the mathematical community itself, deep discussions are taking place concerning the following problems (for a general discussion, see Hersh (1997, Part 1, Chapter 4)):

(i) Is a proof that involves the use of computers to be considered a proof? The most interesting case that started the whole discussion was, of course, the four-colour theorem (or conjecture?). As part of the proof consists of a computer program,[5] to a number of mathematicians the proof does not deserve to be called such. The debate is still running. See Tymoczko (1986) for more details.

(ii) What is the value of probabilistic proofs? Are these to be considered as proofs? The answer to the latter question has to be yes. After all, one does prove statements such as 'If test T, involving a choice of k numbers, is performed on a given number n, and the answer is yes, then the number n is prime, with a probability of $1 - 1/4^k$'. The more intriguing question is the former one: what does a proof like that tell us? Is it interesting to know that a number is *very likely* a prime number? See, e.g., Ribenboim (1989, 107–128), for a clear presentation.

(iii) What is the value of a 'video-proof'? I have to add here that, although it is claimed that video-proofs introduce an entirely new and novel way of doing mathematics, video-proofs are nothing but modern technological man's version of proofs-by-looking that were mentioned above. Perhaps the question should be phrased more generally: Can there be 'experimental' proofs? See my 1990b for some discussion and examples.

To conclude this paragraph, the general and hence not too detailed picture is that mathematics can be viewed as a network of research projects, whereby larger parts of the network form research traditions. The function of a project is to generate, firstly, problems, conjectures, i.e., work that needs to be done, and, secondly, an agreement on the methods and standards to be used to handle the problems. This is, roughly speaking, the environment wherein a mathematician performs his or her daily task.

4. A Day in the Life of a Mathematician

On the micro-level, individual mathematicians set out to prove theorems, to formulate conjectures, to check proofs or theorems proved by other mathematicians, to search for counter-examples to disprove a statement, and so on. The basic question to be asked is: how do they do it? Given a statement A, how do you go about it to find (or construct) a proof? In short, the question of heuristics has to be dealt with.

As one might expect, there are several suggestions, ideas, and proposals. No doubt, the most familiar one is Lakatos' *method of proofs and refutations*. In his own words:

> '*Rule 1.* If you have a conjecture, set out to prove it and to refute it. Inspect the proof carefully to prepare a list of non-trivial lemmas (proof-analysis); find counterexamples both to the conjecture (global counterexamples) and to the suspect lemmas (local counterexamples).
>
> *Rule 2.* If you have a global counterexample discard your conjecture, add to your proof-analysis a suitable lemma that will be refuted by the counterexample, and replace the discarded conjecture by an improved one that incorporates that lemma as a condition. Do not allow a refutation to be dismissed as a monster. Try to make all 'hidden lemmas' explicit.
>
> *Rule 3.* If you have a local counterexample, check to see whether it is not also a global counterexample. If it is, you can easily apply Rule 2.' (Lakatos 1976, 50)

However, this cannot be the whole story. What, for instance, is a mathematician supposed to do if no proof is to be found in the first place? As an example, let me briefly summarize some aspects of the history of *Fermat's Last Theorem* (FLT) (for more details, see my 1987).

4.1. Fermat's Last Theorem

In a first phase, instead of tackling the general problem, proofs were found for particular cases: thus, it was shown that $x^n + y^n = z^n$ did not have integer solutions for $n = 3, 4, 5, 7, 14$. In these proofs the method of infinite descent was crucial (see above). One has to wait for Sophie Germain who, in 1823, showed the following. To formulate the theorem it is necessary to make the following distinctions:

– the equation is restricted to prime numbers p, thus:

$$x^p + y^p = z^p,$$

– the first case of FLT states that there are no x, y, z, such that p does not divide $x.y.z$ and $x^p + y^p = z^p$ (the second case is, obviously, the one where p does divide $x.y.z$).

Germain's theorem says the following:

For every odd prime p such that $2.p + 1$ is also a prime, the first case holds.

With some additional theorems, Germain and Legendre were able to deal with all prime numbers <100. As one might expect, the problem is to determine how many primes p there are such that $2.p + 1$ is also a prime, a very difficult problem indeed. I must add here that the technique to split up a theorem in subcases and to treat these separately, is an almost continuous characteristic of the history of FLT. For example, in the next breakthrough by Eduard Kummer, because he had transposed the problem to the domain of complex numbers, another division was made into regular primes and irregular primes. The theorem Kummer arrived at stated boldly: If the prime p is regular, then FLT holds.

The point of importance to note here is that FLT has now entered a new domain: it is no longer a problem in number theory but a problem in complex number theory. This is a second constant phenomenon to be observed: a problem such as FLT remains in a specific domain and 'migrates' to another domain if no interesting results are found. However, deciding whether a prime p is regular is about as difficult as deciding whether it is such that $2.p + 1$ is also prime. Nevertheless, it would enable mathematicians in the years to follow to raise the upper bound to 125.000.

But FLT did not remain in this domain, another 'migration' took place. Rewriting the equation as follows:

$$(x/z)^p + (y/z)^p = 1, \ : \text{or}$$
$$X^p + Y^p = 1,$$

FLT says that this curve does not go through rational points. We enter the domain of algebraic number fields and from this domain, the area of elliptic curves, modular forms, where finally a proof would be found (see Wiles (1995)). Note that top mathematicians involved such as Gerd Faltings, Gerhard Frey, Ken Ribet, even

Andrew Wiles,[6] were not really working on FLT, but on other problems that as a corollary would prove FLT.

What I want to emphasize is that the search for a proof – the initial step in Lakatos' model – apparently happens in a methodical way (or methodical ways). Thus, it should be possible to set up rules and guidelines. Some of these rules will be rather evident – splitting up your problem into different cases, was already advised by Polya (see further) – but the suggestion to look for 'translations' of your problem in other domains and fields, thereby encouraging a 'migration' is perhaps less trivial.

I will not go into any details, but similar stories can be told about other open problems in mathematics. I refer the reader to Echeverria (1996) for a beautiful treatment of Goldbach's conjecture.

But even that cannot be the whole story. Proofs are curious things. It is perhaps trivial to say that it takes a mathematician to see one if there happens to be one, but they definitely use more criteria than mere formal correctness, as the summarized account of the following case shows (for more details, see my (1988)).

4.2. Apéry and the Riemann Zeta Function

Suppose you attend a seminar where a mathematician presents a proof to some of his colleagues. Suppose further that what he is proving is an important mathematical statement. Now the following happens: as the mathematician proceeds, his audience is amazed at first, then becomes angry and finally ends up disturbing the lecture (some walk out, some laugh, ...). Nevertheless, the proof is formally speaking (nearly) correct. What has happened?

Roger Apéry investigated the Riemann Zeta Function, $Z(s) = \Sigma_n 1/n^s$, where s is a complex number and n goes from 1 to infinity (the same function in Hilbert's eight problem, see above). There is no doubt that this is an important subject in the mathematical community. More specifically, he was interested in the integer values, $Z(n)$. Some results were known, such as:

$$\text{For } n = 2k,\ Z(2k) = (-1)^{k-1}(2\pi)^{2k} B_{2k}/(2.(2k)!),$$

where B_{2k} is the $2k$-th Bernoulli number, i.e., the $2k$-th coefficient in the equation:

$$x/(e^x - 1) = \Sigma B_i x^i/k!.$$

Example: For $n = 2$, $k = 1$ and, given that $B_2 = 1/6$, we find that:

$$\begin{aligned}\Sigma_n 1/n^2 = Z(2) &= (-1)^0(2\pi)^2/6.2.2!\\ &= 1.4.\pi^2/6.2.2\\ &= \pi^2/6,\end{aligned}$$

a well-known result.

However, much less is known about the odd values. Are $Z(3)$, $Z(5)$, ..., in general, $Z(2n+1)$, rational or irrational? The problem was known to Euler but neither Euler nor mathematicians after him managed to handle the problem. In June 1978, Roger Apéry presented a proof of the irrationality of $Z(3)$. It is this proof that provoked the strange reaction of his colleagues.

If the question does not sound too silly: what was wrong with Apéry's formally correct proof? Mathematicians gave the following comments:

(i) The proof was 'mysterious' and consisted of a series of 'miracles'. Thus, e.g., Apéry uses the following series, defined recursively:

$$n^3 u_n = (34n^3 - 51n^2 + 27n - 5)u_{n-1} - (n-1)^3 u_{n-2}.$$

Apéry claimed the following: if one starts with $u_0 = 1$ and $u_1 = 5$, then all u_n are integers! This is indeed very surprising as each u_n is of the form A/n^3. Therefore the right-hand-side must be divisible by n^3, for all n.[7]

(ii) The proof offers no clues at all for other values of $Z(s)$ for $s = 2n+1$. Apparently, mathematicians consider proofs that do not have this property as proofs of low quality.

(iii) Part of the disbelief in Apéry's proof had to do with the fact that he did not use any 'new' methods. In short, the proof could have been found by Euler. So why did Euler not find the proof or anybody soon after that?

After rewriting the proof (done by other mathematicians) the result has now been accepted and generalizations have been found (see Apéry (1996, 58)).[8]

In summary, as one might expect, the processes that lead to new mathematical results and insights, are complicated, diversified, hard to understand and to grasp and, quite simply, tricky. Therefore, although for some limited sets of problems, a Polya set of heuristics may be helpful, it does not really address the hard issues.

To some extent, the same can be said for work being done in the field of psychology. The main focus is still on the development and growth of mathematical concepts in children and, occasionally, something is said about adults and about professional mathematicians, e.g., in Tall (1991). It does seem odd that the Hadamard book is still being referred to, although it dates from 1945 (strangely enough, also the publication year of Polya's *How to Solve It* and that too is still being referred to). That procedures such as generalization or reducing the problem to simpler cases can be extremely helpful and fruitful, does not really need any comment. But to move beyond that, is the core problem. Without going into any details (once more, I am afraid), it may seem rather ironic that one of the oldest approaches to the problem is still doing very well: *the method of analysis and synthesis*. For a recent overview, see Otte and Panza (1997).

4.3. MATH WORLD OR MAD WORLD?

Perhaps this is the right moment to return to one of my initial claims, viz. the fact that one's philosophical view of mathematics – both ontological and epistemological – will co-determine one's ideas about the growth and development of mathematics. Mathematical realists will happily compare the universe of numbers, sets and geometrical figures with the material world we are part of, but the comparison has to be treated extremely carefully. For, if the mathematical universe is a real universe, then it is a funny one, to say the least. In our material universe, it is pretty safe to generalize from time to time. After all, ravens do turn out to be black, and birds, notwithstanding Tweety and his friends, do tend to fly. However, math world is a mad world. Here are a few examples. (a) and (b) are well-known within the mathematical world, whereas (c) is a rather more general observation.

(a) *Approximation of the Distribution of Prime Numbers.* Let $\pi(n)$ be the function that counts the number of prime numbers $\leq n$. Let $Li(n)$ – the logarithmic integral – be the function:

$$\int_2^n (1/\ln(x))dx, \text{ where } \ln(x) \text{ is the natural logarithm of } x.$$

The prime-number theorem says that $Li(n)$ is an extremely good approximation of $\pi(n)$ (to be precise: $Li(n)$ is asymptotically equal to $\pi(n)$).

For finite values for n, one notes, by direct calculation, that, although the difference is small, $Li(n) > \pi(n)$. Calculations up to 10^9 showed that this is the case. It seemed more than reasonable to conclude that this is always the case. Which it is not. Littlewood has shown that the difference $Li(n) - \pi(n)$ changes sign *infinitely many* times! The first estimate for what value of n this is supposed to happen, was given by Skewes. He arrived at the impressive number

$$10\text{\textasciicircum}(10\text{\textasciicircum}(10\text{\textasciicircum}34))),$$

meaning that a change of sign has to take place before this number. This upper bound was improved to $6{,}69.10^{370}$, still a quite impressive number. See Devlin (1988, 207–213) for more details.

(b) *Mertens Conjecture.* If n is a natural number, then either n is divisible by the square of a prime, p^2, or not. In the latter case, we call n square-free. Now we define a function $m(n)$ as follows:

– if n is not square-free: $m(n) = 0$
– if n is square-free and the number of primes in n is even: $m(n) = 1$
– if n is square-free and the number of primes in n is odd: $m(n) = -1$.

Example: $m(6) = m(2.3) = 1$, $m(9) = 0$, $m(11) = -1$.

Finally, define the function $M(n)$ as follows:

$$M(n) = m(1) + m(2) + \ldots + m(n).$$

Mertens Conjecture claims that:

$$|M(n)| < \sqrt{n}.$$

Straightforward checking reveals that the inequality is satisfied for values of n into the billions. However, there is a counter-example for a value of n in the order of 10^{65}.

(c) To a certain extent, one expects to be 'fooled' almost all of the time. Think of the real numbers, that, classically speaking, come in two sorts: the algebraic reals and the transcendental reals. The former ones can be defined in terms of polynomials of a certain degree – e.g., $\sqrt{2}$ is one of the solutions of $x^2 - 2 = 0$ – whereas the latter ones are not so definable. Therefore to show that a specific real number is transcendental is not an easy task.

However, one of the criteria to be used is this:

> Given a number r, if there exists an infinite sequence of distinct rational numbers p_i/q_i, such that $|r - p_i/q_i| < 1/q_i^{ni}$ where n_i goes to infinity as i does, then r is transcendental.

This means that some transcendental numbers can be approximated almost arbitrarily close to rational numbers. Hence, to make the distinction will be extremely difficult and very often one expects that numerical calculations will lead one into an entirely wrong direction.

Another stunningly nice example, involving the two best known transcendental number, viz., π and e, is this (from Borwein and Borwein (1992, 827)). The following formula gives the correct value for π up to 42 billion digits and only then do things go wrong:

$$\pi = [(1/10^5)[\Sigma_e^{-n^2/10^\wedge 10}]]^2, \text{ where } n \text{ goes from } -\infty \text{ to } \infty.$$

What is the point of these examples and comments? Even if we were to find a high quality set of heuristics that are able to deal with a mass of mathematical problems, one must still be prepared for the odd and queer thing every now and then. In other words, it is true that heuristics are not supposed to guarantee success in all cases, but, if one fails occasionally, one hopes for the best. In the math universe, the best thing to do is to fear for the worst.

I will not explore this line of approach any further and instead I will look at an alternative that has been developed in recent years: computer programs that prove theorems for you.

5. Success and Failure of Automated Reasoning

Although the focus of this paragraph will be on automated reasoning, I will throw a brief look at some other approaches, again not aiming at being exhaustive. No doubt, one of the most famous programs, written by Douglas Lenat, is *Automatic Mathematician* (AM). This program does not prove theorems, it operates at a deeper level, namely the generation of new concepts on the basis of given concepts and the formulation of possibly interesting conjectures.

5.1. Artificial Mathematician

The basic structure of AM is fairly simple: a small collection of basic, rather general notions and an extensive set of heuristics to apply to these concepts. Some examples:

(a) Suppose that two sets A and B are given, as well as a function $f\colon A \times A \to B$. Thus $f(a, b) = c$. In this case, an interesting heuristic is to see what happens if the two arguments are identified, thus obtaining a function $g\colon A \to B$. If, e.g., f is multiplication, $a.b = c$, then g is the square function $a^2 = c$.

(b) Given any function $f\colon A \to B$, see what happens if f is applied repeatedly (if such is possible, of course), say $f^n = f^\circ f^\circ f \ldots^\circ f$ (n times). If, e.g., f is addition, thus $f\colon N \times N \to N$ and the previous heuristic is applied, then we have the function $g\colon N \to N$, such that $g(a) = 2a$. Repeated applications of g, produce functions that map a onto $2a, 3a, \ldots, na$, in other words a basic multiplication appears.

(c) Given any function $f\colon A \to B$, see what happens with the inverse function, if it exists. Thus, if multiplication is defined, $a.b = c$, division will be produced by this heuristic, taking into account the impossible case a/b, where $b = 0$.

(d) Given any function $f\colon A \to B$, look at extreme cases, i.e., if some concept or other takes on values in a given range, look at the end-values to see what happens. Thus, e.g., if the notion of divisor is available, then one can construct the function d that maps numbers onto the number of divisors of that number (this function actually exists in number theory and is a quite fundamental function). One extreme case is to look for the minimum of $d(n)$. Clearly the lowest value is 2, thus those n for which $d(n) = 2$ are special. In fact, they are nothing else but the prime numbers.

(e) Given any function $f\colon A \to B$, and a set of specific values of f, look for a pattern and formulate the conjecture that all values confirm to the proposed pattern. Example: continuing with the function d in the above example, suppose we look at the n, such that $d(n) = 3$. If AM generates a number of examples for this function, arguments such as 4, 25, 121, ... will come up. These are all squares, hence the conjecture: 'If $d(n) = 3$, then $n = m^2$, for some m' (which happens to be the case). There is actually an even stronger conjecture possible: '$d(n) = 3$ if and only if $n = m^2$, where m is a prime'.

It is not easy to evaluate the values and shortcomings of AM in a few lines. I refer the reader to Boden (1990, 206–209), for a balanced judgment. Let me just mention that AM does not escape the horror of all computer scientists: combinatorial explosion. Unless additional metaheuristics are fed into the system, AM will generate concept after concept, conjecture after conjecture, all things interesting and also all things uninteresting. But that, of course, is not a particular critique of AM, but of programs, intended to model a creative process, in general.

A lot of attention has been given to programs that are capable of checking existing proofs. No doubt, one of the most famous is the AUTOMATH program, developed by N. G. de Bruijn. But equally impressive are programs such as Mathematics Understander (MU) developed by Edmund Furse, and ONTIC, developed by David A. McAllester (see his (1989)). In the same range, are programs such as MACSYMA, REDUCE, MATHEMATICA, and so many others. An interesting overview is presented in Johnson et al. (1994). I will not discuss these programs but, instead, focus my attention on the underlying ideas of automated reasoning.[9]

5.2. Automated Reasoning

One of the major advantages of automated reasoning (AR) is that the *basic* ideas are extremely easy to explain. However, spelling out the basics, is only a minor part of the whole undertaking of AR. To illustrate the basis, I will first say a few things about classical propositional logic (PC) and then about classical first-order predicate logic (PL).[10]

1. *Automated Reasoning in Propositional Calculus*

Classical logic has the extremely nice property that every formula A can be rewritten in a standard format A-*cnf*, the conjunctive normal form. A formula in A-*cnf* format consists of a series of conjunctions (possibly empty), each conjunct itself is a series of disjunctions (possibly empty), and the members of the disjunctions are either letters $p, q, r, \ldots$ or negated letters. Thus, e.g., the formula

$$(p \supset q) \supset ({\sim}q \supset {\sim}p)$$

becomes

$$(p \vee q \vee {\sim}p) \,\&\, ({\sim}q \vee q \vee {\sim}p).$$

If we drop the conjunctions, then we are left with clausal forms:

(i) $p \vee q \vee {\sim}p$
(ii) ${\sim}q \vee q \vee {\sim}p$.

Why is this interesting? Because one can show that all logical *rules* can be reduced to one *single* rule, the so-called resolution rule:

Suppose that two formulas are given in *cnf* format and such that:

(i) $A_1 \vee A_2 \vee \ldots \vee p \vee \ldots \vee A_n$
(ii) $B_1 \vee B_2 \vee \ldots \vee {\sim}p \vee \ldots \vee B_m$,

the one can conclude to:

(iii) $A_1 \vee A_2 \vee \ldots \vee A_n \vee B_1 \vee B_2 \vee \ldots \vee B_m$.

In words: if in two formulas one has an occurrence of a letter p and the same letter with negation, then the two clauses can be joined into a new clause deleting p and the negation of p.

With this material at hand, it becomes easy to prove theorems. One of the standard ways is through refutation. If one has to show that B follows from a set of premisses $A_1, A_2, \ldots, A_n$, rewrite all premisses and the negation of the conclusion in *cnf* format and try to find a contradiction – indicated by the *empty* clause, f – by successive applications of the resolutions rule.

EXAMPLE. Show that $p \supset s$ follows from $(p \vee q) \supset r$ and $r \supset s$. The translation in *cnf* format give the following clauses:

1.$\sim p \vee r$
2.$\sim q \vee r$
3.$\sim r \vee s$
4.p
5.$\sim s$

Resolution applied to 3 and 5: 6. $\sim r$
Resolution applied to 1 and 6: 7. $\sim p$
Resolution applied to 4 and 7: 8. f. contradiction.

Of course, since PC is decidable (although NP-hard), we know that this method will always produce correct answers. For PL the situation is more interesting.

2. *Automated Reasoning in Predicate Logic*

First the good news. As in PC it is possible to rewrite any formula in a standard format. I will not go into details, but generally speaking the standard format, the so-called prenex normal form (*pnf*) has all the quantifiers in front and then a quantifier-free expression in *cnf* format. Just as in PC it is possible to work with (as good as) one single rule, the full resolution rule. So it seems that we can do the same things as in PC.

However, as we know, PL is not decidable, therefore repeated applications of the rule does not necessarily lead to certain success. Mathematics needs at least PL, so there is the challenge. Basically, two options are open:

(a) *Find Restricted Cases.* There are parts of PL that are decidable, hence for these cases an algorithm can be formulated.

(b) *Search for Heuristics.* Find additional "rules" that can help to give guidance to the search for the empty clause.

It would be extremely unfair to take one page or half a page in order to evaluate the virtues and faults of AR. I will list a few of the successes for the simple reason that most of mathematicians, logicians and philosophers are deeply convinced that the general value of AR is close to zero. And it has to be said, there are some nice results.

(i) One of the really impressive and extremely recent results is the solution to the problem of *Robbins algebras*. The question is quite simple. Given the following axioms, that define a Robbins algebra:

(R1) : $(\forall x)(\forall y)(x + y = y + x)$,
(R2) : $(\forall x)(\forall y)(\forall z))((x + y) + z = x + (y + z))$,
(R3) : $(\forall x)(\forall y)(-(-(x + y) + -(x + -y)) = x)$,

show that you have a Boolean algebra.

As the other way is easy to show, the question comes down to showing that Robbins algebras are the same as Booleans algebras. I refer the reader to McCune (1997), for an overview of this problem and how the solution was found. I will limit myself to some general remarks. The basic approach has been to find additional statements A such that the Robbins axioms together with A produce a proof of the equivalence with a Boolean algebra. The problem was then reduced to proving A from the axioms (R1), (R2), and (R3). This history in itself is quite intriguing. Here is a list of some of these formulas:

(A1) $(\forall x)(--x = x)$,
(A2) $(\exists y)(\forall x)(y + x = x)$,
(A3) $(\exists y)(\forall x)(1.x = x)$,
(A4) $(\forall x)(x + x = x)$,
(A5) $(\exists x)(x + x = x)$,
(A6) $(\exists x)(\exists y)(x + y = y)$.

Especially (A4) through (A6) are fascinating, as each one is weaker than its predecessor. Note that this practice of looking for intermediate statements is standard practice among human mathematicians.

(ii) AR programs are extremely good at generating counter-examples and finite models. One might perhaps think that this is trivial, but it is not. For the simple reason that a blind generation of all possibilities, even if finite, is exponentially difficult. Thus a guided search is needed. Using this technique, it has been possible to answer some questions in the theory of finite semi-groups (see Wos et al.

(1992, 320–323)). Looking outside of the domain of mathematics, this technique has proven its worth in formal logic.

(iii) Inspired by AR and other programs, some curious results have appeared. Bailey et al., 1997, discuss diverse methods for calculating the decimals of π and observe that most identities converge far too slow.[11] Thus, better identities are needed. On the one hand, the work of Ramanujan has been a source of inspiration, and, on the other hand, for the purpose of calculating individual digits, a computer method (called ‘PSLQ’) was used to generate new identities, such as (where i runs from 0 to infinity):

$$\pi = \Sigma_i (1/16^i)[4/(8i+1) - 2/(8i+4) - 1/(8i+5) - 1/(8i+6)]$$

Apart from the fact that this sort of identity makes one think of Apéry’s proof, it is important to realize that the program looks for identities on the basis of numerical identity and then a proof was searched for.

For similarly inspired work, see Wilf and Zeilberger (1990). They present a general method for generating identities where the proof can be automatically checked.

Finally, I might add that AR gives a nice formal idea of *reasoning by analogy*. Suppose that a proof of a statement A follows a particular route, selected by the heuristics applied, then you obtain a proof schema. This schema can be used as the proof frame for a proposition similar to A.

EXAMPLE. Think of proof by infinite descent (see above). This is indeed a proof schema that can be applied to any Diophantine problem as a heuristic. For a nice and interesting example, see Melis (1998).

5.3. Proofs From the Unexpected

What goes for humans, goes for machines. At least in this case. It is obvious that automated reasoning does shed new light on what the search for and the nature of a m athematical proof is. At the same time, I wish to repeat my comment made before that the mathematical universe, if there is any such thing at all, is a strange place. The same goes for proofs. I do not doubt that many proofs are standard and do not involve anything strange or bizarre, but, nevertheless, occasionally one must wonder. To end this section, let me present a few of such proofs.[12]

EXAMPLE 1. The following definition is given. For n a natural number, define $S(d)$ as the sum of its divisors. Then three cases are possible:

(i) $S(d) = 2n$, the number is perfect,
(ii) $S(d) > 2n$, the number is abundant,
(iii) $S(d) < 2n$, the number is deficient.

Prove that every even number greater than 46 can be expressed as the sum of two abundant numbers. (Honsberger 1970, Essay Fourteen).

Confronted with this problem for the first time, it seems a reasonable strategy to take two abundant numbers a and b and to wonder what properties their sum $a+b$ must have. Although this might perhaps be successful, a rather direct solution is given through proving the following lemma:

If a number n is perfect or abundant, then its multiples are abundant. (I leave the quite simple proof to the reader).

The next step is to write the number n as $6k+m$, where $m = 0, 2$ or 4. If $m = 0$, then $n = 6k = 6k' + 6k''$, and, as 6 is a perfect number, n is abundant. If $m = 2$, then $n = 6k' + 20$, and, as 20 is abundant, so is n. If $m = 4$, then $n = 6k' + 40$, and, as 40 is abundant, so is n. QED.

EXAMPLE 2. Consider 18 consecutive natural numbers, smaller than 1.000. Show that at least one of these numbers is divisible by the sum of its digits.

The shortest proof I know relies on the simple fact that if the number is abc then $a + b + c \leq 27$. Exclude 999 (no problem for 999 is divisible by 27), then $a+b+c < 27$. In a row of 18 numbers, there is at least one multiple of 18. That number is divisible by 9 and by 2, hence the sum of its digits is divisible by 9, thus $a+b+c = 9$ or 18. QED

EXAMPLE 3. This problem is truly my favourite, because things cannot get any simpler than this. Consider a real function $f\colon R \to R$. A real function f is *symmetric* if $f(x) = f(-x)$, *anti-symmetric* if $f(x) = -f(-x)$. Show that any real function is the sum of a symmetric and anti-symmetric function.

Probably one would tend to 'subtract' from f a symmetric function g and then try to show that $f - g$ is an anti-symmetric function under certain conditions. Whereas the answer is just this:

$$f(x) = [f(x) + f(-x)]/2 + [f(x) - f(-x)]/2.$$

Obviously $f(x)+f(-x) = f(-x)+f(x)$ and $f(x)-f(-x) = -[f(-x)-f(x)]$. QED

Generally speaking, it is this sense of unexpectedness that seems quite difficult to be captured by AR. But do note at the same time that human mathematicians consider these proofs to be ingenious as well. Freely translated, this means that they themselves did not expect a proof of this kind. So, once again human and machine meet.

6. Afterthought

The contents of this paper are of an almost entirely descriptive nature. I have tried to bring together some elements that must be part of such a description, if it claims to be representative of mathematical practice as we know it. All this being said and done however, there is a subsequent question to consider: is mathematical practice, as it is, the best we have? Is there room for improvement? Is it possible that not all aspects of the proof idea have been explored? I have no other choice than to reiterate a comment made several times in the course of this text: to answer these questions, one's philosophical views enter into the picture. If, e.g., one believes that there is such a thing as the ideal proof, and one believes that this ideal is humanly reachable, then there will be a moment where things can improve no further.[13] If, however, one believes that mathematics is (nothing but) a human product, then, on the basis of this description, it does leave room for further reflection and it opens the possibility of 'planning' mathematics itself in a particular direction.

This last idea is not ludicrous at all. I end this paper by mentioning the interesting but not enough mentioned work done by van Gasteren in her 1990, where she proposes that mathematicians should try to present their proofs in such a way that maximum clarity is achieved. One of her motives is that such proofs are easier to check using automated reasoning programs and thus a certain division of labour within the mathematical community can be installed. After all, whether one likes it or not, mathematics is a social phenomenon from this perspective as well.

Notes

[1] The expression "proof by looking" is actually an entry in David Wells, 1991 and I quote: "Many simple arithmetical facts can be proved 'at sight', by examining a suitable figure" (198). If Koetsier is right (see his 1991, 188–190), one might just as well leave out the "simple", for he claims that, according to Oskar Becker, there is a proof by looking of this arithmetical fact: any number of the form $2^n.(1 + 2 + 2^2 + \ldots + 2^n)$ such that $p = 1 + 2 + 2^2 + \ldots + 2^n$ is a prime, is perfect.

[2] Thus, to a structuralist, the local structures can be viewed as substructures of the general large-scale structure. Note that one other problem I am completely ignoring, is whether the 'impetus' of the research projects or the substructures derives from the projects and/or structures themselves, and, if not, whether it comes from individual mathematicians, groups of mathematicians, and, if that is not the end of the story, whether other non-mathematical individuals and/or groups enter into the picture.

[3] These confusions and difficulties are reponsible for the immense, diversified, and amusing literature on paradoxes in probability theory, see, e.g., Northrop (1978), especially chapter eight.

[4] An example may help. One needs a singular distribution function to solve this problem: On the kth toss of a fair coin a gambler receives 0 if it is a tail and $(2/3)^k$ if it is a head. Let X be the total gain of the gambler after an infinite sequence of tosses of the coin. The problem is solved in Grimmett and Welsh (1994, 102–104).

[5] To be a bit more precise: the proof comes in two parts. The first part – a classical mathematical masterpiece – shows that the set of all maps to be coloured can be reduced to a finite set of maps, such that if the finite set can be coloured, so can all of them. The second part consists of a computer program that actually colours every map in the finite set. One therefore has no other choice than to

run the program and see if the final answer is yes or no. A rather amazing situation. See Appel and Haken (1989).

[6] In the case of Andrew Wiles there is even a curious twist to the story. Already as a child he wanted to prove FLT, but in his mathematics study he was strongly advised not to waste time on this problem. Rather he should use his talents for important problems in elliptic curves and modular forms. However, after Ribet's theorem that showed the connection between the two, Wiles realized he was after all working on his dream project. It is worth noting that in the famous Wiles' paper of 1995, though FLT is mentioned in the title, FLT itself is only referred to in the introduction.

[7] There is an intriguing historical remark to be made. The inspiration for this type of series, Apéry had found in the work of Ramanujan, the famous Indian mathematician. However, for the latter, the term 'miracle' is precisely used as a positive qualification.

[8] The recognition afterwards for his result has apparently wiped out the bad memories of the occasion itself. François Apéry, his son, makes no mention of the incident, but writes that: 'The proudest moment of his career was his proving, at more than 60 years of age, the irrationality of $Z(3)$' (Apéry 1996, 58).

[9] There is one thing I must mention. In many of these approaches, the author or authors emphasize the double use of bottom-up and top-down strategies. In terms of looking for proofs, this translates into: (i) start from the axioms and derive as much as you can keeping in mind the conclusion you want to reach, and (ii) start with the conclusion and reason backwards keeping in mind what your axioms are. If there is an overlap somewhere, you will have the backbone of a proof. The thing worth mentioning is that this procedure is nothing but a modern translation of the well-known method of analysis and synthesis. One is tempted to say: *nihil novi sub sole*.

[10] The ideas about AR here presented are taken from Wos et al. (1992), and from Bundy (1983).

[11] If someone happens to be interested, according to the paper of Bailey et al. (1997), the current 'record' is 6.442.450.938 decimals. However, now (= June 1998) the correct number is close to $51{,}5.10^9$ decimals. This impressive result has been achieved by Yasumasa Kanada.

[12] It is almost inevitable that the example should be given of the chess board with two opposite corners removed and the problem to solve is to show that the board cannot be covered with bricks that cover exactly two squares. As everybody gives this example, one might have the impression that this is the only example. Hence, I present here three different examples.

[13] In my (1993) I have presented a sketch of this ideal mathematical community (IMC). Its basic characteristics are: (a) in the IMC, all members are equal, (b) the IMC is relatively isolated from the rest of society, (c) all members of the IMC share the same idea of the existence of a unique mathematical universe U (independent of the question whether this is actually the case) and the task of mathematics is the search for a complete description of U, (d) all members share the idea that there is a unique or preferred language L wherein this description is formulated, (e) for any mathematical statement, if there is a proof, then it can, in principle, be found by any mathematician, (f) any proof of any statement can be checked by any mathematician, and, finally, though optional, (g) how the proof is to be found is mostly a matter of some kind of innate capabilities. Do note that the description of the IMC is a lot poorer than the real(istic) community.

References

Note. The list of references is quite extensive. It is not meant to impress the reader but it is the consequence of the set-up of this paper. I have tried to cover as many areas as possible, each time with indications where to proceed therefrom if the discussion in this paper is unsatisfactory for the reader.

Alexandrov, P. S. (ed.): 1971, *Die Hilbertschen Probleme*, Leipzig, Akademische Verlagsgesellschaft, Geest & Portig.

Apéry, François: 1996, 'Roger Apéry, 1916–1994: A Radical Mathematician', *The Mathematical Intelligencer* **18**(2), 54–61.

Appel, Kenneth and Wolfgang Haken: 1989, *Every Planar Map is Four Colorable*, Providence, AMS (Contemporary Mathematics, vol. 98).

Bailey, D. H., J. M. Borwein, P. B. Borwein and S. Plouffe: 1997, 'The Quest for Pi', *Mathematical Intelligencer* **19**(1), 50–57.

Bell, John L. (guest ed.): 1994, *Categories in the Foundations of Mathematics and Language*, Special issue of *Philosophia Mathematica*, vol. 2, first third.

Boden, Margaret: 1990, *The Creative Mind*, London, Sphere Books.

Borwein, Jonathan and Peter Borwein: 1992, 'Some Observations on Computer Aided Analysis', *Notices of the AMS* **39**(8), 825–829.

Browder, Felix E. (ed.): 1976, *Mathematical Developments Arising from Hilbert Problems,* Providence, AMS, (Proceedings of Symposia in Pure Mathematics, vol. 28).

Bundy, Alan: 1983, *The Computer Modelling of Mathematical Reasoning*, New York, Academic Press.

Cipra, Barry: 1996, *What's Happening in the Mathematical Sciences, 1995–1996*, Vol. 3, Providence: AMS.

Corry, Leo: 1992, 'Nicolas Bourbaki and the Concept of Mathematical Structure', *Synthese* **92**, 315–348.

Crowe, Michael: 1992, 'Ten "Laws" Concerning Patterns of Change in the History of Mathematics', in Donald Gillies (ed.), pp. 15–20 (originally published in 1975).

Dauben, Joseph: 1992, 'Conceptual Revolutions and the History of Mathematics: Two Studies in the Growth of Knowledge', in Donald Gillies (ed.), pp. 49–71 (originally published in 1984).

Devlin, Keith: 1988, *Mathematics: The New Golden Age*, Harmondsworth, Penguin.

Dieudonné, Jean: 1987, *Pour l'honneur de l'esprit humain. Les mathématiques aujourd'hui*, Paris, Hachette.

Duda, Roman : 1997, 'Mathematics: Essential Tensions', *Foundations of Science* **2**(1), 11–19.

Dunham, William: 1990, *Journey Through Genius. The Great Theorems of Mathematics*, New York, John Wiley & Sons.

Echeverria, Javier: 1996, 'Empirical Methods in Mathematics. A Case Study: Goldbach's Conjecture', in Gonzalo Munévar (ed.), *Spanish Studies in the Philosophy of Science*, Dordrecht: Kluwer Academic, pp. 19–55.

Gonzalo Gelbart, Stephen: 1984, 'An Elementary Introduction to the Langlands Program', *Bulletin (New Series) of the American Mathematical Society* **10**(2), 177–219.

Gillies, Donald (ed.): 1992, *Revolutions in Mathematics*, Oxford, Clarendon Press.

Glas, Eduard: 1981, *Wiskunde en samenleving in historisch perspectief*, Muiderberg, Coutinho.

Glas, Eduard: 1991a, 'Lakatos Revisited', *Kennis en Methode* **XV**(3), 307–311.

Glas, Eduard: 1991b, 'Koetsiers verfijnde metamethodologie – een repliek', *Kennis en Methode* **XV**(4), 404–405.

Gorenstein, Daniel: 1986, 'Classifying the Finite Simple Groups', *Bulletin (New Series) of the AMS* **14**, 1–98.

Graham, L. A. : 1959, *Ingenious Mathematical Problems and Methods*, New York, Dover Publications.

Grattan-Guinness, Ivor: 1997, *The Fontana History of the Mathematical Sciences*, London, Fontana Press.

Grimmett, Geoffrey and Dominic Welsh: 1994, *Probability. An Introduction*, Oxford, Clarendon Press.

Hersh, Reuben: 1997, *What is Mathematics, Really?*, London, Jonathan Cape.

Honsberger, Ross: 1970, *Ingenuity in Mathematics*, Washington, New Mathematical Library, MAA.

Johnson, Jeffrey, Sean McKee and Alfred Vella (eds.): 1994, *Artificial Intelligence in Mathematics*, Oxford, Clarendon Press.

King, Jerry P.: 1992. *The Art of Mathematics*, New York, Plenum Press.

Kitcher, Philip: 1983, *The Nature of Mathematical Knowledge*, Oxford, Oxford University Press.

Koetsier, Teun 1991, *Lakatos' Philosophy of Mathematics. A Historical Approach*, New York/Amsterdam, North-Holland (Studies in the History and Philosophy of Mathematics, volume 3).

Kuhn, Thomas: 1977, *The Essential Tension. Selected Studies in Scientific Tradition and Change*, Chicago, University of Chicago Press.

Lakatos, Imre: 1976, *Proofs and Refutations*, Cambridge, Cambridge University Press.

Langley, Pat, Herbert A. Simon, Gary L. Bradshaw and Jan M. Zytkow: 1987, *Scientific Discovery. Computational Explorations of the Creative Processes*, Cambridge, MA, MIT.

Lenat, D. B.: 1980, 'AM: Discovery in Mathematics as Heuristic Search', in R. Davis and D. B. Lenat (eds.), *Knowledge-Based Systems in Artificial Intelligence*, New York, McGraw-Hill, pp. 3–228.

MacLane, Saunders: 1986, *Mathematics. Form and Function*, Heidelberg, Springer.

McAllester, David A.: 1989, *ONTIC. A Knowledge Representation System for Mathematics*, Cambridge, MA, MIT.

McCune, William: 1997, 'Solution of the Robbins Problem', *Journal of Automated Reasoning* **19**(3), 263–276.

Melis, Erica: 1998, 'The Heine-Borel Challenge Problem. In Honor of Woody Bledsoe', *Journal of Automated Reasoning* **20**(3), 255–282.

Munévar, Gonzalo (ed.): 1996, *Spanish Studies in the Philosophy of Science*, Dordrecht, Kluwer Academic (BSPS volume 186).

Northrop, Eugene P.: 1978, *Riddles in Mathematics*, Harmondsworth, Penguin.

Otte, M. and M. Panza (eds.): 1997, *Analysis and Synthesis in Mathematics*, Dordrecht, Kluwer.

Polya, Georg: 1945, *How to Solve It. A New Aspect of Mathematical Method*, Princeton, Princeton University Press . (New York: Doubleday Anchor Books, 1957.)

Rav, Yehuda: 1993, 'Philosophical Problems of Mathematics in the Light of Evolutionary Epistemology', in Sal Restivo et al. (eds.), *Math Worlds: New Directions in the Social Studies and Philosophy of Mathematics*, New York, State University New York Press, pp. 80–109.

Restivo, Sal: 1983, *The Social Relations of Physics, Mysticism, and Mathematics*, Dordrecht, Reidel.

Restivo, Sal: 1992, *Mathematics in Society and History*, Dordrecht, Kluwer Academic.

Restivo, Sal, Jean Paul Van Bendegem and Roland Fischer (eds.): 1993, *Math Worlds: New Directions in the Social Studies and Philosophy of Mathematics*, New York, State University New York Press.

Ribenboim, Paulo: 1989, *The Book of Prime Number Records*, Heidelberg, Springer.

Schoenfeld, Alan H.: 1985, *Mathematical Problem Solving*, New York, Academic Press.

Stewart, Ian: 1987, *The Problems of Mathematics*, Oxford, Oxford University Press.

Tall, David (ed.): 1991, *Advanced Mathematical Thinking*, Dordrecht, Kluwer (Mathematics Education Library).

Taylor, Richard and Andrew Wiles: 1995, 'Ring-theoretic Properties of Certain Hecke Algebras', *Annals of Mathematics*, Second Series, **141**(3), 553–572.

Tymoczko, Thomas (ed.): 1986, *New Directions in the Philosophy of Mathematics*, Stuttgart, Birkhäuser.

Van Bendegem, Jean Paul: 1987, 'Fermat's Last Theorem seen as an Exercise in Evolutionary Epistemology', in Werner Callebaut and Rik Pinxten (eds.), *Evolutionary Epistemology*, Dordrecht, Kluwer, pp. 337–363.

Van Bendegem, Jean Paul: 1988, 'Non-Formal Properties of Real Mathematical Proofs', in Arthur Fine and Jarrett Leplin (eds.), *PSA 1988*, Volume One, East Lansing, PSA, pp. 249–254.

Van Bendegem, Jean Paul: 1990a, 'Characteristics of Real Mathematical Proofs', in A. Diaz, J. Echeverria and A. Ibarra (eds.), *Structures in Mathematical Theories*, San Sebastian, Servicio Editorial Universidad del Pais Vasco, pp. 333–337.

Van Bendegem, Jean Paul: 1993, 'Foundations of Mathematics or Mathematical Practice: Is One Forced to Choose?', in S. Restivo, J. P. Van Bendegem and R. Fischer (eds.), pp. 21–38.

Van Bendegem, Jean Paul: 1996, 'Mathematical Experiments and Mathematical Pictures', in Igor Douven and Leon Horsten (eds.), *Realism in the Sciences. Proceedings of the Ernan McMullin Symposium Leuven 1995*, Louvain Philosophical Studies 10. Leuven, Leuven University Press, pp. 203–216.

van Gasteren, A. J. M.: 1990, *On the Shape of Mathematical Arguments*, Heidelberg, Springer.

Wells, David: 1991, *The Penguin Dictionary of Curious and Interesting Geometry*, Harmondsworth, Penguin Books.

Wilder, Raymond L.: 1981, *Mathematics as a Cultural System*, Oxford, Pergamon Press.

Wiles, Andrew: 1995, 'Modular Elliptic Curves and Fermat's Last Theorem', *Annals of Mathematics*, Second Series, **141**(3), 443–551.

Wilf, Herbert S. and Doron Zeilberger: 1990, 'Towards Computerized Proofs of Identities', *Bulletin (New Series) of the AMS* **23**(1), 77–83.

Wos, Larry; Ross Overbeek; Ewing Lusk and Jim Boyle: 1992, *Automated Reasoning. Introduction and Applications*, New York, McGraw-Hill.

QUANTUM LOGIC AND THE UNITY OF SCIENCE

JOHN WOODS[1] and KENT A. PEACOCK[2]
[1]*The Abductive Systems Group, The University of British Columbia, 1866 Main Mall, E370 Buchanan Building, Vancouver, British Columbia, V6T 1Z1, Canada, E-mail:* jhwoods@interchange.ubc.ca [2]*Department of Philosophy University of Lethbridge, 4401 University Drive, Lethbridge, Alberta, T1K 3M4, Canada E-mail:* kent.peacock@uleth.ca

Abstract. This paper is an exploratory prolegomenon to the construction of a quantum logic that could shed some light on the thesis of the unity of science. We attempt to take account of the following factors, among others: the difficulty of saying just what a logic *is*, the startlingly *simple* queerness of quantum mechanics from the classical point of view, the consequences of the breakdown of bivalence and individuation in quantum mechanics, and the implications of recent work in quantum computation for quantum logic. We tentatively endorse modal interpretations of quantum mechanics, and suggest that quantum computation points to ways in which quantum logic could be extended beyond the traditional Birkhoff-von Neumann lattice theoretic approach.

1. Motivating Quantum Logic

The unity of science is the idea that all of the sciences are partially ordered by a set of reduction relations. This notion has been found attractive to thinkers at least as far back as Descartes, who posited the familiar metaphor of the tree of knowledge, with metaphysics as its roots and physics as its trunk.[1]

Since its heyday under the sway of logical positivism, the thesis of the unity of science has growingly been judged plausible in highly localized junctures of the universe of science – which is just to say that the thesis itself has not been found very convincing.

Nor has every local candidate for reduction (logicism in mathematics is a case in point) returned the promise of its youth. Notwithstanding the general discouragements that have fallen on it, the idea of unity in the sciences retains some minimal credibility in two main respects. One is that all of science is driven by essentially the same methodology. The other is that there is one science, namely logic, to which all of science must answer, a suggestion which we examine here. It would be natural to suppose that it is classical logic (or some near thing) that lays rightful claim on the status of this *ur*-science. Not everyone takes this view, however. Ambitious claims have been made by Putnam (1975a, b) and Deutsch (1985, 1997) (in different ways) to the effect that classical (Boolean) logic is a *special case* of quantum logic. If such a claim is correct, then the roots of Descartes' tree must be the logic of quantum mechanics.

S. Rahman et al. (eds.), Logic, Epistemology, and the Unity of Science, pp. 257–287.

This is an exploratory paper that will be limited to three related questions which bear on the challenge posed by Deutsch and Putnam. One is whether quantum physics in any way discredits classical logic. For example, does the failure of distributivity in the quantum domain (when statements about noncommuting observables are involved) show that the classical principle of distributivity is invalid? The second issue is whether quantum physics requires or allows for a logic of its own. And, third, would this be one that gave the unity of science thesis any encouragement?

We want to lay some emphasis on the preliminaries of our intentions in this chapter. We do not have a smoothly coherent view to propose. This will come in later work, now in progress. Our task here is to assemble the bits and pieces that a smoothly coherent positive account will have to take into account.

Possible answers to our questions cannot help being heavily conditionalized by a rather substantial *pluralism* in present-day logical theory. Our answers must include the rider: "It depends on what you mean by 'logic' ". We intend to expose enough of this pluralism to help determine whether there is a *respectable* notion of logic which meets the following conditions: that there are features of quantum mechanics (*QM*) which require that there is a logic in this respectable sense; that this logic is nonclassical; and that it is indispensable or especially conducive to the logical business of *QM* itself.

With appropriate tentativeness we call this logic or possible logic *QL*, appropriating the acronym from Birkhoff and von Neumann (1936) and giving it a more comprehensive use than it had originally. We are asking therefore whether there is reason to suppose that *QL* exists and that it meets the conditions we have just set out; and whether it instantiates a respectable notion of logic.

Fully identifying the logical concerns of *QM* (or any other empirical science, for that matter) is a bigger task than we have time for here. Even so, certain undertakings stand out as fairly obvious. For example, it is certainly part of the logic of an empirical science to articulate a principled distinction between its valid and its invalid deductions and its correct and incorrect entailments.

Logicians have long held that the laws of logic are prior to those of the other disciplines. In one way of representing this priority, logic is seen – as Wittgenstein saw it in the *Tractatus* – as an account or model of the logical structure of the world. There are two ways of reading this notion, one silly and the other interesting. On the silly view every truth about the world instantiates a law of logic. On the interesting view, no truth about the world contradicts a law of logic (minimal unity of science, again).

If, for example, a logic must accommodate strong logicism, then nothing counts as a logic unless it accommodates everything that counts as mathematics, including the mathematics peculiar to *QM*.

Here are some further questions that will guide our enquiry. If logic is, or is in part, a theory of certain properties of linguistic structures, and if the type of language that is appropriate to such an investigation is an ideal or artificial language, how can it be the case that the logical laws of such structures have any bearing on natural language structures, such as deductions in *QM*? (We return to this important question in Section 5.2 below.) Another question is that if logic is, or is in part, the theory of certain properties of reasoning, is it necessary to restrict the notion of reasoning to what human beings do? Or is there room for a logic of reasoning for any type of system for which a concept of higher order computation is definable?

Answers to these questions assist the would-be quantum logician in a non-trivial ways. If strong logicism is true and if its accommodation is a requirement of anything aspiring to be a logic, then *QL* must exist and must be very different from classical logic. Strong logicism is the view that all of mathematics can be reproduced without relevant loss in logic. *QM* contains some mathematics that classical logic cannot accommodate. Hence, on the present assumption, *QL* would be a needed alternative to classical logic in the quantum domain. If, on the other hand, logic is a theory of reasoning, or of aspects of reasoning, for any system for which the concept of higher order computation is definable, then there is reason to suppose that accounts of *quantum computation* may embed a nonclassical logic.

Examples could be multiplied, but we shall forbear for the present. We have exposed enough of the structure of logic's pluralism to be getting on with here.

Here is a point on which we wish to lay some emphasis. It is one thing to show that a purpose-built *QL* is needed or deserved by *QM*. It is another thing entirely as to what such a logic would look like, apart from its not looking classical. In this chapter we concentrate on the first issue. Concerning the second we leave a promissory note.[2]

2. Feynman's Problem

We can now say something more definite about what our starting position is. If it is necessary or justified to posit a *QL*, the following conditions must be met:

1. The positing of *QL* is driven by particular traits of the *physics* of *QM*.
2. Whether or not these physical traits constitute a refutation of classical logic, *QL* itself must be nonclassical.
3. The objective of *QL* is to discharge the *logical* requirements of *QM*; e.g., it must provide an account of validity which honours the distinction between valid and invalid *QM*-deductions.

What, then, is it about the *physics* of *QM* that calls out for a purpose-built nonclassical logic? A loose and informal (and very common) answer is the utter queerness of *QM*.

What is so special about quantum mechanics? What makes it "utterly queer"? And is it *these* considerations that call for a special logic? Of course, one automatically thinks of startling phenomena such as nonlocality or Bose-Einstein condensation; but Richard Feynman has argued that the deepest puzzle is that we can't see why anything so terribly *simple* is true.

QM *simple*? How can this be? Aren't the elaborations of quantum theory often of great mathematical complexity? But, as Feynman points out, the basic rules are terribly indeed simple and can be grasped by anyone with a slight familiarity with *probabilities*, *complex numbers*, and the most elementary parts of *linear algebra*. Here we review just enough of the structure of quantum theory to make Feynman's point clear. It should also be enough to satisfy condition (1) of three paragraphs ago.

The basic notion is that we think of a physical system as being capable of passing from some initial prepared state, via some definite procedure acting on the system, to a final output state. Input (or preparation) states are represented by so-called *kets* of the form $|\psi\rangle$; these are column vectors in a Hilbert space, which is a linear space of complex-valued vectors. There are many possible Hilbert spaces with different dimensionalities, depending on the number of states that the system can assume. By saying that a Hilbert space is linear, we mean that the superposition principle holds: any linear combination of allowable state vectors is an allowable state vector. (There are some special restrictions on superposition, called superselection rules, but we need not be concerned with them here.)

Output (or outcome) states are represented by *bras* of the form $\langle\phi|$, which are row vectors; the components of a bra are the complex conjugates of the corresponding ket. The scalar product of a bra and a ket is represented by a *probability* or *transition amplitude*, written as a 'bra-ket' of the form $\langle\phi|\psi\rangle$. Although Dirac notation can be manipulated with great ease and read in any direction, if the transition amplitude is read from right to left (like Hebrew) it has a natural interpretation as tracing the evolution of the experimental set-up from preparation to outcome. Such transition amplitudes are complex numbers of the form $e^{i\theta}$, where θ is a phase angle. All the predictions of *QM* come directly or indirectly from phase relationships between transition amplitudes.

State vectors (bras and kets) can be transformed into other state vectors by linear operators, represented by square matrices. If a linear operator $\hat{O}$ is such that

$$\hat{O}|\alpha\rangle = \alpha|\alpha\rangle, \tag{1}$$

where α is a real number, we say that the operator is Hermitian, and the state $|\alpha\rangle$ is said to be in an eigenstate of $\hat{O}$ with eigenvalue α. In the conventional reading of *QM*, Hermitian operators represent possible observations on the system, and are called *observables*; the spectrum of eigenvalues are the possible results that one could obtain. (In some versions of the modal interpretation, discussed below, this eigenvalue-eigenstate rule is modified.) Some eigenvalue spectra are discrete (such as spin states, or the energy states of bound systems); others are presumed to be continuous, such as the energy of open systems. (It may eventually turn out that the appearance of continuity for some observables is an artifact of fine-graining.)

Observables can also be thought of as definite procedures applied to prepared systems. The probability amplitude to move from input $|\psi\rangle$ to output $\langle\phi|$ by means of procedure $\hat{O}$ is

$$\langle\phi|\hat{O}|\psi\rangle. \tag{2}$$

The *probability* that the system will undergo this transition is given by

$$|\langle\phi|\hat{O}|\psi\rangle|^2. \tag{3}$$

This is a real number, and the description can be normalized such that it comes out in the interval [0, 1]. In other words, to get from probability amplitudes to probabilities, we square up (i.e., take the modulus) of the complex amplitude to get a real number. Or to put it another way, in *QM* probabilities have complex-valued square roots, a fact that has no counterpart in classical probability theory. This notion (which remains to be defined more precisely) of 'square root of a classical concept' may have quite wide applicability, and points toward a way in which the quantum view is a natural generalization of the classical view.

Suppose now that the system can be taken from an initial state $|\psi\rangle$ to a final state $\langle\phi|$ via two possible operations $\hat{O}_1$ and $\hat{O}_2$. Let us suppose that these two routes are mutually exclusive, and – most important – that when the system arrives in its final state we cannot tell, without further experimentation, which route the system took. Then the probability that the system will get from the initial to the final state will be of the form

$$|\langle\phi|\hat{O}_1|\psi\rangle + \langle\phi|\hat{O}_2|\psi\rangle|^2. \tag{4}$$

We add the amplitudes *first* and then square up to get the probability. Now, had the systems been classical (meaning in important part that it would be possible in principle to distinguish which paths the system took *by independent means*, without disrupting the system), we would find the probability that a system followed either of two mutually exclusive paths by directly adding the probabilities for each path. In other words, in the computation of probabilities in quantum systems there is an *extra* step (i.e., summing up the amplitudes). This step makes all the difference in the world, for transition amplitudes are complex exponentials, and if they are not perfectly in phase there will be interference terms that have no counterpart in the classical realm.

As Feynman et al. say, that's it; all of quantum theory is merely an elaboration upon these simple rules.[3] But these rules are as deep as anyone has been able to go: 'We have no ideas about a more basic mechanism from which these results can be deduced' (Feynman et al. 1965, p. 1–10). And this strikes us (and has struck others) as the deepest mystery about *QM*, nonlocality notwithstanding: why can we not see the reason for a mathematical structure that is so utterly simple, and yet so widely applicable? It should be *obvious* why something as basic as this is the way to go – but it is not. In the following we shall call this the Feynman Problem.

Feynman et al., state correctly that we get interference so long as we do not have a way of distinguishing the paths in the experiment; but how would you distinguish the paths that the system can take? The answer is that, one way or another, you would have to end up measuring observables that fail to commute with the first observables you thought you were measuring. That's what it takes to destroy interference. For instance, if you try to pin down the trajectories of the particles as they zip through a double-slit apparatus, an apparatus which measures the positions of the particles as they hit a detector plate, you have to measure their momentum vectors; this destroys the interference because position and momentum fail to commute.

So to explain the mathematical structure we must take account of the *physical* fact of non-commutativity. However, we still don't have a complete story of why non-commutativity forces upon us the Hilbert space structure. And, then of course, there is the wholly central question as to whether the contradictions implied by commutativity assumptions (such as the Kochen-Specker paradox; see Bub 1997) are *fruitfully* construable as logical.

Very well, then. We have this m athematical structure to take note of. It is in a number of ways odd, even though it is not as mathematically complex as it appears to be conceptually. Is there any feature of it that tells us that we must have a purpose-built **QL**? If there is, it is far from obvious that there is. So we shall keep looking.

3. Lattice Theory: A Mined-out Vein?

Historically, quantum logic originated in the 1936 paper of Birkhoff and von Neumann. It arose as one of several rival interpretations of *QM*. Birkhoff and von Neumann's essential idea was that we try to read a logic off from the mathematics of quantum theory. They did not presume to offer a deeper analysis in terms of which *QM* could be explained; rather, they seem to have realized from the outset that they were simply *redescribing* the formalism of *QM* in a way that would, they hoped, be more perspicuous. And they showed us that while a classical, or more precisely a Boolean logic has the structure of an orthocomplemented distributive lattice, quantum lattices are nondistributive – a mathematical condition that reflects or encodes the physical fact of non-commutativity, the fact that observable quantities come in conjugate pairs.

So conceived, the distinction between quantum and Boolean lattices is essentially topological; they connect in different ways. That is why a quantum lattice cannot be mapped homomorphically onto Z_2, the simplest Boolean lattice. This is the same thing as saying that it is not possible to evaluate all possible observation-claims about a quantum system at once. It is not that they are not all known at once, but rather that it is contradictory to even suppose that they all have a truth value at once! (We are not aware whether anyone has done any studies of quantum lattice theory from an explicitly topological point of view, but it seems that this approach could be fruitful.)

This non-Booleanity was anticipated by Schrödinger in the same paper of 1935 in which he annunciated his cat paradox:

> At no moment does there exist an ensemble of classical states of the model [the quantum mechanical system in consideration] that squares with the totality of quantum mechanical statements of this moment. ... if I wish to ascribe to the model at each moment a definite (merely not exactly known to me) state, or (which is the same) to *all* determining parts definite ([not] merely not exactly known to me) numerical values, then there is no supposition as to these numerical values *to be imagined* that would not conflict with some portion of quantum theoretical assertions. (Schrodinger 1983(1935), 156)

It is this apparently irreducible non-Booleanity that is behind the several 'no-go' theorems of quantum mechanics, the most central of which is the Kochen-Specker Theorem. (See Bub 1997 for extensive review.) A no-go theorem places limitations on our attempts to reproduce the predictions of quantum theory with a Boolean model. It is often said that such a model would be expressed in terms of so-called 'hidden variables' – although this is a misnomer since the most promising hidden variable theory, the Bohm-de Broglie causal interpretation, is based simply on position as a determinate variable. What really counts for the 'ontological' aspirations that motivated Bohm (and other authors such as J. S. Bell who sought to interpret *QM* in terms of 'beables') is whether or not the statistics of *QM* can be underpinned by a Boolean structure (Bell 1987; Bohm and Hiley 1993; Bub 1997).

It is undoubtedly too soon to say that Bohm's project must be accounted a failure, although we must go on record as conjecturing that it will be very unlikely that anyone can find a way to make the non-Booleanity of *QM* 'go away.' As Bub says, 'The really essential thing about a quantum world is the irreducible indeterminism associated with non-commutativity or non-Booleanity' (Bub 1997, 240). Does this mean that Bohm's efforts were wasted? No; the really interesting thing that Bohm showed, almost in spite of himself, was the pervasive *nonlocality of dynamics* in *QM*. But it would take us too far afield to justify this claim here.[4]

In summary: it would be nice if the Birkhoff-von Neumann logic could give us insights into why the quantum world is the way it is. However, in the end, we simply read this logic off the physics. This gives us an interestingly different way of looking at the physics, but it could not be convincing as any sort of explanation of why quantum phenomena are the way they are unless quantum logic had some sort of *independent* motivation. And this is what we utterly lack; the facts of *QM* were forced by Nature upon more or less unwilling but open-minded physicists such as Planck. Nature said, if you want to get good predictions, these are the sorts of calculations you have to do. Compare this with classical statistical mechanics, which depends upon the classical laws of probability, themselves having an intuitive justification in terms of elementary set theory and Boolean reasoning. There is nothing like that for *QM*.

We now have a clear question to ask. Classical propositional logic is Boolean. *QM* is non-Boolean. Doesn't this mean that, at a minimum, *QM* requires a non-Boolean propositional logic? It is true that the vertices of the lattices can be interpreted as propositions of a sort, and the meet and join operations have a formal resemblance to the meet and join of classical logic; but it is not clear that we do any discursive *reasoning* in the language of quantum mechanics – we just calculate with it. If you look through the pages of any paper or text using quantum mechanics, you will find quantum-mechanical calculations set within text written in perfectly ordinary classical language, with the reasoning done by means of classical natural deduction, and with the predictions of the theory being interpreted and described classically. No physicists, to our knowledge, actually use quantum logic in their day-to-day work. Furthermore, one can point out that quantum logic has no obvious *semantics* (unless we follow Dickson (2001) and just say that the entire lattice structure itself is the semantics). There is a very clear syntax, but we are not at all sure how to interpret it. (Indeed, this is another way of expressing the Feynman problem.) Is a logic without a semantics a logic at all?

This discussion harks back to important conceptual questions about the relationships between mathematics and logic. Is all of mathematics a form of logic, as Russell and Frege hoped? Or is logic a form of applied mathematics? What would we say, for example, if it turned out that the *mathematics* of quantum theory permitted interestingly expanded methods of reasoning? So, again, if logic must accommodate strong logicism, a *QL must* be deployed. And if logic is a theory of reasoning or computational systems, a *QL might* also have to be. (We return to this point in Section 5.8 below.)

But we have left a question dangling. What if a theory – any theory – has a non-Boolean propositional structure. Doesn't that show decisively that its propositional logic must be (at least) non-Boolean? Let us see.

4. Does Quantum Physics Discredit or Undermine Classical Logic?

QM is formulable in any scientifically mature natural language. This is important. Logicians long since have wondered what the logic of a natural language is, and this has given rise to some peppery contentions. If we know how to settle this more general question, this might help us box our compass with respect to those parts of natural languages in which *QM* is formulated.

Classical logic is the extensional logic of truth functions extended to a theory of quantification. There is a wide (though hardly unanimous) consensus that in such a logic quantification is restricted to domains of individuals; hence classical logic is first order logic.

Classical logic is a theory of certain properties definable for linguistic structures. These latter are constructions of elements of a designated formal language, which is a language whose atomic sentences are uninterpreted and whose molecular sentences are constructions of atomic sentences attached to connectives and/or quantificational symbols, both of which bear weak interpretations by way of truth conditions. The net effect is that the sentences of classical logic are devoid of *propositional* content.

The target properties of classical logic are properties such as consequence, truth in a model, entailment, consistency, and so on. These properties are either properties of linguistic structures alone, or of linguistic structures in virtue of relations they bear to non-linguistic set theoretic structures.

4.1. Classical Logic

Consider a logical truth S of classical logic, e.g., "$q \supset (p \vee q)$". To say that S is a truth of logic is to say that S is true for all valuations. Valuations are functions taking truth values into truth values. In our example, if 'p' and 'q' are both true, then so is S itself. Consider now any pair of true sentences, E and E^*, from an empirical theory. Suppose that we assign E and E^* to p and q in S. Are there *any* such substitutions in S that produce a false sentence? The answer is No. Any such interpretations of p and q in S will either have a classical truth value or not. If the former, the resulting interpretation of S will be a logical truth by the classical definition of logical truth. If the latter, the resulting interpretation of S is defective. If either E or E^* is not classically truth-valued, then it is not an admissible interpretation of p (or of q, as the case may be), since it is a requirement of p and q that they be classically truth-valued. So there is no admissible empirical interpretation of a logical truth that itself is other than true.

Similar considerations apply to the quantificational component of classical logic.

4.2. Quantum Logic

Here is the nub of the problem at hand. There are quantum states that are complexes of simpler structures. Some of these states are describable by sentences in the form '$X \oplus (Y \otimes Z)$', in which '$\oplus$' denotes an operation which set theorists call *join* and '$\otimes$' denotes an operation which they call *meet*. Now join is a kind of disjunction and meet a kind of conjunction.[5] So, *informally speaking*, it would not be wrong to read our sentence as 'X or (Y and Z)'.

Of course, classical logic sanctions the law of distributivity. Expressed in purely natural language terms, this law provides that if 'X and (Y or Z)' is true, so necessarily is '(X and Y) or (X and Z)' also true. But in quantum mechanics, if 'X and (Y or Z)' is true it does *not* follow that '(X and Y) or (X and Z)' is true. Indeed both 'X and Y' and 'X and Z' are contradictions in *QM*, if X and Y (X and Z) assert the joint measurement of non-commuting observables. This leads some theorists to the view that the validity of the classical distributivity principle is overturned by certain established facts about the quantum domain.

It is easy to see that this case against classical logic turns on an elementary mistake. It is an error that involves a fundamental misconception about formalization. The claim in question is that there are true quantum physical sentences of the form '$X \otimes (Y \oplus Z)$'. It is demonstrable that there are cases in which a sentence of the form '$X \otimes (Y \oplus Z)$' is true and yet '$(X \otimes Y) \oplus (X \otimes Z)$' is not true. It follows from this that for sentences 'X', 'Y' and 'Z' and for operators '$\oplus$' and '$\otimes$', the distributivity rule fails in the micro-domain. The question is whether this shows that it also fails in classical logic. If so, there will be valuations which make

$$A \wedge (B \vee C)$$

true, and which make

$$(A \wedge B) \vee (A \wedge C)$$

false, where 'A', 'B' and 'C' formalize classical sentences and '$\wedge$' and '$\vee$' are the connectives for truth functional conjunction and disjunction. It take no more than a simple review of the truth table definitions of '$\wedge$' and '$\vee$' to see that no such valuation exist. There *is* no valuation which makes '$A \wedge (B \vee C)$' true and '$(A \wedge B) \vee (A \wedge C)$' false. The distributivity law is valid in classical logic.

Of course, in *QM* 'and' and 'or' are interpreted as $\otimes$ and $\oplus$, for which distributivity fails, whereas in classical logic 'and' and 'or' are interpreted as $\wedge$ and $\vee$, for which distributivity does not fail. But it is no part of the empirical adequacy of *QM* that $\otimes$ and $\oplus$ are correct or even plausible interpretations of the meaning of the English connectives 'and' and 'or'. Nor is it a condition of the completeness and soundness of classical logic that $\wedge$ and $\vee$ capture the ordinary meanings of 'and' and 'or'. It is true, of course, the $\{\otimes, \oplus\}$-pair and the $\{\wedge, \vee\}$-pair are incompatible interpretations of 'and' and 'or'. But since the $\{\otimes, \oplus\}$-pair have no occurrence in classical logic, there is no treatment of 'and' and 'or' in classical logic in which distributivity *could* fail.

We note in passing Hilary Putnam's bold claim that since distributivity fails for the English connectives 'and' and 'or', *QM* gets them right and classical logic gets them wrong (Putnam 1975a). This may be so, but it is of no mind. Neither *QM* nor classical logic is a linguistics for 'and' and 'or' in English.

We conclude, therefore, that there is nothing good to be said for "the fundamental claim" of quantum logic:

> *QL* claims that quantum logic is the 'true' logic. It plays the role traditionally played by logic, the normative theory of right-reasoning. Hence the distributive law is wrong. It is not wrong 'for quantum systems' or 'in the context of physical theories' or anything of the sort. It is just wrong, in the same way that '(p or q) implies p' is wrong. It is a logical mistake, and any argument that relies on distributivity is not logically valid ... (Dickson 2001, S275)

We note that it is a consequence of the constraints that Putnam places on $\neg$ and $\vee$ that the classical rule, disjunctive syllogism, also fails. This suffices to suppress *ex falso quolibet*, the classical theorem that establishes the equivalence of negation-inconsistency and absolute inconsistency. This makes Putnam's system a paraconsistent logic. In some ways, this is an attractive outcome, in as much as it innoculates the purported logic of *QM* against any of the contradiction-making paradoxes. (Another way of saying this is that a paraconsistent logic is not intrinsically hostile to paradox. For an exploration of old quantum mechanics from a paraconsistent viewpoint, see Brown (1992).)

On the debit side is the complexity of systems that lack laws such as disjunctive syllogism. To take just one example, the propositional system *LR* is a system of relevant propositional logic of the Anderson and Belnap kind (Anderson and Belnap 1975), except that the distributivity principle fails in *LR*. This makes the decision problem for *LR at best* ESPACE hard (Tennant 1993, 7).

4.3. THE QUANTUM QUESTION AGAIN

We now turn to the question of the kind of fit that can be constructed between *QM* and the semantics of **Pred**, the predicate calculus. For example, if the basic entities of *QM* are *not* representable as arbitrary elements in an arbitrarily large domain *D* of individuals, then the issue is settled. The sentences of *QM* could not have classical models. Hence they could do no violence to the sentences of **Pred**, which do have classical models. On the other hand, if the quanta of the quantum domain are legitimately construable as first-order individuals, then they are classical objects model-theoretically.

It is also necessary to determine whether the predicates of *QM* are representable as n-tuples of model-theoretically classical objects from D. If they are not, then the issue closes negatively. If, contrariwise, they are, then the properties of *QM* have classical representations.

But what of quantifiers? In a classical (Tarski-Henkin) semantics, a universally quantified sentence $\ulcorner \forall \alpha_k(\Psi) \urcorner$ is satisfied by a countably infinite sequence σ of objects in D if and only if every countably infinite sequence of such objects that differs from σ at most in its kth element satisfies Ψ. We note a particularly important condition on sequences: Sequences of individuals must be enumerable; i.e., they must stand in a one-to-one correspondence with the the natural line. It must be possible to refer unambiguously and with definiteness to the kth element of every sequence. That is to say, the members of D must satisfy strict conditions on *individuation*.

Some logicians are of the view that since the basic entities of the quantum world are intrinsically stochastic, they lack the determinancy required for individuation. To say that objects are not well-individuated is, among other things, to say that there is no function that takes them to the natural line. If this is right, then two consequences fall out. One is that the quantified sentences of *QM* can't be modelled classically. The other, relatedly, is that it is a mistake to suppose that the basic entities of *QM* are in the classical sense members of *D*. If so, *QM* lacks a classical semantics.

It is important, by the way, that we emphasize the distinction between a theory's ontology and a theory's model. Ontologies are structures that facilitate the distribution of truth values over a theory's interpreted sentences. Models are set theoretic structures which facilitate the distribution of target logical properties over (sets of) its uninterpreted sentences. From the point of view of model theory, a model is all the ontology that an uninterpreted theory can have. By these lights, then, the original *QL* of Birkhoff and von Neumann may qualify as an ontology for *QM* but not as a model for it.

5. The Breakdown of Bivalence; or *ex superpondendo quolibet*?

We shall have no more to say about whether *QM* contradicts any law of classical logic. Our position is that it does not. And that, as they say in a presently popular TV game-show, is our 'final answer'. Much more interesting is whether, granting its consistency with *QM*, classical logic meets our condition (of section 3); i.e., takes care of the *logical* business of *QM* itself. Well, what *is* the logical business of *QM*?

One of the most logically dramatic features of *QM* is the apparent collapse of bivalence. Not every sentence about quantum states will be either true or false without exception. This again follows from the 'no-go' results such as the Kochen-Specker Theorem, noted above. As Bub has it, 'we can take propositions corresponding to the properties in the property state as true, but we can't take the propositions that correspond to all the properties that are not in the property state as false, for this will involve a contradiction' (Bub 1997, 31).

It also appears that if we must say that Schrödinger's cat is both alive and dead at the same time, we are open to all the consequences of a classical contradiction – in particular, that any proposition whatsoever could be deduced. Let us call this phenomenon *quantum detonation*. But this sort of logical detonation takes a special form in quantum mechanics.

In classical logic we have the principle *ex falso quolibet*: from a classical contradiction we can deduce any proposition whatsoever. Quantum mechanics, however, does not exactly allow us to infer from the supposed contradiction of the cat being in a superposition of live and dead states to the conclusion that any state whatsoever is allowable; rather, it says that any superposition (linear combination) of allowable states is an allowable state. But there are often a lot of ways in which a number of given states can be linearly combined, including states that violate locality and other classical expectations. Hence we might say, with tongue in cheek, that in quantum mechanics we should replace the classical *ex falso* with *ex superponendo quolibet.*[6] If it is not strictly accurate to say that *anything* can happen in quantum mechanics, it is certainly true that some very strange things can happen. Strictly speaking, for instance, it is only highly improbable, but *not* impossible, that the next time Harry steps out of his office at his University that he should find himself in the court of King Henry VIII. We have something in QM that is analogous to what happens in statistical mechanics, according to which it is possible (though *highly* improbable) that a pot of water might freeze solid the next time it is put on a gas flame. One can hardly say that there are no constraints in nature, but the few truly global constraints (such as mass-energy conservation) that do seem to stand up tend to be of an extremely general nature; and within those broad constraints the laws of probability are free to work their magic. Sam Treiman's sardonic remark comes to mind: 'Impossible things usually don't happen' (Benford 1998).

Thus in quantum mechanics we really do stand, as Pitowsky aptly puts it, "on the edge of a contradiction" (Pitowsky 1994). But, as noted above, it is still not enough to discredit Boolean reasoning *where Boolean reasoning applies* (technically, because meet and join are defined differently for quantum and Boolean lattices). There is a discontinuity between the Boolean and the quantum world (indicated, as we have noted, by the essential discontinuity between Boolean and quantum lattices) that acts as a kind of protective membrane sealing off the classical world from the contagion of logical paradox. Does this really mean that we live in two metaphysically distinct worlds? No thinker with an instinctive taste for the unity of science will find this a comfortable conclusion. As we shall note below, however, there is a way out of this colossal dilemma, which is to consider the possibility that the distinction between quantum and classical talk is *modal*. We return to this point in Section 5.4.

5.1. Putnam's Analogy With Riemannian Geometry

Hilary Putnam (1975a, b)) has suggested that a very attractive analogy may stand between the relationship between quantum and Boolean logics, on the one hand, and Riemannian and Euclidean geometry on the other. Gauss and Riemann famously showed that ordinary flat Euclidean geometry (which Kant thought was *a priori*) can be seen simply as a special case of more general curvilinear geometries; namely, the special case in which the curvature is zero. Einstein then found that in order to describe gravitation in a way that accords with the general principle of relativity (no preferred frames, accelerated or otherwise), one must presume that the geometry of space-time is, in general, Riemannian. So Putnam has argued that just as Nature tells us that Riemannian geometry is the appropriate generalization of Euclidean geometry, Nature is also telling us that quantum logic is the appropriate generalization of Boolean logic.

There is one difficulty with this appealing analogy. There is a smooth transition between a curved and a flat space; the curvature can go continuously to zero. However, as noted above, the difference between quantum and Boolean lattices has to do with how they connect, and there is no *smooth* transition between structures with different connectivity. (This difference between quantum and classical lattices is a consequence of non-commutativity and the existence of a finite quantum of action.) This is *not* to suggest that it is impossible to embed Boolean logic within a more general quantal scheme; as we shall see, the truth is quite the contrary. But Putnam's analogy must break down. The best we can do to salvage it is to note that there is frequently (but by no means inevitably) a *statistical* transition from quantum to classical behaviour; for instance, quantum interference phenomena will often be washed out in systems with large numbers of particles, so that the behavior of such systems will tend to *numerically approximate* the behavior of purely classical systems.

5.2. Trivializing Validity in Quantum Contexts

As we said at the beginning, an important question for any logic is whether it has interesting external *applications*. There are two main ways in which this question receives an affirmative answer.

1. A logic has an interesting application to natural language structures when, or to the extent that, for certain target properties, the natural language structure instantiates that property in virtue of *logical forms* recognized by the logic.

   EXAMPLE: Arguments in English that are *valid* in virtue of having a valid form in first order logic.
2. A logic has an interesting application to natural language structures when, or to the extent that, for certain target properties the natural language structure instantiates that property in virtue of its satisfying the logic's *definition* of that property.

   EXAMPLE: The argument
   (a) The shirt is red
   (b) Therefore, the shirt is coloured
   has no valid form in first order logic; but it satisfies that logic's definition of validity, namely, that any valuation making (i) true also makes (ii) true.[7]

How, then, do these matters bear on nonbivalence?

The short answer is that nonbivalent sentences have no formalization in classical logic. So deductions from nonbivalent premisses cannot, even if valid, be so in virtue of having valid classical forms.

Beyond that, let **I** be a class of nonbivalent English sentences. Let D be any deduction of a sentence of English from a set Σ of **I**-sentences and let $\mathbf{I}_{qm}$ be a subset of **I** containing nonbivalent quantum sentences. It is easy to see that all such deductions are classically valid. Since there is no valuation that makes the sentences of Σ true, Σ is classically inconsistent. But any statement is classically deducible from an inconsistent set of premisses.

We have it, then, that *all* Ds in *QM* satisfy the classical definition of validity, and therewith is lost the essential distinction between valid and invalid quantum deductions. It is obvious that if *QM* has any logical business to perform, it is to show a certain favoritism for valid, rather than invalid deductions. The classical account of validity cannot serve that end. So *QM* has a stake in someone's providing an account of validity that *can* serve that end; and this, we may suppose, is a fundamental task of *QL*. However, for this to be the case, it is not in slightest degree necessary that any law of classical logic fail.

It is well to be clear about what motivates the quest by quantum logic for quantum validity. If classical validity were the only validity, then every deduction from $\mathbf{I}_{qm}$-sentences would be valid. This alone would dispossess validity of any reasonable standing as a deductive target for quantum physics. It follows, then, that either validity sets no applicable or defensible deductive standard for quantum physics or that the validity of quantum deductions is non-classical.

So long as we retain the assumption that quantum theory contains **I**-sentences indispensably, then the set of *QM* sentences will fail to have the Lindenbaum property. That is, there will be non-contradictory sentences that cannot be extended to a consistent and complete set of sentences K such that for any wff Φ, either $\ulcorner\Phi\urcorner \epsilon K$ or $\ulcorner\neg\Phi\urcorner \epsilon K$. Informally, the K of a theory T is the set of all its (logically) true sentences. If T's every sentence is such that either it or its negation is in K, the sentences of K must without exception have the classical truth values. Accordingly, were a theory to contain an **I**-sentence, it cannot be consistently extended to a K that is in the requisite sense complete.

So far, we have been considering the consequences of supposing that some sentences of *QM* are indispensably members of a set of **I**-sentences. The loss of classical bivalence is deeply enough consequential to motivate a quantum logic even at the level of sentential logic. Since the story of the classical truth values T and F is in turn laid out in the semantical theory of the quantifiers, we may expect some corresponding deviation from classical norms in the semantic theory of *QL*'s non-classical truth values. Another way of saying the same thing is this. The classical logic of quantification gives a model theoretic account of the logical truth values. It is easy to see that no such account can be a wholly accurate story of the *non-classical* truth values. We may expect in turn that the model theory that delivers the goods for the nonclassical truth values will also make nonclassical provision for the quantifiers themselves.[8] This expectation is met in the failure of the Existential Instantiation rule. For we may have it that it is provable that some individual (i.e., particle) satisfies an open sentence $\ulcorner\Phi\upsilon\urcorner$ but no individual α satisfies $\ulcorner\Phi\upsilon\urcorner$. For this to be a coherent outcome we need a model theory in which quantification is possible even when individuation is not. In classical model theory, quantification and individuation go hand in hand; 'no entity without identity', in Quine's nice quip.

Just as we had it that the physics of *QM* gave a principled reason to deny classical bivalence to at least some of its sentences, so too would we expect that the failure of the classical quantifier law is occasioned wholly by the physical provisions of *QM*. For this to be so, it is essential that the ontology of physics not be that of classical individuals and the requisite set theoretical constructions on them.

A final conjecture regarding the breakdown of bivalence in *QM*: just as probabilities in *QM* have complex-valued square roots (i.e., the corresponding probability amplitudes), perhaps it will be possible to define something like a square root of a truth value. But we shall have to leave this as a conjecture for now.

5.2.1. *Further Conditions on a QL*

We now find ourselves in a position to say something more about what a *QL* should be. Given classical nonbivalence, we have reason to postulate a *QL* and to mandate it to produce a validity relation that will facilitate the logical work of *QM*. What else might we say of *QL*? We might propose that *QL* is a bona fide logic for *QM* only if

1. for some deductions in *QM* it is essential to their correctness in *QM* that they employ a nonclassical consequence relation, C, supplied by *QL*
2. C is not a physical relation.

Similarly, *QL* is a *bona fide* logic for *QM* only if

1. for some truths of *QM* it is necessary that they be expressed nonclassically, i.e., they be in the domain or counterdomain of a nonclassical truth-predicate T_{QM} (perhaps involving 'square roots of truth' as hinted at above) supplied by the model theory of *QL*;
2. T_{QM} does not denote a physical property.

To establish that *QL* is a real or respectable candidate for logic, it suffices to find a deduction in *QM* or a truth of *QM* that meets these conditions. For example, suppose that a given deduction D is up for scrutiny. D passes the test if and only if it is provable that D deploys a nonclassical, non-physical consequence relation in delivering the deductive goods intended by *D* and that no classical relation of consequence can deliver those same goods.

5.3. Quantum Individuation

There is a large literature that bears directly on identity in *QM*.[9] Several authors argue that the intuitive idea of numerical distinctness simply does not apply to quantum objects. Paul Teller offers an interesting analogy. Suppose – adapting Teller – that you transfer from a bank in, say, Vancouver to your dollar account in a bank in London 100 dollars, and that later in the same day you transfer another hundred from Vancouver to London. In the week following, you find yourself in London and in need of some cash. You present yourself at your bank, and ask not only for a hundred dollars but the very hundred dollars that was your first deposit last week. Of course, there is no such thing for the bank to give you. The two hundred dollar sums are not numerically distinct. The familiar metaphysical distinction between qualitative and numerical or strict identity breaks down in such contexts (Teller 1998, 115).

It is widely supposed that *QM* presents situations that resemble our banking example, or at least resembles it enough to generate the collapse of the distinction between qualitative and numerical identity. Many writers ascribe a common cause to this collapse. Quanta, unlike classical particles, fail the principle of the indiscernibility of identicals (see e.g., van Fraassen 1998), which means that quanta cannot instantiate the relation of numerical identity. Again, the distinction between qualitative and numerical identity topples in the quantum domain.

The fact (or the appearance) of the damage done this distinction by quantum objects presents the would-be quantum logician with a twofold task. His negative task is to show that the loss of strict identity carries implications for any model for quantum discourse that preclude classical workings-up of those properties in which *QM* has a legitimate *logical* interest – for example, validity as a desired characteristic of quantum deductions. But the interested logician also has a positive task. He must describe very carefully the sorts of thing quantum models are, and he must do so in such a way as to show that how they are – especially in their model theoretically nonclassical details – is how *QM* requires them to be. It would be an easy thing to rig models whose basic entities are electronic movements of sums of money. Such could not be classical models, of course; the collapse of identity would see to that. But neither could they be quantum models, even when these electronic entities were treated with considerable abstraction.

The would-be quantum logician is thus faced with a possible problem. Given the present state of our knowledge of the quantum domain, he may be able to perform his negative task with precise and accurate reference to the quantum realities that bring identity to heel; but he might also find it very difficult to find in those quantum details what it is that requires us to jig quantum model theory in the required ways, that is, in ways that both honour the physical imperatives and yet provide the wherewithal to account for something recognizable as logical properties. It is easy to see that the negative task is more easily performed than the positive. Showing the inapplicability of classical models requires no more than giving up on numerical identity. Anyone satisfied that the physics overturns the indiscernibility principle can say with perfect accuracy that his rejection of classical models flows directly from how the *physics* of the quantum world goes. But in offering this, that, or the other nonclassical model as a model for quantum discourse, it is not enough that, in so choosing, the would-be logician honours the physics; for he must also "honour the logic" as well. To do this latter thing it is essential that, in his proferred model theory for quantum discourse, he not merely redescribe *QM* in a new formalism, as Birkhoff and von Neumann did. What the quantum logician wants (or should do) is not an analysis of physical states, but rather an analysis of logical properties of discourse about physical states. He wants, that is to say, an account of *quantum validity* and *quantum truth*.

In discharging his negative and positive responsibilities, it is essential that the quantum logician not fall into the trap of taking every peculiarity or downright oddity of the quantum domain as logically significant, as a second thought experiment, also due to Teller, will show (Teller 1998, 114–116). Imagine two quantitatively identical particles roaming about in a closed box, whose right and left sides are of equal volume. Suppose also that the particles are independent in their motion and small enough to avoid collision. Where at a given time would we find these particles? Intuitively, there are four equiprobable possibilities: they are both on the right side; they are both on the left side; one is on the right side and the other on the left side; or vice versa. However, as Teller observes, there are situations galore in which in quantum contexts these are simply the wrong numbers. The probabilities, instead, are identical for *three* possibilities, not four: Both particles are on the right; both are on the left; and one on each side. One way of capturing the difference between the intuitively expected options and the experimentally indicated options is to say that quantum setups such as this one don't obey classical statistics, and obey instead something we dignify as *quantum statistics*.

Imagine now a logician who reasons as follows: Since the physics of such setups is queer enough to defeat classical statistics and to honour instead quantum statistics, the physics in question is *very* queer indeed. So it will require a purpose-built logic of its own.

Unless we are ready to plead that mathematics just is logic, at least in part, this is a mistake. The mathematics involved amply attests to the queerness of quantum physics. But queerness is not what counts. What counts is that in the quantum setup it is impossible to discern the distinction between the third and fourth options which classical statistics is obliged to posit. In the quantum setup, there is no discernible difference between these same particles exchanging these positions. Not everything that explains this indiscernibility is logically significant. But if it is explained by the fact or apparent fact that there *is* no numerical distinctness between the two particles, then they cannot be the individuals required by classical models. That, of course, would make the indiscernibility logically significant.

It is possible, but far from perfectly obvious, that quantum sentences owe their nonbivalence directly to the fact that *QM*'s basic entities lack the property of numberability.[10] Teller makes an interesting case for saying that what unnumberable objects lack is the haecceititic 'property' of *thisness*. If this is right, there is no quantum body that is this body, hence no situation in which this quantum body differs from that quantum body. Equally, there is no quantum body to which, for any, this is identical to it. This gives unnumberability, right enough. But does it give nonbivalence? It would seem not. If it did, bivalence would then imply numberability. But (reverting now to Teller's bank example), it would seem that the sentences that record the deposits and withdrawals of that example are straightforwardly bivalent without its being the case that for some clump of a hundred dollars it is this clump that was transferred from the chequing account to, say, the Water Company's account. If the Teller example is a good enough analogue of quantum unnumberability, we have a case for supposing that nonbivalence and unnumberability are logically independent properties in quantum contexts.

Still, we are not entirely home-free. We admit to some lingering doubts. We strongly suspect that if the view that particles can be continuously existent entities with continuous trajectories could be formalized in just the right way, it would lead to a Bell Inequality – one that would almost certainly be violated by actual quantum predictions.

5.4. Modal Interpretations: Is Quantum Logic a Modal Logic?

One of the most interesting lines of investigation is the modal viewpoint advocated by Bub (1997), and first put forward by van Fraassen (1981). It is that the state vector does not actually describe a putative reality, but a possibility. This solves one of the central interpretive puzzles. Consider Schrödinger's cat, which exists in a superposition of states before we open the box and look at it. No real cat can be simultaneously alive and dead. It is very hard to interpret the state vector for the cat realistically. However, the possibilities for incompatible outcomes can easily co-exist. It is therefore very natural to think of quantum mechanics as a calculus of the evolution of *possibilia*. And quantum logic must therefore be a species of modal logic. And, in fact, quite a lot of work has already been done to interpret quantum logic as a modal logic. (For review, see della Chiara 1986.)

On the modal view, we can interpret kets as possible input states – states whose possibility is consistent with the way we know the system was prepared, while bras are possible output states. The mysterious extra step in the *QM* algorithm is therefore a move from possibility to probability, a distinction that is entirely collapsed in the classical, Boolean picture.

It may be that Bub is on the right track in insisting that this insight has to be taken very seriously. However, much work done on the modal interpretation has possibly been side-tracked into the pursuit of a red herring. It focuses on the problem of finding a way of assigning a so-called determinate variable, a quantity that can be presumed to be definite, and thus represent the closest approach we can make to a classical realism, throughout the course of a physical investigation – *despite* the modality of quantum talk.

This work has culminated in a recent 'uniqueness' theorem by Jeffrey Bub and Rob Clifton (see Bub 1997). It shows precisely *how definite* one can be in the face of quantum indeterminacy. It is always possible to assign some observable as 'determinate' in the course of a physical investigation. But, at the same time, we find ourselves right back where Schrödinger warned us we would be: it is, in general, impossible to make *all* the variables pertaining to a given system determinate at once. We can move quantum indeterminacy around, but we can't make it go away.

The modal interpretation therefore poses a philosophical challenge: if our very best physical theory (best in the sense of having a great predictive power over the widest range of phenomena) can do no better than describe the world in terms of *possibilia* and *probabilia*, if the very notion of something that can *be* is at best a large-scale approximation and at worst a mathematical contradiction, then perhaps we have to simply give up Einstein's hope of finding a theory that can successfully map one to one with the deepest 'elements of physical reality', whatever those elements might be. Perhaps physical theory in the last analysis is simply a highly sophisticated *descriptive schema*, or an elaborate form of applied mathematics. Perhaps some will say that it is not so much that *QM* obeys or expresses a modal logic, but that quantum mechanics *is* a quantitative modal logic in the spirit of the original *QL* of Birkhoff and von Neumann (1936).

The apparent collapse of physical theory as a means for describing what *is* does not imply that there is no such thing as physical reality; we are not advocating some species of idealism, solipsism, or social constructivism here. We merely suggest that (essentially because of non-commutativity) that we cannot hope to faithfully *map* that reality all at once. The notion of the incompletability of physical theory is entirely consistent with the notion of a larger reality that can be partially or wholly independent of human consciousness and observation. We really do not think that the Great Nebula in Andromeda has been shimmering in its glory for billions of years precisely because some anthropoids on a minor planet in a neighbouring galaxy finally noticed its existence (although we are certainly prepared to suppose that human astronomers may have interacted in interesting ways with the wave functions of a few photons from that vast Galaxy).

We said at the beginning "It depends on what you mean by 'logic'." We also said that a logic cannot be a *physical* system, that, in effect, the models of a logical language shouldn't be confused with the ontology of a physical language. If this is right, the modal suggestion presently under review will suggest not a logic, but rather a modal ontology for *QM*. It depends on what you mean by 'logic.'

5.5. Quantum Paradoxes

It is proposed by those, among others, who favour the quantum logical interpretation of *QM*, that a further way in which to attempt to motivate a purpose-built nonclassical logic for *QM* would be to show that the so-called paradoxes of *QM* are solvable if classical logic is replaced by a suitable nonclassical logic. For this to have any real chance of being brought off, it must be demonstrated that at least the following four conditions are met:

1. classical logic has an irreducible role in the generation of the quantum paradoxes;
2. displacement of those classical principles that abet the paradoxes by quantum logical principles meets two important conditions:
   (a) the paradoxes disappear

(b) no damage is done to the physical integrity of *QM*;

3. it is independently ascertainable that the principles of a putative quantum logic are logical principles, and not just principles of *QM* dressed up in a new formalism; and
4. the quantum paradoxes must give genuinely logical offence rather than just being queer.

This is all very problematic. If, as condition (4) requires, quantum paradoxes must be genuinely illogical, it is necessary to ask "By the standards of what logic?" If we were to decide that the relevant standards are those of classical logic, then, if (4) is satisfied, the *QM* paradoxes are classical contradictions or logical falsehoods. But, since the classical rules are provably consistent and sound, the fault must be that of *QM*, not of classical logic. On the other hand, if the paradoxes violate the standards of some other logic, it would have to be shown *independently* that this logic is a better bet for *QM* than classical logic. So it cannot be a sufficient motivation of quantum logic merely that conditions (1) to (4) are discharged.

Here is one such possibility. If *QM* sentences are nonbivalent, then, as we have said, classical validity utterly fails to serve the logical purpose of distinguishing in a principled way between valid and invalid quantum deductions. There is an important similarity between this situation and the *QM* paradoxes. From the point of view of classical logic, every quantum deduction is valid, and, from that same point of view, every sentence of the language of *QM* is true. Just as classical logic can be said to have nothing to offer the quantum theorist with regard to his interest marking the difference between validity and invalidity, neither does it offer anything valuable as regards the *QM* theorist's interest in the distinction between truth and falsehood.

Suppose now that it comes to pass that in the quantum logic, *QL*, that *does* serve the theorist's interest in validity, the paradoxes are underivable, i.e., the deductions of *QM* of which the paradoxes are their respective conclusions are invalid in *QL*. Finally, suppose that nothing else in the *physics* is deranged by this blockage in *QL* of the quantum paradoxes. Under these conditions, not only would we have a way of blocking the paradoxes, it would be a principled way. It would be a consequence of a theory of quantum validity which is already needed because of the collapse of classical validity in quantum contexts.

Even so, this leaves the fact that the quantum paradoxes are *classical contradictions* wholly undealt with. A further step is needed. One possibility is that the quantum logician will insist that classical logic is defective, that sentences of the form $\ulcorner\Phi$ and $\neg\Phi\urcorner$ are not always logically false in *QM*. But this overlooks the fact that the classical logical falsehoods are false under every semantic variation which they are capable of acknowledging. With this in mind, a second possibility is that the quantum logician will insist that the *metalogic* of *QL* cannot be classical. But this turns out to be a costly solution for the quantum logician; for it denies him his argument that the nonbivalence of *QM* occasions the collapse of classical validity in quantum contexts (which is an argument worked up in wholly classical way).

A third possibility is that the state of affairs described by our present assumptions carries only one consequence for classical logic, which is that classical logic has nothing to do with the quantum paradoxes (for recall, quantum paradoxes have no formalization in classical logic). This will be so, however, only if no paradoxical consequence of *QM* has a classical truth value. Supposing this to be so, the quantum logician still retains the considerable burden of showing that our present assumptions are indeed met.

5.6. Further Reflections on *QL*

When, as in the case of *QM*, conditions like these are satisfied, we shall say that the physics itself requires a nonclassical logic. This meets a fundamental constraint on a quantum logic. The physics must be such as to show that it cannot be served by a classical logic with regard to the basic logical properties (such as validity) in which it has a nontrivial stake.

It is one thing to say that a scientific theory such as *QM* cannot have a classical logic. It is another thing to say in appropriate detail what kind of non-classical theory it does or must have. Two things stand out as basic necessities. A quantum logic must display models in which it is ascertainable in a principled way that the "right" sentences came out classically non-truth-valued, and it must develop a theory of nonclassical validity such that the valid deduction of *QM* are valid according to the model's account of validity, and its invalid deductions are invalid in the same way. In this we *may* come to see that the deductions of *QM* are intrinsically mathematical (e.g., set theoretic). This being so, the model theory would have no chance of characterizing a concept of validity that applied in a principled way to the valid and non-valid deductions of *QM*, *unless* it made validity intrinsically mathematical in the model. We could then say that, whereas there is a principled reason to distinguish mathematics from classical logic, the opposite is true of quantum logic and mathematics. The sixty-four dollar question, of course, is what such a notion of validity would look like, and – flowing from it – what is it about the *physics* that makes it so?

Here is a possibility that seems to us interesting enough to introduce now. Given that it is a requirement on any system that aspires to the name of quantum logic that it produce an account of quantum validity, a standard enough way of producing this *kind* of analysis is via a description of the consequence relation. In Engesser and Gabbay (2002), it is proposed that for each ray of a Hilbert space H it is possible to associate one-to-one a non-monotonic consequence relation. This enables us to conceive of the projectors in H as revision operators of a certain kind. It can then be shown that the lattice of closed subspaces of H is a natural generalization of the classical notion of a Lindenbaum algebra. It also emerges that the quantum consequence relations reflect, to a rather surprising extent, their metatheory at the object level. The logic of this approach involves a tight partnership between non-monotonic logic and Hilbert space theory. Without here judging the success of this approach in any detail, we can say that the Engesser-Gabbay account meets our two main criteria. It gives an account of quantum validity and it does so in a way that is motivated by and enabled by peculiarities of quantum physics. So whether ultimately a success in all its details or not, this is the right *kind* of approach.

Much is made of the sheer conceptual difficulty of *QM*, by its queerness, as we were saying earlier on. Some would go further, with harsher verdicts such as incoherence. Often the failure of Existential Instantiation is taken as evidence of these difficulties. This is a mistake. If someone reports a dream in which there were some flowers in a vase on a table but, for no colour, reported that the flowers were that colour, then the flowers of his dream would have been coloured but of no particular colour. (To say that, the dreamer would have overcome this indeterminacy had he been more attentive simply misjudges what it is to report a dream.) Dreams are utterly common experiences, concerning which something like the failure of Existential Instantiation is also common. Then, too, the mathematically-minded seem to be reasonably at home with the ω-inconsistency of first order number theory, in which for some property Φ it is provable that some number lacks Φ even though each individual number has it.

There are still further examples from quite ordinary English to which EI seems rather obviously inapplicable. One whole class of these is mass terms, such as "snow", which take the quantifier and yet have no instances. So 'Some snow fell overnight' is perfectly all right, but there is no name that names *a* snow, hence no candidate for an instantiation of this quantification. This is not to say that quantification over quantum states is quantification of mass terms. But if it is not, what is it, pray?

It is also true that, from 'Some snow fell overnight', there must be inferrable something like '*This* snow fell overnight'. This is interesting. "Snow" is a mass term, but it also takes indexicals. Indexicals particularize occurrences of snow without naming them. This matters in two ways. One is that EI doesn't instantiate to indexicals. The other is that, even if it did, it would give us nonclassical quantification, since indexicals have no occurrence in classical logic.

5.7. Complementarity

One of the most important tasks of the logician is to take the measure of complementarity, considered as a property of physical states prior to the collapse of the wave function. It does little good to for the logician-theorist to consult his "intuitions" about what is and what is not real negation. What counts logically is the bearing that such complementarities have on the truth values of the formulae that describe them. Here are some possibilities to consider.

1. Complementary states contradict one another. In that case, they are fit candidates for a dialethic logic.

2. Complementary states don't contradict one another, even though one at most is realizable. In that case, we might take the states as representing two different but logically incompatible physical possibilities, one of which and only one is made actual by measurement. Again, this might indicate the suitability of a modal logic, in which $\ulcorner M\Phi \urcorner$ and $\ulcorner M\neg\Phi \urcorner$ can both be true (or have a designated value) and yet Φ and $\ulcorner \neg\Phi \urcorner$ must have opposite or conflicting truth values.

We keep saying that the fundamental question for the logician is whether the statements of *QM* take the classical truth values. If so, this works an enormous simplification into the question of the extent, if any, to which *QM* requires or deserves a *QL*. If not, then *QM* is straightaway a many-valued logic. The trouble is that there are more many-valued logics than one can shake a stick at. Which, then, is the one that *QM* demands to have? A related difficulty is sorting out which of these values to designate and which to contradesignate. A further problem is that a logic with an even number of truth values tends to operate classically,[11] whereas a logic with an odd number of truth values can make it hard to get negation (or complementarity) right. (See here Rescher 1968, 89–90.)

5.8. Convention T and Quantum Computation

Consider again our putative set, $\mathbf{I}_{qm}$, of *QM* sentences which fail to have a classical truth value. Informally, the lesson that we are invited to learn from $\mathbf{I}_{qm}$ is not that its sentences *say nothing* (which would be why interrogative sentences lack truth values), but rather that they say nothing *determinate*. Ordinary English has lots of such sentences; e.g.,

"Bill is somewhat angry"

"Sue is rather tall"

"Harry is kind of angry"

Notice, however, that in each of these cases, indeterminacy is no bar to bivalence; for each of these sentences satisfies Convention T. So what is it about the $\mathbf{I}_{qm}$ sentences that is unlike this? What is it that requires us to code up the quantum indeterminacies in the *metalanguage*?

There has been much ado of late about the peculiarities of quantum calculation (e.g., Deutsch et al. 1999, Nielsen and Chuang 2000). This might lead us to suppose that since calculating is a kind of reasoning, and since logic is the science of reasoning, logic must take these quantum peculiarities into account in a principled way.

One of the fascinating consequences of quantum computation is that we have to give up the idea that we should look for the kind of definiteness of result that

Descartes insisted upon. The results will often be merely probable – although if the circuit designer knows what she is doing, she can arrange phase factors such that the circuit will almost certainly compute the answer we want. As Deutsch et al. say,

> The basic idea of quantum computation is to use quantum interference to amplify the correct outcomes and to suppress the incorrect outcomes of computations (Deutsch et al. 1999).

Furthermore, quantum computations, if mapped into sequences of parallel classical computations, would be intractably complex; the human mind must give up all hope of being able to simply follow and thereby check all the steps of the calculation. The price we pay for the greater computational power of a q uantum computer is the near-total loss of any sort of Cartesian 'clear and distinct' grasp of why the result is what it is.

Despite these difficulties, we are drawn to Deutsch's argument that this must change our whole picture of how computation, indeed all reasoning, works. In particular, we think that this carries the implication (which will not make everyone happy) that *all logicians, especially those interested in computation, decidability, and similar meta-issues, must learn quantum mechanics*. This seems almost too much to ask of the chronically-overworked logician. But we don't see how one could get around this. Consider, for instance, the fact that Deutsch (1985) has apparently succeeded in *proving* the Church-Turing thesis (subject to some very natural constraints) in terms of a generalized quantum Turing machine. How can anyone from now on discuss such things as completeness or provability without taking this into account?

Closely following from the last point is that it may be more helpful to think of quantum logical operations in terms of linear operators (represented by matrices), rather than as a problem in lattice theory. The lattice theory vein may well be played out, for now at least. One sees this possibility from the construction of the square root of 'not' as a certain kind of matrix (representing the transformations of a state in a Mach-Zehnder interferometer). (Nielsen and Chuang 2000) The theory of quantum logical circuits seems to represent a natural generalization of classical Boolean circuit theory; the quantum theory can represent all of the Boolean logical operations that can be represented by the classical theory of logic gates, but it can also represent other operations (such as $\sqrt{\text{NOT}}$) that have no classical Boolean meaning. There is, therefore, a clear sense in which Boolean logic can be embedded in a more general non-Boolean theory; the question is how we are to interpret these peculiar 'square roots' of classical concepts. If something like the modal interpretation is correct, then could we say that $\sqrt{\text{NOT}}$ is to NOT as possibility is to actuality?

5.9. Too Many Worlds?

Mention of quantum computation makes it is necessary to venture a very brief comment about the many-worlds interpretation of *QM*, even though Abner Shimony has sagely warned us that 'discussions of the Many-Worlds interpretation tend to take up all the time allotted to them... ' (Shimony 1986).

Having said this, we note that Deutsch has put forward a very challenging point in favour of the Many-Worlds theory, or 'multiverse' theory as he calls it. Quantum computers can sometimes perform calculations very much faster than any possible classical computer, and the way they do this, in effect, is that there is a separate amplitude for each possible state of the circuit and the computation is carried out in all of these Schrödinger-cat circuit-states at once; one has, in effect, massive parallelism. Now, Deutsch says that the fact that such a computer can produce a result, and even out-perform a classical computer, shows that all these parallel computations must be actually taking place *somewhere*; and if they are not taking place in our space-time, they must be taking place in real parallel space-times.

Deutsch, in other words bites the ontological bullet posed by Schrödinger's cat paradox. The cat really is both alive and dead. But we avoid quantum detonation because the alive and dead states are correlated to two distinct observer states, in one of which the observer sees an alive cat; the other a dead cat. These two branching states define distinct worlds which appear classical to the observers in them; there is no logical detonation within any given world, although there is an on-going ontological detonation on a colossal scale in the multiverse.

We have not seen any comment by Deutsch on the modal interpretation of *QM*, but we presume that he would not accept it; the sheer fact that quantum computation works, he would say, argues for an outright realism about state functions. Hugh Everett, founder of the MWT, was quite explicit that there is no such thing in *QM* as the transition from the possible to the actual; the wave function is all actual, always. (Everett 1957, 459–460; cited in Bub 1997, 224; see also papers by Everett and others in de Witt and Graham (eds.), 1973.).

We tend to reject any sort of modal realism – either of the Lewisian variety, or the many-worlds theory which is now so popular with quantum computationalists. It seems to us that modal realism simply fails to accurately construe possibility-talk. The entire point of possibility-talk is to be able to discuss things without ontological commitment. (There could be room, however, for reifying potentia, along Aristotelian lines, perhaps.) But Deutsch argues that the many-worlds (or multiverse) theory is the only conceivable explanation of how quantum computers can carry out computations that are vastly more complex than any classical Turing-machine-equivalent computer can do in the same time. Any sort of logical or mathematical operation is something that has to take place on a physical device, not in Plato's heaven. Any critic of the multiverse theory, Deutsch insists (1997, Chapter 9), must provide an alternate *physical* explanation for the power of quantum computation; *where*, pray tell, does all that computation actually take place?

There are, perhaps, alternative explanations. In a Bohm-de Broglie pilot-wave model, for instance, we suspect that one could write a consistent story according to which the computations are carried out within the pilot wave itself. (Deutsch (1997) rejects this possibility, for unclear reasons.) Another candidate location for quantum computation is the quantum vacuum itself, which probably contains quite enough complexity to account for all the computations a quantum computer can perform, without having to move to alternate space-times. But much work remains to articulate these possibilities in detail.

5.10. THE PROMISE OF QUANTUM COMPUTATION

Certainly, we must concede that quantum computation is a particular challenge for modal interpretations of *QM*; certainly, also, regardless of what interpretation proves to be the most effective in the long run, we are in for some fairly radical ontological readjustments.

Quantum computation is an interesting idea, but care needs to be taken. There are at least two respects in which this is so.

Respect number 1. Mainstream classical logic makes no claim whatever to be a science of actual, everyday reasoning. (See Barwise (1977), Kleene (1967), and Shoenfield (1967)). If quantum peculiarities do enter the reasoning picture, no logic need or should take account of it. *Or* we should change our notion of logic across the board. That is, we should insist that *anything* deserving of the name of logic is required to say something about at least certain aspects of the operation of reasoners or cognitive systems. (See Gabbay and Woods 2001.) In which case, there is nothing special about *QM* in necessitating a logic of this kind.

Respect number 2. "The calculations are different". This could mean that *QM*'s *mathematics* is nonclassical; in which case, it is indeed nonclassical. Or it could mean that there is a certain kind of reasoning whose description must have a *QM* component. If so, then *QM* and cognitive science come together fruitfully. But this has no impact on *logic* unless we are already pledged to the idea that logic is a description of the reasoning behaviour of cognitive agents or cognitive systems, or some such thing.

6. The Unity of Science, Again

What, then, does this come to? As we warned at the beginning, we would not be able to establish many firm conclusions in this exploratory paper. On the one hand, we might note that the pluralisms inherent in trying to say just what logic (any logic) really *is* tends to undermine a Cartesian tree-of-knowledge model of the unity of science. If there are many logics, with different motivations and different structures, then they might not map into each other with sufficient lack of ambiguity to constitute a genuine root-system for Descartes' tree. On the other hand, we have noted that there is considerable evidence (especially through recent work on quantum computation) that Boolean reasoning can be embedded in a broader quantal structure, so perhaps unity can be salvaged after all. This unity would come, if it comes at all, at a price: the conclusions of a quantal computation are inherently probabilistic, and the steps of most quantum computations are so intractably complex as to be utterly unsurveyable by the human mind. Quantum logic must therefore fail to satisfy two of the conditions that Descartes insisted are essential for any discipline to be called a science: certainty and surveyability.

Despite all of these large difficulties, it seems certain that could Descartes be brought back to our time he would have been thrilled and fascinated by the tremendous predictive power of quantum mechanics – a power we cannot deny even though we still have no satisfying explanation of the grounds of its possibility.

Acknowledgements

An early version of this chapter was presented to the New York Logic Group in February 1999. For incisive and helpful comments we thank Jonathan Adler, Arnold Koslow, Rohit Parikh and Alex Orenstein. A revised version was read at the University of Alberta the following March, and additional valuable advice was forthcoming from Bernard Linsky and David Sharp, for which our thanks. The present version reflects the fruitful stimulation provided by participants in Kent Peacock's course on Deviant Logic in Lethbridge in the Fall of 2000, and John Woods' Group on Quantum Logic at the University of Groningen in the Spring of 2001. Our thanks to David Atkinson and Jan Willem Romeyn. A special word of gratitude to our colleague, Peter Alward, for helpful advice. We thank the Social Sciences and Humanities Research Council of Canada, the Engineering and Physical Sciences Research Council of the United Kingdom, and Professor Christopher Nicol, Dean of Arts and Science of the University of Lethbridge, for financial support.

Notes

1 To be more precise, Descartes himself believed that the roots of the tree of knowledge would be metaphysics, or first philosophy; (see, e.g., *Principles of Philosophy*, especially the Letter to Abbé Picot, in Wilson (ed.) 1969). But Descartes' own attempts to construct a rationalistic theology were never convincing, and such projects have been largely abandoned. Instead, we tend, perhaps too uncritically, to think of logic itself has having a kind of metaphysical import, at least insofar as it governs the forms of *possible* worlds; and to this extent the modern version of the Cartesian view is that it is mathematical logic itself that should be at the root of the tree.

[2] To be redeemed in a future work of our. We hardly mean to suggest that we would be the first to develop a quantum logic. Apart from the pioneering work of Birkhoff and von Neumann themselves (1936), one can point to, among other notable attempts, J. L. Bell (1986), Pitowsky (1989), and Gibbins (1987).

[3] Feynman really should have added one more rule, which would be a statement of the value of Planck's fundamental constant of action. If action were not quantized, there would be no uniquely quantum phenomena, and the existence of interference and Bohm's quantum potential (another *sine qua non* of *QM*) must be somehow deeply connected with the basis of Planck's constant. But at present we have absolutely no idea why action must be quantized, much less how to calculate the value of the constant of action.

[4] There exists an offshoot of Bohm's pilot-wave theory called 'Bohmian Mechanics', which has been advocated with missionary zeal by Sheldon Goldstein and co-workers. (See Dürr et al. 1995; Cushing et al. 1996.) These authors downplay the nonlocality of quantum dynamics, and proclaim that 'particles can still be particles'; by which they apparently mean that the Uncertainty Relations are only statistical constraints because position and momentum really *do* commute, after all. It is beyond the scope of this paper to carry out an adequate critique of Goldstein's Bohmian mechanics. We can only point to the abundant (though arguably not completely conclusive) evidence that without fundamental non-commutativity there would be no interference phenomena and thus no quantum mechanics.

[5] To be precise, in a quantum lattice the meet of two vectors is their intersection, while their join is the subspace they span. See Hughes (1981, 1989) for very clear expositions.

[6] We thank Chris Epplett for his dipomatic reminder that we should use the gerund in this expression.

[7] For a more detailed discussion see (Woods 2004).

[8] It may appear, then, that the lack of numberability does give rise to nonbivalence, contrary to what we say late in section 5.3. Again, it all depends. If by bivalence we mean that every sentence is either true or false, bivalence need not be a casualty of a nonstandard theory of truth just because it gives a nonstandard notion of the truth values. But if, as we have already remarked, by bivalence we mean that every sentence takes the *classical* truth values, then we have it trivially that a model theory that gives a nonstandard notion of truth is nonbivalent even if every sentence is, by the lights of that account, either true or false.

[9] See, for example, (Castellani 1998).

[10] Numberability is not to be confused with denumerability, which is a necessary but not sufficient condition of it. Items count as numberable when they can be identified and *reidentified* in experimental settings.

[11] More precisely, if L is a many-valued Kleene-regular logic with equal numbers of designated and contradesignated values, then L is isomorphic to classical logic.

References

Anderson, A. R. and N. D. Belnap: 1975, *Entailment: The Logic of Relevance and Necessity*, Vol. I, Princeton, Princeton University Press.

Barwise, J.: 1977, *Handbook of Mathematical Logic*, Amsterdam, North Holland.

Bell, J. L.: 1986, 'A New Approach to Quantum Logic', *British Journal for the Philosophy of Science* **37**, 83–99.

Bell, J. S.: 1987, *Speakable and Unspeakable in Quantum Mechanics*, Cambridge, Cambridge University Press.

Beltrametti, E. G. and B. C. van Fraassen (eds.): 1981, *Current Issues in Quantum Logic*, New York, Plenum.

Benford, Greg: 1998, *Cosm*, New York, Avon Books.

Bohm, David, and Basil Hiley: 1993, *The Undivided Universe: An Ontological Interpretation of Quantum Theory*, London, Routledge.

Birkhoff, G. and J. von Neumann: 1936, 'The Logic of Quantum Mechanics', *Annals of Mathematics* **37**, 823–843; reprinted in Hooker (ed.) 1975, pp. 1–26.

Bohm, D. and B. J. Hiley: 1993, *The Undivided Universe: An Ontological Interpretation of Quantum Theory*, London and New York, Routledge.

Brown, Bryson: 1992, 'Old Quantum Theory: A Paraconsistent Approach', in *Proceedings of the Biennial Meeting of the Philosophy of Science Association* Vol. II, pp. 397–411.

Bub, J.: 1997, *Interpreting the Quantum World*, Cambridge, Cambridge University Press.

Catellani, E. (ed.): 1998, *Interpreting Bodies: Classical and Quantum Objects in Modern Physics*, Princeton, Princeton University Press.

Cushing, J. T., A. Fine and S. Goldstein (eds.): 1996, *Bohmian Mechanics and Quantum Theory: An Appraisal*, Dordrecht, Kluwer.

della Chiara, M. L.: 1986, in Gabbay and Guenther (eds.), *Quantum Logic*, pp. 427–467.

Deutsch, D.: 1985. 'Quantum Theory, the Church-Turing Principle and the Universal Quantum Computer', *Proceedings of the Royal Society of London A* **400**, 97–117.

Deutsch, David: 1997, *The Fabric of Reality*, London, Penguin Books.

Deutsch, D., A. Ekert and R. Lupacchini: 1999, 'Machines, Logic, and Quantum Physics', `http://xxx.lanl.gov/abs/math.HO/9911150`.

de Witt, B. S. and N. Graham (eds.): 1973, *The Many-Worlds Interpretation of Quantum Mechanics*, Princeton, Princeton University Press.

Dickson, M.: 2001, 'Quantum Logic is Alive $\wedge$ (It is True $\vee$ It is False)', *Philosophy of Science* **68**, S274–S287.

Dürr, D., S. Goldstein and N. Zanghì: 1995, 'Quantum Physics without Quantum Philosophy', *Studies in the History and Philosophy of Modern Physics* **26**(2), 137–149.

Engesser, K. and D. Gabbay: forthcoming, 'Quantum Logic, Hilbert Space, and Revision Theory'.

Everett, H.: 1957, '"Relative State" Formulation of Quantum Mechanics', *Physical Review* **29**(3), 454–462.

Feynman, R. P., R. B. Leighton and M. Sands: 1965, *The Feynman Lectures on Physics* Vol. III, Reading, MA, Addison-Wesley.

Gabbay, D. and F. Guenther (eds.): 1986, *Handbook of Philosophical Logic*, Vol. III: *Alternatives to Classical Logic*, Dordrecht, Kluwer.

Gabbay, D. and J. Woods: 2001, 'The New Logic', *Logic Journal of the IGPL* **9**, 157–190.

Gibbins, P.: 1987, *Particles and Paradoxes: The Limitations of Quantum Logic*, Cambridge, Cambridge University Press.

Hooker, C. A. (ed.): 1975, *The Logico-Algebraic Approach to Quantum Mechanics*, Vol. I, Dordrecht, Reidel.

Hughes, R. I. G.: 1981, 'Quantum Logic', *Scientific American* **245**(4), 202–213.

Hughes, R. I. G.: 1989, *The Structure and Interpretation of Quantum Mechanics*, Cambridge, MA, Harvard University Press.

Kleene, S.: 1967, *Mathematical Logic*, New York, John Wiley and Sons.

Nielsen, M. A. and I. L. Chuang: *Quantum Computation and Quantum Information*, Cambridge, Cambridge University Press.

Penrose, R. and C. Isham (eds.): 1986, *Quantum Processes in Space and Time*, Oxford, Clarendon Press.

Pitowsky, I.: 1984. 'Goerge Boole's "Conditions of Possible Experience" and the Quantum Puzzle', *British Journal for the Philosophy of Science* **45**, 95–125.

Pitowsky, I.: 1989, *Quantum Probability – Quantum Logic*, Berlin, Springer-Verlag.

Putnam, Hilary: 1975a, 'A Philosopher Looks at Quantum Mechanics', in *Mathematics, Matter and Method*, Cambridge, Cambridge University Press, pp. 130–158.

Putnam, Hilary: 1975b, 'The Logic of Quantum Mechanics', in *Mathematics, Matter and Method*, Cambridge, Cambridge University Press, pp. 174–197; first published as 'Is logic Empirical?', in R. Cohen and M. Wartofsky (eds.), *Boston Studies in the Philosophy of Science 5*, Dordrecht, D. Reidel, 1968.

Rescher, N.: 1968, *Topics in Philosophical Logic*, Dordrecht, D. Reidel.

Schrödinger, E.: 1983 (1935), 'The Present Situation in Quantum Mechanics', translation by J. D. Trimmer of 'Die gegenwärtige Situation in der Quantenmechanik', *Naturwissenschaften* **23**, 807–812, 823–828, 844–849; reprinted in Wheeler and Zurek (eds.): 1983, pp. 152–167.

Shimony, A.: 1986, in Penrose and Isham (eds.), *Events and Processes in the Quantum World*, pp. 182–203.

Shoenfield, J. R.: 1967, *Mathematical Logic*, Reading, MA, Addison-Wesley.

Teller, P.: 1998, in Castellani (ed.), *Quantum Mechanics and Haecceities*, pp. 114–141.

Tennant, N.: 1993, *Autologic*, Edinburgh, Edinburgh University Press.

van Fraassen, B. C.: 1981, in Beltrametti and van Fraassen (eds.), *A Modal Interpretation of Quantum Mechanics*, pp. 229–258.

van Fraassen, B. C.: 1988, in Castellani (ed.), *The Problem of Indistinguishable Particles*, pp. 73–92.

Wheeler, J. A. and W. H. Zurek (eds.): 1983, *Quantum Theory and Measurement*, Princeton, Princeton University Press.

Wilson, M. D. (ed.): 1969, *The Essential Descartes*, New York, New American Library.

Woods, J.: 2004, *The Death of Argument: Fallacies in Agent-Based Reasoning*, Dordrecht and Boston, Kluwer.

PART IV
THE LOGIC OF THE KNOWLEDGE-SEEKING ACTIVITIES

BELIEF CONTRACTION, ANTI-FORMULAE AND RESOURCE OVERDRAFT: PART II DELETION IN RESOURCE UNBOUNDED LOGICS

DOV M. GABBAY and ODINALDO RODRIGUES
Department of Computer Science, King's College London, London WC2R 2LS

JOHN WOODS
Department of Computer Science, King's College London, London WC2R 2LS
Department of Philosophy, University of British Columbia, Vancouver, B.C. Canada V6T 1Z4

Abstract. The operation of deletion plays an important role in many areas of applied logic. However, there are a number of logical difficulties relating to the removal of *elements* from a database. These difficulties are usually handled by executing the operation of deletion on the meta-level.

In Gabbay et al. 2002, we argued that bringing the operation of deletion to the *object* level was a useful exercise and we analysed how this can be achieved in resource bounded logics such as linear logic.

In this paper, we continue the investigation by analysing ways of effecting object level deletion for logics which do not have a resource bound on the number of times a formula can be used.

1. Introduction

In this paper we continue our investigation of the logic of deletion. In Gabbay et al. 2002, we argued that bringing the operation of deletion to the object level was a useful exercise. We started our investigation with the formulation of object level deletion for resource bounded logics and we now proceed by considering resource unbounded logics. Let us briefly recap our main ideas.

Deletion can always be made at the meta-level. If we have $\Delta \cup \{A\} \vdash B$, we can physically take A out and possibly end up with $\Delta \nvdash B$. In Gabbay et al. 2002, we discussed object level deletion of formulae from databases Δ of a given logic $\mathbf{L}$. We were looking for ways of effecting d eletion in the object level, i.e., by logical means. We distinguished two kinds of object level deletions.

1. *Logical deletion in the object level*: This means that if $\Delta \vdash_{\mathbf{L}} B$, we want to add some formula $\bigcirc(B)$ such that $\Delta + \bigcirc(B) \nvdash B$.
2. *Physical deletion at the object level*: This means that if $B' \in \Delta$, then we want to add some formula $\bigcirc(B')$ such that $\Delta + \bigcirc(B')$ is equivalent to $\Delta - B'$. By equivalent we mean $\Delta \equiv \Gamma$ iff (def.) $\forall Y(\Delta \vdash Y \text{ iff } \Gamma \vdash Y)$.

To make a concrete example, let Δ be $\{B', B' \Rightarrow B\}$, we can physically delete B' but can only logically delete B.

S. Rahman et al. (eds.), Logic, Epistemology, and the Unity of Science, pp. 291–326.

A solution for object level deletion of a formula B was put forward for resource bounded logics by using the idea of diverting B towards proving nothing. To explain the idea briefly, consider the case of linear logic. In this logic we have $\Delta \vdash B$ iff each element of Δ is correctly used in the proof of B *exactly once*. Linear logic contains a constant **e** which is a multiplicative unit (Girard calls it **1**). **e** is basically equivalent to any theorem of the logic (it becomes $\top$ in stronger logics) and such that essentially in the logic, $X \Rightarrow (\mathbf{e} \Rightarrow Y)$ is the equivalent to $X \Rightarrow Y$ for all X, Y. If we let $\bigcirc(B) = B \Rightarrow \mathbf{e}$, then to delete B we simply add to the data $B \Rightarrow \mathbf{e}$. B is now diverted to proving **e** which is nothing. In other words, B is deleted. So going back to our example Δ, we can add $B' \Rightarrow \mathbf{e}$ for object level physical deletion or $B \Rightarrow \mathbf{e}$ for object level logical deletion. $B' \Rightarrow \mathbf{e}$ will delete B' (same effect as physical deletion) and $B \Rightarrow \mathbf{e}$ will delete/divert B the minute it is derived (logical deletion). This idea works provided there is a resource bound on the number of times a formula can be used. If there is no such bound, as in intuitionistic logic, other (object level) means for deleting are to be found. This is the task of this paper.

There are two main ways we shall consider. We illustrate by examples

EXAMPLE 1.1. Let Δ be

1. A
2. $(B \Rightarrow A) \Rightarrow (A \Rightarrow (A \Rightarrow C))$

Clearly, $\Delta \vdash C$ in intuitionistic logic. We want to delete A.

The first method is to use negation as failure in a goal directed formulation of the logic, see Section 3 below. Let n_1, n_2 be new propositional constants naming (1) and (2) an let $\neg$ be negation as failure. Let Δ^* be

1^* $\neg n_1 \rightarrow (1)$

2^* $\neg n_2 \rightarrow (2)$

Δ^* is equivalent to Δ as long as n_1, n_2 are not in Δ^*. We refer to n_1, n_2 as names of (1) and (2) because we are using them to write (1^*), (2^*) respectively. To delete (1^*) we add n_1. To delete (2^*) we add n_2.

To use $\neg$ for deletion we need to use names systematically.

EXAMPLE 1.2. Consider Δ of the previous example (still in intuitionistic logic) and turn the logic into a labelled deductive system (LDS) allowing each formula to name (i.e., to be a label acting as a name for) itself. Our database $\Delta^\sharp$ becomes

$1^\sharp$ (1):(1)

$2^\sharp$ (2):(2)

where (1) and (2) are the formulae in Δ.

The labels are multisets and keep track of the use of the resources. In particular, we use the rules (see next section):

$\Rightarrow E$:

$$\frac{t : A; s : A \Rightarrow B}{st : B}$$

and

$\Rightarrow I$:

$t : B \Rightarrow A$ can be shown at line X, if we can start a subcomputation at line $(X.1)$ with some arbitrary label x; at line $(X.Y)$ we obtain the formula $\gamma(x) : B \Rightarrow A$; and an *exit* condition ε on $\gamma(x)$ holds at line $(X.Y{+}1)$. The label t will be the result of an *exit* function on $(x, \gamma(x))$:

$X.1$	$x : B$	assumption (where x is an arbitrary label)
	$\vdots$	
$X.Y$	$\gamma(x) : A$	from $(1^\sharp)$
$X.Y+1$	Exit condition ε holds	

Exit t: $B \Rightarrow A$

In the case of linear logic, the exit condition ε is that $\gamma(x)$ contains exactly one x. In the case of intuitionistic logic, $\gamma(x)$ may contain 0 or more occurrences of x. The exit function in both cases deletes all occurrences of x from $\gamma(x)$.

Now, let us see how we prove C from $\Delta^\sharp$:

$3^\sharp$ (1): $B \Rightarrow A$ from subcomputation $(3.1^\sharp)$–$(3.3^\sharp)$:

$3.1^\sharp$	$x : B$	assumption
	x	arbitrary label
$3.2^\sharp$	$(1) : A$	from $(1^\sharp)$
$3.3^\sharp$	Exit condition holds (trivially)	

Exit (1): $B \Rightarrow A$

Now by $(\Rightarrow E)$ using $(1^\sharp)$ twice and $(2^\sharp)$, $(3^\sharp)$ once, we get

$4^\sharp$ $(2)(1)(1)(1) : C$

We get a proof of C with a label indicating exactly which formulae of $\Delta^\sharp$ were used in the proof and how many times.

It is a logical option in the LDS context to put conditions on the kinds of labels allowed in acceptable proofs. (In fact we have to put such conditions if we want the deduction theorem to hold. See Example 2.3 below.) In the above special LDS formulation of our logic suppose we have shown that

$$\Gamma \vdash t : C$$

where t itself is a database in, say, linear logic or any resource logic $\mathbf{L}$ where the databases are multisets. Suppose we stipulate that the proof of $t : C$ from Γ is accepted provided $t \vdash_{\mathbf{L}} C$.

Thus $\Delta \vdash_\sharp C$ iff $\Delta^\sharp \vdash t : C$ and $t \vdash_{\mathbf{L}} C$.

If $\mathbf{L}$ is a resource logic, and we want to logically delete $t : C$, then we can delete from t instead of from Δ, and we know how to do that.

In our example, change the label of $A : A$ (i.e., of item $(1^\sharp)$ of $\Delta^\sharp$) into $\{A, \bigcirc(A)\} : A$. If we now proceed with the same proof as before, we will end up with $t \cup \{\bigcirc(A)\} \nvdash_{\mathbf{L}} C$.

Of course, we have to make sure that deletion from t is done in LDS object level and not just physical metalevel deletion.

This paper will consider both options.

2. Logical Deletion in LDS

This section will use an LDS formulation for intuitionistic $\Rightarrow$ and show how to affect logical deletion using the labels.

First let us recall our problem in the case of intuitionistic $\Rightarrow$.

Suppose we have $A \vdash A$. Here $\Delta = \{A\}$ and we have $\Delta \vdash A$.

This holds in any logic **L**. Say we want to delete A from Δ. In linear logic it is easy to do, as we have seen in Part 1 of this paper. Add $\bigcirc(A)$ to delete A. Thus let $\Delta - A = \Delta \cup \{\bigcirc(A)\} = \{A, \bigcirc(A)\}$ which is equivalent to $\varnothing$. In fact we have seen that $\bigcirc(A)$ can be defined as $\bigcirc(A) = (A \Rightarrow \mathbf{e})$. In intuitionistic logic we cannot do that because even if databases are taken as multisets, we still have

$$\{A\} \equiv \{A, A\}.$$

Hence we would expect

$$\varnothing \equiv \{A, \bigcirc(A)\} \equiv \{A, \bigcirc(A), A\} \equiv \{A\}.$$

so we cannot just use $\bigcirc(A)$ in the same way.

Our proposed solution to this problem is done in two steps:

Step 1: Give an equivalent formulation of intuitionistic logic as an LDS.

Step 2: Define object level deletion in the LDS formulation.

The following is a series of definitions setting the scene for the LDS mechanisms for deletion.

To fix our notation, let **I** be the logic of intuitionistic $\Rightarrow$ and let $\mathbf{L_e}$ be linear logic for $\Rightarrow$ and **e**, as defined in Gabbay et al. (2002, Definition 2.1).

For the readers' convenience we list the axioms of $\mathbf{L_e}$, the linear logic with $\Rightarrow$ and **e**:

DEFINITION 2.1 (Axioms and rules for $\mathbf{L_e}$).

1. $A \Rightarrow A$
2. $(A \Rightarrow B) \Rightarrow ((C \Rightarrow A) \Rightarrow (C \Rightarrow B))$
3. $(A \Rightarrow (B \Rightarrow C)) \Rightarrow ((A \Rightarrow B) \Rightarrow (A \Rightarrow C))$
4. $\dfrac{\vdash A \Rightarrow B}{\vdash (B \Rightarrow C) \Rightarrow (A \Rightarrow C)}$
5. $\dfrac{\vdash A, \vdash A \Rightarrow B}{\vdash B}$
6. $\dfrac{\vdash A}{\vdash (A \Rightarrow B) \Rightarrow B}$
7. $\vdash \mathbf{e}$

For a multiset Γ of formulae, $\Gamma = \{A_1, \ldots, A_n\}$ let $\Gamma \Rightarrow A$ be $A_1 \Rightarrow \ldots \Rightarrow (A_n \Rightarrow A)\ldots)$. By axiom (2) this is independent of the order. Using this notation we have the following axiom.

8. $\vdash \Gamma \Rightarrow \mathbf{e}$ implies that $\vdash \Delta \Rightarrow B$ iff $\vdash \Delta \cup \Gamma \Rightarrow B$.
9. One can prove that $\mathbf{e}$ is equivalent to any B such that $\vdash B$.

Thus $\mathbf{e}$ is a constant standing for any theorem.

We now define the kind of LDS we need.

DEFINITION 2.2 (Resource LDS for $\Rightarrow$). Let $\mathbb{L}$ be a language with atomic propositions $\{q_1, q_2, \ldots\}$ and a binary $\Rightarrow$. Let $\mathcal{A}$ be a set of atomic labels. Let α be a function giving for each label variable x and a formula X a new label $\alpha(x, X)$. Let f be a binary function from multisets of labels to multisets of labels. Let $\gamma(x)$ be a label containing a variable x. Let *Exit*$(x, \gamma(x))$ be a partial function giving a label δ not containing x when defined. Let φ be a binary predicate on multisets of labels.

Define the following concepts:

1. A declarative unit has the form $t : A$ where t is a multiset of atomic labels and A a formula.
2. A database Δ is a set of declarative units, together with a notion $\Delta + (t : A)$ of input of declarative units $t : A$ into Δ..
3. The $\Rightarrow E$ rule has the form:

$$\frac{t : A \Rightarrow B;\ s : A;\ \varphi(t, s)}{f(t, s) : B}$$

4. The $\Rightarrow I$ rule has the form:
   To show $t : A \Rightarrow B$ assume $\alpha(x, A) : A$ and further assume $\forall y \varphi(y, \alpha(x, A))$, where x is a new variable for atomic label and then show $\gamma(x) : B$, where *Exit*$(x, \gamma(x)) = t$.

EXAMPLE 2.3. The following example explains the various components of Definition 2.2. Consider the implication $\Rightarrow$ of intuitionistic logic. This can be presented using the two natural deduction rules:

$\Rightarrow E$ $\quad \frac{A, A \Rightarrow B}{B}$

and

$\Rightarrow I$ $\quad$ To show $A \Rightarrow B$ assume A and show B.

Thus to show

$$A \Rightarrow (A \Rightarrow B) \vdash A \Rightarrow B$$

in intuitionistic logic we assume A and then use A twice in modus ponens with $A \Rightarrow (A \Rightarrow B)$ and get B.

Similarly to show $A \vdash B \Rightarrow A$, we assume B and then show A, even without using B because we already have A.

Linear implication $\Rightarrow$ requires that all assumptions be used each exactly once. Thus $A \Rightarrow (A \Rightarrow B) \not\vdash A \Rightarrow B$ because A needs to be used twice and $A \not\vdash B \Rightarrow A$ because B is not used.

A general resource implication may have all kinds of conditions on the proofs. The best way to express them is to label the assumptions (label all the data and whenever you use the $\Rightarrow I$ rule, use a new label to label the new assumption) and propagate the labels. The functions and predicates α, φ, f and Exit help us control and express what we want.

DEFINITION 2.4 (Consequence). Let Δ be an *LDS* theory as in Definition 2.2 and let $t : A$ be a declarative unit. We define the consequence $\Delta \vdash_{\mathbb{C}} t : A$ subject to a condition $\mathbb{C}(t, \Delta)$ a follows.

First define $\vdash_{m,n}$, m, n non-negative integers. m counts the total number of uses of $\Rightarrow E$ rule in the proof and n counts the maximal number of nested uses of $\Rightarrow I$ in the proof.

1. $\Delta \vdash_{0,0} t : A$, if $t : A \in \Delta$.
2. If $\Delta \vdash_{m_1,n_1} t : A$ and $\Delta \vdash_{m_2,n_2} s : A \Rightarrow B$ and $\varphi(s, t)$ holds then $\Delta \vdash_{m,n} f(s, t) : B$, where

$$n = \max(n_1, n_2) \text{ and}$$
$$m = m_1 + m_2 + 1$$

Note that $\varphi(s, t)$ may be used to control the order in which the assumptions are used. For example, suppose we have a data item of the form $x : A \Rightarrow A$ and suppose we also have just proved $t : A$. We can use φ to force an immediate $\Rightarrow E$ step with $x : A \Rightarrow A$ and get $f(x, t) : A$. This has the effect of 'sticking' the label x into t the 'moment' A is proved (with label t). See Example 2.13
3. $\Delta \vdash_{m,n+1} t : A \Rightarrow B$, if for some new variable x we have $\Delta + \{\alpha(x, A) : A\} \vdash_{m.n} \gamma(x) : B$ and *Exit*$(x, \gamma(x))$ is defined and is equal to t.
4. Let $\Delta \vdash_{\mathbb{C}} t : A$ hold if for some m, n $\Delta \vdash_{m,n} t : A$ and condition $\mathbb{C}(t, \Delta)$ holds.

EXAMPLE 2.5. We now explain the role of the condition $\mathbb{C}$ appearing in Definition 2.4.

Consider again

$$A \Rightarrow (A \Rightarrow B) \vdash^? A \Rightarrow B.$$

Using labels we get

$$t : A \Rightarrow (A \Rightarrow B) \vdash^? t : A \Rightarrow B.$$

We assume $x : A$ with arbitrary x and prove $\gamma(x) = txx : B$. The label $\gamma(x) = txx$ tells us that A was used twice in the proof of B. The function *Exit*$(x, \gamma(x))$ wants to exit with label t. This is allowed if x appears exactly once and then we exit with $\gamma(x) - x$. So Exit is used to control the resource conditions.

Now consider the equivalent proof problem (if we want the deduction theorem to hold) of

$$t_1 : A \Rightarrow (A \Rightarrow B), t_2 : A \vdash^? B.$$

In this case we can prove $t_1 t_2 t_2 : B$. There is no use of Exit and so how do we block this proof? We need another predicate $\mathbb{C}$ for such a case. $\mathbb{C}$ will really be defined using iterations of the Exit condition.

EXAMPLE 2.6. We show how to get linear $\Rightarrow$ as an LDS of this form. Let $\mathcal{A}$ be a set of atomic labels. A database has the form $\{t_1 : A_1, \ldots, t_n : A_n\}$ where all t_i are pairwise disjoint and we regard 't' as a unit multiset $\{t\}$. Let $\alpha(x, A)$ be x and let $f(t, s) = t \cup s$, where t, s are multisets. Let $\varphi(x, y) = \top$. Let $Exit(x, \gamma(x))$ be defined if x occurs in $\gamma(x)$ exactly once and in this case $Exit(x, \gamma(x)) = \gamma(x) - x$ (i.e., take x out of $\gamma(x)$).

Let $\mathbb{C}(t, \{t_1 : A_1, \ldots, t_n : A_n\})$ be the condition $t = \{t_1, \ldots, t_n\}$.

We obviously have that $\Delta \vdash_{\mathbb{C}} t : A$ iff A is proved from Δ using each assumption exactly once!

DEFINITION 2.7 ($\mathbf{L_e}$: **I**-logic). We define an LDS version of intuitionistic $\Rightarrow$ where the labels come from linear logic with anti-formulae (i.e., $\Rightarrow$ and **e** logic), as defined in Section 2 of Gabbay et al. (2002).

1. A label is any multiset of $\Rightarrow$, **e** formulae.
2. A declarative unit is a labelled $\Rightarrow$ formula.
3. The $\Rightarrow E$ rule has the form

$$\frac{t : A; s : A \Rightarrow B}{t \cup s : B}$$

   where $\cup$ is the multiset union.
4. The $\Rightarrow I$ rule has the form: to show $\Delta \vdash t : A \Rightarrow B$ assume $\{x\} : A$, for x new atomic label and show $\gamma(x) : B$. The exit condition is $\gamma(x/A) \vdash_{\mathbf{L_e}} B$ and $t = \gamma(x) \cup \{\bigcirc(x)\}(x/A)$, where x/A is the result of substituting A for x.[1]
5. Let a database Δ be a multiset of labelled formulae of the form $\{\Theta_1 : A_1, \ldots, \Theta_n : A_n\}$ where A_i are formulae with $\Rightarrow$ and Θ_i are multisets of formulae with $\Rightarrow$ and **e**.
6. Define $\Delta \vdash_{m,n} \Theta : B$ as follows:
   6.1. $\Delta \vdash_{0,0} \Theta : B$ if $\Theta : B \in \Delta$.
   6.2. If $\Delta \vdash_{m_1,n_1} t : A$ and $\Delta \vdash_{m_2,n_2} s : A \Rightarrow B$ then $\Delta \vdash_{m,n} s \cup t : B$, where $n = \max(n_1, n_2)$ and $m = m_1 + m_2 + 1$.
   6.3. $\Delta \vdash_{m,n+1} t : A \Rightarrow B$ if $\Delta \cup \{\{x\} : A\} \vdash_{m,n} \gamma : B$ and $\gamma(x/A) \vdash_{\mathbf{L_e}} B$ and $t = \gamma \cup \{\bigcirc(A)\}$ if x is not in γ and $t = \gamma(A) \cup \{\bigcirc(A)\}$ otherwise.
   6.4. $\Delta \vdash_{\mathbb{C}} t : A$ if for some m, n, $\Delta \vdash_{m,n} t : A$ and $\mathbb{C}(\Delta, t : A)$ holds.

7. Let $\mathbb{C}(\Delta, t : A)$ be $t \vdash_{\mathbf{L_e}} A$.

EXAMPLE 2.8.

(1) In intuitionistic logic we have

$$A \Rightarrow (A \Rightarrow B) \vdash A \Rightarrow B$$

The derivation is as follows:

1. $A \Rightarrow (A \Rightarrow B)$ data
2. Show $A \Rightarrow B$ from box subcomputation

2.1	A assumption
2.2	$A \Rightarrow B$, using $\Rightarrow E$, (1) and (2.1)
2.3	B, using $\Rightarrow E$ (2.2) and (2.1).

Exit $A \Rightarrow B$

(2) The same proof can be done in $\mathbf{L_e} : \mathbf{I}$ logic as follows:

1* $\{A \Rightarrow (A \Rightarrow B)\} : A \Rightarrow (A \Rightarrow B)$ data

2* Show $\{\bigcirc(A), A, A, A \Rightarrow (A \Rightarrow B)\} : A \Rightarrow B$ from box subcomputation

2.1*	$\{x\} : A$ assumption
2.2*	$\{x, A \Rightarrow (A \Rightarrow B)\} : A \Rightarrow B$, using $\Rightarrow E$, (1*) and (2.1*)
2.3*	$\gamma(x) = \{x, x, A \Rightarrow (A \Rightarrow B)\} : B$, using $\Rightarrow E$ (2.2*) and (2.1*).
2.4*	The exit condition $\gamma(A) \vdash_{\mathbf{Le}} B$ holds.

Exit with $\{\bigcirc(A), A, A, (A \Rightarrow (A \Rightarrow B))\} : B$

3* The proof of (2*) is acceptable because in linear logic with $\Rightarrow$ and $\mathbf{e}$ (where $\bigcirc(X)$) is $X \Rightarrow \mathbf{e}$), we have $\{\bigcirc(A), A, A, A \Rightarrow (A \Rightarrow B)\} \vdash A \Rightarrow B$.

EXAMPLE 2.9.

(1) Let us look at the proof of $A \vdash_{\mathbf{I}} B \Rightarrow A$.

1. A assumption
2. $B \Rightarrow A$, from box subcomputation

2.1	B assumption
2.2	A, from (1)

Exit $B \Rightarrow A$

(2) Let's do the same in $\mathbf{L_e} : \mathbf{I}$

1*. $\{A\} : A$ assumption

2*. Show $\{\bigcirc(B), A\} : B \Rightarrow A$, from box

2.1*	$\{x\} : B$ assumption
2.2*	$\{A\} : A$ from (1*)
2.3*	Exit condition holds

Exit $\{A, \bigcirc(B)\} : B \Rightarrow A$

We have defined the $\mathbf{L_e} : \mathbf{I}$ logic and have given some examples of proofs in this logic. We need to show now that it is equivalent to intuitionistic logic **I**. This will complete Step 1 of our strategy.

For the next theorem it is convenient to consider the databases for intuitionistic logic as multisets of formulae. This allows us to compare provability in **I** to that of $\mathbf{L_e}$ and $\mathbf{L_e} : \mathbf{I}$.

Let Δ be a multiset $\Delta = \{A_1, \ldots, A_n\}$, let Δ' be $\{\{A_1\} : A_1, \ldots, \{A_n\} : A_n\}$. Δ' is a labelled database of $\mathbf{L_e} : \mathbf{I}$.

We can state our theorem.

THEOREM 2.10. Let $\Delta = \{A_1, \ldots, A_n\}$ be a multiset. Let $\Delta' = \{\{A_1\} : A_1, \ldots, \{A_n\} : A_n\}$. Then we have for any B that (1) is equivalent to (2):

1. $\Delta \vdash_{m,n} B$ in **I**.
2. For some t we have $\Delta' \vdash_{m,n} t : B$ in $\mathbf{L_e} : I$ with $t \vdash_{\mathbf{L_e}} B$.

Proof. Any proof in $\mathbf{L_e} : \mathbf{I}$ is a valid proof in **I**, if we ignore the labels. Thus one direction of our theorem holds. We show the other direction by induction on the index (m, n) of the proof of $\Delta \vdash_{m,n} B$.

Case (0, 0)
This means that for some i, $B = A_i$. Then $\{A_i\} : A_i \in \Delta'$ and since $\{A_i\} \vdash_{\mathbf{L_e}} A$ we get $\Delta' \vdash_{\mathbf{L_e:I}} \{A_i\} : A_i$.

Case (m, n), $\Rightarrow$ *E rule:*
This means that for some k_1, k_2 such that $1 + k_1 + k_2 = m$ and for some n_1, n_2 such that $\max(n_1, n_2) = n$ and for some A we have $\Delta \vdash_{k_1,n_1} A$ and $\Delta \vdash_{k_2,n_2} A \Rightarrow B$ in **I**.

By the induction hypothesis we have t_1, t_2 such that $\Delta' \vdash_{k_1,n_1} t_1 : A$ and $\Delta' \vdash_{k_1,n_2} t_2 : A \Rightarrow B$ in $\mathbf{L_e} : \mathbf{I}$, with $t_1 \vdash_{\mathbf{L_e}} A$ and $t_2 \vdash_{\mathbf{L_e}} A \Rightarrow B$. Hence $\Delta' \vdash_{m,n} t_1 \cup t_2 : B$ and $t_1 \cup t_2 \vdash_{\mathbf{L_e}} B$.

Case $(m, n + 1)$, $\Rightarrow$ *I rule:*
This is the case where we used the $\Rightarrow I$ rule. This means that B has the form $B = (B_1 \Rightarrow B_2)$ and that we assumed B_1 and from $\Delta \cup \{B_1\}$ we proved B_2 with index (m, n). By the induction hypothesis for the database $\Delta' \cup \{x : B_1\}$ (for the substitution $x = B_1$), we have $\Delta \vdash_{m,n} \gamma(B_1) : B_2$ and $\gamma(B_1) \vdash_{\mathbf{L_e}} B_2$.

From the deduction theorem for $\mathbf{L_e}$[2] we have $\gamma \vdash_{\mathbf{L_e}} B_2$ iff $\gamma \cup \{\bigcirc(B_1)\} \vdash_{\mathbf{L_e}} B_1 \Rightarrow B_2$.

By the rules $\mathbf{L_e} : \mathbf{I}$ we can exit with $\gamma(B_1) \cup \{\bigcirc(B_1)\}$ and hence $\Delta' \vdash_{m,n+1} \gamma \cup \{\bigcirc(B_1)\} : B$ with $\gamma \cup \{\bigcirc(B_1)\} \vdash_{\mathbf{L_e}} B$.

This completes the proof of the theorem. ■

We now have the machinery to do deletion. We are ready for Step 2 of our strategy, namely we work in $\mathbf{L_e} : \mathbf{I}$ instead of **I** and do deletion there. Theorem 2.10 allows us to do that. The question is how to execute deletion. Assume $\{A\} : A \in \Delta$

needs to be deleted, one's first attempt is simply to add $\bigcirc(A)$ to the label of A, obtaining $\{\bigcirc(A), A\} : A$ and thus deleting A.

The following example illustrates our point.

EXAMPLE 2.11. Let us see how to delete A in the $\mathbf{L_e} : \mathbf{I}$ version of the proof of Example 2.8. To delete A add $\{\bigcirc(A)\} : A$.

If we do that the database becomes

1**. $\bigcirc(A) : A, \{A \Rightarrow (A \Rightarrow B)\} : A \Rightarrow (A \Rightarrow B)$.

The box proof of $A \Rightarrow B$ becomes

2.1**. $\{\bigcirc(A), A\} : A$ assumption

(we are inputting into the database $\{A\} : A$. Since $\{\bigcirc(A)\}LA$ is already there, the overall label is the multiset union, i.e., $\{\bigcirc(A), A\} : A$.

If we continue the proof we get

2.2**. $\{\bigcirc(A), A, A \Rightarrow (A \Rightarrow B)\} : A \Rightarrow B$

2.3**. $\{\bigcirc(A), A, A, A \Rightarrow (A \Rightarrow B)\} : B$

2.4**. If the exit condition is satisfied, we exit with $\{\bigcirc(A), \bigcirc(A), A, A, A \Rightarrow (A \Rightarrow B)\} : A \Rightarrow B$.

Line 2.4** does not work because in linear logic the label $\{\bigcirc(A), \bigcirc(A), A, A, A \Rightarrow (A \Rightarrow B)\}$ is equivalent to $\{A \Rightarrow (A \Rightarrow B)\}$ and it does not prove $A \Rightarrow B$.

The problem with the previous example is that it is not object level deletion. Even in *LDS*, changing a label in a database is meta- level. We can add $\{\bigcirc(A)\} : A$ to Δ, but unless we have a rule for aggregating labels, it will not help. We still have $\{\{\bigcirc(A)\} : A, \{A\} : A\} \vdash \{A\} : A$.

Somehow we want to put something in the database of the form $x : X$ (this is a logical move) and force it to interact with $\{A\} : A$ and thus delete it. The answer is to put $\{\bigcirc(A)\} : A \Rightarrow A$ in the database. If we use modus ponens

$$\frac{\{\bigcirc(A)\} : A \Rightarrow A, \{A\} : A}{\{\bigcirc(A), A\} : A}$$

Can we force these two to interact? The answer is yes if we change the logic slightly. Recall that the difference between linear logic and intuitionistic logic is that linear logic wants each assumption to be used exactly once while intuitionistic logic does not care. We can have a compromise between the two. A database has the form $\Delta_1 \cup \Delta_2$. The elements of the multiset Δ_1 have all to be used exactly once and the elements of Δ_2 we do not care. Now we can put $\{\bigcirc : A \Rightarrow A$ into Δ_1. It has to be used and when used it will add $\{\bigcirc\}$ to the label of A. This of course allows us to do logical deletion as well. We do not care if A is in the database or not. This solution might have some technical problems. For instance, suppose $\Delta \nvdash A$ from Δ. If we delete A, i.e., Add $\{\bigcirc : A\}$ to Δ then it *cannot* be used and the new database cannot prove anything. Put differently, we *do not* have the metalevel rule for vacuous deletion:

– If $\Delta \nvdash A$ then $\mathbf{Delete}(\Delta, A) = \Delta$, see Example 6.1.

Note that we have the same problem in linear logic itself. In Part 1 of this paper we saw that to delete A from Δ we add $\bigcirc(A) = A \Rightarrow \mathbf{e}$. But if $\Delta \not\vdash A$ then we are stuck with $\bigcirc(A)$ in the database. So what are we to do? We are going to address this in Part 3 of this paper. But let us look at some examples.

EXAMPLE 2.12 (Logical deletion).

(1) In intuitionistic logic we have that

1. $C \Rightarrow A$
2. C
3. $D \Rightarrow A$
4. D
5. $A \Rightarrow B$

prove B. The same holds in $\mathbf{L_e} : \mathbf{I}$ logic. The proofs will be the same except that in $\mathbf{L_e} : \mathbf{I}$ logic the multiset label will record what assumptions have been used and how many times. So using (1), (2) and (5), we can prove B. In $\mathbf{L_e} : \mathbf{I}$ logic we prove $\{(1), (2), (5)\} : B$ and of course $\{(1), (2), (5)\} \vdash B$ in linear logic.

If we delete D, we do this by adding in $\mathbf{L_e} : \mathbf{I}$ logic $\{\bigcirc(D)\} : D \Rightarrow D$, and force modus ponens with $\{D\} : D$; so the label of D becomes $\{\bigcirc(D), D\} : D$. This will not affect the above proof of B from (1), (2), (5).

If we also delete C in a similar way, the label of C will become $\{\bigcirc(C), C\} : C$ and the above proof will end up with a label $\{\bigcirc(C), C, C \Rightarrow A, A \Rightarrow B\}$.

In linear logic with $\Rightarrow$, $\mathbf{e}$, this database does not prove B and hence the $\mathbf{L_e} : \mathbf{I}$ proof of B is blocked.

Let us now delete A. This means we add to the database $\{\bigcirc(A)\} : A \Rightarrow A$. However, since A is not in the database, we really want to do *logical deletion*: the minute A is proved, it must be deleted. So we should have the following sequence:

Step 1: From $C \Rightarrow A$ and C get A

Step 2: Delete A

Let us see what the labels do:

Step 1*: From $\{C\} : C$ and $\{C \Rightarrow A\} : C \Rightarrow A$ get $\{C, C \Rightarrow A\} : A$.

Step 2*: Since $\{\bigcirc(A)\} : A \Rightarrow A$ is in the database, if we force modus ponens at this point then the current label of A becomes $\{\bigcirc(A), C, C \Rightarrow A\}$. This label does not prove A in linear logic.

The problem is that using (3) and (4) we can get another copy of A and thus rescue the proof and we must remember that $\{\bigcirc(A)\} : A \Rightarrow A$ has already been spent, i.e., one copy of A has already been deleted!

We can make the problem more severe. Why not use (1) and (2) again to get another copy of A? In intuitionistic logic we can use the data as many times as we want. Hence we can get A with the label $\{C, C \Rightarrow A, C, C \Rightarrow A\}$. Adding $\bigcirc(A)$ will leave us with one copy of A!

Obviously our thinking of our resource management is not clear cut enough.

Let us look more closely at the problem mentioned in Example 2.12. $\mathbf{L_e}$ proof theory works as follows:

To have $\Delta \vdash A$, we must be able to decompose Δ into (possibly empty) subsets Δ_i, Γ_j such that $\Delta = \bigcup_i \Delta_i \cup \bigcup_j \Gamma_j$ such that $\Delta_i \vdash A_i, \Gamma_j \vdash \mathbf{e}$ and $\Theta = \{A_i\} \vdash B$.

Note that we know how to delete A_1 from Θ if it is written in Θ explicitly but how do we logically delete A_1 from Δ? The way we do it in $\mathbf{L_e}$ is that we add $\bigcirc(A_1)$ to Δ. When A_1 is proved then $\bigcirc(A_1)$ will cancel it. However in $\mathbf{L_e}$: **I**, Δ is accumulated dynamically during the proof so how and where are we going to get $\bigcirc(A_1)$ as a label?

The solution is simple. We add the item of data $\{\bigcirc(A_1)\} : A_1 \Rightarrow A_1$.

The minute $A_1$is prov ed with a label $t : A_1$ if we perform modus ponens with $\{\bigcirc(A_1)\} : A_1 \Rightarrow A_1$, then this will add $\bigcirc(A_1)$ to the label t.

Our problem is now how to force the proof procedure to use this new clause the minute A_1 is proved? Here we use φ. It is best explained by example:

EXAMPLE 2.13.
(1) $A_1 : A_1$
(2) $A_2 : A_2$
(3) $(A_1 \Rightarrow B) : A_1 \Rightarrow B$
(4) $(A_2 \Rightarrow B) : A_2 \Rightarrow B$
This proves B. How do we logically delete B? We add
(5) $\bigcirc(B) : B \Rightarrow B$
Let $\varphi(s, t)$ be $(t \not\vdash B \vee s \vdash \bigcirc(B))$.

Now we can do modus ponens with (1), (3) or (2), (4) because B is not provable from the label. But once B is provable, the only modus ponens we can do (because of φ) is with $\bigcirc(B) : B \Rightarrow B$, which will delete B from the label and we can continue our modus ponens.

There is still a problem with this proposed solution. To delete B we need to change φ and this changing operation is also metalevel.

Another option for logical deletion is to compute in the metalevel what physical assumptions we need to physically delete in order to affect the logical deletion and we know how to do that! But this is not as satisfactory as the proof theoretical method.

3. Introducing N-Prolog with Negation as Failure

We can give meaning to anti-formulae in intuitionistic logic by translation into N-Prolog with negation as failure. N-Prolog is intuitionistic implication augmented by negation as failure $\neg$. In the N-Prolog system we can perform deletion by addition. This N-Prolog deletion will satisfy that if $\Delta \not\vdash A$, then the result of deleting A from Δ yields Δ. This section will introduce N-Prolog. The definitions are as follows.

DEFINITION 3.1. Consider a propositional language with $\Rightarrow$, $\bot$ and $\neg$. Define the notions of *literal, data-clause* and *goal- clause* as follows.

1. An atom q or $\bot$ or $\neg q$ or $\neg\bot$ are literals.[3] q and $\bot$ are positive and $\neg q$ and $\neg\bot$ are negative.
2. A positive literal is a data clause and a literal is a goal clause.
3. If $A_1, \ldots, A_n$ are goal clauses then $A_1 \Rightarrow (\ldots \Rightarrow (A_n \Rightarrow q)\ldots)$ is a data clause, where q is a positive literal. We say q is the head of the clause.
4. If $A_1, \ldots, A_n$ are data clauses and q is a literal then $A_1 \Rightarrow (\ldots \Rightarrow (A_n \Rightarrow q)\ldots)$ is a goal clause, with head q.

DEFINITION 3.2 (Success for N-Prolog).

1. *Immediate success case*
   a) $Success(\Delta, q) = 1$ if $q \in \Delta$ for q a positive literal.
   b) $Success(\Delta, \neg q) = 1$ if q is not the head of any clause in Δ, for q a positive literal.
2. *Implication Case*
   $Success(\Delta, B \Rightarrow C) = x$ if $Success(\Delta \cup \{B\}, C) = x$.
3. *Immediate failure case*
   a) $Success(\Delta, q) = 0$ if q is not the head of any clause in Δ, for q a positive literal.
   b) $Success(\Delta, \neg q) = 0$ if $q \in \Delta$, q a positive literal.
4. *Cut reduction case*
   Success$(\Delta, q) = 1$ (resp. $Success(\Delta, q) = 0$), for q a positive literal if for some (resp. all) clauses of the form $A_1 \Rightarrow (\ldots \Rightarrow (A_n \Rightarrow q')\ldots)$ in Δ with $q' = q$ or $q' = \bot$ we have that for all (resp. some) $1 \leq i \leq n$, we have $Success(\Delta, A_i) = 1$ (resp. $Success(\Delta, A_i) = 0$).
5. *Negation as failure case*
   $Success(\Delta, \neg q) = x$ iff $Success(\Delta, q) = 1 - x$.
6. *Consequence*
   We have $\Delta \vdash A$ iff $Success(\Delta, A) = 1$.

The above (1)–(6) define $\{\Rightarrow, \bot, \neg\}$ as N-Prolog.

Note that for the language without $\neg$, with $\Rightarrow, \bot$ only we have completeness: $\Delta \vdash G$ in intuitionistic logic iff $Success(\Delta, G) = 1$.

If we do not allow for embedded implications and not allow for $\bot$ then we get the usual Prolog clauses with negation as failure written in the form $q_1 \Rightarrow (\ldots \Rightarrow (q_n \Rightarrow q)\ldots)$ where q_i, $1 \leq i \leq n$ are literals and q a positive literal.

REMARK 3.3. *We saw that the language of N-Prolog is obtained from that of intuitionistic implication (with $\bot$) by adding the negation as failure connective $\neg$. The meaning of $\neg A$ in the goal directed computation of the previous Definition 3.2 is that A fails, i.e.,*

$$Success(\Delta, \neg A) = x \text{ iff } Success(\Delta, A) = 1 - x.$$

Because of the deduction theorem (the implication case 2 of Definition 3.2) we have that

$$\neg(A_1 \Rightarrow \ldots (A_n \Rightarrow q)\ldots)$$

is equivalent to

$$A_1 \Rightarrow (\ldots \Rightarrow (A_n \Rightarrow \neg q)\ldots)$$

and so it is sufficient to allow for the $\neg$ connective to apply to atoms only.

We must take care that $\neg q$ never occurs as a head of clause in databases, only in goals. This is because we do not have a meaning for $\neg q$ being an element of a database. We can of course try to give it some (integrity constraint?) meaning but, this is another story.[4]

Hence we can accept a clause such as

$$a \Rightarrow \neg q$$

as a goal clause but not as a data clause.

However

$$(a \Rightarrow \neg q) \Rightarrow r$$

is acceptable as a data clause. Similarly we cannot accept $\neg q \Rightarrow a$ as a goal clause or $(\neg q \Rightarrow r) \Rightarrow r$ as a data clause, since either case would force us to put $\neg q$ in the database. This explains the rationale of the definition 3.1.

N-Prolog is a pretty powerful system. It allows us to delete through addition and to perform many metalevel operations in the object level. Let us illustrate its properties through some examples and then give general definitions.

EXAMPLE 3.4 (Properties of negation as failure). In ordinary Prolog, the database Δ does not change. Thus given a clause in Δ with negation $\neg a$ in it, the meaning of $\neg a$ is 'a fails from Δ', where Δ is the program.

In *N*-Prolog, the program Δ changes throughout the computation and hence the meaning of $\neg a$ is dynamic; it means 'a fails from the current database'. Consider Δ with

1. $\neg a \Rightarrow q$
2. $(b \Rightarrow q) \Rightarrow q$

A query is represented as $?q$. If we use (1) first, $\neg a$ means a failure from (1) and (2). If we use (2) first, we ask for $b \Rightarrow q$, and therefore we add

3. b

to the database and ask $?q$ and then using (1) we get that $\neg a$ means 'failure from (1), (2) and (3)'.

N-Prolog with negation as failure does not satisfy Cut. We have, for example. that

$$\neg n \Rightarrow x \vdash x$$

and that

$$\neg n \Rightarrow x, x \vdash n \Rightarrow x$$

but

$$\neg n \Rightarrow x \nvdash n \Rightarrow x$$

These features make it more difficult to give semantics for $\neg$. Nevertheless, semantics for N-Prolog with negation as failure was given by Olivetti and Terracini (1992) in a very long paper.

The goal directed process allows us to define the deletion. Assume we have $\Delta \vdash A$ and we are seeking a suitable $\Delta' \subset \Delta$ such that $\Delta' \nvdash A$. We look at $\mathit{Success}(\Delta, A) = 1$. We follow the computation until we get to steps of immediate success (clause (1) in Definition 3.2). At that step some q is indeed in the then current theory Δ_1. We spoil the computation by taking q out of Δ_1.

The following example explains our options.

EXAMPLE 3.5. Let Δ_0 be $\{q \Rightarrow a, r_0 \Rightarrow (r_1 \Rightarrow a), q, r_0, r_1\}$. We have $\Delta_0 \vdash a$. We are looking for $\Delta'_0 \subset \Delta$ such that $\Delta'_0 \nvdash a$. We follow the computation. There are two ways a can be proved: either from q or from $\{r_0, r_1\}$. So to spoil the success we need to take out either $\{q, r_0\}$ or $\{q, r_1\}$. Let $\mathit{Abduce}^-(\Delta, A)$, a metapredicate to be defined below, give us all the options for sets to *delete*, to take out to spoil $\Delta \vdash A$ (if $\Delta \nvdash A$ the result is $\varnothing$). Then in our case $\mathit{Abduce}^-(\Delta_0, a) = \{\{q, r_1\}, \{q, r_0\}\} =$ Delete$(\{q, r_1\})$ or Delete$(\{q, r_0\})$.

Notice that this process removes only atoms. Thus $\Delta_0^i = \{q \Rightarrow a, r_0 \Rightarrow (r_1 \Rightarrow a), r_{1-i}\}$ for $i = 0, 1$ and $\Delta_0^i \nvdash a$.

We could have removed $q \Rightarrow a$ and $r_0 \Rightarrow (r_1 \Rightarrow a)$ and got a Δ' such that $\Delta' \nvdash a$ but our *deletion* process does not do that.[5]

EXAMPLE 3.6. This example shows the need for anti-formulae $\bigcirc(A)$ which in our context we also write as $\mathit{Delete}(A)$.[6]

We have

$$Success(\Delta, A \Rightarrow B) = Success(\Delta \cup \{A\}, B).$$

Therefore we want something like

$$Abduce^{-}(\Delta, A \Rightarrow B) = A \Rightarrow Abduce^{-}(\Delta \cup \{A\}, B)$$

So $Abduce^{-}(\varnothing, q \Rightarrow q)$ should equal $q \Rightarrow Abduce^{-}(\{q\}, q)$, but $Abduce^{-}(\{q\}, q) = Delete(q)$. Thus the theory we need is $q \Rightarrow Delete(q)$. Whereupon

$$q \Rightarrow Delete(q) \nvdash q \Rightarrow q.$$

The following is a formal definition of $Abduce^{-}$ using the operator *Delete*. The problem is that we do not have a logic for *Delete*. This is what we are looking for.

DEFINITION 3.7 (Abduce^{-} for intuitionistic logic). $Abduce^{-}(\Delta, Q)$ is a family of sets of pseudo formulae of the form $A_1 \Rightarrow \ldots \Rightarrow (A_n \Rightarrow Delete(q))\ldots)$ where *Delete* is a metapredicate and q is atomic.

1. $Abduce^{-}(\Delta, Q) = \{\varnothing\}$ if $\Delta ? Q = 0$
2. $Abduce^{-}(\Delta, q) = \{Delete(q)\}$ if $q \in \Delta$, q atomic and is the only clause in Δ with head q.
3. $Abduce^{-}(\Delta, A_1 \Rightarrow (A_2 \Rightarrow \ldots (A_n \Rightarrow q)\ldots) = \{A_1 \Rightarrow (A_2 \Rightarrow \ldots (A_n \Rightarrow X)\ldots) | X \in Abduce^{-}(\Delta \cup \{A_1, \ldots, A_n\}, q)\}$ and where $A_1 \Rightarrow \ldots \Rightarrow (A_n \Rightarrow X)\ldots) = \{A_1 \Rightarrow \ldots \Rightarrow (A_n \Rightarrow y)\ldots) | y \in X\}$

For clause (4) below, we need to assume that

$$B^j = (B^1_1 \Rightarrow \ldots \Rightarrow (B^j_{n(j)} \Rightarrow q)\ldots), j = 1, \ldots, m,$$

lists all clauses of heads q in Δ.

We also need the notion of a choice function $\mathbf{c}$. For each $1 \leq j \leq m$, $\mathbf{c}(j)$ is a theory $\Gamma = \mathbf{c}(j)$ such that for some $1 \leq i \leq n(j)$, $\Gamma \in Abduce^{-}(\Delta, B^j)$.

4. $Abduce^{-}(\Delta, q) = \{\Gamma_{\mathbf{c}} | \mathbf{c}$ is a choice function as explained above and $\Gamma_{\mathbf{c}} = \bigcup_{j=1}^{m} \Gamma_{\mathbf{c}(j)}\}$.

The explanation for clause (4) is the following: for each clause B^j as above, we want to choose an $1 \leq i \leq n(j)$ such that $\Delta ? B^j_i = 0$. To ensure that we look to $Abduce^{-}(\Delta, B^j_i)$.

Our choice functions are functions $\mathbf{c}$ choosing for each $j = 1, \ldots, m$ and $1 \leq i \leq j$ and a theory $\mathbf{c}(j) \in Abduce^{-}(\Delta, B^j_i)$.

Let $\Gamma_{\mathbf{c}} = \bigcup_{j=1}^{m} \Gamma_{\mathbf{c}(j)}$.

Note that Definition 3.7 gives a connection between $\bigcirc(X)$ (or $Delete(X)$) and contraction in intuitionistic logic.

$Abduce^{-}(\Delta, A)$ gives us formulae with $Delete(X)$ in them. Add these to Δ and in the suitable logical extension, which we are still trying to formulate; the addition does the job of deletion.

EXAMPLE 3.8 (Deletion by addition).

1. Consider the database Δ of the preceding example and suppose that we want to delete clause (1). How do we do it? We want to do it by addition! Let n_1, n_2 be two new atoms. Consider the database Δ^* with

   (1*) $\neg n_1 \Rightarrow (\neg a \Rightarrow q)$

   (2*) $\neg n_2 \Rightarrow ((b \Rightarrow q) \Rightarrow q)$

   Clearly since n_1, n_2 are new atoms which are not heads of any clause in Δ^*, we have for any z
   – $Success(\Delta, z) = Success(\Delta^*, z)$.
   Now we can effectively delete (1*) from Δ^* by adding n_1, and delete (2*) from Δ^* by adding n_2.
2. We can also make sure we can delete, for example, all clauses in the database with head q. To achieve this let f be unary function giving for each atom q of the language a new atom $f(q)$ (thus we also have $ff(q)$, $f^3(q), \ldots$). Let Δ be a database and let $B = A_1 \Rightarrow (\ldots \Rightarrow (A_n \Rightarrow q)\ldots)$ be any clause with head q. Rewrite Δ to $\Delta^\sharp$ by replacing each B above by $\neg f(q) \Rightarrow B = \neg f(q) \Rightarrow (A_1 \Rightarrow \ldots \Rightarrow (A_n \Rightarrow q)\ldots)$.
   Clearly, by adding $f(q)$ to $\Delta^\sharp$ we delete all clauses with head q.

Let us now do this naming in a systematic way.

We first show how to do this operationally. Start with a language $\mathbb{L}_0$. Let $n_1^1, n_2^1, \ldots$ be an infinite sequence of new names not in $\mathbb{L}_0$. Form $\mathbb{L}_1$ using these names. Get new names $n_1^2, n_2^2, \ldots$ and form $\mathbb{L}_2$, and go on to form $\mathbb{L}_k$, $k = 1, 2 \ldots$. We can now assume every formula φ has a unique name n_φ.

Starting with a database Δ of the form $\Delta = \{A_1, \ldots, A_k\}$ replace it by the database $\Delta' = \{\neg n_1 \Rightarrow A_1, \ldots, \neg n_k \rightarrow A_k\}$ where n_i is the name of A_i.

Make a note, for future use, of which name names which items in the database. Whenever we add a formula B to the database, add it with a new name, i.e., add $\neg n_r \Rightarrow B$, where n_r is the unique name of B. This operational rule also applied to deletion. So to delete B, we add n_r. However, since n_r is to be added, we follow our procedure and actually add $\neg n_s \Rightarrow n_r$, where n_s is the unique name of n_r. The best way to do this is to record the steps in the computation and use the step as an index. Now to add B back after it has been deleted, we can delete the deletion of B, i.e., we add $\neg m_2 \Rightarrow n_s$, where m_2 is the unique name of n_s.

So, comparing with our $\lambda X \bigcirc (X)$ notation we have the following correspondence:

–formula B : $\neg n_r \Rightarrow B$

–formula $\bigcirc(B)$: $\neg n_s \Rightarrow n_r$

–formula $\bigcirc(\bigcirc(B))$: $\neg m_2 \Rightarrow n_s$

etc.

EXAMPLE 3.9. Let us do Example 2.13 again, using this method:

Data:
1. $\neg n_1 \Rightarrow A_1$
2. $\neg n_2 \Rightarrow A_2$
3. $\neg n_3 \Rightarrow (A_1 \Rightarrow B)$
4. $\neg n_4 \Rightarrow (A_2 \Rightarrow B)$.

This proves B. How do we logically delete B? In the N-Prolog method cannot do that. We can only calculate what items of data the proof of B depends on and delete them. So we need to add for example

$$\neg n_5 \Rightarrow n_1$$

and

$$\neg n_6 \Rightarrow n_2.$$

REMARK 3.10. *Note that the N-Prolog approach does satisfy that if $A \notin \Delta$, then deleting A fromΔ does yield Δ. To achieve this we need to systematically give names to formulae.*

What we cannot do is to delete formulae B that can be proved from Δ but are not physically in Δ. To delete such a B we need to find which sets of assumptions from Δ can be used to prove B and then delete from Δ enough assumptions to destroy all possible proofs. Here we need the assistance of the metapredicate Abduce$^-$. *It will identify such assumptions. This will be addressed in Section 4. Note that we might need to do conditional deletion, i.e., introduce expressions of the form* $q \Rightarrow$ **Delete**(r).

Actually we can use a device to delete such a B. Let r_B be a special name for B; different from B's standard name. Add $B \Rightarrow r_B$ to the database and try to prove r_B. This will succeed only if B can be proved. So if we ask for r_B whenever we want B, then we can delete r_B by deleting $B \Rightarrow r_B$. We shall address this in part 3 of this series of papers. See also Section 6 and remark 6.3.

DEFINITION 3.11 (N-Prolog sublanguage allowing systematic deletion).
1. Let $Q_i = \{q^i_j | j = 1, 2, 3, \ldots\}, i = 1, 2, 3\ldots$ be pairwise disjoint sets of atoms. Let f be a function symbol creating new atoms. Let $\mathbb{L}_0$ be the N-Prolog language with $\Rightarrow$, $\bot$ and $\neg$ based on the atoms $Q_0 = \{q_1, q_2, \ldots\}$ and let $\mathbb{L}_n$ be the language based on the atoms in $\bigcup_{i=0}^{n}(Q_i \cup \bigcup_{m=1}^{\infty}\{f^m(x)|x \in Q_i\})$. Assume that for atoms $x \in \bigcup_n Q_n$, $f^m(x), m = 1, 2, \ldots$ are all pairwise different.
2. For each formula A of $\mathbb{L}_m$, let $\mathbf{n}(m+1, A)$ be a unique atom from Q_{m+1} associated with A, acting as its name. Thus for $A \neq B$ we have $\mathbf{n}(m+1, A) \neq \mathbf{n}(m+1, B)$.
3. Let A be a formula of $\mathbb{L}_m$. Let $A_d^{[m+1]}$ and $A_g^{[m+1]}$ be the formulae of $\mathbb{L}_{m+1}$ being the result of naming all subformulae of A for data or for goal respectively. We define these formulae by structural induction based on Definition 3.1 as follows

– $q_d^{[m+1]} = \neg f(q) \Rightarrow (\neg \mathbf{n}(m+1, q) \Rightarrow q)^7$

$q_g^{[m+1]} = q$

for q atomic or $\perp$ or of the form $f^m(x)$, x atomic,

– $(\neg q)_d^{[m+1]} = 1 = f(q)^8$

$\neg q_g^{[m+1]} = \neg q$

for q atomic or $\perp$ or $f^m(x)$, x atomic.

– If $B = (A_1 \Rightarrow \ldots \Rightarrow (A_n \Rightarrow q)\ldots)$ is a data clause then A_i are goal clauses and we let
$B_d^{[m+1]} = \neg f(q) \Rightarrow \neg \mathbf{n}(m+1, B) \Rightarrow (A_{1,g}^{[m+1]} \Rightarrow \ldots \Rightarrow (A_{n,g}^{[m+1]} \Rightarrow q)\ldots))$

– If $B = A_1 \Rightarrow \ldots \Rightarrow (A_n \Rightarrow q)\ldots)$ is a goal clause, then A_i are data clauses, and we let

$$B_g^{[m+1]} = A_{1,d}^{[m+1]} \Rightarrow (\ldots \Rightarrow (A_{n,d}^{[m+1]} \Rightarrow q)\ldots).$$

LEMMA 3.12. Let Δ be a database of formulae in $\mathbb{L}_m$ and let $\Delta_d^{[m+1]} = \{A_d^{[m+1]} | A \in \Delta\}$. Then for any z, $Success(\Delta, z) = Success(\Delta_d^{[m+1]}, z_g^{[m+1]})$.

Proof. Proved by induction on the computation and follows from the fact that none of the new names are heads in $\Delta_d^{[m+1]}$.

We follow the inductive steps of Definition 3.2

1. *Immediate success/failure case*

   a) $Success(\Delta, q) = 1$ if $q \in \Delta$. But $q \in \Delta$ iff $\neg f(q) \Rightarrow (\neg \mathbf{n}(m+1, q) \Rightarrow q)$ is in $\Delta_d^{[m+1]}$. Hence $Success(\Delta_d^{[m+1]}, q) = 1$ since $\mathbf{n}(m+1, q)$ is not a head in $\Delta_d^{[m+1]}$.
   b) $Success(\Delta, q) = 0$ if q is not a head in Δ. But then q is not a head in $\Delta_d^{[m+1]}$ and hence $Success(\Delta_d^{[m+1]}, q) = 0$
   c) If $Success(\Delta_d^{[m+1]}, q)$ is 1 using one step (not counting $?\neg f(q)$ and $?\neg \mathbf{n}$ kind of steps) then $\neg f(q) \Rightarrow (\neg \mathbf{n}(m+1, q) \Rightarrow q)$ must be in $\Delta_d^{[m+1]}$ and hence $q \in \Delta$. Hence $Success(\Delta, q) = 1$ in one step.
   d) Clearly if $Success(\Delta_d^{[m+1]}, q)$ is 0 in one step then q is not a head in Δ and $Success(\Delta, q) = 0$.

2. *Implication case*
   $Success(\Delta, A_1 \Rightarrow \ldots \Rightarrow (A_n \Rightarrow q)\ldots) = Success(\Delta \cup \{A_i\}, q) =$ (by induction) $success(\Delta_d^{[m+1]} \cup \{A_{i,d}^{[m+1]}\}, q) = Success(\Delta_d^{[m+1]}, A_1 \Rightarrow \ldots A_n \Rightarrow q)\ldots)_g^{[m+1]})$.

3. *Cut rule case*
   $Success(\Delta, q) = 1$ (resp 0) iff for all (resp. some) $A = (A_1 \Rightarrow \ldots \Rightarrow (A_n \Rightarrow q)\ldots))$ in Δ we have that for all (resp. some) $Success(\Delta, A_i) = 1$ (resp. 0) iff (by induction) for all (reps. some) $A_d^{[m+1]} \in \Delta_d^{[m+1]}$ we have that for all

(resp. some) $A_{i,g}^{[m+1]}$, $Success(\Delta_d^{[m+1]}, A_{i,g}^{[m+1]}) = 1$ (resp. 0). Note that $\neg f(q)$ and $\neg\mathbf{n}(m+1, A)$ succeed from $\Delta_d^{[m+1]}$ and hence can be disregarded. We continue, iff $Success(\Delta_d^{[m+1]}, q) = 1$ (resp. 0). ∎

LEMMA 3.13.

1. Let Δ be a database of $\mathbb{L}_m$ and let $A \in \Delta$ with head x. Then $(\Delta - \{A\})_d^{[m+1]}$ is N-Prolog equivalent to $\Delta_d^{[m+1]} \cup \{\mathbf{n})(m+1, A)\}$. In other words, for any q, $Success(\Delta - \{A\}, q) = Success((\Delta - \{A\})_d^{[m+1]}, q) = Success(\Delta_d^{[m+1]} \cup \{\mathbf{n}(m+1, A\}, q)$.
2. Let $A_i \in \Delta$ list all clauses in Δ with head q, then $(\Delta - \{A_i\})_d^{[m+1]}$ is N-Prolog equivalent to $\Delta_d^{[m+1]} \cup \{f(q)\}$.

Proof. The proof is clear since $A_d^{[m+1]}$ has the form $\neg f(x) \Rightarrow (\neg\mathbf{n}(m+1, A) \Rightarrow A')$. ∎

The last lemma gives deletion by addition. Let us see how it works.

EXAMPLE 3.14. Consider

$$p \vdash p$$

We want to make p not provable. Obviously we need to delete p.

The above is equivalent (after inserting all the appropriate names) to $\neg f(p) \Rightarrow (\neg\mathbf{n}(p) \Rightarrow p) \vdash p$.

Hence we delete p by adding $\mathbf{n}(p)$ We get the database

$$\{\neg f(p) \Rightarrow (\neg\mathbf{n}(p) \Rightarrow p), \mathbf{n}(p)\}$$

and this database is equivalent to $\varnothing$.

4. Exploring Deletion via Addition

We saw in Remark 3.10 that we need the *Abduce*$^-$ metapredicate to enable us to make logical deletion by deleting from the data enough assumption that destroy all possible proofs. This section studies this metapredicate.

The aim of this section is to check whether, within the framework of N-Prolog with negation as failure and our naming methodology, we can give meaning to the deletion metapredicate *Abduce*$^-$ of Definition 3.7, i.e., effectively add to the database expressions which mean $A \Rightarrow Delete(X)$. Here $Delete(X)$ is our $\bigcirc(X)$.

Let us see what we can do so far. Consider again the problem of stopping by deletion the fact that $p \vdash p$. Obviously we should delete p and get $\varnothing \nvdash p$. The problem of $p \vdash p$ is equivalent to the problem of $\varnothing \vdash p \Rightarrow p$. So what do we delete from $\varnothing$ to stop $p \Rightarrow p$ being provable? Does it make sense to want a tautology not to be provable? Well, let us examine the equivalent problem, using Lemma 3.12. The problem becomes $\neg f(p) \Rightarrow (\neg \mathbf{n}(p) \Rightarrow p) \vdash p$, which is equivalent to $\varnothing \vdash (\neg f(p) \Rightarrow (\neg \mathbf{n}(p) \Rightarrow p)) \Rightarrow p$. To delete p we add $\mathbf{n}(p)$. Thus $\mathbf{n}(p) \nvdash (\neg f(p) \Rightarrow (\neg \mathbf{n}(p) \Rightarrow p)) \Rightarrow p$.

In fact, $\mathbf{n}(p)$ is the anti-formula for $\neg f(p) \Rightarrow (\neg \mathbf{n}(p) \Rightarrow p)$, and $\mathbf{n}(p)$ deletes it. There are still some points to check in the next example. In the previous example, we performed the abduction in the language of $\varnothing \vdash p \Rightarrow p$. We discovered we need to add $p \Rightarrow \mathit{Delete}(p)$, then we moved to the translation of the problem above in the language with names, namely to $\varnothing \vdash (\neg f(p) \Rightarrow (\neg \mathbf{n}(p) \Rightarrow p)) \Rightarrow p$ and added the translation of $p \Rightarrow \mathit{Delete}(p)$, namely $(\neg f(p) \Rightarrow (\mathbf{n}(p) \Rightarrow p)) \Rightarrow \mathbf{n}(p)$.

The question is what will we find if we apply the abduction process directly in the translation language?

EXAMPLE 4.1. Let us apply our abduction process of Definition 3.7 to the database and query

$$\varnothing \vdash p \Rightarrow p.$$

We consider the equivalent problem of

$$\varnothing \vdash (\neg f(p) \Rightarrow (\neg \mathbf{n}(p) \Rightarrow p)) \Rightarrow p.$$

We have it that

$$\mathit{Abduce}^{-}(\varnothing, (\neg f(p) \Rightarrow (\neg \mathbf{n}(p) \Rightarrow p)) \Rightarrow p)$$

is equal to

$$(\neg f(p) \Rightarrow (\neg \mathbf{n}(p) \Rightarrow p)) \Rightarrow \mathit{Abduce}^{-}(\neg f(p) \Rightarrow (\neg \mathbf{n}(p) \Rightarrow p), p).$$

We continue:

$$\mathit{Abduce}^{-}(\neg f(p) \Rightarrow (\neg \mathbf{n}(p) \Rightarrow p), p)$$

is equal to

$$\mathit{Abduce}^{-}(\neg f(p) \Rightarrow (\neg \mathbf{n}(p) \Rightarrow p), \neg f(p)) \text{ union with}$$
$$\mathit{Abduce}^{-}(\neg f(p) \Rightarrow (\neg \mathbf{n}(p) \Rightarrow p), \neg \mathbf{n}(p))$$

and this is equal to[9]

$$\mathit{Abduce}^{+}(\neg f(p) \Rightarrow (\neg \mathbf{n}(p) \Rightarrow p), f(p)) \text{ union with}$$
$$\mathit{Abduce}^{+}(\neg f(p) \Rightarrow (\neg \mathbf{n}(p) \Rightarrow p), \mathbf{n}(p))$$

which is equal to

$$\{\{\mathbf{n}(p)\}, \{f(p)\}\}$$

Thus the abduced formulae are either

$$(\neg f(\mathbf{n}(p)) \Rightarrow (\neg f(p) \Rightarrow (\neg \mathbf{n}(p) \Rightarrow p)) \Rightarrow \mathbf{n}(p))$$

or

$$\neg ff(p) \Rightarrow ((\neg f(p) \Rightarrow (\neg \mathbf{n}(p) \Rightarrow p)) \Rightarrow f(p)).$$

We can agree to abduce only on names of the form $\mathbf{n}(A)$, A a formula, because that was the original intention. In fact, we abduce only on names of the form $\mathbf{n}(q)$, q atomic because our original abduction process either added or deleted atoms.

The clause $f(p)$ is intended to delete when necessary (i.e., when $\neg p$ is supposed to be added into the database, all clauses with head p. So it does not participate in the abduction. Its purpose is different.

The above example shows we may have a problem. The clause abduced in the example is a goal clause, not a data clause. If we put it in the database and ask for $?\mathbf{n}(p)$ we need to ask for $\neg\mathbf{n}(p) \Rightarrow p$, and so we need to put $\neg\mathbf{n}(p)$ in the database. We thus get the new database and query $\Delta = \{\neg\mathbf{n}(p), \neg f(\mathbf{n}(p)) \Rightarrow (\neg f(p) \Rightarrow (\neg\mathbf{n}(p) \Rightarrow p)) \Rightarrow \mathbf{n}(p))\}?p$.

We have not said what putting $\neg\mathbf{n}(p)$ in a database means, i.e., N-Prolog with negation as failure does not tell us what it means to put a literal $\neg q$ into a data base. In fact, the language makes the distinctions of data clauses and goal clauses in order that the problem of putting a $\neg q$ into databases will never arise! However we have already made provisions in the translation of Definition 3.11 that the translation of $\neg q_d^{[m+1]}$ is $f(q)$, thus to add $\neg\mathbf{n}(p)$ means to add $f(\mathbf{n}(p))$. This indeed will kill any clause with head $\mathbf{n}(p)$.

We seem to have two options for solving our abduction problem:

Option 1 Use a *modified N-Prolog*.
This option will allow for inserting literal $\neg q$ into databases and allow also for clauses of the form $\neg q \Rightarrow A$ to be goal clauses provided A is a goal clause. We need to say what it means to have $\neg q$ in a database Δ. Let us agree that we view $\neg q$ as an integrity constraint, *cancelling* all clauses in Δ with head q.[10]

Thus if Δ contains $\neg q$ then no clause of the form $A_1 \Rightarrow \ldots \Rightarrow (A_n \Rightarrow q)\ldots)$ can be used in any computation in Δ. (More specifically items (1a), (3b) and (4) of Definition 3.2 must be qualified by the phrase 'and $\neg q$ is not in Δ' and items (1b), (3b) and (4) must be qualified by the phrase 'or $\neg q$ is in Δ').

The reader may note that N-Prolog actually knows how to do deletions, and so really that this modification can be implemented within N-Prolog itself, through the translation of Definition 3.11.

Option 2 Change the definition of $Abduce^-$

We eliminate the clause in the $Abduce^-$ definition (Definition 3.7) which puts us in this undesirable situation, namely item 3 of Definition 3.7

(*) $Abduce^-(\Delta, A_1 \Rightarrow \ldots (A_n \Rightarrow q)\ldots) = A_1 \Rightarrow \ldots (A_n \Rightarrow Abduce^-(\Delta \cup \{A_i\}, q))\ldots)$ replacing it by the rule

(**) $Abduce^-(\Delta, A_1 \Rightarrow \ldots (A_n \Rightarrow q)\ldots)) = Abduce^-(\Delta \cup \{A_i\}, q)$

(**) will mean that we add $\mathbf{n}(p)$ to our database.

Our initial position may be that we are reluctant to adopt Option 1, even though N-Prolog with negation as failure is a well known system with good semantics. We want to explore other options first. Unfortunately Option 2 is unacceptable because it would give unintuitive results, as the next example shows:

EXAMPLE 4.2. Consider the database Γ containing:

1. $p \Rightarrow (r \Rightarrow q)$
2. $r \Rightarrow a$
3. r

This database can prove a and it can also prove $p \Rightarrow q$.

Suppose we want to make sure, using $Abduce^-$, that $p \Rightarrow q$ does not follow. Then (*) gives us

$$Abduce^-(\Gamma, p \Rightarrow q) = p \Rightarrow Abduce^-(\Gamma \cup \{p\}, q),$$

which gives us two possibilities for abduction, which we will loosely write as $p \Rightarrow Delete(p)$ and $p \Rightarrow Delete(r)$.

If we use (**) we get the two options $Delete(p)$ and $Delete(r)$.

Let us choose the option of deleting r in some way.

Using (*) means that our database is primed to delete r the minute p is introduced into it. But as long as p has not been introduced into Γ, our database does have r in it and can therefore prove a.

Using (**), on the other hand, deletes r immediately. So even if p never arrives, a cannot be proved.

This seems counter-intuitive. $A \Rightarrow B$ has the interpretation that whenever A happens B must be true. Thus $p \Rightarrow Delete(r)$ may mean 'in case of emergency relax budgetary restrictions on spending'. We do not want to delete r now – only when p happens!

So it looks as if we are stuck without any new options. What shall we do? Shall we now adopt option 1?

Let us check one more angle; we tried to use (**) in $Abduce^-(\Gamma, p \Rightarrow q)$ and got counter-intuitive results. Let us ask, does the problem arise in the context of the naming scheme we introduced? In other words, if we use (**) in $Abduce^-(\Gamma_d^{[m+1]}, (p \Rightarrow q)_g^{[m+1]})$ do we still get a problem?

EXAMPLE 4.3. Let us see what happens to our database when we use names.

Our database becomes (call it Γ_d for short):[11]

(1*) $\neg n_1 \Rightarrow (r \Rightarrow (p \Rightarrow q))$
(2*) $\neg n_2 \Rightarrow (r \Rightarrow a)$
(3*) $\neg n_3 \Rightarrow r$
Our goal becomes $(\neg n_4 \Rightarrow p) \Rightarrow q$.

We want the goal to fail.

If we use (*) we get that we have two possibilities for deletion via addition: we either add

(4*) $(\neg n_4 \Rightarrow p) \Rightarrow n_4$
corresponding to $p \Rightarrow Delete(p)$
or we add
(5*) $(\neg n_4 \Rightarrow p) \Rightarrow n_3$ corresponding to $p \Rightarrow Delete(r)$.
corresponding to $p \Rightarrow Delete(r)$.

If we use (**) we get that we need to either add n_4 (corresponding to $Delete(p)$) or add n_3 (corresponding to $Delete(r)$).

We seem to have the same problem as before.

So we are still in a search for a third option.

It seems that we were either operating in the language $\mathbb{L}$ of Γ in which case $Abduce^-(\Gamma, p \Rightarrow q)$ gave us $p \Rightarrow Delete(r)$ which is not part of the Γ language, or we were operating in the language of N-Prolog, into which we translated Γ (to Γ_d) and $p \Rightarrow q$ to $(p \Rightarrow q)_g$ and we applied $Abduce^-(\Gamma_d, (p \Rightarrow q)_g)$ and got a new kind of difficulty. What we have not yet considered is the following:

Option 3 Mixed option
This option is a compromise on languages. We first execute the abduction in the language of Γ and get the abduced set (i.e., $p \Rightarrow Delete(r)$ and $p \Rightarrow Delete(p)$ in our example) and then *translate* the abduced set into N-Prolog and add it to Γ_d. This way the abduction for Γ_d is done in the original language (of Γ) and then translated and is not directly executed in the translation (i.e., in N-Prolog).

Will Option 3 solve our difficulties? Our experience so far, in Example 4.3 and the discussion preceding is encouraging.

Let us check.

EXAMPLE 4.4. The translation of Γ is Γ_d below (again we omit the use of the f function).

(1*) $\neg n_1 \Rightarrow (r \Rightarrow (p \Rightarrow q))$
(2*) $\neg n_2 \Rightarrow (r \Rightarrow a)$
(3*) $\neg n_3 \Rightarrow r$.

If we do the $Abduce^-(\Gamma, p \Rightarrow q)$ we get, as we have seen before, two possibilities for abduction, $p \Rightarrow Delete(r)$ and $p \Rightarrow Delete(p)$. Let us examine what happens with each choice in turn.

1. Case of $p \Rightarrow Delete(r)$:
   In this case r is in the database so it has a name n_3. So let us agree to
   – Translate '$Delete(X)$' as the *name of* X, '$\mathbf{n}(X)$'.

So we add to Γ_d the clause

(6*) $p \Rightarrow n_3$.

Let us see now whether $\Gamma_d + (6*) \vdash p \Rightarrow q$?
The answer is no because $\Gamma_d + (6*) + p \nvdash q$, because r is no longer provable. In fact, even if we ask as a goal the translation of $p \Rightarrow q$, namely $(\neg n_4 \Rightarrow p) \Rightarrow q$ we still get that $\Gamma_d + (6*) + (\neg n_4 \Rightarrow p) \nvdash q$ because p is still provable and hence r is thrown out.

2. Case of $p \Rightarrow \mathit{Delete}(p)$
The first problem in this case is that p does not have a name. The answer is that we have given clauses their names in a systematic way so p does indeed have a name, it is $\mathbf{n}(p) = n_4$ and so we can add

(7*) $p \Rightarrow n_4$

to Γ_d in this case.
To check whether

$$\Gamma_d + (7*) \vdash ?p \Rightarrow q$$

we have to be careful and require that the computation be done in N-Prolog with names. We know this choice works for case (1): so we must ask not $p \Rightarrow q$ but its translation $(p \Rightarrow q)_g = (\neg n_4 \Rightarrow p) \Rightarrow q$.
This gives:

$$\Gamma_d + (7*) + \neg n_4 \Rightarrow p \vdash ?q$$

and of course we loop. We have the two clauses

$$p \Rightarrow n_4$$
$$\neg n_4 \Rightarrow p$$

and we need to ask $?p$, hoping it would fail, but it actually loops. This is no cause for alarm. These two clauses are actually clauses of ordinary Prolog with negation as failure and a lot is known about looping in ordinary Prolog. We just want a device which will fail p. The trouble is that this is a genuine loop. If p fails then n_4 must succeed from the second clause and therefore p must succeed from the first clause. Any technical device which may work for this example may not work to our satisfaction in more complex examples.

EXAMPLE 4.5 (Problems with loops). Continuing the previous example, we know, however, that we are dealing with a specific task of finding a way of executing deletion by addition and that we are not concerned with the general problem of resolving nasty loops in ordinary Prolog. We may therefore utilise any specific features of our problem towards the success of our task. We note first that N-Prolog is based on intuitionistic logic, which is complete for a goal directed computation with diminishing resource, where whenever a clause is used it is immediately deleted. See Gabbay and Olivetti (2000, Section 3.1). The policy of diminishing resource eliminates loops. Can this help us?

Let us look at our loop again. The database is

$$p \Rightarrow n_4$$
$$\neg n_4 \Rightarrow p.$$

If we ask for $?p = 1$ we get

$\{p \Rightarrow n_4, \neg n_4 \Rightarrow p\}?p = 1$

if

$\{p \Rightarrow n_4\}?\neg n_4 = 1$

if

$\{p \Rightarrow n_4\}?n_4 = 0$

if

$\varnothing ?n_4 = 0$

if

Success.

Now ask for $?n_4 = 1$.

$\{p \Rightarrow n_4, \neg n_4 \Rightarrow p\}?n_4 = 1$

if

$\{\neg n_4 \Rightarrow p\}?p = 1$

if

$\varnothing ?\neg n_4 = 1$

if

$\varnothing ?n_4 = 0$

if

Success.

We get it that both n_4 and p succeed. This is nonsense. The diminishing resource policy does not work in the presence of negation as failure.

However, we still may be able to save the situation. We recall that anti-formulae are to be used only once. This means that the data item $p \Rightarrow n_4$, which represents the black hole should be used only once, while the data item $\neg n_4 \Rightarrow p$, which represents p is a genuine data item and can be used as many times as needed.

Let us reconsider our loop with this in mind:

$\{p \Rightarrow n, \neg n_4 \Rightarrow p\}?p = 1$

if

$\{p \Rightarrow n_4, \neg n_4 \Rightarrow p\}?\neg n_4 = 1$

if $\{p \Rightarrow n_4, \neg n_4 \Rightarrow p\}?n_4 = 0$

if

$\{\neg n_4 \Rightarrow p\}?p = 0$

if

$\varnothing ?\neg n_4 = 0$

if

$\varnothing ?n_4 = 1$

if

failure.

So p fails. This is good. How about $?n_4 = 1$. This should succeed.

$\{\Rightarrow n_4, \neg n_4 \Rightarrow p\}?n_4 = 1$

if

$\{\neg n_4 \Rightarrow p\}?p = 1$

if

$\varnothing ?\neg n_4 = 1$

if

$\varnothing ?n_4 = 0$

if

success.

It looks as if we have made it. n_4 succeeds and p fails.[12]

We have one more item to check. We need to check Example 4.4 where we chose the possibility of $p \Rightarrow Delete(r)$ and added

(6*) $p \Rightarrow n_3$

does it still work with (6*) being a once only clause? The answer is yes. We use (6*) only once in this example.

It is time now to move to the next section and give a formal definition of how we are to deal with $Abduce^-$ of Definition 3.7.

But before we do that, why don't we go back to Option 1 and see how it fares with our example?

EXAMPLE 4.6. According to Option 1 we have the following database, after $Abduce^-(\varnothing,$
$(\neg f(p) \Rightarrow (\neg \mathbf{n}(p) \Rightarrow p)) \Rightarrow p)$ of Example 4.3 has been executed, and the result of the abduction added to $\varnothing$:

$$\Delta^* = \{(\neg f(\mathbf{n}(p) \Rightarrow (\neg \mathbf{n}(p) \Rightarrow p))) \Rightarrow \mathbf{n}(p)\}.$$

Let us ask whether $?(\neg f(p) \Rightarrow (\neg \mathbf{n}(p) \Rightarrow p)) \Rightarrow p = 1$,
from this database.

We get the above reducing to

$$\{\neg f(\mathbf{n}(p)) \Rightarrow (\neg \mathbf{n}(p) \Rightarrow p)) \Rightarrow \mathbf{n}(p), \neg f(p) \Rightarrow (\neg \mathbf{n}(p) \Rightarrow p)\}?p = 1$$

which reduces to, (since $?\neg f(p) = 1$),

$$\{\neg f(\mathbf{n}(p)) \Rightarrow (\neg \mathbf{n}(p) \Rightarrow p)) \Rightarrow \mathbf{n}(p), \neg f(p) \Rightarrow (\neg \mathbf{n}(p) \Rightarrow p)\}?\mathbf{n}(p) = 0$$

which reduces to, (since $?\neg f(\mathbf{n}(p)) = 1$),

$$\{\neg f(\mathbf{n}(p)) \Rightarrow ((\neg\mathbf{n}(p) \Rightarrow p)) \Rightarrow \mathbf{n}(p), \neg f(p) \Rightarrow (\neg\mathbf{n}(p) \Rightarrow p), f(\mathbf{n}(p))\}?p = 0.$$

The presence of $f(\mathbf{n}(p))$ cancels all clauses with head $\mathbf{n}(p)$, and so our problem reduces to[13]

$$\{\neg f(p) \Rightarrow (\neg\mathbf{n}(p) \Rightarrow p)\}?p = 0$$

which reduces to

$$\{\neg f(p) \Rightarrow (\neg\mathbf{n}(p) \Rightarrow p)\}?\mathbf{n}(p) = 1$$

which fails, as required.

Let us now check whether $?\mathbf{n}(p) = 1$ succeeds.

$$\{\neg f(\mathbf{n}(p) \Rightarrow ((\neg\mathbf{n}(p) \Rightarrow p)) \Rightarrow \mathbf{n}(p), \neg f(p) \Rightarrow (\neg\mathbf{n}(p) \Rightarrow p)\}?\mathbf{n}(p) = 1$$

reduces to, (since $?\neg f(\mathbf{n}(p)) = 1$ at this stage),

$$\{\neg f(\mathbf{n}(p)) \Rightarrow ((\neg\mathbf{n}(p) \Rightarrow p)) \Rightarrow \mathbf{n}(p), \neg f(p) \Rightarrow (\neg\mathbf{n}(p) \Rightarrow p), f(\mathbf{n}(p))\}?p = 1$$

and this reduces to

$$\{\neg f(p) \Rightarrow (\neg\mathbf{n}(p) \Rightarrow p)\}?p = 1$$

which reduces to

$$\{\neg f(p) \Rightarrow (\neg\mathbf{n}(p) \Rightarrow p)\}?\mathbf{n}(p) = 0$$

which succeeds.

So all is well.

Let us also check Example 4.2 according to Option 1.

EXAMPLE 4.7. Our database is (we omit the use of f):
(1*) $\neg n_1 = \neg(r \Rightarrow (p \Rightarrow q))$
(2*) $\neg n_2 \Rightarrow (r \Rightarrow a)$
(3*) $\neg n_3 \Rightarrow r$.
The goal to fail is

$$(\neg n_4 \Rightarrow p) \Rightarrow q.$$

The two possible abduced sentences are
(4*) $(\neg n_4 \Rightarrow p) \Rightarrow n_4$
corresponding to $p \Rightarrow \mathit{Delete}(p)$ and
(5*) $(\neg n_4 \Rightarrow p) \Rightarrow n_3$

corresponding to $p \Rightarrow Delete(r)$.

Let us check if the job gets done[14]

First we check

$$\{(1*) - (3*), (4*)\}?(\neg n_4 \Rightarrow p) \Rightarrow q = 1$$

This reduces to

$$\{(1*) - (3*), (4*), \neg n_4 \Rightarrow p\}?q = 1.$$

From (1*), and using (3*), this reduces to

$$\{(1*) - (3*), (4*), \neg n_4 \Rightarrow p\}?p = 1$$

which reduces to

$$\{(1*) - (3*), (4*), \neg n_4 \Rightarrow p\}?n_4 = 0$$

which reduces to (from (4*))

$$\{(1*) - (3*), (4*), \neg n_4 \Rightarrow p, \neg n_4\}?p = 0$$

which reduces to (by using $\neg n_4 \Rightarrow p$ and simplifying, since $\neg n_4$ cancels (4*)

$$\{(1*) - (3*), \neg n_4 \Rightarrow p\}?n_4 = 1$$

which fails.

5. A Formal System for Deletion via Addition

We begin with a methodological remark. The discussion in the previous section gave us hope that $Abduce^-$ can be given respectable meaning and semantics by translating it into N-Prolog with negation as failure. The task of this section is to show how it can be done formally. We basically need to give meaning to expressions like $p \Rightarrow Delete(q)$.

The methodological point we want to make is that once we succeed in our task, we no longer need to actually work in N-Prolog. We can extend our language of intuitionistic $\Rightarrow$ and $\bot$ with a $Delete(X)$ predicate or $\bigcirc(X)$, and work out the local rules via the translation into N-Prolog. This should be our final elegant aim.

First let us examine whether our intuitive properties for anti- formulae discussed in Section 3 of Part 1 of this paper hold for the N-Prolog translation.

1. *Annihilation*

This rule says that

$$\Delta + X + \bigcirc(X) = \Delta$$

If $*$ is the translation into N-Prolog, then we need to check whether

$$\Delta^* + \neg\mathbf{n}(X) \Rightarrow X + \mathbf{n}(X) = \Delta^*$$

where $\Gamma_1 = \Gamma_2$ means $\Gamma_1?Y = \Gamma_2?Y$ for every Y, i.e., Γ_1, Γ_2 give the same answers. This is clearly true.

2. *Black hole deduction theorem*
The deduction theorem says

$$\Delta \vdash A \text{ iff } \Delta + \bigcirc(X) \vdash X \Rightarrow A$$

where

$$\Delta, A, X$$

are arbitrary. Translated it becomes

$$\Delta_d \vdash A_g \text{ iff } \Delta_d + \mathbf{n}(X) \vdash (\neg\mathbf{n}(X) \Rightarrow X) \Rightarrow A_g.$$

We have already noticed that we need the *once only* rule, because without it we could have the wrong results. Consider

$$X \Rightarrow A \vdash X \Rightarrow A$$

hence by the black hole deduction theorem for X we get

$$X \Rightarrow A + \bigcirc(X) \vdash X \Rightarrow (X \Rightarrow A)$$

hence

$$X \Rightarrow A + \bigcirc(X) + X + X \vdash A$$

hence

$$X \Rightarrow A \vdash A.$$

Let us see what happens in the translation.
Let

$$n_1 = \mathbf{n}(X \Rightarrow A)$$
$$n_2 = \mathbf{n}(X)$$

The translation of $X \Rightarrow A \vdash X \Rightarrow A$ becomes $(X \Rightarrow A)_d \vdash (X \Rightarrow A)_g$, namely

$$\neg n_1 \Rightarrow (X \Rightarrow A) \vdash (\neg n_2 \Rightarrow X) \Rightarrow A$$

hence by the black hole deduction theorem for X

$$\neg n_1 \Rightarrow (X \Rightarrow A) + n_2 \vdash (\neg n_2 \Rightarrow X) \Rightarrow (\neg n_2 \Rightarrow X) \Rightarrow A$$

which is equivalent to

$$\neg n_1 \Rightarrow (X \Rightarrow A) + n_2 + (\neg n_2 \Rightarrow X) + (\neg n_2 \Rightarrow X) \vdash A.$$

It is clear that we need a once only use of n_2, otherwise we get $\neg n_1 \Rightarrow (X \Rightarrow A) \vdash A$.

Consider now a database with $\{X, \bigcirc(X), \circledcirc(x)\}$. What would be its translation? The obvious choice is to take $\{\neg\mathbf{n}(X) \Rightarrow X, \neg\mathbf{n}(\mathbf{n}(X) \Rightarrow \mathbf{n}(X), \mathbf{n}(\mathbf{n}(X)\}$.

We may think, however, that we are not being consistent in naming. We usually name 'X' by '$\mathbf{n}(X)$' and put $\neg\mathbf{n}(X) \Rightarrow X$' in the database as the translation of 'X'. But if '$\mathbf{n}(X)$' is also named by '$\mathbf{n}(\mathbf{n}(X))$' then if we put '$\neg\mathbf{n}(\mathbf{n}(X) \Rightarrow \mathbf{n}(X)$' in the database as the translation of '$\mathbf{n}(X)$', then to be consistent, '$\neg\mathbf{n}(X) \Rightarrow X$' should be translated as $\neg(\neg\mathbf{n}(\mathbf{n}(X) \Rightarrow \mathbf{n}(X)) \Rightarrow X$ which is equivalent to $(\neg\mathbf{n}(\mathbf{n}(X)) \Rightarrow \neg\mathbf{n}(X)) \Rightarrow X$.

We therefore might consider the translation:

$$(X)_d = (\neg\mathbf{n}(\mathbf{n}(x) \Rightarrow \neg\mathbf{n}(X)) \Rightarrow X$$
$$\bigcirc(X) = \neg\mathbf{n}(\mathbf{n}(X)) \Rightarrow \mathbf{n}(X)$$

and

$$\circledcirc(X) = \mathbf{n}(\mathbf{n}(X))$$

We shall see that this translation does not give the correct results. Let us check what can be proved from the database containing all three. We have

$$\Delta = \{X, \bigcirc(X), \circledcirc(X)\}_d$$
$$= \{(\neg\mathbf{n}(\mathbf{n}(X)) \Rightarrow \neg\mathbf{n}(X)) \Rightarrow X, \neg\mathbf{n}(\mathbf{n}(X)) \Rightarrow \mathbf{n}(X), \mathbf{n}(\mathbf{n}(X))\}$$

We ask

$$\Delta?X = 1$$

if

$$\Delta \cup \{\neg\mathbf{n}(\mathbf{n}(X))\}?\neg\mathbf{n}(X) = 1$$

if

$$\Delta \cup \{\neg\mathbf{n}(\mathbf{n}(X))\}?\mathbf{n}(X) = 0$$

if

$$\Delta \cup \{\neg\mathbf{n}(\mathbf{n}(X)\}?\mathbf{n}(\mathbf{n}(X)) = 1$$

if

fail.

Similarly

$$\Delta?\mathbf{n}(X) = 0$$

succeeds,

and

$\Delta?\mathbf{n}(\mathbf{n}(X)) = 1$

succeeds.

Since the name of X fails, $?X$ must succeed; but this is not the case!

We should have translated $(X)_d$ as $\neg\mathbf{n}(X) \Rightarrow X$ and the database $\{X, \bigcirc(X), \circledcirc(X)\}_d$ should be $\{\neg\mathbf{n}(X) \Rightarrow X, \neg\mathbf{n}(\mathbf{n}(X)) \Rightarrow \mathbf{n}(X), \mathbf{n}(\mathbf{n}(X))\}$.

This database does give the right answers.

Notice that $X \neq \circledcirc(X)$ in this translation. In fact $\{\circledcirc(X), \bigcirc(X)\} = \varnothing$ and annihilation is done in the order from higher anti-formulae to lower ones.

6. Discussion and Future Work

We presented in this paper two methods for object level deletion in intuitionistic logic. We also saw some limitations.

This section discusses some problems left open. We hope to tackle the issues in a forthcoming paper. We want to address the problem of what happens when we want to delete A from Δ, in the case where $\Delta \nvdash A$. We also want to see how to improve the N-Prolog model and be able to logically delete formulae A from Δ in case $A \notin \Delta$ but $\Delta \vdash A$.

The following examples illustrate what can happen.

EXAMPLE 6.1 (Vacuous Deletion). Suppose $\Delta \nvdash A$. We want to execute the operation of deletion of A from Δ, which we denote as $\Delta' = \mathbf{Delete}(\Delta, A)$. What do we get in this case? Shall we expect that $\Delta' = \Delta$? Using the methods of Part 1 of this paper, we have $\mathbf{Delete}(\Delta, A) = \Delta \cup \{A \Rightarrow \mathbf{e}\}$ and so for example $\mathbf{Delete}(\{B\}, A) = \{B, A \Rightarrow \mathbf{e}\} = \Delta'$. Now $\Delta' \nvdash B$ because$A \Rightarrow \mathbf{e}$ cannot be used. This may be undesirable! We might try to solve the problem by allowing the system to ignore $A \Rightarrow \mathbf{e}$, that is, no need to use it. We get some form of variation of linear logic, but this is not satisfactory because using or not using $A \Rightarrow \mathbf{e}$ should be context independent of whether A is among the data or not.

EXAMPLE 6.2 (Conditional Deletion). The problem of vacuous deletion can still be troublesome. To explain the difficulty, consider the following database:

1. B
2. $A \Rightarrow (B \Rightarrow C)$

This database proves $A \Rightarrow C$. Suppose we want to delete A. Using the methods of Part 1 we add

3. $(A \Rightarrow \mathbf{e})$

Since A is not in the database nor is it provable from it, we would expect to be able to still prove $A \Rightarrow C$. How do we prove $A \Rightarrow C$? We need to assume A and prove C so let us do so. We get

1. B
2. $A \Rightarrow (B \Rightarrow C)$
3. $A \Rightarrow \mathbf{e}$
4. A

From this database C cannot be proved because A is deleted.

To avoid this we must indicate that

3. $A \Rightarrow \mathbf{e}$

is not accessible. This gives proof theory similar to that of strict implication, where $A, A \Rightarrow B \nvdash B$, when A is atomic to be used when A is assumed. So we must use box proof theory as follows:

1.	B	data
2.	$A \Rightarrow (B \Rightarrow C)$	data
3.	$A \Rightarrow \mathbf{e}$	local deletion
4.	$A \Rightarrow C$	from Box

3.1	A assumption
3.2	$B \Rightarrow C$, from (2) and (3.1)
3.3	C, from (3.2) and (1).
	Note that (3) is not accessible and so 3.1 cannot be deleted.

EXAMPLE 6.3. As an example of logical deletion in N-Prolog, let us reconsider Example 3.9. The data in this example proves B but we have no means to preform logical deletion. We now show how to do it in N-Prolog.

With each formula X associate three names

1. n_X name for deleting X if it is among the data. This means we put X in as $\neg n_X \Rightarrow X$.
2. r_X and q_X, two names to use when X is a goal (to be proved). We agree to prove $\gamma(X) = ((\neg q_X \wedge X \Rightarrow r_X) \Rightarrow r_X)$ instead of proving X.

Let us see what happens in this case. To show $\gamma(X)$, we add to the data

$$\neg q_X \wedge X \Rightarrow r_X$$

and ask for r_X.

The only way for r_X to succeed is that q_X fails and X succeeds. If q_X and r_X are new names, then r_X succeeds iff X succeeds and q_X was not added deliberately to stop X from succeeding.

So in our example, we ask for $\gamma(B)$ whenever we want to prove B. To delete B logically, add q_B.

The above is just an idea and needs to be investigated. For suppose we also have in our database a clause $B \Rightarrow Y$. Adding q_B deletes B for the purpose of asking B because we need to ask $\gamma(B)$. But if we ask for Y (i.e., ask for $\gamma(Y)$, the query will succeed because we can still get B and by using $B \Rightarrow Y$ we can get Y.

We postpone addressing this problem until the third part of this investigation.

Notes

[1] To explain the definition of t, assume A was used in the proof of B k times, $k \geq 0$. For $k = 0$, $t = \gamma \cup \{\bigcirc(A)\}$ and otherwise t has in it the number of copies of A required to facilitate the proof of B. Thus we would expect $t \vdash_{\mathbf{L_e}} A \Rightarrow B$.

[2] The deduction theorem for $\mathbf{L_e}$ says that for any γ, X and B (no matter whether X is in γ or not), we have $\gamma \vdash B$ iff $\gamma \cup \{\bigcirc(X)\} \vdash X \Rightarrow B$. This gives the usual deduction theorem as a special case because $\gamma \cup \{X\} \vdash B$ iff $\gamma \cup \{X, \bigcirc(X) \vdash X \Rightarrow B)$ iff $\gamma \vdash X \Rightarrow B$ because $\{\bigcirc(X), X\} \equiv \mathbf{e}$ and $\gamma \cup \{\mathbf{e}\} \equiv \gamma$.

[3] $\neg\bot$ is actually $\top$.

[4] We shall see later that it may be convenient to allow goals of the form $\neg q \Rightarrow a$, provided we say what it means to 'add $\neg q$ to the data'. We may interpret this as 'Delete all clauses with head q' from the database, and keep deleting such clauses as long as $\neg q$ is in the database.

[5] This is a serious commitment, which may not be intuitive in some applications. Consider a political Party commitment to disclose (D) all cash gifts (G) to the Party chairperson. This can be formalised as $G \Rightarrow D$. Suppose we want to delete this. Using our method, to achieve $G \Rightarrow D \not\vdash G \Rightarrow D$, we consider the equivalent problem of $G, G \Rightarrow D \not\vdash D$ and delete G. This means a policy of denying all gifts, rather than changing the rule.

[6] We thus change notation and write *Delete*(A) for $\bigcirc(A)$. It is easier to use for complex expressions.

[7] $\neg f(q)$ is the negation as failure of the 'name' of the clause head 'q'. It is put in so that we can delete all clauses with head q in one action by adding '$f(q)$'. $\mathbf{n}(m+1, q)$ is the name of this particular atomic clause. We put in $\neg\mathbf{n}(m+1, q)$ so that we can delete this particular clause by adding its name. Thus any clause $A \Rightarrow q$ becomes $\neg f(q) \wedge \neg\mathbf{n}(m+1, A) \wedge A \Rightarrow q$ or if written with $\Rightarrow$ alone it becomes $\neg f(q) \Rightarrow |(\neg\mathbf{n}(m+1, A) \Rightarrow (A \Rightarrow q))$.

[8] The significance of this definition becomes apparent in light of Example 3.4

[9] *Abduce*$^+(\Delta, q)$ is what is needed to add to Δ to make q succeed. An inductive definition for *Abduce*$^+$ can be given, similar to *Abduce*$^-$. But in our case it is clear what needs to be added.

[10] Adding $f(q)$ will continue to delete for ever any clause with head q.

[11] We omit to add, for the sake of simplicity, the names $\neg f(q)$, $\neg f(a)$ and $\neg f(r)$ in the respective clauses.

[12] Think of our database classically: $(\neg n_4 \Rightarrow p) \wedge (p \Rightarrow n_4)$ is equivalent to $(\neg(p \Rightarrow n_4) \wedge (p \Rightarrow n_4)$, which is equivalent to n_4. It is not surprising therefore that n_4 succeeds and p fails.

[13] Notice that the minute $(\neg\mathbf{n}(p) \Rightarrow \neg p) \Rightarrow \mathbf{n}(p)$ was used, $\neg\mathbf{n}(p)$ was supposed to be added to the database, and thus $f(\mathbf{n}(p))$ was added and thus cancelled the use of the clauses, *making it a use once clause only*. This is fully compatible with our view of black holes! to add $\neg\mathbf{n}(p)$ we added $f(\mathbf{n}(p))$ as required in item 3 of Definition 3.8.

[14] Notice again that any clause of the form $(\neg x \Rightarrow q) \Rightarrow x$ is a use once only clause. The minute it is used with $?x$, it immediately adds $\neg x$ to the database and continues with $?q$. Thus immediately cancelling itself.

In fact any clause of the form $x \Rightarrow q$ can be made a once only clause by 'equivalently' replacing it by $(\neg x \Rightarrow q) \Rightarrow x$, (provided no other clauses $A \Rightarrow x$ exist).

References

Alchourrón, C. A. and D. Makinson: 1982, 'On the Logic of Theory Change: Contraction Functions and their Associated Revision Functions', *Theoria* **48**, 14–37.

Alchourrón, C. A. and D. Makinson: 1985, 'The Logic of Theory Change: Safe Contraction', *Studia Logica* **44**, 405–422.

Alchourrón, C. A., P. Gärdenfors and D. Makinson: 1985, 'On the Logic of Theory Change: Partial Meet Contraction and Revision Functions', *The Journal of Symbolic Logic* **50**, 510–530.

Anderson, A. R. and N. D. Belnap: 1975, *Entailment*, Vol. 1, Princeton University Press.

Connolly, T., C. Begg and A. Strachan: 1999, *Database Systems: A Practical Approach to Design, Implementation and Management*, 2nd edn, Addison-Wesley.

Darwiche, A. and J. Pearl: 1994, 'On the Logic of Iterated Belief Revision', in Ronald Fagin (ed.), *Proceedings of the 5th International Conference on Principles of Knowledge Representation and Reasoning*, Pacific Grove, CA, Morgan Kaufmann, pp. 5–23.

Darwiche, A. and J. Pearl: 1997, 'On the Logic of Iterated Belief Revision', *Artificial Intelligence* **89**, 1–29.

Date, C. J. and H. Darwen: 1997, *A Guide to the SQL Standard: A User's Guide to the Standard Database Language*, 4th edn, Addison-Wesley Longman, Inc.

Freund, M. and D. Lehmann: 1994, 'Belief Revision and Rational Inference', Technical Report TR 94-16, The Leibniz Center for Research in Computer Science, Institute of Computer Science, Hebrew University.

Gabbay, D. M.: 1996, *Labelled Deductive Systems*, Oxford University Press.

Gabbay, D. M. and O. Rodrigues: 1997, 'Structured Belief Bases: A Practical Approach to Prioritised Base Revision', in Dov M. Gabbay, Rudolf Kruse, Andreas Nonnengart, and Hans Jürgen Ohlbach (eds.), *Proceedings of First International Joint Conference on Qualitative and Quantitative Practical Reasoning*, Springer-Verlag, pp. 267–281.

Gabbay, D. M.: 1998, *Elementary Logic: A Procedural Perspective*, Prentice Hall.

Gabbay, D. M.: 1999, 'Compromise Update and Revision: A Position Paper', in B. Fronhoffer and R. Pareschi (eds.), *Dynamic Worlds*, Kluwer, pp. 111–148.

Gabbay, D. M. and N. Olivetti: 2000, *Goal Directed Proof Theory*, Kluwer.

Gabbay, D. M., O. Rodrigues and A. Russo: 2000, 'Revision by Translation', in B. Bouchon-Meunier, R. Yager and L. A. Zadeh (eds.), *Information, Uncertainty and Fusion: Proceedings of IPMU 98*, Kluwer, pp. 3–32.

Gabbay, D. M. and J. Woods: 2004, *The Reach of Abduction: Insight and Trial, A Practical Logic of Cognitive Systems*, Vol. 2, Elsevier.

Gabbay, D. M., O. Rodrigues and J. Woods: 2002, 'Belief Contraction, Anti-formulae and Resource Overdraft: Part I Deletion in Resource Bounded Logics', *Logic Journal of the IGPL* **10**, 601–652.

Gabbay, D. M., O. Rodrigues and J. Woods: in preparation, 'Belief Contraction, Anti-formulae and Resource Overdraft: Part III Fine Tuning of the Deletion Process'.

Gabbay, D. M., O. Rodrigues and J. Woods: in preparation, 'Existence and Anti-existence in Non-classical Logics'.

Gärdenfors, P.: 1978, 'Conditionals and Changes of Belief', *Acta Philosophica Fennica* **30**, 381–404.

Gärdenfors, P.: 1982, 'Rules for Rational Changes of Belief', in T. Pauli (ed.), *Philosophical Essays Dedicated to Lannart Åqvist on His Fiftieth Birthday*, Vol. 34 of *Philosophical Studies*, Philosophical Society and the Department of Philosophy, Uppsala University, pp. 88–101.

Gärdenfors, P.: 1988, *Knowledge in Flux: Modeling the Dynamics of Epistemic States*, Cambridge, MA, London, England, A Bradford Book – The MIT Press.

Girard, J. Y.: 1998, 'Light Linear Logic, Information and Computation', **143**(2), 175–204.

Groff, J. R. and P. N. Weinberg: 1999, *SQL: The Complete Reference*, Osborne/McGraw-Hill.

Grosof, B. N.: 1992, *Updating and Structure in Non-montonic Theories*, Ph.D. Thesis.

Hansson, S. O.: 1999, *A Textbook of Belief Dynamics: Theory Change and Database Updating*, Dodrdrecht, Kluwer Academic Publishers.

Kurosh, A. G.: 1963, *General Algebra*, Chelsea Publishing Commpany.

Lehmann, D.: 1995, 'Belief Revision, Revised',, in *Proceedings of the 14th International Joint Conference of Artificial Intelligence (IJCAI-95)*, pp. 1534–1540.

Olivetti, N. and L. Terracini: 1992, 'N-prolog and Equivalence of Logic Programs', *Journal of Logic, Language and Information* **1**, 253–340.

Restall, G.: 2000, *An Introduction to Substructural Logics*, New York, Routledge, ISBN 0-415-21534-X.

Rodrigues, O.: 1998, *A Methodology for Iterated Information Change*, Ph.D. thesis, Department of Computing, Imperial College.

Woods, J.: 1974, *The Logic of Fiction: Philosophical Soundings of Deviant Logic*, The Hague and Paris, Mouton.

REASONING ABOUT KNOWLEDGE IN LINEAR LOGIC: MODALITIES AND COMPLEXITY

MATHIEU MARION and MEHRNOUCHE SADRZADEH
Département de philosophie, Université du Québec à Montréal, C.P. 8888, Succursale Centre-Ville, Montréal, Québec, Canada H3C 3P8

Abstract. In this paper, we briefly argue, following ideas set forth by Jacques Dubucs, for a radical version of anti-realism and claim that it leads to the adoption of a 'substructural' logic, linear logic. We further argue that, in order to avoids problems such as that of 'omniscience', one should develop an epistemic linear logic, which would be weak enough so that the agents could still be described as omniscient, while this would not be problematic anymore. We then examine two possible ways to develop an epistemic linear logic, and eliminate one. We conclude on some remarks about complexity. The paper contains a coding in *Coq* of fragments of modal linear logic and a proof of the 'wise men' puzzle.

1. Introduction

In a recent paper, Jean-Yves Girard commented that "it has been a long time since philosophy has stopped intereacting with logic" (Girard 1998). Actually, it has not been such a "long" time since, e.g., Dag Prawitz and Michael Dummett developed philosophical arguments within the paradigm of Gentzen-style systems in favour of the adoption of intuitionistic logic. But recent developments within logic have left philosophers far behind. Prawitz's timely book *Natural Deduction* (Prawitz 1965), along with a key result obtained at around the same time, the Curry-Howard isomorphism (Howard 1980), initiated deep changes within logic; to wit, the development of the substructural logics (Schroeder-Heister and Dosen 1993; Restall 2000). Dummett developed within proof-theoretical semantics a mature version of his anti-realist challenge, closely allied to Prawitz's own philosophical considerations. But within the debate generated since by Dummett's anti-realism, philosophers have for the most part not paid much attention to the developments within logic since the 1970s. The almost complete absence of any discussion of linear logic within this context should convince any one that there is something seriously amiss here. The present paper is related to an attempt by Jacques Dubucs to provide a new impetus to the anti-realism debate by providing a radical anti-realist line of argument that also ties the debate more closely to issues of concern within substructural logics (Dubucs 1997, 2002; Dubucs and Marion 2003). We shall only make brief remarks about his argument in Section 2. Our intention is to

S. Rahman et al. (eds.), Logic, Epistemology, and the Unity of Science, pp. 327–350.

push it a few steps further and explore the possibilities for a radically anti-realist epistemic logic.

In a nutshell, it has been suggested that a radical form of anti-realism should force one to look at structural rules within Gentzen-style systems that are responsible for the idealizations of the full structural logic. By 'idealization' we merely mean features of structural logic which allow for infinity to creep in, so to speak, and which should not go without notice within the interpretation of proofs as actions that has dominated much of the thinking about proof-theoretical semantics since the days of Prawitz and Dummett (e.g., in the work of Girard or of Martin-Löf). Linear logic, by forbidding contraction and weakening as structural rules and by simultaneously introducing the exponentials ! and ? to recover them in some way, appears at first blush to be a promising candidate. Since its beginning, epistemic logic has been plagued with a serious case of idealization, the problem of logical omniscience. We shall propose here one new avenue for coping with this problem. Here again we build on ideas set forth by Dubucs (1998). This new approach requires that one develops a modal linear logic. In Section 3, we shall discuss only two candidates for epistemic linear logic and eliminate one. The radical anti-realism advocated for here requires that issues concerning complexity should not be ignored, as they have been traditionally, precisely because, when discussing idealizations, complexity is part of the diagnosis. This is a serious problem for epistemic logic, where one seems not to be able to escape exponential complexity, especially in multi-agent systems. In the concluding section of our paper, we shall suggest one way out, namely the encoding of proofs in the proof assistant *Coq* (Coquand and Huet 1988).

The following considerations are of an exploratory and programmatic (in other words: philosophical) nature. There are no new results, except a coding in *Coq* of two different fragments of modal linear logic and the proof of the puzzle known as the 'wise men' or 'King, 3 wise men, and 5 hats' puzzle, which is a well-known version of the 'muddy children' puzzle. This proof is given in the appendix. This paper is rather an informal discussion which should help to orient and to motivate future research. Although we shall go over some elementary points in order to make our philosophical points, especially in the next section, some basic knowledge is presupposed.

2. Radical Anti-Realism and Linear Logic

The stance adopted here is that of *radical anti-realism*. It is the result of a radicalization of the anti-realist philosophy of Michael Dummett, on the basis of an argument already presented in other places (Dubucs 1997, 2002; Dubucs and Marion 2003). This argument certainly needs to be butressed by further philosophical considerations but this is not the place to do so. We should like merely to insist here on a few consequences from this argument, since they provide the initial impetus for the following. In these papers, it is argued that the traditional form of anti-realism propounded by Dummett is not satisfactory because, in a nutshell, it relies on the notion of assertability-conditions, where assertability is claimed to be effective *in principle*. This very notion is claimed to be as obscure as the realist notion of truth-conditions (which transcend our cognitive capacities) and it is argued that one should replace effectivity in principle by the notion of *feasibility* in practice. The argument is in essence as follows.

According to the definition of assertability-conditions, a statement is assertable if there exists an effective proof of it, that is a finite sequence of statements of which it is the last and of which every statement follow another as the result of an application of a rule of inference (there is only a finite number of such rules). Such a definition does not allow for an hypothetical being whose cognitive capacities would be such that it could, say, recognize the truth of a universal statement by inspection of an infinity of particular cases. The realist could still point out, however, that when the anti-realist admits of finite proofs that can be carried out merely in principle he does not fare much better than someone who admits of truth-conditions which transcend our cognitive capacities. Therefore, the definition must be such that our cognitive capacities must allow us always to recognize a sentence as assertable when it is, i.e., that one must be able to recognize the object $Pr(s)$ which is an effective proof of s, when there is one. This statement is ambiguous, since one may understand this either, as Dummett did, as the weaker claim that one has to be able recognize a proof of s when presented with one or as the stronger claim that one must be able to produce or reproduce the object $Pr(s)$. For the anti-realist really to distinguish his position from that of the realist on this rather crucial point, he must claim not only that circumstances in which an assertion is justified must be such that we should recognize them when we are in a position to do so, he must also claim that we must always be able in practice to put ourselves in such a position whenever such circumstances exist. Otherwise, it would be open for the realist to admit there should always exist circumstances under which we would recognize that an assertion is justified and merely to deny that we should always have the practical capacity to put ourselves in that position. To repeat, the weaker claim that one has to be able recognize a proof of s when presented with one won't do, because there may simply be situations where we could recognize a proof when presented with one, but we would never be able in practice to put ourselves in such a position. Therefore, in order to develop a coherent alternative to the realist, the anti-realist must develop a notion of assertability-conditions based on the fact that our own cognitive capacities must allow us not only always to recognize a sentence as assertable when it is, i.e., that one must be able to recognize the object $Pr(s)$ which is an effective proof of s, when there is one, but also to be able in practice to produce or construct the object $Pr(s)$. (In the jargon of computer scientists, once an answer to a problem is obtained, one may further produce a polynomial-time certificate while the algorithm had an exponential-time worst-case running time.)

There have already been arguments in favour of a radical form of constructivism known as 'strict finitism', e.g., (Wright 1993), but Dubucs' proposal differs from these in two fundamental ways. First, the discussion is not anymore conducted in terms of the Hilbert-Style systems to which the notion of 'effectivity' is associated. It is conducted in terms of Gentzen-style systems, more precisely: it is about sequent calculi and their structural rules (and not even about introduction and elimination rules for natural deduction systems, as it was for Dummett and Prawitz). Secondly, Dubucs argues for feasibility in a more principled way, i.e., by looking at a weakening of structural rules as opposed to, say, a mere bounding of the length of computations. In other words, bounds should remain hidden, i.e., the logic for radical anti-realism should reflect limitations to human cognitive capacities in a 'structural' fashion. Furthermore, the key to the whole argument is seriously to take into account the *physical cost* of the proof, which is precisely the initial motivation for the development of linear logic, according to Girard (see, e.g., Girard (1995b)).

The optional discharge in the case of the introduction rule for implication has the structural rule of weakening on the left as its counterpart. Relevant logic and linear logic both reject it, as opposed to intuitionistic logic, which merely distinguishes itself from the full structural logic by its restriction on weakening on the right. However, the counterpart of obligatory discharge is contraction on the left, which still holds within relevant logic but not within linear logic, where discharge is obligatory but not multiple. Now, once the focus is on strutural rules, a radical anti-realist may argue that there are no specific reasons to adopt intuitionistic logic, for which traditional anti-realists such as Dummett had argued. The relevance of relevant logic is not clear either, since the cause of idealizations, from the radical anti-realist point of view, is the rules of contraction, here shown with weakening:

$$\textit{Contraction} \qquad \frac{\Gamma, A, A \vdash \Delta}{\Gamma, A \vdash \Delta}\ \textit{Left} \qquad\qquad \frac{\Gamma \vdash A, A, \Delta}{\Gamma \vdash A, \Delta}\ \textit{Right}$$

$$\textit{Weakening} \qquad \frac{\Gamma \vdash \Delta}{\Gamma, A \vdash \Delta}\ \textit{Left} \qquad\qquad \frac{\Gamma \vdash \Delta}{\Gamma \vdash A, \Delta}\ \textit{Right}$$

But relevant logic consists in rejecting weakening (both left and right) while keeping contraction. Moreover, in absence of weakening, one needs to introduce *ad hoc* distributivity rules in order to keep to classical logic. In linear logic, rules for contraction do not disappear entirely (the resulting system would not be expressive enough); they reappear as the rules for special connectives, the exponentials. The sequent rules for exponentials are:

$$\text{Of Course} \qquad \frac{\Gamma, A \vdash \Delta}{\Gamma, !A \vdash \Delta}\ !L \qquad\qquad \frac{!\Gamma \vdash B, ?\Delta}{!\Gamma \vdash !A, ?\Delta}\ !R$$

$$\text{Why Not} \qquad \frac{!\Gamma, A \vdash ?\Delta}{!\Gamma, ?A \vdash ?\Delta}\ ?L \qquad\qquad \frac{\Gamma \vdash A, \Delta}{\Gamma \vdash ?A, \Delta}\ ?R$$

Linear *contraction* and *weakening* are:

Contraction $\dfrac{\Gamma, !A, !A \vdash \Delta}{\Gamma, !A \vdash \Delta}\ Left$ $\dfrac{\Gamma \vdash !A, !A, \Delta}{\Gamma \vdash !A, \Delta}\ Right$

Weakening $\dfrac{\Gamma \vdash \Delta}{\Gamma, !A \vdash \Delta}\ Left$ $\dfrac{\Gamma \vdash \Delta}{\Gamma \vdash !A, \Delta}\ Right$

Classical *contraction* enables us to use a formula infinitely many times in a proof (this is an idealization). Classical *weakening*, on the other hand, allows us to bring unused hypotheses into our proofs, with that we may be left with unrelated hypotheses. In linear logic, the exponentials are used to control the structural rules of contraction and weakening. An infinite resource, i.e., a resource that can be consumed more than once is handled using the linear exponentials $!A$ and its De Morgan dual $?A$. The idea behind this move is to *control* the use of contraction, e.g., the length of proof search or of normalization procedures. This ability to control contraction should be a prime topic for investigation from our radical anti-realist standpoint.

We should wrap up this section with some basic remarks about other connectives, which will turn out useful in the following section. A sequent of the form $\Gamma \vdash \Delta$ in linear logic means that resources presented by Γ are to be consumed to yield resources Δ. This makes linear logic a *resource-sensitive* logic. This property of linear logic makes the conjunction and disjunction of classical logic ambiguous. For example, we can use $A \wedge B$ both for producing A and also $A \wedge B$ itself (see Girard (1987) for a more detailed discussion). To overcome these ambiguities, linear logic introduces two distinct connectives for each of conjunction and disjunction, resp., the multiplicatives and the additives. We write $A \oplus B$ and $A \& B$ for the two additives, and $A \otimes B$ and $A \parr B$ for the two multiplicatives. Negation is defined by means of the following rules:

Negation $\dfrac{\Gamma \vdash A, \Delta}{\Gamma, A^{\perp} \vdash \Delta}\ Left$ $\dfrac{\Gamma \vdash A, \Delta}{\Gamma \vdash A^{\perp}, \Delta}\ Right$

One should note that according to the following sequent rules, the two multiplicatives are De Morgan duals of each other. The same is true for the two additives. Linear implication will be the same as linear deduction and will be denoted by $A \multimap B$. Multiplicatives, additives and linear implication have the following left and right sequent rules:

Times $\dfrac{\Gamma, A, B \vdash \Delta}{\Gamma, A \otimes B \vdash \Delta}\ \otimes L$ $\dfrac{\Gamma \vdash A, \Delta 1 \quad \Gamma \vdash B, \Delta 2}{\Gamma \vdash A \otimes B, \Delta 1, \Delta 2}\ \otimes R$

Par $\dfrac{\Gamma 1, A \vdash C \quad \Gamma 2, A \vdash D}{\Gamma 1, \Gamma 2, A \parr B \vdash C, D}\ \parr L$ $\dfrac{\Gamma \vdash A, B, \Delta}{\Gamma \vdash A \parr B, \Delta}\ \parr R$

Plus $\dfrac{\Gamma, A \vdash \Delta \quad \Gamma, B \vdash \Delta}{\Gamma, A \oplus B \vdash \Delta} \oplus L \quad \dfrac{\Gamma \vdash A, \Delta}{\Gamma \vdash A \oplus B, \Delta} \oplus R1 \quad \dfrac{\Gamma \vdash B, \Delta}{\Gamma \vdash A \oplus B, \Delta} \oplus R2$

With $\dfrac{\Gamma, A \vdash \Delta}{\Gamma, A\&B \vdash \Delta} \&L1 \quad \dfrac{\Gamma, B \vdash \Delta}{\Gamma, A\&B \vdash \Delta} \&L2 \quad \dfrac{\Gamma \vdash A, \Delta 1 \quad \Gamma \vdash B, \Delta 2}{\Gamma \vdash A\&B, \Delta 1, \Delta 2} \&R$

Implies $\dfrac{\Gamma 1, B \vdash \Delta \quad \Gamma 2 \vdash A}{\Gamma 1, \Gamma 2, A \multimap B \vdash \Delta} \multimap L \qquad \dfrac{\Gamma, A \vdash B, \Delta}{\Gamma \vdash A \multimap B, \Delta} \multimap R$

3. Epistemic Logic and Modal Linear Logic

Reasoning about knowledge is one of the many areas where problems about computational complexity cannot be eluded. The problem of logical omniscience in epistemic logic is a perfect case of an idealization in the above sense. It is usually presented in Hintikka's original Hilbert-style system (Hintikka 1962). We shall present a Gentzen-style version below. Informally the problem is this: if an agent knows that p and knows that 'p implies q', deductive closure requires that the agent also knows that q. This is obviously not the case for real agents: for example, one does not know all the consequences of the axioms of elementary arithmetic. Nor would it be true of a computer because the resources necessary for the knowledge of q might not be available, if, for example, the computation involves exponential complexity. Here too, philosophers have not been able to engage with the issues raised by the logicians. Joseph Halpern's remark still stands today:

> ... reasoning about knowledge has found applications in such diverse fields as economics, linguistics, artificial intelligence, and computer science. While researchers in these areas have tended to look to philosophy for their initial inspiration, it has also been the case that their more practical concerns, which often centred around more computational issues such as the difficulty of computing knowledge, have not been treated in the philosophical literature (Halpern 1986, 2).

The literature contains many attempted solutions to the problem of logical omniscience, from Hintikka's own ideas about 'impossible possible worlds' (Hintikka 1975), based on Rantala's 'urn' models (Rantala 1978, 1982), to the syntactical solutions of (Eberle 1974; Konolige 1986), Parikh's 'knowledge algorithms' (Parikh 1987, 1995), and the logic of 'awareness' (Fagin and Halpern 1988). This is not the place for a critical evaluation of these alternatives. It should be pointed out, however, that number of these solutions can be characterized by the wish to adhere come what may to the full structural logic: a sort of superstructure – one is tempted to say: epicycle – is then added to it within which one could talk about agents reasoning about knowledge without conceiving of them as omniscient. There are no reasons, except philosophical prejudice, not to explore the substructural world. We prefer here to follow Dubucs (1998) and try and look for *a weaker logic relatively to which agents can be said to be omniscient, without this omniscience being problematic*. In that paper, Dubucs only discussed intuitionistic logic as a posssible alternative but in (1991) linear logic is discussed. Intuitionistic logic is not an interesting candidate for us precisely because it does not keep contraction and weakening under control and, in that sense it is no more likely than classical logic (as the full structural logic) to be the weaker logic that we are looking for, relatively to which agents can be said to be omniscient. Further reasons for its inadequacy in the epistemic context will surface below.

The only attempt that we know of at developing an epistemic logic by using a substructural logic is by Hector Levesque (Levesque 1984a, b), who used relevant logic. Levesque's approach is original in many respects. First, he distinguishes between 'implicit' and 'explicit' knowledge. According to Levesque, the possible worlds semantics is an idealization because it is not about what is known by an agent, but what is true given what is known. One must distinguish between what is known in this sense from what is *explicitely* known by the agent. Implicit knowledge is therefore defined as something that is true in all the worlds that an agent considers as possible and explicit knowledge is what is known as true for that agent. Levesque introduces the operator I and E as, resp., '$E_i\phi$ is true if ϕ is explicitely known' and '$I_i\phi$ is true if ϕ is implicitely in what is known'. (It is on the basis of this distinction that Fagin and Halpern introduced a logic of 'awareness' (Fagin and Halpern 1988), where an agent knows explicitely ϕ if she is aware that ϕ and knows implicitely that ϕ.) Secondly, Levesque uses the situation semantics of Barwise and Perry (1983) to deal with the explicit knowledge of agents. In a given situation, some formulas will have a truth-value assigned to them but it is possible that some other formulas can have no truth-value. Levesque also uses the notion of an 'incoherent' situation, which is not compatible with any possible worlds. In such sitations some formulas can be seen as both true and false. Levesque identifies explicit knowledge as a set of situations and gives a semantics and a proof theory for it. The proof theory consists of propositional tautologies, modus ponens and axioms of propositional logic. He adds axioms for E and I which include closure under implicit knowledge. He also proves the following theorem:

$$\models (E_i\phi \supset E_i\psi) \quad \text{iff} \quad \phi \quad \textit{entails} \quad \psi$$

The proof of this theorem can be found in Levesque (1984b). This theorem allows Levesque to complete his axiomatization by using the axioms of entailment for explicit knowledge E.

From our radical anti-realist point of view, relevant logic is not an interesting alternative to the full structural logic. It may go further than intuitionistic logic in rejecting weakening on both right and left sides but it leaves contraction untouched and the resulting system is unappealing from a computational point of view. On top of this, Levesque's approach suffers from many defects of its own. To begin with, it is limited to only one agent and there seems to be no room for the multi-agent case, which is needed for a full epistemic logic. Secondly, a clear philosophical motivation for the distinction between 'implicit' and 'explicit' knowledge is lacking. Thirdly, the notion of an 'incoherent' situation needs to be clearly distinguished from that of impossible possible worlds. These last two defects would require a fuller philosophical discussion for which this is not the appropriate forum. It remains, however, that in absence of control over contraction within relevant logic, it is not clear that the idea that agents remain omniscient for their explicit knowledge really solves the problem of omniscience. For these reasons, it is worth looking at an *epistemic linear logic*. In order to deal with the logic of knowledge and belief, we must extend linear logic by introducing modalities. In this paper, we cannot do more that merely explore various ways of doing so. Semantical considerations cannot be truly dealt with at this stage. Jaakko Hintikka's seminal discussion in Hintikka (1962) is a model of clarity and elegance that has had no equivalent and we are certainly not yet in a position to present our own. But the following considerations should help us to select a candidate for further semantical and philosophical elaboration.

There are two main strategies for introducing modalities within linear logic. First, one could interpret exponentials as modalities, as in Avron (1988), the system would be a form of linear S4 with indexed modalities. However, the resulting connectives would both control contraction and weakening and serve as modalities. They would appear in both structural and modal rules; a very bizarre cocktail. Secondly, one could add modalities to fragments of linear logic. Some such combinations and their semantics have been studied in Martin (2002). In what follows, we shall assess only two of them, namely the multi-modal linear logic or MMLL of Kobayashi et al. (1999), which has been inspired by and shown to have applications in reasoning about location-dependent distributed network processes, and the system $KDT4_{Lin}$ developed in Martin (2002). In what follows we first give a brief overview of both of these logics. (Some familiarity with both linear and modal logics is assumed.) We then give examples of the application of these logics to the problem of logical omniscience and the wise men puzzle. The proofs of these problems will be compared with each other and a short presentation of our encoding of modal linear logic in *Coq* will be given (the full proof tree and *Coq* code are in the Appendix). Finally, we shall give our reasons for choosing $KDT4_{lin}$, which is classical, over MMLL, which is intuitionistic.

MMLL uses the proof-search-as-computation paradigm where formulas of linear logic are seen as processes. The concurrency aspects and location of processes in the network are dealt by using the added modality L. The semantics is based on the resource-indexed model of Milner et al. (1992), it is of a set of located resources with different combinations of these resources. These combinations correspond to the additive and multiplicative connectives of linear logic. An algebra of resources and a resource structure on this algebra are defined and the semantics is proved to be sound and complete in Kobayashi et al. (1999). The BNF of the logic is:

$$A ::= a | !A | L_i A | A \otimes A | A \& A | A \multimap A | 1$$

where a is an atomic formula; 1 is the unit for tensor ($(A \otimes 1) = A$); L is the modal operator; i is a natural number ranging over a denumerably infinite set;

$L_i A$ intuitively means that resource A is available at location i. $\otimes$ and & are the multiplicative and additive connectives of linear logic. (Note that this system does not contain negation or dual operators.) The sequent rules for this logic are the same as for linear logic except for the following two cases that are used for the congruences:

$$\frac{\Gamma, A' \vdash D \quad A \equiv A'}{\Gamma, A \vdash D} \equiv L \qquad\qquad \frac{\Gamma \vdash A \quad A \equiv A'}{\Gamma, \vdash A'} \equiv R$$

This system is intuitionistic, i.e., its sequents have a single formula on their right-hand side. There is no sequent rule for introduction and elimination of modalities. The only way to work with modalities is the following congruences over formulas:

1. $L_i(A \otimes B) \equiv L_i A \otimes L_i B$
2. $L_i(A \& B) \equiv L_i A \;\&\; L_i B$
3. $L_i(A \multimap B) \equiv L_i A \multimap L_i B$
4. $L_i!A \equiv !L_i A$
5. $L_i 1 \equiv 1$
6. $L_i L_j A \equiv L_j A$

The system MMLL has some advantages. First, it has the same contraction rules as linear logic and thus avoids idealizations. Secondly, as we can see from the syntax, this system has indexed modalities that make it a multi-modal system; one that can be applied to multi-agent systems. But the indexed modalities of this system are problematic, as opposed to those in $KDT4_{lin}$, which will be discussed below, because of the existence of the last congruence in the above list. This congruence allows for connections between the knowledge of different agents. In one direction it says that if agent 1 knows that agent 2 knows A, then agent 2 himself knows A. In the other direction, it says that if agent 1 one knows A, then there is another agent, say, agent 2, who knows that agent 1 knows A. This is a bit counterintuitive. One should note that it is not necessary that whenever I know A, there exists someone else that knows that I know A, but the existence of that other agent is not impossible. (In other words, there are no explicit quantifiers over the indexed modalities, this congruence is not terribly counterintuitive.) Had it not been so counterintuitive, this congruence would have made MMLL of interest not only for reasoning about knowledge, but also for a discussion of the intersubjectivity in knowledge; it is a sort of iteration principle. However for reasons that will apprear below, MMLL is at any rate not suited for some crucial epistemic purposes.

$KDT4_{lin}$ has an algebraic semantics and it has been proven to be sound and complete in Martin (2002). The BNF of this logic is shown below:

$$A ::= a|!A|?A|K_i A|A \otimes A|A \parr A|A \oplus A|A\&A|A \multimap A|A^{\perp}|1|0|\top|\perp$$

where a is an atomic formula; $1, \perp, \top$, and 0 are the units for $\otimes$, $\parr$, &, and $\oplus$ respectively; K is the modal operator; i is a natural number ranging over a denumerably infinite set; $K_i A$ intuitively means that i knows A.

The sequent rules of this logic are the same as the full propositional classical linear logic, in which we take sequents $\Gamma \vdash \Delta$ for lists of formulas. These lists, together with the *Exchange* rule, will have the properties of multisets. The sequent calculus will also have three modal sequent rules shown below. These rules correspond to T, KD, and S4 axioms of the classical Hilbert-style modal logics:

$$\textit{TRules} \qquad \frac{\Gamma, A \vdash B, \Delta}{K_i\Gamma, K_iA \vdash B, \Delta}\, Left \qquad\qquad \frac{\Gamma, A \vdash B}{K_i\Gamma, K_iA \vdash K_iB}\, Right$$

$$\textit{KDRules} \qquad \frac{\Gamma, A \vdash 0}{K_i\Gamma, K_iA \vdash 0}\, Left \qquad\qquad \frac{\Gamma, A \vdash B}{K_i\Gamma, K_iA \vdash K_iB}\, Right$$

$$\textit{S4Rules} \qquad \frac{\Gamma, A \vdash B, \Delta}{\Gamma, K_iA \vdash B, \Delta}\, Left \qquad\qquad \frac{K_i\Gamma, K_iA \vdash B}{K_i\Gamma, K_iA \vdash K_iB}\, Right$$

Where Γ is a multiset of formulas and $K_i\Gamma$ is the multiset $\{K_iA | A \in \Gamma\}$. (Note that *Trule-Right* and *KDrule-Right* are the same, because we have chosen to work on one modality K_i as opposed to dual modalities K_i and B_i; the dismissed modality B_i would distinguish between the two rules.)

One should observe that all the connectives input linear propositions, but K_i takes as input a list of linear propositions. The modality sequent rules need our modality to operates over a list of formulas rather than a single formula. The modality is also an indexed modality, *making our logic a multi-modal linear logic* where the modality K_i expresses the knowledge of agent i. For example K_1D intuitively means that *agent* 1 *knows that* D, i.e., he knows all of the formulas of the list D. The modality operator can be seen as a binary operator with two operands: an integer and a list of formulas.

The system $KDT4_{lin}$ has many advantages. It avoids idealization, as it keeps the exponentials and also all the structural rules of linear logic. Hence, it is capable of controlling the resources by marking them with exponentials. It also fares better for epistemic purposes than Levesque's system in (Levesque 1984a, b) because it is *multi-modal*. There is also no trace of the problematic congruence that we found in MMLL.

We shall now encode both MMLL and $KDT4_{lin}$ in the proof assistant *Coq*, developped in Cornes et al. (1999, 20003) on the basis of the Calculus of Constructions (*CC*) of (Coquand and Huet 1988). Systems can be encoded in *Coq*'s higher-order logic; these encodings allows us to state and prove theorems using facilities of this proof assistant. Intuitionistic linear logic has been previously encoded in *Coq*, in Powers and Webster (1999), by associating the constructs of *Coq* together with linear logic proofs. Modal logic, too, has been previously encoded in *Coq* in Lescanne (2001). Our encoding method will be based on that of Felty (2002), in which the system to be encoded is treated as the *object logic* and *Coq*'s Calculus of Constructions (CC) as the *metalogic*.

We are here encoding for the first time modalities in classical linear logic. (Our sequents are classical in the sense that we are not limiting ourselves to sequents with single formulas on the right hand side.) The encoding has been done in two steps: (i) defining modal linear logic formulas and (ii) modal linear logic sequent rules inductively, using the set of inductive datatypes of *Coq*, and (iii) proving some lemmas to work with lists.

In the first phase, we define inductively a set of linear logic propositions: *MLinProp*, which stands for *Modal LINear PROPosition*. The smallest formulas of our modal linear logic will be the different cases of induction. The definition in *Coq* is:

```
Inductive  MLinProp  :  Set  :=
| Implies : (MLinProp) → (MLinProp) → MLinProp
| Times : (MLinProp) → (MLinProp) → MLinProp
| Par : (MLinProp) → (MLinProp) → MLinProp
| Plus : (MLinProp) → (MLinProp) → MLinProp
| With : (MLinProp) → (MLinProp) → MLinProp
| OfCourse : (MLinProp) → MLinProp
| WhyNot : (MLinProp) → MLinProp
| Box : (nat) → (list MLinProp)→(list MLinProp)
| Negation : (MLinProp) → MLinProp
| One : MLinProp
| Zero : MLinProp
|⊥ : MLinProp
|⊤ : MLinProp
```

Now we can use our *MLinProp* as a *Coq* type. We can define variables of this type. For example we can define *A* and *B* as modal linear propositions, and *D* as a list of Modal linear proposition:

```
Variable A, B : MLinProp. Variable D : (list MLinProp).
```

We can also define predicates over this type. For example red is a 1-ary modal linear predicate:

```
Variable red: nat → MLinProp.
```

Using *Coq*'s syntax definition and pretty-printing facilities, we can give a notation to each of our modal linear connectives. This will allow us to infix and prefix our connectives. The *Coq* code for Bang, Times, and modality is given below. We are augmenting the grammar rules and giving pretty-printing rules to represent Bang as "!", Times as "*", and modality as "*K*".The reader is assumed to be familiar with the syntax of these *Coq* commands (see Section 6.7.3 and 6.7.4 in Cornes et al. 1999–2003).

```
Grammar command command2 :=
OfCourse ["!" command2($c)] →[⟨⟨(OfCourse $c)⟩⟩].
Syntax constr level 2:
[⟨⟨(OfCourse $c)⟩⟩]→["!" $c].
Grammar command command6 :=
```

```
Times [command5($c1) "*" command6($c2)] → [⟨⟨(Times $c1 $c2)⟩⟩].
Box [command5($c1) "K" command6($c2)] → [⟨⟨(Box $c1 $c2)⟩⟩].
Syntax constr level 6:
PTimes [⟨⟨(Times $c1 $c2)⟩⟩] → [ $c1:L "*" $c2:E ].
Syntax constr level 6:
PBox [⟨⟨(Box $c1 $c2)⟩⟩]→ [ $c1:L "K" $c2:E ].
```

The notation for all of our modal linear connectives is given in this table for further reference:

Connective	Symbol	Syntax in Coq	Example
Times	$\otimes$	**	$A**B$
Par	⅋	%	$A\%\%B$
Plus	$\oplus$	$\oplus$	$A++B$
With	&	&	$A\&B$
Box	K	K	¡KD
OfCourse	!	!	$!A$
Implies	$\multimap$	$\multimap$	$A \multimap B$

In the second phase of our encoding, we will implement the sequent calculus of our modal linear logic. The sequent rules are defined inductively. The induction is made on the linear sequent relation $\Gamma \vdash \Delta$. The sequent relation *LinCons* has been represented as a 2-ary function. It takes two arguments as input: the hypothesis Γ and the conclusion Δ. Remember that Γ and Δ are implemented as lists of formulas. These lists together with the exchange and permutation rules will act as multisets. The output of the function, which is either true or false, is defined as a *Coq* proposition *Prop*. The *Coq* code for *LinCons* is:

```
Inductive LinCons : (list MLinProp) → (list MLinProp) →
Prop :=
```

The connective "⊢" is defined as a binary operator with a low precedence using the *Coq* Syntax and pretty-printing commands:

```
Grammar command command9 :=
LinCons [command8($t1) "⊢" command9($t2)] → [⟨⟨(LinCons $t1
$t2)⟩⟩].
Syntax constr level 9:
PLinCons [⟨⟨(LinCons $t1 $t2)⟩⟩]→ [ $t1 "⊢" $t2 ].
```

The axiom and the sequent rules of the modal linear logic will be the cases of the induction. They are added individually. For example the axiom *Identity* is added as follows:

```
Identity :
(A : MLinProp)
('A ⊢ 'A)
```

The sequents of our system are of the form $D1 \frown `A \vdash D2 \frown `B$, where $D1$ and $D2$ are lists of formulas of type *MLinProp*, and A and B are formulas of the type *MLinProp*. Note that we have lists on both sides of the sequent. Following the encoding of Powers and Webster (1999), two symbols $\frown$ and ‘ are used to work with lists in *Coq*; $\frown$ is used to concatenate two lists and ‘ presents a singleton list. For example, $D1 \frown `A$ concatenates list $D1$ and the singleton A. The empty list will be shown as *Empty*. Logical and structural rules of modal linear logic are added next. These rules are coded using *Coq*'s implication → for deduction. For example the *Cut* rule:

$$\frac{\Gamma_1 \vdash A, \Delta_1 \quad \Gamma_2, A \vdash \Delta_2}{\Gamma_1, \Gamma_2 \vdash \Delta_1, \Delta_2}\ Cut$$

is coded as:

```
| Cut :
(A, B : MLinProp)(D1, D2, D3, D4 : (list MLinProp))
((D1 ⊢ D3 ^ `A) → (D2 ^ `A ⊢ D4) → (D1 ^ D2 ⊢ D3, D4))
```

As examples of logical rules, consider the *Coq* code for Par Left and Times Right:

```
| ParLeft :
(A, B, C1 , C2 : MLinProp)(D1, D2 : (list MLinProp))
((D1 ^ `A ⊢ ` C1) → (D2 ^ `B ⊢ `C2) → (D1 ^ D2 ^
`(A%%B) ⊢ `C1 ^ `C2))
|TimesRight :
(A, B : MLinProp)(D1, D2 , D3 , D4: (list MLinProp))
((D1 ⊢ `A ^ D3) → (D2 ⊢ `B ^ D4) → (D1 ^ D2 ⊢ `(A **
B) ^ D3 ^ D4))
```

The modal sequent rules are KD, T, and S4. The difference is that the modal operator has two operands: an index i and a list of formulas D. $K_i D$ will be shown as $i K D$ in *Coq*. For example the KD rule below:

$$\frac{\Gamma, A \vdash B}{iK\Gamma, iKA \vdash iKB}\ KD$$

will be coded as:

```
| KDRule :
(i : nat)(A, B : MLinProp)(D : (list MLinProp))
((D ^ `A ⊢ `B) → (`(iKD) ^ `(iK`A) ⊢ `(iK`B)))
```

In the third phase of our encoding we will deal with some lemmas to work with lists. Our sequent rules have lists to the right and left of the sequents. That will cause difficulty while working with the sequents that do not contain lists on one side or on both sides. For example, sequents of the form $A \vdash A$ or $A \vdash A \ominus B$ are not accepted in our encoding. Moreover, a deduction using these sequents, which is an acceptable deduction, will not be accepted in our system. One such deduction would be:

$$\frac{A \vdash A}{A \vdash A \oplus B} \oplus R1$$

To solve the problem we will have to make lists out of single formulas. This will be done by adding Nil lists to the left hand side of them. Applying these changes to the above deduction makes it look like:

$$\frac{Empty, A \vdash Empty, A}{Empty, A \vdash Empty, A \oplus B} \oplus R1$$

This will be done using two lemmas: *AddNilLeft* and *AddNilRight*. *AddNilFront* is shown below:

`Lemma AddNilLeft` : $(D1, D2 :$ `(list MLinProp)`)

((`Empty` $\frown$ $D1 \vdash D2) \rightarrow (D1 \vdash D2)$).

Each of these lemmas has a dual to eliminate the added Nils. This is necessary because we are working with sequents with distinguished formulas. So we need to have a list and a single formula on both sides of the sequent. By adding Nil we will have sequents without distinguished formulas. So we have to eliminates the nils that we added before. Eliminating Nils will be done using *ElimNilLeft* and *ElimNilRight* lemmas. *ElimNilRight* is shown below.

`Lemma ElimNilRight` : $(D1, D2 :$ `(list MLinProp)`)

$((D1 \vdash D2) \rightarrow (D1 \vdash Empty \frown D2))$.

List concatenation and singleton lists are dealt with the same way as the encoding of Powers and Webster (1999).

As examples of encoding, we shall state the problem of logical omniscience and the wise men puzzle as theorems and then prove them, in both MMLL and $KDT4_{lin}$. Lescanne (2001) has already encoded in *Coq* the latter along with the muddy children puzzle. But those are in Hilbert-style classical modal logic.

The sequent calculus version of logical omniscience is the following:

$$K_1 A, K_1(A \multimap B) \vdash K_1 B$$

The proof tree for the $KDT4_{lin}$is:

$$\frac{\dfrac{A \vdash A \quad B \vdash B}{A, (A \multimap B) \vdash B} \multimap L}{K_1 A, K_1 A \multimap B \vdash K_1 B} \text{KDRule}$$

The proof in *Coq* is done as in the above proof tree, using the sequent rules encoded in *Coq*. Some extra work has to be done while working with lists of formulas, namely adding *Nil* to the left and right of our sequents to be able to apply the *ImpliesLeft* and *ImpliesRight* rules and the *Identity* axiom. The *Coq* code is thus:

```
Intros.
Apply KDRule.
Apply AddNilLeft.
Apply ImpliesLeft.
Apply AddNilLeft.
Apply Identity.
Apply Identity.
```

The proof tree in MMLL is:

$$\dfrac{\dfrac{L_1 A \vdash L_1 A \quad L_1 B \vdash L_1 B}{L_1 A, L_1 A \multimap L_1 B \vdash L_1 B}\multimap L}{L_1 A, L_1(A \multimap B) \vdash L_1 B}\text{3d congruence}$$

One of the standard puzzles for multi-modal epistemic logic is 'wise men' or 'King, three wise men and 5 hats' puzzle (see Fagin et al. (1995 p. 12)): *a king has three wise men and 5 hats: 2 green and 3 red. He asks the wise men to close their eyes and puts a hat on the head of each of them. He then asks them to open their eyes and poses a question to each of them in order. He asks the first man: 'Do you know the colour of your hat?' He answers: 'No'. The same question is asked from the second man and he, too, answers: 'No'. But when the third man is asked the same question, he answers: 'Yes! The colour of my hat is red'. How this is possible?* This conclusion is based on the information provided by the answers of previous wise men, together with the fact that each agent knows the color of the hats of the other agents except for himself. In more formal terms we have: if agent 3 knows that agent 1 does not know the colour of his hat, and he knows that agent 2 does not know the colour of his hat and moreover he knows that agent 2 knows that agent 1 does not know the colour of his hat, he will know the colour of his own hat. Therefore, agent 3 knows three things that help him, together with a good number of assumptions and some lemmas, to reach a conclusion about the colour of his own hat (red). These three things are:

1. Agent 1 does not know the color of his hat.
2. Agent 2 does not know the color of his hat.
3. Agent 2 knows that agent 1 does not know the color of his hat.

These three pieces of information will help agent 3 to conclude that the colour of his own hat is red. From (1) it can be concluded that at least one of the agents 2 and 3 wear a red hat. Indeed, if both of them had green hats, since we only have two green hats, agent 1 would know the colour of his hat. So a corollary of (1) is that agents 2 and 3 both know the following fact: at least one of agents 2 or 3 wears a red hat (or both of them do) This fact, together with (2) and (3) above help agent 3 to conclude that his hat is red. The fact that agent 2 does not know the colour of his hat shows that agent 3 is not wearing a green hat. Because if this were the case, agent 2, who knows that at least one of them is wearing a red hat, would have easily concluded the color of his own hat.

In order to prove this theorem in *Coq*, we need three agents, two colour predicates and one definition:

1. Three agents:

   ```
   agent1, agent2, agent3 : nat.
   ```

2. Two color predicates:
   - `(red i): the color of the hat of ith agent is red`
   - `(green i): the color of the hat of ith agent is green`

3. Definition

   When each agent knows the color of his hat, it means he knows whether it is red or green. This can be shown using the additive ⊕ because it expresses a choice between two cases, where both of the cases cannot happen at the same time.

   (Lhat *i*): agent *i* knows that his hat is either red or green.

   or in *Coq* terms:

   ```
   Definition Lhat := [i: nat](K_i `(red i)) ⊕ (K_i `(green i)).
   ```

We will use the proof method in Lescanne (2001), with linear logic axioms:

1. AOne:Each hat is either red or green. This can again be shown using the additive ⊕ because (green i) and (red i) cannot both happen at the same time, i.e., each hat cannot be both red and green at the same time.

   ```
   (i:nat)(D : (list MLinProp))( D⊢ `((green i)⊕ (red i))).
   ```

2. ATwo: this says that if two agents wear a green hat then the third one wears a red one. In this axiom, as opposed to the previous one, we want to be able to express that two cases happen at the same time, i.e., both agents wear a green hat. A multiplicative connector is called for and we are going to use ⅋.

   ```
   Axiom ATwo : (`((green agent2) ⅋ (green agent3)) ⊢ `(red agent1)).
   ```

3. AThree: If agent 2 has a green hat, then agent one knows it. The reason is obvious because he is seeing the hat of agent 2.

   ```
   (`(green agent2) ⊢ `(agent1 K `(green agent2))).
   ```

4. AFour: If agent 3 has a green hat, then agent 1 knows it. The reason is obvious because he is seeing the hat of agent 3.

   ```
   (`(green agent3) ⊢ `(agent1 K `(green agent3))).
   ```

5. AFive : If an agent is wearing a red hat, then he is not wearing a green one.

   ```
   (`(red i) ⊢ `(Not (green i))).
   ```

6. ASix : If an agent is wearing a green hat, then he is not wearing a red one.

   ```
   (`(green i)) ⊢ `(Not (red i))).
   ```

The theorem to be proved in sequent calculus is:

(agent2 K (Not (Lhat agent1))), (Not (Lhat agent2)) ⊢ (red agent3)

Or in *Coq* terms:

```
Theorem ThirdKnows :
('(Not (Lhat agent2)) ^ ('(agent2 K '(Not(Lhat agent1)))) ⊢ '(red
agent3)).
```

The proof is done mostly with cuts. The proof tree and the *Coq* code are given in the Appendix.

An attempt at this encoding with MMLL will fail. This system is intuitionistic so the first encoding of our logic has lists only in the left-hand side of sequents and only single formulas on the right-hand side. Therefore, this fragment does not have all the connectives of linear logic. It misses ⅋, the dual of ⊗, because the sequent rules for ⅋ are not intuitionistic:

$$\frac{\Gamma \vdash A, B}{\Gamma \vdash A \parr B}\, R \qquad\qquad \frac{\Gamma_1, A \vdash C \quad \Gamma_2, B \vdash D}{\Gamma_1, \Gamma_2, A \parr B \vdash C, D}\, L$$

The problem with this fragment is that in the proof of the puzzle we need at one stage the dual of ⊗. We had to prove the following sequent:

```
Not ((red 1) ⊗ (red 2)) ⊢ (green 1) ⅋ (green 2)
```

Thus, the puzzle cannot be proved in the fragment without ⅋. This does not settle the matter entirely, as one could attempt to re-phrase the puzzle (without any loss of meaning) so that it could be expressed in MMLL and use the intuitionistic version of ⅋ introduced in Brauner and de Paiva (1996), but this is mere speculation. Therefore, in order to be able to solve the puzzle, we had to work with the *full* modal linear logic. So we had to add lists to both sides of our sequents:

$$D1 \,\hat{}\; `A \vdash D2 \,\hat{}\; `B$$

One very important consequence of this is that there seem to be no real prospect for an intuitionistic epistemic linear logic.

4. Complexity

From our radical anti-realist point of view, computational complexity is to be taken seriously. Furthermore, the issue of complexity cannot be avoided when dealing with practical applications, e.g, in case of epistemic logic, applications in the domain of cognitive science or artificial intelligence. The key idea here is that the brain can be analysed as a computational system and one would propose models for cognitive activities in terms of computational tasks. But, to put it crudely, such tasks can hardly have an exponential lower bound, because one would not know how this task or cognitive activity is physically possible: idealizations must be avoided. Although the topic was hardly dealt with a mere 25 years ago, computational complexity is by now a well-studied phenomenon, with proof-theoretical measures of complexity, see, e.g., the survey in Urquhart (1995). One obvious proposal here would be to limit oneself to polynomial time. (This has been suggested, e.g., in Levesque (1988) for epistemic logic, because of applications to cognitive science.) There are candidates for this in linear logic, such as the Bounded Linear Logic (BLL) of Girard et al. (1992), in which the use of exponentials is bounded in advance, or the more recent Light Linear Logic (LLL) of Girard (1998), that has a (locally) polynomial-time cut-elimination. Both BLL and LLL are strong enough to represent all polynomial-time functions. We would like to suggest, however, that this is not, *prima facie*, the right approach. We shall give here, in very brief outline, three arguments. These will hardly settle the question but we hope to initiate a discussion.

First, it seems that epistemic logic is a hopeless case from the point of view of complexity, especially if we deal with multi-agent systems. At any rate, MMLL and $KDT4_{lin}$ are mere variants on the full classical linear logic, which is exponential. Moreover, it is not clear if the system resulting from an hypothetical addition of modalities to BLL or LLL would remain polynomial-time. Secondly, there is a conflict, so to speak, between theory and practice. This can be illustrated by considering a well-known algorithm, the simplex method in linear programming. This is a perfectly constructive method: when optimal solutions to linear programming problems exist, it gives us an algorithm to compute them. However, this algorithm is exponential-time. Still, in practice it outperformed a polynomial-time algorithm and cases where the simplex run for an exponential amount of time hardly ever occur. Thirdly, on a more philosophical note, the issue of complexity is, from a proof-theoretical standpoint, rather paradoxical. Indeed, use of cut, which corresponds to the use of lemmas in ordinary mathematics, allows for proofs that can be taken in. Now, if we keep in mind applications to domains such as reasoning about knowledge, the proof-theoretical approach creates a paradoxical situation:

> The idea of lengths of proofs is quite amusing from the perspective of reasoning. It suggests that there are some statements that are true that we cannot understand in practice because it would take too long, and that there are statements which we can understand if we permit ourselves to use cuts and not otherwise (Carbone and Semmes 1997, 137).

It is for these reasons, which need to be argued for in a more substantial manner (counterarguments easily spring to mind), that we chose to deal with the problem from a computational viewpoint and we chose to have an encoding in *Coq*. The computational approach to linear logic initiated in Abramsky (1993) links it to functional programming languages and the key here is an extension the Curry-Howard isomorphism (Howard 1980), which establishes a correspondence between linear logic proofs and computer programs and allows us to see cut-elimination as computation. This is a powerful paradigm that has taken over proof theory but, although it has been argued for in, e.g., Martin-Löf (1984), it has hardly been noticed by philosophers. The Curry-Howard isomorphism is extended in Abramsky (1993) to classical as well as to the intuitionistic fragment of linear logic. Through this computational interpretation we are able to reduce the complexity of our proofs to the complexity of programs and we think that a first step here should be to limit programs to constuctive programs. In order to be able to get constructive programs out of proofs we need tools and one such tool is the proof assistant *Coq*, which is a higher-order logic based on the calculus of constructions, whose ancestors are to be found in de Bruijn's Automath, Girard's system F (Girard 1970) and Martin-Löf's intuitionistic type theory (Martin-Löf 1984). The calculus of constructions enables us constructively to encode other logics in *Coq*. These logics are treated as object logics vs the metalogic, which is the higher-order logic of *Coq*. The key point here is that, once theorem-proving in these logics becomes automated in *Coq*, then one has constructive programmes (Caldwell 1998). Hence, the encoding of our two modal linear logics in *Coq* provides us with automated proofs which lead to constructive programs. Proof automation has not been examined in this paper but, as mentioned before, our encoding is similar to that of Powers and Webster (1999), where issues related to proof automation are discussed, with a context-handling system and a general proof strategy. Guidelines for context-handling are mentioned and used succesfully in Powers and Webster (1999). The general proof strategy can be found in the linear logic programming approach in Hodas (1992). Once proofs are automated, *Coq* provides a mechanism for construction of programmes out of automated theorem proving. As we just pointed out, automated proofs in *Coq* are constructed proofs that are correct-by-construction. This also provides us with a decision algorithm. Although the issue about complexity is hardly settled, this suggests encoding in *Coq* as a step towards the right direction.

5. Appendix: Proving the King, Three Wisemen, and Five Hats Puzzle

5.1. Proof Tree

$$\dfrac{\dfrac{\Pi_1}{K_2(Not(Lhat1)) \vdash (red2)\otimes(red3)} \qquad \dfrac{\dfrac{Identity}{Not(Lhat2), (red2)\otimes(red3) \vdash (red2)\otimes(red3)}\,\text{Identity} \qquad \dfrac{\dfrac{Identity}{(red2)\otimes(red3) \vdash (red3)}\,\text{Identity}}{(red2)\otimes(red3) \vdash (red3)}\,\otimes\text{L}}{Not(Lhat2), (red2)\otimes(red3) \vdash (red3)}\,\text{CUT}}{K_2(Not(Lhat1)), Not(Lhat2) \vdash (red3)}\,\text{CUT}$$

Π_1 :

$$\dfrac{\dfrac{\dfrac{\dfrac{\dfrac{\dfrac{Identity}{(red1) \vdash (red1)}\,\text{Identity}}{(red1) \vdash (red1)\oplus(green1)}\,\oplus\text{R}}{K_1(red1) \vdash K_1((red1)\oplus(green1))}\,\text{KD}}{K_1(red1) \vdash (Lhat1)}\,\text{Unfold}}{Not(Lhat1) \vdash Not(K_1(red1)}\,\text{Negation} \qquad \dfrac{\Pi_2}{Not(K_1(red1) \vdash (red2)\otimes(red3)}}{\dfrac{Not(Lhat1) \vdash (red2)\otimes(red3)}{K_2(Not(Lhat1)) \vdash (red2)\otimes(red3)}\,\text{S4}}\,\text{CUT}$$

Π_2 :

$$
\dfrac{
\dfrac{
\dfrac{
\dfrac{
\dfrac{\text{A3}}{(green2) \vdash K_1(green2)} \quad \dfrac{\text{A4}}{(green3) \vdash K_1(green3)}
}{(green2) \parr (green3) \vdash K_1(green2), K_1(green3)} \parr \text{L}
}{(green2) \parr (green3) \vdash K_1(green2) \parr K_1(green3)} \parr \text{R}
\quad
\dfrac{
\dfrac{\text{A2}}{(green2) \parr (green3) \vdash (red1)}
}{K_1(green2) \parr K_1(green3) \vdash K_1(red1)} \text{T}
}{(green2) \parr (green3) \vdash K_1(red1)} \text{CUT}
}{Not(K_1(red1)) \vdash Not((green2) \parr (green3))} \text{Negation}
\quad
\dfrac{
\dfrac{
\dfrac{
\dfrac{
\dfrac{\text{A5}}{(red2) \vdash Not(green2)}
}{(red2), (green2) \vdash \text{Empty}} \text{Negation}
\quad
\dfrac{
\dfrac{\text{A5}}{(red3) \vdash Not(green3)}
}{(red3), (green3) \vdash \text{Empty}} \text{Negation}
}{(red2), (red3), (green2), (green3) \vdash \text{Empty}} \text{CUT}
}{(red2), (red3) \vdash Not((green2) \parr (green3))} \text{Negation}
}{(red2) \otimes (red3) \vdash Not((green2) \parr (green3))} \otimes\text{L}
}{Not(K_1(red1)) \vdash (red2) \otimes (red3)} \text{CUT}
$$

5.2. Coq Code

```
Section Hats.
Load MALL.
Variables red, green : nat → MLinProp.
Variables agent1, agent2, agent3 : nat.
Definition Lhat := [i:nat](i K `(red i)) ++ (i K `(green i)).
Axiom AOne :
(i:nat)(D : (list MLinProp))
( D ⊢ ` ((green i) %% (red i))).
Axiom ATwo :
(`((green agent2) %% (green agent3)) ⊢ `(agent1 K `(red agent1))).
Axiom AThree :
(`(green agent2) ⊢ `(agent1 K `(green agent2))).
Axiom AFour :
(`(green agent3) ⊢ `(agent1 K `(green agent3))).
Axiom AFive :
(i : nat)(`(red i) ⊢ `(Not (green i))).
Axiom ASix :
(i : nat)(`(green i) ⊢ `(Not (red i))).
Lemma Duals :
(i , j : nat)
(`(Not ((green i)Proof.
Apply TimesLeft.
Apply AddNilRight.
Apply NegationRight.
Apply NegationRight.
Apply ParLeft.
Apply NegationRight.
Apply AFive.
Apply NegationRight.
Apply AFive.
Qed.
(* Main Theorem *)
Theorem ThirdKnows :
(`(Not (Lhat agent2))  ^  (`(agent2 K `(Not(Lhat agent1))))⊢
`(red agent3)).
(* Proof *)
Intros.
Apply Cut with (Times (red agent2) (red agent3)).
Apply AddNilLeft.
Apply S4Rule1.
Apply Cut with (Negation (agent1 K `(red agent1))).
```

```
Apply AddNilLeft.
Apply NegationLeft.
Apply AddNilRight.
Apply ExchangeRight.
Apply ElimNilRight.
Apply NegationRight.
Unfold Lhat.
Apply PlusRight1.
Apply Identity.
Apply Cut with (Negation (Par (green agent2) (green agent3))).
Apply AddNilLeft.
Apply NegationLeft.
Apply AddNilRight.
Apply ExchangeRight.
Apply ElimNilRight.
Apply NegationRight.
Apply ElimNilLeft.
Apply ATwo.
Apply ElimNilLeft.
Apply Duals.
Apply Cut with (red agent3).
Apply AddNilLeft.
Apply TimesLeft.
Apply ElimNilLeft.
Apply Identity.
Apply Identity.
End Hats.
```

References

Abramsky, S.: 1993, 'Computational Interpretations of Linear Logic', *Theoretical Computer Science* **111**, 3–57.

Avron, A.: 1988, 'Syntax and Semantics of Linear Logic', *Theoretical Computer Science* **57**, 161–184

Barwise, J. and J. Perry: 1983, *Situations and Attitudes*, Cambridge MA, MIT Press.

Brauner, T. and V. de Paiva: 1996, 'Cut-Elimination for Full Intuitionistic Linear Logic', Technical Report 395, Computer Laboratory University of Cambridge and BRICKS, Denmark.

Caldwell, J.: 1998, 'Decidability Extracted: Synthesizing Correct-by-Construction Decision Procedure from Constructive Proofs', Ph.D. thesis, Cornell University.

Carbone, A. and S. Semmes: 1997, 'Making Proofs without Modus Ponens: An Introduction to the Combinatorics and Complexity of Cut Elimination', *Bulletin of the American Mathematical Society* **34**, 131–159.

Coquand, T. and G. Huet: 1988, 'The Calculus of Constructions', *Information and Computation* **76**, 95–120.

Cornes, C., J. Courant, J-C. Filliatre, G. Huet, P. Manoury, C. Murioz, C. Murthy, C. Parent, C. Paulin-Mohring, A. Saibi and B. Werber: 1999–2003, The Coq Proof Assistant Reference Manual, Version 7.4, Rapport Technique 177, INRIA.

Dubucs, J.: 1997, 'Logique, effectivité et faisabilité', *Dialogue* **36**, 45–68.

Dubucs, J.: 1998, 'Hintikka et la question de l'omniscience logique', in E. Rigal (ed.), *Jaakko Hintikka. Questions de logique et de phénoménologie*, Paris, Vrin, pp. 141–148.

Dubucs, J.: 2002, 'Feasibility in Logic', *Synthese* **132**, 213–237.

Dubucs, J. and M. Marion: 2003, 'Radical Anti-Realism and Substructural Logics', in A. Rojszczak, J. Cachro and G. Kurczewski (eds.), *Philosophical Dimensions of Logic and Science. Selected Contributed Papers from the 11th International Congress of Logic, Methodology, and the Philosophy of Science*, Krakòw, 1999, Dordrecht, Kluwer, pp. 235–249.

Eberle, R. A.: 1974, 'A Logic of Believing, Knowing and Inferring', *Synthese* **26**, 356–382.

Fagin, R., J. Y. Halpern, Y. Moses and M. Y. Vardi: 1995, *Reasoning about Knowledge*, Cambridge MA, MIT Press.

Felty, A.: 2002, 'Two-Level Meta-Reasoning in Coq', *Fifteenth International Conference on Theorem Proving in Higher Order Logics*, Springer-Verlag LNCS 2410.

Girard, J.-Y.: 1970, 'Une extension de l'interprétation de Gödel a l'analyse et son application a l'elimination des coupures dans l'analyse et la théorie des types', in J.-E. Fenstad (ed.), *Proceedings of the Second Scandinavian Logic Symposium*, Amsterdam, North-Holland, pp. 63–92.

Girard, J-Y.: 1987, 'Linear Logic', *Theoretical Computer Science* **50**, 1–102.

Girard, J-Y.: 1995a, 'Light Linear Logic', in D. Leivant (ed.). *Logic and Computational Complexity*, Berlin, Springer, pp. 145–176.

Girard, J-Y.: 1995b, 'Linear Logic: Its Syntax and Semantics', in J-Y. Girard, Y. Lafont and L. Regnier (eds.), *Advances in Linear Logic*, Cambridge, Cambridge University Press, pp. 1–42.

Girard, J-Y.: 1998, 'On the Meaning of Logical Rules I: Syntax vs. Semantics', `http://iml.univ-mrs.fr/ girard/Articles.html`.

Girard, J-Y., A. Scedrov and P. Scott: 1992, 'Bounded Linear Logic, A Modular Approach to Polynomial-Time Computability', *Theoretical Computer Science* **97**, 1–66.

Halpern, J. Y.: 1986, 'Reasoning about Knowledge: An Overview', in J. Y. Halpern (ed.), *Theoretical Aspects of Reasoning about Knowledge. Proceedings of the 1986 Conference*, San Francisco, Morgan Kaufmann, pp. 1–17.

Hintikka, J.: 1962, *Knowledge and Belief, An Introduction to the Logic of Two Notions*, New York, Cornell University Press.

Hintikka, J.: 1975, 'Impossible Possible Worlds Vindicated', *Journal of Philosophical Logic* **4**, 475–484.

Hodas, J. S.: 1992, 'Lolli: An Extension of Lambda Prolog with Linear Logic Context Management', *Workshop on the lambda Prolog Programming Language*, Philadelphia, 159–168.

Howard, W. A.: 1980, 'The Formulae-as-Types Notion of Construction', in J. P. Seldin and J.R. Hindley (eds.), *To H. B. Curry: Essays on Combinatory Logic, Lambda Calculus and Formalism*, London, Academic Press, pp. 479–490.

Kobayashi, N., T. Shimizu and A. Yonezawa: 1999, 'Distributed Concurrent Linear Logic Programming', *Theoretical Computer Science* **227**, 185–220.

Konolige, K.: 1986, *A Deduction Model of Belief*, San Franscisco, Morgan Kaufmann.

Lescanne, P.: 2001, 'Epistemic Logic in Higher Order Logic', `http://www.ens-lyon.fr/LIP/Pub/rr2001.html`.

Levesque, H.: 1984a, 'A Logic of Implicit and Explicit Belief', *Proceedings of the National Conference on Artificial Intelligence*, AAAI-84, pp. 192–202.

Levesque, H.: 1984b, 'A Logic of Implicit and Explicit Belief', Fairchild Laboratory for Artificial Intelligence Research, Technical Report.

Levesque, H.: 1988, 'Logic and the Complexity of Reasoning', *Journal of Philosophical Logic* **17**, 355–389.

Martin, A.: 2001, *Modal and Fixpoint Linear Logic*, Masters Thesis, Department of Mathematics and Statistics, University of Ottawa.

Martin-Löf, P.: 1984, *Intuitionistic Type Theory*, Naples, Bibliopolis.

Milner, R., J. Parrow and D. Walker: 1992, 'A Calculus of Mobile Processes I, II', *Information and Computation* **100**, 1–77.

Powers, J. and C. Webster: 1999, 'Working with Linear Logic in Coq', 12th International Conference on Theorem Proving in Higher Order Logics.

Parikh, R.: 1987, 'Knowledge and the Problem of Logical Omniscience', in Z. W. Ras and M. Zemankova (eds.), *Methodologies for Intelligent Systems*, The Hague, Elsevier, pp. 432–439.

Parikh, R.: 1995, 'Logical Omniscience', in D. Leivant (ed.), *Logic and Computational Complexity*, Berlin, Springer, pp. 22–29.

Prawitz, D.: 1965, *Natural Deduction. A Proof-Theoretical Study*, Stockholm, Almqvist & Wicksell.

Rantala, V.: 1978, 'Urn Models: A New Kind of Non-Standard Model for First-Order Logic', in E. Saarinen (ed.), *Game-Theoretical Semantics*, Dordrecht, D. Reidel, pp. 347–366.

Rantala, V.: 1982, 'Impossible Worlds Semantics and Logical Omniscience', in I. Niiniluoto and E. Saarinen (ed.), *Intensional Logic: Theory and Applications*, *Acta Philosophica Fennica*, pp. 1–9.

Restall, G.: 2000, *An Introduction to Substructural Logics*, London & New York, Routledge.

Schroeder-Heister P. and K. Dosen (eds.): 1993, *Substructural Logics*, Oxford, Clarendon Press.

Urquhart, A.: 1995, 'The Complexity of Propositional Proofs', *Bulletin of Symbolic Logic* **1**, 425–467.

Wright, C.: 1993, 'Strict Finitism', in C. Wright, *Realism, Meaning and Truth*, 2nd edn, Oxford, Blackwell, pp. 107–175.

A SOLUTION TO FITCH'S PARADOX OF KNOWABILITY

HELGE RÜCKERT
Department of Philosophy University of Mannheim, Germany, E-mail: helge_rueckert@web.de

Abstract. There is an argument (first presented by Fitch), which tries to show by formal means that the anti-realistic thesis that every truth might possibly be known, is equivalent to the unacceptable thesis that every truth actually is known (at some time in the past, present or future). First, the argument is presented and some proposals for the solution of Fitch's Paradox are briefly discussed. Then, by using Wehmeier's modal logic with subjunctive marker (S5*), it is shown how the derivation can be blocked if one respects adequately the distinction between the indicative and the subjunctive mood. Essentially, this proposal amounts to the one by Edgington which was formulated with the help of the actuality-operator. Finally it is shown how the criticisms by Williamson against Edgington can be answered by the formulation of a new conception of possible knowledge that α (thereby α being in the indicative mood and thus referring to the actual world). This conception is based on the concept of same *de re* knowledge in different possible worlds.

Anti-realists such as Dummett and Wright claim that linguistic meaning is intimately related to the use of relevant expressions by the human linguistic community. What is expressed by a certain sentence thus depends essentially on how it is used. According to this position it is not possible that the states of affairs that are expressed might be in principle independent from the corresponding contexts of use that may arise in the linguistic community:

> The meaning of a mathematical statement determines and is exhaustively determined by its *use*. The meaning of such a statement cannot be, or contain as an ingredient, anything which is not manifest in the use made of it, lying solely in the mind of the individual who apprehends that meaning: if two individuals agree completely about the use to be made of the statement, then they agree about its meaning. (Dummett 1978, 216)[1]

Thus, it is plausible to accept the thesis that there are no states of affairs that may be expressed by linguistic means and that are nevertheless in principle inaccessible to the members of the linguistic community. This means for the anti-realistic concept of truth that it is epistemically constrained: Truths have to be epistemically accessible to the members of the linguistic community in principle. In its strongest reading this condition says that every truth might under certain (possibly counterfactual) circumstances also be known.[2] Now, an ordinary language formulation of the anti-realistic thesis (ART) can be given:

(ART) *Every truth might possibly be known.*

S. Rahman et al. (eds.), Logic, Epistemology, and the Unity of Science, pp. 351–380.

1. The Derivation of Fitch's Paradox

If one wants to give (ART) a formal reading, the following formula (which reflects the redundancy of the truth predicate for the purposes at stake and which uses schematic letters instead of explicit quantification over propositions) seems to be the first choice:[3]

$$\alpha \rightarrow \Diamond K\alpha \tag{1}$$

K is the knowledge operator meaning something like "it is known by somebody at some point in time, that...", and the concept of modality at issue is a metaphysical one and not a mere epistemic one.

If anti-realism is understood according to thesis (1), it has to face a serious problem: There is a simple argument – it was firstly presented by (Fitch 1963),[4] and has become much more popular through (Hart 1979) – that aims to show, that (1) implies the much more problematic thesis that every truth at some point in time actually is known, that there are no unknown truths.

For the derivation two more premises are needed. First, the unproblematic assumption about the concept of knowledge which says that it is a necessary condition of knowledge that it is factive. If something is known then it has to be true:

$$\Box(K\alpha \rightarrow \alpha) \tag{2}$$

Also it is assumed that knowing a conjunction necessarily implies that both conjuncts are known:

$$\Box(K(\alpha \wedge \beta) \rightarrow (K\alpha \wedge K\beta)) \tag{3}$$

From (2) and (3) it follows logically, that it is impossible to know that something is a never known truth. Assume this would be possible: $\Diamond K(\alpha \wedge \neg K\alpha)$. Because of (3) it follows that $\Diamond(K\alpha \wedge K\neg K\alpha)$. By help of (2) again we get $\Diamond(K\alpha \wedge \neg\alpha)$, a contradiction. Thus:

$$\neg\Diamond K(\alpha \wedge \neg K\alpha) \tag{4}$$

Now, by substituting in (1) $(\alpha \wedge \neg K\alpha)$ for α we get to

$$(\alpha \wedge \neg K\alpha) \rightarrow \Diamond K(\alpha \wedge \neg K\alpha). \tag{5}$$

From (4) and (5) follows (6):

$$\neg(\alpha \wedge \neg K\alpha). \tag{6}$$

And (6) again is (at least in classical logic) equivalent to

$$\alpha \rightarrow K\alpha. \tag{7}$$

The derivation of Fitch's Paradox is completed. Apparently it has been shown that the thesis that every truth might possibly be known implies the thesis that every truth actually is known at some time or other. From (1) to (7) a modal collapse happened: the possibility operator has disappeared. And even if anti-realists want to accept thesis (1) they usually don't want to accept (7). Because to claim that everything which is true also is known at some time or other seems to be at least very problematic if not absurd.

Even if what (4) says is correct, namely that of no proposition is it possible to know that it is true and also that it is never known, one nevertheless wants to accept the existential claim that there are propositions which are true but actually (because of contingent reasons) never known. For example, it is impossible for us to know of something lying outside our light cone (and that it lies outside of our light cone is a contingent fact), because – if Einstein's theory of relativity is correct in this respect – we can't get into causal contact with such facts.

Thus, the anti-realist has a problem. How should he react to Fitch's Paradox? But, in my opinion, also his enemy, the realist, should be interested in an adequate reaction, because even if he rejects (ART), it seems dubious that the dispute about an apparently interesting difficult philosophical thesis might be decided with the help of simple logical means by showing that (ART) implies the unacceptable (7). Also to the realist an intuitive understanding of (ART) can be assigned, that shows (ART) to be less strong and problematic than (7).[5]

At the very least the anti-realist should have a good response to Fitch's Paradox at his disposal in order to defend his philosophical position. Several different starting points for an anti-realistic reaction are intelligible. Some of them will be discussed shortly in the following. On the reaction to Fitch it also depends how the respective anti-realistic position has to be formulated in more detail. Certainly, one has to agree with Williamson when he writes:

> A diffuse philosophical tendency cannot be refuted once and for all by a single rigorous argument. Nevertheless, such an argument can severely constrain the forms in which the tendency is expressed. (Williamson 2000, 99)

2. Possible Reactions to Fitch's Paradox

Confronted with a philosophical-logical argument of the present kind, four different strategies might be followed:

(i) One accepts the conclusion. But then, one has to show that the conclusion is, against the first impression, nevertheless acceptable or compatible with the defended position. This would mean for the anti-realist, who is surprisingly willed to accept (7), that he has to show that it is, according to his position, plausible to hold the claim that every truth actually is known at some time or other.

(ii) One doubts that the derivation of the conclusion from the premises is correct. To approach the problem in this way, one has to show that one or more argumentative steps haven't been well-founded.

(iii) One rejects one or more of the premises. For the following the principles (2) and (3), which seem to be highly plausible assumptions about the concept of knowing, will not be called into question. This leaves us with the anti-realistic thesis. Here, a distinction should be drawn:

(A) One could defend that (ART) represents adequately anti-realism but that its formal representation with (1) is not correct, and (1) should be replaced by another formula of the underlying logic.

(B) The other possibility consists in claiming that the acceptance of already (ART) is not forced by a correctly understood anti-realistic position.

(iv) One doubts the suitability of the used instruments. In our case the used instruments are standard modal logic and the knowledge operator K. At first sight, this strategy seems to be hopeless, because standard modal logic is very well established and only very few and unproblematic assumptions concerning the K-operator are made use of. But who knows?

In the following, several responses (reflecting all four strategies) to Fitch's Paradox will be discussed. In this section we will discuss proposals that can be related to strategies (i) to (iii). In the subsequent section the main proposal of this paper is presented, performing a combination of strategies (iii)(A) and (iv).[6]

2.1. The Ideal Epistemic Subject: A Way Out?

First, let's discuss the consequent acceptance of thesis (7). Following a terminological proposal by Tennant (1997, 261), we will call an anti-realist, who is ready to accept (7) a '*very hard anti-realist*'. Suppose, as suggested above, that the totality of human knowledge never can comprise the wealth of actually existing truths (and a healthy amount of modesty as well as a realistic appraisal of the limitations of our cognitive capacities seem to lead to this conclusion), which way out is still available not to be forced to withdraw from (7)?

One might think of a similar (theological) move as (Berkeley 1710) had imagined in order to stick with the *esse est percipi* of his radical subjective idealism without having to draw the absurd consequence that objects cease to exist as soon as they are no longer perceived by human beings. In analogy to the objects' always being perceived by God in Berkeley the very hard anti-realist could retain thesis (7) by claiming that even if not every truth is known (at some time or other) by a human being, every truth is known by a postulated ideal epistemic subject or an omniscient God. And thus, (7) would be true nevertheless.[7]

What to think about this theological way out? Leaving aside any problems the hypothesis of an omniscient God or a similar conception might bring with it and whether one's own attitude towards such a doctrine is positive or negative, it is clear that this proposal completely misses the point of the debate that concerns us. The original anti-realistic argument referred to a (our) human linguistic community and aimed to show, that every truth might possibly be known by a (possible) member of this community. This means that the K-operator contains an implicit double existential quantification, one about points of time (in the past, present or future) and the other about the members of the communication community. Let's make this explicit by using Mx as an abbreviation for "x is a member of the linguistic community", Tx as an abbreviation for "x is a point of time", and the doubled indexed knowledge operator $K_{xy}\alpha$ meaning "x knows at y, that α":

$$K\alpha \Leftrightarrow_{\text{def}} \exists_x \exists_y (Mx \wedge Ty \wedge K_{xy}\alpha)$$

By help of this more detailed definition of the K-operator it is now obvious that the strategic move, the idea that God knows all truths, is no promising defence of thesis (7), because, so to speak, God is not accepted as a potentially relevant knowing subject (he doesn't fulfil the predicate M).[8]

2.2. INTUITIONISTIC LOGIC

The philosophical position of anti-realism arose when intuitionistic and constructivist tendencies in the philosophy of mathematics were generalised and applied to other areas.[9] In the field of mathematics the movement coincided with a replacement of classical logic by intuitionistic logic. Thus, it would be quite consequent for the anti-realist to use intuitionistic logic also in his generalised project. Here, we will discuss this proposal very briefly, because the debates in the newer literature about Fitch's Paradox mainly deal with the possibilities for a solution resulting from the acceptance of intuitionistic logic (sometimes accompanied by a weakening of thesis (1)). And also, the main aim of this paper is to propose a different solution.[10]

First, using intuitionistic logic seems to offer another way of defending the acceptability of thesis (7) which was derived by Fitch's argument. Consequently, if one uses intuitionistic logic, one also should use one of the semantics developed for intuitionistic logic. Thus, one can give the conditional in (7) an intuitionistic reading. One such reading says that an intuitionistic conditional is correct if everyone who *knows* the antecedent is thus in a position to know the consequent, too. Understood in this way, (7) seems to be fully acceptable.[11]

The principle problem for this strategy (leaving aside the question whether an consistent formulation of the main thesis without unwanted side-effects can be given[12]) is that if the anti-realist wants to give thesis (7) a special intuitionistic meaning of the mentioned kind (by giving an intuitionistic reading to the conditional), (7) does no longer serve as a translation of the acceptable ordinary-language proposition that not all truths are known by human beings at some time or other, and the anti-realist needs to present another formula of his intuitionistic logical language to express this proposition. Because a position which doesn't allow to do so seems to be at least very dubious.

A further starting point for a possible solution of Fitch's Paradox amounts from the question whether all the steps in its derivation are also intuitionistically valid and not only classically (as known, intuitionistic logic is weaker than classical logic). Indeed, the derivation till

$$\neg(\alpha \wedge \neg K\alpha) \tag{6}$$

is completely correct, also in an intuitionistic setting. But the remaining step leading to

$$\alpha \rightarrow K\alpha \tag{7}$$

is only acceptable from a classical point of view. For the anti-realist using intuitionistic logic this fact opens the possibility to stick with (ART) and (1), but to withdraw from (7), because (7) has been derived by intuitionistically unacceptable means. But then still he has to accept (6) as well as the thesis (8) that follows intuitionistically from (6):

$$\neg K\alpha \rightarrow \neg\alpha \tag{8}$$

To come again back to Tennant's terminology we will call an anti-realist willing to accept (6) and thus (8) as correct, but not (7), a '*moderately hard anti-realist*'.[13]

Translated into ordinary language (8) says something like that everything that is not known at some time or other also is not true. And at first sight, this seems to be almost as absurd as what is said by (7). Once again, one might try to justify (8) by giving a special intuitionistic interpretation to the logical particles contained in (8), for example by giving the conditional one of the readings discussed above. This strategy leads to the problem that such a reading not only allows to justify the acceptance of (8), but that of (7) as well, and thus moderately hard anti-realism turns out to be a very hard anti-realism. This is not the place to engage in a more detailed discussion of moderately hard anti-realism, but it seems to me that this position is not very natural with regard to the ideas that in the beginning did lead to anti-realism (cf. Dummett's argument from the beginning of this paper). Thus, an anti-realist should not be happy with moderately hard anti-realism unless there is no other consistent and useful alternative at his disposition.

The aim of Sections 3 and 4 of this paper is to show that '*soft anti-realism*', accepting (ART) but neither (7) or (8), can be successfully defended against Fitch's argument.[14]

2.3. Considering Time-Parameters

According to the analysis of the K-operator given above using K involves an implicit existential quantification over the members of the linguistic community as well as an implicit existential quantification over all points of time (in the past, present or future). Propositions containing the K-operator thus also refer to the future. Already Aristotle (1975, 18a–19b), at hand of a sentence saying that there would be a see battle the next day, discussed the position that (contingent) sentences about the future don't have a truth-value yet, but that they get one as soon as the point of time they talk about arises. Thus, sentences about the future are no real propositions which are either true or false.[15] And indeed, the thesis that sentences about the future already now have a definite truth-value seems to presuppose the correctness of the metaphysical thesis of determinism, which is called into question by current interpretations of modern quantum physics.

Let's examine whether the adoption of the Aristotelian position leads to a solution of Fitch's Paradox. Instead of using K we introduce a knowledge operator K_t, which says something like "there is a point in time $t' \leq t$, and at t' it is known that…", where t may lie in the past or the present but not in the future. Furthermore, $K_t\alpha$ may only be true if α does not refer to a point in time lying in the future with respect to t, in which case α would not already have a truth-value at t. Thus, we introduce a time index: α_t says that α is true at t. Also let the present point of time with respect to our world be named by t_0.

After these preparations we are now able to give the anti-realistic thesis the following formulation which takes time-parameters into consideration:

$$\alpha_{t_1} \rightarrow \Diamond K_{t_2}\alpha_{t_1} \quad \text{where there is at least one such } t_2 \text{ with } t_2 \geq t_1 \tag{9}$$

But, is this new formulation safe from Fitch? Let's chose α = "$\alpha_{t_0} \wedge \neg K_{t_0}\alpha_{t_0}$". Substitution in schema (9) results in:

$$(\alpha_{t_0} \wedge \neg K_{t_0}\alpha_{t_0})_{t_1} \rightarrow \Diamond K_{t_2}(\alpha_{t_0} \wedge \neg K_{t_0}\alpha_{t_0})_{t_1} \tag{10}$$

On the basis of the assumptions made, it is immediately clear that it has to be $t_1 = t_0$, and thus apparently also $t_2 = t_0$.[16] It seems as if (10) results in (11).

$$(\alpha_{t_0} \wedge \neg K_{t_0}\alpha_{t_0})_{t_0} \rightarrow \Diamond K_{t_0}(\alpha_{t_0} \wedge \neg K_{t_0}\alpha_{t_0})_{t_0} \tag{11}$$

It is easy to check whether starting from (11) Fitch's Paradox can be derived in the usual way (all time-indices coincide and thus no step in the argument will be blocked with the help of them). Do we have to conclude that the consideration of time-parameters does not lead to a way out?

No, we don't have to draw this conclusion. But first, some more modifications are necessary. In the argument leading us from (10) to (11) at the stage that it apparently has to be that $t_2 = t_0$ we used the condition that t_2 must not lie in the future,[17] but that it has to be that $t_2 \geq t_1$. But now, if $t_1 = t_0$, then $t_2 = t_0$. But the expression $K_{t_2}(\alpha_{t_0} \wedge \neg K_{t_0}\alpha_{t_0})_{t_1}$ lies within the scope of a modal operator and thus has to be semantically evaluated always with respect to the corresponding possible worlds at stake. Thus, it results only necessarily that $t_2 = t_0$ if the present point of time with respect to each possible world is identical with t_0.

What still needs to be done in order to block Fitch's derivation with the help of the consideration of time-parameters is a determination of the concept "possible world" where this does not have to be so. Let for each possible world w_i the corresponding present point in time be named by t_{wi}. Then, a world w_m is a possible alternative with respect to w_n[18] if and only if the following conditions are fulfilled:

(C1) $t_{w_m} \geq t_{w_n}$

(C2) $\{\alpha \mid \alpha \text{ is true at } w_n\} \subseteq \{\alpha \mid \alpha \text{ is true at } w_m\}$

(C3) w_n and w_m have to fulfil the usual conditions for possible worlds (in particular, the sets of truths of the different worlds have to be consistent)

(C1) and (C2) together amount to the fact that a possible alternative to a given possible world always has to comprise the complete course of this world (that is the class of true propositions in this world), and (apart from the limit case that the starting world and the possible alternative are identical) develops it further to a certain future point in time (the present point in time with respect to that possible alternative). (C3) demands the further condition of internal consistency of all possible worlds. For example, it excludes that in a world w somebody knows α, but α isn't even true (this would contradict the factive character of knowledge).

The modal logic resulting from the above considerations obviously is of the *S4*-type, because the accessibility relation between possible worlds R has to be reflexive and transitive (but isn't symmetric as soon as the frame contains more than one possible world).[19] What remains to be shown is that the anti-realistic thesis (9) is now no longer subject to an argument *à la* Fitch. Let's consider again (10), which resulted from (9) by the substitution, that is characteristic for Fitch's Paradox:

$$(\alpha_{t_0} \wedge \neg K_{t_0}\alpha_{t_0})_{t_1} \rightarrow \Diamond K_{t_2}(\alpha_{t_0} \wedge \neg K_{t_0}\alpha_{t_0})_{t_1} \tag{10}$$

Because t_0 is the present point of time with respect to the real world and because sentences about the future don't have a truth-value yet, it follows that it has to be that $t_1 = t_0$:

$$(\alpha_{t_0} \wedge \neg K_{t_0}\alpha_{t_0})_{t_0} \rightarrow \Diamond K_{t_2}(\alpha_{t_0} \wedge \neg K_{t_0}\alpha_{t_0})_{t_0} \tag{12}$$

Would $t_2 = t_0$ Fitch's Paradox could again be derived, because for no member of the linguistic community is it possible to know at t_0 that the problematic conjunction is true at the same moment (the second conjunct says that no member of the linguistic community knows the first conjunct at t_0). But the problem is resolved in case of $t_2 > t_0$, because then a member of the linguistic community might know at a future point in time t_2, that the conjunction was true at t_0, which includes

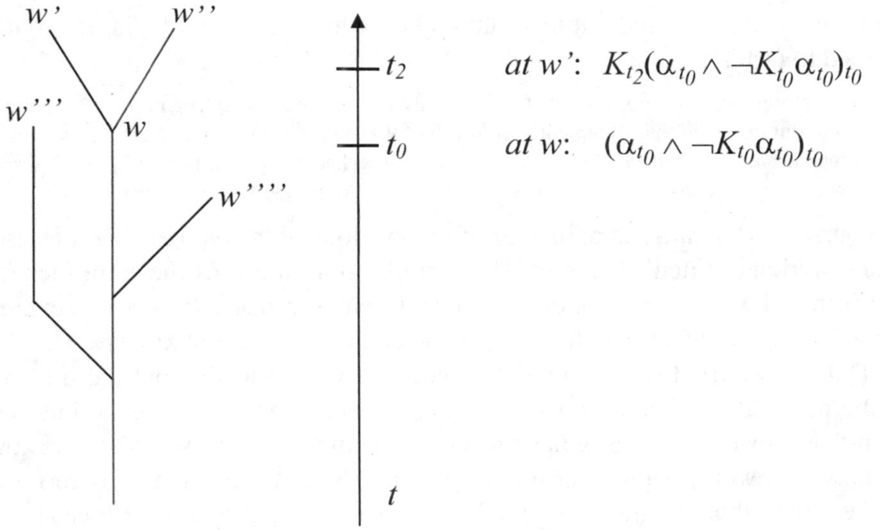

Figure 1.

knowledge of the fact that nobody (himself included) knew at t_0 that α was true at t_0 (see Figure 1).

Now, it should be clear that the assumptions made about the relation between truth-values and time-parameters as well as the given concept of possible world, allow to deal successfully with Fitch's Paradox. But the assumptions one has to concede in order to go this way are not unproblematic. First of all one has to adopt the Aristotelian view that (contingent) sentences about the future don't have a truth-value yet and thus are no real propositions. This leads to some linguistic hardship cases because one is forced, for example, to declare the reply "Yes, that's true" to the assertion "Tomorrow Man.U. plays in the Champion's League", which might arise very naturally in ordinary discourse, to be not fully correct.

Thus, even if the assumptions made might be defended, the proposed analysis does not represent a real alternative for anti-realists like Wright, because they would have to give up the so-called '*timelessness of truth*'-thesis[20] and would have to accept several restrictions of the usual modal logic framework, whereas modern anti-realists tend to be closer to realistic positions with respect to these matters.

2.4. MELIA'S LIBERALISATION OF (ART)

In a very short note (Melia 1991) has argued that the anti-realistic thesis (1) as well as its ordinary language formulation (ART) shouldn't be accepted by the anti-realist, because the claims thus expressed were too strong, and therefore could be exploited to derive Fitch's Paradox. According to Melia's analysis there is an improper assumption (about the independence of the truth-value of a proposition from the attempts to find out this truth-value) which creeps in the initial argument that led to (ART):

> This argument presupposes that we can always discover a statement's truth value without affecting that statement's truth value. But this is not so: there exist statements which are true, yet which would have been false had we performed the procedures necessary to discover that statement's truth value. (Melia 1991, 341)

Of course, Melia thinks in particular of propositions of the type $\alpha \wedge \neg K\alpha$ as used in the derivation of Fitch's Paradox. These propositions are constructed in such a way that if they are true they cannot be known, because in one conjunct it is expressed that the other conjunct (and thus the whole conjunction) is not known.

Thus, if we tried (in an alternative possible world) to find out the truth-value of the proposition in question, we certainly would start by trying to find out the truth-value of α. Two cases have to be distinguished here: Would α turn out to be false, we would know that the conjunction is false, too. Would α turn out to be true, the whole conjunction would turn out to be false as well (because then the second conjunct would be false) and we would have reached a case, called 'pathological' by Melia, in which only the counterfactual circumstances resulting from the fact that it is tried to find out the truth-value of the conjunction, would lead to a change of this truth-value from 'true' (in the real world) to 'false' (in the alternative possible world).

Considerations of this kind lead Melia to the following conclusions concerning the position of anti-realism:

> For the anti-realist can surely distinguish between *(i)* statements whose truth value changes whenever we try to verify them, and *(ii)* statements whose truth value cannot be discovered no matter how we try to verify them. The anti-realist takes exception to statements belonging to class *(ii)*, but I see no reason why that should make him reject statements in class *(i)*. (Melia 1991, 342)

Melia considers the demands resulting from (ART) to be unjustified, and claims that the anti-realist should withdraw from (ART) and replace it with a more liberal thesis.[21] Even if Melia does not formulate explicitly such a thesis of his own, his remarks justify ascribing the following liberalised anti-realistic thesis (LART) to him:

(LART) *For every proposition there are (possible) circumstances under which it is known which truth-value this proposition has under these circumstances*

An anti-realism resulting from a substitution of (ART) by a weaker thesis like (LART) we call a '*very soft anti-realism*', thus expanding Tennant's terminology. (LART) can be adequately rendered by formal means as:

$$\Diamond(K\alpha \vee K\neg\alpha) \tag{13}$$

Now, the question arises whether it is possible (similar to Fitch's Paradox) to derive unpleasant consequences from (13) by clever substitutions for α together with logical means alone. There are propositions that make the first disjunct false, for example those of the form $\alpha \wedge \neg K\alpha$ (as used in the derivation of Fitch's Paradox), as well as propositions that make the second disjunct false, for example those of the form $\neg\alpha \vee K\alpha$. In both cases only $\Diamond(\alpha \rightarrow K\alpha)$ can be derived immediately, but this formula seems to be perfectly acceptable. Also it is not possible by truth-functional compositions of such propositions to get to propositions which make the first and the second disjunct false, thus showing the schema (13) to be incorrect.

Let's assume that Melia's proposal is immune from logical constructions.[22] How is it to be judged then? The liberalised anti-realistic thesis (LART) is much weaker than (ART), because it is no longer claimed that every aspect of the world as it really is, might in principle be epistemically accessible. Now, the very soft anti-realist would even concede to the realist that the actual world might be constituted in a way that some aspects of it might be in principle unknowable.

Were there no other acceptable proposals to deal with Fitch's Paradox, Melia's proposal would maybe be an interesting alternative for the anti-realist, who then would have to weaken his claim concerning the relation between truth and knowability. In the following it is shown that (ART) can be retained, and thus Melia's proposal loses its attraction for anti-realists. Because, if there are no logical arguments against (ART), the discussion as regards content might start from the stronger anti-realistic thesis.[23]

3. Solving Fitch's Paradox with the Help of the Distinction between the Indicative and the Subjunctive Mood

The result of the last section was that there are several possible ways of dealing with Fitch's Paradox. By accepting intuitionistic logic one can try to defend a hard or a moderately hard anti-realism. An Aristotelian neutralism concerning the future also allows for a solution of the difficulty. And finally, as proposed by Melia, one might weaken the anti-realistic thesis (ART), and for example replace it by (LART).

But up to now we still lack (apart from the Aristotelian proposal) a conception which allows to defend what I believe to be the most compelling form of anti-realism, namely soft anti-realism, against Fitch's Paradox. A soft anti-realist wants to retain (ART), but not accept either (7) or (8). Aim of the following two sections is to show how soft anti-realism can be successfully defended against Fitch. It is clear that then (ART) can no longer be expressed by the formula (1), because (1) logically implies (7) and thus excludes soft anti-realism. Thus, we will have to give the natural language thesis (ART) a modified formal translation. In order to prepare this move, we will first present Wehmeier's modal logic with subjunctive marker (*S5**).

3.1. Wehmeier's Modal Logic with Subjunctive Marker

Recently, Wehmeier has criticised Kripke's modal argument (cf. Kripke (1980)) against the so-called Frege and Russell-view of proper names along with standard modal logic, which is implicitly used by Kripke (see (Wehmeier (2003a) and Wehmeier (2003b)). According to his diagnosis the argument does not consider carefully enough the semantically relevant distinction between the modes. While predicates standing in the indicative mood[24] always have to be evaluated with respect to the real world, predicates standing in the subjunctive mood have to be evaluated with respect to the possible worlds at stake. That the difference between the indicative and the subjunctive mood is semantically relevant can easily be shown with two examples:

(a) *Under certain counterfactual circumstances the man who would have taught Alexander would not have taught Alexander.*

(b) *Under certain counterfactual circumstances the man who taught Alexander would not have taught Alexander.*

Both expressions only differ in the fact that the subjunctive "would have taught" in (a) is replaced by the indicative "taught" in (b). This difference is semantically relevant: (a) is false, but (b) expresses a truth. Because (a) says that there are possible circumstances under which the one who taught Alexander (under these circumstances) did not teach Alexander (under these circumstances), a logical contradiction. On the other hand, (b) says that there are possible circumstances under which the one who actually taught Alexander, namely Aristotle, would not have taught Alexander (under these circumstances). And that's perfectly imaginable, because it might have been that Alexander would have been taught by another person than Aristotle, for example by his own father.

But, one may ask, why does this example already speak against standard modal logic? Because standard modal logic has means to deal with the difference between the indicative and the subjunctive mood: If, in a natural language example, a predicate is in the subjunctive mood, this only shows that the proposition has to be rendered formally in such a way that the corresponding predicate of the logical formula stands inside the scope of a modal operator, whereas predicates in the indicative mood have to be rendered by the logical analysis in such a way that they stand outside any modal scope. This conception suggests that the modes of natural language are only means which help to determine the scopes of modal contexts. And as different scopes can easily be dealt with by the help of parentheses it seems not to be necessary to make the difference between the indicative and the subjunctive mood any more explicit in our logical notation.

What to think about this solution to the problem posed by the examples above? Is it always possible to reduce differences in the modes to differences concerning the placement of the parentheses? True, it is easy to explain the difference between (a) and (b) by this manoeuvre. But, there are also propositions that do not allow for such a move:

(c) *Under certain counterfactual circumstances everyone who actually has flown to the moon would not have flown to the moon.*

$\forall_x(Fx \rightarrow \Diamond\neg Fx)$ as the formal rendering of (c) is not correct because according to (c) there must be counterfactual circumstances under which no one of those, who have actually flown to the moon, would have flown to the moon. But this is not required by the formula because it says that for everyone who actually has flown to the moon there have to be circumstances under which he would not have flown to the moon, and these counterfactual circumstances need not to be the same for everyone. Also, no other formula of standard modal logic does the job.[25]

Thus, it has been shown that usual modal logic has a serious problem, at least if it is conceived of as a tool to deal with the modal part of ordinary discourse. It is not able to provide a formula for all natural language propositions one might expect one. Its expressibility is not high enough to even deal with all simple cases of everyday modal discourse.

To repair these defects, Wehmeier proposes to change the formal language and make the difference between the indicative and the subjunctive mood explicit.[26] For our purposes it is not necessary to present the modal logic evolving from this idea in detail. It should be sufficient to discuss the main ideas, which should also give an impression of the exact techniques.

Syntactically, *S5** behaves like *S5* with the only difference that we add subjunctive versions of all predicates and quantifiers, which are symbolised by an attached subjunctive marker "*".[27] Semantically, predicates and quantifiers in the indicative mood are always evaluated with respect to a specially characterised world w^* (the real world), even if they stand inside the scope of a modal operator. Predicates and quantifiers in the subjunctive mood have to be evaluated, as usual, with respect to those possible worlds as determined by the modal operators in whose scopes they stand.

Essentially the subjunctive marker works as a variable over possible worlds. Thus, every occurrence of "*" has to be bound by a modal operator. For the sake of illustration let's give the translations of all our three examples into *S5 **:

(a*) $\Diamond\neg T^*(\iota^* x)(T^* x)$

(b*) $\Diamond\neg T^*(\iota x)(T x)$

(c*) $\Diamond\forall_x(Fx \rightarrow \neg F^* x)$

3.2. The Reformulation of Fitch's Argument Within S5*

After briefly presenting Wehmeier's modal logic with subjunctive marker, we will now examine whether Fitch's argument turns out to be valid also in the *S5** framework or whether the derivation will be blocked by the explicit distinction between the indicative and the subjunctive mood.

In *S5** thesis (1) changes to:

$$\alpha \rightarrow \Diamond K^*\alpha \qquad (1^*)$$

(ART) only requires that what is true, namely α, might possibly be known, and not that it is known. That's why the K-operator has to be in subjunctive mood and thus refer to the respective possible world in question and not (necessarily) to the real world. It should already be mentioned that the second occurrence of α, although standing inside the scope of a m odal operator, does not contain any free subjunctive markers to be bound by this modal operator, because the first occurrence of α does not stand in any modal context and thus all subjunctive markers that might occur in α already have to be bound.

The two necessary conditions about knowledge now look like:

$$\Box(K^*\alpha \to \alpha) \tag{2*}$$

$$\Box(K^*(\alpha \wedge \beta) \to (K^*\alpha \wedge K^*\beta)) \tag{3*}$$

Here, it makes no difference whether α and β contain any free subjunctive markers that might be bound by the necessity operator, or not. From (2*) and (3*) it follows:

$$\neg\Diamond K^*(\alpha \wedge \neg K^*\alpha) \tag{4*}$$

If we now, as in the usual derivation of Fitch's Paradox, substitute $(\alpha \wedge \neg W\alpha)$ for α in (1*), we get:

$$(\alpha \wedge \neg K\alpha) \to \Diamond K^*(\alpha \wedge \neg K\alpha) \tag{5*}$$

But here (4*) is no longer identical with the negation of the consequent of (5*), because the respective second occurrence of the K-operator once is in subjunctive mood, and once in indicative mood. That's why the derivation is blocked and (6) is no longer implied.

Thus, we have now with (1*) a formal rendering of (ART) at hand, that does not imply (6) or even (7), and a consistent formulation of soft anti-realism seems to be in sight.

3.3. Modal Logic with Subjunctive Marker vs. Modal Logic with Actuality-Operator

Wehmeier's modal logic with subjunctive marker is intimately related to usual modal logic with a supplementary so-called *actuality*-operator A.[28] The use of indicative versions of predicates, quantifiers and operators produces, that they always have to be evaluated with respect to the real world, no matter whether they stand in the scope of a modal operator, or not. The A-operator has a similar effect: If there is an occurrence of the form $A\alpha$, then α gets evaluated with respect to the real world, even if the whole expression $A\alpha$ stands in the scope of a modal operator. Thus, outside any modal context $A\alpha$ and α are equivalent, but inside modal scope the A-operator causes, so to say, that α gets extracted from the modal context. The resulting connections are listed in the following table:

Modal logic with A		*Modal logic with subjunctive marker*
α outside modal scope	$\approx$	α in indicative mood
α inside modal scope	$\approx$	α^* in subjunctive mood
$A\alpha$ outside modal scope	$\approx$	α in indicative mood
$A\alpha$ inside modal scope	$\approx$	α in indicative mood

Because of these connections the conjecture that the proposal for a solution of Fitch's Paradox just formulated within the framework of Wehmeier's logic could be reformulated by using standard modal logic plus *actuality*-operator too, suggests itself.

And indeed (Edgington 1985) has made exactly this proposal.[29] The anti-realistic starting thesis gets the following translation by Edgington:

$$A\alpha \rightarrow \Diamond K A\alpha \qquad (1')$$

Here, the derivation of Fitch's Paradox fails for analogous reasons as in the case of (1*), and that needs no further demonstration.

As already Edgington knew, the Achilles' heel of her (and my) proposal is the sequence of signs $\Diamond K A\alpha$. Because, at first sight, it is not at all obvious, what it should mean to be a merely possible knowledge of $A\alpha$ (which refers to the actual world). Edgington's solution (as well as the analogous one in *S5* *) thus depends on whether a sensible conception of someone in a possible world knowing something that actually is the case can be formulated. This is the point where most criticisms, formulated at different places, start from.[30] But the counter-arguments against Edgington's attempts to justify the problematic construction by the help of analogies, especially the analogy between "actually" and the indexical "now" put forward in (Williamson 1987) seem to me to be the clearest and most serious ones. Thus, we will examine Williamson's criticisms in some more detail next.

3.4. The Counter-Arguments by Williamson

The first of Williamson's objections against the proposal to give (ART) the reading (1′) goes as follows:

> In one respect, (4) [i.e., (1′); H.R.] is a surprisingly weak form of verificationism. For, as Edgington notes, 'Ap' always entails '$\Box Ap$', where '$\Box$' abbreviates 'it is necessary that'. Thus, the only knowledge that (4) requires is of necessary truths. One might expect a robust form of verificationism to insist that at least some contingent truths are knowable. (Williamson 1987, 257)

But, outside modal scope $A\alpha$ is equivalent to α. So, why not translate the anti-realistic starting thesis by (1″)?

$$\alpha \rightarrow \Diamond K A\alpha \qquad (1'')$$

The natural objection against (1″) is probably that according to (ART) the same thing that is true should be knowable, and that thus the antecedent of the anti-realistic thesis should be identical with the expression in the scope of the knowledge operator.

But, let us again have a look at the anti-realistic thesis corresponding to (1′), or (1″) respectively, formulated within the framework of *S5**:

$$\alpha \rightarrow \Diamond K^{*}\alpha \qquad (1^{*})$$

Here, Williamson's objection becomes pointless, because the schema is no longer only about necessary propositions and the antecedent is identical to the expression in the scope of the K-operator. Obviously it depends on the chosen formal language in which cases what stands inside the scope of a modal operator is 'the same' as outside, and in which cases not. To defend his own modal logic *S5** in this respect against standard modal logic with *actuality*-operator, we give the word to Wehmeier:

> An obvious constraint on any language of modal predicate logic is that its non-modal part should simply be the language of ordinary predicate logic. Therefore, indicative predicates are to be expressed by the ordinary predicate symbols of non-modal predicate logic (which, after all, formalises ordinary, indicative discourse), and it is the subjunctive for which we need to introduce a new notation. This constraint rules out the standard solution to the expressiveness problem, viz. the introduction of an "actually" operator "**A**". For consider the following example:
>
> (40) Someone has flown to the moon, but under certain counterfactual circumstances, everyone who has flown to the moon would not have flown to the moon.
>
> With an actuality operator, it would have to be formalised as
>
> (41) $\exists x Fx \;\&\; \Diamond\forall x(\mathbf{A}(Fx) \rightarrow \neg Fx)$,
>
> but this is clearly not a faithful representation of the ordinary language sentence (40): In (40), there are two occurrences of the indicative predicate "has flown to the moon", and on both of these occurrences, the predicate has exactly the same semantic function, viz., to refer to how things stand in the real world. There is one occurrence of "would have flown to the moon", which is syntactically distinguished by being subjunctive, and semantically distinguished by referring not to the real, but to some counterfactual situation. In (41), however, the first occurrence of "has flown to the moon" in (40) is modelled by "Fx", and so is the occurrence of "would have flown to the moon", whereas the second occurrence of "has flown to the moon" corresponds to "$\mathbf{A}(Fx)$". This does not appear to be a transparent logical analysis – why should the two occurrences of the predicate "has flown to the moon", which function in precisely the same way semantically, be modelled in *two* typographically distinct ways, as "Fx" and "$\mathbf{A}(Fx)$"? And why is the subjunctive predicate "would have flown to the moon", referring to some counterfactual world, represented in exactly the same way as the indicative predicate "has flown to the moon" (as it occurs first in (40)), which refers to the actual world? (Wehmeier 2003a, 11–12)

As one can easily see the defects of the *actuality*-operator version mentioned in this very clear citation disappear if the example is translated into Wehmeier's *S5**: Occurrences of identical predicates in the natural language formulation are represented identically in the formal language, whereas occurrences of predicates which differ with respect to the indicative/subjunctive-difference are represented differently.

Obviously, against (1*) Williamson's criticism does not get through. But, apart from that, is his criticism sound if it is thought of to be directed only against (1')? I think, it is at least doubtful whether the criterion for necessary statements used by Williamson, namely that α is a necessary truth iff $\Box\alpha$ is a truth, should be accepted. Because in cases like $\Box A\alpha$ the necessity operator is freewheeling, so to say, because of the *actuality*-operator, as the quantifier in $\forall x\alpha$ is without effect, supposed x does not appear freely in α.[31] For a discussion of other criteria, in the context of two-dimensional semantics, see Davies and Humberstone (1980).

But the claim that schema (1′) would only refer to necessary propositions is not the main criticism by Williamson against Edgington's analysis. His main point concerns the problematic conception of non-actual knowledge of something referring to the real world, already mentioned above. Specially, Williamson remarks that Edgington's analogies between modal and temporal discourse don't go through, because there might be causal relations between events that take place at different times, but between events that take place in different possible worlds causal relations are principally excluded (cf. Williamson (1987, 257–258)).

But how then can an expression of the form $\Diamond KA\alpha$ be understood? For it to be true there has to be an alternative possible world, in which somebody knows at some point in time that $A\alpha$. In the easiest case this possible world is the real world itself. Then, there are no bigger problems. But in most cases, as in Fitch's Paradox, it is necessary to consider a possible alternative that is different from the real world, and in which $A\alpha$ is known. Then, the problem arises how the non-actual knowing subject may express his knowledge,[32] because it is obvious that it can't do so by using $A\alpha$ or α (then the knowing subject would only refer to its own world and not to the real one). To give the problem a more general formulation: How can a knowledge that refers to another possible world w_2 be expressed in a possible world w_1?

The only possibility Williamson seems to consider is that the knowing subject in w_1 uses an expression of the kind "in w_2, α". Thus, the knowing subject first has to specify or name world w_2, and then assert about this so specified world, that in it α holds. Williamson discusses four candidates that might perhaps allow for a specification of a different possible world:[33]

(i) by necessary and sufficient conditions
(ii) by counterfactuals
(iii) by space-time-coordinates
(iv) by ostension

Clearly, (iii) and (iv) won't do the job, because with the help of them one is not able to exceed the realm of one's own world. Williamson's arguments concerning the candidates (i) and (ii) run similarly and are not much different. Therefore, we will give in the following a reconstruction of his rather technical and difficult argument against (i), which surprisingly amounts to the thesis that knowledge of the kind "in w, α" consists of mere trivial logical knowledge as soon as w has been specified by a necessary and sufficient condition (we will call this condition β in the following).

After having convincingly explained that "in w, α" amounts to the same as $\Box$ (w obtains $\rightarrow \alpha$), Williamson presents his argument in a rather compact form:[34]

> Assume first that, in knowing that in *s*, *p*, one can specify *s* in way (i). Thus, for some value of '*q*', necessarily, s obtains if and only if *q*: moreover, the knowledge that, necessarily, if *q* then *p* counts as knowledge that, in *s*, *p*. Now it is easy to show that, necessarily, *s* obtains if and only if both *p* and *q*. Thus the condition that both *p* and *q* specifies *s* in way (i) just as well as the condition that *q* does. It would therefore be quite *ad hoc* not to permit its use in *de re* knowledge about s, since the latter condition may be so used. Hence the knowledge that, necessarily, if both *p* and *q* then *r* counts as knowledge that, in *s*, *r*. In particular, the knowledge that, necessarily, if both *p* and *q* then *p* counts as knowledge that, in *s*, *p*. Thus, given the assumption about (i), the knowledge that, in *s*, *p* requires no more than knowledge of a trivial logical truth. (Williamson 1987, 259)

The argument is difficult and a detailed reconstruction will be helpful. Let's assume that α is really true in w, thus

$$\Box(w \text{ obtains} \rightarrow \alpha), \tag{14}$$

and ask what a knowledge of this fact must consist of. According to the reflections above (specially variant (i)), the knowing subject first has to specify w by a necessary and sufficient condition β, and further has to know that this condition necessarily implies α:

$$\Box(w \text{ obtains} \leftrightarrow \beta) \tag{15}$$

$$K\Box(\beta \rightarrow \alpha). \tag{16}$$

But, if α is true in w and if β is a necessary and sufficient condition for the obtaining of w, then:

$$\Box(\beta \leftrightarrow (\beta \wedge \alpha)) \tag{17}$$

And thus, also $\beta \wedge \alpha$ is a necessary and sufficient condition for the obtaining of w:

$$\Box(w \text{ obtains} \leftrightarrow (\beta \wedge \alpha)) \tag{18}$$

And consequently also $\beta \wedge \alpha$ can be used in both clauses for knowledge of "in w, α":

$$\Box(w \text{ obtains} \leftrightarrow (\beta \wedge \alpha)) \tag{15$'$}$$

$$K\Box((\beta \wedge \alpha) \rightarrow \alpha). \tag{16$'$}$$

The knowledge required by (16′) is mere trivial logical knowledge, and thus knowing "in w, α" reduces to being able to specify w by a necessary and sufficient condition. But then, the knowledge that in w, α can no longer be distinguished from the knowledge that in w, γ (for any γ different from α and holding in w). As soon as one is able to specify a possible world (by a necessary and sufficient condition) one knows all the truths holding in this world.

How to make this surprising consequence comprehensible? In order to specify a world by a necessary and sufficient condition, this condition has to guarantee that the world can be distinguished from all other possible worlds at hand of this condition. But this can only be done by a condition that comprises all truths of this world (metaphorically such a necessary and sufficient condition for the obtaining of w can be thought of as something like a big, maybe endless, conjunction of all truths of w). Because, if the condition β would not determine for a certain (contingent) proposition α whether it is true or false in the possible world to be specified, there would be two metaphysically possible worlds meeting the condition β, and the crucial proposition α being true in one world and false in the other.

Williamson's argument shows that up to now the sequence of symbols $\Diamond KA\alpha$ (or $\Diamond K^*\alpha$ in *S5**, respectively) has not yet been given a sensible interpretation. Because we still lack a convincing answer to the question how the non-actual knowing subject might even express its knowledge: By α and $A\alpha$ it refers to its own non-actual world, and the expression of its knowledge with a construction like "in w, α" (where w designates the real world) fails, because such a knowledge reduces to trivial logical knowledge, an unacceptable consequence. Williamson draws the conclusion that as long as no justifiable conception of non-actual knowledge of $A\alpha$ (which also has to determine how the non-actual knowing subject can express its knowledge) has been given the scheme (1′) (or (1*) respectively) "should be treated as uninterpreted formalism" (Williamson 1987, 261).

4. A New Conception of Possible Knowledge

First, let's come back to the standard understanding of (ART) via thesis (1). What (1) requires is that if α is true in the real world there has to be a possible alternative w' in which α is also true and someone knows it (see Figure 2).

Concerning necessary *a priori* knowable truths this picture is perfectly acceptable, because then the α in the one possible world and the α in the other possible world don't differ with respect to what is required from the respective inhabitants of the corresponding worlds in order to know α.[35] Things look different when we are concerned with contingent truths (or necessary, but only *a posteriori* knowable truths), because then the knowing subjects have to know something (for example via experience) that might be different between the two possible worlds.

An example: Take α to be "The President of the United States is male". In our real world in order to know α one has to have had experiences that are somehow

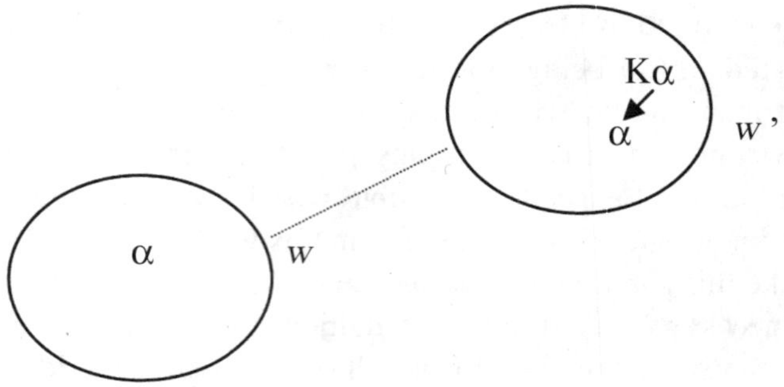

Figure 2.

related to George Bush (to know α is knowing something of Bush), whereas in another possible world, in which Al Gore would have become president, one needs experiences that are somehow related to Al Gore in order to know α (to know α is to know something of Gore).

That's the deeper reason why we rejected schema (1) in the beginning and replaced it with schema (1*): It is not at all clear that someone in one possible world having knowledge which he expresses by α and someone in another possible world having knowledge which he expresses by α, know 'the same'. As the example above shows, their knowledge might be of different objects, a very good reason for assuming that they are not knowing 'the same'.

On the other hand there was a problem with (1*), too. As Williamson's argument shows, the schema can't be understood in a way, such that what is required is the knowledge that α is true in another possible world (see Figure 3).

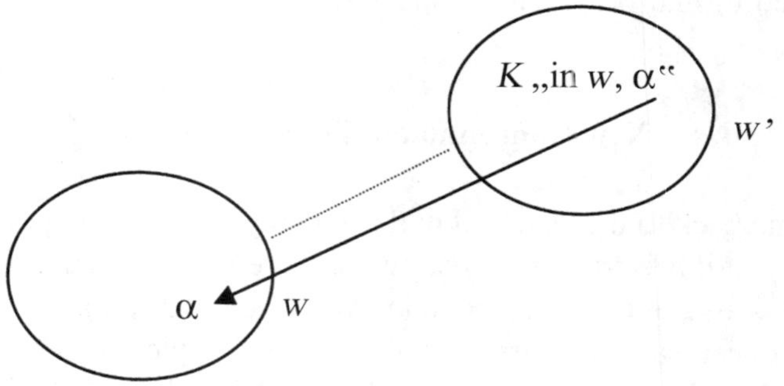

Figure 3.

What we are still in need of is a conception of how to understand non-actual knowledge of α, where α refers to the actual world. The main idea of the proposal I want to make can be seen by means of the following picture: a non-actual knowing subject knows α (α referring to the real world), if it has knowledge (about its own world) that it can express by β and α (with respect to w) and β (with respect to w')

express 'the same'. To determine more exactly what it is that is 'the same' we need some more preparations (see Figure 4).

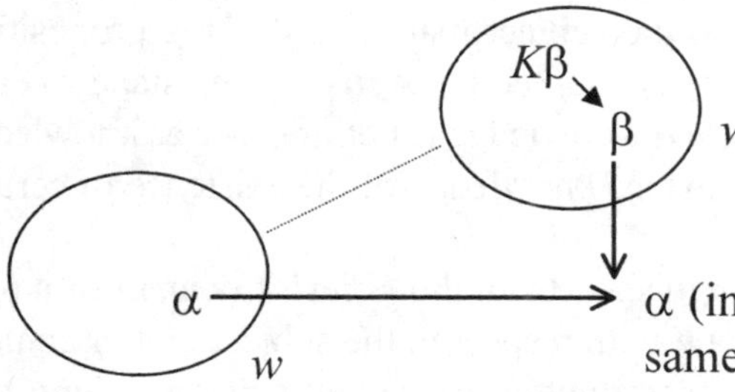

Figure 4.

4.1. KNOWLEDGE DE RE AND KNOWLEDGE DE DICTO

Suppose that the American basketball fan Bill knows that in the game between the Dallas Mavericks and the Portland Trailblazers Dirk Nowitzki was the top-scorer. In fact, Dirk Nowitzki is the best German player in the NBA. But, Bill does not know that (let's say, Bill thinks that Nowitzki is a Russian, because of the name, and thus still considers Detlef Schrempf to be the only German NBA-player, even if Schrempf has retired meanwhile and Shawn Bradley got a German passport: well, Bill is kind of an uninformed Basketball fan). What about Bill's knowledge that the best German NBA-player scored the most in that game? In one sense of the word he has that knowledge because he knows that Nowitzki has been the highest scorer and Nowitzki is in fact identical with the best German NBA-player. But in another sense, Bill does not have the knowledge in question (ask him, whether the best German player has scored most, and his answer will be "No"), because he does not know that the proper name "Dirk Nowitzki" and the definite description "the best German NBA-player" denote the same person.

Knowledge in the first sense is called *de re* knowledge, knowledge in the second sense *de dicto* knowledge.[36] If we apply this terminology to our example, it follows that Bill has the *de dicto* knowledge that Dirk Nowitzki was the top-scorer but he has not the *de dicto* knowledge that the best German NBA-player was the top-scorer. But, Bill has *de re* knowledge of the best NBA-player (who is Dirk Nowitzki) to have been the top-scorer.

In order to determine more exactly the notions of *de dicto* and *de re* knowledge, let's consider first only knowledge concerning elementary propositions. Take an elementary proposition of the form $Fa_1, a_2, \ldots, a_n$ containing a predicate F and n singular terms $a_1, a_2, \ldots, a_n$. We will apply some terminological proposals from Perry. He defines the *subject matter* of such a proposition as follows:

> The **subject matter** of a sentence. This is the objects (or conditions) designated by the terms in the sentence, and the condition designated by the condition word in the sentence. (Perry 2001b, 147)[37]

Thus, the subject matter of $Fa_1, a_2, \ldots, a_n$ is the set containing the relation designated by F and the objects named by $a_1, a_2, \ldots, a_n$. Now, we can develop the concept of *subject matter content*. The subject matter content of a proposition says what has to be the case with respect to the subject matter so that the proposition is true. Namely in our case, that the objects named by $a_1, a_2, \ldots, a_n$ stand to each other in the relation designated by F. *De re* knowledge is nothing else as knowledge of the subject matter content, and this means knowledge of the existence of certain states of affairs or facts.[38]

The subject matter content is distinguished from the *reflexive content* of a proposition by Perry.[39] While it is irrelevant with respect to the subject matter content *how* for example the objects of the subject matter are named (the only thing that matters is which objects are named), this relation between language and world is decisive for the reflexive content. Because it determines the truth-conditions concerning the relation between language and world that have to be fulfilled so that the proposition is true. The reflexive content of for example $Fa_1, a_2, \ldots, a_n$ reads as follows: "The expression 'a_1' names an object d_1, the expression 'a_2' names an object $d_2, \ldots$, the expression 'a_n' names an object d_n, and 'F' designates a relation R in such a way that the objects $d_1, d_2, \ldots, d_n$ stand in the relation R."

De re knowledge can be conceived of as knowledge of the subject matter content of a proposition and *de dicto* knowledge can be conceived of as knowledge of the reflexive content of a proposition. But what about the relation between *de re* and *de dicto* knowledge? A person has *de re* knowledge that α if and only if there is a β having the same subject matter content as α and the person has the *de dicto* knowledge that β. A person has no *de re* knowledge that α if and only if there is no such β.

A few words remain to be said regarding *de re* and *de dicto* knowledge concerning complex propositions. In order to avoid problematic entities like complex states of affairs we simply define *de re* knowledge as above via *de dicto* knowledge, and the subject matter content of complex propositions is determined in the usual way via the subject matter contents of their contained elementary propositions. For example, the subject matter content of a conjunction $\alpha \wedge \beta$ says that concerning the subject matter of α it has to be as α says and concerning the subject matter of β it is to be as β says for the whole conjunction to be true.

4.2. Quantifiers

It is common usage in logic to work with unrestricted quantifiers, that range over the whole universe of discourse, only. This is so because a formula containing a restricted quantifier ranging over a certain specific domain D is always logically equivalent to a formula containing an unrestricted quantifier instead, for example:

$$\forall_{x \in D} \alpha \leftrightarrow \forall_x (x \in D \rightarrow \alpha)$$

But for our purpose of defining *de re* knowledge in terms of *de dicto* knowledge of propositions containing quantifiers it is necessary to use restricted quantifiers again. The subject matter of a general proposition of the form $\forall_{x \in D}\alpha$ is constituted by the subject matter of α plus the set that is the domain of the quantifier, here D. And the subject matter content of $\forall_{x \in D}\alpha$ says that for every object $d \in D$ named by τ it has to be true that $\alpha_{[\tau/x]}$.[40]

To sum up our conception of *de re* knowledge we can say that it is build by the construction of equivalence classes of *de dicto* knowledges, and a certain equivalence class is constituted of those *de dicto* knowledges that are indifferent with respect to the subject matter content.

Examples: If Bill knows that Dirk Nowitzki was the top-scorer and if I know that the best German basketball player was the top-scorer, two different *de dicto* knowledges are involved; but nevertheless we have the same *de re* knowledge. With ordinary language examples concerning knowledge of quantified propositions we have to be careful:

All philosophy students passed their logic exam. (E)

Suppose that Bill knows (E) in the *de dicto* sense. Then it depends on our logical analysis of (E) which *de re* knowledge we ascribe to Bill. If we analyse (E) as $\forall_{Sx} Px$ (with Sx standing for "x is a philosophy student" and Px abbreviating "x passed his exam") we attribute to Bill that he knows *de re* of all philosophy students that they passed their exam, but if we analyse (E) as $\forall_x(Sx \rightarrow Px)$ we ascribe to Bill the *de re* knowledge of everything that if it is a philosophy student it passed its exam. The first alternative (as in almost all similar cases) seems to be the more natural one.

4.3. Same de re Knowledge in Different Possible Worlds

The question arises what kind of knowledge is at stake in Fitch's Paradox. As already argued above, to claim that two persons in two different possible worlds know 'the same' if they have the same *de dicto* knowledge is a little bit problematic because it might be that their knowledge is even of different objects (as in the Bush/Gore-example). I propose that the knowledge operator in Fitch's Paradox should be understood as expressing *de re* knowledge (with respect to *a posteriori* truths):

$$\alpha \rightarrow \Diamond k_{de\ re}{}^{*}\alpha \qquad (1^*)'$$

Finally, a very simple example will illustrate the conception and at the same time give a hint how Fitch's Paradox is now resolved: Suppose that there are only two objects in the real world w, namely the two persons Tom and Bob, and that both are stupid (to be stupid means here to know nothing at all). Thus, in w the following proposition is true:

All are stupid. (α)

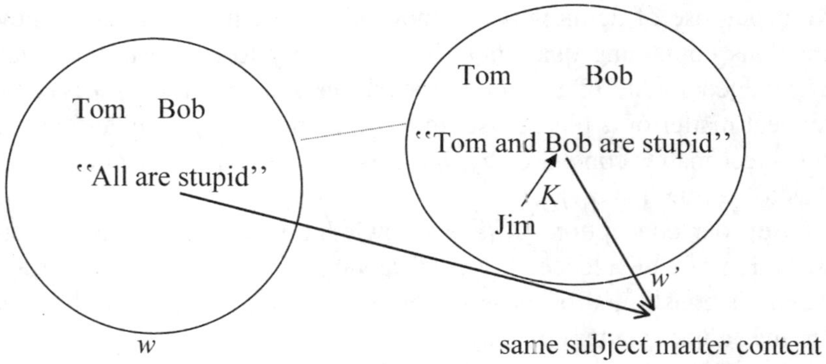

Figure 5.

Certainly, α isn't known in w, because Tom and Bob are both stupid, and thus know nothing at all. And there is also no possible alternative in which α is knowable *de dicto*. Nevertheless there is a possible world in which somebody knows *de re* what α expresses with respect to the real world, namely that all inhabitants of w (Tom and Bob) are stupid. Take for example the world w' with its only three objects Tom, Bob and Jim. Tom and Bob, again, are stupid, but Jim is not, because he knows (at least) that Tom and Bob are stupid. For example he might express his knowledge by "All but me are stupid" or "Tom and Bob are stupid". Thus, Jim in world w' has the *de re* knowledge that α (α referring to the real world), even if he himself is not able to express his knowledge by α (see Figure 5).

5. Conclusion

Aim of this paper has not been to answer the question whether all truths might possibly be known. Whether the thesis is correct cannot be decided given the arguments presented here, alone. But I hope to have shown that Fitch's argument is not able to decide the question against (ART), either. The position of soft anti-realism stays untouched by Fitch's Paradox if the anti-realist is willed to give (ART) the reading (1*) within *S5** (or (1′) within modal logic with *actuality*-operator, respectively) and to understand the kind of knowledge involved in the thesis as *de re* knowledge in the sense developed here in order to deal with non-actual knowledge of α (α referring to the actual world).[41]

Notes

[1] This citation only talks about the meaning of a mathematical proposition, but a little later Dummett withdraws from this restriction:

> The argument involved only certain considerations within the theory of meaning of a high degree of generality, and could, therefore, just as well have been applied to any statements whatever, in whatever area of language. (Dummett 1978, 226)

[2] The epistemic constraint that every truth must be knowable under adequate (counterfactual) circumstances may be too strong and should be replaced by a weaker one claiming that for every truth there must possibly be good reasons to believe it or that there must possibly be accessible evidences for every truth. But with respect to the following arguments the differences between stronger and weaker formulations of the anti-realistic thesis in this sense should be of no relevance.

[3] Here, as in the following, for the sake of simplicity the anti-realistic thesis is not formulated as a necessity claim, which could easily be done by adding a necessity operator: $\Box(\alpha \to \Diamond K\alpha)$. Even if such a modal rendering is adequate (for the anti-realist the epistemic constraint of the concept of truth is a necessary one), the chosen simplification should have no effect on the following arguments.

[4] The argument really originates with a referee of an unpublished paper by Fitch in 1945. Thus the name "Fitch's Paradox" does not seem to be fully justified, but the real originator is unknown (at least to me).

[5] It should be conceded that there exist certainly lots of full-blooded realists who consider already the anti-realistic thesis to be (almost) absurd, and who take Fitch's Paradox only as a further confirmation of their anyhow existing opinion.

[6] A short note on other approaches: (Wansing 2002) proposes to use a modal epistemic logic based on Nelson's constructive propositional logic with strong, constructive negation. In such a logic *modus tollens* fails and thus the step from (5) to (6) in the derivation of Fitch's Paradox is no longer justified.

(Dummett 2001) argues that the anti-realistic thesis should be restricted to basic statements, because the property of knowability is not transmitted from atomic to complex formulas: Even if α is knowable and if β is knowable $\alpha \wedge \beta$ doesn't need to be so.

Proposals with some relation to the one presented later on in this paper (although not the same) can be found in (Kvanvig 1995), (Lindström 1997) and (Rabinowicz and Segerberg 1994).

A number of issues coming up in my paper are discussed in Williamson (2000a, 270–301).

[7] Sure, there might be several non-human subjects which, so to say, could distribute the knowledge of all truths among each other. But then, according to a result by (Humberstone 1985) the class of knowing subjects needs to be infinite, because if for a finite group of individuals every truth is known by someone in the group, then there is someone in the group who knows every truth. And thus, in the case of an finite group, again there would be at least one omniscient subject.

[8] Probably some theists might insist that God actually is a (possible) member of our communication community (think of prayers and the like). But I can't and I don't want to engage in such a debate here.

[9] Thus, (Usberti 1995) proposes to restrict thesis (1) as referring only to mathematical propositions. Then it is no more possible that the knowledge operator K appears in α, and thus constructions as the one used in the derivation of Fitch's Paradox are automatically excluded. But, a restriction of its theses to mathematics is no longer in the spirit of modern anti-realism, a position which tries to generalise ideas that had been developed in the field of the philosophy of mathematics to other areas. The drastic restriction proposed by Usberti concerning the range of α in thesis (1) amounts, in the end, to give up the position of anti-realism.

[10] Williamson, the most eager defender of Fitch's Paradox against anti-realistic proposals, seems to think that the possibilities arising from a replacement of classical logic by intuitionistic logic may be the only rescue for the anti-realistic position:

> How [...] might a verificationist escape from Fitch's argument? One way would be to substitute intuitionistic for classical logic. It may even be the only way. (Williamson 1987, 261)

[11] The more general conception of the intuitionistic conditional saying that the assertion of a conditional $\alpha \to \beta$ is justified whenever the justified assertability of α entails the justified assertability of β, leads to similar consequences.

A detailed discussion of the relevant consequences arising from a further conception of the intuitionistic conditional – namely the one which says that in an intuitionistic conditional $\alpha \to \beta$ a prove of α has to be transformable into a proof of β (cf. for example Dummett (1977, 12–13) – can be found in Williamson (1988, 429–431).

[12] For example, it could be argued that the domain of the general quantifier of the semantic condition for the intuitionistic conditional does not necessarily have to coincide with the domain of the existential quantifier implicit in the K-operator (ranging over the members of the linguistic community). Because then, knowledge of α would not automatically entail knowledge of $K\alpha$, because under these circumstances $K\alpha$ even could be false.

[13] Concerning the question how the position of moderately hard anti-realism can be further developed, see the proposals made in Williamson (1982, 1988, 1992).

[14] It should be mentioned that within the framework of intuitionistic logic there have been made other proposals dealing with restrictions of thesis (1). For example, (Tennant 1997), using his intuitionistic relevance logic IR, proposes to restrict the range of thesis (1) to 'cartesian' propositions. And a proposition α is cartesian if and only if no contradiction follows from $K\alpha$. (Tennant 2001) answers the critics of (Hand and Kvanvig 1999) that the proposed restriction being *ad hoc*, but (Williamson 2000b) sharp-wittedly shows that Tennant's proposal does not succeed in avoiding the problematic consequences it wants to get rid of.

[15] For the sake of simplicity we suppose that the sentences in question are not necessary truths (or falsities, respectively). Of course, concerning necessary truths (or falsities, respectively) it can be supposed that they are true (or false, respectively) at all points of time.

[16] Explanation: It is not possible that $t_1 > t_0$ because t_0 is the present point in time and thus in case of $t_1 > t_0$ the antecedent would lack a truth-value. On the other hand, $t_0 > t_1$ isn't possible either, because the formula inside the parentheses refers to t_0, and thus it does not have a truth-value before t_0.

[17] Because if t_2 would lie in the future $K_{t_2}(\alpha_{t_0} \wedge \neg K_{t_0}\alpha_{t_0})_{t_1}$ would not have a truth-value yet.

[18] That means, the accessibility relation R is such that $w_n R w_m$.

[19] Such *S4*-structures as the one defined by (C1), (C2) and (C3) seem to be very natural when dealing with temporal issues.

[20] For a detailed, but in parts rather difficult, analysis of this thesis as well as of further questions concerning the relation between truth-values and time-parameters that arise from an anti-realistic point of view, see the chapter *Anti-Realism, Timeless Truth and Nineteen Eighty-Four* in Wright (1993, 176–203). A classical source for the discussion of these and similar problems is the paper *The Reality of the Past* in Dummett (1978, 358–374).

[21] The discussion of Melia's analyses and arguments in the main text is maybe too goodwilling, because the formulations in the two citations are very problematic. On the one hand, I think, Melia makes a mistake when he demands that one only has to *try* to find out the truth-value of certain propositions to cause a 'change' of it. This demand certainly is too weak because in the problematic cases (as for example the one with the Fitch formula) the truth-value necessarily 'changes' only if these efforts are successful. But, what should it mean to 'find out' something that is 'changed' by this 'finding out' itself? It seems to me that the talk of the 'change' of truth-values in this context doesn't amount to more than the fact that the same (contingent) proposition may have two different truth-values with respect to two different possible worlds (but that's trivial). And as there are propositions, as is shown by Fitch's Paradox, that cannot be known when they are true, their truth-value can only be known in worlds in which they are false.

[22] A proof that it is impossible to show the falsity of (13) by pure logical means can be provided using modal and epistemic possible worlds models in which some world w is modally accessible from every world but only w is epistemically accessible from w.

[23] Naturally, it is not excluded that there might be philosophical (non-logical) arguments that might favour a replacement of (ART) by (LART), or even philosophical (non-logical) arguments that might discredit the philosophical position of anti-realism altogether.

[24] In ordinary language, the semantic role that is typically played by the subjunctive mood is sometimes articulated by other linguistic means. For a discussion of this point see Wehmeier (2003a).
[25] For a detailed analysis of the example see Wehmeier (2003a, 7-8). Sure, that there are propositions that cannot be expressed within standard modal logic is well-known for quite a long time (for example, see Hazen (1976) and Crossley and Humberstone (1977)).
[26] The standard way to deal with the expressibility problem is to introduce a so-called *actuality*-operator. For a discussion of the relation between Wehmeier's proposal and modal logic with *actuality*-operator, see below.
[27] Should the formal language be enriched by supplementary operators, there has to be an indicative as well as a subjunctive version also of those, as for example of the knowledge operator (K and K^*). It is possible to work with several subjunctive markers in order to distinguish different binding relations in encapsulated modal contexts (but such subtleties are of no relevance for the aim of this paper). Furthermore, Wehmeier limits his discussion to *S5**, in which all possible worlds are accessible to each other. For corresponding versions of other modal logic systems that pose weaker conditions on the accessibility relation R (as for example T^* or *S4**), there have to be indicative and subjunctive versions of the modal operators themselves.
[28] One might wonder why the solution proposed to block the derivation of Fitch's Paradox has been presented first in the new framework of *S5** and not in the framework of the much better known modal logic with *actuality*-operator. But, I consider Wehmeier's modal logic with subjunctive marker to be the better and more natural formalism compared to modal logic with actuality-operator for the reasons given later on and in Wehmeier (2003a).
[29] As with Fitch's Paradox, also here the identity of the first one to have made this proposal for a solution is unclear. Because, in a footnote Edgington reports that Lloyd Humberstone had independently reached essentially the same solution, and Williamson (1987) writes in the first footnote that he had been confronted with such a proposal already in 1982 by an anonymous referee.

At the same time Humberstone has published a paper (cf. Humberstone (1982)), in which he suggests a modal logic, which respects the semantically relevant differences between the indicative and the subjunctive mood by the help of a subjunctive operator. A discussion of the similarities and differences between this modal logic and his own can be found in Wehmeier (2003a, 12).

Finally, it should be mentioned that I discovered the presented blocking of the derivation of Fitch's Paradox in *S5** before having been acquainted with Edgington's paper and the subsequent literature.
[30] See for example Sorensen (1988) and Wright (1993).
[31] The argument and the analogy between necessity propositions and general propositions is inspired by Wehmeier (2003a, 24–25). But, I think, his further thesis that propositions of the form $\Box(a = b)$, or $\Box(a =^* b)$ in *S5** respectively, are no real necessity statements, and thus identity statements are no potential candidates for necessary *a posteriori* truths, is misleading because in the scope of the modal operator there is a predicate that might be in subjunctive mood. That the difference between the indicative and the subjunctive mood does not really matter here depends on the peculiarities of the identity relation, I think, and not on the logical form of the proposition.
[32] Here, I suppose that knowledge is principally mediated by language and that it is part of the truth-conditions of a proposition like "S knows α" that for the subject S it is at least in principle possible to express this knowledge from his perspective within his language.
[33] Those acquainted with so-called *hybrid logic* (cf. for example Blackburn (2001) and Blackburn et al. (2001)) might think that using this kind of logic solves the problem of the specification of other possible worlds straightforwardly, because hybrid logic is essentially modal logic enriched with so-called *nominals*, formulas that are true in exactly one possible world and thus name possible worlds in a certain sense. The problem with this idea is that the hybrid logic used by the logician or philosopher in order to model counterfactual situations and modal discourse from an outside perspective is not necessarily available to the non-actual knowing subject, too. The knowing subject should be able to express its knowledge with the means available to it, mainly ordinary language, and I doubt whether

there is something in ordinary language corresponding to the nominals of hybrid logic (at least when hybrid logic is used to deal with metaphysical modality).

34 The s in the citation corresponds to our w, the p to our α, and the q to our β.

35 With respect to necessary *a priori* truths the anti-realistic thesis seems to be much less problematic, and maybe even uncontroversial. Thus, we will not consider such truths anymore in the rest of the paper.

36 The distinction between *de re* and *de dicto* knowledge and thus between *de re* and *de dicto* beliefs is heavily discussed in the literature. Often it is thought of to be a genuine distinction: a certain belief is either *de re* or *de dicto* (see for example Burge (1977)). I will not use the distinction that way. According to my conception of the distinction it is a matter of different perspectives: From the perspective of the knowing person each knowledge has to be formulated in a certain way, and thus each knowledge is *de dicto* in the first place. *De re* knowledge is a construction from an outside perspective, if one abstracts from how the objects, properties and relations the knowledge is about are designated in the *de dicto* knowledge. Thus, no *de re* knowledge without *de dicto* knowledge. True, there are problems if indexicals as for example "here" are involved because then what is known depends on the context. But, I think, this does not speak immediately against the conception that assumes that all knowledge is in the first place *de dicto* (for an attempt to deal with these difficulties see McDowell (1984)). Because these problems with indexicals are of no importance for the aim of this paper, I will assume for the following that no indexicals are involved. A classical source for the discussion of the difficulties that might evolve concerning the relation between *de re* and *de dicto* beliefs and indexicals is Perry (1979).

37 I have to thank Albert Newen (Bonn) for sending me Perry's paper already before its publication.

38 Instead of applying Perry's conception of subject matter content in order to specify *de re* knowledge it would have been possible to work with so-called *Russellian propositions*, too (cf. for example Barwise and Etchemendy (1987)).

39 The concept of reflexive content was introduced in Barwise and Perry (1983). (Perry 2001a) makes use of it to refute several counter-arguments against the thesis of the identity between the physical and the mental.

40 Similar considerations hold in the case of existential quantifiers.

41 I understand this paper as a tentative, initial presentation of a new idea how to solve Fitch's Paradox of knowability. Certainly, some conceptions have to be examined in much more detail, especially the exact way of constructing equivalence classes out of items of *de dicto* knowledge in order to determine the *de re* knowledge in the case of the involvement of quantifiers. I have to thank Ulrich Nortmann (Saarbrücken), Shahid Rahman (Lille) and Kai Wehmeier (Irvine) for stimulating discussions. Mark Siebel (Leipzig), John Symons (El Paso), Heinrich Wansing (Dresden) and Timothy Williamson (Oxford) read an earlier version and provided many criticisms and remarks that lead to an improvement of the paper. The main ideas of this paper were presented at the philosophical colloquium of the University of Saarbrücken (December 2000) and at the *Logic and Logical Philosophy 2001* workshop at Dresden (March 2001).

References

Aristotle: 1975, *Categories and De Interpretatione*, translated with notes by J. L. Ackrill, Oxford, Clarendon Press.

Barwise, J. and J. Etchemendy: 1987, *The Liar. An Essay on Truth and Circularity*, New York, Oxford, Oxford University Press.

Barwise, J. and J. Perry: 1983, *Situations and Attitudes*, Cambridge, Bradford-MIT

Berkeley, G.: 1710, *Treatise Concerning the Principles of Human Knowledge*, Dublin.

Blackburn, P.: 2001, 'Modal Logic as Dialogical Logic', in S. Rahman and H. Rückert (eds.), *New Perspectives in Dialogical Logic, Synthese* **127**(1/2), 57–93.

Blackburn, P., M. de Rijke and Y. Venema (eds.): 2001, *Modal Logic*, Cambridge, Cambridge University Press.

Burge, T.: 1977, 'Belief *de re*', *Journal of Philosophy* **74**, 317–338.

Crossley, J. and L. Humberstone: 1977, 'The Logic of "Actually" ', *Reports on Mathematical Logic* **8**, 11–29.

Davies, M. and L. Humberstone: 1980, 'Two Notions of Necessity', *Philosophical Studies* **38**, 1–30.

Dummett, M.: 1977, *Elements of Intuitionism*, Oxford, Clarendon Press.

Dummett, M.: 1978, *Truth and Other Enigmas*, London, Duckworth.

Dummett, M.: 2001, 'Victor's Error', *Analysis* **61**, 1–2.

Edgington, D.: 1985, 'The Paradox of Knowability', *Mind* **94**, 557–568.

Fitch, F.: 1963, 'A Logical Analysis of Some Value Concepts', *Journal of Symbolic Logic* **28**, 135–142.

Hand, M. and J. Kvanvig: 1999, 'Tennant on Knowability', *Australasian Journal of Philsopophy*, **77**, 422–428.

Hart, W.: 1979, 'The Epistemology of Abstract Objects', *Proceedings of the Aristotelian Society Supplementary Volume*, pp. 164–165.

Hazen, A.: 1976, 'Expressive Completeness in Modal Logic', *Journal of Philosophical Logic* **5**, 25–46.

Humberstone, L.: 1982, 'Scope and Subjunctivity', *Philosophia* **12**, 99–126.

Humberstone, L.: 1985, 'The Formalities of Collective Omniscience', *Philosophical Studies* **48**, 401–423.

Kripke, S.: 1980, *Naming and Necessity*, Cambridge, Harvard University Press.

Kvanvig, J.: 1995, 'The Knowability Paradox and the Prospects for Anti-Realism', *Noûs* **29**, 481–500.

Lindström, S.: 1997, 'Situations, Truth and Knowability: A Situation-Theoretic Analysis of a Paradox by Fitch', in E. Ejerhed and S. Lindström (eds.), *Logic, Action and Cognition: Essays in Philosophical Logic*, Dordrecht, Kluwer, pp. 183–209.

McDowell, J.: 1984, '*De re* senses', *Philosophical Quarterly* **34**, 283–294.

Melia, J.: 1991, 'Anti-Realism Untouched', *Mind* **100**, 341–342.

Perry, J.: 1979, 'The Problem of the Essential Indexical', *Noûs* **13**, 3–21.

Perry, J.: 2001a, *Knowledge, Possibility and Consciousness: the 1999 Jean Nicod Lectures*, MIT Press.

Perry, J.: 2001b, 'Frege on Identity, Cognitive Value, and Subject Matter', in A. Newen, U. Nortmann and R. Stuhlmann-Laeisz (eds.), *Building on Frege. New Essays about Sense, Content and Concept*, CSLI publications, pp. 141–158.

Rabinowicz, W. and K. Segerberg: 1994, 'Actual Truth, Possible Knowledge', *Topoi* **13**, 101–115.

Sorensen, R.: 1988, *Blindspots*, Oxford, Oxford University Press.

Tennant, N.: 1997, *The Taming of the True*, Oxford, Clarendon Press.

Tennant, N.: 2001, 'Is every Truth Knowable? Reply to Hand and Kvanvig', *Australasian Journal of Philosophy* **79**, 107–113.

Usberti, G.: 1995, *Significato e Conoscenza: Per una Critica del Neoverificazionismo*, Milano, Guerini Scientifica.

Van Fraassen, B.: 1966, 'Singular Terms, Truthvalue Gaps and Free Logic', *Journal of Philosophy* **63**, 481–494.

Wansing, H.: 2002, 'Diamonds are a Philosopher's Best Friends. The Knowability Paradox and Modal Epistemic Relevance Logic', *Journal of Philosophical Logic* **31**, 591–612

Wehmeier, K.: 2003a, 'Modality, Mood, and Descriptions', unpublished manuscript, R. Kahle (ed.), *Intensionality – An Interdisciplinary Discussion*, AK Peters (Lecture Notes in Logic).

Wehmeier, K.: 2003b, 'World Travelling and Mood Swings', in B. Löwe, T. Räsch and W. Malzkorn (eds.), *Foundations of the Formal Sciences II*, Kluwer (Trends in Logic).
Williamson, T.: 1987, 'On the Paradox of Knowability', *Mind* **96**, 256–261.
Williamson, T.: 1988, 'Knowability and Constructivism', *The Philosophical Quarterly* **38**, 422–432.
Williamson, T.: 1992, 'On Intuitionistic Modal Epistemic Logic', *Journal of Philosophical Logic* **21**, 63–89.
Williamson, T.: 2000a, *Knowledge and its Limits*, Oxford University Press.
Williamson, T.: 2000b, 'Tennant on Knowable Truth', *Ratio*, **XIII**(2), 99–114.
Wright, C.: 1993, *Realism, Meaning and Truth*, 2nd edn, Oxford, Blackwell.

THEORIES OF KNOWLEDGE AND IGNORANCE

WIEBE VAN DER HOEK[1], JAN JASPARS[2] and ELIAS THIJSSE[3]
[1] *Computer Science, The University of Liverpool, United Kingdom,*
E-mail: wiebe@csc.liv.ac.uk; [2] *Free Lance Logician, Amsterdam, The Netherlands,*
E-mail: jaspars@science.uva.nl; [3] *Computational Linguistics & AI, Tilburg University, Tilburg, The Netherlands, E-mail:* thysse@uvt.nl

Abstract. What does it mean to say that an agent only knows a particular fact, i.e., knowing that fact and not more than that? The problem of describing so-called minimal knowledge has been discussed in the literature since 1985, when Halpern and Moses published a paper on Knowledge and Ignorance. The present chapter reviews a considerable part of the most important proposals for only knowing and provides a number of generalizations over these proposals. The focus of this study is on a theoretical understanding of the subject, but several (possible) applications are indicated too.

The proposals for solving the problem of minimal knowledge vary along some dimensions. Most of these proposals are restricted to a single agent, whereas a few deal with the multi-agent case. Also, most deal with the problem of minimal knowledge on the level of meta-language, by formulating inferential conditions, semantic constraints on verifying models, or rules for establishing belief sets; a few suggest an explicit operator for only knowing in the object language. Moreover, the majority of proposals employ specific modal systems which, e.g., point out whether the agent is (fully) introspective, i.e., knows that she knows (and knows that and what she does not know); the authors, however, suggest and discuss general modal approaches too. Finally, the advantages of 'going partial' (using models that may leave the truth value of certain propositions undefined) are demonstrated.

1. Introduction

Imagine a rich uncle, Scroogy, who owns a huge safe in his office. He adopts the following policy regarding security of this safe:

> I know that I do not trust private access to my office by non-relatives. As to relatives, as long I do not know that they are trust-worthy, I do not trust them. I moreover know the facts that Brigitta is not a relative, where Donald is.

Surely, we can derive that Scroogy knows he does not trust Brigitta alone in his office. What about Donald? We are tempted to conclude that Scroogy does not trust him either, but this does not follow from our premisses. For that, we have to stipulate that the knowledge we ascribed to Scroogy above is *all that* Scroogy knows, so that, in particular, we can derive that Scroogy *does not know* that Donald is trust-worthy.

This mirrors a common phenomenon in default reasoning, and, more in general, nonmonotonic reasoning: some conclusions (I do not trust Donald) are drawn on the basis of ignorance of certain preconditions (I do not know that this relative is

S. Rahman et al. (eds.), Logic, Epistemology, and the Unity of Science, pp. 381–418.

trust-worthy), and this ignorance is often not explicitly stored. Thus, there should be some way to derive this ignorance from a knowledge base, or from a set of statements about the knowledge and ignorance of the subject under consideration. Thus, there should be some way to treat a set of sentences as 'all that is known'. This appears to be not a trivial problem. For instance, when formalizing the last sentence of the premisses with $\Box_S$ denoting 'Scroogy knows that', one has a choice between $\Box_S(\neg r_B \wedge r_D)$ and $(\Box_S \neg r_B \wedge \Box_S r_D)$, which, for all the logics that we consider in this paper, are equivalent. The notion of 'only knowing' should also respect this equivalence: even if S claims to 'only know $(\neg r_B \wedge r_D)$', it should follow that he also knows $\neg r_B$.

To further the example, suppose that Scroogy decides to have two keys k_1 and k_2 for the safe: in order to gain access, both keys are needed. Scroogy reveals the first key to Donald which we, slightly abusing notation, denote as $\Box_D k_1$: Donald knows what the key k_1 is. He intends to give the second key to Gladstone Gander, but at that very moment Scroogy does remember that he gave one of the keys to Donald, but does not recall which one, which we could abbreviate as: $O_S(\Box_D k_1 \vee \Box_D k_2)$, where O_S denotes 'all that S knows is that'. Focussing on the identity of the keys, this assertion perfectly makes sense, but one could compare this to the statement that Donald looks for an excuse not to reveal the key k_1 to his nephews, and he declares, upon their continuous bagging, $O_D(\Box_D k_1 \vee \Box_D k_2)$: "I only know that I know the first key or that I know the second". Obviously, his nephews will note the absurdity of this claim.

To finish the story, suppose Scroogy gets back to his senses, and gives the second key to Gladstone Gander: $\Box_D k_1 \wedge \Box_G k_2$. Since Scroogy distributed the keys, we also have $\Box_S(\Box_D k_1 \wedge \Box_G k_2)$. Our uncle, planning his first holiday trip, has asked one of his nephews, Huey, to guard his office. He freezes at the moment Huey tells him, without hesitation, that he knows the secret k_1, after which $\Box_S \Box_H k_1$. Would Scroogy go on holiday, without concern? Definitely not, he wants to ensure $\Box_S(\Box_H k_1 \wedge \neg\Box_H k_2)$. And even that may not be enough: he may want to ensure himself that Huey does not know that Gladstone Gander knows k_2: Scroogy wants to convert the situation in which $\Box_S \Box_H k_1$ into one in which $\Box_S O_H k_1$. More realistically, Scroogy's aim would be to circumscribe Huey's knowledge in some formula φ_H such that $\Box_H \psi$ iff ψ somehow follows from φ_H, and with this φ_H at his disposal, he can plan his strategy.

This story, in a nutshell, exemplifies the study of this paper: given some knowledge of an agent, we want to apply some kind of closure on it in order to derive what the agent does not know, and we may want to use an explicit operator for only knowing for this. Also, the story demonstrates that not for all sentences ψ, it makes sense to claim that one 'only knows ψ'. In other words, this paper also investigates so-called *honest formulas*: formulas that are candidates for a complete description of an agent's knowledge. The story already indicated that this can become rather involved, even for the one-agent case: whereas for Donald, it makes sense to claim that he only knows (the value of) k_1 or k_2, written as $O_D(k_1 \vee k_2)$, it does not make sense for him to state that $\Box_D k_1 \vee \Box_D k_2$ is all he knows, since from it we could infer that the Donald does not know k_1: $\neg\Box_D k_1$, and in the same fashion, we would infer that he does not know that k_2: $\neg\Box_D k_2$. However, these two simple inferences together contradict Donald's initial knowledge. This demonstrates the dishonesty of the formula $\Box_D k_1 \vee \Box_D k_2$.

The relevance of the problem of circumscribing knowledge is hopefully clear by now. A satisfactory logical analysis of 'only knowing' is of essential importance for knowledge representation and inference, in analogy with closed world assumptions in the field of database theory and logic programming. Formulas which can be 'only known', so-called *honest* formulas, can be used as a precise description of an agent's state of knowledge. Moreover, drawing 'common sense' inferences from a knowledge base, describing the agent's knowledge, often relies on the assumption that this knowledge base represents the agent's only knowledge.

Another important example is the quantity maxim in the logic of conversation (Grice 1975). This principle of cooperation roughly says that the information conveyed by utterances should be the only information that a speaker has about the subject. The quantity maxim explains the anomaly of a speaker saying e.g., "I know whether it is raining" in a dialogue about the weather. For this conversational maxim would lead to the inference that the speaker only knows $\Box p \vee \Box\neg p$, which is impossible since this formula is dishonest (cf. the earlier formula describing Donald's knowledge of the keys). In fact Grice's analysis goes beyond this, since *if* a speaker utters such a sentence, the hearer can infer the apparent violation of the quantity maxim, and since the speaker also knows this, the intention of the seemingly self-defeating message may be to signal unwillingness to exchange the information to the hearer.

There are many situations where knowledge of other people's ignorance is crucial in strategic decision making: in security and authorization protocols, one typically wants to derive that intruders to the protocol *do not know* certain sensitive information. Also, one needs a way to circumscribe the agents' knowledge, when verifying the claim that one agent 'knows *more than*' another. But also in the setting of a single agent, we want to compare his different states of knowledge – over time, or during a sequence of moves in a game, for example. A central issue here is to obtain a formal description of the agent's knowledge containing *not more* than the information conveyed by some formula φ, that is, the case in which φ is the agent's *only* knowledge. In many games then, finding moves that maximize the mover's information state and at the same time minimize the opponents' information state, is at stake. As a final example, recently (cf. van Ditmarsch (1999)), there is an interest in *modelling particular (card-)games*, where the initial situation of such a game is assigned one particular model. This presupposes that there is some 'minimal information state', and that the agents know the initial description, *and not more*.

The rest of this survey paper is organized as follows. In Section 2 we give some technical preliminaries, fixing the basic modal language, systems and models.

Then, in Section 3, we present systems that study minimal knowledge in the one agent case. In its first subsection, Section 3.1, we present systems that formalize an explicit operator 'O' for only knowing in the object language, suitable to do default-like reasoning as in our example. In subsequent subsections, we do not assume to have this operator in the language: in Section 3.2, we present several approaches, all for different logical systems, to only knowing in the one agent case. Section 4 resumes the enterprise of Section 3 for many knowers. All these approaches thus far have in common that 'knowing little' semantically corresponds with large models, representing 'many possibilities'. In Section 5 we show how moving to a partial modal logic in many ways gives a more natural approach to finding 'minimal models' for knowledge. Finally, in Section 6 we round off.

2. Technical Preliminaries

Language. Our basic language $\mathcal{L}$ is that of modal logic, with the constant $\top$ (true), the modal operator $\Box$ (necessity, here interpreted as 'knowledge') and the connectives $\neg$ (negation), $\wedge$ (conjunction) and $\vee$ (disjunction). Unless stated otherwise, we assume that formulas φ and ψ in $\mathcal{L}$ are composed from a finite set of propositional atoms $\mathcal{P} = \{p, q, r \ldots\}$, using these operators. Other operators are introduced as abbreviations: $\Diamond$ (possibility) is defined by $\Diamond\varphi = \neg\Box\neg\varphi$, $\rightarrow$ (implication) by $\varphi \rightarrow \psi = \neg\varphi \vee \psi$, $\leftrightarrow$ (equivalence) by $\varphi \leftrightarrow \psi = (\varphi \wedge \psi) \vee \neg(\varphi \vee \psi)$ and $\bot$ (false) by $\bot = \neg\top$. This general format only needs a little expansion for multi-modal systems. One expansion is for bimodal systems (cf. Section 3.1) where $\Box$ is accompanied by a second, similar 'necessity-type' operator; another is for many agents systems where one finds a family of operators $\Box_i$ indexed for the agent. For the classical (i.e., not partial) systems studied in the following two sections, we can define $\top$ as $p \vee \neg p$. A formula without modal operators is called *objective*; if every atomic subformula is in the scope of a $\Box$-operator, the formula is called *subjective*.

The function $d : \mathcal{L} \rightarrow \mathbb{N}$ calculates the *modal depth* of formulas as follows: $d(p) = d(\top) = d(\bot) = 0\ (p \in \mathcal{P})$, $d(\neg\varphi) = d(\varphi)$, $d(\varphi \star \psi) = max\{d(\varphi), d(\psi)\}$ for $\star = \wedge, \vee, \rightarrow, \leftrightarrow$, and, finally, $d(\Box\varphi) = d(\Diamond\varphi) = 1 + d(\varphi)$. Some properties to be presented are relative to a given subset $\mathcal{L}' \subseteq \mathcal{L}$. An example of such a sublanguage of $\mathcal{L}$ is $\mathcal{L}_{(n)} = \{\varphi \in \mathcal{L} \mid d(\varphi) \leq n\}$. For a unary operator $\triangle = \neg, \Diamond, \Box$ and language $\mathcal{L}' \subseteq \mathcal{L}$, $\triangle\mathcal{L}' = \{\triangle\varphi \mid \varphi \in \mathcal{L}'\}$. We also use an 'inverse' $\triangle^{-}\mathcal{L}'$ denoting $\{\alpha \mid \triangle\alpha \in \mathcal{L}'\}$.

Semantics. We use Kripke models $M = \langle W, R, V\rangle$ for a standard interpretation of the modal language, where W is the set of worlds, accessibility R is a relation on W (or a family of such relations in the multi-modal case) and V is a mapping from W to the set of assignments or propositional valuations ($\{0, 1\}^{\mathcal{P}}$). Instead of wRv, we also write $v \in R[w]$, where w and v denote worlds in M. Here, the key notion is the pair $\langle M, w\rangle$ (often written as M, w), called a *state*, in which each modal formula φ receives its standard interpretation with the typical modal case: $M, w \models \Box\varphi$ iff for all $v \in R[w]$, one has $M, v \models \varphi$. For any state $\langle M, w\rangle$, its theory is $\mathrm{Th}(M, w) = \{\varphi \mid M, w \models \varphi\}$. For $\Gamma \subseteq \mathcal{L}$, $M, w \models \Gamma$ ('$\langle M, w\rangle$ verifies Γ') means that $\Gamma \subseteq \mathrm{Th}(M, w)$. Relative to a given set of models $\mathcal{S}$, consequence is defined by $\Gamma \models_{\mathcal{S}} \varphi$ iff for all $M \in \mathcal{S}$, $w \in W$: $M, w \models \Gamma$ implies $M, w \models \varphi$.

A major issue in this paper is how truth is preserved by moving from one state to another. Let $\mathcal{L}^* \subseteq \mathcal{L}$ and let S be a set of states. We say that an order $\leq$ on S *preserves* the sublanguage $\mathcal{L}^*$, or that $\mathcal{L}^*$ is *persistent* over $\leq$ iff for all states $\langle M, w\rangle$ and $\langle M', w'\rangle$ in S:

$$M, w \leq M', w' \;\Rightarrow\; \text{for all } \varphi \in \mathcal{L}^* : (M, w \models \varphi \Rightarrow M', w' \models \varphi).$$

If the overall converse holds, we say that $\mathcal{L}^*$ *characterizes* $\leq$ on S.

Inference and Modal Systems. We discuss several logical systems **S** on top of the minimal modal system **K**. For a set of premises Γ, we write $\Gamma \vdash_{\mathbf{S}} \varphi$ if there is a derivation of φ (without applications of necessitation to the premises) from Γ in **S**. When **S** is clear from context, or when the particular normal system is not relevant, we will simply drop it. For a single premise we omit parentheses, i.e., $\varphi \vdash \psi$ means $\{\varphi\} \vdash \psi$; for derivability without premises we write $\vdash \varphi$ instead of $\emptyset \vdash \varphi$. The set $\mathrm{Cn}_{\mathbf{S}}(\varphi) = \{\psi \mid \varphi \vdash_{\mathbf{S}} \psi\}$ contains all formulas that can be derived from φ in **S**.

The formulas φ and ψ are *equivalent in* **S**, or **S**-*equivalent*, if both $\varphi \vdash_{\mathbf{S}} \psi$ and $\psi \vdash_{\mathbf{S}} \varphi$. The logic **S** is called *finitary* for a sublanguage $\mathcal{L}^*$ if it induces finitely many **S**-equivalence classes in $\mathcal{L}^*$. Notice that such an $\mathcal{L}^*$ is not in general finite. As an immediate consequence of our assumption that $\mathcal{P}$ is finite, we have that every **S** we will consider is *finitary layered*, in the sense that for each $n \in \mathbb{N}$, **S** is finitary for $\mathcal{L}_{(n)}$.

The minimal system **K** contains the rule of necessitation ($\vdash \varphi \;\Rightarrow \vdash \Box\varphi$) and the modal axiom K: $\Box(\varphi \rightarrow \psi) \rightarrow (\Box\varphi \rightarrow \Box\psi)$ on top of any Hilbert-style axiomatization of propositional logic. Normal systems **S** are obtained by adding modal axioms to **K**. One such extension, **T**, is obtained by adding the axiom T: $\Box\varphi \rightarrow \varphi$ to **K**. The system **KD** is named after the axiom that distinguishes it from **K**, which is D: $\Diamond\top$ or, equivalently, $\Box\varphi \rightarrow \Diamond\varphi$. In epistemic logic, systems that include axiom 4: $\Box\varphi \rightarrow \Box\Box\varphi$ are called *positively introspective,* those that have axiom 5: $\neg\Box\varphi \rightarrow \Box\neg\Box\varphi$ are called *negatively introspective.* Two other axioms worth mentioning are B: $\varphi \rightarrow \Box\Diamond\varphi$ and G: $\Diamond\Box\varphi \rightarrow \Box\Diamond\varphi$. Any combination of the axioms mentioned is called an *epistemic logic*, here. Typical examples of such systems are **S4** (**T** + 4), **S5** (**S4** + 5) and **S4.2** (**S4** + G). When $\Box$ is interpreted as belief, the axiom T is often replaced by D, which gives rise to systems that obtain their name directly from the constituting axioms: **KD**, **KD4**, **KD45**, etc.

Each of the systems just mentioned and most of those discussed here is a *Geach logic* (Chellas 1980), i.e., a normal system which contains (in addition to **K**) axioms of the form:

$$\Diamond^k \Box^l \varphi \rightarrow \Box^m \Diamond^n \varphi \text{ with } k, l, m, n \in \mathbb{N}, \tag{1}$$

where $\Diamond^k$ is defined recursively: $\Diamond^0 \varphi = \varphi$ and $\Diamond^{k+1} \varphi = \Diamond \Diamond^k \varphi$. For example, **KD45** has three Geach axioms: D ($k = m = 0, l = n = 1$), 4 ($k = n = 0, l = 1, m = 2$) and 5 ($k = m = n = 1, l = 0$).

Given a logic **S**, a set of formulas Γ is **S**-*consistent* if $\Gamma \nvdash_{\mathbf{S}} \bot$. We say that $\Gamma \subseteq \mathcal{L}$ satisfies the **S**-*disjunction property* (**S**-DP) over a sublanguage $\mathcal{L}^*$, if Γ is **S**-consistent and for every $\psi_1, \psi_2, \ldots, \psi_k \in \mathcal{L}^*$:

$$\Gamma \vdash_{\mathbf{S}} (\psi_1 \vee \cdots \vee \psi_k) \;\Rightarrow\; \text{for some } i \leq k : \Gamma \vdash_{\mathbf{S}} \psi_i.$$

A state verifies a logic **S** if it verifies all the theorems of **S**, i.e., $\{\varphi \mid \vdash_{\mathbf{S}} \varphi\}$. The set of states verifying **S** (**S**-states for short) is called STATE$_{\mathbf{S}}$. An **S**-model is a model of which all states verify the logic **S**.

Completeness and Stability. A set of formulas Γ is *maximal* **S**-*consistent* (**S**-m.c.) if it is **S**-*consistent* and moreover contains all the formulas φ for which $\Gamma \cup \{\varphi\}$ is consistent. Such sets are used for proving *completeness* of **S** w.r.t. a class of models $\mathcal{S}$: $\Gamma \models_{\mathcal{S}} \varphi \Rightarrow \Gamma \vdash_{\mathbf{S}} \varphi$; the converse is called *soundness*.

Many classes of models $\mathcal{S}$ have been identified that are sound and complete with respect to the systems **S** that were mentioned above, see Chellas (1980). Most significant are those classes of models that are determined by a property of their accessibility relation. For instance, for **KD** one takes the serial (i.e., $\forall x \exists y\, xRy$) Kripke models, for **T** the accessibility relation has to be reflexive, in **KD4** it is serial and transitive, and in **S5** it is an equivalence relation. In general, the Geach axiom (1) corresponds to a confluency condition: $\forall x y y' : x R^k y \,\&\, x R^m y' \Rightarrow \exists z (y R^l z \,\&\, y' R^n z)$.

Maximal consistent sets also play an important role in characterizing minimal knowledge. Following Jaspars (1991), we generalize the notion of a *stable set* as introduced in Moore (1985) to identify the knowledge contained in an m.c. set: Σ is stable if $\Sigma = \Box^- \Gamma$ for some **S**-m.c. Γ. If $\varphi \in \Sigma$ and Σ is stable, Σ is called a *stable expansion* of φ.

3. One Agent Minimal Knowledge

3.1. FORMALIZING ONLY KNOWING WITH ONE AGENT

In Levesque (1990), Levesque introduces an explicit operator O for only knowing in the object language. His motivation for this is to obtain a logic for knowledge or belief that can be useful in consistency-based approaches to non-monotonic reasoning. Logics of belief are relevant for consistency-based nonmonotonicity, by considering a 'current theory' as a set of beliefs, and interpreting the consistency test for a sentence and a theory as the question whether the negation of that sentence is disbelieved. For instance, the default that birds typically fly then relates to the belief that any bird that is not believed to be flightless, can fly. However, taking this route, one still has to give an account on deriving such disbeliefs, which is not provided by ordinary modal doxastic or epistemic logics. To repeat an example from Levesque (1990), consider the premisses

Tweety is a bird (a)
If a bird can be consistently believed to fly, it flies (b)

From these two premisses, one cannot conclude that Tweety flies (c). In order to derive (c), we need the prerequisite of (b):

It can be consistenly believed that Tweety flies (p)

However, there seems to be no justification for (p). Clearly, such a justification is not given by (a), rather, it is the fact that, besides (b), there *is* nothing believed else than (a):

This is *all* that is known (about Tweety) (d)

Since now (p) seems to follow from (a) and (d), we can conclude eventually that Tweety flies, using (a), (d) and (b).

The problem that is addressed in Levesque (1990) now is to express a property like (d) within the object language, rather than on a meta-level characterizing sets of beliefs (stable expansions) where (d) intuitively might be said to hold. As far as we know, Levesque's approach is also the only one that gives an account of only knowing in the context of first order logic, rather than just propositional logic. Although Levesque uses the notions of 'knowledge' and 'belief' interchangeably, we will, in the formal analysis, restrict ourselves to the use of the verb 'to know'.

A System for Knowledge. In order to do so, we need two modal operators: $\Box$ and O, where $\Box\alpha$ means 'α is known', and $O\alpha$ that 'α is *all* that is known', or '*only* α is known'. The full language for only knowing, $\mathcal{ONL}(\Phi)$, is then obtained by adding to the basic language predicate symbols of any arity from an infinite set Φ, quantifiers and the special two-place identity symbol, as well as an infinite number of *standard names* $n_1, n_2, \ldots$. Sentences without O operators are called *basic*. Also, α^x_n denotes α, with the name n substituted for the variable x. We first give the semantics of basic sentences:

DEFINITION 3.1. An *assignment* w is a function from atomic sentences to $\{0, 1\}$. Let W be a set of assignments. Then:

$W, w \models_{\text{Le}} \varphi$	iff	$w(\varphi) = 1$	for atomic φ
$W, w \models_{\text{Le}} (n_i = n_j)$	iff	$i = j$	
$W, w \models_{\text{Le}} \neg\alpha$	iff	$W, w \not\models_{\text{Le}} \alpha$	
$W, w \models_{\text{Le}} \alpha \wedge \beta$	iff	$W, w \models_{\text{Le}} \alpha$ and $W, w \models_{\text{Le}} \beta$	
$W, w \models_{\text{Le}} \exists x\alpha$	iff	$W, w \models_{\text{Le}} \alpha^x_n$	for some n
$W, w \models_{\text{Le}} \Box\alpha$	iff	$W, w' \models_{\text{Le}} \alpha$	for every $w' \in W$

The logic for knowledge thus obtained is 'weak **S5**' which is **K45**. Moreover, we have the Barcan formula $\forall x \Box \alpha \rightarrow \Box \forall x \alpha$. In the case of subjective sentences φ, we often write $W \models_{\text{Le}} \varphi$.

Only Knowing. To motivate the definition of the operator O, one has to put the intuition to work that the more (objective) formulas are known, the smaller the set W must be, and vice versa. Saying that one *at least* knows α, $\Box\alpha$, then amounts to requiring that $W \subseteq [\![\alpha]\!]$, where $[\![\alpha]\!] = \{w \mid W, w \models_{\text{Le}} \alpha\}$: one may in fact know more so that some of the worlds w that satisfy α or not present in W as well. Then, claiming that one *at most* knows α corresponds to $[\![\alpha]\!] \subseteq W$: one may in fact even know less than α so that also worlds that falsify α are present in W. The conclusion is that stating that one *exactly*, or *only* knows α comes down to requiring $[\![\alpha]\!] = W$. This gives us:

DEFINITION 3.2. (Truth Definition of Only Knowing)
$W, w \models_{\text{Le}} O\alpha$ iff for every w' : $w' \in W \Leftrightarrow W, w' \models_{\text{Le}} \alpha$

This definition relates $\Box$ and O in the following way:

$$W, w \models_{\text{Le}} O\alpha \quad \text{iff} \quad W, w \models_{\text{Le}} \Box\alpha \text{ and for every } w'\text{: if } W, w' \models_{\text{Le}} \alpha \text{ then } w' \in W \tag{2}$$

It may be illuminating to relate our motivational paragraph which explained *only knowing* in terms of knowing *at least* and *at most*, to the formal definition. For that, we need an operator A, where the intended meaning of $A\alpha$ is 'at most α is known'. The truth-definition for this operator is easily obtained: $W, w \models_{\text{Le}} A\alpha$ iff for every w', if $W, w' \models_{\text{Le}} \alpha$ then $w' \in W$. Instead of introducing this operator A directly, Levesque introduces $N\alpha = A\neg\alpha$ which he reads as 'α is at most known to be false', and which, by rewriting the above clause for A, has the following truth-definition:

$$W, w \models_{\text{Le}} N\alpha \quad \text{iff} \quad \text{for every } w'\text{, if } w' \notin W \text{ then } W, w' \models_{\text{Le}} \alpha \tag{3}$$

Writing $w \in W^c$ for $w \notin W$, one can perceive N as just another necessity operator defined with respect to another set of accessible worlds. This implies that the notions $\Box$ and N, and hence also A and O, can both be defined in other modal systems using ordinary accessibility relations $R_\Box$ for $\Box$ and R_N for N, with the structural requirement that $R_\Box = R_N^c$, which gives a straightforward axiomatization.

Before going into that, we have to make precise the notion of validity and satisfiability, which involves some technical detail. Let W be the set of assignments that make p true. Then we have $W \models_{\text{Le}} Op$ and $W \models_{\text{Le}} \Box p$. Now the issue is whether the truth of Op should be determined by all known and unknown basic sentences. Let

$$\Sigma(W) = \{\Box\alpha \mid \alpha \text{ is basic and } W \models_{\text{Le}} \Box\alpha\} \cup \{\neg\Box\alpha \mid \alpha \text{ is basic and } W \models_{\text{Le}} \neg\Box\alpha\}$$

and let $W^* = W \setminus \{w\}$, where w is an arbitrary assignment making p true. Then, given the fact that we have infinitely many atomic propositions (or predicates, in the first order case), it follows that $\Sigma(W) = \Sigma(W^*)$, but $W^* \not\models_{\text{Le}} Op$, so $\Sigma(W)$ would not imply Op in the usual sense! In other words, although the same basic formulas are known in W and W^*, they differ in the verification of only knowing p. One would expect that if one only knows p in W, but not in W^*, there should be some additional basic statement that is known in W^*, which is not the case. This is the reason that Levesque defines the essential semantic relations in terms of *maximal sets* of assignments. Let us call two sets of assignments W_1 and W_2 *equivalent* if for every basic α, $W_1 \models_{\text{Le}} \Box\alpha$ iff $W_2 \models_{\text{Le}} \Box\alpha$. Now, the following theorem justifies that we take a straightforward representative in every class:

THEOREM 3.3. For every set of assignments W there is a unique largest superset W^+ that is equivalent to W. (W^+ is called *maximal* then.)

Now, a set of formulas Γ is *satisfiable* if there is a *maximal* set W^+ and an assignment w such that $W^+, w \models_{\text{Le}} \gamma$ for every $\gamma \in \Gamma$. Moreover, Γ is said to *imply* φ (written $\Gamma \models_{\text{Le}} \varphi$), iff $\Gamma \cup \{\neg\varphi\}$ is not satisfiable, hence, iff for every maximal set W^+ and any assignment w: if $W^+, w \models_{\text{Le}} \Gamma$ then $W^+, w \models_{\text{Le}} \varphi$. Formula φ is called *valid* iff it is implied by the empty set.

Levesque notices that this restriction to maximal sets is harmless as long as we restrict ourselves to basic sentences: 'the satisfiable and valid sentences do not change' (Levesque 1990, 274). Although in using maximal sets the notion of semantic implication may seem to have some built-in circumscriptive features, it can be shown that, restricted to basic formulas, Levesque's notion of consequence amounts to the usual one, and therefore by itself does not result in nonmonotonic behaviour. In other words, to distinguish Levesque's consequence relation ($\models_{\text{Le}}$) from the usual consequence relation ($\models$) either one of the premises or the conclusion must be non-basic. This is exemplified in the earlier example, where $\Sigma(W) \not\models Op$ yet $\Sigma(W) \models_{\text{Le}} Op$. Finally, also note that $O\alpha$ is satisfiable and $\Box\alpha \wedge N\alpha$ a valid formula, for every basic sentence α that is valid.

DEFINITION 3.4. *(Axiom system* **Le***)*
Let axiom Fol denote that every instance of a theorem in FOL is also a theorem of **Le**. The system **Le** is then obtained from Fol and the axiom K for both $\Box$ and N, which also both satisfy necessitation (Nec). It adds to this the Barcan formula Bf, and the definition of O, O and, finally, Ign that allows one to derive statements about ignorance. In that rule, $\vdash_{\mathbf{Fol}}$ denotes FOL-derivability.

Int	$\sigma \to (\Box\sigma \wedge N\sigma)$, for subjective σ
Bf	$(\forall x\Box\alpha \to \Box\forall x\alpha) \wedge (\forall x N\alpha \to N\forall x\alpha)$
O	$O\alpha \equiv (\Box\alpha \wedge N\neg\alpha)$
Ign	if $\not\vdash_{\mathbf{Fol}} \varphi$, then $\vdash_{\mathbf{Le}} N\varphi \to \neg\Box\varphi$, for objective φ

Note that Int makes the system fully introspective: we not only have positive and negative introspection for both $\Box$ and N, but also properties like $\neg\Box\alpha \rightarrow N\neg\Box\alpha$. Mirroring an earlier observation, Ign can be rephrased as: an objective α is FOL-valid if $\Box\alpha \wedge N\alpha$ is satisfiable. In Levesque (1990), our rule Ign is phrased as an axiom. Levesque then claims that **Le** is sound with respect to the notion of validity that is defined above. He also provides a partial completeness result:

THEOREM 3.5. Any valid sentence without quantifiers is a theorem.

As the title of the paper *Levesque's Axiomatization of Only Knowing is Incomplete* (Halpern and Lakemeyer 1995) already suggests, **Le** is not complete for our notion of validity. Moreover, whereas Levesque already observed that his axiomatization is not recursively enumerable, in Halpern and Lakemeyer (1995) it is observed that a complete axiomatization of validity of sentences in $\mathcal{ONL}$ *cannot* be recursive, since, if there would be one, say **Ax**, then we could recursively enumerate all falsifiable objective formulas by just generating all objective formulas α for which $N\alpha \rightarrow \neg\Box\alpha$ is provable from **Ax**. However, we know that the set of all falsifiable objective formulas is not recursively enumerable.

Before linking Levesque's approach to meta-theoretical approaches, we first demonstrate his solution to the introductory Tweety example. Let γ be the following default rule

$$\forall x[(Bird(x) \wedge \neg\Box\neg Fly(x)) \rightarrow Fly(x)]$$

and let the knowledge base $\text{KB} = \{Bird(tweety)\}$. Then we can derive the following, using **Le**:

EXAMPLE 3.6.
(1) $O(\text{KB} \wedge \neg Fly(tweety) \wedge \gamma) \rightarrow \Box\neg Fly(tweety)$
(2) $O(\text{KB} \wedge Fly(tweety) \wedge \gamma) \rightarrow \Box Fly(tweety)$
(3) $O(\text{KB} \wedge \gamma) \rightarrow \Box Fly(tweety)$

Property (3) is obtained by first deriving $\Box(\text{KB} \wedge \gamma)$ from the antecedent, from which one concludes the 'normal default' $\neg\Box\neg Fly(tweety) \rightarrow \Box Fly(tweety)$ (a). The antecedent of (3) also yields $N(\text{KB} \rightarrow \neg\gamma)$, and hence $N(\text{KB} \rightarrow \exists x \neg Fly(x))$. Applying Ign to this yields $\neg\Box(\text{KB} \rightarrow \exists x\neg Fly(x))$, from which one concludes $\neg\Box\neg Fly(tweety)(b)$, and (a) and (b) give us (3).

Stable Sets and Expansions. In Levesque (1990), stable sets are generalized to the language including the operator O. A generalized stable set Γ is then a set of sentences that is closed under consequence, and the constraints $\alpha \in \Gamma \Rightarrow \Box\alpha \in \Gamma$ and $\alpha \notin \Gamma \Rightarrow \neg\Box\alpha \in \Gamma$. Alternatively, a generalized belief set for W is the set of sentences α for which $W \models_{\text{Le}} \alpha$. The notions appear to be the same: Γ is a generalized belief set iff it is a generalized stable set.

Where for stable sets in the propositional **K45** case it is known that they are determined by their objective content, in the quantified language this is not the case (Levesque 1990, Theorem 3.6). We finally mention the following result that links *stable expansions* to only knowing. The set Γ is stable expansion of a set of sentences Δ if it satisfies the following fixed-point equation:

$$\Gamma = \mathrm{Cn}_{\mathbf{Le}}(\Delta \cup \Box\Gamma \cup \neg\Box\Gamma^c)$$

Now, for any $\varphi \in \mathcal{ONL}$ and every maximal set of assignments W, $W \models_{\mathrm{Le}} O\varphi$ iff the generalized belief set of W is a generalized expansion of $\{\varphi\}$.

Finally, let us consider how this approach deals with dishonest formulas. What happens if we 'feed' a dishonest formula to the operator O? What are the consequences, if any, of $O(\Box p \vee \Box q)$? To determine whether this formula is satisfiable, we ask ourselves whether there is a maximal set W and an assignment w such that $M, w \models_{\mathrm{Le}} O(\Box p \vee \Box q)$. That is, a maximal W with the property that for all w': $w' \in W$ iff $W, w' \models_{\mathrm{Le}} (\Box p \vee \Box q)$. But we have $W, w' \models_{\mathrm{Le}} (\Box p \vee \Box q)$ iff for all v: $W, v \models_{\mathrm{Le}} (\Box p \vee \Box q)$, so that the only candidate for the maximal set is that of all assignments, which, on the other hand, does not make $(\Box p \vee \Box q)$ true. Thus, there is no largest W for $O(\Box p \vee \Box q)$, hence the formula is not satifiable, and we have $O(\Box p \vee \Box q) \models_{\mathrm{Le}} \alpha$, for every α.

Generalizing Levesque's Approach. The point of departure for Halpern and Lakemeyer (2001) is the truth-definition (3) as given earlier this section. This definition has the following important features:

- The set of possibilities W remains fixed in that definition. In fact, this guarantees the validity of axiom Int in Definition 3.4.
- Also, the set of *conceivable worlds* that is, the set W together with its complement, is fixed, and independent of the *situation* (W, w). This property corresponds to the rule Ign in the definition of **Le**.
- Finally, for every set of conceivable worlds, there is a model where that set is precisely the set of worlds that the agent considers possible. This property ensures that every objective formula can be "all you know".

The first insight that Halperns and Lakeymeyer observe is that the completeness result Theorem 3.5 only holds if the set of atoms in the language is infinite. The axiom system is not complete if this set is finite: consider the case where the only atoms are p and q. Then we would have $\models \neg\Box\neg(p \wedge \neg q) \rightarrow N\neg(p \wedge \neg q)$: if the agent considers a world $w_{p\wedge\neg q}$ possible in which $p \wedge \neg q$ holds, meaning that $w_{p\wedge\neg q} \in W$, then every world $v \in W^c$ must satisfy $\neg(p \wedge \neg q)$, and hence we have $N\neg(p \wedge \neg q)$. Two remedies are now offered in Halpern and Lakemeyer (2001) for this incompleteness of Levesque's logic: an alternative axiomatization that *is* complete w.r.t. Levesque's semantics for finite sets of atomic propositions, and also, an alternative semantics is provided, for which the original axiomatization **Le** is complete.

To demonstrate the former, let us denote the set of atomic propositions by Φ. If Φ is finite, we can identify a world with the literals it verifies, like we did for $w_{p\wedge\neg q}$. Then, for an objective formula α, we write $W_{\alpha,\Phi}$ for the set of assignments over Φ that satisfy α. The axiom system $\mathbf{Ax}(\Phi)$ is then obtained by replacing Ign in Levesque's system by

$$\mathsf{Ign}(\Phi) \quad \text{if } \not\vdash_{\mathbf{Fol}} \alpha, \text{ then } \vdash_{\mathbf{Ax}(\Phi)} N\alpha \leftrightarrow \bigwedge_{w\in W_{\neg\alpha,\Phi}} \neg\Box\neg w, \text{ for objective } \alpha$$

THEOREM 3.7. If Φ is finite, then for every $\varphi \in \mathcal{ONL}(\Phi)$, $\vdash_{\mathbf{Ax}(\Phi)} \varphi$ iff $\models_{\mathrm{Le}} \varphi$

Let us now discuss the alternative semantics for Levesque's language. This semantics takes the observation serious that we made after displaying (3) as truth-definition for N, which said that each of the weak **S5**-operators $\Box$ and N could be interpreted with respect to its own 'balloon' of worlds. To do this, Halpern and Lakemeyer introduce so-called *extended situations* $(W_\Box, W_N, w)$, where $W_\Box \cup W_N$ comprises the set of all truth assignments. This gives rise to a satisfaction relation $\models_{\mathrm{Ex}}$ which is exactly like $\models_{\mathrm{Le}}$, except for the modal formulas:

$$(W_\Box, W_N, w) \models_{\mathrm{Ex}} \Box\varphi \quad \text{iff} \quad (W_\Box, W_N, w') \models_{\mathrm{Ex}} \varphi \text{ for all } w' \in W_\Box$$
$$(W_\Box, W_N, w) \models_{\mathrm{Ex}} N\varphi \quad \text{iff} \quad (W_\Box, W_N, w') \models_{\mathrm{Ex}} \varphi \text{ for all } w' \in W_N$$

This semantics generalizes Levesque's $\models_{\mathrm{Le}}$ to the case of arbitrary Φ:

THEOREM 3.8. For all Φ, and all $\varphi \in \mathcal{ONL}(\Phi)$: $\vdash_{\mathbf{Le}} \varphi$ iff $\models_{\mathrm{Ex}} \varphi$

3.2. Particular Systems with Minimal Knowledge Models

We now move to approaches in which, 'only knowing' is not formalized using a special epistemic operator, i.e., it is not represented at the object level. It is fair to say that this meta-level approach is the dominant one among proposals dealing with the minimal knowledge problem. So 'only knowing' is characterized by basic 'knowing' on object level, combined with 'higher order' properties such as inferential conditions, extensions to certain infinite sets of formulas, and relations between models.

Halpern and Moses. The seminal paper in this area is Halpern and Moses (1985). Much of the terminology and methodology of the subject can be traced back to the work of Halpern and Moses (H&M, henceforth). As demonstrated in the introduction, key notions as *honesty* and *only knowing* were suggested by H&M. Here we focus on some of the formal aspects of their approach.

So, the language of this epistemic logic is that of unimodal logic where $\Box\varphi$ can be interpreted as 'φ is known'. The modal system of this logic is **S5**. Intuitively, a formula φ is *honest* if an agent can be in a knowledge state in which she can (sincerely) claim to only know φ. Formally, honesty can be characterized in a number of ways. It is shown that φ is honest iff one of the following equivalent conditions holds:

- φ can be extended to an **S5**-stable set whose propositional subset is (a unique) minimum;
- φ is preserved by union of universal models;
- the set of consequences of the recursively defined knowledge and ignorance of subformulas of $\Box\varphi$ is consistent;
- if $\Box\varphi \vdash \Box\alpha_1 \vee \cdots \vee \Box\alpha_n$ for propositional formulas $\alpha_1, \ldots, \alpha_n$, then for some i: $\Box\varphi \vdash \Box\alpha_i$ (i.e., $\Box\varphi$ satisfies the propositional disjunction property).

Schwarz and Truszczyński. In Schwarz and Truszczyński (1994), it is argued that **S5** and the resulting notions of honesty may not be ideal. Their main motivation lies in the fact that the H&M notions of honesty are not conservative with respect to adding *explicit definitions*. To appreciate this observation, first note that there is a natural generalisation from honest formulas to honest sets of formulas: a finite honest set $\mathcal{T}$ corresponds to the honest formula $\bigwedge \mathcal{T}$. In particular, $\emptyset$ ("zero knowledge" or "full ignorance") should be honest according to any standard. Yet, adding the definition $q \leftrightarrow \Box p$ renders the set H&M-dishonest, since $q \leftrightarrow \Box p$ is itself H&M-dishonest: e.g., $\Box(q \leftrightarrow \Box p) \vdash \Box(p \wedge q) \vee \Box\neg q$, whereas neither disjunct follows itself from the given knowledge.

Basically, Schwarz and Truszczyński (1994) blame the modal system for this unintuitive result. The same problem occurs in the 'weak **S5**' systems **K45** and **KD45**, so they are not of any help here. Schwarz and Truszczyński (S&T) therefore propose the system **S4F**, which is **S4** with axiom F added. Axiom scheme F can be expressed as $(\Diamond\varphi \wedge \Diamond\Box\psi) \rightarrow \Box(\Diamond\varphi \vee \psi)$. In the context of **S4**, it is possible to define the class of corresponding models as those with a simple structure, viz. the ones for which $W \subseteq 2^{\mathcal{P}}$, $W = W_1 \oplus W_2$ and $wRv \Leftrightarrow w \in W_1 \vee v \in W_2$; so W_1 and W_2 are equivalence *clusters* and (all of) W_2 is accessible from (all of) W_1. Therefore, where **S5**-models can be restricted to single clusters, for **S4F** we find double clusters. In fact **S5** models can be seen as **S4F** models with an empty first cluster. The bi-cluster models are now sufficiently rich to separate facts from knowledge, loosely speaking.

DEFINITION 3.9. *M is said to be a minimal-knowledge model for $\mathcal{T}$* iff

1. $M = \langle W, V\rangle$ is a universal **S5** model
2. $M \models \mathcal{T}$
3. for every **S4F** model $N = \langle W^*, W, V^*\rangle$ such that $V = V^*|_W$: if $N \models \mathcal{T}$ then for all $\alpha \in \mathcal{L}_{(0)}$, $M \models \alpha \Rightarrow N \models \alpha$

What seems less satisfying in this definition is that purely modeltheoretic criteria seem mixed with syntactic conditions. We hypothesize that the relation between M and N in the above definition might be expressed in structural terms, e.g., that the W^*-cluster of N is isomorphic to a submodel of M (which by itself reoccurs as the W-cluster of N). Let us use the definition above to characterize *honesty* in terms of minimal-knowledge models.

DEFINITION 3.10. Theory $\mathcal{T}$ is *honest* if it has a unique minimal-knowledge model.

Like Halpern and Moses, S&T now relate this (largely) modeltheoretic condition to a purely syntactic condition, involving stability. Surprisingly, not the expected **S4F**-stability features, but the old notion of **S5**-stability. In particular, the S&T stable sets are closed under 'negative introspection': if $\psi \notin \mathcal{T}$ then $\neg\Box\psi \in \mathcal{T}$. This is theoretically unsatisfactory, but it works.

THEOREM 3.11. M is a minimal-knowledge model for $\mathcal{T}$ iff $\mathcal{S} = \text{Th}(M)$ is an **S4F**-expansion of $\mathcal{T}$, i.e., $\mathcal{S} = \text{Cn}_{\mathbf{S4F}}(\mathcal{T} \cup \neg\Box[\mathcal{L} - \mathcal{S}])$.

One derives from this that a theory $\mathcal{T}$ is *honest* iff there is a unique stable expansion of $\mathcal{T}$. There are no counterparts of the subformula test and the disjunction property for **S4F** honesty in Schwarz and Truszczyński (1994). What this approach *does* have in common with other approaches, is that purely propositional $\mathcal{T}$ are S&T-honest. For standard negative cases, such as $\mathcal{T} = \{\Box p \vee \Box q\}$ the uniqueness condition enforces dishonesty. The interesting distinctive cases are disjunctions of facts and knowledge:

EXAMPLE 3.12. Let $\mathcal{T} = \{\Box p \vee q\}$, $\mathcal{P} = \{p, q\}$. This formula is H&M-dishonest since the propositional disjunction property and locomotive modal system gives: $\Box(\Box p \vee q) \vdash_{\mathbf{S5}} \Box p \vee \Box q$, yet $\Box(\Box p \vee q) \not\vdash_{\mathbf{S5}} \Box p$ and $\Box(\Box p \vee q) \not\vdash_{\mathbf{S5}} \Box q$. But the same formula *is* S&T-honest, since there are two maximal **S5** cluster models, viz. $M\ \boxed{p \bullet q \quad p \bullet \neg q}$ and $M'\ \boxed{p \bullet q \quad \neg p \bullet q}$. However, M does not qualify as a minimal-knowledge model, since $N\ \boxed{\neg p \bullet q} \longrightarrow \boxed{p \bullet q \quad p \bullet \neg q}$ is an **S4F** model disproving condition (3): $M \models p$, $N \not\models p$, whereas $N \models \mathcal{T}$. Inspection of the 16 possible start clusters of the **S4F**-extensions of M' shows that if such an extension globally verifies $\mathcal{T}$, propositional truth is preserved. Therefore M' is the unique minimal-knowledge model of $\Box p \vee q$.

By the same strategy, i.e., by adding worlds (different from those in the end cluster) to the start cluster which preserve the global truth of the theory (but not of some propositional formula), one can prove that the definition $q \leftrightarrow \Box p$ is S&T-honest. It is also shown that the set of consequences of stable sets with definitions added are conservative extensions of the original stable sets, proving transfer of expansions. Another bonus of this approach is that theories consisting of so-called *positively defined clauses*, i.e., formulas of the form $\Box\alpha_1 \wedge \cdots \wedge \Box\alpha_n \rightarrow \Box\alpha_0$ (with $\alpha_0, \ldots, \alpha_n \in \mathcal{L}_{(0)}$), are honest, as desired by Konolige in Konolige (1989). Moreover, **S4F** is in some sense the strongest system with this property.

THEOREM 3.13.

1. If $\mathcal{T}$ consists of positively defined clauses, then it has a unique minimal-knowledge model (i.e., $\mathcal{T}$ is **S4F** honest).
2. If **S** is a modal logic such that **S4F** $\subset$ **S**$\subseteq$ **S5** then there is a theory $\mathcal{T}$ cf positively defined clauses such that $\mathcal{T}$ has several **S**-expansions.

3.3. Arbitrary Systems with Minimal Knowledge Models

Let **S** be an arbitrary modal system and the information order $\leq$ a pre-order (i.e., a reflexive and transitive relation) on $\text{STATE}_{\mathbf{S}}$. A formula φ is called *honest* with respect to **S** and $\leq$, if there exists a *least* **S**-state verifying $\Box\varphi$. More precisely,

DEFINITION 3.14. A formula φ is **S**-*honest* (for $\leq$) iff there is an **S**-state M, w such that:

- $M, w \models \Box\varphi$,
- $M', w' \models \Box\varphi \Rightarrow M, w \leq M', w'$ for all $\langle M', w'\rangle \in \text{STATE}_{\mathbf{S}}$.

The next result relates three main approaches to honesty.

THEOREM 3.15. Let $\mathcal{L}^* \subseteq \mathcal{L}$ be persistent and characterizing for $\leq$. Then the following propositions are equivalent:

- φ is **S**-honest for $\leq$ (i.e., $\Box\varphi$ has a $\leq$-least verifying state)
- $\Box\varphi$ has an $\mathcal{L}^*$-smallest **S**-m.c. expansion
- $\Box\varphi$ has **S**-DP over $\mathcal{L}^*$

In terms of stability (cf. Section 1), the second condition for φ being honest can be rephrased as:

- φ has a $\Box^-\mathcal{L}^*$-smallest stable expansion

Although Theorem 3.15, together with the condition above, solves the problem of alternative characterizations of honesty in an abstract sense, the solution is not entirely satisfactory. Notice that Theorem 3.15 and the resulting minimal information equivalences are based on the assumption that $\mathcal{L}^*$ is both persistent and characterizing for $\leq$. It is unclear, up to this point, whether suitable orders *exist* that enable persistent sublanguages to characterize them. And, if so, we would like to specify them in an independent, insightful way. With this end in view van der Hoek et al. (1996) proposes several specific orders which we now briefly discuss.

As is argued in van der Hoek et al. (1996), unrestricted bisimulation (van Benthem 1983) turns out to be too strong for a proper information order: there are states M, v and M', v' with the same theory that do not bisimulate. Therefore, van der Hoek et al. (1996) propose to use Ehrenfeucht-Fraïssé orders, which are defined by means of underlying, 'layered' pre-orders. The idea is that two states verify the same modal formulas if and only if every finite path originating from one state has a corresponding path arising from the other.

DEFINITION 3.16. Suppose $\leq_n$ is a pre-order on STATE_S for each natural number n ('layer n'). We say that $\leq$ is *induced by* the family $\leq_n$, ($n \in \mathbb{N}$) iff

$$M, w \leq M', w' \Leftrightarrow \forall n \in \mathbb{N}\, \forall v' \in R'[w']\, \exists v \in R[w] : M, v \leq_n M', v'$$

Now, let $\mathcal{L}^*$ be a sublanguage and $\mathcal{L}^*_{(n)} = \mathcal{L}^* \cap \mathcal{L}_{(n)}$ be its subset of formulas of modal depth up to n. The following lemma facilitates to define two important Ehrenfeucht-Fraïssé orders.

LEMMA 3.17. If $\mathcal{L}^*_{(n)}$ is persistent and characterizing for $\leq_n$, and $\mathcal{L}^*$ is closed under $\vee$, then $\Box\mathcal{L}^*$ is persistent and characterizing for $\leq$, i.e.,

$$M, w \leq M', w' \Leftrightarrow \forall \varphi \in \Box\mathcal{L}^* : (M, w \models \varphi \Rightarrow M', w' \models \varphi)$$

General Information Order. In the first Ehrenfeucht-Fraïssé order the underlying, layered order is an equivalence relation.

DEFINITION 3.18. The relation $\simeq_n$ is defined recursively by:
– $M, w \simeq_0 M', w' \Leftrightarrow V(w) = V'(w')$
– $$M, w \simeq_{n+1} M', w' \Leftrightarrow \begin{cases} M, w \simeq_n M', w' \;\& \\ \forall v' \in R'[w']\, \exists v \in R[w] : M, v \simeq_n M', v' \;\; (back) \;\& \\ \forall v \in R[w]\, \exists v' \in R'[w'] : M, v \simeq_n M', v' \;\; (forth) \end{cases}$$

Then it can be shown that two states are $\simeq_n$-equivalent iff they verify the same formulas up to depth n. Now, the *general information order* $\sqsubseteq$ between two states $\langle M, w\rangle$ and $\langle M', w'\rangle$ is defined as being induced by $\simeq_n$. Since $\mathcal{L}_{(n)}$ is persistent and characterizing for $\simeq_n$, Lemma 3.17 shows that $\Box\mathcal{L}$ is both persistent and characterizing for $\sqsubseteq$, and hence, from Theorem 3.15, we know that, for any **S**, the minimal information equivalences hold for $\sqsubseteq$ and $\Box\mathcal{L}$. From now on, let us call honesty with respect to $\sqsubseteq$ *general honesty*.

Positive Information Order. The second order is based on a genuine stratified pre-order. The relation $\preceq_n$ is defined as $\simeq_n$ in Definition 3.18, but without the forth-clause. Furthermore, let the *positive information order* $\preceq$ be the order that is induced by $\preceq_n$. In terms of knowledge, the order preserves positive knowledge. Technically this amounts to persistence of formulas that do not contain boxes in the scope of a negation:

$$\mathcal{L}^+ = \{\varphi \in \mathcal{L} \mid \varphi \text{ contains no } \Box \text{ in the scope of } \neg\}$$

In other words, formulas in $\mathcal{L}^+$ contain no negative occurrence of $\Box$; therefore $\Diamond$ is also not allowed in a positive knowledge formulas. Thus $\mathcal{L}^+$ amounts to the closure under $\wedge$, $\vee$ and $\Box$ of propositional formulas. So $\Box p \vee \Box q$, $\Box\neg p$ and $\Box p \wedge \neg q$ are members of $\mathcal{L}^+$, but $\neg\Box p$ and $\Diamond p \vee \Box q$ are not. Let $\mathcal{L}^+_{(n)}$ be $\mathcal{L}^+ \cap \mathcal{L}_{(n)}$.

THEOREM 3.19. For any **S**, the minimal information equivalences hold for $\preceq$ and $\Box\mathcal{L}^+$.

Let us call honesty with respect to $\preceq$ *positive honesty*. A simple, yet important case of the positive information order is the submodel relation. $\langle M', w\rangle$ is a submodel of $\langle M, w\rangle$ iff $W' \subseteq W$, $R' = R \cap W' \times W'$ and $V'(u) = V(u)$ for all $u \in W'$. A modal formula is preserved under submodels iff it is equivalent to a 'positive knowledge' formula, i.e., a formula in $\mathcal{L}^+$. Since $\Box\mathcal{L}^+ \subseteq \mathcal{L}^+$ this implies that $\supseteq$ preserves $\mathcal{L}^+$ (1) and if $\langle M', w\rangle$ is a submodel of $\langle M, w\rangle$, then $M, w \preceq M', w$ (2). Since the submodel relation is easily established, this provides a convenient tool for proving that two models are related by the positive information order. In Halpern and Moses (1985) the submodel relation has been used for the definition of minimal models for the system **S5**.

3.4. Evaluating Minimality in Epistemic Systems

We now briefly evaluate the two information orders introduced before in the light of our broad class of epistemic systems, as introduced in Section 2. In fact, we will first state a negative result for a large class introduced in the preliminary section, that of *Geach logics*.

General Minimality. The general information order specifies that one world is smaller than another world if and only if the first world represents less knowledge than the second. It turns out that for most systems this order is not appropriate. In most Geach logics this order trivializes the notion of general honesty: either all formulas are honest, or (nearly) all formulas are dishonest.

For weaker modal systems such as **K**, **K4**, **KD** and **KD4**, which may be used for belief, it can be proved by a simple model-theoretic technique that all formulas φ such that $\Box\varphi$ is consistent are generally honest. This technique is called *simple amalgamation*, which is a slight modification of what is called 'amalgamation' in Hughes and Cresswell (1984). For two **S**-states a simple amalgamation is constructed by adding one world from which all worlds are accessible which are accessible from the original two states. The construction is depicted in Figure 1. For every formula φ we obtain: $(M, w \models \Box\varphi \;\&\; M', w' \models \Box\varphi) \Leftrightarrow M^*, w^* \models \Box\varphi$ (†).

Let $\mathcal{S}$ be a class of **S**-models which are characterized by the Geachean relational restrictions corresponding to **S**, and $\mathcal{S}$ be closed under amalgamation —which is the case for each of these weak modal systems. Then one can prove the **S**-disjunction property of any consistent $\Box\varphi$ over $\Box\mathcal{L}$, by using contraposition: Suppose that $\Box\varphi \not\vdash_{\mathbf{S}} \Box\psi_1$, $\Box\varphi \not\vdash_{\mathbf{S}} \Box\psi_2$, then we can find **S**-states with $M, w \models \Box\varphi \wedge \neg\Box\psi_1$ and $M', w' \models \Box\varphi \wedge \neg\Box\psi_2$. By (†) we know that $M^*, w^* \models \Box\varphi$, but obviously also $M^*, w^* \models \neg(\Box\psi_1 \vee \Box\psi_2)$. In other words, $\Box\varphi \not\vdash_{\mathbf{S}} \Box\psi_1 \vee \Box\psi_2$.

For many Geach logics the notion of general honesty deflates seriously. Such systems often contain theorems of the form $\Box\varphi_1 \vee \Box\varphi_2$ where $\Box\varphi_1$ and $\Box\varphi_2$ are not theorems (*). In particular, systems which incorporate Geach axioms with $k, m > 0$

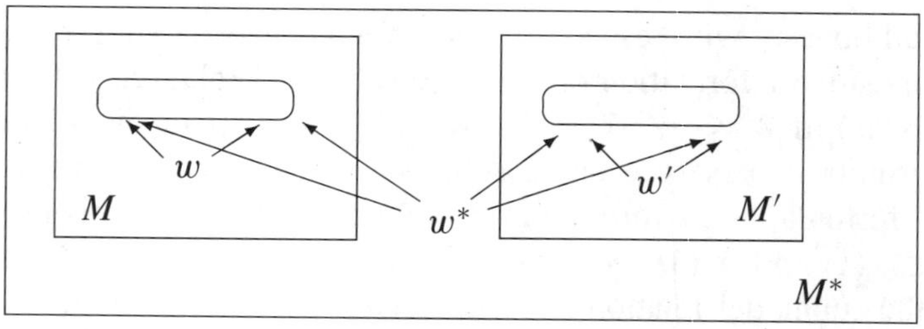

Figure 1. A simple amalgamation of the states $\langle M, w\rangle$ and $\langle M', w'\rangle$.

have the property (*). The disjunction property is then easily violated. In fact, bearing in mind that epistemic logics are defined here to be systems extending **K** with a selection of T, D, 4, 5, B and G, we know from van der Hoek et al. (1996), Theorem 15, that for any epistemic logic **S** except for **T**, **S4**, and the weak systems **K**, **KD**, **K4**, and **KD4**, no **S**-consistent formula is generally honest. Examples of systems without consistent generally honest formulas are **K45**, **KD45** and **S5**, but also those with a milder form of negative introspection such as **S4.2**. The only generally honest formulas in these systems are those φ that are **S**-inconsistent whereas $\Box\varphi$ is **S**-consistent.

Positive Minimality. As we have seen the systems **T** and **S4** permit non-trivial generally honest and dishonest formulas. Even then, there are good reasons to question the feasibility of the notion of general honesty. To begin with, general honesty cannot serve as a paramount notion of honesty, since in many systems this notion trivializes, as we have seen above. Moreover, for epistemic purposes it seems intuitively more sound to exclude formulas which represent ignorance, i.e., formulas of the form $\neg\Box\varphi$, when it comes to minimizing knowledge.

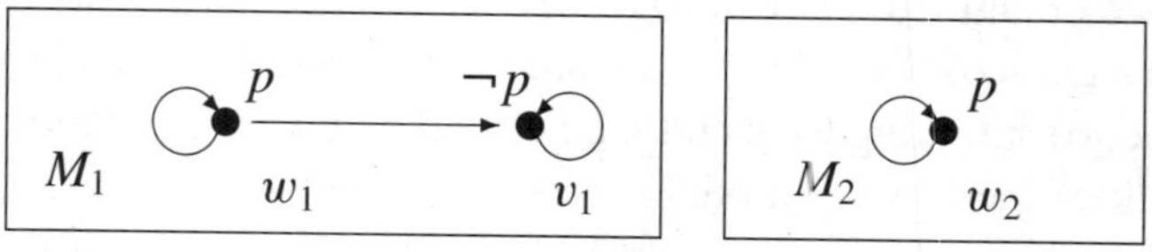

Figure 2. Two **S4**-models, illustrating 'knowing more'

To understand the problem, consider the two models above. Intuitively, one would say that the agent knows more in state $\langle M_2, w_2\rangle$ than in $\langle M_1, w_1\rangle$, since the agent considers less possibilities in $\langle M_2, w_2\rangle$. However, $\langle M_1, w_1\rangle$ is not smaller than $\langle M_2, w_2\rangle$ in terms of the general information order. In the first configuration the agent knows that he does not know that p, while in the second he does not know that he does not know that p, since he knows that p. This shows that the general information order on possible worlds does not fit in with our intuition that 'more knowledge' corresponds to 'less uncertainty'. In the case of the positive information order, however, we *do* obtain $M_1, w_1 \prec M_2, w_2$, i.e., $M_1, w_1 \preceq M_2, w_2$ but $M_2, w_2 \npreceq M_1, w_1$

For the system **S5**, the restriction to positive minimality turns out to be equivalent with the original analysis of honesty in Halpern and Moses (1985). In fact a more restricted version of minimality is given in Halpern and Moses (1985), viz. with respect to the language $\Box\mathcal{L}_{(0)}$ (factual knowledge). However, it can be shown that in the system **S5** the disjunction property with respect to this restricted language is equivalent to the disjunction property with respect to the language of positive knowledge formulas, using the fact that every formula $\Box\psi$ is **S5**-equivalent to a formula of modal depth 1, see e.g., Chellas (1980).

For some modal systems such as **S4**, in which neither general nor positive minimality trivializes, it is interesting to compare the two orders.

THEOREM 3.20. For any normal system **S**, if φ is **S**-honest with respect to $\sqsubseteq$ then φ is also **S**-honest with respect to $\preceq$.

However, a similar transfer between different modal systems (and one kind of honesty) is not easily obtained. It may therefore be illuminating to contrast general and positive honesty for **S4** with positive honesty for **S5.** Table I displays formulas which are honest (√) or dishonest (–) in the indicated sense.

TABLE I

Several (dis-)honest formulas for **S4** and **S5**

Case	*Formula*	**S4**gen	**S4**pos	**S5**pos
1	$p \vee q$	√	√	√
2	$\Box p \vee q$	√	√	–
3	$\Box p \vee \Box\Diamond q$	–	√	√
4	$\Box p \vee \Box\Diamond\Box q$	–	√	–
5	$(\Box(\Box p \vee q) \wedge \neg\Box q) \vee \Box(p \vee r)$	–	–	√
6	$\Box p \vee \Box q$	–	–	–

From Theorem 3.20, we know that there are no (consistent) formulas that are generally honest in **S5**. Also, Theorem 3.20 tells us that that are no formulas which are honest in the sense of **S4**gen, but not in the sense of **S4**pos. Cases 3 and 4 show that Theorem 3.20 cannot be strengthened to an 'if and only if' statement. Finally, note that there is no relationship between positive honesty in **S5** and either general or positive honesty in **S4**.

4. Extensions to Multi-Agent Systems

4.1. Formalising Only Knowing with Many Agents

Lakemeyer (1993) and Halpern (1993) both generalized the approach of Levesque as discussed in Section 3.1 to the multi-agent case. Rather than discussing their approaches separately here, we can build upon a paper co-authored by Halpern and Lakemeyer (2001), in which they synthesize their earlier individual treatments. Their analysis is restricted to the propositional case.

In order to redefine Levesque's **Le** for the multi-agent case, it seems natural to focus on rule Ign of Definition 3.4. There are two ways in which this rule should be adapted: the notion of satisfiability and that of objective formulas has to be generalized. For the one agent case, objective formulas are those he can 'only know', and for weak **S5**, that are the propositional formulas. When there are many agents, more formulas can be rendered objective for agent i, like $\Box_j p$ and even, according to Lakemeyer (1993), $\Box_j \Box_i p$. Following Lakemeyer (1993) here, we call a formula *i-objective*, if it is a Boolean combination of primitive propositions and formulas of the form $\Box_j \varphi$ and $N_j \varphi$, with $j \neq i$. Furthermore, a formula is *basic* if it does not involve any of the operators N_i, and *i-subjective* if it is a Boolean combination of $\Box_i$- and N_i-formulas. Since the system **Le** is a **K45**-system for knowledge, Lakemeyer proposes the following straightforward generalisation of Ign:

Ign′ for every *basic i-objective* formula, if $\not\vdash_{\mathbf{K45}} \alpha$, then $\vdash_{\mathbf{La}} N_i \alpha \rightarrow \neg \Box_i \alpha$

The semantics for this system is inspired by Humberstone (1986), of which the key-idea is

$$M, w \models N_i \varphi \quad \text{iff } M, w' \models \varphi \text{ for all } w' \in R_i^c(w) \tag{4}$$

This definition says that at most $\neg\alpha$ is known at w, if α is true in all worlds that are not accessible from w. However, this definition does not comply with the requirement that the set of conceivable worlds should not depend on w. To overcome this, (Lakemeyer 1993) takes a specific model for $\models_{\text{La}}$, a model that has 'enough worlds': the canonical model for **K45**, where the worlds are maximal **K45**-consistent sets of basic formulas. He then shows that $\vdash_{\mathbf{La}}$, which is $\vdash_{\mathbf{Le}}$ with Ign replaced with Ign′, is sound for this semantics.

To explain the semantics that Halpern (1997) defines for $\mathcal{ONL}$, we first define $obj_i(M, w)$ to be all i-objective basic formulas α that are true at (M, w), and $Obj_i(M, w) = \{obj_i(M, w') | w R_i w'\}$. Then, the truth-definition in $\models_{\text{Ha}}$ for N_i formulas becomes is equivalent to:

$$M, w \models_{\text{Ha}} N_i \varphi \quad \text{iff } M', w' \models \varphi \text{ for all } M', w' \text{ for which}$$
$$Obj_i(M, w) = Obj_i(M', w') \text{ and } obj_i(M', w') \notin Obj_i(M, w)$$

Technically speaking, when comparing the notions of $\models_{\mathrm{La}}$ and $\models_{\mathrm{Ha}}$, we first mention a similarity. Let $\mathcal{ONL}_n^{-i}$ denote the sublanguage of $\mathcal{ONL}_n$ in which no N_j occurs in the scope of an N_i or $\Box_i$, for any $j \neq i$.

THEOREM 4.1. For all formulas $\varphi \in \mathcal{ONL}_n^{-i}$:
- $\models_{\mathrm{La}} \varphi$ iff $\models_{\mathrm{Ha}} \varphi$
- $\models_{\mathrm{La}} \varphi$ iff $\vdash_{\mathbf{La}} \varphi$

For arbitrary formulas, the two semantics differ. In that of Lakemeyer, there are, in some sense, 'not enough worlds' in the canonical model for the basic objective formulas to make $O_j O_i p$ true. Hence, $\neg O_j \neg O_i p$ is valid under $\models_{\mathrm{La}}$. Under Halpern's semantics, we face another problem for formulas in $\mathcal{ONL}_n \setminus \mathcal{ONL}_n^{-i}$: rule Ign′ is not valid for them, since, as Halpern (1993) demonstrates, the formula $N_i \neg O_j p \wedge \Box_i \neg O_j p$ is $\models_{\mathrm{Ha}}$-satisfiable. In the joint paper Halpern and Lakemeyer (2001), the authors try to remedy this, by first extending the language with operators *Val* and *Sat* to reason about validity and satisfiability of formulas, respectively. Then, the proper generalisation of Ign becomes:

Ign″ for i-objective formulas α, $\vdash_{\mathbf{Hl}} Sat(\neg\alpha \rightarrow (N_i\alpha \rightarrow \neg\Box_i\alpha))$

System **Hl** has the following rules to reason about *Val* and *Sat*:

V1 $\vdash_{\mathbf{Hl}} Val(\varphi \rightarrow \psi) \rightarrow (Val(\varphi) \rightarrow Val(\psi))$

V2 if φ is a satisfiable propositional formula, then $\vdash_{\mathbf{Hl}} Sat(\varphi)$

V3 if $\alpha, \beta_1 \dots \beta_k, \gamma, \delta_1 \dots \delta_m$ are i-objective, then
$\vdash_{\mathbf{Hl}} (Sat(\alpha \wedge \beta_1) \wedge \dots \wedge Sat(\alpha \wedge \beta_k)$
$\wedge Sat(\gamma \wedge \delta_1) \wedge \dots \wedge Sat(\gamma \wedge \delta_m) \wedge Val(\alpha \vee \gamma)) \rightarrow$
$Sat(\Box_i\alpha \wedge \neg\Box_i\neg\beta_1 \wedge \dots \wedge \neg\Box_i\neg\beta_k \wedge N_i\gamma \wedge \neg N_i\neg\delta_1 \wedge \dots \wedge \neg N_i\neg\delta_m)$

V4 for i-objective α and i-subjective β:
$\vdash_{\mathbf{Hl}} (Sat(\alpha) \wedge Sat(\beta) \rightarrow Sat(\alpha \wedge \beta)$

Nec(v) if $\vdash_{\mathbf{Hl}} \varphi$ then $\vdash_{\mathbf{Hl}} Val(\varphi)$

Demonstrating even soundness of these rules would take too much rescurces from this paper. However, let **Hl** be the system comprised of V1 – V4, Nec(v), and **Le**, with Ign replaced by Ign″. Then it appears that *Sat* and *Val* behave as intended: for every $\varphi \in \mathcal{ONL}_n$, if φ is provable in **Hl** then so is $Val(\varphi)$, and if φ is not provable in **Hl** then $\neg Val(\varphi)$ *is* provable. The semantics for **Hl** is obtained by looking at the canonical model $\mathcal{M}$, where the worlds are all maximal consistent sets of $\mathcal{ONL}_n^+$, which is the extension of $\mathcal{ONL}_n$ with the *Val*-operator. Furthermore, we conceive this as an extended canonical model, where we have canonical relations for $\Box_i$ and for N_i. Finally, the truth of $Val(\varphi)$ in a world in the canonical model is defined as φ being true in all the worlds of this model. The following nice result now states completeness:

THEOREM 4.2. For every formula $\varphi \in \mathcal{ONL}_n^+$:

- there is a formula $\varphi' \in \mathcal{ONL}$ that is provably equivalent
- $\vdash_{\mathbf{Hl}} \varphi$ iff $\mathcal{M}, w \models \varphi$ for every world w in the canonical model $\mathcal{M}$.

4.2. PARTICULAR MULTI-AGENT MINIMAL KNOWLEDGE SYSTEMS

In their paper, Halpern and Moses (1985) briefly considered some problems in generalizing their approach to many agents. The most obvious generalization of **S5**-honesty to many agents was believed to involve what was later called *a-objective* formulas, i.e., formulas in which each $\Box_a$ is in the scope of a $\Box_b$ operator, with $b \neq a$. This was shown to be insufficient, due to the fact that any consistent formula, say $\Box_a p$, does not have a stable expansion with minimum a-objective subset. Instead of proving this, it is easier to observe that in multi-**S5**, $\Box_a p \vdash \Box_a q \vee \Box_a \neg\Box_b\Box_a q$ whereas neither disjunct is implied by $\Box_a p$; since both q and $\neg\Box_b\Box_a q$ are a-objective this shows that even p would not be honest, clearly a very indesirable result. In the next (sub)sections we will have have a closer look at some proposals to overcome this difficulty. During the subsequent discussion, $\mathbf{S_m}$ indicates the multi-agent system where each operator $\Box_i$ respects the axioms of modal system **S**.

Parikh. Despite the claim in the beginning of Parikh's paper that "the more powerful logical and model-theoretic methods developed for dynamic logic and game logic provide us with better tools for studying both monotonic and non-monotonic logics of knowledge", these methods are hardly used in his paper and can easily be replaced by standard modal techniques. Parikh considers the interpretation of modal completeness (a theory $\mathcal{T}$ is modally complete if $\mathcal{T} \nvdash \Box_a\varphi \Rightarrow \mathcal{T} \vdash \neg\Box_a\varphi$, which boils down to McCarthy's nonmonotonic rule of inference M) as the key to the problem of $\mathbf{S5_m}$ honesty. No doubt the contemplation on the constructive effect of this rule is meaningful, yet the rule does not play a major role in his characterization.

As Vardi and others did, Parikh uses a special kind of Kripke model, viz. a tree structure. A noticeable difference is that Parikh's trees are *finite*. These trees have a nice and easy inductive definition.

DEFINITION 4.3. A *tree of height h* is a structure $\langle W, R, V, w_0\rangle$ (the designated world w_0 is called the *root* of the tree) which is recursively defined by:

1. a tree of height 0 is the singleton structure $\langle\{w\}, \emptyset, V, w\rangle$ for some $w \in 2^{\mathcal{P}}$
2. a tree of height $h+1$ is a structure $\langle W, R, V, w_0\rangle$ with root $w_0 \in 2^{\mathcal{P}}$, from which, for each agent i, there is at most one $w \in W$ such that $w_0 R_i w$ is a link to (essentially the submodel generated by w) a copy of a tree $\langle W_w, R', V', w\rangle$ of height h, where W_w is iteratively generated by w, $R' = R|_{W_w}$, and $V' = V|_{W_w}$.

Notice that in the 'growth step', since there are finitely many (m) agents and finitely many (a) atomic propositions, by induction only finitely many trees of finite size can be obtained. Loosely speaking, growing trees are booming both in number and size, but still finite. Also note that these trees can be obtained by the standard modal model-theoretic techniques unravelling and filtration over subformulas.

Suppose we have two trees M and M' of height h with the same truth assignment w as root. Subordination $M \subseteq M'$ is then defined by: M' has each link wR_iv to a subtree of height $h-1$ that M has. The sum $M + M'$ is simply the smallest tree subordinating M and M', which amounts to glueing M and M' in w and removing duplicate subtrees linked to this root (cf. the amalgamation technique of Section 3.4). Then M is a tree model for φ if φ is true in the root of the tree. We call a formula φ to be *adhesive* if φ is propositionally complete (i.e., φ entails either p or $\neg p$ for each atom p) and the sum operation preserves φ. Gluing models then in the limit characterizes honesty:

DEFINITION 4.4. A propositionally complete formula φ is *honest* if it has a largest tree model of height at least $d(\varphi)$. If φ is honest, then $\varphi \mid\!\sim \psi$ if ψ is true in the root of the largest tree model for φ of height at least $d(\varphi \wedge \psi)$.

It follows that the adhesive formulas are honest, but the converse need not hold. Also, honesty thus defined is decidable, as desired. Moreover, if φ is honest, then each nonmonotonic entailment ψ can be obtained by a non-monotonic proof which enforces restricted modal completeness through a Henkin construction: i.e., by constructing a series of theories $\mathcal{T}_0, \mathcal{T}_1, \ldots, \mathcal{T}_l$ where $\mathcal{T}_0 = \mathrm{Cn}_{\mathbf{S5_m}}(\varphi)$, $\mathcal{T}_{k+1} = \mathrm{Cn}_{\mathbf{S5_m}}(\mathcal{T}_k \cup \{\Diamond_i \xi\})$ such that $\Box_i \neg\xi \notin \mathcal{T}_k$ and for each subformula $\Box_i \eta$ of ξ: either $\Box_i \eta \in \mathcal{T}_k$ or $\Diamond_i \neg\eta \in \mathcal{T}_k$ (notice this formulation presupposes all formulas to be in *negation normal form*, i.e., $\Box$, $\Diamond$, $\wedge$, $\vee$ compounds of literals).

Although a nonmonotonic entailment $\mid\!\sim$ is thus determined, this characterization only works for honest formulas. And honesty is only described in model-theoretic terms, i.e., a syntactic or deductive counterpart is missing. In fact, there is but one sufficient syntactic condition for honesty: a consistent conjunction of (negations of) a formula and its subformulas is shown to be adhesive, and therefore honest. Moreover, Definition 4.4 merely applies to propositionally complete formulas φ. This is quite a strict condition and a rather brute way to avoid the problem with the $\mathbf{S5_m}$ derivable disjunction involving a-objective formulas, as noticed in the introduction to this subsection: for propositionally complete φ we cannot find such a q. But the price is too high: partial knowledge cannot even be honest anymore. Another remarkable feature of this approach is that there is no dependence on the agent. Now we could restrict ourselves to honest formulas $\Box_a \varphi$, but even then the notion itself would not be related to a in any way.

We believe it is fair to say that Parikh (1991) solved the problem of $\mathbf{S5_m}$ honesty to a limited extent, and since we are not aware of any subsequent progress in this line, we should consider rival proposals. We will now turn to one of the few others we know.

Halpern. Halpern (1997) defines 'only knowing' in the multi-agent case as a generalization of the single agent analysis for **S5** as defined in Halpern and Moses (1985). The truth condition of 'all an agent a knows is φ' means that the agent knows that φ and that the state representing this knowledge contains a *maximum* set of *possibilities*, that is, this state is *a-minimal* for φ. A maximum of possibilities represents maximal ignorance, and therefore, a minimum of knowledge. Halpern focuses in his paper on the formalization of a proper notion of possibility $Poss_a^{\mathbf{S}}(M, w)$ of an agent a for an **S**-state $\langle M, w\rangle$ so that it yields a satisfactory definition of 'only knowing' and honest formulas.

Halpern formulates the two following conditions for candidate definitions of possibility in order to judge their appropriateness.

Determination. If two states contain the *same* set of possibilities for an agent a then they contain the *same* knowledge of a.

$$Poss_a^{\mathbf{S}}(M, w) = Poss_a^{\mathbf{S}}(M', w') \Rightarrow \text{ for all } \varphi : M, w \models \Box_a\varphi \text{ iff } M', w' \models \Box_a\varphi.$$

Union. For every pair of states there must exist a third state which contains at least all the possibilities contained by the first two.

For all pairs of states $\langle M, w\rangle$ and $\langle M', w'\rangle$ there exists $\langle N, v\rangle$ such that:

$$Poss_a^{\mathbf{S}}(N, v) \supseteq Poss_a^{\mathbf{S}}(M, w) \cup Poss_a^{\mathbf{S}}(M', w').$$

The first condition seems to be a minimal condition: possibility determines knowledge. The union condition provides a way of joining the ignorance as represented by two different states, i.e., two different possibility sets are never mutually **S**-inconsistent.

The definition of knowledge state in terms of worlds in Kripke models does not give us a canonical definition of what a possibility is. Halpern uses a translation of worlds into infinite trees as a canonical representation of a state. If $\langle M, v\rangle$ is a state then the tree which corresponds to $\langle M, v\rangle$ consists of a root which coincides with $V(v)$ and a set of labeled edges, of which every label represent an agent a. The set of a-edges connect the root with the trees that correspond with the possible worlds in $R_a[v]$. This procedure gives a tree of which the branches are just representations of the accessibility paths in the model M starting from w. The illustration below

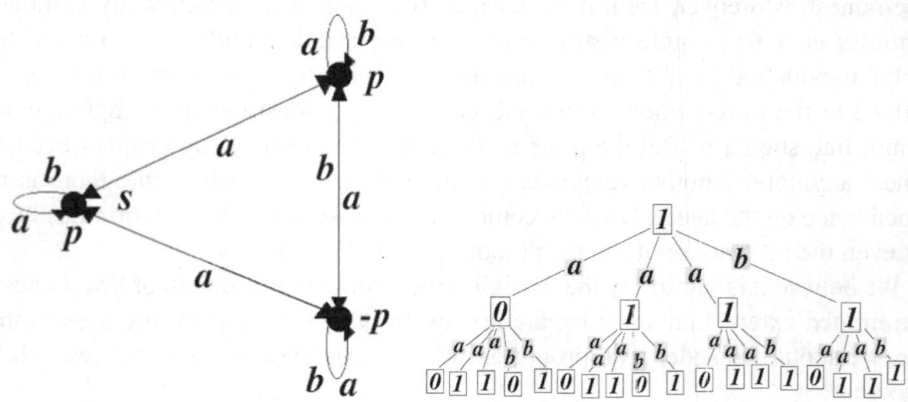

shows the partial result up to length 2 for a simple $\mathbf{S5_2}$-state s in a three world model with a single proposition letter p. The nodes in the tree only contain the truth-value of p.

Every tree $T_{M,v}$ obtained in this way can be translated back to a unique world within the model of all trees M^t such that $M^t, T_{M,v} \models \varphi$ iff $M, v \models \varphi$ for all formulas φ. This technique yields a canonical representation of possibilities in the way that two different trees obtained by the translation mentioned above contain different information: $\text{Th}(M^t, T) = \text{Th}(M^t, T')$ iff $T = T'$.

The most obvious candidate definition of the set of possibilities for a state $\langle M, w\rangle$ in terms of trees is the set of trees which correspond to the accessible states of w in M.

$$Poss_a^{\mathbf{S}}(M, w) = \{T_{M,v} \mid v \in R_a[w]\}$$

A formula φ is then called $\mathbf{S}$-a-honest iff there exists an $\mathbf{S}$-state $\langle M, w\rangle$ with a maximal set of possibilities: $\langle M, w\rangle \models \Box_a\varphi$ and for every M', w' with $M', w' \models \Box_a\varphi$ we have $Poss_a^{\mathbf{S}}(M', w') \subseteq Poss_a^{\mathbf{S}}(M, w)$.

If a formula has such a minimal state $\langle M, w\rangle$, then its $\mathcal{L}$-smallest $\mathbf{S}$-a-stable expansion is simply the set of formulas φ such that $M, w \models \Box_a\varphi$. The definition of honesty above also establishes the *general* $\mathbf{S}$-disjunction property for φ along the same lines as exposed in Theorem 3.15. In other words, ordinary inclusion between possibility sets coincides with the general information order over states as defined for the single agent case in the paragraph on *General minimality* in Section 3.4. As Halpern observes, this information order over possibilities only works for epistemic logics without negative introspection: the determination and union condition are both satisfied. Validity of determination is trivial, and for proving the union condition we can use the amalgamation technique as has been demonstrated in Figure 1.

In the case of logics such as $\mathbf{K_m}$, $\mathbf{K4_m}$, $\mathbf{KD_m}$ and $\mathbf{KD4_m}$ the strong union condition also holds since the corresponding model classes of these logics are preserved under simple amalgamation. The notion of honesty trivializes since simple amalgamation provides an a-minimal model for every formula φ whenever $\Box_a\varphi$ is $\mathbf{S}$-consistent.

For the logics $\mathbf{T_m}$ and $\mathbf{S4_m}$ the straightforward definition of possibility above yields a non-trivial notion of honesty, just like for the single agent case as explained in Section 3.4. The strengthening of the union condition is no longer possible, since the amalgamation technique requires for these two logics reflexivity of the root state. Despite the fact that honesty is not trivial for $\mathbf{T_m}$ and $\mathbf{S4_m}$, the ordering of possibility sets on a general basis seems counterintuitive, which was shown in Figure 2. It is not hard to verify that $Poss_a^{\mathbf{S4_m}}(M_1, w_1) \not\subseteq Poss_a^{\mathbf{S4_m}}(M_2, w_2)$ for the two simple $\mathbf{S4}$-states in Figure 2. The first possibility set is a simple tree consisting of a single branch of p-valuations. This tree does not belong to the second set of possibilities.

For fully introspective logics the definition of possibility above does not work, essentially for the same reason as why the general information order is not satisfactory for these logics in the single agent case of Section 3.4. No formula is honest if we would use this notion of possibility. The solution which Halpern provides for the logics $\mathbf{K45_m}$ and $\mathbf{KD45_m}$ is to exclude the agent's knowledge about her own knowledge from the definition of possibility. In terms of trees as possibities, the a-edges reaching from the roots are to be ignored. We assemble the accessibility paths which do not start with an a-step as branches of the tree. The tree we obtain in this way from a state $\langle M, w\rangle$ is written as $T^a_{M,w}$ and the set of possibilities is defined as

$$Poss^{\mathbf{S}}_a(M, w) = \{T^a_{M,v} \mid v \in R_a[w]\} \text{ if } \mathbf{S} = \mathbf{K45_m}, \mathbf{KD45_m}$$

Despite the restriction on the definition of possibility, the determination and union condition hold. The essential idea behind the succesfulness of this definition is that every formula $\Box_a\varphi$ is equivalent to a Boolean combination of formulas of the form $\Box_a\psi$ with ψ being a-objective. Halpern provides a special amalgamation technique to prove that this possibility definition has the strong union property. This special technique is needed to take care that euclidicity of the accessibility relations is preserved. The analysis fits with the honesty diagram of Figure 2 with $\mathcal{L}^*$ being the sublanguage of formulas $\Box_a\varphi$ with φ being a-objective.

In the case of $\mathbf{S5_m}$ the restriction to objective languages does not work. The combination of negative introspection and the $\mathbf{T}$-axiom entails a problematic sort of disjunctive $\mathbf{S5_m}$-theorems. False formulas cannot be known, and therefore every agent in the case of $\mathbf{S5_m}$ knows that she does not know false formulas by negative introspection. This yields $\mathbf{S5_m}$-theorems like $\Box_a p \vee \Box_a\Box_b\neg\Box_b\Box_a p$: disjunctions of knowledge of a-objective information of which none of the disjunctions is a theorem. Implementation of the objective notion of possibility would then cause a collapse of what honesty is: all formulas are dishonest.

On the level of states and trees we can strengthen the restriction technique which has been used for the weaker fully introspective systems. Besides the a-accessibilities the real world is also excluded in order to disarm the $\mathbf{T}$-axiom. Halpern introduces a special sort of trees with empty nodes so that when a state $\langle M, w\rangle$ is transformed into tree form the content of $\langle M, w\rangle$ is represented as an empty node at each stage of the transformation. In this way the real world is factored out. $Poss^{\mathbf{S5_m}}_a(M, w)$ is then defined as the set of trees of this limited form which correspond to the a-accessible worlds at $\langle M, w\rangle$.

The stronger union condition holds for this notion of possibility. The amalgamation technique for the weaker negative introspective logic can be used here as well. The only adaptation that has to be made is that the root of the amalgamation has to be reflexive. By the exclusion of the real world in the definition of $\mathbf{S5_m}$-possibility the stronger union condition is established in exactly the same manner.

Unfortunately, there is no appropriate sublanguage such that a determination condition can be proved. Moreover, minimality of stable sets and a definition of a restricted disjunction property is impossible. The epistemic language is not expressive enough to capture the factoring out of reality as a sublanguage restriction.

To overcome this expressiveness problem, Halpern introduces an extension of the language with special dyadic modal operators. This introduction facilitates the proof of determination with respect to the enriched language. The definition of minimality of states and stable sets, and also a disjunction property can then be recaptured by an appropriate sublanguage restriction of the enriched language. The formal and conceptual status of the new operators remain vague in Halpern's paper: no axiomatization of the underlying enriched logic is given, and moreover, their introduction seems to be purely technically motivated.

In the next section an alternative for minimal knowledge analysis in the case of $\mathbf{S5_m}$ is introduced on the basis of a further restriction of the objective sublanguage of the original language. An enrichment of the epistemic language is then no longer required.

4.3. Arbitrary Multi-agent Minimal Knowledge Systems

This section is a generalisation of the discussion of Section 3.3 to the multi-agent case. Most of the results are from van der Hoek and Thijsse (2002). Here, the aim is to describe what a particular agent a can 'only know' in a multi-agent context. Thus, the notions that are relevant for honesty are now relativized with respect to one particular agent $a \in \mathcal{A}$. A formula φ is *a-honest* with respect to $\mathbf{S}$ and $\leq^a$ iff there is an $\mathbf{S}$-state $\langle M, w\rangle$ that is $\leq^a$-minimal among the states verifying $\Box\varphi$.

Thus, we will reason about $\leq^a$-least verifying $\mathbf{S}$-states, and relate them to $\mathcal{L}^a$-smallest $\mathbf{S}$-m.c. expansions, to $\Box_a^- \mathcal{L}^a$-smallest a-stable expansions and also to the $\mathbf{S}$-Disjunction Property with respect to $\mathcal{L}^a$. In van der Hoek and Thijsse (2002), the authors provide a generalisation of Theorem 3.15 from Section 3.3:

THEOREM 4.5. Let $\mathcal{L}^a$ be a characteristic persistent sublanguage of $\mathcal{L}$ with respect to $\leq^a$. Then φ is a-honest with respect to $\leq^a$ and $\mathbf{S}$ iff (all statements are equivalent):

- $\Box_a\varphi$ has a $\leq^a$-least $\mathbf{S}$-state
- $\Box_a\varphi$ has an $\mathcal{L}^a$-smallest $\mathbf{S}$-m.c. expansion
- φ has a $\Box_a^- \mathcal{L}^a$-smallest a-stable expansion
- $\Box_a\varphi$ has $\mathbf{S}$-DP over $\mathcal{L}^a$.

When the equivalent notions of Theorem 4.5 hold for a particular order $\leq^a$, system $\mathbf{S}$ and language $\mathcal{L}^a$, we say that $\leq^a$ and $\mathcal{L}^a$ *determine* a notion of honesty in $\mathbf{S}$. We can also link up the semantic definition of honesty with deduction, providing a perhaps even more intuitive characterization:

COROLLARY 4.6. Let $\mathcal{L}^a$ be persistent and characterizing for $\leq^a$. Then φ is a-honest with respect to $\leq^a$ and $\mathbf{S}$ iff there is an $\mathbf{S}$-state $\langle M, w\rangle$ such that:

$$\forall\psi \in \mathcal{L}^a : M, w \models \psi \Leftrightarrow \Box_a\varphi \vdash_{\mathbf{S}} \psi$$

This insight was triggered by a question from Arnis Vilks, and it is in fact close to the approach that Pratt-Hartmann takes in Pratt-Hartmann (2000), where only knowing φ holds at a world w iff $\Box\varphi$ holds, and for all objective ψ (hence, $\mathcal{L}^a$ is the objective language) that do not hold in w, one has $\not\models \Box\varphi \to \psi$: one does not know anything stronger than φ.

Layered Information Orders. An information order and its characterizing persistent language can be obtained along fairly general patterns from the underlying layered orders and their characterizing persistent languages. This is a very convenient tool for many orders to follow, since we can restrict attention to one simple layer at the time.

DEFINITION 4.7. Suppose $\leq_n^a$ is a pre-order on the set of model-world pairs for each natural number n ('layer n'). From now on, assuming $M = \langle W, \vec{R}, V\rangle$ and $M' = \langle W', \vec{R}', V'\rangle$, the base case will be defined as $M, w \leq_0^a M', w' \Leftrightarrow V(w) = V'(w')$. Then we define $\leq^a$ for any layered order $\leq_n^a$ by:

$$M, w \leq^a M', w' \Leftrightarrow \forall n \in \mathbb{N}\, \forall v' \in R'_a[w']\, \exists v \in R_a[w] : M, v \leq_n^a M', v'.$$

We say that $\leq^a$ is *induced by* $\leq_n^a$ if the above equivalence holds.

Now, let $\mathcal{L}^a$ be a sublanguage and $\mathcal{L}^a_{(n)} = \mathcal{L}^a \cap \mathcal{L}_{(n)}$ be its subset of formulas of modal depth up to n. The following lemma explains how a persistence and characterization result for layered orders and languages with finite depth can be lifted to the full language and the induced order.

LEMMA 4.8. If $\mathcal{L}^a_{(n)}$ is persistent and characterizing for $\leq_n^a$, and $\mathcal{L}^a$ is closed under $\vee$, then $\Box_a\mathcal{L}^a$ is persistent and characterizing for $\leq^a$.

We now inspect orders inspired by Ehrenfeucht-Fraïssé games. In van der Hoek and Thijsse (2002), the authors start out by defining the so-called *general information order* $\sqsubseteq^a$, for which the full language $\mathcal{L}^{\mathcal{A}}$ is characteristic and persistent, where $\mathcal{L}^{\mathcal{A}}$ is defined as:

$$\mathcal{L}^{\mathcal{A}}: \quad \varphi ::= p\ (p \in \mathcal{P}) \mid \neg\varphi \mid \varphi \wedge \varphi \mid \Box_i\varphi\ (i \in \mathcal{A})$$

However, as we noticed in Section 3.3, this notion of honesty is, though technically correct, intuitively a rather poor one. Therefore we here focus on generalizations of the positive order as defined in Section 3.3.

In a positive information order we merely want to preserve positive knowledge of one or more agents. In other words, we disregard negative knowledge, i.e., knowledge of not knowing. We, typically, encounter some form of back simulation on the underlying layers, but usually no forth simulation, or only a restricted form of forth simulation.

It is not *a priori* clear which notion of positive information order is involved. We will discuss several options in what follows, of which the last is the more general one, using both underlying layers from the general and from the so-called 'objective' information order. We start by discussing a rather straightforward generalization from the single agent case.

Positive Honesty. The positive information order only preserves positive knowledge of agent a. It is the most obvious generalization of one-agent positive honesty. The formulas of the characterizing language do not have negative occurrences of $\Box_a$, and so, by definition, no $\Diamond_a$ as well. Formally, let $\mathcal{L}^{+a}$ consist of those $\varphi \in \mathcal{L}$ for which φ does not contain $\Box_a$ in the scope of $\neg$. Formulas in $\mathcal{L}^{+a}$ are called *a-positive*. So, $\Box_a p \vee \neg\Box_b q$, $\Box_a \neg p$ and $\Box_a p \wedge \neg q$ are members of $\mathcal{L}^{+a}$, but $\neg\Box_a p$ and $\Diamond_a p \vee \Box_b q$ are not. Formally, in BNF-notation:

$$\mathcal{L}^{+a}: \quad \varphi_0 ::= \varphi \ (\varphi \in \mathcal{L}^{\mathcal{A}\setminus\{a\}}) \mid \varphi_0 \wedge \varphi_0 \mid \varphi_0 \vee \varphi_0 \mid \Box_i \varphi_0 \ (i \in \mathcal{A})$$

Now consider $\mathcal{L}^a = \Box_a \mathcal{L}^{+a}$. This is a correct generalization of the single agent positive language, which by itself is a generalization of the language of so-called objective one-agent formulas which suit **S5**. We will call the elements of $\Box_a \mathcal{L}^{+a}$ *a-positive knowledge formulas*. What is the corresponding $\leq^a$? Essentially, the underlying order displays the back direction of the EF-equivalence for all agents, operating on a-positive formulas until subformulas are reached that are $\Box_a$-free, where full EF-equivalence for all agents except a takes over. Then, $M, w \leq^{+a}_{n+1} M', w'$ iff:

- $M, w \simeq^{\mathcal{A}\setminus\{a\}}_{n+1} M', w'$ &
- $\forall i \in \mathcal{A}\, \forall v' \in R'_i[w']\, \exists v \in R_i[w] : M, v \leq^{+a}_n M', v'$ (*back*)

Let the *positive information order* $\leq^{+a}$ be induced by $\leq^{+a}_n$. Then $\mathcal{L}^{+a}_{(n)}$ is characteristic and persistent for $\leq^{+a}_n$, so Lemma 4.8 guarantees that $\Box_a \mathcal{L}^{+a}$ is persistent and characterizing with respect to $\leq^{+a}$. Thus, we obtain:

THEOREM 4.9. The minimal information equivalences hold for the order $\leq^{+a}$ and the language $\Box_a \mathcal{L}^{+a}$.

Now, φ is called *positively a-honest* if $\Box_a \varphi$ has a $\leq^{+a}$-least model. Thus, we have that $\leq^{+a}$ and $\Box_a \mathcal{L}^{+a}$ determine a notion of a-honesty in **S**, for any system **S**. So, the notion of positive honesty is technically sound, that is, there is a persistent language that characterizes the positive information order, and it seems a proper extension of the unimodal case. It avoids problems objective honesty encounters, such as the one noticed by Halpern that we already mentioned in Section 4.2, and which holds for extensions of $\mathbf{KB4_m}$: suppose p is some fact totally unrelated to a formula φ (for example, p may not occur in φ), then $\Box_a \varphi \vdash \Box_a p \vee \Box_a \Box_b \neg \Box_b \Box_a p$. It is clear, however, that each of the disjuncts itself does not follow from $\Box_a \varphi$. Yet φ may constitute innocent knowledge, e.g., $\Box_b q$. But for our notion of positive honesty, this counter-example to the DP test is avoided by the restriction that the disjuncts should be in the a-positive knowledge language; here, obviously, $\Box_b \neg \Box_b \Box_a p \notin \mathcal{L}^{+a}$.

Yet we do not want to exclude other possible notions of honesty *a priori*, and therefore now turn to one studied earlier.

Objective Honesty. To make a different start in formalizing multi-agent positive honesty, we return to Halpern's (1997) definition of a-objective formulas and the notion of honesty connected to it. Halpern reserves the notion objective honesty for the two strong doxastic systems $\mathbf{K45_m}$ and $\mathbf{KD45_m}$. This seems harmless for these two systems. Our main concern is that developing a whole apparatus for just two modal systems, and again different ones for others, leads to an approach which lacks generality and in fact conceals much of the general pattern. In fact, in Halpern's approach it is not clear *why* a-objective formulas might be suitable for the two systems mentioned. We think that we can in fact explain some of the reasons for its feasibility.

The idea of a-objective knowledge is that agent a only has knowledge of information 'outside' of a, i.e., knowledge of facts and other agents' knowledge. Such other agents' knowledge may again involve a's knowledge, but still counts as external for a. This is easily formalized when we start with the *a-objective* (that is, wide scope a-operator-free) formulas: let $\mathcal{L}^{-a}$ consist of those $\varphi \in \mathcal{L}$ for which φ does not contain wide scope $\Box_a$. In other words, in an a-objective formula, every $\Box_a$ and $\Diamond_a$ has to be in the scope of a $\Box_b$ or $\Diamond_b$ ($b \neq a$). Examples: $\Box_a p \vee \Box_b q$, $\Box_a \neg p$ are not in $\mathcal{L}^{-a}$, but $\neg \Box_b p$ and $\Diamond_b(p \vee \neg \Box_a q)$ are.

$$\mathcal{L}^{-a}: \quad \varphi_1 ::= p \ (p \in \mathcal{P}) \mid \neg \varphi_1 \mid \varphi_1 \wedge \varphi_1 \mid \Box_i \varphi \ (i \in \mathcal{A} - \{a\})$$

So, where does the agent a's knowledge enter the story? Here it is: consider $\mathcal{L}^a = \Box_a \mathcal{L}^{-a}$. A formula is then called an *a-objective knowledge formula* if it is of the form $\Box_a \varphi$ with $\varphi \in \mathcal{L}^{-a}$.

One can find an order $\leq^{-a}$ such that the minimal information equivalences hold for $\leq^{-a}$ and the language $\Box_a \mathcal{L}^{-a}$, we refer the reader to van der Hoek and Thijsse (2002). This again implies that $\leq^{-a}$ and $\Box_a \mathcal{L}^{-a}$ determine a notion of (objective) a-honesty in $\mathbf{S}$, for any system $\mathbf{S}$. As can been seen from the format of the a-objective formulas, agent a's knowledge is not taken into account. For fully introspective systems this is unproblematic in the one-agent case: there one can show that positive knowledge formulas can be reduced to disjunctions of objective knowledge formulas, which implies that for each system containing $\mathbf{K45}$, objective honesty amounts to positive honesty.

It should be emphasized that this equivalence only holds for the one agent case and full introspective knowledge. For more agents there is no such reduction, since an objective knowledge formula need not be (equivalent to) a positive one, e.g., $\Box_a \Box_b \neg \Box_a p$ is an a-objective knowledge formula which is not related to any a-positive knowledge formula whatsoever. If we want to generalize this equivalence to fully introspective multi-agent systems, we have to relax the notion of positive formula somewhat, as will be done in the next subsection.

Positive-Objective Honesty. We want to generalize objective knowledge to what we consider to be a more adequate notion of multi-modal honesty. The *a-positive-objective* formulas can, roughly, be characterized as having no wide scope negative occurrence of $\Box_a$ operators. Again assume for simplicity's sake that we only consider formulas where every $\Diamond_i$ is replaced by $\neg\Box_i\neg$. Let $\mathcal{L}^{\pm a}$ consist of those $\varphi \in \mathcal{L}$ for which every $\Box_a$ in φ in the scope of $\neg$ is also in the scope of a $\Box_i$ with $i \neq a$. Thus, $\mathcal{L}^{\pm a}$ can also be regarded as the closure of $\mathcal{L}^{-a}$ under the operations $\wedge$, $\vee$ and $\Box_a$. Examples: $\Box_a p \vee \Box_b q$, $\Box_a p \wedge \neg\Box_b q$ and $\Box_a\Box_b\neg\Box_a p$ are members of $\mathcal{L}^{\pm a}$, but $\neg\Box_a p$ and $\Box_a\neg\Box_a\neg p \vee \Box_b q$ are not.

$$\mathcal{L}^{\pm a}: \quad \varphi_2 ::= \varphi_1(\varphi_1 \in \mathcal{L}^{-a}) \mid \varphi_2 \wedge \varphi_2 \mid \varphi_2 \vee \varphi_2 \mid \Box_a\varphi_2$$

Once again, for the corresponding $\leq^{\pm a}$ we refer to van der Hoek and Thijsse (2002). There, it is also demonstrated that the minimal information equivalences hold for the order $\leq^{\pm a}$ and the language $\Box_a\mathcal{L}^{\pm a}$. This implies that $\leq^{\pm a}$ and $\Box_a\mathcal{L}^{\pm a}$ determine a notion of a-honesty in **S**. We can show that for fully introspective systems the extension from objective to positive-objective formulas is immaterial, since then again a's positive-objective knowledge can be reduced to a-objective knowledge. So, objective honesty and positive-objective honesty coincide for $\mathbf{K45_m}$, $\mathbf{KD45_m}$, and $\mathbf{S5_m}$. Although positive, objective, and positive-objective honesty agree on (one agent) **S5**, they, surprisingly, do not on $\mathbf{S5_m}$ $(m > 1)$. Since $\Box_a p \vee \Box_a\Box_b\neg\Box_b\Box_a p$ is derivable in $\mathbf{S5_m}$, there are virtually no (positive)-objectively honest formulas in this system. However, we have already seen that for $\mathbf{S5_m}$, the positive information order seems correct.

4.4. Evaluation and Tools

Which of the notions presented here constitutes a feasible notion of multi-agent honesty, independent of the epistemic system? To evaluate these notions, we need to determine the (dis)honesty of a number of key examples.

In van der Hoek and Thijsse (2002), the authors elaborate on ways to check the honesty of a given formula φ. In general, *dishonesty* can be shown fairly easily by using the relevant DP, but it may be harder to show *honesty* more or less directly. It is not *prima facie* clear how to prove honesty, since DP then has to be checked for an infinite set of formulas (although in many cases, see van der Hoek and Thijsse (2002), one only has to inspect disjunctions of length 2). Also, minimality of stable expansions encounters similar problems and finding the least model may be non-trivial, which is related to the complexity of the information orders. Presumably, for many relevant multi-modal systems these intricate orders have simple counterparts.

Despite the name of minimal models, such models can be hard to construct or find, since they have to demonstrate a *maximum* amount of ignorance. Similar to

	$\mathbf{S4_m}$	$\mathbf{K45_m}$	$\mathbf{S5_m}$
gen	a	-	-
pob	b	e	-
obj	c	e	-
pos	d	f	g

	$\mathbf{S4_m}$	$\mathbf{K45_m}$	$\mathbf{S5_m}$
gen	+	-	-
pob	+	-	-
obj	+	-	-
pos	+	-	-

Figure 3. A general and a common pattern of honesty.

Corollary 4.6 we find that a minimal state for $\Box_a\varphi$ has to verify every $\neg\psi$ for which $\psi \in \mathcal{L}^a$ and $\Box_a\varphi \not\vdash_{\mathbf{S}} \psi$. Finding an explicit description of such a minimal state appears to be much harder than in the one agent case.

To sum up the techniques used in van der Hoek and Thijsse (2002) to decide (dis-)honesty, the authors look at generalized submodels, and at (reflexive) amalgamation and (reflexive) rootability, which are sufficient conditions to conclude general honesty in ($\mathbf{T_m}$, $\mathbf{S4_m}$) $\mathbf{KD4_m}$ and the like. Then, (reflexive) embeddings are discussed, which are sufficient to conclude positive honesty (for the same group of systems). Then, they discuss the technique of clustering, to decide positive honesty in $\mathbf{S5_m}$.

Relating Types of Honesty. The types of honesty distinguished in this section are ordered as indicated Observation 4.10. This hierarchy easily follows from DP, using the fact that $\Box_a\mathcal{L}^{+a} \cup \Box_a\mathcal{L}^{-a} \subseteq \Box_a\mathcal{L}^{\pm a} \subseteq \Box_a\mathcal{L}$.

OBSERVATION 4.10. General honesty implies positive-objective honesty, and the latter implies both positive and objective honesty.

We now like to inspect the four types of honesty studied in this section for the modal systems $\mathbf{S4_m}$, $\mathbf{K45_m}$ and $\mathbf{S5_m}$. To keep trace of the amount of 'honesty' displayed in a particular example, we list its pattern in a 4×3 array. Due to Observation 4.10 not all 2^{12} patterns are indeed possible, only 36 survive: ("+" means honest for the listed type, "-" dishonest, and "a,…,g" can be either + or -).

To further constrain the left table of Figure 3 by Observation 4.10, a=+ $\Rightarrow$ b=+, b=+ $\Rightarrow$ c=d=+, and e=+ $\Rightarrow$ f=+. Apart from some exceptional cases which we have left out of the scheme, there are no generally honest formulas for $\mathbf{K45_m}$. As noticed in one of the text corollaries, for $\mathbf{K45_m}$ objective and positive-objective honesty is one and the same thing (e). For $\mathbf{S5_m}$ there is also overall dishonesty. For (positive)-objective honesty (which again amount to the same), there is also very general failure of DP due to the notorious $\mathbf{S5_m}$-theorem $\Box_a\psi \vee \Box_a\Box_b\neg\Box_b\Box_a\psi$ for some a-objective ψ.

Given these constraints, the maximally honest pattern is obtained by choosing a '+' for all the positions in Figure 3, that are not occupied by a '-', yet. This pattern manifests itself in many formulas that are also intuitively honest for agent

a: p, $\Box_i p$, ... The most challenging cases are disjunctions of (negated) knowledge formulas. Whether or not they are intuitively honest largely depends on the agency of the knowing subject. So, also the following innocent formulas are indeed maximally honest: $\Box_b p \vee \Box_b q$, $\Box_a p \vee \Diamond_a q$, $\Box_b p \vee \Diamond_a q$, $\Box_b p \vee \Diamond_b q$ and $p \vee q$.

The other extreme are the obviously totally dishonest formulas displaying the pattern obtained by filling all places in Figure 3 with a minus. The paradigm for this is $\Box_a p \vee \Box_a q$. There are many (34) intermediate cases. A very common pattern here is the one in which honesty only depends on the amount of introspection attributed to the agents, witnessed by the pattern displayed in the second table of Figure 3. Some samples of these are $\Box_a p \vee \Box_b q$, $\Box_a p \vee \Diamond_b q$, and $\Box_a p \vee q$.

To get more complicated patterns we need formulas with nested knowledge. In particular, we suggest the following two types. The formula $\Box_a p \vee \Box_a \Diamond_a q$ seems to have an 'almost maximally honest' pattern: only the 'a' in Figure 3 becomes a '-'. Whereas a dishonest counterpart of this is $\Box_a p \vee \Box_a \Diamond_a \Box_a q$ for which only the 'd' and 'f' in the table are a '+'. As a final example we inspect the formula $\Box_a p \vee \Box_a \Box_b \Diamond_a q$, displaying a pattern where honesty merely depends on the type and not on the modal systems under inspection: this formula only obtains a '+' at the positions 'd' and 'f' in Figure 3.

Summarizing, we have given generalizations of information orders for multi-agent only knowing, which apply to arbitrary modal systems and ordinary Kripke models. Using a general theorem relating information orders and their corresponding (sub-) languages, we were able to identify several equivalent characterizations of honesty. In particular, we have explored the general information order and some positive and objective information orders. So-called positive honesty seems the intuitively correct notion here.

5. Ignorance and Doubt

A wide variety of non-classical logics have been proposed as alternatives to classical propositional logic as a basis of epistemic logic. One of the most prominent alternatives is *partial* logic since the introduction of situation theory in formal semantics of natural language by Barwise and Perry (1983). Situation theory is more than only a partial variant of classical logic, but in the much smaller context of propositional epistemic logic, a situation can be seen as a *part* of a possible world. In terms of Kripke models the change is relatively moderate: the valuation function is no longer required to be total.

In van der Hoek et al. (1996) a partial variant of **S5** has been introduced as an epistemic logic. The paper investigates the notion of minimal knowledge and honesty in this context. The most important advantage of partial logic in terms of modeling knowledge is that two different notions of 'not knowing' come on stage: *doubt* and *mere ignorance*. An agent doubts a proposition φ if she thinks that φ may be false, whereas ignorance with respect to φ means that the epistemic possibilities of an agent does not assign a definite truth-value to that proposition. This differentiation in interpretation of ‘not knowing’ makes a partial variant of negative introspection plausible. When an agent doubts a proposition she also knows that she doubts that proposition. The semantic condition on Kripke models for fully introspective partial logic turns out to be the same as for the classical case: the epistemic alternatives of an agent are mutually accessible. An agent is omniscient with respect to her own uncertainties.

The difference of the logic introduced in van der Hoek et al. (1996) with classical **S5** when it comes to the epistemic axioms is the T-axiom. The axiom stating that knowledge is always true is accepted, but not its contraposition saying that every false proposition is doubted. Partial logic does not contain contraposition, and in this way it is possible to avoid this unwanted classical side effect of T. The corresponding constraint on Kripke models is a weakening of reflexivity: among the epistemic alternatives of an agent there is always one possibility which is a part of the real world. It turns out that the kind of problem of modeling minimal knowledge in the case of $\mathbf{S5_m}$ that Halpern addresses disappears.

When it comes to minimal knowledge partial logic offers another attractive advantage. Knowledge is no longer inversely proportional to epistemic possibility. ‘Knowing nothing’ in classical logic means taking all possible worlds into account, whereas in partial logic ‘no knowledge’ is modelled as a single possible world in which no propositional variable is assigned a definite truth-value: the *tabula rasa*. In the case of classical logic a completely ignorant agent knows only tautologies, whereas in partial logic complete ignorance really reduces to a ‘tabula rasa’, even tautologies are unknown. In van der Hoek et al. (1996) a minimization of epistemic possibility is proposed in addition to minimizing knowledge.

5.1. Partial Modal Logic

A partial Kripke model consists of a classical accessibility frame $\langle W, R\rangle$ augmented with a partial valuation function $V : W \longrightarrow (\mathcal{P} \rightharpoonup \{0, 1\})$ where $\mathcal{P} \rightharpoonup \{0, 1\}$ is the set of all partial truth-value assignments. A partial state is a pair $\langle M, w\rangle$ where M is a partial Kripke model and w in M. As with the proposition letters, the whole language is evaluated in a partial way by such a pair $\langle M, w\rangle$. This means we need to distinguish false formulas and formulas that are not true. We write $M, w \Dashv \varphi$ in the case that a proposition φ is false at $\langle M, w\rangle$. Truth and falsity are defined according to the following inductive clauses:

$$M, w \models p \Leftrightarrow V(w)(p) = 1 \quad M, w \models \varphi \wedge \psi \Leftrightarrow M, w \models \varphi \text{ and } M, w \models \psi$$
$$M, w \models \neg\varphi \Leftrightarrow M, w \Dashv \varphi \quad M, w \models \Box\varphi \Leftrightarrow M, v \models \varphi \text{ for all } v \in R[w]$$

$$M, w \Dashv p \Leftrightarrow V(w)(p) = 0 \quad M, w \Dashv \varphi \wedge \psi \Leftrightarrow M, w \Dashv \varphi \text{ or } M, w \Dashv \psi$$
$$M, w \Dashv \neg\varphi \Leftrightarrow M, w \models \varphi \quad M, w \Dashv \Box\varphi \Leftrightarrow M, v \Dashv \varphi \text{ for certain } v \in R[w]$$

In addition we can extend the system with a weak negation as opposed to the strong negation used in the table above. Strong negation says that the argument is false, whereas weak negation expresses that the argument is not true.

$$M, w \models {\sim}\varphi \Leftrightarrow M, w \not\models \varphi \qquad M, w \;=\!\!\mid {\sim}\varphi \Leftrightarrow M, w \models \varphi$$

In van der Hoek et al. (1996) weak negation is not used, but it is needed for a logical description of ignorance as described in the introduction of this section. A complete axiomatization of partial modal logic can be found in Jaspars and Thijsse (1996). The essential difference with classical logic is that $\neg$-introduction on consequences is absent: $\Gamma, \varphi \vdash \Delta \not\Rightarrow \Gamma \vdash \neg\varphi, \Delta$. In particular, contraposition is invalid.

The following additional axioms for the epistemic part are from van der Hoek et al. (1996):

$$\Box\varphi \vdash \varphi$$

$$\Box\varphi \vdash \Box\Box\varphi \quad \Diamond\Diamond\varphi \vdash \Diamond\varphi$$

$$\Diamond\varphi \vdash \Box\Diamond\varphi \quad \Diamond\Box\varphi \vdash \Box\varphi$$

Just as for classical logic **S5** we can give a very simple class of models for this logic. Instead of using an accessibility relation the class of models for the logic can be described as so-called *balloon models*. A balloon model M consists of a non-empty set of worlds W, a root g and a partial global valuation function $V : W \cup \{g\} \longrightarrow (\mathcal{P} \rightharpoonup \{0, 1\})$ such that there exist a world $v \in W$ with $V(v)(p) = V(g)(p)$ whenever $V(v)(p) \in \{0, 1\}$. The last requirement settles the weaker version of the T-axiom as mentioned in the introduction of this section.

Evaluation for the propositional connectives is the same as in the case of regular partial Kripke models. Evaluation of knowledge formulas is arranged in the following way:

$$\begin{array}{lcl} M, g \models \Box\varphi & \Leftrightarrow & M^w, w \models \varphi \text{ for all } w \in W \\ M, g \;=\!\!\mid \Box\varphi & \Leftrightarrow & M^w, w \models \varphi \text{ for certain } w \in W \end{array}$$

The model M^w is the same as M with the root g replaced by w.

A simple balloon M consisting of a p-root g and a single epistemic alternative at which no proposition variable has a definite truth-value is a simple countermodel for the contraposition of T: $M, g \models p$ but $M, g \not\models \Diamond p$.

5.2. Minimal Knowledge in Partial Epistemic Logic

In partial logic bisimulation-style orders can be implemented quite straightforwardly. If we use the language without weak negation it is even possible to extend the theories of possible worlds.

A partial truth-value assignment V is an *extension* of a second partial truth-value assignment V' if $V(p) = V'(p)$ for all p such that $V'(p) \in \{0, 1\}$. We write

$V' \sqsubseteq V$. The epistemic version for partial **S5** of this simple extension relation can be obtained by a bisimulation remake for balloons:

$$\langle M, g\rangle \sqsubseteq \langle M', g'\rangle \Leftrightarrow \begin{cases} 1.\ V(g) \sqsubseteq V'(g) \\ 2.\ \forall w \in W\, \exists w' \in W' : V(w) \sqsubseteq V'(w') \\ 3.\ \forall w' \in W'\, \exists w \in W : V(w) \sqsubseteq V'(w') \end{cases}$$

It is not hard to verify that $\langle M, g\rangle \sqsubseteq \langle M', g'\rangle$ iff $\mathrm{Th}(M, g) \subseteq \mathrm{Th}(M', g')$.

A positive order can be instantiated by using only the third requirement. In that case we write $\langle M, g\rangle \sqsubseteq^{\Box} \langle M', g'\rangle$. Honesty over this order has been called *weak* honesty in van der Hoek et al. (1996), and it fits within the scheme of Theorem 3.15 with $\mathcal{L}^*$ being the knowledge formulas $\Box\varphi$ with φ being a positive formula. In fact, the language can be compressed without loss of the honesty result by replacing the positive language by the objective language. In this way, a truly partial variant of Halpern and Moses (1985) has been defined.

The difference with the total **S5**-analysis of Halpern and Moses (1985) is that $\Diamond$-formulas are no longer representing ignorance. This pops up when we use minimal non-monotonic entailment over $\sqsubseteq^{\Box}$. For classical **S5** we get $p \mid\!\sim \Diamond q$. A minimal classical **S5**-model consists of all possible p-worlds and therefore q is considered to be possible by someone who only knows that p. A single p-world with all other propositions left undefined is also a minimal balloon model in the case of partial **S5** and therefore $\Diamond q$ cannot be inferred from only knowing p in partial **S5**. Still, minimal entailment is non-monotonic for partial **S5**: $p \vee q \mid\!\sim \Diamond q$ whereas $p \vee q, \neg q \not\mid\!\sim \Diamond q$.

In van der Hoek et al. (1996) a stronger notion of honesty has been defined. The reason is that partial logic makes it possible to minimize $\Diamond$-information as well. Utterances like "I doubt p or I doubt q" can be considered to be as dishonest as the knowledge examples as given in the introduction of this paper. Moreover, by minimizing possibilities we end up with really small minimal models.

If the second (forth) bisimulation requirement in the definition of balloon extension holds between a pair $\langle M, g\rangle$ and $\langle M', g'\rangle$, we write $\langle M, g\rangle \sqsubseteq^{\Diamond} \langle M', g'\rangle$. A formula is then called *strongly honest* if it is weakly honest *and* among the minimal models over $\sqsubseteq^{\Box}$ there exists a model which is minimal with respect to $\sqsubseteq^{\Diamond}$. This settles strong dishonesty for formulas such as $\Box p \vee \Diamond q$ and $\Diamond p \vee \Diamond q$.

6. Conclusion

We demonstrated various ways to capture minimal knowledge of agents. The approaches varied in a number of dimensions. Firstly, there is the choice of modeling only knowing with an explicit operator in the object language. Then, since properties like introspection play a dominant role in the behaviour of only knowing, one has to determine the underlying background epistemic logic. Finally, in order to pinpoint what objective statements are for a given agent, one has to distinguish between one agent and multi-agent systems.

Due to space limitations, we could not cover all contributions to this subject. For instance, we did not consider time- or space-complexity for reasoning about or with minimal knowledge. The paper Halpern and Lakemeyer (2001) provides a result about satisfiability in the system **Hl** presented in Section 4.1. Rosati (1997) considers complexity of only knowing for the propositional case, and, maybe surprisingly, comparing his results with known computational properties of non-monotonic formalisms (see e.g., Gottlob (1992)), he concludes that, given some well-accepted assumptions on the polynomial hierarchy, reasoning using minimal models is harder than for systems with an explicit operator for only knowing.

There are many ways in which the various approaches reported here may be extended or combined in future research. Just to hint upon three directions: it is interesting to give an account of the minimal model approach by using an explicit knowledge operator. Giving constructive ways to obtain a minimal model for an honest formula has both theoretical and practical challenges, and, last but not least, we think that the partial approach to (minimal) knowledge might be revived by, for instance, extending it to the multi-agent case.

References

Barwise, J. and J. Perry: 1983, *Situations and Attitudes*, Cambridge, MA, MIT Press.

Chellas, B. F.: 1980, *Modal Logic. An Introduction*, Cambridge University Press.

Gottlob, G.: 1992, 'Complexity Results for Nonmonotonic Logics', *Journal of Logic and Computation* **2**(3), 397–425.

Grice, P.: 1975, 'Logic and Conversation', in P. Cole and J. Morgan (eds), *Speech Acts, Syntax and Semantics III*, New York, Academic Press, pp. 41–58.

Halpern, J. Y.: 1993, 'Reasoning about Only Knowing with Many Agents', in *Proceedings National Conference on Artificial Intelligence (AAAI'93)*, pp. 655–661.

Halpern, J. Y.: 1997, 'Theory of Knowledge and Ignorance for Many Agents', in *Journal of Logic and Computation*, **7**(1), 79–108.

Halpern, J. Y. and G. Lakemeyer: 1995, 'Levesque's Axiomatization of Only Knowing is Incomplete', *Artificial Intelligence*, **74**, 381–387.

Halpern, J. Y. and G. Lakemeyer: 2001, 'Multi-agent Only Knowing', *Journal of Logic and Computation* **11**(1), 41–70.

Halpern, J. Y. and Y. Moses: 1985, 'Towards a Theory of Knowledge and Ignorance', in Kr. Apt (ed.), *Logics and Models of Concurrent Systems*, Berlin, Springer-Verlag.

Hughes G. and M. Cresswell: 1984, *A Companion to Modal Logic*, London, Methuen.

Humberstone, I. L.: 1986, 'A More Discriminating Approach to Modal Logic', *Journal of Symbolic Logic* **51**(2), 503–504.

Jaspars, J. O. M.: 1991, 'A Generalization of Stability and Its Application to Circumscription of Positive Introspective Knowledge', *Proceedings of the Ninth Workshop on Computer Science Logic* (CSL'90), Berlin, Springer–Verlag.

Jaspars, J. and E. G. C. Thijsse: 1996, 'Fundamentals of Partial Modal Logic', in P. Doherty (ed.), *Partiality, Modality, Nonmonotonicity*, Stanford, CSLI Publications, Studies in Logic, Language and Information, pp. 111–141.

Konolige, K.: 1989, 'On the Relation between Default and Autoepistemic Logic', *Artificial Intelligence* **35**, 343–382.

Lakemeyer, G.: 1993, 'All They Know: A Study in Multi-agent Auto Epistemic Reasoning', IJCAI'93, pp. 376–381.

Levesque, H. J.: 1990, 'All I Know: A Study in Auto-epistemic Logic', *Artificial Intelligence* **42**(3), 263–309.

Moore, R. C.: 1985, 'Semantical Considerations on Non-monotonic Logic', *Artificial Intelligence* **25**, 75–94.

Parikh, R.: 1991, 'Monotonic and Nonmonotonic Logics of Knowledge', *Fundamenta Informaticae* **15**, 255–274.

Pratt-Hartmann, I.: 2000, 'Total Knowledge', In AAAI-00, pp. 423–428.

Rosati, R.: 1997, 'Complexity of Only Knowing: The Propositional Case', in *Proceedings of LPNMR'97*, LNAI 1265, Springer-Verlag, pp. 76–91.

Schwarz G. and M. Truszczyński: 1994, 'Minimal Knowledge Problem: A New Approach', *Artificial Intelligence* **67**, 113–141.

van Benthem, J.: 1983, *Modal Logic and Classical Logic*, Napoli, Bibliopolis.

van Ditmarsch, H.: 1999, 'The Logic of Knowledge Games: Showing a Card', *Proceedings of BNAIC 1999*, pp. 34–42.

van der Hoek, W., J. O. M. Jaspars and E. G. C. Thijsse: 1996, 'Honesty in Partial Logic', *Studia Logica* **56**(3), 323–360.

van der Hoek, W., J. O. M. Jaspars and E. G. C. Thijsse: 2000, 'Persistence and Minimality in Epistemic Logic', *Annals of Mathematics and Artificial Inteliigence* **27** (1999), 25–47.

van der Hoek W. and E. G. C. Thijsse: 2002, 'A General Approach to Multi-Agent Minimal Knowledge: With Tools and Samples', *Studia Logica* **72**, 61–84.

ACTION-THEORETIC ASPECTS OF THEORY CHOICE

HEINRICH WANSING
Institute of Philosophy, Dresden University of Technology, 01062 Dresden, Germany,
E-mail: wansing@mailbox.tu-dresden.de

Abstract. In the controversy about doxastic voluntarism, epistemological and logical issues meet with fundamental considerations in the philosophy of science. The common interface is the logic of theory choice, and the present paper aims at making a contribution to understanding theory choices as concrete actions of belief formation.

Introduction

According to William James (1896, 101),

> [t]hroughout the breadth of physical nature facts are what they are quite independently of us, and seldom is there any such hurry about them that the risk of being duped by believing a premature theory need be faced. The questions here are always trivial options, the hypotheses are hardly living (at any rate not for us spectators), the choice between believing truth or falsehood is seldom forced.

In contrast to James's view, in post-Kuhnian philosophy of science, the choice between competing theories or theory complexes is often characterized as possibly pressing, and in any case the idea of such a choice plays a central role in explaining scientific rationality. In Laudan's opinion, for example, rationality is parasitic upon the progressiveness of science and "*consists in making the most progressive theory choices*" (Laudan 1977, 6). The question I intend to address in this paper is:

What does it mean to acquire (or choose) a theory in a justified way?

In the most simple cases, a theory is, or may be represented, just as a declarative sentence. In mathematics, for instance, Zermelo-Fraenkel set theory with the Axiom of Choice (ZFC) consists of a finite number of axioms forming the theory base together with everything derivable from these axioms using first-order predicate logic. ZFC is therefore representable as the finite conjunction of its finite theory base, and in general a finitely axiomatized theory can be represented as the conjunction of its theory base, with the understanding that this base is deductively closed under some consequence relation, which need not be the classical one. In such a simple case, theory acquisition amounts to acquiring a belief expressed by the conjunction of the finite theory base. I decide to believe ZFC, if I acquire the belief expressed

S. Rahman et al. (eds.), Logic, Epistemology, and the Unity of Science, pp. 419–435

by the conjunction of its theory base. Starting from this simple notion of a theory, I shall first address the following question:

What does it mean that a doxastic subject acquires a belief?

The starting point for this inquiry is an epistemological position that is known as *doxastic voluntarism*. Put roughly and imprecisely, the thesis of doxastic voluntarism is that the acquisition of beliefs is subject to the will. Doxastic voluntarism thus makes a claim about the nature of our human cognitive faculties, and it is not surprising that doxastic voluntarism is highly controversial in epistemology. Among the voluntarists we can find Thomas Aquinas, René Descartes, John Locke, Søren Kierkegaard, William James, Roderick Chisholm, and Bas van Fraassen. Also anti-voluntarism has been given a prominent voice by, for instance, David Hume, Bernard Williams, Louis Pojman, Jonathan Bennett, and Robert Audi.

In Section 1, I shall have a somewhat closer look at the voluntarist's thesis. A number of versions of doxastic voluntarism will be distinguished; in particular, a distinction will be drawn between two varieties of doxastic voluntarism which I shall call possibilistic and factual voluntarism. Whereas possibilistic voluntarism makes a claim about the possibility of acquiring beliefs at will, factual voluntarism makes a claim about beliefs we already have. One version of *possibilistic* voluntarism has been criticized by Bernard Williams (1973). Williams presents an argument purporting to show that the version of doxastic voluntarism in question is necessarily false, that it is mistaken for conceptual reasons. After having drawn my conceptual and terminological distinctions, in Section 2 I shall first briefly discuss both the argument presented by Williams and another anti-voluntaristic argument due to Louis Pojman (1985). The result of this discussion will be negative. Williams does not prove the inconsistency of doxastic voluntarism, nor can Pojman show that doxastic voluntarism gives rise to sentences that are 'pragmatically inconsistent' in the way Moore sentences are.

The failure of these attempted refutations of doxastic voluntarism does not, of course, show that doxastic voluntarism in fact has models and hence is consistent. In Section 3, a positive contribution is made by developing models for a particular version of *factual* voluntarism. My proposal is to interpret ascriptions of belief acquisition in the framework of a certain modal theory of concrete actions, namely a version of the *Seeing-to-it-that* theory (or short stit theory) developed at Pittsburgh University by Nuel Belnap, Michael Perloff and Ming Xu, see (Belnap et al. 2001) and references therein. Independently of the Pittsburgh group, this theory of deliberatively seeing to it that was suggested by von Kutschera (1986) and also by Horty (1989).

Not every theory and none of the developed theories of any empirical science is of the simple kind of ZFC. Elaborated theories in the natural sciences and the humanities have a much richer internal structure. Moreover, according to the structuralist conception of scientific theories (Balzer et al. 1987), a theory should not be seen as a syntactic entity, but rather as a complex model-theoretic entity. Although a theory-net in the structuralist sense determines, up to logical equivalence, a sentence expressing its empirical contents, this so-called Ramsey sentence is not to be identified with the theory-net. Furthermore, the relevant type of justified theory acquisition is not that of a single doxastic subject deciding to believe a theory, but that of a group of scientists, a scientific community, choosing between theories. In order to elucidate what it means that a scientific community sees to it that it believes a theory, an explication of joint agency is needed. This will be the topic of the fourth and final section.

1. What is the Claim of Doxastic Voluntarism?

A manifesto-like statement of doxastic voluntarism is van Fraassen's (1984) claim: "Belief is a matter of the will." Other typical formulations are slightly more informative; according to Heil (1984), for example, doxastic voluntarism is "the view that beliefs are formed through an act of the will". We shall not discuss here what is meant by 'the will', but we shall assume that what is at stake is deciding to believe or acquiring beliefs. What is a belief? Usually (or at least often), philosophers assume that a belief is a psychological state of a doxastic subject.[1] Deciding to believe thus means deciding to enter a certain psychological state. It is also assumed that a belief has a content which is a proposition.[2] Propositions are expressed by declarative sentences, and a proposition can be true or false either by itself or on the strength of a sentence expressing the proposition. Usually, these assumptions are not controversial. But then, is it not outright absurd to suppose that beliefs are subject to the will? And do not beliefs come quite involuntarily to our minds? Do not the facts impose themselves upon us? Do not, for example, visual observations evoke beliefs of doxastic subjects without any mediating decisions to believe? To clarify questions of this kind, it is useful to consider first of all what exactly the voluntarists claim.[3] Contending that belief is a matter of the will admits many different possible readings, and the thesis of doxastic voluntarism may become a complicated statement if modalities are introduced. The following readings have been distinguished in (Wansing 2000):

1. It is possible that one voluntarily acquires arbitrary beliefs in full consciousness. (Universal possibilistic voluntarism UPV)
2. It is possible that one voluntarily acquires some beliefs in full consciousness. (Existential possibilistic voluntarism EPV)

3.1 For all beliefs one acquires it holds true that one voluntarily acquires these beliefs. (Universal weak factual voluntarism UWFV)

3.2 For all beliefs one acquires it holds true that one voluntarily acquires these beliefs in full consciousness. (Universal strong factual voluntarism USFV)

4.1 For some beliefs one acquires it holds true that one voluntarily acquires these beliefs. (Existential weak factual voluntarism EWFV)

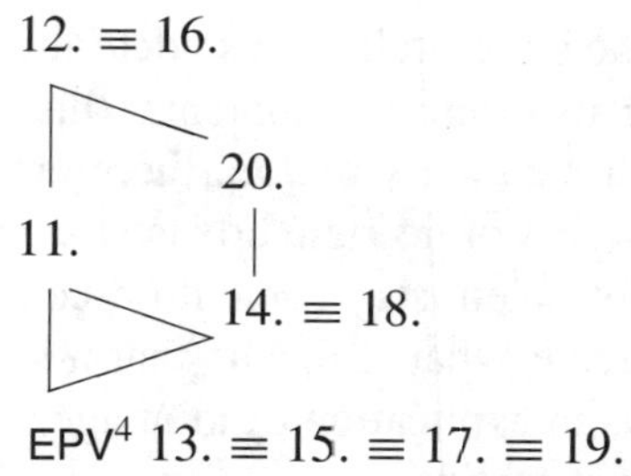

Figure 1. Some claims in the vicinity of possibilistic voluntarism.

4.2 For some beliefs one acquires it holds true that one voluntarily acquires these beliefs in full consciousness. (Existential strong factual voluntarism ESFV)

But this list is just a first step toward disentangling versions of doxastic voluntarism, because also some of these claims support different readings. A translation into formal notation may help. I shall use the letters x, y (α, β) as variables (constants) for doxastic subjects and A, B as variables for formulas expressing the contents of beliefs. Furthermore, I shall use some abbreviations, namely: *x ab A* ("*x* acquires the belief that *A*"), *x vab A* ("*x* voluntarily acquires the belief that *A*"), and *x vabc A* ("*x* voluntarily acquires in full consciousness the belief that *A*"). Let us draw some distinctions within possibilistic voluntarism:

1.	$\Diamond\forall x\forall A\ x\ vabc\ A$	11.	$\Diamond\forall x\exists A\ x\ vabc\ A$
2.	$\Diamond\exists x\forall A\ x\ vabc\ A$	12.	$\Diamond\exists A\forall x\ x\ vabc\ A$
3.	$\Diamond\forall A\exists x\ x\ vabc\ A$	13.	$\Diamond\exists x\exists A\ x\ vabc\ A$
4.	$\forall x\Diamond\forall A\ x\ vabc\ A$	14.	$\forall x\Diamond\exists A\ x\ vabc\ A$
5.	$\exists x\Diamond\forall A\ x\ vabc\ A$	15.	$\exists x\Diamond\exists A\ x\ vabc\ A$
6.	$\forall A\Diamond\forall x\ x\ vabc\ A$	16.	$\exists A\Diamond\forall x\ x\ vabc\ A$
7.	$\forall A\Diamond\exists x\ x\ vabc\ A$	17.	$\exists A\Diamond\exists x\ x\ vabc\ A$
8.	$\forall x\forall A\Diamond\ x\ vabc\ A$	18.	$\forall x\exists A\Diamond\ x\ vabc\ A$
9.	$\exists x\forall A\Diamond\ x\ vabc\ A$	19.	$\exists x\exists A\Diamond\ x\ vabc\ A$
10.	$\forall A\exists x\Diamond\ x\ vabc\ A$	20.	$\exists A\forall x\Diamond\ x\ vabc\ A$

This list can be simplified if we assume that we are working in the modal logic **S5**, so that the Barcan-Formula holds, and the following equivalence is valid: $\Diamond\exists x A(x) \equiv \exists x\Diamond A(x)$. For the right column, we then obtain Figure 1, where the arrangement of formulas reflects the obvious entailment patterns. Another complication arises from the possible readings of the m odality $\Diamond$. Are we talking about logical or psychological possibility? Williams (1973) claimed to have shown that 13. $\Diamond\exists x\exists A\ x\ vabc\ A$ is false, where $\Diamond$ is logical possibility.

Whereas possibilistic voluntarists assume that the voluntary acquisition of beliefs is possible, proponents of factual voluntarism claim of already acquired beliefs that they have been acquired at will. Possibilistic voluntarism usually comes with the assumption that the content of beliefs is presented to an agent, who deliberatively acquires or discards the beliefs in question or not. Since in such cases the doxastic subject is aware of the fact that it is in a choice situation, drawing a distinction between voluntary belief acquisition simpliciter and voluntary belief acquisition in full consciousness (of making a decision) does not make much sense, although the distinction can be drawn, of course. For factual voluntarism this distinction does make sense and leads to differentiating between weak and strong factual voluntarism. We obtain the following readings:[5]

UWFV $\forall A \forall x\ (x\ ab\ A \supset x\ vab\ A)$,
USFV $\forall A \forall x\ (x\ ab\ A \supset x\ vabc\ A)$,
EWFV $\exists A \forall x\ (x\ ab\ A \supset x\ vab\ A)$,
ESFW $\exists A \forall x\ (x\ ab\ A \supset x\ vabc\ A)$.

The thesis of universal possibilistic voluntarism, UPV, whether it is understood as 1. $\Diamond \forall x \forall A\ x\ vabc\ A$ or 2. $\Diamond \exists x \forall A\ x\ vabc\ A$ (or as 4. or 5.), is not a serious variant of voluntarism. According to James (1896), voluntary belief acquisition lawfully occurs if a doxastic subject is confronted with a *genuine option*. An option in the sense of James is a decision between a pair of hypotheses. An option between *A* and *B* is genuine for subject *x* at a certain moment if it satisfies the following three conditions:[6]

- Both *A* and *B* are *live for x* (and hence the option is (*living*)), meaning that both sentences are serious contenders for being the contents of *x*'s belief (at the given moment);
- The option is *forced* in the sense that *A* and *B* can neither both be true nor both be false;
- The choice between believing *A* or *B* is *momentous*; it is a unique irrevocable opportunity for *x* which has important consequences for the subject.

Since an agent as a finitary entity has only a finite number of options, it is clearly logically impossible that for a given agent and every sentence *A* (of a language with infinitely many sentences), '*A* or not *A*' is a genuine option for the agent. Since the liveness of a hypothesis is a matter of degree, even many of the propositions an agent is aware of or considering will fail to be vivid enough to turn up in the agent's forced and momentous options. The class of semantical models to be presented contains models in which for every subject *x* and at least for all *A* such that *A* is neither logically true nor logically false, *x vab A* is satisfiable. In this (restricted) sense we obtain models of UWFV. If a belief is acquired in full consciousness, it seems unlikely that this feature of belief acquisition can be explicated without reference to pragmatic parameters such as intentions. But intentions are difficult to model, and it seems more promising to look for general models enabling the interpretation of ascriptions of voluntary belief acquisition, even if the beliefs in question are not acquired in full consciousness. Models of USFV will then have to be obtained by imposing additional pragmatic conditions. Although every concrete action is voluntary insofar as it requires choices of the agents involved, the agents are not always conscious of a choice when they choose; they just act. As Hoyler (1983, 275 f.) explains, "[u]nconscious choices are certainly an expression of the will and we should surely come to a distorted view of human agency (the will) and human responsibility if we ignored them." Voluntarily acquiring a belief means "that there are ways in which our beliefs could be different as a direct result of our own agency."

2. Two Arguments against Doxastic Voluntarism

2.1. INCONSISTENCY

Bernard Williams's argument against EPV has been cited and discussed many times. I shall repeat it here again together with an earlier critique, because there still seems to be no consensus that the argument fails, and, moreover, this is an opportunity to reject a recent approval of Williams's argument. Here is what Williams (1973, 148) says:

> If I could acquire a belief at will, I could acquire it whether it was true or not; moreover I would know that I could acquire it whether it was true or not. If in full consciousness I could will to acquire a 'belief' irrespective of its truth, it is unclear that before the event I could seriously think of it as a belief, i.e., as something purporting to represent reality. At the very least, there must be a restriction on what is the case after the event; since I could not then, in full consciousness, regard this as a belief of mine, i.e., something I take to be true, and also know that I acquired it at will. With regard to no belief could I know – or, if all this is to be done in full consciousness, even suspect – that I had acquired it at will. But if I can acquire beliefs at will, I must know that I am able to do this; and could I know that I was capable of this feat, if with regard to every feat of this kind which I had performed I necessarily had to believe that it had not taken place?

A philologically detailed reconstruction of the argument has been presented by Winters (1979). She comes to the conclusion that the argument fails in the first place, because it rests on the false assumption that if an agent voluntarily acquires a belief, then after this event it is impossible that the agent takes the content of the belief to be true and believes that she or he acquired the belief at will. Winters remarks that the assumption that an agent voluntarily acquired a belief in full consciousness does not preclude at a later moment the agent taking the content of the belief to be true "for reasons other than those involved in the original acquisition" (Winters 1979, 253). Against the background of this observation, Winters tries to make plausible that voluntary belief *sustainment* is impossible.

In formal notation, the reconstruction of Williams's argument presented in (Wansing 2000) is as follows, where $T_x A$ abbreviates "x takes A to be true", and $K_x A$ abbreviates "x knows that A":

$\langle 1\rangle$	$\neg\Diamond\exists x\exists A\,(T_x A \wedge K_x\Diamond x\ vabc\ A)$	assumption
$\langle 2\rangle$	$\forall x\forall A\,(\Diamond x\ vabc\ A \supset K_x\Diamond x\ vabc\ A)$	assumption
$\langle 3\rangle$	$\forall x\forall A\,(K_x\Diamond\, x\ vabc\ A \supset \neg T_x A)$	$\langle 1\rangle$
$\langle 4\rangle$	$\forall x\forall A\,(\Diamond x\ vabc\ A \supset \neg T_x A)$	$\langle 2\rangle$, $\langle 3\rangle$, transitivity of $\supset$
$\langle 5\rangle$	$\forall x\forall A\,(T_x A \supset \neg\Diamond x\ vabc\ A)$	$\langle 4\rangle$, contraposition

Since Williams assumes that beliefs are taken to be true, from $\langle 5\rangle$ it follows that every belief content is such that it cannot be voluntarily acquired. Winters points out that in general, if it is possible for an agent to perform a basic (generic) action, it does not follow that the agent knows that this is possible. That is true enough. Winters's example of such a basic action is lowering the rate of one's heartbeat. But this is something different from *deciding to enter a mental state* of a certain kind. Anti-voluntarists also normally grant that there are agents x such that for some A, x can decide to imagine that A. If I can decide to imagine that A while being fully conscious of making a decision, it is plausible to assume that I know that I can decide to imagine that A. In this respect belief states are similar to imagination states. If we assume that $\langle 2\rangle$ is unproblematic, then $\langle 1\rangle$ implies the negation of EPV. But the converse holds, too. If $\langle 1\rangle$ is false, then $\Diamond\exists x\exists A\,(T_x A \wedge K_x\Diamond\, x\ vabc\ A)$ and since $K_x B$ entails B, $\Diamond\exists x\exists A\,(T_x A \wedge \Diamond\, x\ vabc\ A)$ and hence $\Diamond\exists x\exists A\Diamond\, x\ vabc\ A$. In **S5**, this is equivalent with $\Diamond\exists x\exists A\, x\ vabc\ A$. The reconstructed argument thus emerges as circular, but perhaps it is not what Williams has in mind.

For Engel (1999, 17) "[t]he argument seems to work". Engel explains that

> if a person were to have a belief merely at will, irrespective of its truth or falsity and irrespective of the evidence one has for them, the state in question would not be a belief. It would not be a belief, because it would violate the constraints (1)–(5) proposed ... above. They specify what a rational belief is according to criteria which are epistemic or cognitive. And the argument claims that a state which does not obey such criteria is not a genuine belief.

Constraint (1) is this: "Beliefs are involuntary and not normally subject to direct voluntary control" (Engel 1999, 10). Hence, under Engel's reading, Williams's argument is concerned with rational belief. But then direct voluntarism (or volitionism, as Engel calls this doctrine) is 'refuted' trivially by pointing out that by definition voluntary belief fails to be rational.

2.2. INCOHERENCE

Another prominent objection against EPV, presented by Pojman (1985), is a claimed analogy to the Moore paradox. The Moore paradox consists in the observation that serious utterances of sentences of a certain type are strange or display a certain incoherence. Moore sentences are sentences of the following form:

(Moore sentences) (a) *A* and I do not believe that *A*.
(b) ¬*A* and I believe that *A*.

If the assumption is made that "the most straightforward or elementary expression of my belief that *p*, is the assertion *that p*" (Williams 1973, 137), then an agent who seriously utters a Moore sentence of type (a) claims both to believe and not to believe a certain proposition. And if an agent seriously utters a Moore sentence of type (b), then the agent claims to believe both a certain proposition and its negation, i.e., to have an inconsistent belief. Instead of Moore sentences, Pojman considers sentences of the following shape:[7]

(Pojman sentences) *A* and I believe that *A* for other than truth considerations.

Surprisingly, Pojman sentences are not about the voluntary acquisition of the belief that *A* but about the belief that *A*. Indeed, Pojman loses sight of belief acquisition when he claims that "a person ... cannot knowingly or in full consciousness acquire a belief or *continue to hold a belief* simply by willing to believe a proposition" (Pojman 1985, 49 (my emphasis)). If belief *formation* is debated, one ought to consider sentences of the following form:

(Pojman* sentences) *A* and I have voluntarily acquired the belief that *A* in full consciousness.

Like Williams and other anti-voluntarists, Pojman assumes that the idea of voluntary belief acquisition implies that the belief acquisition of an agent is independent of any considerations of the agent with respect to the truth or falsity of the belief content in question. (Therefore Pojman* sentences could also be specified as having the form '*A* and I have voluntarily acquired the belief that *A* in full consciousness for other than truth considerations'.) A person who seriously utters a Pojman* sentence thus claims to take a certain belief content *A* to be true, and to have acquired the belief that *A* independently of any considerations with respect to the truth or falsity of *A*. However, a serious utterance of a Pojman* sentence fails to be incoherent or strange in a way that is analogous to a serious utterance of a Moore sentence. The assumptions that a serious utterance of *A* expresses the utterer's belief that *A* and that voluntary belief acquisition is independent of truth considerations of doxastic subjects do not jointly imply that a person who seriously utters 'I believe that A, and I have voluntarily acquired the belief that *A* in full consciousness' claims both to believe that *A* and not to believe that *A*.[8] If an agent has acquired the belief that *A* independently of truth considerations, she or he may nevertheless believe that *A* and have reasons for taking *A* to be true. There is no analogy to the Moore paradox.[9]

The assumption that the voluntary acquisition of beliefs is independent of any considerations agents undertake with respect to the truth or falsity of their belief content is also shared by Winters (1979), who regards this assumption as a

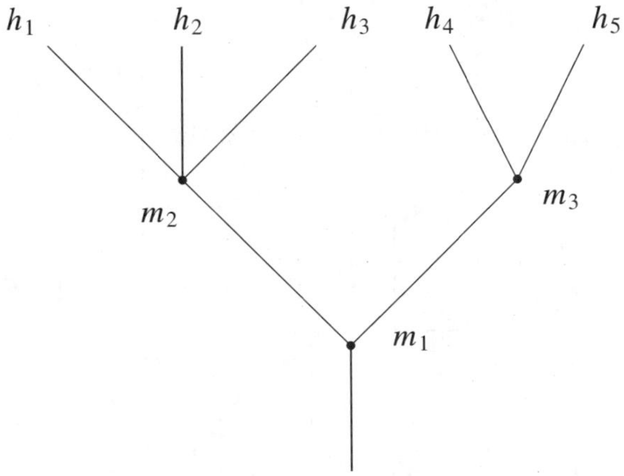

Figure 2. A branching tree of moments of time.

necessary condition for voluntary belief acquisition. This condition can be easily misunderstood, if it is interpreted as in (Pojman 1985, 40) to mean that voluntarily acquired beliefs are underdetermined by the evidence. It is conceivable that an agent – in full consciousness of making a decision or not– decides to believe that A and could have done otherwise, also if the agent classifies the evidence for A as overwhelming.

A spectrum of failed attempts to refute EPV can be found in (Bennett 1990).

3. Models of Doxastic Voluntarism

The understanding of the models for interpreting ascriptions of voluntary belief acquisition is based on the assumption of so-called branching time, see (Belnap et al. 2001), (Thomason 1984) and references therein. The basic idea is that moments of time have a tree-like structure. Such a tree branches toward the future, and this reflects the openness or indeterminacy of the future. Moreover, it is assumed that there is no branching toward the past, and this assumption is meant to reflect the determinacy of the past. Linearly ordered sets of moments from a tree are called histories, if they are maximal, i.e., if they are not contained in any larger linearly ordered set of moments from the tree. Intuitively, these histories may be seen as complete possible temporal developments of the world. In other words, a history represents a possible development the world might take. If one considers a moment that belongs to a history, one may also say that this history passes through the moment, or that the moment occurs in the history. Since it is assumed that the future is open, branching may occur and more than one history may pass through a given moment. These ideas can be nicely depicted, see Figure 2.

There are good reasons for interpreting sentences not just as true or false at moments of a tree, but as true or false at moment/history-pairs (m, h). The semantics of future contingencies makes this clear. Suppose that in Figure 2 A is

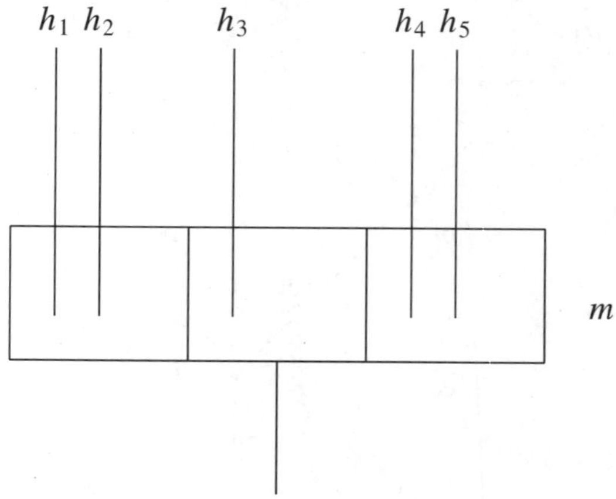

Figure 3. Choice-cells of an agent at moment m.

true at moment m_3 and false at moments m_1 and m_2. What can be said about the truth or falsity of 'Sometimes in the future A' at m_1? With respect to histories h_4 and h_5 the claim is true but with respect to histories $h_1 - h_3$ it is not. But if future contingencies are to be evaluated at moment/history-pairs, then it makes sense to evaluate every sentence at such complexes.

I shall now present a proposal for interpreting ascriptions of voluntary belief acquisition. A doxastic subject is also supposed to be an agent who by her or his actions can influence the further development of the world. In stit theory this idea is accounted for by assuming that for every individual agent, the histories passing through a moment are partitioned into sets of histories choice-equivalent for the agent. If two histories h and h' are choice-equivalent for an agent x at moment m, then x cannot discriminate by her or his actions at m between h or h' as the development of the world. The sets of histories choice-equivalent for an agent at a moment m represent the 'choice-cells' of the agent at m. A very natural postulate then is that for every agent x, histories that pass through a moment m and divide only at a later moment must be choice-equivalent for x at m. The idea of choice-cells can also be nicely graphically represented, see Figure 3.

In a first step, the dstit theory developed by von Kutschera (1986) (and, independently, by Horty (1989)) provides a simple semantics for ascriptions of concrete actions. Here are the definitions from dstit theory that make the above informal semantical remarks precise. A pair $\langle T, \leq \rangle$, is called a *branching temporal frame* if T is a non-empty set (of moments), and $\leq$ is a partial order on T satisfying *historical connectedness* $(\forall m_1 \forall m_2 \exists m (m \leq m_1 \wedge m \leq m_2))$ and *no backward branching* $(\forall m \forall m_1 \forall m_2 ((m_1 \leq m \wedge m_2 \leq m) \supset (m_1 \leq m_2 \vee m_2 \leq m_1)))$. A *history in T* is a maximal set of moments (in T) linearly ordered by $<$, where $m < m'$ iff $m \leq m'$ and $m \neq m'$. The set of histories passing through moment m, H_m, is defined as $\{h \mid h \text{ is history and } m \in h\}$. If $\langle \mathrm{T}, \leq \rangle$ is a branching temporal frame, then $\langle \mathrm{T}, \leq, Agent, Choice \rangle$ is called a *dstit frame*, if *Agent* is a nonempty set (of agents) and *Choice* is a function mapping every agent/moment-pair (x, m) to a partition of H_m (the histories *choice-equivalent for x at m*) satisfying *no choice between undivided histories* $(\forall x \in Agent)$ $(\forall H \in Choice(x, m))$ $\forall h \forall h'[(h \in H \wedge \exists m'(m < m' \wedge m' \in h \cap h')) \supset h' \in H]$. If $h \in H_m$, then $Choice^m_\alpha(h)$ is the particular choice in $Choice(\alpha, m)$ containing h. A *dstit model* is a structure $\langle T, \leq, Agent, Choice, v\rangle$, where $\langle T, \leq, Agent, Choice\rangle$ is a dstit frame, and v is a valuation function that interprets atomic formulas by sets of moment/history-pairs. A complete axiomatization of dstit logic has been presented in (Xu 1998). The truth definition for [α *dstit*: A] ("α deliberatively sees to it that A") is as follows.

DEFINITION 1. [α dstit : A] is true in $\langle T, \leq, Agent, Choice, v\rangle$ at (m, h) iff (i) $\forall h' \in Choice^m_\alpha(h)$ A is true at (m, h'), and (ii) $\exists h' \in H_m$ such that A is not true at (m, h').

There seems to be an agreement that the propositional logic of implicit belief, the belief of completely rational, logically omniscient agents, is the polymodal logic **KD45**, see (Fagin et al. 1995). If so, then for every agent x, 'x implicitly believes that A' may be adequately translated as $B_x A$, where B_x is a **KD45** necessity operator. Intuitively, $B_x A$ is true at a moment/history-pair (m, h) if and only if A is true at every doxastic alternative for x at (m, h), that is, at every moment/history-pair compatible with what x believes at (m, h). Such ascriptions of implicit belief are non-agentive, because A being true at every state compatible with what x believes at (m, h) clearly fails to describe a concrete action. The idea now is to combine the dstit semantics for 'seeing to it that' with the semantics for **KD45**. My suggestion is to read 'x voluntarily acquires the (implicit) belief that A' ('x forms the (implicit) belief that A') as [α *dstit*: $B_\alpha A$].

A *doxastic dstit model* is a structure $\langle T, \leq, Agent, Choice, R, v\rangle$, where $\langle T, \leq, Agent, Choice, v\rangle$ is a dstit model and $R = \{R^m_\alpha \mid \alpha \in Agent, m \in T, R^m_\alpha \subseteq H_m \times H_m\}$ is a set of serial, transitive and Euclidean relations. We then obtain the following truth definitions for [α *dstit* : $B_\alpha A$] ("α deliberatively sees to it that α implicitly believes that A"):

DEFINITION 2. [α dstit: $B_\alpha A$] is true in $\langle T, \leq, Agent, Choice, R, v\rangle$ at (m, h) iff (i) $\forall h' \in Choice^m_\alpha(h)$ $\forall h'' \in H_m$, if $(m, h')R^m_\alpha(m, h'')$ then A is true at (m, h''), and (ii) $\exists h', h'' \in H_m$ such that $(m, h')R^m_\alpha(m, h'')$ and A is not true at (m, h'').

In Figure 4, the moment m is partitioned into three choice cells, for α. Moreover, the R^m_α alternatives to the histories passing through m are depicted by annotated arrows. The formula A is true at the moment/history-pairs (m, h_1), (m, h_3) and (m, h_4). In this simple example, at (m, h_2) it is true that α sees to it that α implicitly believes that A.

In Definitions 1 and 2, (i) is called the positive condition and (ii) the negative condition. Note that the negative condition in Definition 2 prevents the formation

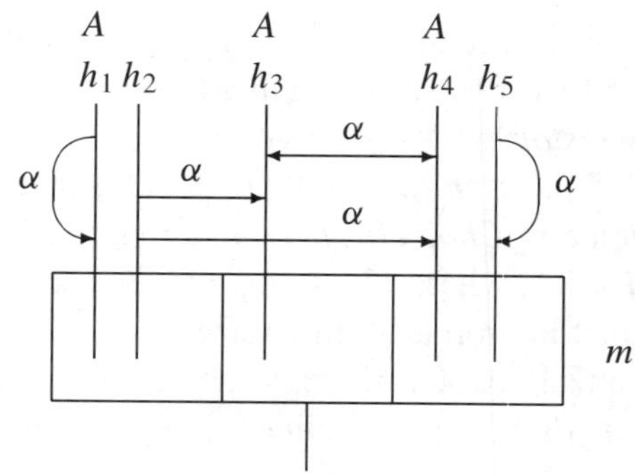

Figure 4. An example illustrating Definition 2.

of implicit beliefs from closure under logical consequence: if an agent voluntarily acquires the belief that A, and A logically implies B, then it does not follow that the agent also acquires the belief that B. Since for every agent x, $B_x\top$ is valid, no agent can *decide* to implicitly believe a valid formula $\top$; valid formulas are already implicitly believed. Since for every agent x and every moment m, the relation R_x^m is serial, $B_x\bot$ is false at every moment/history-pair and hence no agent can decide to implicitly believe a falsehood $\bot$.

If ascriptions of voluntary belief acquisition are interpreted as ascriptions of seeing to it that an agent has an implicit belief, then they have semantical models, namely doxastic dstit models. And in this sense then UWFV has also models.

4. Models of Justified Theory Choice

The choice between competing scientific theories has been a much debated issue in the philosophy of science ever since Kuhn's (1962). Because the interpretation of ascriptions of concrete actions of single agents in models of branching time can be naturally extended to a semantics for ascriptions of collective agency, this semantics suggests itself as a natural starting point for explicating the *collective acquisition* of a theory by a scientific community.

We shall not enter into a discussion of Kuhn's incommensurability thesis, semantic holism, conventionalism, or questions of translatability between theories. We shall assume, however, that there is a language allowing one to talk about the theories from a set $\mathcal{T} = \{T_1, \ldots, T_n\}$ of competing theories. Since we are interested in justified theory choice, we also assume that there is a set of accepted parameters $\mathcal{P} = \{P_1, \ldots, P_r\}$ with respect to which the theories from $\mathcal{T}$ are compared with each other. Such parameters may include, for example:

- the richness of the assumed ontology,
- the predictability of certain experimental results,
- the realization of important methodological ideas,
- the presence or absence of certain paradoxes or other anomalies,

- a theory's internal progressiveness rate in the sense of Laudan (1977).

In general, the accepted parameters for theory comparison will have different weights. Since it is assumed that the competing theories from $\mathcal{T}$ can be pairwise compared with respect to the accepted and weighted parameters from $\mathcal{P}$, we may consider claims of the following form:

Theory T_j is at least as good as theory T_k with respect to parameter P_i.
(In symbols: $T_j \geq_{P_i} T_k$)

We define that a theory T_j is designated among the theories competing with T_j if and only if for every theory T_k competing with T_j, the sum of the weights of those parameters with respect to which T_j is at least as good as T_k is greater than the sum of the weights of those parameters with respect to which T_k is at least as good as T_j. In other words: 'T_j is designated among the theories competing with T_j' (in symbols $d(T_j, \mathcal{T})$) is true at (m, h) iff for all theories T_k competing with T_j:

$$\sum_{i=1}^{r}\{w_i \mid w_i \text{ is the weight of } P_i;\ T_j \geq_{P_i} T_k \text{ at } (m,h)\} >$$
$$\sum_{i=1}^{r}\{w_i \mid w_i \text{ is the weight of } P_i;\ T_k \geq_{P_i} T_j \text{ at } (m,h)\}$$

If not merely a single doxastic subject but a group of agents Δ, a scientific community is considered, we need a semantics for ascriptions of collective belief acquisition:

(∗) Δ sees to it that Δ implicitly believes that A.

The embedded clause 'Δ implicitly believes that A' is unproblematic, since implicit, logically omniscient group belief is normally understood as implicit belief of all group members: $B_\Delta A$ if and only if for every $x \in \Delta$, $B_x A$. Therefore, a semantics for $[\Delta\ \mathit{dstit} : B_\Delta A]$ ("Δ voluntarily acquires the implicit belief that A") is available if a semantics for $[\Delta\ \mathit{dstit} : B_\alpha A]$ is available. The latter is, however, straightforward, because there already exists a formal semantics for $[\Delta\ \mathit{dstit} : A]$. If $\Gamma \subseteq \mathit{Agent}$ and h is a history passing through $m \in T$, the set $\mathit{Choice}^m_\Gamma(h)$ of histories choice-equivalent with h for Γ at moment m is defined as $\{h' \mid (\forall x \in \Gamma)\, h' \in \mathit{Choice}^m_x(h)\}$.

DEFINITION 3. $[\Gamma\ \mathit{dstit}\colon B_\beta A]$ is true in $\langle T, \leq, \mathit{Agent}, \mathit{Choice}, R, v\rangle$ at (m,h) iff (i) $\forall h' \in \mathit{Choice}^m_\Gamma(h)$, $\forall h'' \in H_m$, if $(m,h')R^m_\beta(m,h'')$ then A is true at (m,h''), and (ii) $\exists h', h'' \in H_m$ such that $(m,h')R^m_\beta(m,h'')$ and A is not true at (m,h'').

We can now observe that the formation of implicit group beliefs is neither closed under logical consequence nor closed under group membership. In other words, if $\Delta\ \mathit{vab}\ A$ and $\alpha \in \Delta$, it does not follow that $\alpha\ \mathit{vab}\ A$. Let s_m be any mapping from Agent into the the powerset of H_m such that $s_m(x) \in \mathit{Choice}(x, m)$. It has

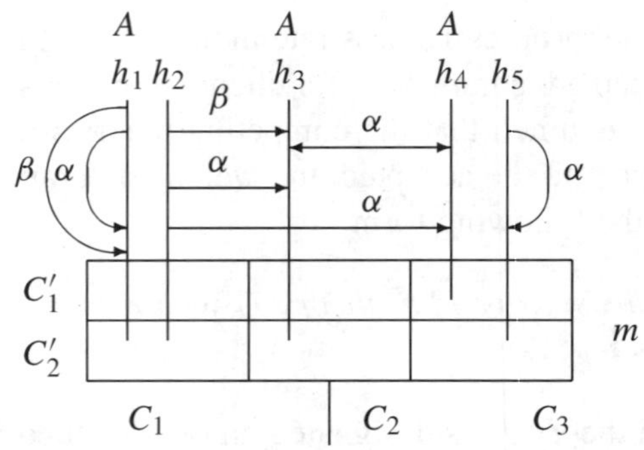

Figure 5. An example illustrating Definition 3.

been suggested in stit theory that the independence of agents can be captured by requiring that for every function s_m,

$$\bigcap_{x \in Agent} s_m(x) \neq \emptyset.$$

For a discussion of this plausible postulate see (Belnap et al. 2001, Section 10B).[10] In Figure 5, moment m is divided into six cells, for $\Gamma = \{\alpha, \beta\}$, where the vertical lines separate β's choice cells $C_1 - C_3$ and the horizontal line separates α's choice cells C'_1 and C'_2. The doxastic alternative relations of α and β are depicted by annotated arrows. In this example $[\Gamma\ dstit : B_\Gamma A]$ is true at moment/history-pair (m, h_2). The agents α and β do not act independently of each other. If α acts so as to realize h_4 and β acts so as to realize h_3, their joint action realizes *no* history.

My proposal for a semantics of ascriptions of theory acquisition is this: 'Group Γ voluntarily acquires the (implicit) belief that T' (understood in the sense that Γ chooses T from the set of competing theories $\mathcal{T}$) is true in a moment/history-pair (m, h) iff

1. $\forall h' \in \text{Choice}^m_\Gamma(h)$, $B_\Gamma d(T, \mathcal{M})$ is true at (m, h'), and
2. $\exists h' \in \text{Choice}^m_\Gamma(h)$, $B_\Gamma d(T, \mathcal{M})$ is not true at (m, h').

Among the many aspects of theory choice, its action-theoretic aspects and the problem of modeling what scientists *do* when they choose a theory have received little attention. In the present paper action-theoretic aspects of theory choice have been highlighted first by reference to the discussion of doxastic voluntarism and second by making a proposal for an application of dstit theory. The result is a semantics for ascriptions of voluntary belief acquisition and theory choice. A semantics for ascriptions of implicit as well as explicit belief along these lines is defined in (Wansing 2002).[11]

Notes

1 We neglect here the idea that a subject might be a society of minds and hence might be in more than one psychological state at a given moment or in a certain situation. We shall, however, consider collective doxastic subjects, without making any particular assumptions about 'collective psychological states'.

2 There is also the view that a belief is a disposition. H. H. Price (1954, 15) writes:

> Believing a proposition is, I think, a disposition and not an occurrence or "mental act", though the disposition is not necessarily a very long-lived one and may only last a few seconds.

For Bennett (1990) such a disposition is a function from intentions and other beliefs of a subject to actions of the subject.

3 Audi (1999) distinguishes between behavioral and genetic voluntarism, where behavioral voluntarism is the view that believing is an action type and genetic voluntarism is the view that *belief formation* is an action type. According to Audi (1999, 100), behavioral voluntarism "is a clear failure", and in the opinion of Engel (1999) it is absurd. Engel (1999) draws a distinction between voluntarism and volitionism, where voluntarism is a normative view and "volitionism", a term suggested by Pojman (1985), denotes the "psychological-descriptive thesis . . . that we can form beliefs "at will" or that we can "decide to believe" ", (1999, 11).

4 If "It is possible that one voluntarily acquires some beliefs in full consciousness" is ambiguous between 11. and 13., we resolve the ambiguity in favour of 13.

5 These readings should be uncontroversial, given the standard understanding of so-called donkey sentences, see (Geach 1962) and (Groenendijk and Stokhof 1991).

6 A discussion of these criteria may be found in (Gale 1999).

7 Actually, Pojman (1985, 48) writes "One cannot believe in full consciousness '*p* and I believe *p* for other than truth considerations.' The statement has a similar incoherence as Moore's paradox 'I believe that *p*, but *p* is false' ". But he also explains that "saying, 'I believe that *p*, but I believe it only because I want to believe it,' has the same incoherence attached to it as Moore's paradox" (1985, 49).

8 If we assume Pojman's version of the Moore paradox, namely serious utterances of 'I believe that *p*, but *p* is false', there are two assumptions under which a serious utterance of a Moore sentence is strange.

(a) If an agent *x* seriously utters " '*A*' is false", then *x* believes that '*A*' is false.

(b) If an agent *x* seriously utters " '*A*' is false", then *x* does not believe that *A*.

If (a) holds, then *x* seriously claims to believe a proposition and disbelieves it, if *x* seriously utters a Moore sentence. If (b) holds, then *x* seriously claims to believe a proposition and does not believe it, if *x* seriously utters a Moore sentence. But the assumption that voluntary belief acquisition is independent of truth considerations of doxastic subjects, does not imply that a person who seriously utters 'I believe that A, and I have voluntarily acquired the belief that *A* in full consciousness' believes that *A* is false. Thus, it is not justified to conclude in analogy with the Moore paradox that a person who seriously utters a Pojman* sentence both seriously claims to believe that *A* and disbelieves that *A*. Moreover, it also does not follow that a person seriously claims to believe that *A* but does not believe that *A*, when she or he seriously utters a Pojman* sentence.

9 Engel (1999, 19) holds that a voluntary believer is in a situation similar to the situation exemplified by a serious utterer of a Moore sentence of type (a). Engel characterizes the psychological situation of a voluntary believer as follows:

> "I believe that P (as a result of self inducement), but I do not believe that P (since self induced beliefs are not typical beliefs)"

But this description presupposes a distinction between typical and untypical beliefs and a classification of voluntarily acquired beliefs as untypical.

[10] Note that the statement of the postulate in Wansing (2000) is erroneous.

[11] In (Wansing 2000) it has been suggested to introduce for every doxastic subject $x \in Agent$ and every formula A a propositional constant $E_{x,A}$ to be understood intuitively as "x is mistaken with respect to A". Moreover, $[\alpha\ vab{:}A]$ ("α voluntarily acquires the belief that A") was defined by stipulating that $[\alpha\ vab{:}A]$ is true in a dstit model $\langle T, \leq, Agent, Choice, v\rangle$ at moment/history-pair (m, h) iff (i) $\forall h' \in Choice^m_\alpha(h)$, $((\neg A \supset E_{\alpha,A}) \wedge (A \supset \neg E_{\alpha,A}))$ is true at (m, h') and (ii) $\exists h'$ such that $m \in h'$ and $((\neg A \supset E_{\alpha,A}) \wedge (A \supset \neg E_{\alpha,A}))$ fails to be true at (m, h'). The present relational approach appears to be more flexible.

Acknowledgement

The author would like to thank two anonymous referees for their thoughtful comments.

References

Audi, R.: 1999, 'Doxastic Voluntarism and the Ethics of Belief', *Facta Philosophica* **1**, 87–109.

Balzer, W., J. D. Sneed and C. U. Moulines: 1987, *An Architectonic for Science. The Structuralist Program*, Reidel, Dordrecht.

Belnap, N., M. Perloff and M. Xu: 2001, *Facing the Future: Agents and Choices in our Indeterminist World*, Oxford, Oxford University Press.

Bennett, J.: 1990, 'Why Belief is Involuntary', *Analysis* **50**, 87–107.

Engel, P.: 1999, 'Volitionism and Voluntarism about Belief', in A. Meijers (ed.), *Belief, Cognition and the Will*, Tilburg, Tilburg University Press, pp. 9–25.

Fagin, R., Y. Halpern, Y. Moses and M. Vardi: 1995, *Reasoning about Knowledge*, Cambridge, MS, MIT Press.

van Fraassen, B.: 1984, 'Belief and the Will', *The Journal of Philosophy* **81**, 235–257.

Gale, R.: 1999, *The Divided Self of William James*, Cambdridge, Cambridge University Press.

Geach, P.: 1962, *Reference and Generality*, Ithaca, New York, Cornell University Press.

Groenendijk J. and M. Stokhof: 1991, 'Dynamic Predicate Logic', *Linguistics and Philosophy* **14**, 39–100.

Heil, J.: 1984, 'Doxastic Incontinence', *Mind* **93**, 56–70.

Horty, J.: 1989, 'An Alternative Stit-operator', Manuscript, Philosophy Department, University of Maryland.

Hoyler, R.: 1983, 'Belief and Will Revisited', *Dialogue* **22**, 273–290.

James, W.: 1896, 'The Will to Believe', in *Essays in Pragmatism*, New York, Hafner Press, 1969, pp. 88–105.

Kuhn, T.: 1962, *The Structure of Scientific Revolutions*, University of Chicago Press, Chicago.

von Kutschera, F.: 1986, 'Bewirken', *Erkenntnis* **24**, 253–281.

Laudan, L.: 1977, *Progress and its Problems*, University of California Press, Berkeley.

Pojman, L.: 1985, 'Believing and Willing', *Canadian Journal of Philosophy* **15**, 37–55.

Price, H. H.: 1954, 'Belief and the Will', *Proceedings of the Aristotelian Society*, Suppl. **28**, 1–26.

Thomason, R.: 1984, 'Combinations of Tense and Modality', in D. Gabbay and F. Guenthner (eds.), *Handbook of Philosophical Logic*, Vol. II, Dordrecht, Reidel, pp. 135–165.

Wansing, H.: 2000, 'A Reduction of Doxastic Logic to Action Logic', *Erkenntnis* **53**, 267–283.

Wansing, H.: 2002, 'Seeing to it that an Agent Forms a Belief', *Logic and Logical Philosophy* **10**, 185–197.

Williams, B.: 1973, 'Deciding to Believe', in *Problems of the Self*, New York, Cambridge University Press, pp. 136–151.

Winters, B.: 1979, 'Believing at Will', *Journal of Philosophy* **76**, 243–256.

Xu, M.: 1998, 'Axioms for Deliberative *Stit*', *Journal of Philosophical Logic* **27**, 505–552.

SOME COMPUTATIONAL CONSTRAINTS IN EPISTEMIC LOGIC

TIMOTHY WILLIAMSON
University of Oxford, New College, Oxford OX1 3BN, U.K.,
E-mail: Timothy.Williamson@philosophy.ox.ac.uk

Abstract. Some systems of modal logic, such as S5, which are often used as epistemic logics with the 'necessity' operator read as 'the agent knows that' are problematic as general epistemic logics for agents whose computational capacity does not exceed that of a Turing machine because they impose unwarranted constraints on the agent's theory of non-epistemic aspects of the world, for example by requiring the theory to be decidable rather than merely recursively axiomatizable. To generalize this idea, two constraints on an epistemic logic are formulated: *r.e. conservativeness*, that any recursively enumerable theory R in the sublanguage without the epistemic operator is conservatively extended by some recursively enumerable theory in the language with the epistemic operator which is permitted by the logic to be the agent's overall theory; the weaker requirement of *r.e. quasi-conservativeness* is similar except for applying only when R is consistent. The logic S5 is not even r.e. quasi-conservative; this result is generalized to many other modal logics. However, it is also proved that the modal logics S4, Grz and KDE are r.e. quasi-conservative and that K4, KE and the provability logic GLS are r.e. conservative. Finally, r.e. conservativeness and r.e. quasi-conservativeness are compared with related non-computational constraints.

1. Introduction

This paper concerns limits that some epistemic logics impose on the complexity of an epistemic agent's reasoning, rather than limits on the complexity of the epistemic logic itself.

As an epistemic agent, one theorizes about a world which contains the theorizing of epistemic agents, including oneself. Epistemic logicians theorize about the abstract structure of epistemic agents' theorizing. This paper concerns the comparatively simple special case of epistemic logic in which only one agent is considered. Such an epistemic agent theorizes about a world which contains that agent's theorizing. One has knowledge about one's own knowledge, or beliefs about one's own beliefs. The considerations of this paper can be generalized to multi-agent epistemic logic, but that will not be done here. Formally, single-agent epistemic logic is just standard monomodal logic; we call it 'epistemic' in view of the envisaged applications.

In epistemic logic, we typically abstract away from some practical computational limitations of all real epistemic agents. For example, we are not concerned with their failure to infer from a proposition q the disjunction $q \vee r$ for every unrelated proposition r. What matters is that if some propositions do in fact follow

S. Rahman et al. (eds.), Logic, Epistemology, and the Unity of Science, pp. 437–456.

from the agent's theory (from what the agent knows, or believes), then so too do all their logical consequences. For ease of exposition, we may idealize epistemic agents and describe them as knowing whatever follows from what they know, or as believing whatever follows from what they believe, but we could equally well redescribe the matter in less contentious terms by substituting '*p* follows from what one knows' for 'one knows *p*' or '*p* follows from what one believes' for 'one believes *p*' throughout the informal renderings of formulas, at the cost only of some clumsiness. Thus, if we so wish, we can make what looks like the notorious assumption of logical omniscience true by definition of the relevant epistemic operators. On suitable readings, it is a triviality rather than an idealization. It does not follow that no computational constraints are of any concern to epistemic logic. For if one's knowledge is logically closed by definition, that makes it computationally all the harder to know that one does *not* know something: in the standard jargon, logical omniscience poses a new threat to negative introspection. That threat is one of the phenomena to be investigated in this paper.

In a recursively axiomatizable epistemic logic, logical omniscience amounts to closure under a recursively axiomatizable system of inferences. Thus all the inferences in question can in principle be carried out by a single Turing machine, an idealized computer. Epistemic logicians do not usually want to make assumptions which would require an epistemic agent to exceed every Turing machine in computational power. In particular, such a requirement would presumably defeat the purpose of the many current applications of epistemic logic in computer science. By extension, epistemic logicians might prefer not to make assumptions which would permit an epistemic agent not to exceed every Turing machine in computational power only under highly restrictive conditions. Of course, such assumptions might be perfectly appropriate in special applications of epistemic logic to cases in which those restrictive conditions may be treated as met. But they would not be appropriate in more general theoretical uses of epistemic logic.

As an example, let us consider the so-called axiom of *negative introspection* alluded to above. It may be read as the claim that if one does not know *p* then one knows that one does not know *p*, or that if one does not believe *p* then one believes that one does not believe *p*. In terms of theories: if one's theory does not entail *p*, then one's theory entails that one's theory does not entail *p*. That assumption is acceptable in special cases for special values of '*p*'. However, for a theory to be consistent is in effect for there to be some *p* which it does not entail. On this reading, negative introspection implies that if one's theory is consistent then it entails its own consistency. But, by Gödel's second incompleteness theorem, if one's theory is recursively axiomatizable and includes Peano arithmetic, then it entails its own consistency only if it is inconsistent. Thus, combined with the incompleteness theorem, negative introspection implies that if one's theory is recursively axiomatizable then it includes Peano arithmetic only if it is inconsistent. Yet, in a wide range of interesting cases, the output of a Turing machine, or the theory of an epistemic agent of equal computational power, is a consistent recursively axiomatizable theory which includes Peano arithmetic. Thus, except in special circumstances, the negative introspection axiom imposes an unwarranted constraint on the computational power of epistemic agents.

Naturally, such an argument must be made more rigorous before we can place much confidence in it. That will be done below. The problem for the negative introspection axiom turns out to be rather general: it arises not just for extensions of Peano arithmetic but for any undecidable recursively axiomatizable theory, that is, for any theory which is the output of some Turing machine while its complement is not. It is very natural to consider epistemic agents whose theories are of that kind.

The aim of this paper is not primarily to criticize the negative introspection axiom. Rather, it is to generalize the problem to which that axiom gives rise, to formulate precisely the conditions which a system of epistemic logic must satisfy in order not to be susceptible to such problems, and to investigate which systems satisfy those conditions. The conditions in question will be called *r.e. conservativeness* and *r.e. quasi-conservativeness*. Very roughly indeed, a system satisfies these conditions if it has a wide enough variety of models in which the epistemic agent is computationally constrained. Such models appear to be among the intended models on various applications of epistemic logic. As already noted, systems of epistemic logic which do not satisfy the conditions may be appropriate for other applications. But it is time to be more precise.

2. Elementary Epistemic Logic

Let L be the language consisting of countably many propositional variables $p_0, p_1, p_2, \ldots$ (p and q represent arbitrary distinct variables), the falsity constant $\bot$ and the material conditional $\supset$. Other operators are treated as metalinguistic abbreviations in the usual way. We expand L to the language $L_\Box$ of propositional modal logic by adding the operator $\Box$. $\Diamond\alpha$ abbreviates $\neg\Box\neg\alpha$. Unless otherwise specified, the metalinguistic variables $\alpha, \beta, \gamma, \ldots$ range over all formulas of $L_\Box$. We use the necessity symbol $\Box$ from modal logic to make various formulas and formal systems look familiar, without prejudice to its interpretation. We reinterpret $\Box$ as something like 'I know that' or 'I believe that'. To generalize over reinterpretations, we use the neutral verb 'cognize' for $\Box$ in informal renditions of formulas.

A *theory* in $L_\Box$ is a subset of $L_\Box$ containing all truth-functional tautologies and closed under modus ponens for $\supset$ (MP). A *model* M of $L_\Box$ induces a function M() : $L_\Box \to \{0, 1\}$ where $M(\bot) = 0$ and $M(\alpha \supset \beta) = 1$ if and only if $M(\alpha) \leq M(\beta)$. Intuitively, $M(\alpha) = 1$ if and only if α is true in M; $M(\alpha) = 0$ if and only if α is false in M. An *application* of epistemic logic determines a class of its intended models. The logic of the application is the set of formulas α such that $M(\alpha) = 1$ for every intended model M; thus the logic is a theory in $L_\Box$. Of course, we can also define a relation of logical consequence on the models, but for present purposes it is simpler to identify a logic with the set of its theorems.

Since atomic sentences are treated simply as propositional variables, we may substitute complex formulas for them. More precisely, we assume that for each intended model M and uniform substitution σ there is an intended model M^σ such that for every α $M^\sigma(\alpha) = M(\sigma\alpha)$. Thus the logic of the application is closed under uniform substitution (US).

A *modal logic* is a theory in $L_\Box$ closed under US. The logic of an application is a modal logic. The smallest modal logic is PC, the set of all truth-functional tautologies. If Σ is a modal logic, we write $\vdash_\Sigma \alpha$ when $\alpha \in \Sigma$. For any $X \subseteq L_\Box$, we define $X \vdash_\Sigma \alpha$ if and only if $\vdash_\Sigma \bigwedge X_0 \supset \alpha$ for some finite $X_0 \subseteq X$($\bigwedge X_0$ and $\bigvee X_0$ are the conjunction and disjunction respectively of X_0 on a fixed ordering of the language). X is Σ-consistent unless $X \vdash_\Sigma \bot$. A maximal Σ-consistent set is a Σ-consistent set not properly included in any Σ-consistent set.

If M is a model, let $\Box^{-1}M = \{\alpha : M(\Box\alpha) = 1\}$. Thus $\Box^{-1}M$ expresses what the agent cognizes in M. If Σ is the logic of an application on which $\Box^{-1}M$ is a theory in $L_\Box$ for every intended model M, then for all formulas α and β, $\vdash_\Sigma \Box(\alpha \supset \beta) \supset (\Box\alpha \supset \Box\beta)$ (axiom schema K) and if $\vdash_{PC} \alpha$ then $\vdash_\Sigma \Box\alpha$ (rule RN_{PC}). A modal logic satisfying RN_{PC} and K is *prenormal.* If cognizing is knowing or believing, then prenormality is an extreme idealization, a form of logical omniscience. But if cognizing is the closure of knowing or believing under at least truth-functional consequence, then prenormality is innocuous. The rule RN_{PC} differs from the stronger and better-known rule RN (necessitation or epistemization): if $\vdash_\Sigma \alpha$ then $\vdash_\Sigma \Box\alpha$. A modal logic Σ satisfying RN and K is *normal.* Unlike RN_{PC}, RN requires the agent to cognize all theorems of the epistemic logic, not just all truth-functional tautologies. For instance, $\Box\top$ is a theorem of every prenormal logic by RN_{PC}, but since it is not a theorem of PC we cannot iterate the rule; $\Box\Box\top$ is not a theorem of the smallest prenormal logic. By contrast, we can iterate RN, and $\Box\Box\top$ is a theorem of every normal modal logic. Prenormality does not imply that agents cognize their own cognizing. It merely implies that they can formulate propositions about cognizing, for since $\Box\alpha \supset \Box\alpha$ is a truth-functional tautology, $\Box(\Box\alpha \supset \Box\alpha)$ is a theorem of every prenormal logic. Since normality entails prenormality, results about all prenormal logics apply to all normal modal logics. Every logic extending a prenormal logic is prenormal; by contrast, some nonnormal logics extend normal logics, although any extension of a normal logic is at least prenormal.

Any normal logic Σ has a possible worlds semantics where $\Box\alpha$ is true at a world w in a model M if and only if α is true at every world in M to which w has the accessibility relation of M. Intuitively, a world x is accessible from w if and only if what the agent cognizes at w is true at x. In other words, one world is accessible from another if and only if for all one cognizes in the latter one is in the former. The formulas α such that $\Box\alpha$ is true at w express what the agent cognizes at w. For every normal logic Σ there is a class C of models such that Σ consists of exactly the formulas true at every world in every model in C.

Many authors require the accessibility relation to be an equivalence relation (reflexive, symmetric and transitive) for every intended model of their application. A common attitude is expressed by the authors of a standard text, who write that the postulate 'seems reasonable for many applications we have in mind',but 'we can certainly imagine other possibilities' (Fagin et al. 1995, 33). For example, if x is accessible from w if and only if appearances to the agent are identical in x and w, then accessibility is an equivalence relation because identity in any given respect is an equivalence relation. The logic of the class of all possible worlds models in which accessibility is an equivalence relation is the modal system known as S5: $\vdash_{S5} \alpha$ if and only if α is true in every model for which accessibility is an equivalence relation. Since equivalence relations correspond to partitions of the set of worlds, S5 is also known as the logic of the *partitional* conception of knowledge. S5 is the smallest normal modal logic with the theorem schemas T ($\Box\alpha \supset \alpha$) and E ($\neg\Box\alpha \supset \Box\neg\Box\alpha$). T (*truthfulness*) says that the agent cognizes only truths; it is appropriate for applications on which one cognizes only what follows from what one knows. T corresponds to the condition that accessibility be reflexive. For applications on which one cognizes what follows from what one *believes*, T would have to be dropped, perhaps replaced by the weaker principle D ($\Box\alpha \supset \Diamond\alpha$). D requires cognition to be consistent in the sense that an agent who cognizes something does not also cognize its negation. D corresponds to the condition that accessibility be serial (from every world some world is accessible). E is the principle of *negative introspection*: cognition records its omissions in the sense that agents who do not cognize something cognize that they do not cognize it. E corresponds to the condition that accessibility be euclidean (worlds accessible from a given world are accessible from each other). In S5 we can derive the principle of *positive introspection* 4 ($\Box\alpha \supset \Box\Box\alpha$), that cognition records its contents in the sense that agents who cognize something cognize that they cognize it. 4 corresponds to the condition that accessibility be transitive. If T is dropped or weakened to D then 4 is no longer derivable from E, so 4 might be added as an independent schema. Accessibility is reflexive (T) and euclidean (E) if and only if it is an equivalence relation.

3. Computational Constraints

To formulate computational constraints, we generalize concepts from recursion theory to $L_\Box$ using a standard intuitively computable coding procedure. A model M is r.e. if and only if $\Box^{-1}$M (which expresses what the agent cognizes in M) is an r.e. (recursively enumerable) theory in $L_\Box$. In that sense, the agent's cognition in an r.e. model does not exceed the computational capacity of a sufficiently powerful Turing machine.

Consider the restriction of $\Box^{-1}$M to the $\Box$-free sublanguage L, L $\cap$ $\Box^{-1}$M. Let $\Box^{-1}$M be an r.e. theory in $L_\Box$. Thus L $\cap$ $\Box^{-1}$M is an r.e. theory in L. It is the part of the agent's overall theory in M which is not specifically epistemic. From the standpoint of general epistemic logic, can we reasonably impose any further constraints on $L \cap \Box^{-1}M$ beyond recursive enumerability?

If $\Box^{-1}M$ is required to be consistent, $L \cap \Box^{-1}M$ is consistent too. Can we limit the possible values of $L \cap \Box^{-1}M$ still further? For many applications we cannot. $L \cap \Box^{-1}M$ simply expresses what the agent cognizes in M about some aspect of reality. The agent can store any r.e. theory in L as a recursive axiomatization (Craig 1953). If the agent might cognize that aspect of reality simply by having learned a theory about it on the testimony of a teacher, any (consistent) r.e. theory in L is possible. In particular, we can interpret the propositional variables as mutually independent. For example, given a black box which may or may not flash a light on input of a symbol for a natural number, we can read p_i as 'The light flashes on input i'. Then any (consistent) r.e. theory in L could exhaust everything expressible in L which the agent (with only the computational power of a Turing machine) has learned about the black box. Such situations seem quite reasonable. If we want an epistemic logic to have a generality beyond some local application, it should apply to them: such situations should correspond to intended models. Now any application which has all those intended models thereby satisfies (*) or ($*_{con}$), depending on whether the epistemic agent's theory is required to be consistent:

(*) For every r.e. theory R in L, $L \cap \Box^{-1}M = R$ for some r.e. intended model M.

($*_{con}$) For every consistent r.e. theory R in L, $L \cap \Box^{-1}M = R$ for some r.e. intended model M.

(*) is appropriate for readings of $\Box$ like 'It follows from what I believe that ...', if the agent is not required to be consistent. For readings of $\Box$ like 'It follows from what I know that ...', only ($*_{con}$) is appropriate, for one can know only truths and any set of truths is consistent. We can define corresponding constraints on a modal logic Σ without reference to models:

Σ is *r.e. conservative* if and only if for every r.e. theory R in L, there is a maximal Σ-consistent set X such that $\Box^{-1}X$ is r.e. and $L \cap \Box^{-1}X = R$.

Σ is *r.e. quasi-conservative* if and only if for every consistent r.e. theory R in L, there is a maximal Σ-consistent set X such that $\Box^{-1}X$ is r.e. and $L \cap \Box^{-1}X = R$.

Here $\Box^{-1}X = \{\alpha \in L_\Box : \Box\alpha \in X\}$. Roughly, if Σ is r.e. (quasi-)conservative then every (consistent) r.e. theory in the language without $\Box$ is conservatively extended by an r.e. theory in the language with $\Box$ such that it is consistent in Σ for R to be exactly what the agent cognizes in the language without $\Box$ while what the agent cognizes in the language with $\Box$ constitutes an r.e. theory. If an application satisfies (*), its logic is r.e. conservative, for X can be the set of formulas true in M. Conversely, any r.e. conservative logic is the logic of some application which satisfies (*), for some appropriate kind of intended model. The same relationships hold between ($*_{\text{con}}$) and r.e. quasi-conservativeness. For many applications of epistemic logic, the class of intended models is quite restricted and even ($*_{\text{con}}$) does not hold. But if the application interprets $\Box$ as something like 'It follows from what I believe/know that', without special restrictions on the epistemic subject, then situations of the kind described above will correspond to intended models and the logic of the application will be r.e. [quasi-] conservative. In this paper we do not attempt to determine which informally presented applications of epistemic logic satisfy (*) or ($*_{\text{con}}$). We simply investigate which logics are r.e. [quasi-] conservative.

Trivially, every r.e. conservative modal logic is r.e. quasi-conservative. Examples will be given below of r.e. quasi-conservative normal modal logics which are not r.e. conservative. For prenormal modal logics, r.e. conservativeness can be characterized in terms of r.e. quasi-conservativeness in a simple way which allows us to transfer results about one to the other:

PROPOSITION 1. Let Σ be a prenormal modal logic. Then Σ is r.e. conservative if and only if Σ is r.e. quasi-conservative and not $\vdash_\Sigma \Diamond\top$.

Proof. Let $\Box\text{L} = \{\Box\alpha : \alpha \in \text{L}\}$. ($\Rightarrow$) Trivially, Σ is r.e. quasi-conservative if r.e. conservative. Suppose that $\vdash_\Sigma \Diamond\top$. Since $\Box\text{L} \vdash_\Sigma \Box\neg\top$, $\Box\text{L}$ is Σ-inconsistent. Thus $\text{L} \cap \Box^{-1}\text{X} = \text{L}$ for no Σ-consistent set X. Since L is an r.e. theory in L, Σ is not r.e. conservative. ($\Leftarrow$) Suppose that Σ is r.e. quasi-conservative but not r.e. conservative. Since L is the only inconsistent theory in L, there is no maximal Σ-consistent set X such that $\Box^{-1}\text{X}$ is r.e. and $\text{L} \cap \Box^{-1}\text{X} = \text{L}$. If $\Box\text{L}$ is Σ-consistent, then some maximal Σ-consistent set X extends $\Box\text{L}$, so $\text{L} \cap \Box^{-1}\text{X} = \text{L}$; but for $\alpha \in \text{L}_\Box \vdash_\Sigma \bot \supset \alpha$, so $\vdash_\Sigma \Box\bot \supset \Box\alpha$ by prenormality, so $\Box^{-1}\text{X} = \text{L}_\Box$ because $\Box\bot \in X$, so $\Box^{-1}\text{X}$ is r.e. Thus $\Box\text{L}$ is Σ-inconsistent, i.e., for some $\alpha_0, \ldots, \alpha_m \in \text{L}$, $\vdash_\Sigma \neg\bigwedge\{\Box\alpha_i : i \leq m\}$. But for $i \leq m$, $\vdash_\Sigma \Box\neg\top \supset \Box\alpha_i$ by prenormality, so $\vdash_\Sigma \neg\Box\neg\top$.

Examination of the proof shows that the prenormality condition can be weakened to this: if $\vdash_{\text{PC}} \alpha \supset \beta$ then $\vdash_\Sigma \Box\alpha \supset \Box\beta$. An example of a reading of $\Box$ which verifies this weaker condition but falsifies prenormality is 'There is a subjective probability of at least x that', where $0 < x < 1$, for prenormality implies that $\vdash_\Sigma (\Box p \wedge \Box q) \supset \Box(p \wedge q)$, whereas this reading invalidates that formula. Prenormality can be weakened in similar ways for subsequent propositions.

R.e. conservativeness and r.e. quasi-conservativeness do not state upper or lower bounds on the epistemic agent's computational capacity. Rather, they state upper bounds on the strength of the epistemic logic itself; evidently a modal logic with an r.e. [quasi-] conservative extension is itself r.e. [quasi-] conservative. But too strong a logic can impose unwarranted restrictions on the agent's theory of the world given an upper bound on the agent's computational capacity.

4. Some Non-R.e. Quasi-Conservative Logics

Which modal logics are not r.e. [quasi-] conservative? Obviously, since $\vdash_{S5} \Diamond\top$, the logic S5 is not r.e. conservative. Since S5 is decidable, this does not result from non-recursiveness in S5 itself. More significantly:

PROPOSITION 2. S5 is not r.e. quasi-conservative.

Proof. (Skyrms 1978, 377 and Shin and Williamson 1994, Proposition 3 have similar proofs of related facts about S5): Let R be a non-recursive r.e. theory in L; R is consistent. Suppose that $\Box^{-1}X$ is r.e. and $L \cap \Box^{-1}X = R$ for some maximal S5-consistent set X. Now $L - R = \{\alpha : \Box\neg\Box\alpha \in X\} \cap L$. For if $\alpha \in L - R$ then $\Box\alpha \notin X$, so $\neg\Box\alpha \in X$; but $\vdash_{S5} \neg\Box\alpha \supset \Box\neg\Box\alpha$, so $\Box\neg\Box\alpha \in X$ since X is maximal S5-consistent. Conversely, if $\Box\neg\Box\alpha \in X$ then $\neg\Box\alpha \in X$ since $\vdash_{S5} \Box\neg\Box\alpha \supset \neg\Box\alpha$, so $\Box\alpha \notin X$, so $\alpha \notin R$ since $L \cap \Box^{-1}X = R$. Since $\Box^{-1}X$ is r.e., so is $\{\alpha : \Box\neg\Box\alpha \in X\} \cap L$, i.e. $L - R$. Contradiction.

Thus the partitional conception of knowledge prevents a subject with the computational capacity of a Turing machine from having as the restriction of its theory to the $\Box$-free language any non-recursive r.e. theory (for other problems with the S5 schema in epistemic logic and further references see Williamson (2000, 23–24, 166–167, 226–228, 316–317)). Thus S5 is unsuitable as a general epistemic logic for Turing machines.

The proof of Proposition 2 depends on the existence of an r.e. set whose complement is not r.e. By contrast, the complement of any recursive set is itself recursive; decidability, unlike semi-decidability, is symmetric between positive and negative answers. The analogue of Proposition 2 for a notion like r.e. quasi-conservativeness but defined in terms of recursiveness rather than recursive enumerability would be false. For it is not hard to show that if R is a consistent recursive theory in L, then there is a maximal S5-consistent set X in $L_\Box$ such that $\Box^{-1}X$ is recursive and $L \cap \Box^{-1}X = R$. Thus S5 imposes computational constraints not on very clever agents (whose theories need not be r.e.) or on very stupid agents (whose theories must be recursive) but on half-clever agents (whose theories must be r.e. but need not be recursive).

Proposition 2 is the rigorous version of the argument sketched in the introduction. Can we generalize it? The next result provides a rather unintuitive necessary condition for r.e. quasi-conservativeness which nevertheless has many applications.

THEOREM 3. Let Σ be a modal logic such that for some formulas $\alpha_0, \ldots, \alpha_n \in L_\Box$ and $\beta_0, \ldots, \beta_n \in L$, $\vdash_\Sigma \bigvee\{\Box\alpha_i : i \leq n\}$ and, for each $i \leq n$, $\vdash_\Sigma (\Box\alpha_i \wedge \Box\beta_i) \supset \Box\bot$ and not $\vdash_{PC} \neg\beta_i$. Then Σ is not r.e. quasi-conservative.

Proof. There are pairwise disjoint r.e. subsets $I_0, I_1, I_2, \ldots$ of the natural numbers $\mathbb{N}$ such that for every total recursive function f, $i \in I_{f(i)}$ for some $i \in \mathbb{N}$. For let $f[0], f[1], f[2], \ldots$ be a recursive enumeration of all partial and total recursive functions on $\mathbb{N}$ and set $I_i = \{j : f[j](j)$ is defined and $= i\}$; then $j \in I_{f[j](j)}$ whenever $f[j]$ is total, I_i is r.e. and $I_i \cap I_j = \{\}$ whenever $i \neq j$. Now suppose that (i) $\vdash_\Sigma \bigvee\{\Box\alpha_i : i \leq n\}$; (ii) $\vdash_\Sigma(\Box\alpha_i \wedge \Box\beta_i) \supset \Box\bot$ for each $i \leq n$; (iii) $\vdash_{PC} \neg\beta_i$ for no $i \leq n$. Let m be the highest subscript on any propositional variable occurring in $\beta_0, \ldots, \beta_n$. For all $i \in \mathbb{N}$, let σ_i and τ_i be substitutions such that $\sigma_i p_j = p_{i(m+1)+j}$ and $\tau_i p_{i(m+1)+j} = p_j$ for all $j \in \mathbb{N}$. Set $U = \{\sigma_i\beta_j : i \in I_j\}$. Since the σ_i are recursive and the I_j are r.e., U is r.e. Now $\vdash_{PC} \neg\sigma_i\beta_j$ for no i, j, otherwise $\vdash_{PC} \neg\tau_i\sigma_i\beta_j$, i.e., $\vdash_{PC} \neg\beta_j$, contrary to (iii). Moreover, if $h \neq i$ then $\sigma_h\beta_j$ and $\sigma_i\beta_k$ have no propositional variable in common. Thus if $h \in I_j$ and $i \in I_k$ and $\sigma_h\beta_j$ has a variable in common with $\sigma_i\beta_k$, then $h = i$, so $j = k$ because the I_j are pairwise disjoint. Hence no two members of U have a propositional variable in common. Thus U is consistent. Let R be the smallest theory in L containing U; R is consistent and r.e. Suppose that for some maximal Σ-consistent set X, $\Box^{-1}$X is r.e. and $L \cap \Box^{-1}X = R$. Let the total recursive function g enumerate $\Box^{-1}$X. Fix $j \in \mathbb{N}$. By (i), $\vdash_\Sigma \bigvee\{\Box\sigma_j\alpha_i : i \leq n\}$ since Σ is closed under US, so $\Box\sigma_j\alpha_i \in Y$ for some $i \leq n$ since Y is maximal Σ-consistent. Thus $g(k) = \sigma_j\alpha_i$ for some k; let $k(j)$ be the least k such that $g(k) \in \{\sigma_j\alpha_i : i \leq n\}$. Let $f(j)$ be the least $i \leq n$ such that $g(k(j)) = \sigma_j\alpha_i$. Since g enumerates $\Box^{-1}$X, $\Box\sigma_j\alpha_{f(j)} \in X$. Since g and σ_j are total recursive, k is total recursive, so f is total recursive. Thus $j \in I_{f(j)}$ for some $j \in \mathbb{N}$, so $\sigma_j\beta_{f(j)} \in U \subseteq R$ since $f(j) \leq n$. Since $L \cap \Box^{-1}X = R$, $\Box\sigma_j\beta_{f(j)} \in X$. By (ii), $\vdash_\Sigma(\Box\alpha_{f(j)} \wedge \Box\beta_{f(j)}) \supset \Box\bot$, so $\vdash_\Sigma(\Box\sigma_j\alpha_{f(j)} \wedge \Box\sigma_j\beta_{f(j)}) \supset \Box\bot$; since X is maximal Σ-consistent, $\Box\bot \in X$. Thus $\bot \in R$, contradicting the consistency of R. Thus no such set as X can exist, so Σ is not r.e. quasi-conservative.

In other words, a necessary condition for Σ to be r.e. quasi-conservative is that for all formulas $\alpha_0, \ldots, \alpha_n \in L_\Box$ and $\beta_0, \ldots, \beta_n \in L$, if $\vdash_\Sigma \bigvee\{\Box\alpha_i : i \leq n\}$ and, for each $i \leq n$, $\vdash_\Sigma(\Box\alpha_i \wedge \Box\beta_i) \supset \Box\bot$ then, for some $i \leq n$, $\vdash_{PC} \neg\beta_i$. Of course, if Σ is prenormal and contains the D axiom (requiring the agent to be consistent) then the condition that $\vdash_\Sigma(\Box\alpha_i \wedge \Box\beta_i) \supset \Box\bot$ can be simplified to the condition that $\vdash_\Sigma \neg\Box(\alpha_i \wedge \beta_i)$.

OPEN PROBLEM. Is the necessary condition for r.e. quasi-conservativeness in Theorem 3 (or some natural generalization of it) also sufficient?

OBSERVATION. The proof of Theorem 3 uses significantly more recursion theory than does the proof of Proposition 2, which relies only on the existence of an r.e. set whose complement is not r.e. Samson Abramsky observed (informal communication) that the proof of Proposition 2 would generalize to a setting in which r.e. sets were replaced by open sets in a suitable topology (in which not all open sets have open complements). It would be interesting to see whether a generalization along such lines yielded a smoother theory. One might then seek an intuitive interpretation of the topology.

To see that Proposition 2 is a special case of Theorem 3, put $n = 1$, $\alpha_0 = \Diamond\neg p$, $\alpha_1 = \Diamond p$, $\beta_0 = p$ and $\beta_1 = \neg p$. Now $\vdash_{S5} \Box\Diamond\neg p \vee \Box\Box p$; $\vdash_{S5} (\Box\Diamond\neg p \wedge \Box p) \supset \Box\bot$ because $\vdash_{S5} \Box p \supset \Box\Box p$ and S5 is normal; likewise $\vdash_{S5} (\Box\Diamond p \wedge \Box\neg p) \supset \Box\bot$; finally, neither $\vdash_{S5} p$ nor $\vdash_{S5} \neg p$. These features of S5 follow easily from the fact that it is a consistent normal extension of K4G_1, the smallest normal logic Σ including both 4 and G_1 ($\Diamond\Box\alpha \supset \Box\Diamond\alpha$). Since the inconsistent logic is certainly not r.e. quasi-conservative, we have this generalization of Proposition 2:

COROLLARY 4. No normal extension of K$G_1$4 is r.e. quasi-conservative.

We can use Corollary 4 to show several familiar weakenings of S5 not to be r.e. quasi-conservative. G_1 corresponds to the condition that accessibility be convergent, in the sense that if x and y are both accessible from w, then some world z is accessible from both x and y. Informally, G_1 says that agents either cognize that they do not cognize α or cognize that they do not cognize $\neg\alpha$. Any normal logic satisfying E also satisfies G_1, so Corollary 4 implies in particular the failure of r.e. quasi-conservativeness for the logics K4E and KD4E. Those two logics are the natural analogues for belief of S5 as a logic for knowledge, since they retain positive and negative introspection while dropping truthfulness altogether (K4E) or weakening it to consistency (KD4E). Thus they are often used as logics of belief. But positive and negative introspection together violate the computational constraint in a normal logic even in the absence of truthfulness. Thus, in a generalized context, K4E or KD4E impose unacceptably strong computational constraints as logics of belief, just as S5 does as a logic of knowledge.

For more examples, consider the schemas B($\alpha \supset \Box\Diamond\alpha$) and D_1($\Box(\Box\alpha \supset \beta) \vee \Box(\Box\beta \supset \alpha)$). B corresponds to the condition that accessibility be symmetric, D_1 to the condition that accessibility be connected, in the sense that if x and y are both accessible from w, then either x is accessible from y or y is accessible from x. Any normal logic satisfying B or D_1 also satisfies G_1, so KB4 and K$D_1$4 are not r.e. quasi-conservative. *A fortiori*, the same holds if one requires the agent to be consistent or truthful by adding D or T respectively. Thus KD4E, KDG14, KT$G_1$4 (= S4.2), KD$D_1$4 and KT$D_1$4 (= S4.3) are also not r.e. quasi-conservative. All these are sublogics of S5; we shall need to weaken S5 considerably to find an r.e. quasi-conservative logic.

Theorem 3 is also applicable to logics without positive introspection. We can use T rather than 4 to derive $(\Box\Diamond\neg p \wedge \Box p) \supset \Box\bot$, so:

CORLLARY 5. No normal extension of KTG_1 is r.e. quasi-conservative.

Again, consider Alt_n ($\bigvee\{\Box(\bigwedge\{p_j : j < i\} \supset p_i) : i \leq n\}$), e.g., Alt_2 is $\Box p_0 \vee \Box(p_0 \supset p_1) \vee \Box((p_0 \wedge p_1) \supset p_2)$. Alt_n corresponds to the condition that from each world at most n worlds be accessible; informally, the agent rules out all but n

specific possibilities. Setting $\alpha_i = \bigwedge\{p_j : j < i\} \supset p_i$ and $\beta_i = \neg\alpha_i$ in Theorem 3 gives:

COROLLARY 6. For any n, no r.e. quasi-conservative prenormal modal logic contains Alt_n.

An epistemic logic which imposes an upper bound on how many possibilities the agent can countenance thereby excludes the agent from having some consistent r.e. theories about the black box.

5. Some R.e. Conservative Logics

Since every modal logic with an r.e. [quasi-] conservative extension is itself r.e. [quasi-] conservative, an efficient strategy is to seek very strong r.e. [quasi-] conservative logics, even if they are implausibly strong for most epistemic applications, because we can note that the weaker and perhaps more plausible logics which they extend will also be r.e.[quasi-] conservative.

A large class of r.e. conservative logics arises as follows. Let Σ be any epistemic logic. The agent might cognize each theorem of Σ. Moreover, an epistemic logic Σ^* may imply this, in that $\vdash_{\Sigma^*} \Box\alpha$ whenever $\vdash_\Sigma \alpha$. Σ and Σ^* may be distinct, even incompatible. For example, let Ver be the smallest normal modal logic containing $\Box\bot$. Interpreted epistemically, Ver implies that the agent is inconsistent; but Ver itself is consistent. An epistemic theory consisting just of Ver falsely but consistently self-attributes inconsistency, and an epistemic logic may report that the agent self-attributes inconsistency without itself attributing inconsistency to the agent. Thus Ver* may contain $\Box\Box\bot$ without $\Box\bot$. Similarly, let Triv be the smallest normal modal logic containing all theorems of the form $\alpha \equiv \Box\alpha$. Interpreted epistemically, Triv implies that the agent cognizes that his beliefs contain all and only truths; but Triv itself does not contain all and only truths (neither $\vdash_{\mathrm{Triv}} p$ nor $\vdash_{\mathrm{Triv}} \neg p$). Thus Triv* may contain $\Box(p \equiv \Box p)$ without $p \equiv \Box p$. To be more precise, for any modal logics Λ and Σ let $\Lambda\Box\Sigma$ be the smallest normal extension of Λ containing $\{\Box\alpha : \vdash_\Sigma \alpha\}$. We will prove that if Σ is consistent and normal then $\mathrm{K}\Box\Sigma$ is r.e. conservative. $\mathrm{K}\Box\Sigma$ is an epistemic logic for theorizing about theories that incorporate the epistemic logic Σ. R.e. conservativeness implies no constraint on what epistemic logic the agent uses beyond consistency (if Σ is inconsistent, then $\mathrm{K}\Box\Sigma$ contains Alt_0 and so is not even r.e. quasi-conservative). In particular, the smallest normal logic K itself is r.e. conservative. Moreover, if Σ is consistent and normal, then $\mathrm{K4}\Box\Sigma$ is r.e. conservative; that is, we can add positive introspection. In particular, K4 itself is r.e. conservative. We prove this by proving that K$\Box$Ver and K$\Box$Triv are r.e. conservative. Since K$\Box$Ver and K$\Box$Triv contain $\Box\Box\bot$ and $\Box(p \equiv \Box p)$ respectively, they are too strong to be useful epistemic logics themselves, but equally they are strong enough to contain many other logics of epistemic interest, all of which must also be r.e. conservative. By contrast, Ver and Triv are not themselves even r.e. quasi-conservative, for $\vdash_{\mathrm{Ver}} \mathrm{Alt}_0$ and $\vdash_{\mathrm{Triv}} \mathrm{Alt}_1$.

For future reference, call a mapping ϕ from $\mathrm{L}_\Box$ into $\mathrm{L}_\Box$ *respectful* if and only if $\phi p = p$ for all propositional variables p, $\phi\bot = \bot$ and $\phi(\alpha \supset \beta) = \phi\alpha \supset \phi\beta$ for all formulas α and β.

LEMMA 7. K□Triv is r.e. conservative.

Proof. Let R be an r.e. theory in L. Let δ and κ be respectful mappings from $\mathrm{L}_\Box$ to L such that $\delta\Box\alpha = \delta\alpha$; $\kappa\Box\alpha = \top$ if $\mathrm{R} \vdash_{\mathrm{PC}} \delta\alpha$ and $\kappa\Box\alpha = \bot$ otherwise for all formulas α. (i) Axiomatize Triv with all truth-functional tautologies and formulas of the form $\alpha \equiv \Box\alpha$ as the axioms and MP as the only rule of inference (schema K and rule RN are easily derivable). By an easy induction on the length of proofs, $\vdash_{\mathrm{Triv}} \alpha$ only if $\vdash_{\mathrm{PC}} \delta\alpha$. (ii) Axiomatize K□Triv with all truth-functional tautologies and formulas of the forms $\Box(\alpha \supset \beta) \supset (\Box\alpha \supset \Box\beta)$ and $\Box\gamma$ whenever $\vdash_{\mathrm{Triv}} \gamma$ as the axioms and MP as the only rule of inference (RN is a derived rule; its conclusion is always an axiom because the logic so defined is a sublogic of Triv). We show by induction on the length of proofs that $\vdash_{\mathrm{K\Box Triv}} \alpha$ only if $\vdash_{\mathrm{PC}} \kappa\alpha$. Basis: If $\vdash_{\mathrm{PC}} \alpha$, $\vdash_{\mathrm{PC}} \kappa\alpha$. If $\kappa\Box(\alpha \supset \beta) = \top$ and $\kappa\Box\alpha = \top$ then $\mathrm{R} \vdash_{\mathrm{PC}} \delta\alpha \supset \delta\beta$ and $\mathrm{R} \vdash_{\mathrm{PC}} \delta\alpha$, so $\mathrm{R} \vdash_{\mathrm{PC}} \delta\beta$, so $\kappa\Box\beta = \top$, so $\mathrm{R} \vdash_{\mathrm{PC}} \kappa(\Box(\alpha \supset \beta) \supset (\Box\alpha \supset \Box\beta))$; otherwise $\kappa\Box(\alpha \supset \beta) = \bot$ or $\kappa\Box\alpha = \bot$ and again $\mathrm{R} \vdash_{\mathrm{PC}} \kappa(\Box(\alpha \supset \beta) \supset (\Box\alpha \supset \Box\beta))$. If $\vdash_{\mathrm{Triv}} \gamma$ then $\vdash_{\mathrm{PC}} \delta\gamma$ by (i), so $\mathrm{R} \vdash_{\mathrm{PC}} \delta\gamma$, so $\kappa\Box\gamma = \top$, so $\vdash_{\mathrm{PC}} \kappa\Box\gamma$. Induction step: trivial. (iii) Put $Y = \{\Box\alpha \in \mathrm{L}_\Box : \mathrm{R} \vdash_{\mathrm{PC}} \delta\alpha\} \cup \{\neg\Box\alpha \in \mathrm{L}_\Box : \text{not } \mathrm{R} \vdash_{\mathrm{PC}} \delta\alpha\}$. Y is K□Triv-consistent, for if $\mathrm{Y}_0 \subseteq \mathrm{Y}$ is finite and $\vdash_{\mathrm{K\Box Triv}} \bigwedge \mathrm{Y}_0 \supset \bot$, then $\vdash_{\mathrm{PC}} \kappa(\bigwedge \mathrm{Y}_0 \supset \bot)$ by (ii), i.e. $\vdash_{\mathrm{PC}} \bigwedge\{\kappa\alpha : \alpha \in \mathrm{Y}_0\} \supset \bot$, which is impossible since $\{\kappa\alpha : \alpha \in \mathrm{Y}\} \subseteq \{\top, \neg\bot\}$. Let X be a maximal K□Triv-consistent extension of Y. By definition of Y, $\Box^{-1}\mathrm{X} = \{\alpha : \mathrm{R} \vdash_{\mathrm{PC}} \delta\alpha\}$, which is r.e. because R is r.e. and δ is recursive (although κ need not be). If $\alpha \in \mathrm{L}$, $\delta\alpha = \alpha$, so $\Box\alpha \in \mathrm{X}$ if and only if $\mathrm{R} \vdash_{\mathrm{PC}} \alpha$, i.e., if and only if $\alpha \in \mathrm{R}$ because R is a theory; thus $\mathrm{L} \cap \Box^{-1}\mathrm{X} = \mathrm{R}$. Hence K□Triv is r.e. conservative.

LEMMA 8. K□Ver is r.e. conservative.

Proof. Like Lemma 7, but in place of δ use a respectful mapping λ such that $\lambda\Box\alpha = \top$.

A notable sublogic of K□Ver is GL, the smallest normal modal logic including $\Box(\Box\alpha \supset \alpha) \supset \Box\alpha$. Thus a corollary of Lemma 8 is that GL is r.e. conservative. GL is in a precise sense the logic of what is provable in Peano arithmetic (PA) about provability in PA (Boolos 1993 has exposition and references). More generally, if R is an ω-consistent r.e. extension of PA, then GL is the logic of what is provable in R about provability in R. Since a Turing machine's theory of arithmetic is presumably at best an ω-consistent r.e. extension of PA, GL is therefore a salient epistemic logic for Turing machines, and its r.e. conservativeness is not surprising.

Caution. We must be careful in our informal renderings of results about provability logic. A provability operator creates an intensional context within which the substitution of coextensive but not provably coextensive descriptions can alter the truth-value of the whole sentence; this point applies in particular to descriptions of agents or their theories. On a provability interpretation of $\Box$, occurrences of $\Box$ within the scope of other occurrences of $\Box$ in effect involve just such occurrences of descriptions of agents or their theories in an intensional context, so which logic is validated can depend on the manner in which a given agent or theory is described. The validity of GL as an epistemic logic is relative to a special kind of descriptive self-presentation of the theory T in the interpretation of $\Box$, by a coding of its axioms and rules of inference. GL is not valid relative to some extensionally equivalent but intensionally distinct interpretations of $\Box$, e.g. the indexical reading 'I can prove that' as uttered by an epistemic subject with the computational capacity of a Turing machine (Shin and Williamson 1994, Williamson 1996 and 1998).

PROPOSITION 9. If Σ is a consistent normal modal logic, $K\Box\Sigma$ and $K4\Box\Sigma$ are r.e. conservative.

Proof. By Makinson (1971), either $\Sigma \subseteq$ Triv or $\Sigma \subseteq$ Ver. Hence either $K\Box\Sigma \subseteq K\Box$Triv or $K\Box\Sigma \subseteq K\Box$Ver. But Schema 4 is easily derivable in both $K\Box$Triv and $K\Box$Ver, so $K4\Box\Sigma \subseteq K\Box$Triv or $K4\Box\Sigma \subseteq K\Box$Ver. By Lemmas 7 and 8, $K\Box$Triv and $K\Box$Ver are r.e. conservative, so $K4\Box\Sigma$ is.

All the logics salient in this paper are decidable, and therefore r.e., but we should note that an epistemic logic need not be r.e. to be r.e. conservative:

COROLLARY 10. Not all r.e. conservative normal modal logics are r.e.

Proof. (i) We show that for any normal modal logic Σ, $\vdash_\Sigma \alpha$ if and only if $\vdash_{K\Box\Sigma} \Box\alpha$. Only the $\Leftarrow$ direction needs proving. Axiomatize $K\Box\Sigma$ with all truth-functional tautologies and formulas of the forms $\Box(\alpha \supset \beta) \supset (\Box\alpha \supset \Box\beta)$ and $\Box\gamma$ whenever $\vdash_\Sigma \gamma$ as the axioms and MP as the only rule of inference (RN is a derived rule; its conclusion is always an axiom because the logic so defined is a sublogic of Σ). Let η be a respectful mapping from $L_\Box$ to $L_\Box$ such that $\eta\Box\alpha = \alpha$ for all formulas α (η is distinct from δ in the proof of Lemma 7 since $\eta\Box\Box p = \Box p$ whereas $\delta\Box\Box p = p$). By induction on the length of proofs, $\vdash_{K\Box\Sigma} \alpha$ only if $\vdash_\Sigma \eta\alpha$. Hence $\vdash_{K\Box\Sigma} \Box\alpha$ only if $\vdash_\Sigma \alpha$. (ii) By (i), for any normal modal logics Σ_1 and Σ_2, $K\Box\Sigma_1 = K\Box\Sigma_2$ if and only if $\Sigma_1 = \Sigma_2$. But there are continuum many consistent normal modal logics (Blok 1980 has much more on these lines). Hence there are continuum many corresponding logics of the form $K\Box\Sigma$; all are r.e. conservative by Proposition 9. Since only countably many modal logics are r.e., some of them are not r.e.

One limitation of Proposition 9 is that $K\Box\Sigma$ and $K4\Box\Sigma$ never contain the consistency schema D. In a sense this limitation is easily repaired. For any modal logic

Σ, let $\Sigma[D]$ be the smallest extension of Σ containing D; thus $\vdash_{\Sigma[D]} \alpha$ just in case $\vdash_{\Sigma} \Diamond\top \supset \alpha$.

PROPOSITION 11. For any r.e. conservative modal logic Σ, Σ[D] is r.e. quasi-conservative.

Proof. For any consistent theory R, any maximal Σ-consistent set X such that $L \cap \Box^{-1}X = R$ is Σ[D]-consistent because $\Diamond\top \in X$.

COROLLARY 12. If Σ is a consistent normal modal logic, (K$\Box\Sigma$)[D] and (K4$\Box\Sigma$)[D] are r.e. quasi-conservative.

Proof. By Propositions 9 and 11.

Although Σ[D] is always prenormal, it may not be normal, even if Σ is normal; sometimes not $\vdash_{\Sigma[D]} \Box\Diamond\top$. But we can also consider epistemic interpretations of normal logics with the D schema, e.g., KD and KD4. Such logics contain $\Box\Diamond\top$; they require agents to cognize their own consistency. By Gödel's second incompleteness theorem, this condition cannot be met relative to a Gödelian manner of representing the theory in itself; no consistent normal extension of the provability logic GL contains D. But $\Box\Diamond\top$ is true on other epistemic interpretations; for example, we know that our knowledge (as opposed to our beliefs) does not imply a contradiction. Since GL $\subseteq$ K$\Box$Ver, Proposition 9 does not generalize to the r.e. quasi-conservativeness of KD$\Box\Sigma$. But we can generalize Lemma 7 thus:

PROPOSITION 13. If $\Sigma \subseteq$ Triv then KD$\Box\Sigma$ and KD4$\Box\Sigma$ are r.e. quasi-conservative.

Proof. It suffices to prove that KD$\Box$Triv (=KD4$\Box$Triv) is r.e. quasi-conservative. Argue as for Lemma 7, adding $\Diamond\top$ as an axiom for KD$\Box$Triv and noting that if R is c onsistent then $\kappa\Box\neg\top = \bot$, so $\vdash_{PC} \kappa\Diamond\top$.

In particular, KD and KD4 are themselves r.e. quasi-conservative; they are our first examples of r.e. quasi-conservative logics which are not r.e. conservative.

We now return to systems with the T schema. Since T implies D, only r.e. quasi-conservativeness is at issue. That constraint was motivated by the idea that any consistent r.e. theory in the non-modal language might be exactly the restriction of the agent's total r.e. theory to the non-modal language. On many epistemic interpretations, it is in the spirit of this idea that the agent's total theory might be true in the envisaged situation (for example, the agent's theory about the black box might be true, having been derived from a reliable witness). To require an epistemic logic Σ to leave open these possibilities is to require that Σ[T] be r.e. quasi-conservative, where Σ[T] is the smallest extension of Σ containing all instances of T. As with Σ[D], Σ[T] need not be normal even when Σ is; sometimes not $\vdash_{\Sigma[T]} \Box(\Box\alpha \supset \alpha)$ (Williamson 1998, 113–116 discusses logics of the form Σ[T]). Agents may not cognize that they cognize only truths. Nevertheless, particularly when $\Box$ is interpreted in terms of knowledge, one might want an epistemic logic such as KT containing $\Box(\Box\alpha \supset \alpha)$.

Proposition 11 and Corollary 12 have no analogues for T in place of D. For any modal logic Σ, if $\vdash_{\Sigma} \alpha$ then $\vdash_{(K\Box\Sigma)[T]} \Box\alpha$, but $\vdash_{(K\Box\Sigma)[T]} \Box\alpha \supset \alpha$, so $\vdash_{(K\Box\Sigma)[T]} \alpha$; thus (K$\Box\Sigma$)[T] extends Σ and is r.e. quasi-conservative only if Σ is. Similarly, Proposition 13 would be false with T in place of D (counterexample: $\Sigma =$ S5). Therefore, needing a different approach, we start with the system GL[T]. GL[T] has intrinsic interest, for it is the provability logic GLS introduced by Solovay and shown by him to be the logic of what is true (rather than provable) about provability in PA; more generally, it is the logic of what is true about provability in an ω-consistent r.e. extension of PA. GLS is therefore a salient epistemic logic for Turing machines, and its r.e. quasi-conservativeness is not surprising. Although GLS is not normal and has no consistent normal extension, we can use its r.e. quasi-conservativeness to establish that of normal logics containing T.

PROPOSITION 14. GLS is r.e. quasi-conservative.

Proof. Let R be an consistent r.e. theory in L. Axiomatize a theory R+ in $L_{\Box}$ with all members of R, truth-functional tautologies and formulas of the forms $\Box(\alpha \supset \beta) \supset (\Box\alpha \supset \Box\beta)$ and $\Box(\Box\alpha \supset \alpha) \supset \Box\alpha$ as the axioms and MP and RN as the rules of inference. Since R is r.e., so is R+. Let λ be the respectful mapping such that $\lambda\Box\alpha = \top$ for all formulas α. By an easy induction on the length of proofs, if $\vdash_{R+} \alpha$ then $R \vdash_{PC} \lambda\alpha$. But if $\alpha \in L$ then $\lambda\alpha = \alpha$, so $\vdash_{R+} \alpha$ only if $R \vdash_{PC} \alpha$, i.e., $\alpha \in R$; conversely, if $\alpha \in R$ then $\vdash_{R+} \alpha$; thus $L \cap R+ = R$. Let $Y \subseteq L$ be a maximal consistent extension of R. Define a set $X \subseteq L_{\Box}$ inductively: $p_i \in\Leftrightarrow p_i \in Y$; $\bot \notin X$; $\alpha \supset \beta \in X \Leftrightarrow \alpha \notin X$ or $\beta \in X$; $\Box\alpha \in X \Leftrightarrow \vdash_{R+} \alpha$. For $\alpha \in L_{\Box}$, either $\alpha \in X$ or $\neg\alpha \in X$. We show by induction on the length of proofs that if $\vdash_{R+} \alpha$ then $\alpha \in X$. Basis: If $\alpha \in R$ then $\alpha \in Y \subseteq X$. If $\Box(\alpha \supset \beta) \in X$ and $\Box\alpha \in X$ then $\vdash_{R+} \alpha \supset \beta$ and $\vdash_{R+} \alpha$, so $\vdash_{R+} \beta$, so $\Box\beta \in X$; thus $\Box(\alpha \supset \beta) \supset (\Box\alpha \supset \Box\beta) \in X$. If $\Box(\Box\alpha \supset \alpha) \in X$ then $\vdash_{R+} \Box\alpha \supset \alpha$, so $\vdash_{R+} \Box(\Box\alpha \supset \alpha)$ because R+ is closed under RN; but $\vdash_{R+} \Box(\Box\alpha \supset \alpha) \supset \Box\alpha$, so $\vdash_{R+} \Box\alpha$, so $\vdash_{R+} \alpha$, so $\Box\alpha \in X$; thus $\Box(\Box\alpha \supset \alpha) \supset \Box\alpha \in X$. Induction step: Trivial. Now axiomatize GLS with all theorems of GL and formulas of the form $\Box\alpha \supset \alpha$ as the axioms and MP as the only rule of inference. We show by induction on the length of proofs that, for all formulas α, if $\vdash_{GLS} \alpha$ then $\alpha \in X$. Basis: If $\vdash_{GL} \alpha$ then $\vdash_{R+} \alpha$ because $GL \subseteq R+$, so $\alpha \in X$ by the previous induction. If $\Box\alpha \in X$ then $\vdash_{R+} \alpha$, so again $\alpha \in X$; thus $\Box\alpha \supset \alpha \in X$. Induction step: Trivial. Hence $GLS \subseteq X$, so X is maximal GLS-consistent. Now $L \cap \Box^{-1}X = L \cap R+ = R$ and $\Box^{-1}X = R+$ is r.e. Thus GLS is r.e. quasi-conservative.

We can extend Proposition 14 to another system of interest in relation to provability logic. Grz is the smallest normal modal logic containing all formulas of the form $\Box(\Box(\alpha \supset \Box\alpha) \supset \alpha) \supset \alpha$. Grz turns out to be in a precise sense the logic of what is both provable and true in PA (Boolos 1993, 155–161 has all the facts about Grz used here). Grz is intimately related to GLS in a way which allows us to extend the r.e. quasi-conservativeness of GLS to Grz:

PROPOSITION 15. Grz is r.e. quasi-conservative.

Proof. Let R be a consistent theory in L. By Proposition 14, for some maximal GLS-consistent X, $L \cap \Box^{-1}X = R$ and $\Box^{-1}X$ is r.e. Let τ be the respectful mapping from $L_\Box$ to $L_\Box$ such that $\tau\Box\alpha = \Box\tau\alpha \wedge \tau\alpha$ for all formulas α. Put $\tau^{-1}X = \{\alpha : \tau\alpha \in X\}$. Now $Grz \subseteq \tau^{-1}X$, for $\vdash_{Grz} \alpha$ if and only if $\vdash_{GLS} \tau\alpha$ (Boolos 1993, 156), so $\tau\alpha \in X$ since X is maximal GLS-consistent, so $\alpha \in \tau^{-1}X$. Since X is maximal GLS-consistent, $\tau^{-1}X$ is maximal Grz-consistent. Suppose $\alpha \in L$, so $\tau\alpha = \alpha$, so $\tau\Box\alpha = \Box\alpha \wedge \alpha$; if $\Box\alpha \in X$ then $\alpha \in X$ because $\vdash_{GLS} \Box\alpha \supset \alpha$, so $\tau\Box\alpha \in X$, so $\Box\alpha \in \tau^{-1}X$; conversely, if $\Box\alpha \in \tau^{-1}X$ then $\tau\Box\alpha \in X$, so $\Box\alpha \in X$. Thus $L \cap \Box^{-1}\tau^{-1}X = L \cap \Box^{-1}X = R$. Moreover, $\Box^{-1}\tau^{-1}X$ is r.e. because X is r.e. and τ is recursive. Thus Grz is r.e. quasi-conservative.

Grz is not plausible as the logic of other epistemic applications. It is not a sublogic of S5 and $\vdash_{Grz} \neg\Box(\Box p \wedge \Diamond\neg p)$, which in effect forbids agents to cognize that they do not cognize whether p is true. Yet you can know that what you know neither entails that the coin came down heads nor entails that it did not. However, since Grz extends the epistemically more plausible *S*4, the smallest normal modal logic including both the T and 4 schemas, its r.e. quasi-conservativeness entails that of S4. Truthfulness and positive introspection are together consistent with r.e. quasi-conservativeness.

COROLLARY 16. [Compare Shin and Williamson 1994 Proposition 4.] S4 is r.e. quasi-conservative.

Since S4 is r.e. quasi-conservative while S5, its extension by E, is not, and K4 is r.e. conservative while K4E is not, one might be tempted to blame E for the failure to satisfy the constraints, and to suppose that no normal logics with E is r.e. quasi-conservative. That would be a mistake; the next two propositions show that E is harmless when not combined with 4.

PROPOSITION 17. KDE is r.e. quasi-conservative.

Proof. Let R be a consistent r.e. theory in L. Let μ and θ be respectful mappings from $L_\Box$ to L such that for all formulas α, $\theta\Box\alpha = \top$ if $\vdash_{PC} \theta\alpha$ and $\theta\Box\alpha = \bot$ otherwise; $\mu\Box\alpha = \top$ if $R \vdash_{PC} \theta\alpha$ and $\mu\Box\alpha = \bot$ otherwise. Axiomatize KDE with all truth-functional tautologies and formulas of the forms $\Box(\alpha \supset \beta) \supset (\Box\alpha \supset \Box\beta)$, $\neg\Box\bot$ and $\neg\Box\alpha \supset \Box\neg\Box\alpha$ as the axioms and MP and RN as the rules of inference. We show by induction on the length of proofs that for all formulas α, $\vdash_{KDE} \alpha$ only if $\vdash_{PC} \theta\alpha$ and $\vdash_{PC} \mu\alpha$. Basis: If $R \vdash_{PC} \theta\alpha$, then $\mu\Box\alpha = \top$, so $\mu(\neg\Box\alpha \supset \Box\neg\Box\alpha) = \neg\top \supset \mu\Box\neg\Box\alpha$; if not, then not $\vdash_{PC} \theta\alpha$, so $\theta\Box\alpha = \bot$, so $\theta\neg\Box\alpha = \neg\bot$, so $R \vdash_{PC} \theta\neg\Box\alpha$, so $\mu\Box\neg\Box\alpha = \top$, so

$\mu(\neg\Box\alpha \supset \Box\neg\Box\alpha) = \neg\bot \supset \top$; either way, $\vdash_{PC} \mu(\neg\Box\alpha \supset \Box\neg\Box\alpha)$. The rest of the induction is by now routine. The rest of the proof is like that of Lemma 7, with θ and μ in place of δ and κ respectively.

COROLLARY 18. KE is r.e. conservative.

Proof. KE is r.e. quasi-conservative by Proposition 17. Since not $\vdash_{KE} \Diamond\top$, KE is r.e. conservative by Proposition 1.

Although both positive and negative introspection are individually consistent with r.e. [quasi-] conservativeness, their conjunction is not. Part of the explanation is this: without positive introspection, an r.e. but non-recursive theory R can count as satisfying negative introspection by falsely equating the agent's theory with a recursive subtheory of R; the idea behind the clause for $\theta\Box\alpha$ in the proof of Proposition 17 is to use PC as such a subtheory. That R satisfies negative introspection by making false equations is crucial, for KE[T] is S5 itself. Although both negative introspection and truthfulness are individually consistent with r.e. [quasi-] conservativeness, their conjunction is not.

6. Related Non-Computational Constraints

Although r.e. conservativeness and r.e. quasi-conservativeness are defined in computational terms, something remains when the computational element is eliminated. For given any [consistent] theory R in L, r.e. or not, we might require an epistemic logic to leave open the possibility that R is exactly the restriction of the agent's theory to L. On this view, an epistemic logic should impose no constraint beyond consistency on the agent's non-epistemic theorizing. Thus we define a modal logic Σ to be *conservative* if and only if for every theory R in L, $L \cap \Box^{-1}X = R$ for some maximal Σ-consistent set X. Σ is *quasi-conservative* if and only if for every consistent theory R in L, $L \cap \Box^{-1}X = R$ for some maximal Σ-consistent set X. Equivalently, Σ is [quasi-] conservative if and only if for every [consistent] theory R in L, $\{\Box\alpha : \alpha \in R\} \cup \{\neg\Box\alpha : \alpha \in L - R\}$ is Σ-consistent. We can assess how far r.e. conservativeness and r.e. quasi-conservativeness are specifically computational constraints by comparing them with conservativeness and quasi-conservativeness respectively.

THEOREM 19. A prenormal modal logic Σ is quasi-conservative if and only if for no $n \vdash_\Sigma \mathrm{Alt}_n$.

Proof. ($\Rightarrow$) Suppose that $\vdash_\Sigma \mathrm{Alt}_n$. Put $X = \{\Box\alpha : \alpha \in \mathrm{PC}\} \cup \{\neg\Box\alpha : \alpha \in L - \mathrm{PC}\}$. For all $i \leq n$, not $\vdash_{PC} \bigwedge\{p_j : j < i\} \supset p_i$, so $\neg\Box(\bigwedge\{p_j : j < i\} \supset p_i) \in X$. Hence $X \vdash_\Sigma \neg\mathrm{Alt}_n$, so $X \vdash_\Sigma \bot$. Since PC is a theory in L, Σ is not quasi-conservative. ($\Leftarrow$) Suppose that R is a consistent theory in L and $\{\Box\alpha : \alpha \in R\} \cup \{\neg\Box\alpha : \alpha \in L - R\}$ is not Σ-consistent. Thus for some $\alpha_0, \ldots, \alpha_m \in R$ and

$\beta_0, \ldots, \beta_n \in \mathrm{L} - \mathrm{R}$ (such β_i exist because R is consistent), $\vdash_\Sigma \bigwedge\{\Box\alpha_i : i \le m\} \supset \bigvee\{\Box\beta_i : i \le n\}$. Let $i \le n$; since $\alpha_0, \ldots, \alpha_m \in \mathrm{R}$, $\beta_i \in \mathrm{L} - \mathrm{R}$ and R is a theory, it follows that for some valuation v_i of L onto $\{0, 1\}$ (where $v_i(\bot) = \bot$ and $v_i(\gamma_1 \supset \gamma_2) = 1$ just in case $v_i(\gamma_1) \le v_i(\gamma_2)$), $v_i(\alpha_j) = 1$ for all $j \le m$ and $v_i(\beta_i) = 0$. Put $v_{n+1} = v_0$. Set $\delta_i = \bigwedge\{p_j : j < i\} \wedge \neg p_i$ for $i \le n$ and $\delta_{n+1} = \bigwedge\{p_j : j \le n\}$. Let σ be the substitution such that for all $j, \sigma p_j = \bigvee\{\delta_i : v_i(p_j) = 1, i \le n+1\}$. Since Σ is closed under US, $\vdash_\Sigma \bigwedge\{\Box\sigma\alpha_i : i \le m\} \supset \bigvee\{\Box\sigma\beta_i : i \le n\}$. We can prove by induction on the complexity of γ that for all $\gamma \in \mathrm{L}$ and $i \le n+1$, if $v_i(\gamma) = 1$ then $\vdash_{\mathrm{PC}} \delta_i \supset \sigma\gamma$ and if $v_i(\gamma) = 0$ then $\vdash_{\mathrm{PC}} \delta_i \supset \neg\sigma\gamma$. Basis: Immediate by definition of σ, for $\vdash_{\mathrm{PC}} \delta_i \supset \neg\delta_k$ whenever $i \ne k$. Induction step: Routine. Now for $i \le n+1$ and $j \le m$, $v_i(\alpha_j) = 1$, so $\vdash_{\mathrm{PC}} \delta_i \supset \sigma\alpha_j$; since $\vdash_{\mathrm{PC}} \bigvee\{\delta_i : i \le n+1\}$, $\vdash_{\mathrm{PC}} \sigma\alpha_j$, so $\vdash_{\mathrm{PC}} \top \supset \sigma\alpha_j$. Hence by prenormality $\vdash_\Sigma \Box\top \supset \Box\sigma\alpha_j$ and so $\vdash_\Sigma \Box\sigma\alpha_j$. Thus $\vdash_\Sigma \bigvee\{\Box\sigma\beta_i : i \le n\}$. Moreover, for each $i \le n$, $v_i(\beta_i) = 0$, so $\vdash_{\mathrm{PC}} \delta_i \supset \neg\sigma\beta_i$, so $\vdash_{\mathrm{PC}} \sigma\beta_i \supset (\bigwedge\{p_j : j < i\} \supset p_i)$, so $\vdash_\Sigma \Box\sigma\beta_i \supset \Box(\bigwedge\{p_j : j < i\} \supset p_i)$. Thus $\vdash_\Sigma \mathrm{Alt}_n$.

PROPOSITION 20. A prenormal modal logic Σ is conservative if and only if Σ is quasi-conservative and not $\vdash_\Sigma \Diamond\top$.

Proof. Like Proposition 1 with 'r.e.' omitted.

Thus S5 is a quasi-conservative normal modal logic which is not r.e. quasi-conservative; K4E is a conservative normal modal logic which is not r.e. conservative. Most of the examples given above of logics which are not r.e. [quasi-] conservative are [quasi-] conservative. It is the distinctively computational requirements of r.e. quasi-conservativeness and r.e. conservativeness which those logics fail to meet.

COROLLARY 21. Every r.e. quasi-conservative prenormal modal logic is quasi-conservative; every r.e. conservative prenormal modal logic is conservative.

Proof. From Proposition 1, Corollary 6, Theorem 19 and Proposition 20.

Although quasi-conservativeness exceeds r.e. quasi-conservativeness in requiring an epistemic logic to leave open the possibility that the restriction of the subject's theory to the language L is any given non-r.e. theory in L, this requirement is met by any epistemic logic which leaves open the corresponding possibility for every consistent r.e. theory in L.

7. Conclusion

Our investigation has uncovered part of a complex picture. The line between those modal logics weak enough to be r.e. conservative or r.e. quasi-conservative and those that are too strong appears not to coincide with any more familiar distinction between classes of modal logics, although a solution to the problem left open in Section 4 about the converse of Theorem 3 might bring clarification. What we have seen is that some decidable modal logics in general use as logics of knowledge (such as S5) or belief (such as KD45 and K45) when applied in generalized settings impose constraints on epistemic agents that require them to exceed every Turing machine in computational power. For many interpretations of epistemic logic, such a constraint is unacceptably strong.

The problem is not the same as the issue of logical omniscience, since many epistemic logics (such as S4 and various provability logics) do not impose the unacceptably strong constraints, although they do impose logical omniscience. Interpretations that finesse logical omniscience by building it into the definition of the propositional attitude that interprets the symbol □ do not thereby finesse the computational issue that we have been investigating. Nevertheless, the two questions are related, because the deductive closure of a recursively axiomatised theory is what makes its theorems computationally hard to survey. In particular, it can be computationally hard to check for *non*-theoremhood, which is what negative introspection and similar axioms require. In fact, negative introspection by itself turned out not to impose unacceptable computational requirements (Corollary 18), but its combination with independently more plausible axioms does so. Perhaps the issues raised in this paper will provide a more fruitful context in which to discuss some of the questions raised by the debate on logical omniscience and bounded rationality.

The results proved in the paper also suggest that more consideration should be given to the epistemic use of weaker modal logics that are r.e. conservative or quasi-conservative. The plausibility of correspondingly weaker axioms must be evaluated under suitable epistemic interpretations. Weaker epistemic logics present a more complex picture of the knowing subject, but also a more nuanced one, because they make distinctions that stronger logics erase. We have seen that the more nuanced picture is needed to express the limits in general cognition of creatures whose powers do not exceed those of every Turing machine.

Acknowledgements

Material based on this paper was presented to colloquia of the British Society for the Philosophy of Science and the Computer Science Laboratory at Oxford. I thank participants in both for useful comments.

References

Blok, W. J.: 1980, 'The Lattice of Modal Logics: An Algebraic Investigation', *Journal of Symbolic Logic* **45**, 221–236.

Boolos, G.: 1993, *The Logic of Provability*, Cambridge, Cambridge University Press.

Craig, W.: 1953, 'On Axiomatizability within a System', *Journal of Symbolic Logic* **18**, 30–32.
Fagin, R., J. Halpern, Y. Moses and M. Vardi: 1995, *Reasoning About Knowledge*, Cambridge MA, MIT Press.
Makinson, D. C.: 1971, 'Some Embedding Theorems in Modal Logic', *Notre Dame Journal of Formal Logic* **12**, 252–254.
Shin, H. S. and T. Williamson: 1994, 'Representing the Knowledge of Turing Machines', *Theory and Decision* **37**, 125–146.
Skyrms, B.: 1978, 'An Immaculate Conception of Modality', *The Journal of Philosophy* **75**, 368–387.
Williamson, T.: 1996, 'Self-knowledge and Embedded Operators', *Analysis* **56**, 202–209.
Williamson, T.: 1998, 'Iterated Attitudes', in T. Smiley (ed.), *Philosophical Logic*, Oxford, Oxford University Press for the British Academy, pp. 85–133.
Williamson, T.: 2000, *Knowledge and its Limits*, Oxford, Oxford University Press.

PART V
CONTRIBUTIONS FROM NON-CLASSICAL LOGICS

THE NEED FOR ADAPTIVE LOGICS IN EPISTEMOLOGY*

DIDERIK BATENS
Centre for Logic and Philosophy of Science, Universiteit Gent, Belgium, E-mail: Diderik.Batens@ugent.be

Abstract. After it is argued that philosophers of science have lost interest in logic because they applied the wrong type of logics, examples are given of the forms of dynamic reasoning that are central for philosophy of science and epistemology. Adaptive logics are presented as a means to understand and explicate those forms of reasoning. All members of a specific (large) set of adaptive logics are proved to have a number of properties that warrant their formal decency and their suitability with respect to understanding and explicating dynamic forms of reasoning. Most of the properties extend to other adaptive logics.

1. Aim of this Paper

In the first half of the twentieth century epistemology largely reduced to the philosophy of science and logic played a central role in it. We are here interested in the last half of the previous sentence. It raises at once the question why logic lost its central role in epistemology, including the philosophy of science. We all know when this happened – the Vienna Circle was succeeded by people like Hanson, Kuhn, Lakatos, Feyerabend, and Laudan, just to name a few central ones, who hardly ever mention logic. But *why* did it happen?

Of these philosophers of science, only Feyerabend made some explicit claims on the topic. While arguing, in Feyerabend (1970), that inconsistencies often occur in episodes of the history of the sciences, he remarks that 'logic' cannot handle inconsistencies, and comments that this is a problem for logic, not for the history and philosophy of science. I do not think that this diagnosis is correct. For one thing, logics that can handle inconsistencies had been around for a while in those days – see, for example, da Costa (1963) and many other papers on paraconsistent[1] logics (by da Costa and associates); even some of Jaśkowski's work had been translated into English, for example Jaśkowski's (1969).

The correct diagnosis, it seems to me, is that philosophers of science had been applying the wrong *type* of logic. They had mainly been applying **CL** (Classical Logic). Even the few that recurred to 'non-standard logics' applied modal logic or

* The research for this paper was financed by the Fund for Scientific Research – Flanders, by the Research Fund of Ghent University, and indirectly by the Flemish Minister responsible for Science and Technology (contract BIL98/73). I am indebted to Kristof De Clercq for locating several mistakes and misprints in a former draft of this paper, and to the referees for helpful comments and corrections.

S. Rahman et al. (eds.), Logic, Epistemology, and the Unity of Science, pp. 459–485.

intuitionistic logic – systems that share their central properties with **CL**. I do not claim that there is anything wrong with those logics. I only claim that they had been used for purposes for which they are unfit.

That the wrong type of logics was chosen was not merely an accident. Mainstream Western epistemology has always been foundational. Notwithstanding some occasional remarks concerning the revision of (mainly) theories, for example Neurath's idea that we have to rebuild our ship in the open sea, the Vienna Circle fitted perfectly within this tradition. 'Protokollsätze' provided the absolute basis. The only epistemological role for logic was to relate Protokollsätze to theories, mainly by deriving consequences from theories together with Protokollsätze. Needless to say, **CL** performs that role in an excellent way, especially as the Vienna Circle's view on the relation between observation and theory was simple and one-dimensional: saving the phenomena.[2]

When Hanson, Kuhn, and the others came around, it soon became clear that the most interesting aspects of scientific reasoning do not concern the relation between observations and theories, *and* that the Vienna Circle view on the relation was simplistic and mistaken – see especially Laudan (1977). This led to renewed attention for discovery processes, which was greatly activated by Nickles (1980a, b). Not long thereafter, the study of explanation as a logical relation was replaced by a study of the *process* of explanation: how does one, given an *explanandum* E and a theory, arrive at 'initial conditions' to explain E?[3] And these are just a few examples.

In Section 2, I shall consider some reasoning processes that are essential to epistemology and philosophy of science, and that display an internal and possibly also an external dynamics. Such reasoning clearly cannot be handled by logics of the same type as **CL**. In Section 3, I shall introduce adaptive logics, the type of logics that is able to handle such reasoning processes. Characterizing such logics semantically will enable me, in Section 4, to prove that the consequence relations of these logics have the required properties. Much more important, however, is the dynamic proof theory of adaptive logics, which I shall discuss in Section 5. Unlike the semantics, which characterizes the consequence relations by means of abstract definitions, the dynamic proof theory enables one to explicate the actual reasoning processes. Moreover, the metatheory enables one to show that, notwithstanding their dynamic character, these proofs (i) lead to the correct conclusions in the long run, and (ii) lead to conclusions that provide a basis for decision and action even in the short run.

2. The Problem

The most common mechanism that, in specific situations, leads to general knowledge is inductive generalization. To be more precise, I mean the 'derivation' from data of formulas of the form $\forall A$ (the universal closure of A) and of **CL**-

consequences of the data and the generalizations.[4] It has often been argued that there is no logic of induction. The only argument that has ever been adduced for this claim is that the resulting consequence relation is not monotonic: a generalization that is derivable from a set of data need not be derivable after the set of data is extended. Today, however, many non-monotonic consequence relations have been decently defined and well-studied.

Non-monotonic consequence relations display an *external dynamics*. Suppose that $\Gamma \vdash_{\mathbf{LI}} A$ states that A is an inductive consequence (defined by the logic of induction **LI**) of Γ.[5] If Γ is the set of data available at some point in time, then $\Gamma \vdash_{\mathbf{LI}} A$ enables one to accept A. At a later point in time, the set of available data might be $\Gamma \cup \Delta$ and if $\Gamma \cup \Delta \nvdash_{\mathbf{LI}} A$, one has to give up the conclusion A. This dynamics is external in that it does not derive from the reasoning process. Both $\Gamma \vdash_{\mathbf{LI}} A$ and $\Gamma \cup \Delta \nvdash_{\mathbf{LI}} A$ always were true and always will be true. If one knows them to be true, then one justly accepted A at the point in time where Γ was the set of available data, and one justly rejected A after Γ was extended with Δ.

Suppose that one is only interested in generalizations $\forall A$ in which A does not contain quantifiers or individual constants. The basic mechanism behind (thus restricted) inductive reasoning is joint compatibility. A generalization G is inductively derivable from a set of data iff it holds for all sets Δ of generalizations that $\Delta \cup \{G\}$ is compatible with the data whenever Δ is compatible with the data. Given that the data are singular formulas and given the form of the generalizations, the matter is (effectively) decidable. Now, let us make the picture slightly more realistic and suppose that background theories are available. Suppose moreover, to keep things simple, that the data do not contradict the background theories. The available knowledge now consists of the data, the background theories, and their (**CL**-)consequences. Which set of inductive generalizations is compatible with this knowledge is not in general a decidable matter. Worse, there is no positive test for it.[6]

Given the absence of a positive test, how is it possible that people ever arrive at inductive generalizations? The answer is quite obvious: by reasoning. In specific cases, the reasoning may enable one to arrive at a final judgement. However, the reasoning cannot in general lead to a final judgement, even if the premises (data and background generalizations) remain stable during the reasoning process. It can, however, lead to a good estimate. Some people will arrive at a better estimate than others, and the efficiency of such reasoning processes may be studied.

Even if the reasoning does not result in a final judgement, one may consider its outcome sufficiently reliable for making a decision – one may know that a final judgement is impossible, one may consider it too expensive or time consuming to obtain a better judgement, etc.

The absence of a positive test makes the reasoning process necessarily dynamic. Even when reasoning from a stable set of premises, one will have to consider certain formulas as derived provisionally. In other words, it cannot be avoided that, at some point in the reasoning process, one considers as derived certain formulas that later have to be considered as not derived. This I shall call the *internal dynamics*. It is not caused by the introduction of new premises, but is a property of the reasoning process itself, even if it proceeds from a stable premise set. For example, if background knowledge is present, one cannot avoid deriving certain generalizations that later turn out incompatible with the (stable) available knowledge.

In the subsequent paragraphs, I give some more examples of reasoning processes that display an internal dynamics. However, let me stress at once that the discussed logic of inductive generalization is only a special case of a broader phenomenon. Whenever a new theory is adduced, it is supposed to be compatible[7] with available knowledge (the data and formerly accepted theories, or at least part of them). As there is no positive test for compatibility, any reasoning that leads to accepting the new theory necessarily displays the i nternal dynamics and, except in the specific cases in which a final judgement can be reached, the decision taken as a result of this reasoning is necessarily defeasible and hence provisional.

A recent version of the theory of the process of explanation is presented by Ilpo Halonen and Jaakko Hintikka (to appear). In Section 6, they discuss the conditions on (nonstatistical) explanations (with a number of restrictions). The conditions (I slightly change their notation) concern an explanandum Pb, a background theory T (in which the predicate P occurs) and an initial condition (antecedent condition) I (in which b occurs). Among the six conditions are the following:

(iii) I is not inconsistent ($\nvdash_{\mathbf{CL}} \sim I$).
(iv) The explanandum is not implied by T alone ($T \nvdash_{\mathbf{CL}} Pb$).
(vi) I is compatible with T, i.e. the initial condition does not falsify the background theory ($T \nvdash_{\mathbf{CL}} \sim I$).

Obviously, there is no positive test for any of the three conditions. In other words, no finite reasoning process can (in general) lead to the conclusion that Pb is explained by I and T.

So although the 'logic' appears to be **CL** (see the formal conditions above), it is quite obvious that the reasoning process that leads to the conclusion that I and T together explain Pb cannot possibly be explicated in terms of **CL** (at the object level). The reasoning is *about* **CL**-derivability, and necessarily displays the internal dynamics. This is why it cannot be explicated by a **CL**-proof but only by a dynamic proof.[8]

The logic of questions forms a further example. According to Wiśniewski (1995, 1996), where this problem is studied and solved, a question Q is evoked by a set of declarative statements Γ iff the (prospective) presupposition of Q is derivable from Γ but no direct answer to Q is derivable from Γ. Note that there is no positive test for non-derivability[9] from Γ. Hence, although the definition is itself unobjectionable, only a d ynamic proof may (in general) lead to the conclusion that Q is evoked by Γ.

Another example concerns handling inconsistency. Consider the case in which a scientific (empirical or mathematical) theory T was meant to be consistent and was formulated with **CL** as its underlying logic, but turned out to be inconsistent. As we know from the literature,[10] scientists do not just throw away such a theory. They *reason* from T in search for a consistent replacement. Of course, they do not reason in terms of **CL**, because this is known to lead to triviality. However, they also do not reason in terms of some monotonic paraconsistent logic **PL**. In their reasoning, they want to interpret T *as consistently as possible*. After all, T was meant as a consistent theory.

Let us consider an utterly simplistic but instructive example. Let the 'theory' consist of $\sim p$, $p \vee r$, $\sim q$, $q \vee s$, and p. One obviously should not derive r from $\sim p$ and $p \vee r$ by Disjunctive Syllogism. To do so would lead to triviality: $p \vee r$ is itself a consequence of p, and so is any formula of the form $p \vee A$. However, as the theory was meant to be consistent, one will apply Disjunctive Syllogism to derive s from $\sim q$ and $q \vee s$. Indeed, it is quite obvious that q behaves consistently on this theory. To be more precise, q is consistently false on the theory, for $\sim q$ is obviously derivable whereas q is not, except of course by explicitly or implicitly applying Ex Falso Quodlibet to p and $\sim p$.[11]

So the reasoning from T should proceed in such a way that one obtains T in its full richness, except for the pernicious consequence of its inconsistency. Precisely for this reason, the reasoning cannot proceed in terms of some monotonic paraconsistent logic **PL**. Indeed, **PL** will invalidate certain **PL**-rules, for example Disjunctive Syllogism.[12] However, as we saw from the previous example, the requested reasoning should not invalidate certain rules of inference of **CL**, but only certain *applications* of these rules. Let me express this more precisely. For certain rules, an application should be valid if specific involved formulas behave consistently on the theory, and invalid otherwise. Precisely this proviso causes the reasoning to be internally dynamic: there is no positive test for the consistent behaviour of some formula on a set of premises.[13]

Up to now we have considered forms of reasoning that display an internal dynamics. All of them concerned a single unstructured set of premises. In many cases, however, the premises are structured, usually as a n-tuple of sets. I shall now consider some examples of this type.

Let us return for a moment to inductive generalization. Usually background knowledge is available in addition to the data. Let us restrict the discussion to the case where the background knowledge consists of generalizations in the sense meant before. An obvious complication is that the data may falsify some of the background generalizations. So two forms of dynamics have to be combined. First, we retain the background generalizations in as far as they are not falsified by the data. Next, from the data and the retained background generalizations we obtain new generalizations by the aforementioned logic **LI**. Remark that it is in general impossible to perform the first selection (of background generalizations) before proceeding to the second selection (of new inductive generalizations). This means that both forms of dynamics are necessarily combined in the reasoning process. After deriving some new inductive generalizations, one may be forced to change one's judgement on the compatibility of some background generalization with the data, and this will affect the derivability of new generalizations.[14]

Often not all background generalizations will be considered equally trustworthy. So instead of a set of background generalizations, one confronts a sequence of such sets, each having a different priority. In this case one has to combine a multiplicity of dynamics concerning the background generalizations with the dynamics that pertains to the new generalizations. Moreover, even certain falsified background generalizations may be considered as applying 'normally'. This means that an *instance* of the generalization is considered to hold unless and until proven incompatible with the data. Such 'pragmatic generalizations' may also be ordered by some priority relation. All this leads to more forms of dynamics (which, however, are all of three kinds).

Let us now consider a very different example. A participant in a discussion may change his or her position in view of arguments adduced by other participants. As a result, the interventions of the participant will be mutually incompatible, even if the participant's position is consistent during each intervention. However, the participant will not state his or her full new position whenever there is a change. So, after an intervention, the participant's position has to be reconstructed from the sequence of his or her interventions. In order to do so, one starts with (the consistent part of) the last intervention, to this one adds that part of the previous intervention that is compatible with it, and so on. Remark that, while doing so, one does not select statements that are made during an intervention, but rather their consequences.[15]

Diagnostic reasoning forms a further example in which the premises are prioritized and hence require a multiplicity of dynamics. Reasoning proceeds from data on the one hand and expectancies (that may have varying degrees of trustworthiness) on the other hand. The expectancies, or rather their consequences, are retained (in their order of priority) until and unless proven inconsistent with the data. (See Weber and Provijn (1999) Provijn and Weber (2002) and Batens et al. (2003) for the adaptive logics.)

In all examples mentioned before, the flat ones as well as the prioritized ones, the reasoning displays both the internal dynamics and the external dynamics. It is worth mentioning that, whenever the external dynamics (non-monotonicity) is present, the reasoning necessarily displays the internal dynamics (even if the premises are stable). The converse, however, does not hold. The *Weak* consequence relation, of Rescher and Manor – see, for example, Rescher and Manor (1970) and Benferhat et al. (1997) – is monotonic. Nevertheless, it may be shown that the reasoning from premises to weak consequences requires an internal dynamics.[16] Some consequence relations that are monotonic as well as decidable may even be characterized (in an enlightening and attractive way) by a form of reasoning that displays an internal dynamics – see Batens (2001) for an example.

The preceding paragraphs do by no means contain an exhaustive list of the reasoning mechanisms (or even of the types of reasoning mechanisms) that display an internal dynamics. Nevertheless, the problem should be clear by now. Essential forms of human reasoning, that are common and that are important for understanding the way in which humans arrive at knowledge and revise it, display an internal dynamics.

In order to arrive at a precise theory of knowledge, one needs to explicate such forms of reasoning. To do so requires specific logics: adaptive logics. These unavoidably have some non-standard properties. More important, however, is that they are characterized in a formally stringent way, and that their properties are studied in agreement to the professional standards.

3. What are Adaptive Logics?

In the loose sense of the term, a logic is adaptive iff it adapts itself to the specific premises to which it is applied. This should not be misunderstood. First, I do not mean to say that the consequence set, determined by the logic, depends on the set of premises. This obviously holds for nearly all[17] logics. I mean that the *logic* adapts to the premises in that it depends on properties of the premise set whether some formula is derivable from some of the premises. Next, I really mean that the logic adapts *itself* to the premises. The reasoner does not interfere in this. The logic is defined by a set of rules as well as by a semantics. Both lead to the adaptive effect, independently of any decision of the human or machine that applies the rules.

The previous paragraph describes the underlying idea. I shall also present a more technical characterization. This should not be understood as a definition, but rather as a hypothesis on the properties of all adaptive logics. It relies on present best insights. These may change as more logics are studied or new insights in them are gained. I have a good reason to insert this remark: during the last twenty years the dynamics of the adaptive logics programme forced the Ghent logic group several times to revise the technical characterization.

Flat (Non-prioritized) Adaptive Logics. I start with these because the prioritized ones may be seen as (systematic) combinations of them. An adaptive logic **AL** may be characterized by a triple: the lower limit logic, the set of abnormalities, and the strategy. The *lower limit logic* **LLL** is a monotonic logic for which it holds that $Cn_{\mathbf{LLL}}(\Gamma) = \bigcap\{A \mid \Gamma \cup \Delta \vdash_{\mathbf{AL}} A; \emptyset \subseteq \Delta \subseteq \mathcal{W}\}$, in which $\mathcal{W}$ is the set of all closed formulas of the language.[18] Intuitively, the lower limit logic is the stable part of the adaptive logic, the part that is not subject to any adaptation. From a proof theoretic point of view, the lower limit logic delineates the rules of inference that hold unexceptionally. From a semantic point of view, all adaptive models of Γ are lower limit models of Γ (but not conversely). It follows that $Cn_{\mathbf{LLL}}(\Gamma) \subseteq Cn_{\mathbf{AL}}(\Gamma)$.

Suppose that we are dealing with a context in which **CL** is taken as the standard of deduction. If the lower limit logic of **AL** is **CL** (or, for example, a modal extension of **CL**), it is said that **AL** is *ampliative*. This is the case for inductive generalization (without background knowledge), for compatibility, etc. If the lower limit logic is weaker than **CL**, as in the case of inconsistency-adaptive logics, the adaptive logic is called *corrective* – the theory was intended to be interpreted in terms of **CL**, but turned out to be inconsistent and hence is interpreted as consistently as possible.[19]

If the lower limit logic is a fragment of **CL**, it is wise to extend it with the missing classical logical symbols (by means of explicit definitions or, where this is impossible, by a straightforward extension of the language – see Batens (1999c)). This changes nothing to the way in which the premises are handled, but greatly simplifies the metatheory.

The second component of an adaptive logic is the set of *abnormalities* Ω. These are the formulas that are presupposed to be false, unless and until proven otherwise. In the standard format, Ω is characterized by a metalinguistic formula that may be restricted. In handling inconsistency, the set of abnormalities comprises the formulas of the form $\exists(A \wedge \sim A)$, in which $\exists A$ abbreviates the existential closure of A. The set may be restricted. For some lower limit logics the set is restricted by the requirement that A be a primitive formula (a formula containing no logical symbols except for identity). In the case of an inductive logic, the set of abnormalities may consist of all formulas of the form $\exists A \wedge \exists \sim A$ with the restriction that no individual constants or quantifiers occur in A.

Extending the lower limit logic with the requirement that no abnormality is logically possible results in a monotonic logic, which is called the *upper limit logic*. The effect is presumably most easily seen by considering the semantics. The upper limit logic is characterized by the lower limit logic models that verify no abnormality. If the adaptive logic is corrective, the lower limit logic is weaker than **CL**, and the upper limit logic will usually be (and in all cases studied up to now is) **CL**. If the adaptive logic is ampliative, the lower limit is (in all cases studied so far) **CL** or a modal extension of **CL**, and the upper limit logic is an extension of this.

Some examples are useful to clarify the matter. If the lower limit logic is **CL** and the set of abnormalities comprises all formulas of the form $\exists A \wedge \exists \sim A$ (see two paragraphs ago), then the upper limit logic is **CLU**, obtained by adding to **CL** the axiom $\exists A \supset \forall A$.[20] If, as in the case of an inconsistency-adaptive logic, the lower limit logic is a paraconsistent logic **PL** that is a fragment of **CL**, and the set of abnormalities comprises all formulas of the form $\exists(A \wedge \sim A)$, then the upper limit logic is **CL**. The importance of the set of abnormalities will be obvious once a strategy is chosen – see below. If the premise set does not require any abnormality to obtain, the adaptive logic will deliver the same consequences as the upper limit logic. If the premise set requires some abnormalities to obtain, the adaptive logic will still deliver more consequences than the lower limit logic, viz. all upper limit consequences that are not 'blocked' by those abnormalities.

It became only clear during the last years that the combination of a lower limit logic with different sets of abnormalities may result in the same upper limit logic but in a different adaptive logic. This may be easily exemplified in terms of inconsistency-adaptive logics. Consider a lower limit logic that validates all reductions of negations (double negation, de Morgan laws, the standard negation laws for the quantifiers, etc.). Whether the set of abnormalities Ω comprises all formulas of the form $\exists(A \wedge \sim A)$, or is restricted to the case in which A is primitive, the upper logic is **CL**. However, on both the Reliability strategy and the Minimal Abnormality Strategy, the resulting adaptive logic is very different. The latter set of abnormalities defines an inconsistency-adaptive logic of the usual kind, whereas the former set of abnormalities defines an adaptive logic of a somewhat weird kind, viz. a flip–flop – see below.

A very important matter has to be brought up at this point. For all that was said before, an adaptive logic is obtained by presupposing that all formulas behave normally, except for those that need to behave abnormally in view of the premises. This formulation suggests that there is a well-defined set of formulas that behave abnormally in view of the premises, but this need not be the case. The complication derives from the fact that a set of premises may entail a disjunction of abnormalities (members of Ω) without entailing any of its disjuncts.[21] Let us again consider the adaptive logic of induction and let the premise set be $\{Pa, Qa, Rb, \sim Qb\}$. Even with so small a data set, $(\forall x)(Px \supset Qx)$ and $(\forall x)(Rx \supset \sim Qx)$ are derivable. Suppose next that the premise set is extended to $\{Pa, Qa, Rb, \sim Qb, Pc, Rc\}$. Neither $(\exists x)(Px \wedge Qx) \wedge (\exists x) \sim (Px \wedge Qx)$ nor $(\exists x)(Rx \wedge Qx) \wedge (\exists x) \sim (Rx \wedge Qx)$ is **CL**-derivable from these premises. However, their (classical) disjunction *is* **CL**-derivable from the premises.

Classical disjunctions[22] of abnormalities will be called *Dab-formulas* and will be written as $Dab(\Delta)$, in which $\Delta \subset \Omega$ is finite.[23] The *Dab*-formulas that are derivable by the lower limit logic from the premise set Γ will be called *Dab-consequences* of Γ. If $Dab(\Delta)$ is a *Dab*-consequence of Γ, then so is $Dab(\Delta \cup \Theta)$ for any (finite) Θ. For this reason, the following definition is important. Let **LLL** be the lower limit logic as before.

DEFINITION 1. $Dab(\Delta)$ is a *minimal Dab*-consequence of Γ iff $\Gamma \vdash_{\mathbf{LLL}} Dab(\Delta)$ and there is no $\Theta \subset \Delta$ such that $\Gamma \vdash_{\mathbf{LLL}} Dab(\Theta)$.

That $Dab(\Delta)$ is a minimal *Dab*-consequence of Γ means that it is derivable (by the lower limit logic) from Γ that some member of Δ behaves abnormally, whereas it is not derivable which member of Δ behaves abnormally. Adaptive logics are obtained by interpreting a set of premises 'as normally as possible'. But clearly, this phrase is not unambiguous. This is why we need to disambiguate it by choosing a specific adaptive strategy.

The oldest known *strategy* is *Reliability* from Batens (1989), where it is discussed at the propositional level. Let $U(\Gamma) = \{A \mid A \in \Delta$ for some minimal *Dab*-consequence $Dab(\Delta)$ of $\Gamma\}$ (the set of formulas that are unreliable on Γ). The Reliability strategy considers a formula as behaving abnormally iff it is a member of $U(\Gamma)$. As for the other strategies, the effect of this on the semantics and proof theory will be discussed in subsequent sections.

The *Minimal Abnormality* strategy (presented in Batens (1986) but first fully studied in Batens (1999a)) delivers some more consequences than the Reliability strategy. If, for example, $Dab(\Delta_1), \ldots, Dab(\Delta_n)$ are the minimal *Dab*-consequences of Γ, the Minimal Abnormality strategy takes one member of each Δ_i to behave abnormally, while all other formulas behave normally.[24] Obviously, the Minimal Abnormality strategy does not pick out a specific such combination, but considers all of them. Here is a simple propositional example for an inconsistency-adaptive logic: $\Gamma = \{\sim p, \sim q, p \vee q, p \vee r, q \vee r\}$. If the lower limit logic validates all of full positive logic, $(p \wedge \sim p) \vee (q \wedge \sim q)$ is a minimal *Dab*-consequence of Γ. On the Reliability strategy, both $p \wedge \sim p$ and $q \wedge \sim q$ are unreliable with respect to Γ, and hence r is not an adaptive consequence of Γ. However, if the Minimal Abnormality strategy is chosen, then r is an adaptive consequence of Γ. Indeed, if $p \wedge \sim p$ behaves abnormally, then $q \wedge \sim q$ behaves normally and hence r is true in view of $\sim q$ and $q \vee r$; if $q \wedge \sim q$ behaves abnormally, then $p \wedge \sim p$ behaves normally and hence r is true in view of $\sim p$ and $p \vee r$. In subsequent sections, we shall see that both strategies are simple and perspicuous from a semantic point of view, and that the Reliability strategy leads to simple dynamic proofs, but that the dynamic proofs determined by the Minimal Abnormality strategy are rather complicated. Which strategy is adequate in a specific context of application is obviously a very different matter.

For some specific lower limit logics and sets of abnormalities, any minimal *Dab*-consequence $Dab(\Delta)$ of any premise set is such that Δ is a singleton. In such cases, the Reliability and Minimal Abnormality strategies lead to the same result and coincide with what is called the *Simple* strategy: a formula behaves abnormally just in case the abnormality is derivable from the premise set. Examples may be found in Meheus (2000) and Batens and Meheus (2000a).

Several other strategies have been studied, but seem to have a less general import. Most of them were the result of characterizing an existing consequence relation by an adaptive logic. Examples may be found in Batens (2000b), Batens (to appear), de Clercq (2000) and Verhoeven (to appear).

A different way to characterize most flat adaptive logics is by seeing them as formula-preferential systems. The idea was first presented in Lev (2000) (see also Avron and Lev (to appear)). This means that Ω is taken to be an arbitrary set of formulas. We shall see that the idea is sensible in general for so-called direct formulations of prioritized adaptive logics. I am not sure that it will work for all adaptive logics. Moreover, many sets of formulas do not define a sensible upper limit logic. The idea to interpret a set of premises 'as much as possible' in agreement with the upper limit logic is an attractive feature of the adaptive logic enterprise and is lost if they are rephrased as formula-preferential systems.

Ongoing work concerns an adaptive logic of inductive prediction. If ω-incomplete models are considered, one needs to define the 'abnormal part' of a model in terms of the model itself rather than in terms of the formulas it verifies. The first example of an approach in terms of models was Graham Priest's system $\mathbf{LP}^m$ from Priest (1991) – see Batens (1999b) for a discussion of some mistakes in this construction.

Prioritized Adaptive Logics and Combining Adaptive Logics. Consider $\Sigma = \langle \Gamma_0, \ldots, \Gamma_n \rangle$, in which Γ_0 is a set of data (that are taken to be certain) and $\Gamma_1, \ldots, \Gamma_n$ are sets of expectancies – formulas that are supposed to obtain but may be overruled. The members of Γ_i $(1 \leq i \leq n)$ carry a higher degree of certainty as i is smaller.

One prioritized adaptive logic to handle such n-tuples is obtained as follows. Where $\Diamond^i$ abbreviates a sequence of i diamonds, $\Sigma^\Diamond = \{\Diamond^i A \mid A \in \Gamma_i\}$. The lower limit logic is (for example) the modal logic[25] **T** and the set of abnormalities comprises all formulas of the form $\Diamond^i A \wedge {\sim}A$, in which A is either a primitive formula or the negation of a primitive formula and $1 \leq i \leq n$. The upper limit logic is **Triv**, which is obtained by extending **T** with (for example) the axiom $\Diamond A \supset A$. Finally, one combines the above with either the Reliability or Minimal Abnormality strategy with the following proviso: an abnormality $\Diamond^i A \wedge {\sim}A$ is considered as worse according as i is smaller. This means that, if either an abnormality of level i or an abnormality of level j is unavoidable in view of the premises, then the abnormality of level i is avoided if $i < j$. The results are the nice adaptive logics $\mathbf{T}^{sr}$ and $\mathbf{T}^{sm}$ from Batens et al. (2003). For more examples see Batens and Haesaert (2001), Verhoeven (2003) and Verhoeven (to appear), the latter containing adaptive logics that characterize all prioritized Rescher–Manor consequence relations from Benferhat et al. (1999).

A different way to characterize prioritized adaptive logics is by seeing them as the result of applying a sequence of flat adaptive logics, each of these logics having a (nearly) similar structure. This characterization of prioritized adaptive logics is a special case of a more general mechanism, viz. that adaptive logics, which may have a very different structure, are combined with each other. Consider the search for an explanation of some singular fact, given a theory. If one relies on Hintikka's conditions, one needs an adaptive logic to deal with the conditions for which there is no positive test – see Section 2. Suppose that moreover the involved theory is inconsistent (or that the data are inconsistent) – see Batens (in print(b)) for a discussion of such inconsistencies. In such a case one needs to combine the adaptive logic for the process of explanation with an inconsistency-adaptive logic that interprets the theory as consistently as possible. The result is a sequence of (two) adaptive logics.

The term 'sequence' deserves a comment. The definition of the logic refers to such a sequence, and intuitively one may understand the logic as resulting from applying one adaptive mechanism after the other. However, as there is no positive test for any of the two, it is essential that the dynamic proof theory is able to handle all the adaptive steps in any order.

Applying the above logics $\mathbf{T}^{sr}$ and $\mathbf{T}^{sm}$ requires the transition from the n-tuple Σ to the set $\Sigma^{\Diamond}$.[26] Given such transition, the characterization of the prioritized consequence relation comes to a combination of (similar) flat adaptive logics.

It might be objected that the prioritized adaptive logic does not define the consequence relation itself, but defines it only under some translation. The matter is not as simple as it looks. If A is not a premise but only an expectancy, it seems desirable that this is expressed in the object language. Typically, in Weber and Provijn (2002), which started the work on adaptive logics for diagnosis, $E(A)$ is used to express that A is an expectancy. So, from a philosophical viewpoint, rendering an expectancy of degree i as $\Diamond^i A$ is superior to an approach in terms of a sequence of sets of premises.

Still, if the adaptive logic is intended to characterize an existing prioritized consequence relation, one might be interested in a faithfully covering. In all cases studied up to now, we were able to also articulate direct formulations. These are formulations in terms of the original language. The same applies for the dynamic proofs. It is not completely clear whether these formulations fit within the technical characterization of an adaptive logic. Apparently, more work is required before the matter can be settled. However, it is quite obvious that the direct formulations may be seen as adaptive logics in the sense of formula-preferential logics: the logic selects the models that verify 'as much as possible' the members of each $\Gamma_i \in \Sigma$ (in its order of priority). If adaptive logics are seen as formula preferential, and the set of abnormalities is arbitrary (and not defined by some logical form), the lower limit logic and the requirement that no abnormalities obtain might not together define an upper limit logic. If the priorities are expressed in the object language (for example as $\Diamond^i A$), abnormalities have a specific logical form (in the example $\Diamond^i A \wedge \sim A$ in which A is either a primitive formula or its negation). This matter too requires more study.

Some Further Comments. In the subsequent paragraphs, I mention some miscellaneous findings. Some of them created confusion at a time, others have not yet been settled.

It is not absolutely clear whether all adaptive logics are either flat or prioritized. Several times our research group discovered items that did not seem to fit in either category. Sooner or later, the exception turned out to be apparent only. Often such apparent exceptions led to a better understanding of adaptive logics, and sometimes to broadening the notion.

A very different matter concerns the introduction of new premises. From a traditional point of view, extensions of the premise set are logically uninteresting. An extension of the premise set leads to a distinct premise set, and distinct premise sets may define distinct consequence sets. If the logic is non-monotonic, extending the premise set may result in a consequence set that displays gain as well as loss with respect to the original consequence set. In the context of adaptive logics the matter is slightly more interesting. Often the very reasoning that is performed in terms of an adaptive logic, or in terms of the consequence relation that is explicated by the adaptive logic, leads to the introduction of new premises. I shall mention two simple examples, a straightforward one first, and a somewhat unexpected one next – both concern forthcoming work.

Suppose that one is reasoning in terms of an adaptive logic – say a flat one – and that $Dab\{A_1, \ldots, A_n\}$ appears to be a minimal *Dab*-consequence of the premise set. Often this very fact causes one to search for good reasons to narrow down the minimal *Dab*-consequence and sometimes good reasons are found. So reasoning from the premises may cause one to extend the premises. Of course the new premises do not provide from the reasoning; the aforementioned good reasons are not provided by the reasoning. But the search for specific new premises is.

Let us consider an example. Suppose that one applies an inconsistency-adaptive logic and hence that the abnormalities have the form $\exists(B \wedge \sim B)$. Finding out that $Dab\{A_1, \ldots, A_n\}$ is apparently a minimal disjunction of abnormalities may trigger the question as well as the resulting insight that there are good reasons not to blame specific A_i. For example, these A_i may pertain to well-entrenched theories, or to observational criteria that are considered as unproblematic. In such cases, one may want to posit that those A_i are not abnormal. There are several ways to handle such situation. Each of them involves some complications that fall beyond the scope of the present paper – not because they are difficult, but because they require too much space.[27] This feature is not typical for corrective adaptive logics. For example, in the context of inductive adaptive logics it leads to introducing generalizations that are not justifiable in terms of the data, but possibly in terms of a worldview or some other bias. Typically, the new premises are conjectures, expressing claims that transcend theories and observations, and hence should be handled in such a way that they are revokable in view of the internal dynamics of the reasoning – that is in view of later gained insights in the meaning of the original premises.

I promised a second, somewhat unexpected example. Consider a problem-solving process, even one of a very basic kind, in which empirical data are relevant. Empirical data may be gathered, by observation or experiment, but gathering them may be expensive or time consuming. So some observations and experiments will be postponed until they turn out relevant for the problem-solving process. And indeed, the reasoning in terms of the suitable adaptive logic will indicate that some information is relevant or is possibly relevant.[28] Note that such considerations are quite remote from the traditional logical viewpoint. The problem one is facing is not whether A is derivable from a premise set Γ; the problem is not what is the consequence set of Γ. The problem is, in the simplest case, to settle whether A is true or not in view of the premises *and* of the available empirical means.

A very different type of adaptive logic is most easily illustrated in terms of inconsistency. Suppose that one confronts an inconsistent theory, that the inconsistency is taken to render the theory inadequate, and that one is interested in the 'consistent part' of the theory, which one deems unproblematic. To locate this consistent part (in other words, to obtain consistency by brute force) is a task for logic. Given the absence of a positive test for consistency, it is a task for adaptive logic. The task seems to be a difficult one; so far no adequate adaptive logic has been characterized.

The last comment concerns an amusing phenomenon. After first reading or hearing about adaptive logics, some people think that adaptive logics are (what I like to call) flip–flops. Thus some people think that inconsistency-adaptive logics (i) deliver the classical consequences of consistent premise sets, and (ii) deliver the paraconsistent consequences (defined by the lower limit logic) of inconsistent premise sets. While (i) is correct, (ii) is false for most inconsistency-adaptive logics. Even if the premise set is abnormal, most adaptive logics still interpret the set as normally as possible, and hence deliver more consequences than the lower limit logic. Nevertheless, it was amusing to discover that it is very easy to define flip–flops. These are indeed adaptive logics: they adapt themselves to the premise set, even if only in the crudest possible way.

4. Semantics

The dynamic proof theory of adaptive logics is certainly their most fascinating feature. It was this proof theory that led to the discovery of adaptive logics – see Batens (1989). I nevertheless start by discussing the semantics because this will be easier for most people.

Let us consider an arbitrary flat adaptive logic **AL**, defined from a lower limit logic **LLL**, a set of abnormalities Ω and a strategy. I shall moreover suppose that the following conditions are fulfilled:

C1 **LLL** is a monotonic logic.
C2 **LLL** is left and right compact.[29]
C3 Ω is a set of formulas characterized by a (possibly restricted) logical form.
C4 **AL** is a flat adaptive logic defined from **LLL** and Ω by either the Reliability strategy or the Minimal Abnormality strategy.[30]
C5 All classical logical symbols are present in **LLL** – see Section 3.

Where it matters, I refer to the strategy by the name of the adaptive logic thus: $\mathbf{AL}^r$ and $\mathbf{AL}^m$.

The use of C5 was explained in Section 3.[31] Even for corrective adaptive logics the presence of classical negation, which I shall write as $\neg$, is often not required for extending the axiomatic characterization of the lower limit logic **LLL** into an axiomatic characterization of the upper limit logic **ULL**.[32]

DEFINITION 2. The upper limit logic **ULL** is semantically characterized by the **LLL**-models that verify no member of Ω.

I shall suppose that an adequate semantics for **LLL** is present, whence Definition 2 provides an adequate semantics for **ULL**. In view of this I shall pass freely from the proof theory to the semantics.

The **AL**-models of a premise set Γ are a subset of the **LLL**-models of Γ. How the selection is made depends on the strategy. For both strategies, we need the set of abnormalities verified by the model M: $Ab(M) = \{A \in \Omega \mid M \models A\}$. For the Reliability strategy, we moreover need the minimal *Dab*-consequences of Γ (defined by the semantic counterpart of Definition 1), and $U(\Gamma)$ is defined as in Section 3, viz. as $\{A \mid A \in \Delta$ for some minimal *Dab*-consequence $Dab(\Delta)$ of $\Gamma\}$.

DEFINITION 3. A **LLL**-model M of Γ is *reliable* iff $Ab(M) \subseteq U(\Gamma)$.

DEFINITION 4. $\Gamma \models_{\mathbf{AL}^r} A$ iff A is verified by all reliable models of Γ.

DEFINITION 5. A **LLL**-model M of Γ is *minimal abnormal* iff there is no **LLL**-model M' of Γ such that $Ab(M') \subset Ab(M)$.

DEFINITION 6. $\Gamma \models_{\mathbf{AL}^m} A$ iff A is verified by all minimal abnormal models of Γ.

I shall prove several theorems for all adaptive logics under consideration. To simplify the notation, let $\mathcal{M}_\Gamma^{\mathbf{L}}$ denote the set of **L**-models of Γ. Where the logic is $\mathbf{AL}^r$ or $\mathbf{AL}^m$, I shall write $\mathcal{M}_\Gamma^r$ and $\mathcal{M}_\Gamma^m$ respectively.

Given that the adaptive logics were defined by a selection of lower limit models of the premise set, it is important to prove that this selection has suitable properties. A very strong property is that, for any lower limit model M that is not selected ($M \notin \mathcal{M}_\Gamma^{\mathbf{AL}}$), there is a selected model M' that is less abnormal than M ($Ab(M') \subset Ab(M)$). I call this property Strong Reassurance, Avron calls it Stopperedness, and it is closely related to what is called Smoothness in Kraus et al. (1990). That its absence leads to undesired results is shown, for example, in Batens (2000a). I first prove the property for the Minimal Abnormality strategy. An unqualified "model" will always refer to a **LLL**-model.

THEOREM 1. If $M \in \mathcal{M}_\Gamma^{\mathbf{LLL}} - \mathcal{M}_\Gamma^m$, then there is a $M' \in \mathcal{M}_\Gamma^m$ such that $Ab(M') \subset Ab(M)$. (Strong Reassurance for Minimal Abnormality.)

Proof. The theorem holds (vacuously) if Γ has no **LLL**-models or if $\mathcal{M}_\Gamma^m = \mathcal{M}_\Gamma^{\mathbf{LLL}}$. Consider a $M \in \mathcal{M}_\Gamma^{\mathbf{LLL}} - \mathcal{M}_\Gamma^m$, let $D_1, D_2, \ldots$ be a list of all members of Ω, and define[33]

$$\Delta_0 = \emptyset$$

$$\Delta_{i+1} = \Delta_i \cup \{\neg D_{i+1}\}$$

if there is a model M' of $\Gamma \cup \Delta_i \cup \{\neg D_{i+1}\}$ such that $Ab(M') \subseteq Ab(M)$, and

$$\Delta_{i+1} = \Delta_i$$

otherwise. Finally,

$\Delta = \Delta_0 \cup \Delta_1 \cup \Delta_2 \cup \ldots$

Given the left compactness of **LLL**, $\Gamma \cup \Delta$ has models in view of the construction.

Step 1. I first show that, if M' is a model of $\Gamma \cup \Delta$, then $Ab(M') \subset Ab(M)$. Suppose that there is a $D_j \in \Omega$ such that $D_j \in Ab(M') - Ab(M)$. Let M'' be a model of $\Gamma \cup \Delta_{j-1}$ for which $Ab(M'') \subseteq Ab(M)$. As $D_j \notin Ab(M)$, $D_j \notin Ab(M'')$. Hence M'' is a model of $\Gamma \cup \Delta_{j-1} \cup \{\neg D_j\}$ and $Ab(M'') \subseteq Ab(M)$. So $\neg D_j \in \Delta_j \subseteq \Delta$. As M' is a model of $\Gamma \cup \Delta$, $D_j \notin Ab(M')$. But this contradicts the supposition.

Step 2. I now show that every model of $\Gamma \cup \Delta$ is a minimal abnormal model of Γ. Suppose that M' is a model of $\Gamma \cup \Delta$, but is not a minimal abnormal model of Γ. Hence, by Definition 5, there is a model M'' of Γ for which $Ab(M'') \subset Ab(M')$.

It follows that M'' is a model of $\Gamma \cup \Delta$. If it were not, then, as M'' is a model of Γ, there is a $\neg D_j \in \Delta$ such that M' verifies $\neg D_j$ and M'' falsifies $\neg D_j$. But then M' falsifies D_j and M'' verifies D_j, which is impossible in view of $Ab(M'') \subset Ab(M')$.

Consider any $D_j \in Ab(M') - Ab(M'') \neq \emptyset$. As M'' is a model of $\Gamma \cup \Delta_{j-1}$ that falsifies D_j, it is a model of $\Gamma \cup \Delta_{j-1} \cup \{\neg D_j\}$. As $Ab(M'') \subset Ab(M')$ and $Ab(M') \subseteq Ab(M)$, $Ab(M'') \subset Ab(M)$. It follows that $\Delta_j = \Delta_{j-1} \cup \{\neg D_j\}$ and hence that $\neg D_j \in \Delta$. But then $D_j \notin Ab(M')$. Hence, $Ab(M'') = Ab(M')$. So the supposition leads to a contradiction.

As $\mathcal{M}^m_\Gamma \subseteq \mathcal{M}^r_\Gamma$ (by property 1 of Theorem 3 below), it follows that:

THEOREM 2. If $M \in \mathcal{M}^{\mathbf{LLL}}_\Gamma - \mathcal{M}^r_\Gamma$, then there is a $M' \in \mathcal{M}^r_\Gamma$ such that $Ab(M') \subset Ab(M)$. (Strong Reassurance for Reliability.)

COROLLARY 1. If Γ has lower limit models, then it has minimal abnormal models as well as reliable models. (Reassurance.)

At this point, I need some lemmas that require a specific characterization of the minimal abnormal models. In Batens (1999a), such a characterization is offered in terms of a set Φ_Γ, which is a set of sets of abnormalities.[34] It is shown there that, where M is a minimal abnormal model of some Γ, $Ab(M)$ is characterized by some $\phi \in \Phi_\Gamma$ – all members of $Ab(M)$ are **LLL**-consequences of some $\phi \in \Phi_\Gamma$.

Recently, I found a drastically simpler such characterization, which only has the disadvantage to be less ‘finitistic’. I shall now redefine Φ_Γ. The proofs in Batens (1999a) may be easily modified in view of this change and may be generalized to all adaptive logics considered, but I cannot, in the present paper, spell out the required modifications to the proofs. Let Φ°_Γ comprise all sets that contain a disjunct out of each minimal *Dab*-consequence of Γ and that are **LLL**-closed with respect to Ω.[35] Let Φ_Γ contain all members of Φ°_Γ that are not supersets of other members of Φ°_Γ. Suitably modifying and generalizing the proof of Lemmas 7.2 and 7.3 of Batens (1999a) gives us:

LEMMA 1. M is a minimal abnormal model of Γ iff $M \in \mathcal{M}_{\Gamma}^{\mathbf{LLL}}$ and $Ab(M) \in \Phi_{\Gamma}$.

The strength of this lemma may be seen from the fact that each of the following properties are immediate or nearly immediate consequences of it:[36]

THEOREM 3. Each of the following holds:

1. Every minimal abnormal model of Γ is a reliable model of Γ ($\mathcal{M}_{\Gamma}^{m} \subseteq \mathcal{M}_{\Gamma}^{r}$). Hence $Cn_{\mathbf{AL}^r}(\Gamma) \subseteq Cn_{\mathbf{AL}^m}(\Gamma)$.
2. If $A \in \Omega - U(\Gamma)$, then $\neg A \in Cn_{\mathbf{AL}^r}(\Gamma)$.
3. If $Dab(\Delta)$ is a minimal Dab-consequence of Γ and $A \in \Delta$, then there is a minimal abnormal model M of Γ that verifies A and falsifies all members (if any) of $\Delta - \{A\}$.
4. All minimal abnormal models of Γ are minimal abnormal models of $Cn_{\mathbf{AL}^n}(\Gamma)$ and vice versa ($\mathcal{M}_{\Gamma}^{m} = \mathcal{M}_{Cn_{\mathbf{AL}^m}(\Gamma)}^{m}$) and hence $Cn_{\mathbf{AL}^m}(\Gamma) = Cn_{\mathbf{AL}^m}(Cn_{\mathbf{AL}^m}(\Gamma))$. (Fixed Point.[37])
5. All reliable models of Γ are reliable models of $Cn_{\mathbf{AL}^r}(\Gamma)$ and vice versa ($\mathcal{M}_{\Gamma}^{r} = \mathcal{M}_{Cn_{\mathbf{AL}^r}(\Gamma)}^{r}$) and hence $Cn_{\mathbf{AL}^r}(\Gamma) = Cn_{\mathbf{AL}^r}(Cn_{\mathbf{AL}^r}(\Gamma))$. (Fixed Point.)
6. For all $\Delta \subseteq \Omega$, $Dab(\Delta) \in Cn_{\mathbf{AL}}(\Gamma)$ iff $Dab(\Delta) \in Cn_{\mathbf{LLL}}(\Gamma)$. (Immunity.)
7. If $\Gamma \vDash_{\mathbf{AL}} A$ for every $A \in \Gamma'$, and $\Gamma \cup \Gamma' \vDash_{\mathbf{AL}} B$, then $\Gamma \vDash_{\mathbf{AL}} B$. (Cautious Cut.)
8. If $\Gamma \vDash_{\mathbf{AL}} A$ for every $A \in \Gamma'$, and $\Gamma \vDash_{\mathbf{AL}} B$, then $\Gamma \cup \Gamma' \vDash_{\mathbf{AL}} B$. (Cautious Monotonicity.)

These properties are well-known from the study of non-monotonic logics. They ensure that, although adaptive logics are non-monotonic, they fulfil a set of desirable properties. Thus Immunity warrants that the transition from the lower limit logic to the adaptive logic does not lead to the derivability of new disjunctions of abnormalities (or to the non-derivability of old disjunctions of abnormalities). Cautious Cut warrants that extending a premise set with some of its adaptive consequences does not lead to any gain in consequences, and Cautious Monotonicity warrants that extending a premise set with some of its adaptive consequences does not lead to any loss of consequences. So, if the adaptive logic is applied to a belief set, its consequences may be considered as beliefs themselves.

A premise set Γ will be called normal if $\mathcal{M}_{\Gamma}^{\mathbf{ULL}} \neq \emptyset$; it is called abnormal otherwise. Note that Γ is normal iff $\Omega \cap Cn_{\mathbf{LLL}}(\Gamma) = \emptyset$.

THEOREM 4. Each of the following obtains:

1. $\mathcal{M}_{\Gamma}^{\mathbf{ULL}} \subseteq \mathcal{M}_{\Gamma}^{m} \subseteq \mathcal{M}_{\Gamma}^{r} \subseteq \mathcal{M}_{\Gamma}^{\mathbf{LLL}}$ and hence $Cn_{\mathbf{LLL}}(\Gamma) \subseteq Cn_{\mathbf{AL}^r}(\Gamma) \subseteq Cn_{\mathbf{AL}^m}(\Gamma) \subseteq Cn_{\mathbf{ULL}}(\Gamma)$.
2. If Γ is normal, then $\mathcal{M}_{\Gamma}^{\mathbf{ULL}} = \mathcal{M}_{\Gamma}^{m} = \mathcal{M}_{\Gamma}^{r}$ and hence $Cn_{\mathbf{AL}^r}(\Gamma) = Cn_{\mathbf{AL}^m}(\Gamma) = Cn_{\mathbf{ULL}}(\Gamma)$.
3. If Γ is abnormal and $\mathcal{M}_{\Gamma}^{\mathbf{LLL}} \neq \emptyset$, then $\mathcal{M}_{\Gamma}^{\mathbf{ULL}} \subset \mathcal{M}_{\Gamma}^{m}$ and hence $Cn_{\mathbf{AL}^m}(\Gamma) \subset Cn_{\mathbf{ULL}}(\Gamma)$.[38]

4. $\mathcal{M}^r_\Gamma \subset \mathcal{M}^{\mathbf{LLL}}_\Gamma$ iff $\Gamma \cup \{A\}$ is **LLL**-satisfiable for some $A \in \Omega - U(\Gamma)$.
5. $Cn_{\mathbf{LLL}}(\Gamma) \subset Cn_{\mathbf{AL}^r}(\Gamma)$ iff $\mathcal{M}^r_\Gamma \subset \mathcal{M}^{\mathbf{LLL}}_\Gamma$.
6. $\mathcal{M}^m_\Gamma \subset \mathcal{M}^{\mathbf{LLL}}_\Gamma$ iff there is a (possibly infinite) $\Delta \subseteq \Omega$ such that $\Gamma \cup \Delta$ is **LLL**-satisfiable and there is no $\varphi \in \Phi_\Gamma$ for which $\Delta \subseteq \varphi$.
7. If there are $A_1, \ldots, A_n \in \Omega$ ($n \geq 1$) such that $\Gamma \cup \{A_1, \ldots, A_n\}$ is **LLL**-satisfiable and $\{A_1, \ldots, A_n\} \nsubseteq \varphi$ for every $\varphi \in \Phi_\Gamma$, then $Cn_{\mathbf{LLL}}(\Gamma) \subset Cn_{\mathbf{AL}^m}(\Gamma)$.
8. $Cn_{\mathbf{AL}^m}(\Gamma)$ and $Cn_{\mathbf{AL}^r}(\Gamma)$ are non-trivial iff $\mathcal{M}^{\mathbf{LLL}}_\Gamma \neq \emptyset$.

Proof. Ad 2. If Γ is normal, then $U(\Gamma) = \emptyset$ and only **ULL**-models of Γ are minimal abnormal.

Ad 3. If Γ is abnormal, then $\mathcal{M}^{\mathbf{ULL}}_\Gamma = \emptyset$.

Ad 1. $\mathcal{M}^{\mathbf{ULL}}_\Gamma \subseteq \mathcal{M}^m_\Gamma$ follows from 2 and 3. $\mathcal{M}^r_\Gamma \subseteq \mathcal{M}^{\mathbf{LLL}}_\Gamma$ is immediate in view of the definition of reliable model of Γ. $\mathcal{M}^m_\Gamma \subseteq \mathcal{M}^r_\Gamma$ is item 1 of Theorem 3.

Ad 4. Immediate in view of Definitions 3 and 4.

Ad 5. By 4, there is an $A \in \Omega - U(\Gamma)$ such that all $M \in \mathcal{M}^r_\Gamma$ verify $\neg A$ whereas some $M \in \mathcal{M}^{\mathbf{LLL}}_\Gamma - \mathcal{M}^r_\Gamma$ does not.

Ad 6. Immediate in view of Definitions 5 and 6.

Ad 7. Suppose that the antecedent is true. All $M \in \mathcal{M}^m_\Gamma$ verify $\neg A_1 \vee \ldots \vee \neg A_n$ whereas some $M \in \mathcal{M}^{\mathbf{LLL}}_\Gamma$ (viz. an $M \in \mathcal{M}^{\mathbf{LLL}}_{\Gamma \cup \{A_1, \ldots, A_n\}}$) does not.

Ad 8. Immediate in view of Reassurance (Theorem 1) and the fact that no **LLL**-model is trivial.

This theorem states that adaptive logics are well-behaved. They deliver at least the lower limit consequences and at most the upper limit consequences, as desired (property 1). It moreover specifies the cases in which there is a gain with respect to the lower limit (properties 4–5 and 7 respectively). Property 2 states that all upper limit consequences are delivered if the premise set is normal.

Other known adaptive logics are obtained by combining adaptive logics of the type described above. Usually, it is easy to check that all aforementioned properties extend to them.[39]

5. Dynamic Proof Theory

Just like any other proof, a dynamic proof consists of a sequence of formulas. Annotated proofs consist of a sequence of lines that have five elements: (i) a line number, (ii) the derived formula A, (iii) the line numbers of the formulas from which A is derived, (iv) the rule by which A is derived, and (v) a (possibly empty) 'condition'. The condition specifies which formulas have to behave normally in order for A to be so derivable.

Apart from the fifth element of the lines, the only unusual thing is that lines of a dynamic proof may be marked. The marks may change from one stage of the proof to the next – adding a line to the proof brings the proof to its next stage.[40] The formula (second element) of a line that is marked at stage s is considered as not derived at stage s. Marking is governed by a definition, which depends on the strategy.

What does all this mean? The conditions and marks enable us to control the internal dynamics of the proofs in a formally precise way. The Derivability Adjustment Theorem (Theorem 5 below) is central for this. It states that a formula A is derivable from the premises by the upper limit logic – this is the ideal that we want to approach 'as much as possible' by the adaptive logic – just in case $A \vee Dab(\Delta)$ (the classical disjunction of A and of a classical disjunction of abnormalities) is derivable from the premises by the lower limit logic (for some finite $\Delta \subseteq \Omega$). In other words, whenever $A \vee Dab(\Delta)$ is derivable from the premises (for some Δ) by the lower limit logic, we want A to be derivable by the adaptive logic unless the premises require that the members of Δ behave abnormally.

There is no negative test for "the members of Δ behave abnormally". If $A \vee Dab(\Delta)$ is derivable from the premises by the lower limit logic, we cannot in general find out whether the members of Δ behave abnormally. This is why we derive A *on the condition* Δ. We presume that the members of Δ do not behave abnormally, and hence that A is derivable from the premises, *unless and until* shown otherwise. Technically this is indicated by the fact that the line in which A is derived on the condition Δ is unmarked at a stage of the proof unless it has been shown at that stage that the members of Δ behave abnormally.

Of course one needs to distinguish between derivability at a stage and final derivability – the latter provides the 'final judgement' from Section 2. The matter is discussed in the next to last paragraph of this section.

Let us now move to technical matters. As before, I shall suppose that $\vee$ denotes classical disjunction.

THEOREM 5. $\Gamma \vdash_{\mathbf{ULL}} A$ iff there is a finite $\Delta \subset \Omega$ such that $\Gamma \vdash_{\mathbf{LLL}} A \vee Dab(\Delta)$. (Derivability Adjustment Theorem.)

Proof. For the left–right direction suppose that $\Gamma \vdash_{\mathbf{ULL}} A$. It follows that all **ULL**-models of Γ verify A. All *other* **LLL**-models of Γ verify some member of Ω. Hence, there is a $\Delta' \subseteq \Omega$ such that all **LLL**-models of Γ verify a member of $\Delta' \cup \{A\}$. By the right compactness of **LLL**, there is a finite $\Delta \subseteq \Delta'$ such that all **LLL**-models of Γ verify a member of $\Delta \cup \{A\}$. In other words, all **LLL**-models of Γ' verify $A \vee Dab(\Delta)$.

For the right–left direction suppose that $\Gamma \vdash_{\mathbf{LLL}} A \vee Dab(\Delta)$. As no **ULL**-model verifies any member of Ω, all **ULL**-models of Γ (if any) verify A.

This theorem provides the motor for the dynamic proof theory. I shall list the rules for a proof from Γ in the form of generic rules.[41] Apart from a premise rule, there is an unconditional rule and a conditional rule.

PREM If $A \in \Gamma$, then one may add a line consisting of

(i) the appropriate line number,
(ii) A,
(iii) "–",
(iv) "Prem", and
(v) $\emptyset$.

RU If $B_1, \ldots, B_m \vdash_{\mathbf{LLL}} A$ and $B_1, \ldots, B_m$ occur in the proof with the conditions $\Delta_1, \ldots, \Delta_m$ respectively, then one may add a line consisting of

(i) the appropriate line number,
(ii) A,
(iii) the line numbers of the B_i,
(iv) "RU", and
(v) $\Delta_1 \cup \ldots \cup \Delta_m$.

RC If $B_1, \ldots, B_m \vdash_{\mathbf{LLL}} A \vee Dab(\Theta)$ and $B_1, \ldots, B_m$ occur in the proof with the conditions $\Delta_1, \ldots, \Delta_m$ respectively, then one may add a line consisting of

(i) the appropriate line number,
(ii) A,
(iii) the line numbers of the B_i,
(iv) "RC", and
(v) $\Theta \cup \Delta_1 \cup \ldots \cup \Delta_m$.

To wind up the characterization of the dynamic proofs, I now present the marking definitions for the two discussed strategies. At any stage of the proof, zero or more *Dab*-formulas will be derived. Some of them are minimal (at that stage). Let $U_s(\Gamma)$ be the union of all Δ for which $Dab(\Delta)$ is a minimal *Dab*-formula at stage s. Let $\Phi^\circ_s(\Gamma)$ be the set of all sets that contain one disjunct out of each minimal *Dab*-formula at stage s, and let $\Phi_s(\Gamma)$ contain those members of $\Phi^\circ_s(\Gamma)$ that are not proper supersets of other members of $\Phi^\circ_s(\Gamma)$.[42]

DEFINITION 7. Marking for $\mathbf{AL}^r$: Line i is marked at stage s iff, where Δ is its fifth element, $\Delta \cap U_s(\Gamma) \neq \emptyset$.

DEFINITION 8. Marking for $\mathbf{AL}^m$: Line i is marked at stage s iff, where A is the second element and Δ the fifth element of line i, (i) there is no $\varphi \in \Phi_s(\Gamma)$ such that $\varphi \cap \Delta = \emptyset$, or (ii) for some $\varphi \in \Phi_s(\Gamma)$, there is no line k that has A as its second element and has as its fifth element some Θ such that $\varphi \cap \Theta = \emptyset$.

At this point I can define **AL**-derivability:

DEFINITION 9. A is *derived at stage* s in an **AL**-proof from Γ iff A is the second element of a line that is not marked in the proof (at stage s).

DEFINITION 10. A is *finally derived* on line i of an **AL**-proof (at a stage) from Γ iff (i) A is the second element of line i, (ii) line i is not marked at stage s, and (iii) any extension of the proof in which line i is marked may be further extended in such a way that line i is unmarked.

DEFINITION 11. $\Gamma \vdash_{\mathbf{AL}} A$ (A is *finally derivable* from Γ) iff A is finally derived on some line of an **AL**-proof from Γ.

The marking rules deserve some comments. For both strategies, the minimal *Dab*-consequences of the premises are estimated in terms of the *Dab*-formulas that have been derived on the empty condition at the stage. To see the relation with the semantics, suppose that the estimate is correct. $U_s(\Gamma)$ is then identical to $U(\Gamma)$ and $\Phi_s(\Gamma)$ to $\Phi(\Gamma)$. Consider a line in which A has been derived on a condition Δ, which corresponds to $A \vee Dab(\Delta)$ being verified by all **LLL**-models of the premises. This line is unmarked on the Reliability strategy iff no element of Δ is a member of $U_s(\Gamma)$. The latter corresponds to A being verified by all reliable models of the premises.

On the Minimal Abnormality strategy, a line in which A has been derived on a condition Δ warrants that A is verified by every model that does not verify any element of Δ. So the line is *un*marked on the Minimal Abnormality strategy just in the following case: for every minimal abnormal model of the premises – any such model is characterized by some $\varphi \in \Phi(\Gamma)$ – A has been derived on a condition that is falsified by that model and if Δ is falsified by some minimal abnormal model of the premises. In other words the line witnesses, together with other lines, that every minimal abnormal model of the premises verifies A.

For the specific logics that were studied, the Soundness and Completeness of the dynamic proof theory with respect to the semantics was proved. Apparently, these proofs may be generalized for all adaptive logics under consideration. For examples of dynamic proofs, I refer to the many papers on specific logics.[43]

While Definition 10 may be taken at face value for Reliability, some weird premise sets require that, in the case of Minimal Abnormality, infinite extensions of proofs are considered – see Batens (1999a, 466) for an example.[44]

The following theorem is central for the dynamic proof theory.

THEOREM 6. If $\Gamma \vdash_{\mathbf{AL}} A$, then any proof from Γ can be extended into a proof in which A is finally derived from Γ. (Proof Invariance.)

Proof. Consider any proof from Γ – call it P1. If $\Gamma \vdash_{\mathbf{AL}} A$, there is a proof from Γ – call it P2 – in which A has been finally derived at some line i and that, if extending it with P1 results in line i being marked, may be further extended in such a way that line i is unmarked. Call the last extension E. Definitions 7 and 8 warrant that,[45] if P1 is first extended with P2 and then with E, then the line that had number i in P2 is unmarked.

What about decidability? The propositional fragments (and some other fragments) of most adaptive logics are decidable. This means that the dynamics of the proofs can in principle be avoided by deriving formulas in a suitable order and by not deriving any formulas that are marked in view of formulas that were derived earlier. The full predicative versions of adaptive logics are obviously undecidable and have no positive test for final derivability.

Even in undecidable waters there may be certain criteria that enable one to decide that a specific formula has been finally derived in some line of a dynamic proof from Γ. Some such criteria provide from work on the block approach (see for example Batens (1995)) and from work on tableau methods for adaptive logics (see Batens and Meheus (2000b) and Batens and Meheus (2001b)). Much more efficient criteria derive from goal directed dynamic proofs ((Batens 2002) and work in progress, partly with Dagmar Provijn).

But what if no such criterion applies? It was shown in Batens (1995) – the result may be easily generalized to all considered adaptive logics – that as dynamic proofs proceed, the sets of formulas derived at subsequent stages offer increasingly better estimates of the set of finally derivable formulas. This estimate is not merely a computational approximation, but there is an idea behind it: as the proof proceeds, it provides an increasingly better insight in the premises, and hence in the minimal *Dab*-formulas that are derivable from them. Moreover, the goal directed dynamic proofs provide means to speed up the gain of insight in the premises. The upshot is that dynamic proofs form a sensible basis for decision and action. In this sense, they not only enable one to explicate actual forms of dynamic reasoning, but also justify such forms of reasoning.

I shall be brief on prioritized and combined adaptive logics. The essential point was already mentioned: where different adaptive mechanisms are combined, one obtains dynamic proofs in which the dynamic mechanisms do not operate consecutively but at the same time. As a result, the dynamic proofs obtain their full explicatory and justificatory function.

6. In Conclusion

Several open problems were mentioned in the previous sections and I shall not repeat them here. I shall rather add a final comment concerning the epistemological function of adaptive logics.

Adaptive logics explicate consequence relations for which there is no positive test and that may be fit into the scheme: a lower limit logic, a set of abnormalities and a strategy. We have seen in Section 2 that those consequence relations abound in epistemological contexts and play a central role in them. The dynamic proofs not only provide the logics with a proof theory. With their conditions and marking definitions, they explicate the actual reasoning in terms of such consequence relations. This is extremely important because they thus provide a clear and transparent conceptual analysis for forms of reasoning that were often qualified as mere tinkering or even as logically flawed.

In my own work on epistemology (see for example Batens (1985, 1992)), I have stressed that the dynamics of human knowledge depends essentially on the fact that humans (as individuals or as groups) shift from one context to the other in solving problems. Adaptive logics do not offer an explication for this inter-contextual dynamics. They apply within contexts and explicate the intra-contextual dynamics in a formally precise way. It may be hoped that one will be able to move on to understand the inter-contextual dynamics, to explicate it, and to find means to increase its computational as well as its problem-solving efficiency.[46]

Notes

[1] A logic is paraconsistent iff (if and only if) it does not validate ex falso quodlibet ($A, \sim A \vdash B$).

[2] Logic also had a curative function: to prevent people from talking nonsense. I shall not discuss this here, as the topic has long been settled by now.

[3] See, for example, the work of Hintikka and associates, who were among the few philosophers of science that tried to keep applying logic.

[4] Nearly always, background knowledge plays a role – see below in the text.

[5] Actually, the adaptive logic **LI** is described in Batens (2003) and several further results are forthcoming in papers by Lieven Haesaert and myself, for example Batens and Haesaert (in print). However, one needs not to know those systems in order to follow the argument in the text.

[6] A positive test for a property is a systematic procedure that leads, after finitely many steps, to a "yes" if the property applies, but may go on forever if it does not. See Boolos and Jeffrey (1989) for such matters.

[7] See Batens and Meheus (2000a) for the adaptive logic of compatibility (in the framework of **CL**).

[8] See Batens and Meheus (2001a) for adaptive logics that explicate several forms of reasoning that underly the search for explanations.

[9] I mean non-**CL**-derivability, in agreement with the cited papers, but the matter is the same for any other sensible logic.

[10] See for example Norton (1987, 1993), Smith (1988), Brown (1990), Nersessian (2002), Meheus (1993, 2002). Not all these authors side with me on the required approach, but that is immaterial.

[11] One way to implicitly apply Ex Falso Quodlibet proceeds by first applying Addition to obtain $p \vee q$ from p and next applying Disjunctive Syllogism to obtain q from $\sim p$ and $p \vee q$.

[12] See Meheus (2000) for an exception: the paraconsistent logic **AN** validates Disjunctive Syllogism (and all 'analysing' rules of **CL**) but invalidates Addition (and Irrelevance and similar rules)

[13] Explicating this kind of reasoning was at the origin of the adaptive logic programme – see Batens (1989, 1999), and many other papers.

[14] The complication of falsifiable background generalizations is dealt with in Batens (2003). Most complications discussed in the subsequent paragraph of the text are handled in Batens and Haesaert (in print).

[15] Adaptive logics for this reconstruction are spelled out in Verhoeven (in print) and Batens (in print(a)).

[16] A is a Weak consequence of Γ iff it is a **CL**-consequence of some consistent subset of Γ – remember that there is no positive test for consistency.

[17] The two obvious exceptions are zero logic, according to which nothing is derivable from any premise set (not even the premises themselves) and trivial logic according to which everything is derivable from any premise set. These logics may seem completely uninteresting, but actually zero logic is not. From it, an adaptive logic may be defined, thus making all logical reasoning contingent on specific properties of the premises (put differently: on 'the world'). For example, adaptive zero logic assigns to consistent sets of premises exactly the same consequence set as **CL**. See Batens (1999c) for a study of zero logic and the (most straightforward) adaptive logic definable from it.

[18] $Cn_{\mathbf{L}}(\Gamma)$ abbreviates $\{A \in \mathcal{W} \mid \Gamma \vdash_{\mathbf{L}} A\}$ as usual.

[19] Remark that the lower limit logic may be zero logic – see footnote 17.

[20] Semantically, this logic is characterized by those **CL**-models in which, for every predicate π of adicity i, $v(\pi) \in \{D^i, \emptyset\}$ in which D^i is the i-th Cartesian product of the domain. The name **CLU** refers to the fact that this logic is characterized by **CL**-models that are (completely) uniform: all elements of the domain have the same properties.

[21] This holds for nearly all combinations of lower limit logics and sets of abnormalities. I shall mention exceptions when I introduce the Simple strategy.

[22] Remember that, whenever a logical symbol of the original language is not classical, then the language is extended with the corresponding classical symbol.

[23] Note that $Dab(\Delta)$ is the classical disjunction of the members of Δ. In some previous papers on specific adaptive logics, $Dab(\Delta)$ has a slightly different function.

[24] For some sets of minimal *Dab*-consequences, at least two members of some Δ_i behave abnormally – see Batens (1999a, 468) for an example.

[25] Given the variety of predicative extensions of propositional modal logics, one needs to pick a specific predicative version of **T**. I shall only discuss the propositional case here and refer to Batens et al. (2003) for a predicative version that is adequate for diagnosis logic.

[26] Actually, the situation is similar for several flat adaptive logics that characterize formerly known consequence relations.

[27] The central point is that, in the absence of a positive test for the minimality of the *Dab*-formula, the new premises have to be defeasible – see also Section 5.

[28] Obviously interesting problem-solving processes are not algorithmic. A road that seems attractive at one point may later turn out to be a dead end.

[29] Here are semantic versions. Left compactness: Γ has a model iff every finite $\Gamma' \subseteq \Gamma$ has a model; right compactness: every model of Γ verifies a member of Δ iff every model of Γ verifies a member of some finite $\Delta' \subseteq \Delta$. In the presence of classical negation, left compactness warrants right compactness.

[30] Whenever the Simple strategy is sensible, the theorems below extend to it immediately because both Reliability and Minimal Abnormality come to the Simple strategy in such cases.

[31] In Batens (1999a) the adaptive logics **ACLuN1** and **ACLuN2** are defined and studied in the presence of an object language that does not contain classical negation. In Batens (2000a), Strong Reassurance is proved in the presence of a similar language.

[32] If A is the logical form characterizing Ω, then, in the presence of classical implication, extending **LLL** with the axiom schema $\mathsf{A} \supset B$ delivers a characterization of **ULL**.

[33] Recall that $\neg$ is classical negation.

[34] In Batens (1999a) Φ_Γ is a set of sets of factors of abnormalities, but this modification is inconsequential and factors of abnormalities are a nuisance in the present setup.

[35] By the second half of the requirement I mean that $\varphi = Cn_{\mathbf{LLL}}(\varphi) \cap \Omega$.

[36] As the proof theory is only defined in the following section, expressions such as $Cn_{\mathbf{AL}^r}(\Gamma)$ refer to $\{A \in \mathcal{W} \mid \Gamma \vDash_{\mathbf{AL}^r} A\}$.

[37] The label might suggest that recurrent applications of some closure operation ultimately lead to a fixed point. This, however, is not the case: a single application of the closure operation leads to a fixed point ($Cn_{\mathbf{AL}^m}(\Gamma)$ is a fixed point with respect to $\mathbf{AL}^m$-closure).

[38] If Γ is abnormal, it has no **ULL**-models and $Cn_{\mathbf{ULL}}(\Gamma)$ is trivial.

[39] More properties (see for example Arieli and Avron (2000) and Avron and Lev (to appear)) may be established for the adaptive logics under consideration. For example, Right Cautious Cut (and hence Plausibility) holds for $\mathbf{AL}^m$ (but not for $\mathbf{AL}^r$).

[40] A stage of a proof is obviously a sequence of lines and not a line. The stage that results from adding line n (and applying the marking definitions) comprises line 1 up to line n.

[41] While generic rules are unavoidable in the present setup, they are in general most convenient and transparent for characterizing the proof theory of specific adaptive logics.

[42] The proofs may be made somewhat more efficient by introducing some closing operations in the definitions of $U_s(\Gamma)$ and $\Phi_s(\Gamma)$. In doing so one should take computational matters into account: it should be decidable whether a line is marked or unmarked.

[43] A list is available: `http://logica.ugent.be/adlog/albib.html`.

[44] Extensions of infinite proofs are obtained by inserting formulas in the proof.

[45] All we need is that the order of the lines of a stage is immaterial for the marking definitions.

[46] Unpublished papers in the reference section (and many others) are available from the internet address `http://logica.ugent.be/centrum/writings/`.

References

Arieli, Ofer and Arnon Avron: 2000, 'General Patterns for Nonmonotonic Reasoning: From Basic Entailments to Plausible Relations', *Logic Journal of the IGPL* **8**, 119–148.

Avron, Arnon and Iddo Lev: to appear, 'Formula-preferential Systems for Paraconsistent Non-monotonic Reasoning (An Extended Abstract)'.

Batens, Diderik: 1985 'Meaning, Acceptance, and Dialectics', in Joseph C. Pitt (ed.), *Change and Progress in Modern Science*, Dordrecht, Reidel, pp. 333–360.

Batens, Diderik: 1986, 'Dialectical Dynamics within Formal Logics', *Logique et Analyse* **114**, 161–173.

Batens, Diderik: 1989 'Dynamic Dialectical Logics', in Graham Priest, Richard Routley and Jean Norman (eds.), *Paraconsistent Logic. Essays on the Inconsistent*, München, Philosophia Verlag, pp. 187–217.

Batens, Diderik: 1992, 'Do we Need a Hierarchical Model of Science?', in John Earman (ed.), *Inference, Explanation, and Other Frustrations. Essays in the Philosophy of Science*, University of California Press, pp. 199–215.

Batens, Diderik: 1995, 'Blocks. The Clue to Dynamic Aspects of Logic', *Logique et Analyse* **150–152**, 285–328, appeared in 1997.

Batens, Diderik: 1999a, *Inconsistency-adaptive Logics*, in Orłowska (1999, 445–472).

Batens, Diderik: 1999b, 'Linguistic and Ontological Measures for Comparing the Inconsistent Parts of Models', *Logique et Analyse* **165–166**, 5–33, appeared in 2002.

Batens, Diderik: 1999c, 'Zero Logic Adding up to Classical Logic', *Logical Studies* **2**, 15. (Electronic Journal: `http://www.logic.ru/LogStud/02/LS2.html`).

Batens, Diderik: 2000a, 'Minimally Abnormal Models in Some Adaptive Logics', *Synthese* **125**, 5–18.

Batens, Diderik: 2000b, 'Towards the Unification of Inconsistency Handling Mechanisms', *Logic and Logical Philosophy* **8**, 5–31, appeared in 2002.

Batens, Diderik: 2001, 'A Dynamic Characterization of the Pure Logic of Relevant Implication', *Journal of Philosophical Logic* **30**, 267–280.

Batens, Diderik: 2002, 'On a Partial Decision Method for Dynamic Proofs', in Hendrik Decker, Jürgen Villadsen and Toshiharu Waragai (eds.), *PCL 2002. Paraconsistent Computational Logic*, (= *Datalogiske Skrifter* vol. 95), pp. 91–108. Also available as cs.LO/0207090 at `http://arxiv.org/archive/cs/intro.html`.

Batens, Diderik: 2003, 'On a Logic of Induction', in Roberto Festa, Atocha Aliseda and Jeanne Peijnenburg (eds.), *Confirmation, Empirical Progress, and Truth Approximation. Essays in Debate with Theo Kuipers*, Vol. 1, Poznan Studies in the Philosophy of the Sciences and the Humanities, Amsterdam, Rodopi, in print.

Batens, Diderik: in print(a), 'Aspects of the Dynamics of Discussions and Logics Handling Them', *Logical Studies*.

Batens, Diderik: in print(b), 'The theory of the Process of Explanation Generalized to Include the Inconsistent Case', *Synthese*.

Batens, Diderik: to appear, 'A Strengthening of the Rescher–Manor Consequence Relations', *Logic and logical Philosophy*.

Batens, Diderik and Lieven Haesaert: 2001, 'On Classical Adaptive Logics of Induction', *Logique et Analyse* **173–175**, 255–290, appeared 2003.

Batens, Diderik and Joke Meheus: 2000a, 'The Adaptive Logic of Compatibility', *Studia Logica* **66**, 327–348.

Batens, Diderik and Joke Meheus: 2000b, 'A Tableau Method for Inconsistency-adaptive Logics', in Roy Dyckhoff (ed.), *Automated Reasoning with Analytic Tableaux and Related Methods*, Lecture Notes in Artificial Intelligence Vol. 1847, Springer, pp. 127–142..

Batens Diderik and Joke Meheus: 2001a, 'On the Logic and Pragmatics of the Process of Explanation', in Mika Kiikeri and Petri Ylikoski (eds.), *Explanatory Connections. Electronic Essays Dedicated to Matti Sintonen*, `http://www.valt.helsinki.fi/kfil/matti/`, 22 pp.

Batens, Diderik and Joke Meheus: 2001b, 'Shortcuts and Dynamic Marking in the Tableau Method for Adaptive Logics', *Studia Logica* **69**, 221–248.

Batens, Diderik, Joke Meheus, Dagmar Provijn and Liza Verhoeven: 2003, 'Some Adaptive Logics for Diagnosis', *Logic and Logical Philosophy* **11/12**, 39–65.

Benferhat, Salem, Didier Dubois, and Henri Prade: 1997, 'Some Syntactic Approaches to the Handling of Inconsistent Knowledge Bases: A Comparative Study. Part 1: The Flat Case', *Studia Logica* **58**, 17–45.

Benferhat, Salem, Didier Dubois, and Henri Prade: 1999, 'Some Syntactic Approaches to the Handling of Inconsistent Knowledge Bases: A Comparative Study. Part 2: The Prioritized Case', in Orłowska (1999, pp. 473–511).

Boolos, George S. and Richard J. Jeffrey: 1989, *Computability and Logic*, 3rd edn, Cambridge University Press.

Brown, Bryson: 1990, 'How to be Realistic about Inconsistency in Science', *Studies in History and Philosophy of Science* **21**, 281–294.

da Costa, Newton C. A.: 1963, 'Calculs propositionnels pour les systèmes formels inconsistants', *Comptes rendus de l'Académie des sciences de Paris* **259**, 3790–3792.

De Clercq, Kristof: 2000, 'Two New Strategies for Inconsistency-adaptive Logics', *Logic and Logical Philosophy* **8**, 65–80, appeared in 2002.

Feyerabend, Paul K: 1970, 'Against Method: Outline of an Anarchistic Theory of Lnowledge', in M. Radner and S. Winokur (eds.), *Analyses of Theories and Methods of Physics and Psychology*, Vol. 4 of *Minnesota Studies in the Philosophy of Science*, Minneapolis, University of Minnesota Press, pp. 17–30.

Halonen, Ilpo and Jaakko Hintikka: to appear, 'Toward a Theory of the Process of Explanation', *Synthese*.

Jaśkowski, Stanisław: 1969, 'Propositional Calculus for Contradictory Deductive Systems', *Studia Logica* **24**, 243–257.

Karus, Sarit, Daniel Lehman, and Menachem Magidor: 1990, 'Nonmonotonic Reasoning, Preferential Models and Cumulative Logics', *Artificial Intelligence* **44**, 167–207.

Laudan, Larry: 1977, *Progress and its Problems*, Berkeley, University of California Press.

Lev, Iddo: 2000, 'Preferential Systems for Plausible Non-classical Reasoning', Master's thesis, Department of Computer Science, Tel-Aviv University.

Meheus, Joke: 1993, 'Adaptive Logic in Scientific Discovery: The Case of Clausius', *Logique et Analyse* **143–144**, 359–389, appeared in 1996.

Meheus, Joke: 2000, 'An Extremely Rich Paraconsistent Logic and the Adaptive Logic Based On It', in Diderik Batens, Chris Mortensen, Graham Priest and Jean Paul Van Bendegem (eds.), *Frontiers of Paraconsistent Logic*, Baldock, UK, Research Studies Press, pp. 189–201.

Meheus, Joke: 2002, 'Inconsistencies in Scientific Discovery. Clausius's Remarkable Derivation of Carnot's Theorem', in Helge Krach, Geert Vanpaemel and Pierre Marage (eds.), *History of Modern Physics*, Brepols, Brepols, pp. 143–154.

Nersessian, Nancy: 2002, 'Inconsistency, Generic Modeling, and Conceptual Change in Science', in Joke Meheus (ed.), *Inconsistency in Science*, Dordrecht, Kluwer, pp. 197–211.

Nickles, Thomas (eds.): 1980a, *Scientific Discovery, Logic, and Rationality*, Dordrecht, Reidel.

Nickles, Thomas (eds.): 1980b, *Scientific Discovery: Case Studies*. Dordrecht, Reidel.

Norton, John: 1987, 'The Logical Inconsistency of the Old Quantum Theory of Black Body Radiation', *Philosophy of Science* **54**, 327–350.

Norton, John: 1993, 'A Paradox in Newtonian Gravitation Theory', *PSA 1992* **2**, 421–420.

Orłowska, Ewa (ed.): 1999, *Logic at Work. Essays Dedicated to the Memory of Helena Rasiowa*, Heidelberg, New York, Physica Verlag (Springer).

Priest, Graham: 1991, 'Minimally Inconsistent **LP**', *Studia Logica* **50**, 321–331.

Provijn, Dagmar and Erik Weber: 2002, 'Two Adaptive Logics for Non-explanatory and Explanatory Diagnostic Reasoning', in Lorenzo Magnani, Nancy J. Nersessian and Claudio Pizzi (eds.), *Logical and Computational Aspects of Model-Based Reasoning*, Dordrecht, Kluwer, pp. 117–142.

Rescher, Nicholas and Ruth Manor: 1970, 'On Inference from Inconsistent Premises', *Theory and Decision* **1**, 179–217.

Smith, Joel: 1988, 'Inconsistency and Scientific Reasoning', *Studies in History and Philosophy of Science* **19**, 429–445.

Verhoeven, Liza: 2003, 'Changing One's Position in Discussions. Some Adaptive Approaches', *Logic and Logical Philosophy* **11/12**, 277–297.

Verhoeven, Liza: to appear, 'Proof Theories for Some Prioritized Consequence Relations'.

Weber, Erik and Dagmar Provijn: 1999, 'A Formal Analysis of Diagnosis and Diagnostic Reasoning', *Logique et Analyse* **165–166**, 161–180, appeared in 2002.

Wiśniewski, Andrzej: 1995, *The Posing of Questions. Logical Foundations of Erotetic Inferences*, Dordrecht, Kluwer.

Wiśniewski, Andrzej: 1996, 'The Logic of Questions as a Theory of Erotetic Arguments', *Synthese* **109**, 1–25.

LOGICS FOR QUALITATIVE REASONING

PAULO A. S. VELOSO[1] and WALTER A. CARNIELLI[2]
[1]*Institute of Mathematics and COPPE, UFRJ, Rio de Janeiro Praça Eugênio Jardim, 6/apt. 501, 22061-040 Rio de Janeiro, RJ, Brazil, E-mail:* srmv@bridge.com.br; [2] *CLE, Unicamp, Campinas Caia Postal 6133, 13083-970 Campinas, SP, Brazil, E-mail:* carniell@cle.unicamp.br

Abstract. Assertions and arguments involving vague notions occur often both in ordinary language and in many branches of science. The vagueness may be plainly expressed by "modifiers", such as 'generally', 'rarely', 'most', 'many', etc., or, less obviously, conveyed by objects termed 'representative', 'typical' or 'generic'. A precise treatment of such ideas has been a basic motivation for logics of qualitative reasoning. Here, we present some logical systems with generalized quantifiers for these modifiers, also handling 'generic' reasoning. Other possible applications for these and related logics for qualitative reasoning are indicated. These (monotonic) generalized logics, with simple sound and complete deductive calculi, are proper conservative extensions of classical first-order logic, with which they share various properties. For generic reasoning, special individuals can be introduced by means of 'generally', and internalized as representative constants, thereby producing conservative extensions where one can reason about generic objects as intended. Some interesting situations, however, require such assertions to be relative to various universes, which cannot be captured by relativization. Thus we extend our generalized logics to sorted versions, with qualitative notions relative to the universes, which can also be compared.

1. Introduction

In this introductory section, we will indicate some motivations for the idea of logics for qualitative reasoning about vague notions and then outline the structure of this chapter.

1.1. Motivation

We will initially examine some motivations for vague notions and logics for qualitative reasoning about them.

Assertions and arguments involving some vague notions occur often, not only in ordinary language, but also in some branches of science. The vagueness may be given by "modifiers", such as 'generally', 'rarely', 'most', 'many', etc., or apparent from objects termed 'representative', 'typical' or 'generic'. A precise treatment of such ideas has been a basic motivation underlying logics for qualitative reasoning. Some logical systems with generalized quantifiers able to express such notions and to reason about them will be examined below.

For instance, one often encounters assertions such as "Bodies 'generally' expand when heated", "Birds 'generally' fly" and "Metals 'rarely' are liquid un-

S. Rahman et al. (eds.), Logic, Epistemology, and the Unity of Science, pp. 487–526.

der ordinary conditions". Somewhat vague terms such as 'likely', 'prone', etc., are frequently used in everyday language. More elaborate expressions involving 'propensity' are often used as well. For instance, a physician may say that a patient's genetic background indicates a certain 'propensity', which makes him or her 'prone' to some ailments. Also, in the familiar "Tweety example" (Reiter 1980) one finds arguments wishing to conclude that "Tweety flies" from the assertions "Birds 'generally' fly" and "Tweety is a 'typical' bird". Such notions may also be useful in reporting experimental set-ups and results.[1] One would be able to reason precisely about them once they receive precise meanings. This is what we intend to provide.

It is our intention to express such assertions and reason about them in a precise manner. In our logics, we wish to express assertions, such as "People 'generally' like chocolate", and reason about them in a formal manner. To *express* such "generalized" assertions formally, we introduce the new operator ∇, and express "People 'generally' like chocolate" by $\nabla v C(v)$. To give precise *meaning* to such assertions we extend the usual notions, by providing a family $\mathcal{K}$ of 'important' sets, and stipulate that $\nabla v C(v)$ means that the set $\{p \in P : C(p)\}$ is in the family $\mathcal{K}$, as a rigorous counterpart for "the set of people that like chocolate is an 'important' set'. In order to *reason* about such assertions in a formal manner, it will be necessary to set up deductive systems by extending (conservatively) the classical first-order predicate calculus. Some properties of such logics will then be examined, their usage illustrated and some applications will be commented.

These logics are related to default logic (Reiter 1980) and its variants (Antoniou 1997; Besnard 1989; Brewka 1991; Brewka et al. 1997; Lukaszewicz 1990; Marek and Truszczyński 1993), as well as to belief revision (Gärdenfors 1988; Makinson and Gärdenfors 1991). Indeed, they do have a large intersection in terms of applications, as indicated by benchmark examples, which was one of the motivations for similar systems (Carnielli and Sette 1994; Schlechta 1995). But, they are quite different logical systems, both technically and in terms of intended interpretations (Carnielli and Veloso 1997).[2] As logics with generalized quantifiers, our systems are indeed connected to extensions of first-order logic (Mostowski 1957; Barwise and Feferman 1985; Keisler 1970).

Ideas concerning these notions have already appeared in the literature. Some traditional square-of-opposition relations among 'few', 'many', and 'most' have been analyzed (Peterson 1979) and a quantifier for 'most' in the sense of majority has been examined (Rescher 1962; Slanley 1988). A logic with various generalized quantifiers, for notions such as 'many', 'few', 'most', etc., has been suggested as appropriate to treat quantified sentences in natural language (Barwise and Cooper 1981). These works are also related to the tradition of analysis and formalization of language (Frege 1879; Tarski 1936; Church 1956; Montague 1974).

1.2. Outline of the Chapter

The structure for the rest of this chapter is outlined as follows: Sections 2 through 4 present the basic ideas and results of our approach to reasoning with 'generally': some intuitions behind these notions, our logics for 'generally' and a few of their basic metamathematical properties. These logics for 'generally', with simple sound and complete deductive calculi, are proper conservative extensions of classical first-order logic, sharing with it several properties.

In Section 2 (Some Notions of 'Generally' and 'Rarely') we shall examine a few intuitions underlying vague notions, such as 'generally' and 'rarely', and indicate how one can capture (some of) them precisely by means of families of sets.

Section 3 (Logics for 'Generally' and 'Rarely') is devoted to presenting our logics for 'generally' and 'rarely': syntax, semantics and axiomatics. These logics add to classical first-order logic (non-standard) generalized quantifiers, giving rise to generalized formulas. The intended interpretation of such a formula holding "generally" is captured by requiring its extension to belong to a given family. We will axiomatize such logics by schemata coding properties of these families.

In Section 4 (Logics with Generalized Assertions) we shall establish some properties of our logic with generalized assertions, including soundness and completeness of their deductive systems with respect to the corresponding semantic consequences. We will also examine some other metamathematical properties of these logics, including deductive and expressive powers.

Sections 5 and 6 present some other concepts and results about our logics with generalized assertions. In these logics, the flexibility of the new generalized quantifiers – with behavior intermediate between those of the classical existential and universal quantifiers – also brings about some problems.

Sections 5 (Generic Reasoning and Generalized Assertions) is devoted to some aspects concerning generic reasoning and inference of generalized assertions. In contrast to the classical universal quantifier, instantiation does not hold for our new generalized quantifiers. To overcome this problem, generic individuals are introduced and internalized as generic constants, thereby producing conservative extensions (with ideal elements) where one can reason about generic objects as intended. Our new generalized quantifiers also share with the classical universal quantifier some problems about inference. We shall consider the question of inferring generalized assertions from experiments on samples.

Some interesting situations, however, require assertions relative to various universes, involving "most birds", "several penguins", and "typical eagle", for instance. Section 6 (Relative Notions) is devoted to showing how our concepts should, and can, be adapted to support reasoning under such relative notions. We shall introduce many-sorted versions of our logics, with qualitative notions relative to the universes, which share various properties, such as supporting generic reasoning, with the original versions. Moreover, some situations require comparing distinct qualitative notions over some universes. Our many-sorted logics for 'generally' can handle such comparisons by means of appropriate transfer assertions.

Section 7 (Concluding Remarks) will reassess our concepts and results and consider some prospects for our logics for 'generally', such as possible developments and applications.

2. Some Notions of 'Generally' and 'Rarely'

Recall that we wish to express assertions involving notions (such as 'generally' and 'rarely') and reason about them in a precise manner. For this purpose, one needs a clear understanding of these notions, which appear to be quite vague. We will now examine some intuitions behind such notions and indicate how one can capture them precisely by means of families of sets.

Various possible interpretations seem to be associated with the rather vague notions of 'generally' and 'rarely'. We shall now consider a few reasonable ones and examine some intuitions underlying them.

Consider assertions of the form "objects 'generally' have φ" or "objects 'rarely' have φ", where φ is a given property. How is one to understand these assertions? What would be the possible grounds for accepting them? We shall now examine some answers to these questions stemming from possible accounts for 'generally' and 'rarely'.

2.1. Numerical Accounts for 'Generally'

The intended meaning of "objects 'generally' have a given property" can be given in terms of the set of those objects having this property. One usually understands "Birds 'generally' fly" as "The flying birds form a 'sizable' set". This view tries to reduce, so to speak, 'generally' to 'sizable', but one still has to explain 'sizable'.

For instance, an assertion such as "Brazilians 'generally' like soccer" may be given the following two accounts. One may say that "the Brazilians that like soccer form a 'likely' portion", with more than, say, 50% of the population, or alternatively, "the Brazilians that like soccer form a 'sizable' set", in the sense that their number is above, say, 80 million.[3]

These two accounts of 'generally' may be termed "metric", trying to reduce it to a measurable aspect, so to speak. Paraphrasing "people generally have property φ" as "several people have φ", they seek to explicate it as "the people having this property φ form a 'likely' (or 'sizable') set", i.e., a set having "high" relative frequency (or cardinality), where 'high' is understood as above a given threshold.

These metric accounts, however, differ in one important aspect, as can be seen by considering the relation of having the same size. On the one hand, the size accounts – cardinality above a given threshold – clearly fail to distinguish sets with the same cardinality: they are all either above or below the threshold. We may say that we have a non-local notion. In contrast, sets with the same size may very well have distinct probabilities.[4] So, the family of "likely" sets (with "high" probability) may fail to be closed under permutations of the universe. Thus, in a probabilistic account of 'generally', the family of "likely" sets, is not necessarily invariant under having the same size. It may be said to correspond to a local notion.

Even though these accounts differ in some aspects, the corresponding notions of sizable sets – those having several elements – seem to share some general properties. Some properties that a family of sizable sets (with several elements) may, or may not, be expected to have are illustrated in the next example.

EXAMPLE [*Brazilians and shaving*]. Consider the universe of Brazilians. Imagine that one accepts the two assertions "Several Brazilians have their beards shaved" and "Several Brazilians shave their legs". In this case, one would probably accept also the assertion "Several Brazilians have their beards shaved *or* sport a moustache". This, however, does not seem to be the case with "Several Brazilians have their beards shaved *and* shave their legs".[5]

This example illustrates the following ideas:

- if B is a subset of M and B has several elements, then M also has several elements;
- even though both B and L have several elements, their intersection $B \cap L$ may fail to have several elements.

So, a family of sizable sets – of those having several elements – is expected to be closed under supersets, but *not* under intersection.

These numerical accounts hinge on assigning a threshold, which may seem somewhat arbitrary. Even though they may suffice for some cases, such approaches do not appear to be appropriate for others, where they may fail to clarify the underlying intuitions. In the sequel, we will examine relaxations of these concepts.

2.2. Relaxed Accounts for 'Generally' and 'Rarely'

The intended meaning of "objects 'generally' have property φ" can also be given by means of the set of exceptions, i.e., those objects failing to have this property φ. One may understand "Birds 'generally' fly" as "Birds 'rarely' fail to fly", in the sense that "The non-flying birds form a 'small' set".

For instance, consider the assertion "Natural numbers generally do not divide twelve". One may paraphrase it as "Most natural numbers do not divide twelve" and explain it by saying that "the divisors of twelve form a 'small' set", where 'small' is understood as finite. Similarly, one would understand the assertion "Real numbers generally are irrational" in terms of its set of exceptions (the rationals) being "small", with 'small' now taken as (at most) denumerable.

This account of 'generally' and 'rarely' is still quantitative, but more relaxed. It tries to explicate "most objects have a property φ" as "the exceptional objects, i.e., those failing to have this property φ, form a 'small' set", under a given sense of 'small' (capturing some idea of "having 'very few' elements"). The next example provides some illustration of the properties that the dual families of large and small sets may, or may not, have.

EXAMPLE [*American males*]. Consider the universe of American males. Imagine that one accepts the three assertions "Most American males like beer", "Most American males like sports", and "Most American males are Democrats or Republicans". In this case, one would probably accept also the two assertions "Most American males like beer *or* wine" and "Most American males like beer *and* sports".[6] On the other hand, neither one of the two assertions "Most American males are Democrats" and "Most American males are Republicans" seems to be equally acceptable.

This example illustrates the following ideas:
- if B is a subset of W and B has most elements, then W has most elements as well;
- if both B and S have most elements, then their complements $B^{\sim}$ and $S^{\sim}$ are small and so will be their union $B^{\sim} \cup S^{\sim}$ small, thus the intersection $B \cap S$ will have most elements;
- a union $D \cup R$ may have most elements, without either D or R having most elements.

So, a family of small sets (those having very few elements) is expected to be closed both under subset and unions, thus being an ideal (Halmos 1963). Dually, a family of large sets – of those having most elements – is expected to be closed both under supersets and intersections, thus being a filter, but *not* necessarily an ultrafilter (Halmos 1963; Bell and Slomson 1971).

2.3. Qualitative Accounts for 'Generally'

The accounts of 'generally' and 'rarely' mentioned so far may be termed "quantitative". Even though they may suffice for various cases, such accounts do not seem to cover some examples, where these notions appear to present a qualitative character.

As an example, consider the assertion "Real numbers generally are rational". How is one to understand this assertion? What would be the possible grounds for accepting it? The rationals do not seem to form a "likely", "sizable" or "large" set of reals in a quantitative sense: there are too few of them.[7] Yet, there seems to be a sense in which one may accept that "Real numbers generally are rational". Indeed, one may say that "the rationals are 'almost everywhere' within the reals". More precisely, the rational reals form a dense set of reals, thus, in any open neighborhood of a real one finds a rational[8] (Kelley 1955). In this sense, the rationals may be said to be "ubiquitous" within the reals (Grácio 1999; Carnielli and Grácio 2000).

This example illustrates a local qualitative notion of 'generally'. One explicates "objects generally have a given property" by saying that "the set of objects having this property is a dense set" in a given topology.

We thus have various distinct notions of 'generally' and 'rarely'. We would like to give them a unified treatment. As more neutral names encompassing these notions, we shall prefer to use 'important' in lieu of 'sizable', 'likely' or 'large' (corresponding to 'generally'), and, accordingly 'negligible' for 'non-sizable', 'unlikely' or 'small' (corresponding to 'rarely'). The previous terms are somewhat vague, the more so with the new ones. Nevertheless, they present some advantages. First, the reliance on a – somewhat arbitrary – threshold is less stringent. Also, they have a wider range of applications, stemming from the liberal interpretation of 'important' as carrying considerable weight or importance.

EXAMPLE [*Interpretations of 'important'*]. Imagine that a socialite visiting Hollywood and eager to attend interesting parties receives the following pieces of advice:

– "Important parties are those attended by the celebrities", and
– "Important parties are those attended by Madonna".

Then, "important" sets of guests are those including the celebrities, for the former advisor, and those where Madonna is, for the latter advisor. In both cases, the family of important sets is closed under supersets and intersections, being filters, and an ultrafilter in the Madonna interpretation.

As these examples suggest, the notions of 'important' and 'negligible' are relative to the situation or person.[9]

We can perhaps distinguish the earlier quantitative accounts from the more flexible qualitative accounts in terms of the properties stressed. They are of a topological nature in the latter, rather than metrical as in the former. We can also see that the earlier quantitative versions can be subsumed under the more flexible qualitative notions.

2.4. Families for 'Generally' and 'Rarely'

We have seen that one has various distinct notions of 'generally' and 'rarely', which may be explicated in terms of families of important and negligible sets, respectively.

Under the light of the preceding considerations, the interpretation of

– "objects 'generally' have property φ" and "objects 'rarely' have property φ";

respectively, as

– "the objects having φ form an 'important' set" and "the objects failing to have φ form a 'negligible' set";

can be seen to amount to

– “the set of objects having φ belongs to a given family $\mathcal{W}$” (of important subsets) and “the set of objects failing to have φ belongs to a given family $\mathcal{N}$” (of negligible subsets) of the universe of discourse.

In this sense, ‘generally’ and ‘rarely’, within a given universe of discourse, can be explained in terms of the families $\mathcal{W}$ of important subsets and $\mathcal{N}$ of negligible subsets of universe V. The relative character of ‘important’ and ‘negligible’ is embodied in these families, which may vary according to the situation. They, however, may be expected to share some properties, if they are to be appropriate for capturing reasonable notions of ‘generally’ (and ‘rarely’), corresponding to ‘several’ (and ‘few’) or ‘most’ (and ‘very few’).

Some general properties can be expected to be shared by all our notions corresponding to ‘generally’ and ‘rarely’.

On the one hand, the idea of exceptions – involved in understanding “Objects ‘generally’ have property φ” as “Objects ‘rarely’ fail to have property φ” – corresponds to the duality between these families: a subset N of the universe V is negligible ($N \in \mathcal{N}$) iff its complement is important ($N^{\sim} \in \mathcal{W}'$).

On the other hand, we wish to deal with non-trivial notions: there should exist negligible and non-negligible sets (as well as important and non-important sets). Now, one would probably regard the empty set as (most) negligible and the universe as non-negligible: the empty set is negligible ($\varnothing \in \mathcal{N}$) and the universe is non-negligible ($V \notin \mathcal{N}$). So, our dual families are proper: $\varnothing \neq \mathcal{N} \neq \wp(V)$ and $\varnothing \neq \mathcal{W} \neq \wp(V)$.

Other properties, seen below, would be expected to be shared only by some notions corresponding to ‘generally’ and ‘rarely’. First, in the interpretation of ‘generally’ in terms of ‘several’ or ‘many’ (cf. Brazilians and shaving in Section 2.1), the family of important sets – those having several elements – is closed under supersets: each superset $Y \supseteq X$ of an important set $X \in \mathcal{W}$ is important ($Y \in \mathcal{W}$). In this case, the family of important set is a proper upward closed family.

In addition, in the interpretation of ‘generally’ in terms of ‘most’ (cf. American males in 2.2), the family of important sets – those having most elements – is also closed under intersection: the intersection $X \cap Y$ of important sets X and $Y (X, Y \in \mathcal{W})$ is important ($X \cap Y \in \mathcal{W}$). In this case, the family of important set is a proper filter.

Also, in other interpretations of ‘generally’ (cf. the Madonna interpretation in Section 2.3), the family of important sets is a proper ultrafilter.

Thus, these interpretations of ‘generally’ give rise to a hierarchy of families. Indeed, the proper family of important subsets of the universe is upward closed in the ‘several’ (or ‘many’) interpretation, a filter in the ‘most’ interpretation, and an ultrafilter in other interpretations.[10]

Our logics for ‘generally’ add to classical first-order logic generalized quantifiers, intended to be interpreted as ranging over given families of important subsets of the universe of discourse.

3. Logics for 'Generally' and 'Rarely'

We now present our logics for 'generally' and 'rarely': syntax, semantics and axiomatics. These logics add generalized quantifiers to classical first-order logic, giving generalized formulas. The intended interpretation of such a formula holding generally as the set of objects satisfying it is important (in a given sense) is captured by requiring the extension of the formula to belong to a given family. We will axiomatize such logics by means of schemata coding properties of these families. We will concentrate on our logic $\mathcal{L}_{\omega\omega}(\rho)^f$ for 'most'. In it, the intended interpretation of a formula holding generally – in the sense that most objects satisfy it – is captured by requiring its extension to belong to a given filter. It can be axiomatized by schemata coding properties of filters. We shall also mention some variants of our logic $\mathcal{L}_{\omega\omega}(\rho)^f$: the logic $\mathcal{L}_{\omega\omega}(\rho)^s$ for 'several' (or 'many') and the ultrafilter logic $\mathcal{L}_{\omega\omega}(\rho)^u$ (Carnielli and Veloso 1997; Sette et al. 1999; Veloso 1998, 1999, 2000).

Consider a fixed denumerably infinite set V of symbols for variables. Given a signature (logical type) ρ, with repertoires of new symbols for predicates, functions and constants, we use $\mathsf{L}(\rho)$ for the usual first-order language (with equality $\equiv$) of signature ρ, closed under the propositional connectives, as well as under the classical quantifiers $\forall$ and $\exists$.

3.1. Syntax of ∇

We now examine the syntax of the generalized quantifier ∇. We use $\mathsf{L}^\nabla(\rho)$ for the extension of the usual first-order language $\mathsf{L}(\rho)$ obtained by adding the new operator ∇.

The formulas of $\mathsf{L}^\nabla(\rho)$ are built by the usual formation rules and the following new, variable-binding, *formation rule*: for each variable $v \in V$, if φ is a formula in $\mathsf{L}^\nabla(\rho)$ then so is $\nabla v\varphi$.

We shall also employ the notation $\varphi(\underline{v}/\underline{t})$ for the result of simultaneously substituting each term t_i for all the free occurrences of variable v_i in formula φ (for given sets $\underline{v}:=\{v_1, \ldots, v_n\}$ of variables and $\underline{t}:=\{t_1, \ldots, t_n\}$ of terms), which we sometimes simplify to $\varphi(\underline{t})$, when safe. Other usual syntactic notions, such as sentence, (free) substitution (Enderton 1972; Shoenfield 1967), can be appropriately adapted.

We can now express generalized assertions, such as "Birds generally fly" and "Metals generally are solid". The next example illustrates the expressive power of such languages.

EXAMPLE [*Expressive power of* ∇]. Consider a signature λ consisting of the binary predicate L, for a relation between persons.

a. First, let $L(x, y)$ stand for "x loves y". We can express some assertions by means of purely first-order sentences. Some assertions expressed by sentences of $\mathsf{L}^\nabla(\lambda)$ are as follows.

– "People generally love somebody" by $\nabla x \exists y L(x, y)$.
– "Somebody loves people in general" by $\exists x \nabla y L(x, y)$.
– "Everybody loves people in general" by $\forall x \nabla y L(x, y)$.
– "People generally love everybody" by $\nabla x \forall y L(x, y)$.
– "People generally love each other" by $\nabla x \nabla y L(x, y)$.

b. Now, let $L(x, y)$ stand for "y is taller than x". We then can express properties, such as "people generally are taller than x" and "x is taller than people in general" by the formulas $\nabla y L(x, y)$ and $\nabla y L(y, x)$ of $\mathsf{L}^{\nabla}(\lambda)$, respectively.[11]

3.2. Semantics of 'Generally' and 'Rarely'

The semantic interpretation for our generalized logics is provided by extending the usual first-order definition of satisfaction to the new quantifier ∇. For this purpose, we resort to modulated structures: expansions of first-order structures by families of subsets (Grácio 1999).

A *modulated structure* $\mathtt{A}^{\mathcal{K}} = (\mathtt{A}, \mathcal{K})$ for signature ρ consists of a usual (first-order) structure $\mathtt{A}$ for signature ρ together with a family $\mathcal{K}$, called a *complex*, of subsets of the universe A of $\mathtt{A}$.

Now, we extend the familiar definition of *satisfaction* of a formula φ in a structure under an assignment $\mathtt{s} : V \to A$ to variables to generalized formulas as follows.

– For a formula $\nabla v\varphi$, we define $\mathtt{A}^{\mathcal{K}} \models \nabla v\varphi[\mathtt{s}]$ iff the set $\{a \in A : \mathtt{A}^{\mathcal{K}} \models \varphi[s(v \mapsto a)]\}$ belongs to the given complex $\mathcal{K}$;

where $\mathtt{s}(v \mapsto a)$ is the assignment agreeing with s on every variable but v, and $\mathtt{s}(v \mapsto a)(v) = a$.

As usual, satisfaction of a formula depends only on the realizations assigned to its symbols. In particular, satisfaction of a formula without ∇ does not depend on the complex, i.e., for a formula φ of $\mathsf{L}(\rho) \subseteq \mathsf{L}^{\nabla}(\rho) : \mathtt{A}^{\mathcal{K}} \models \varphi[\mathtt{s}]$ iff $\mathtt{A}\, \varphi[\mathtt{s}]$. Also, satisfaction of a formula hinges only on the values assigned to its free variables. So, we can employ the familiar notation $\mathtt{A}^{\mathcal{K}} \models \varphi[\underline{a}]$ (for satisfaction of a formula φ – with at most n free variables – by $\underline{a} \in A^n$); such a formula *defines* an n-ary relation: $\mathtt{A}^{\mathcal{K}}[\varphi] := \{\underline{a} \in A^n : \mathtt{A}^{\mathcal{U}} \models \varphi[\underline{a}]\}$. Similarly, we can introduce the *extension* as $\mathtt{A}^{\mathcal{K}}[\varphi(\underline{a}, v)] := \{b \in A : \mathtt{A}^{\mathcal{K}} \models \varphi(\underline{u}, v)[\underline{a}, b]\}$. With this notation, satisfaction of a generalized formula $\nabla v\varphi(u, v)$ becomes:

$$\mathtt{A}^{\mathcal{K}} \models \nabla v\varphi(\underline{u}, v)[\underline{a}] \text{ iff the extension } \mathtt{A}^{\mathcal{K}}[\varphi(\underline{a}, v)] \text{ belongs to the complex } \mathcal{K}.$$

Other familiar semantic notions, such as reduct, model ($\mathtt{A}^{\mathcal{K}} \models \Gamma$), etc., are as usual (Enderton 1972; Shoenfield 1967). The notion of *filter consequence* is as expected:[12]

$$\Gamma \models^{\mathcal{F}} \tau \text{ iff } \mathtt{A}^{\mathcal{F}} \models \tau \text{ whenever } \mathtt{A}^{\mathcal{F}} \models \Gamma.$$

Similarly, we have *filter validity*: $\models^{\mathcal{F}} \tau$ iff $\varnothing \models^{\mathcal{F}} \tau$.

	Affirmatives		Negatives	
Universal	$\forall v \varphi$	$\xleftrightarrow{\text{contraries}}$	$\forall v \neg\varphi$	(or $\neg\exists v\, \varphi$)
	$\Downarrow$		$\Downarrow$	
Generalized	$\nabla v \varphi$	$\xleftrightarrow{\text{contradictories}}$	$\Delta v \neg\varphi$	(or $\neg\nabla v\, o$)
	$\Downarrow$		$\Downarrow$	
Particular	$\exists v \varphi$	$\xleftrightarrow[\text{sub-contraries}]{}$	$\exists v \neg\varphi$	(or $\neg\forall v \varphi$)

Figure 1. Hexagon of oppositions.

Clearly, the behavior of the new quantifiers is intermediate between those of the classical quantifiers $\forall$ and $\exists$: the formulas $\forall v\varphi \to \nabla v\varphi$ and $\nabla v\varphi \to \exists v\varphi$ are valid (but not the converse implications[13]). The behavior of iterated ∇'s, however, contrasts sharply with the commutativities of each classical $\forall$ and $\exists$: the formulas $\nabla y \nabla x\varphi \to \nabla x \nabla y\varphi$ fail to be valid.

More positive examples of the behavior of the new quantifiers are their transfers over the classical quantifiers $\forall$ and $\exists$: the validity of $\nabla x \forall y\varphi \to \forall y \nabla x\varphi$ and $\exists y \nabla x\varphi \to \nabla x \exists y\varphi$.

One can also introduce a dual generalized quantifier Δ for 'not rarely': $\Delta v\varphi$ as an abbreviation for $\neg\nabla v\neg\varphi$. Then, the classical square of oppositions becomes a hexagon[14] (see Figure 1).

3.3. Axiomatics of 'Generally'

We will now formulate deductive systems for our logics of 'generally', by adding schemata (coding properties of the semantic families) to a calculus for classical first-order logic.

To set up deductive system for our logics we can start with a sound and complete deductive calculus for classical first-order logic, with Modus Ponens as the sole inference rule, as in (Enderton 1972). We then extend its set $A(\rho)$ of axiom schemata by adding a set of generalizations of axiom schemata (coding properties of corresponding semantic family), to form a set of schemata for 'generally'.

To introduce the form of these new axioms, we may consider some basic principles as well as the expression of some properties of the families by means of ∇ (for 'generally'). Among the basic principles of such a logic for qualitative notions, one would expect the satisfaction of a generalized formula to hinge only on its extension. These should include invariance under alphabetic variants. We thus consider the following set of formulas:

$[\nabla\alpha] := \{\nabla v\varphi \to \nabla u\varphi(v/u) : \varphi \in \mathsf{L}^{\nabla}(\rho)$, for a new variable u not occurring in $\varphi\}$

As the families are proper and non-empty, consider also

$$[\forall\nabla] := \{\forall v\varphi \to \nabla v\varphi : \varphi \in L^{\nabla}(\rho)\} \text{ and } [\nabla\exists] := \{\nabla v\varphi \to \exists v\varphi : \varphi \in L^{\nabla}(\rho)\}.$$

Now, considering the union $B^i(\rho) := [\nabla\alpha] \cup [\forall\nabla] \cup [\nabla\exists]$, we will form a chain of extensions by adding further schemata, as follows.

- For the upward closed logic $\mathfrak{L}_{\omega\omega}(\rho)^s$ of 'several' (or 'many'), we extend $B^i(\rho)$ to $B^s(\rho) := B^i(\rho) \cup [\to\nabla]$, where $[\to\nabla] := \{\forall v(\psi \to \theta) \to (\nabla v\psi \to \nabla v\theta) : \psi, \theta \in \mathsf{L}^{\nabla}(\rho)\}$.
- For the filter logic $\mathfrak{L}_{\omega\omega}(\rho)^f$ of 'most', we extend $B^s(\rho)$ to the set $B^f(\rho) := B^s(\rho) \cup [\nabla\wedge]$, where $[\nabla\wedge] := \{(\nabla v\psi \wedge \nabla v\theta) \to \nabla v(\psi \wedge \theta) : \psi, \theta \in \mathsf{L}^{\nabla}(\rho)\}$.
- For the ultrafilter logic $\mathfrak{L}_{\omega\omega}(\rho)^u$, we further extend $B^f(\rho)$ to $B^u(\rho) := B^f(\rho) \cup [\neg\nabla]$, where $[\neg\nabla] := \{\neg\nabla v\varphi \to \nabla v\neg\varphi : \varphi \in \mathsf{L}^{\nabla}(\rho)\}$.

Now, take $A^s(\rho)$, $A^f(\rho)$ and $A^u(\rho)$ to consist of the generalizations of the formulas in $B^s(\rho)$, $B^f(\rho)$ and $B^u(\rho)$, respectively.[15] Thus, generalized derivability amounts to first-order derivability from the schemata for 'generally', more precisely:

$$\Gamma \vdash^f \varphi \text{ iff } \Gamma \cup A^k \vdash \varphi. \qquad (\vdash^k)$$

In particular, these deductive systems are monotonic. We also have substitutivity of equivalents: $\nabla v\psi \leftrightarrow \nabla v\theta$ follows from $\psi \leftrightarrow \theta$.

As an example, consider the following facts about a universe of people: "People generally oppose those in conflict with whom they sympathize" and "People generally sympathize with Bill" expressed by $\nabla x\forall y\nabla z[S(x, y) \wedge C(z, y) \to O(x, z)]$ and $\nabla y S(y, b)$, respectively. From them, one can infer the sentence $\nabla x\nabla z[C(z, b) \to O(x, z)]$, expressing "People generally oppose those in conflict with Bill".

Other usual deductive notions, such as (maximal) consistent sets, witnesses, conservative extension (Enderton 1972; Shoenfield 1967), can easily be adapted.

4. Logics with Generalized Assertions

This section is devoted to establishing some properties of our logic with generalized assertions, including soundness and completeness of their deductive systems with respect to the corresponding semantic consequences, and examine their deductive and expressive powers.

4.1. Soundness and Completeness

The soundness of our deductive systems with respect to consequences will be examined first. As usual, soundness is easily established. Indeed, the axioms in each $A^k(\rho)$ code properties of the corresponding class of families, so they hold in all their modulated structures. We thus have soundness of our deductive systems with respect to consequences.

For completeness of our deductive systems with respect to consequences, we can adapt Henkin's well-known proof for classical first-order logic (Henkin 1949; Enderton 1972; Shoenfield 1967). The crucial point is providing an adequate complex, which we can do by means of witnesses. We proceed to outline how this can be done in our logics.

Given a consistent set Γ in $\mathsf{L}^{\nabla}(\rho)$, extend it to a maximal consistent set Σ in $\mathsf{L}^{\nabla}(\rho \cup C)$, with witnesses for the existential sentences of $\mathsf{L}^{\nabla}(\rho \cup C)$ in set C of new constants.[16] Considering the set T of variable-free terms of $\mathsf{L}(\rho \cup C)$, form the canonical structure $\mathtt{H}$, for signature $\rho \cup C$ as usual. It has universe $H := T/{\equiv^{\Sigma}}$ where $t' \equiv^{\Sigma} t''$ iff $\Sigma \vdash^{k} t' \equiv t''$. We provide a complex, by considering the formulas of $\mathsf{L}^{\nabla}(\rho \cup C)$, having a single variable free, as follows.

We consider the set represented within Σ by formula φ of $\mathsf{L}^{\nabla}(\rho \cup C)$ with single free variable v, namely $\varphi^{\Sigma} := \{t/{\equiv^{\Sigma}} \in H : \varphi(v/t) \in \Sigma\}$, and form the family of provably important represented subsets, i.e., ${}_{\nabla}\Sigma := \{\varphi^{\Sigma} \subseteq H : \nabla v\varphi \in \Sigma\}$. In view of our axioms, this family ${}_{\nabla}\Sigma$ has the finite intersection property and can be used to provide an adequate complex.[17] Thus, in each case, we have an appropriate complex $\mathcal{H}$ to expand the canonical structure $\mathtt{H}$ to a modulated structure $\mathtt{H}^{\mathcal{H}} := (\mathtt{H}, \mathcal{H})$ for signature $\rho \cup C$.

It can be now shown, by induction, that $\mathtt{H}^{\mathcal{H}} \models \tau$ iff $\tau \in \Sigma$, for each sentence τ of $\mathsf{L}^{\nabla}(\rho \cup C)$. The inductive step for the new quantifier ∇, namely, for a sentence $\nabla v\varphi$: $\mathtt{H}^{\mathcal{H}} \models \nabla v\varphi$ iff $\nabla v\varphi \in \Sigma$, follows from the crucial property $\varphi^{\Sigma} \in {}_{\nabla}\Sigma$ iff $\varphi^{\Sigma} \in \mathcal{H}$ of the complex $\mathcal{H}$.[18] We thus have a Löwenheim-Skolem Theorem for our logical systems. Löwenheim-Skolem Theorem: Each consistent set Γ of sentences of $\mathsf{L}^{\nabla}(\rho)$ has a modulated model $\mathtt{H}^{\mathcal{H}} := (\mathtt{H}, \mathcal{H})$ with cardinality at most $|\mathsf{L}^{\nabla}(\rho)|(|\mathtt{H}| \leq |\mathsf{L}^{\nabla}(\rho)|)$.

Hence, we have the desired result for our logics for 'generally'.

THEOREM [*Completeness of logics for 'generally'*]. Each deductive system $\vdash^{s}$, $\vdash^{f}$ and $\vdash^{u}$ is complete with respect to the consequence $\models^{\mathcal{S}}$, $\models^{\mathcal{F}}$ and $\models^{\mathcal{U}}$, respectively.

4.2. Other Metamathematical Properties

Other metamathematical properties of our logics $\mathcal{L}_{\omega\omega}(\rho)^{s}$, $\mathcal{L}_{\omega\omega}(\rho)^{m}$ and $\mathcal{L}_{\omega\omega}(\rho)^{u}$ for 'generally' can be obtained as shown below.

We have sound and complete deductive systems for our logics. As usual, such a result transfers the finitary character of derivability to the compactness of the corresponding semantic consequence. Thus, our logics are proper (as we shall see) extensions of classical first-order logic with compactness and Löwenheim-Skolem properties.[19]

Also, our logics for 'generally' have other connections with classical first-order logic $\mathfrak{L}_{\omega\omega}(\rho)$: the pleasing fact that they are conservative extensions of classical first-order logic, as well as the related reduction of simply generalized consequences of a first-order theory to first-order consequences. By a *simply generalized formula* we mean one of the form $\nabla v\varphi$, for some purely first-order formula φ.

PROPOSITION [*Logics for 'generally' and classical logic*].

(a) Conservativeness. For each set $\Delta \cup \{\sigma\}$ of sentences of $\mathsf{L}(\rho)$: $\Delta \vdash \sigma$ iff $\Delta \vdash^{k} \sigma$.

(b) Generalized consequences of first-order theory. Given a set Δ of sentences of $\mathsf{L}(\rho)$, for every formula φ of $\mathsf{L}(\rho)$: $\Delta \vdash^{k} \nabla v\varphi$ iff $\Delta \vdash \forall v\varphi$ and $\Delta \vdash^{k} \neg\nabla v\varphi$ iff $\Delta \vdash \neg\exists v\varphi$.

Proof outline. Any nonempty set is in some ultrafilter, which yields part (a) and one half of part (b), the other half following from the schemata [$\forall\nabla$] and [$\nabla\exists$] in 3.3.

EXAMPLE (Theories of solid metals). Consider consistent theories with information about which metals are solid under ordinary conditions.

a. First, consider a purely first-order theory Δ, with two axioms expressing "Mercury is not solid and is not the only metal" and "Every metal, other than mercury, is solid". In this case, it cannot be decided whether "metals generally are solid".

b. Now, consider a consistent theory Γ extending Δ with the 'generalized' information $\nabla v \neg v \equiv Hg$ for "metals generally are distinct from mercury". Then, one infers that "metals generally are solid", i.e., $\nabla v S(v)$.

We can also reduce to first-order some consequences of an extension of a first-order theory by a simply generalized axiom. By slightly strengthening the preceding argument, we can see that the consequences, in this case too, become somewhat trivialized.

Extension by a simply generalized axiom and classical logic. Consider a set Δ of sentences of $\mathsf{L}(\rho)$ and a formula ψ of $\mathsf{L}(\rho)$.

(a) For every formula θ of $\mathsf{L}(\rho)$: $\Delta \cup \{\nabla v\psi\} \vdash^{k} \nabla v\theta$ iff $\Delta \vdash \forall v(\psi \to \theta)$ and if $\Delta \cup \{\nabla v\psi\} \vdash^{k} \neg\nabla v\theta$ then $\Delta \vdash \neg\exists v(\psi \wedge \theta)$.

(b) For every sentence τ of $\mathsf{L}(\rho)$: $\Delta \cup \{\nabla v\psi\} \vdash^{k} \tau$ iff $\Delta \cup \{\exists v\psi\} \vdash \tau$.

As a simple example, consider purely first-order information about workers in a plant. Assume that one observes that "workers generally are careless", expressed by $\nabla v C(v)$, and asks whether one can then conclude that "workers generally are accident prone", in the sense $\nabla v A(v)$. One can infer this generalized assertion iff the first-order information entails the universal assertion $\forall v[C(v) \to A(v)]$ i.e., "all careless workers are accident prone".

In the case of the stronger logics $\mathcal{L}_{\omega\omega}(\rho)^f$ and $\mathcal{L}_{\omega\omega}(\rho)^u$, the preceding trivialization can be seen to hold for sets of simply generalized axioms. The next result is stated for $\mathcal{L}_{\omega\omega}(\rho)^f$.

Extension by simply generalized axioms and classical logic. Consider a set Δ of sentences of $\mathsf{L}(\rho)$ and a set Γ of simply generalized sentences of $\mathsf{L}^{\nabla}(\rho)$.

(a) For every formula θ of $\mathsf{L}(\rho)$, we have:

$$\Delta \cup \Gamma \vdash^f \nabla v\theta \text{ iff } \Delta \vdash \forall v[\psi_1 \wedge \ldots \wedge \psi_n) \to \theta]$$

$$\Delta \cup \Gamma \vdash^f \neg\nabla v\theta \text{ iff } \Delta \vdash \neg\exists v[\psi_1 \wedge \ldots \wedge \psi_n) \wedge \neg\theta];$$

for some sentences $\nabla v\psi_1, \ldots, \nabla v\psi_n \in \Gamma$.

(b) For every sentence τ of $\mathsf{L}(\rho)$: $\Delta \cup \Gamma \vdash^f \tau$ iff $\Delta \cup \{\exists v(\psi_1 \wedge \ldots \wedge \psi_n)\} \vdash \tau$, for some sentences $\nabla v\psi_1, \ldots, \nabla v\psi_n \in \Gamma$.

These results can be summarized as follows: recall that the behavior of the generalized quantifier ∇ in our logics for 'generally' is intermediate between those of the classical quantifiers $\exists$ and $\forall$ (cf. Section 3). Now, in the context of a classical first-order theory Δ, the single quantifier ∇ in a simply generalized sentence γ behaves as either extreme: as universal $\forall$, when γ is a consequence of Δ, and as existential $\exists$, when γ is added as axiom to Δ.

EXAMPLE [*Birds and flying*]. Consider consistent theories with information about birds.

a. First, consider a consistent purely first-order theory Δ. Assume that one knows that "some birds fly", i.e., $\Delta \vdash \exists vF(v)$, "every bird is a biped with beak", i.e., $\Delta \vdash \forall v[D(v) \wedge K(v)]$, and "flying birds have wings", i.e., $\Delta \vdash \forall v[F(v) \to W(v)]$. Then, one does not know that "birds generally do not fly": $\Delta \nvdash^f \nabla v\neg F(v)$. Also, notice that if one does not know that "all birds fly" ($\Delta \nvdash \forall vF(v)$), then one does not know that "birds generally fly" ($\Delta \nvdash^f \nabla vF(v)$). In this case, we cannot conclude whether birds generally fly or do not fly.
b. Now, consider a consistent theory $\Delta \cup \Gamma$ extending Δ by the set Γ of axioms giving the simply generalized information "birds generally have wings" and "birds generally have feathers", expressed respectively by $\nabla vW(v)$ and $\nabla vT(v)$. Then, one can conclude, from $\Delta \cup \Gamma$, the simply generalized assertion "birds generally fly", i.e., $\Delta \cup \Gamma \vdash^f \nabla vF(v)$, iff one can conclude from first-order theory Δ the universal assertion "all feathered winged birds fly", i.e., $\Delta \vdash \forall v[(W(v) \wedge T(v)) \to F(v)]$.
c. Finally, assume that one also knows that "all normal winged birds fly", expressed by the generalized sentence $\nu : \forall v[(N(v) \wedge W(v)) \to F(v)]$. With this additional information, one can conclude $\Delta \cup \Gamma \cup \{\nu\} \vdash^f \nabla v[F(v) \vee \neg N(v)]$ ("Birds generally fly when normal").

By examining more closely the expressive power of the generalized quantifier in Section 4.3 , we will be able to see that the reduction of consequences to classical logic is restricted to simply generalized sentences, failing for other, more complex, sentences.

4.3. Expressive Power

We shall now examine the power of our logics for 'generally', showing that they are indeed proper extensions of classical first-order logic (Veloso 2000). Concerning deductive powers, it is clear that the extensions of classical first-order logic in our chain (cf. Section 3) have strictly increasing deductive powers.[20]

The expressive powers of our logics for 'generally' will be examined next, showing that they extend properly that of classical first-order logic. We know that satisfaction of a formula with the generalized quantifier ∇ depends on the complex, which is not the case for purely first-order formulas. So, it is to be expected that some formulas with ∇ should not be equivalent to formulas without ∇. It remains to exhibit specific examples of such formulas. For this purpose, we will first characterize a simple class of formulas from which the new generalized quantifier can be eliminated.

As motivation, reconsider our example (Birds and flying) in 4.2. In part (a) of this example, we saw a consistent purely first-order theory Δ expressing some facts about birds. Since we did not know that all birds fly (i.e. $\Delta \nvdash \forall v F(v)$) (or, even more strongly, we knew that not all birds fly, i.e., $\Delta \vdash \neg \forall v F(v)$, as long as Δ is consistent.), we could not conclude whether birds generally fly or do not fly. This example was used to illustrate the reduction of simply generalized consequences of a first-order theory to first-order consequences. We now consider the question of expressing, rather than deducing, formulas with the generalized quantifier ∇. Can we express the simply generalized assertion $\nabla v F(v)$ by an equivalent sentence without ∇? The question appears to have a negative answer. Moreover, the reason for this negative answer will be seen to rest entirely on classical first-order reasoning, namely: $\Delta \nvdash [\exists v F(v) \to \forall v F(v)]$ and $\Delta \nvdash [\exists v \neg F(v) \to \forall v \neg F(v)]$.

It will be clear why the only simply generalized formulas $\nabla v \varphi$ (where φ has no ∇) that can be expressed without ∇ are the trivial ones (in the sense that $\exists v \varphi \to \forall v \varphi$ can be derived). The general question we shall now address concerns the elimination of the generalized quantifier ∇ from formulas. This question of eliminating the quantifier ∇ concerns finding an equivalent formula without ∇. Thus, it is appropriate to consider the context of a theory. We shall give some extra freedom by allowing expanding the signatures.[21] We shall concentrate on the ultrafilter logic $\mathcal{L}_{\omega\omega}(\rho)^u$, as the negative results will transfer to the weaker versions.

We say that theory Δ *eliminates* ∇ from formula φ iff there exists a formula θ without ∇ (in some expanded signature, but with the same free variables) such that $\Delta \models^{\mathcal{U}} (\varphi \leftrightarrow \theta)$.[22]

It can be now shown that a purely first-order theory Δ eliminates ∇ from simply generalized formula $\nabla v \psi$ iff $\Delta \vdash (\exists v \psi \rightarrow \forall v \psi)$. Towards this goal, we first consider each direction of the biconditional in the elimination of ∇ from the simply generalized formula $\nabla v \psi$.

LEMMA [*Conditional theorems with a simply generalized formula*]. Given signatures $\rho \subseteq \rho'$, consider a formula ψ in $\mathsf{L}(\rho)$ and a set Δ of sentences of $\mathsf{L}(\rho')$. For every expansion $\rho'' \subseteq \rho'$ and formula θ of $\mathsf{L}(\rho'')$, with the same free variables as $\nabla v \psi$:
(a) if $\Delta \models^{\mathcal{U}} (\nabla v \psi \rightarrow \theta)$, then $\Delta \vdash (\exists v \psi \rightarrow \theta)$;
(b) if $\Delta \models^{\mathcal{U}} (\theta \rightarrow \nabla v \psi)$, then $\Delta \vdash (\theta \rightarrow \forall v \psi)$.

Proof outline. Any nonempty set belongs to some ultrafilter.

We can now conclude our condition for eliminating a ∇ from a simply generalized formula.

PROPOSITION [*∇-eliminable simply generalized formulas*]. Given a formula ψ in $\mathsf{L}(\rho)$, consider a set Δ of sentences of $\mathsf{L}(\rho')$ (where $\rho \subseteq \rho'$). Then, theory Δ eliminates ∇ from formula $\nabla v \psi$ iff $\Delta \vdash (\exists v \psi \rightarrow \forall v \psi)$.

Proof outline. The preceding lemma yields one half of the equivalence, and the other half follows from the schemata [$\nabla\exists$] and [$\forall\nabla$] (cf. 3.3 in Section 3).

As an example illustrating these ideas by a simply generalized sentence from which ∇ cannot be eliminated, consider a consistent purely first-order theory Δ with information about which metals are solid under ordinary conditions (cf. example: Theories of solid metals in Section 4.2). Assume that theory Δ yields the (reasonable) pieces of information: $\exists v S(v)$ and $\exists v \neg S(v)$ ("some, but not all, metals are solid"). Then, one cannot express the simply generalized sentence $\nabla v S(v)$ ("metals generally are solid") by any equivalent purely first-order sentence (even if we allow some other extra-logical symbols).

The expressive power of our logics for 'generally' will be now more closely analyzed, and it will be shown that they are proper extensions of classical first-order logic. For this purpose, we will first introduce some auxiliary concepts.

Given a cardinal number κ, we will call formula φ *κ-eliminable* iff some purely first-order theory Δ, having models with cardinality κ or above, can eliminate ∇ from formula φ. We shall call formula φ *logically eliminable* iff the empty set $\varnothing$ of sentences eliminates ∇ from it.

We can now rely on our preceding characterization to present some simply generalized formulas that cannot be expressed within classical first-order logic. We shall employ $\varnothing$ for the signature of pure equality, without any extra-logical symbols.

Given distinct variables u and v, the formula $\nabla v\, v{\equiv}$ of $\mathbf{L}^{\nabla}(\varnothing)$ is not 2-eliminable and the sentence $\exists u \nabla v\, v{\equiv}$ of $\mathsf{L}^{\nabla}(\varnothing)$ is not $\aleph_0$-eliminable.[23]

Thus, it can be concluded that within our logics for 'generally' we can express some concepts that cannot be expressed by equivalent sentences of classical first-order logic.

THEOREM [*Powers of logics for 'generally' and classical first-order logic*]. Given distinct variables u and v, the sentences $\exists u \nabla v\, v \equiv u$ and $\forall u \neg \nabla v\, v \equiv u$ of $\mathsf{L}^{\nabla}(\varnothing)$ are not equivalent to any purely first-order sentences (without ∇).

It was mentioned, at the end of Section 4.3, that the reduction of consequences to classical logic is restricted to simply generalized sentences. The above sentences provide examples where such reductions fail.

Consider the sentence $\exists u \nabla v\, v \equiv u$ in $\mathsf{L}^{\nabla}(\varnothing)$, with distinct variables u and v, and a set Δ of purely first-order) sentences of $\mathsf{L}(\rho)$ having infinite models (say $\Delta := \varnothing$). First, in contrast to part (b) of the proposition on generalized logic and classical logic in Section 4.3, we have no sentence σ of $\mathsf{L}(\rho)$, such that $\Delta \vdash^{\mathcal{U}} \exists u \nabla v\, v \equiv u$ iff $\Delta \vdash \sigma$. Also, in contrast to the lemma on extension by a simply generalized axiom and classical logic in Section 4.3, given a sentence τ of $\mathsf{L}(\rho)$, we have no sentence σ of $\mathsf{L}(\rho)$, such that $\Delta \cup \{\exists u \nabla v\, v \equiv u\} \vdash^{u} \tau$ iff $\Delta \cup \{\sigma\} \vdash \tau$.[24]

5. Generic Reasoning and Generalized Assertions

We shall now consider some aspects of inference with generalized assertions, including generic reasoning. In our logics for 'generally', the new generalized quantifiers are intermediate between the classical existential and universal quantifiers, in terms of behavior. This is a source of flexibility but also brings about some problems.

On the one hand, in contrast to the classical universal quantifier, instantiation does not hold for our generalized quantifiers. On the other hand, these generalized quantifiers share with the classical universal quantifier some problems about inference. We shall examine such problems in this section.

5.1. Generic Objects

We now wish to argue that our logics for 'generally' can also support forms of reasoning with notions that may be termed 'generic'.[25] The basic ideas of "generic" objects will be first introduced; they will then be internalized as constants, making possible to reason about them.

The familiar "Tweety example" (Reiter 1980) may be used to convey the main ideas underlying our approach. From the assertions "Most birds fly" and "Tweety is a typical bird", one wishes to conclude "Tweety does fly". Our approach involves two steps. First, formulating 'generally' by means of ∇, and, next, regarding 'typical' as a version of 'representative'. The former – formulating "Most birds fly" as

$\nabla v F(v)$ – looks quite natural, in view of our interpretation of ∇ as "holding almost universally". The latter may require some explanation.

How would one imagine a "typical" bird? One would probably visualize (or draw) a picture of a winged, feathered biped with beaks. One may not be too clear about other features, such as its flying status.[26] We propose to interpret a 'typical' bird as "a bird that exhibits the properties that *most* birds exhibit". (So, it is one giving correct instances of generalized assertions. Notice that "the properties that *all* birds exhibit" would be too strong.)

It remains to give a precise formulation for these ideas in terms of "the properties that objects generally possess". We proceed to explain how this can be done. Our approach can be viewed as a symbolic form of 'generic' reasoning, in that the generalized quantifier ∇ can be used to capture these intended interpretations (Carnielli and Veloso 1997; Veloso 1998).

We first introduce and examine generic objects in a modulated structure. We will be more interested in objects that are generic for a set of formulas, but it may be convenient to begin with the special case of a single formula.

Consider a modulated structure $\mathtt{A}^{\mathcal{K}} = (\mathtt{A}, \mathcal{K})$ for a given signature ρ and an element $a \in A$. With respect to a given generalized sentence $\nabla v \varphi$ of $\mathtt{L}^{\nabla}(\rho)$, we will call the element a *typical* in $\mathtt{A}^{\mathcal{K}}$ iff $\mathtt{A}^{\mathcal{K}} \models \varphi[a]$ or $\mathtt{A}^{\mathcal{K}} \not\models \nabla v \varphi$, an *archetypical* element being one such that $\mathtt{A}^{\mathcal{K}} \models \nabla v \varphi$ iff $\mathtt{A}^{\mathcal{K}} \models \varphi[a]$.[27]

For instance, consider a modulated structure $\mathtt{A}^{\mathcal{K}}$, representing a world of animals where "Animals generally are voracious" and "Animals generally do not fly": $\mathtt{A}^{\mathcal{K}} \models \nabla u V(u)$ and $\mathtt{A}^{\mathcal{K}} \models \nabla u \neg F(u)$. Then, voracious animals are archetypical for general voracity and non-flying animals are archetypical with respect to generally not flying. Only voracious animals are typical for general voracity, but (if $\mathcal{K}$ is a filter) any animal is typical for generally flying.

Now, with respect to a set Ψ of formulas of $\mathtt{L}^{\nabla}(\rho)$, we shall call an element a of $\mathtt{A}^{\mathcal{K}}$ *typical* (or *archetypical*) iff a is typical (respectively, archetypical) for every generalized sentence $\nabla v \varphi$ in Ψ.[28] In particular, by a *typical* (or *archetypical*) object we will mean an element that is so for every generalized sentence of the language $\mathtt{L}^{\nabla}(\rho)$.

For example, consider the generalized sentences $\nabla v F(v)$ (for "Birds generally fly") and $\nabla v \neg S(u)$ (for "Birds generally do not swim"). An archetypical element for these two sentences in a modulated structure $\mathtt{B}^{\mathcal{K}}$ representing a world of birds, where birds generally fly and do not swim, will be any flying bird that does not swim.

These typical and archetypical objects are somewhat reminiscent of Hilbert's ideal elements or of Platonic forms. So, it is not surprising that they are somewhat elusive, being present only in some modulated structures.[29] For instance, in the naturals with zero and successor and a non-principal ultrafilter (containing no finite subset), a typical element, if any, must be non-standard as indicated in Figure 2.[30]

Interestingly enough, theories behave much better with respect to genericity, as we shall have occasion to see in the sequel. The next lemma establishes that *finite*

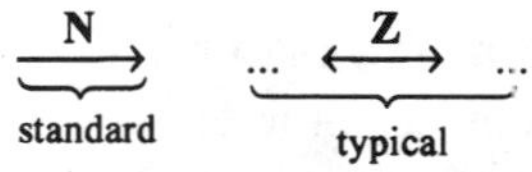

Figure 2. Typical elements and non-standard naturals.

sets of sentences always have typical elements in filter structures and archetypical elements in ultrafilter structures.

LEMMA [*Generic elements for finite sets of sentences in filter structures*]. Consider a modulated structure $\mathtt{A}^{\mathcal{K}} = (\mathtt{A}, \mathcal{K})$ for signature ρ. Then, for each *finite* set Φ of (generalized) sentences of $\mathsf{L}^{\nabla}(\rho)$: the set of its typical elements is in $\mathcal{K}$ if the complex $\mathcal{K}$ is a filter, and the set of its archetypical elements is in $\mathcal{K}$ if the complex $\mathcal{K}$ is an ultrafilter.

Proof outline. The set of typical objects is the finite intersection of the extensions $\mathtt{A}^{\mathcal{K}}[\varphi]$ with $\mathtt{A}^{\mathcal{K}}[\varphi] \in \mathcal{K}$. So, it is in $\mathcal{K}$ if $\mathcal{K}$ is a filter. If the complex $\mathcal{K}$ is an ultrafilter, then the set of archetypical objects is a finite intersection of extensions $\mathtt{A}^{\mathcal{K}}[\varphi]$ in $\mathcal{K}$,[31] thus being in $\mathcal{K}$.

5.2. Reasoning with Generic Constants

We will now internalize the previous ideas in extensions by new constants, which may be regarded as generalized witnesses. We proceed to outline how this can be done in our filter and ultrafilter logics: $\mathcal{L}_{\omega\omega}(\rho)^f$ and $\mathcal{L}_{\omega\omega}(\rho)^u$.

We wish to add a new constant c behaving as a 'generic' witness. Given a signature ρ and a new constant c *not* in ρ, consider the expansion $\rho[c] := \rho \cup \{c\}$ by the new constant c.

Given a sentence $\nabla v\varphi$ of $\mathsf{L}^{\nabla}(\rho)$, we construct the following sentences of $\mathsf{L}^{\nabla}(\rho[c])$ as axioms on the new constant c for $\nabla v\varphi$: $\nabla v\varphi \rightarrow \varphi(v/c)$ as the *typical axiom* $\varpi(\nabla v\varphi \rightarrow c)$, and $\nabla v\varphi \leftrightarrow \varphi(v/c)$ as the *archetypical axiom* $\varpi(\nabla v\varphi \leftrightarrow c)$. Also, the *typical* (or *archetypical*) *axiom schema on c for* a set Σ of sentences of $\mathsf{L}^{\nabla}(\rho)$ is the set $\varpi[\Sigma \rightarrow c] := \{\varpi(\nabla v\varphi \rightarrow c) : \nabla v\varphi \in \Sigma\}$ (respectively $\varpi[\Sigma \leftrightarrow c] := \{\varpi(\nabla v\varphi \leftrightarrow c) : \nabla v\varphi \in \Sigma\}$) of sentences of $\mathsf{L}^{\nabla}(\rho[c])$ consisting of the corresponding axioms on the new constant c for every generalized sentence $\nabla v\varphi$ in Σ. In particular, when Σ is the set of all the generalized sentences of $\mathsf{L}^{\nabla}(\rho)$, we omit it from the notation, using $\varpi[\rightarrow c]$ and $\varpi[\leftrightarrow c]$ for the corresponding *axiom schemata on c*.

These conditions extend conservatively theories in $\mathsf{L}^{\nabla}(\rho)$ to $\mathsf{L}^{\nabla}(\rho[c])$.

PROPOSITION [*Conservative addition of new 'generic' constant*]. Given a set Γ of sentences of $\mathsf{L}^{\nabla}(\rho)$, consider a set Σ of (generalized) sentences of $\mathsf{L}^{\nabla}(\rho)$.

(a) In ultrafilter logic $\mathcal{L}_{\omega\omega}(\rho)^u$: $\Gamma[\Sigma \leftrightarrow c] := \Gamma \cup \varpi[\Sigma \leftrightarrow c]$ is a conservative extension of Γ, such that, for every generalized sentence $\nabla v\varphi \in \Sigma$, $\Gamma \vdash^u \nabla v\varphi$, iff $\Gamma[\Sigma \leftrightarrow c] \vdash^u \varphi(v/c)$.

(b) In filter logic $\mathsf{L}_{\omega\omega}(\rho)^f$ for 'most': $\Gamma[\Sigma \to c] := \Gamma \cup \varpi[\Sigma \to c]$ is a conservative extension of Γ, where $\Gamma[\Sigma \to c] \vdash^f \varphi(v/c)$ whenever $\Gamma \vdash^f \nabla v\varphi$ with $\nabla v\varphi \in \Sigma$.

Proof outline. The lemma yields conservativeness, from which the other assertions follow.

This result establishes the correctness of reasoning with new archetypical or typical constant.[32]

Thus, even though typical objects may fail to exist in particular modulated structures, we may safely use (constants naming) them in theoretical reasoning. Also, reasoning with archetypical constants may quite convenient, as it is easier than manipulating the generalized quantifiers.[33]

The following examples will illustrate these and similar ideas in both filter and ultrafilter logics: $\mathcal{L}_{\omega\omega}(\rho)^f$ and $\mathcal{L}_{\omega\omega}(\rho)^u$.

The next example, similar to flying birds and Tweety, may serve to illustrate some features.

EXAMPLE [*White swans*]. Consider a signature γ with a unary predicate W and theory Γ, over $\mathsf{L}^\nabla(\gamma)$, with single axiom $\nabla v W(v)$ (for "Swans generally are white"). Considering a new constant s (for typical swan), typical extension $\Gamma[\to s]$ (in $\mathsf{L}^\nabla(\gamma[s])$ has the typical axiom $\varpi[\nabla v W(v) \to s]$, i.e., $\nabla v W(v) \to W(s)$. Hence, extension $\Gamma[\to s]$ entails the following sentence of $\mathsf{L}^\nabla(\gamma[s])$: $W(s)$ (i.e., "A typical swan is white"). Now, if b is (a constant naming) a non-white swan, $\Gamma[\to s] \cup \{\neg W(b)\}$ yields $\neg W(b)$ and $W(s)$. Thus $\neg b \equiv s$ (this non-white swan b is not a typical swan). So, by conservativeness, among the consequences of $\Gamma \cup \{\exists y \neg W(y)\}$, i.e., "Swans generally are white, but there is a non-white swan", we have $\exists v W(v) \wedge \exists u \neg W(u)$ and $\exists v \exists u \neg v \equiv u$.

This example also illustrates the monotonic nature of our logics: we do not have to retract conclusions in view of new facts. Given that "Most swans are white", we conclude that "A typical swan is white", which we may hold even if further evidence reveals non-white swans.

The next example illustrates using several typical constants: assuming that "Generally movie stars like authors", one can conclude "A typical movie star likes a typical author".

EXAMPLE [*Movie stars and authors*]. Consider a signature η having a binary predicate L (with $L(x, y)$ standing for x likes y), as well as unary predicates M and A (standing, respectively, for 'is a movie star' and 'is an author').

Assume that theory Λ has as its axiom the following sentence of $\mathsf{L}^\nabla(\eta)$:

$$\nabla x \nabla y[M(x) \wedge A(y) \to L(x, y) \text{ (for "Generally movie stars like authors")}.$$

With a new constant a (for typical movie author), typical extension $\Lambda[\rightarrow a]$ (in $\mathsf{L}^\nabla(\eta[a])$) has the typical axiom for $\nabla x \nabla y[M(x) \wedge A(y) \rightarrow L(x, y)]$. Thus, as a consequence of $\Lambda[\rightarrow]$, we have $\nabla x[M(x) \wedge A(a) \rightarrow L(x, a)]$ ("Movie stars generally like typical authors"). With another new constant m (for typical movie star)), we form extension $\Lambda[\rightarrow a][\rightarrow m]$ over $\eta[a][m] := \eta \cup \{a, m\}$, which has the typical axiom (in language $\mathsf{L}^\nabla(\eta[a][m])$) for sentence $\nabla x[M(x) \wedge A(a) \rightarrow L(x, a)]$ of $\mathsf{L}^\nabla(\eta[a])$. Hence, among the consequences of $\Lambda[\rightarrow a][\rightarrow m]$, we have $M(m) \wedge A(a) \rightarrow L(m, a)$ ("A typical movie star likes a typical author").

Notice that $\Lambda[\rightarrow a][\rightarrow m]$ does not commit us to the existence of (typical) movie stars or authors; all that it entails is that "typical movie stars, if any, like typical authors, if any". Assuming the existence of such typical people, $L(m, a)$ will follow from $\Lambda[\rightarrow a][\rightarrow m] \cup \{M(m), A(a)\}$.

The kind of reasoning in the preceding example, involving several generic constants, can be introduced by iterating our constructions or by means of special sets and predicates, which consist of tuples of generic elements and constants, respectively.

5.3. Inference of Generalized Assertions and Induction

The preceding development has shown that our logical systems with generalized quantifiers can be used to provide rigorous bases for generic reasoning, i.e., with notions such as 'typical' and 'archetypical'. We will now examine some aspects of inference of generalized assertions.

We recall the idea of 'typical' as a correct instance of a generalized assertion. So, failure at a typical element will be enough to refute the corresponding generalized assertion. For instance, if a typical bird fails to fly, then one can conclude that it is not the case that "Birds generally fly". Now, consider establishing a generalized assertion. For this, holding at an archetypical element would be enough. But, as seen before, such objects tend to be elusive. Indeed, it seems too much to expect from a single bird.

For establishing generalized assertions, it seems more reasonable to resort to the idea underlying opinion polls. For instance, to appraise whether "people generally like chocolate", one examines a sample. Also, if all metals in a "representative" sample turn out to be solid, we seem to be entitled to conclude that "metals generally are solid".

The idea of experiments based on samples can be used to introduce some concepts. Let us take a closer look at what is involved in such ideas. From the experimental evidence

(e) "Everyone in this sample likes dogs",

we wish to be able to infer (g) "People generally like dogs".

We would be able to derive such a conclusion if we knew

(r) "This is a 'representative' sample of people (with respect to liking dogs)".

The question then is "How does one know that a sample is representative?".

The case of program testing may be illustrative.[34] It suggests considering a sample representative when it satisfies the following two conditions:

(d) People generally are similar to those in the sample;

(t) Similarity generally transfers liking dogs.

For an example of similarity, consider information about animals. Assume that animals of the same species generally have the same feeding habits:

$$\nabla v \nabla u[S(u, v) \rightarrow (H(u) \rightarrow H(v))].$$

The relation S, then, provides a good transfer for the property H of being herbivorous.

Conditions (d) and (t) involve (generalized) quantifications, but this appears to be a reasonable approach towards inferring generalized assertion (g) from evidence (e). We will argue that indeed this is so, but there are very strong hidden assumptions. To see this, we will formulate these ideas more precisely.

The next example may serve to introduce some of the underlying ideas.

EXAMPLE [*Green emeralds*]. We wish to know whether "Emeralds generally are green", i.e., whether $\nabla v G(v)$ holds. Towards this goal, we resort to examining a sample e of emeralds.

a. Imagine that we found "every emerald in the sample to be green", i.e., $\forall u[e(u) \rightarrow G(u)]$, which we shall denote by ${}_eG$. Assume also that we have good reason to believe that "emeralds generally resemble those in the sample, with respect to being green", i.e., η: $\nabla v \exists u[e(u) \wedge (G(u) \rightarrow G(v))]$. We are then led to believe in our generalized hypothesis.
b. It is also reasonable to consider that we have examined at least one emerald, i.e., "our sample is nonempty": $\exists u e(u)$. Then, good reasons to believe $\nabla v G(v)$ are good reasons to believe η. Also, with the assumption $\exists u e(u)$, good reasons to believe that "emeralds generally are green" are good reasons to believe that "the sample has a prototypical emerald, in that emeralds generally resemble it", i.e., $\exists u[e(u) \wedge \nabla v(G(u) \rightarrow G(v))]$.

We shall have occasion to analyze these connections in the sequel. We will now formulate these ideas in our logics for 'generally' and examine some conditions for inferring a generalized assertion $\nabla\theta : \nabla v \theta(v)$ from experiments on samples.

We will be considering experimentation with a sample, described by predicate e. We shall use $(\exists w : e)\varphi(w)$ and $(\forall w : e)\varphi(w)$, respectively, for the existential and universal relativization of a formula $\varphi(w)$ to predicate e.[35]

The experimental result on sample e can be expressed by the sentence

$${}_e\theta : (\forall u : e)\theta(u) \qquad \text{(every object in sample } e \text{ has } \theta).$$

We will search for sufficient conditions for inferring the assertion $\nabla\theta : \nabla v\theta(v)$.

Consider the following sentence (expressing that e is generally dense for transferring θ):

$\eta : \nabla v(\exists u : e)[\theta(u) \rightarrow \theta(v)]$ (objects generally resemble those in e for θ).

It is not difficult to see that this sentence η provides a sufficient condition for inferring the assertion $\nabla v\theta(v)$ from the experimental result on sample e : $\{_e\theta\}$ entails $\eta \rightarrow \nabla\theta$.[36]

We will now examine necessary conditions for inferring the assertion $\nabla\theta$: $\nabla v\theta(v)$. Consider first the following formula (expressing a strong property of object u):

$\pi_u(u) : \theta(u) \rightarrow \nabla v\theta(v)$ (objects generally have θ if u does).

This property of object u, which we may call prototypicality with respect to θ, asserts that u reflects objects in general.[37] We then form the following prototypicality sentence:

$\pi : (\exists u : e)\pi_u(u)$ (e has prototypical object for θ).

It is not difficult to see that this prototypicality sentence π is stronger than the preceding resemblance sentence η: the conditional sentence $\pi \rightarrow \eta$ is valid.

Now, consider the following sentence (describing a reasonable property of the experiment):

$\exists e : \exists u e(u)$ (non-trivial e).[38]

In this case, it is again not difficult to see that the prototypicality sentence π is a necessary condition for having the assertion $\nabla\theta$: $\nabla v\theta(v)$ given a non-trivial sample e : $\{\exists e\}$ entails $\nabla\theta \rightarrow \pi$.

Hence, in the presence of experimental evidence $_e\theta$ on non-trivial sample e, the following sentences are equivalent: generalized assertion $\nabla v\theta(v)$, prototypicality π and resemblance η.

We can now interpret these equivalences in the context of experiments and the evidence they may provide.[39] We wish to appraise the generalized assertion $\nabla\underline{v}\theta(\underline{v})$ (abbreviated by $\underline{\nabla}\theta$), where $\theta(\underline{v})$ is a formula with set $\underline{v}$ of, say, n free variables.

We resort to experimentation with a non-trivial sample, described by n-ary predicate $\underline{e}$. Consider our present knowledge expressed by theory Δ over language L. The experiments may also involve auxiliary hypotheses, and the experimental set-up – characterized by the sample predicate $\underline{e}$ and the auxiliary hypotheses – may be defined in the language L, but, more generally, it may be characterized in a conservative extension Δ^* of our present knowledge Δ,[40] where we assume the previous conditions to hold: Δ^* entails both $_e\theta$: $(\forall\underline{u} : \underline{e})\theta(\underline{u})$ and

$\exists \underline{e} : \exists \underline{u} \underline{e}(\underline{u})$. In this situation, we can infer $\underline{\nabla}\theta$ from Δ, or consistently hold it, iff we can do so in Δ^* for the corresponding assertions on sample $\underline{e}$, namely resemblance sentence η: $\nabla \underline{v}(\exists \underline{u} : \underline{e})[\theta(\underline{u}) \to \theta(\underline{v})]$ and prototypicality sentence π: $(\exists \underline{u} : \underline{e})[\theta(\underline{u}) \to \nabla \underline{v}\theta(\underline{v})]$.[41]

The next example, continuing the previous one (Green emeralds), illustrates these ideas.

EXAMPLE [*Blue or green emeralds*]. We wish to know whether "Emeralds generally are green" or "Emeralds generally are blue". Assume that, according to our present knowledge Δ, "emeralds are not both green and blue" (or, even "no emerald is both green and blue").

Consider (perhaps *Gedanken*) experiments consisting of examining nonempty samples e' and e'' of emeralds. Imagine that every emerald in the sample e' is found to be green, whereas every emerald in the sample e'' is found to be blue. What, if anything, are we entitled to conclude?

a. First, imagine that we have good reason to believe that emeralds generally resemble those in the sample e', with respect to being green, i.e., in resemblance, for G, to e'. We are then entitled to decide in favor of $\nabla v G(v)$ ("Emeralds generally are green").
b. Now, imagine that we have good reason to believe that emeralds generally resemble those in the sample e'', with respect to being blue, i.e., in resemblance, for B, to e''. We are then entitled to decide in favor of $\nabla v B(v)$ ("Emeralds generally are blue").
c. Finally, if neither resemblance is believed, then our assertions will be left undecided.

Now, if formula $\theta(\underline{v})$ may fail to be amenable to direct experimentation, we can resort to auxiliary (observable) formulas. So, we test an auxiliary formula $\psi(\underline{u})$ (with set $\underline{u}$ of, say, m free variables) and use an m-ary predicate $\underline{e}$ for the sample submitted to experimentation.[42] The experimental evidence is now expressed by the sentence ${}_{\underline{e}}\psi : (\forall \underline{u} : \underline{e})\psi(\underline{u})$.

Resemblance now becomes transfer of formulas, as follows

$${}_{\psi}\eta_{\theta} : \nabla \underline{v}(\exists \underline{u} : e)[\psi(\underline{u}) \to \theta(\underline{v})] \qquad \text{(general resemblance of } \theta \text{ to } \psi \text{ on } \underline{e}\text{).}$$

Similarly, prototypicality now involves transfer of formulas, as follows

$${}_{\psi}\pi_{\theta} : (\exists \underline{u} : \underline{e})[\psi(\underline{u}) \to \nabla \underline{v}\theta(\underline{v})] \qquad \text{(prototypicality of } \underline{e} \text{ for transfer of } \psi \text{ to } \theta\text{).}$$

Much as before, in the presence of experimental evidence ${}_{\underline{e}}\psi$ on non-trivial sample $\underline{e}$, the following sentences are equivalent: generalized assertion $\nabla \underline{v}\theta(\underline{v})$, prototypicality ${}_{\psi}\pi_{\theta}$ and resemblance transfer ${}_{\psi}\eta_{\theta}$. As a result, for an experiment in a conservative extension Δ^* of our present knowledge Δ, such that Δ^* entails both ${}_{\underline{e}}\psi$ and $\exists \underline{e}$, we can infer $\underline{\nabla}\theta$ from Δ, or consistently hold it, iff we can do so in Δ^* for the corresponding assertions on sample $\underline{e}$, namely resemblance transfer ${}_{\psi}\eta_{\theta}$ and prototypicality sentence ${}_{\psi}\pi_{\theta}$.

The next example, continuing the previous one (Blue or green emeralds), involves the concept of 'grue' (Goodman 1955).

EXAMPLE [*Grue emeralds*]. We wish to know whether "Emeralds generally are grue", where 'grue' means "green up to a certain time t, and blue thereafter".

Assume our present knowledge Δ to be as before and experiments on nonempty samples e' and e'' yielding the previous results: ${}_{e'}G : \forall u[e'(u) \rightarrow G(u)]$ and ${}_{e''}G : \forall u[e''(u) \rightarrow B(u)]$.

a. According to our analysis, the experimental results on nonempty samples e' and e'' provide evidence in favor of "Emeralds generally are grue" iff they do so for the general resemblance: "Emeralds up to time t generally resemble those in the sample e', with respect to being green, and thereafter they will generally resemble those in the sample e'', with respect to being blue".
b. Imagine that we have good reason to believe that emeralds up to time t generally resemble those in the sample e', with respect to being green, and thereafter they will generally resemble those in the sample e'', with respect to being blue. We are then entitled to decide in favor of "Emeralds generally are grue".

6. Relative Notions

We shall now examine the idea of having a notion of 'important' (corresponding to 'generally') relative to a universe: how it arises and is formulated, as well as some related issues.

We will first indicate how the proper expression of 'relative generally' assertions suggests the idea of a notion of 'important' with respect to each universe, leading to its natural formulation in sorted versions of our logics for 'generally'. Then, the need for establishing some connections while blocking others leads to comparing such relative concepts. Finally, these ideas are incorporated into a sorted framework for 'relative generally'.

6.1. The Need for Relative Notions

Our generalized quantifiers may exhibit somewhat unexpected behavior in some cases. We shall now examine these undesirable side-effects and propose a way to overcome this difficulty. The generalized quantifier ∇ is meant to capture the idea of holding generally, i.e., for an important set of objects of the universe. Sometimes we wish to express the idea of holding generally over a given subset of the universe, i.e., for several or most objects of a given sub-universe. We now examine the expression of such relative generally assertions.

Over a universe B of birds, we express "Birds generally fly" by $\nabla v F(v)$. How are we to express relative generalized assertions, like "Several eagles have wings" or "Most penguins have beaks"? By analogy with the classical quantifiers, relativization is an apparently natural suggestion, i.e., expressing "M's generally are N's' by $\nabla v[M(v) \to N(v)]$. Unfortunately, relativization fails to provide an adequate way of expressing 'relative generalized' assertions, due to the behavior of the generalized quantifiers ∇.

EXAMPLE [*Penguins and winged birds*]. Consider expressing facts about birds by relativization:

"All penguins are winged birds" by $\forall v[P(v) \to W(v)]$,

and

"Winged birds generally fly" by $\forall v[W(v) \to F(v)]$.

From these two sentences, one concludes $\nabla v[P(v) \to F(v)]$, which would be read as "Penguins generally fly". Now, the two given premises appear to express reasonable facts. On the other hand, the conclusion, *as read*, does not look so reasonable.[43]

This example indicates that relativization fails to express the intended idea. The reason comes from neglecting the relative aspect.

For a generalized formula $\nabla v[M(v) \to N(v)]$ the reading "most M's are N's" is not appropriate. For, one must bear in mind that what this does assert is "for most *birds b*, if $M(b)$ then $N(b)$". Indeed, given the (classical) meaning of the conditional, sentence $\nabla v[P(v) \to \neg F(v)]$ means that the set $P \cap F$ of flying penguins is a small set of *birds* (rather than of *penguins*). Thus, the change in context (Peterson 1979, 166; Barwise and Cooper 1981, 217) is not reflected.[44]

A natural approach to overcome this problem, and expressing relative generalized assertions, rests on relative notions of 'important': each universe has its own relative notion of 'important' subsets. This idea may be formulated by providing a proper family $\mathcal{C}_V$ (of important subsets) over each given universe V. With such relative notions of 'important', we can express "M's generally are N's" as the $\{m \in M : N(m)\}$ is an important set of M's more precisely by $M \cap N \in \mathcal{C}_M$. Thus, we can also distinguish, say, "Eagles generally fly" from "Penguins generally fly", since the former becomes $E \cap F \in \mathcal{C}_E$, whereas the latter becomes $P \cap F \in \mathcal{C}_P$. Since the universe is important (cf. Section 2.4), we shall sometimes write $M \cap N \in \mathcal{C}_M$ as $M \cap N \approx M$ to suggest the reading "$M \cap N$ 'as important as' M".

6.2. SORTED LOGICS FOR 'GENERALLY'

A many-sorted approach can provide a framework for formulating the idea of distinct notions of 'important' for the universes, where one assigns proper families corresponding to these relative notions of important. We shall now examine sorted versions of our logics for 'generally'. The basic idea is relativizing to sorts the previous (unsorted) concepts.

We consider many-sorted signatures, where the extra-logical symbols, as well as variables, come classified according to sorts (Enderton 1972). Quantifiers are relativized to sorts, as expressed in the formation rules: for each variable v over sort s, if φ is a formula in $\mathsf{L}^{\nabla}(\rho)$, then so are $(\forall v : s)\varphi$, $(\exists v : s)\varphi$ and $(\forall v : s)\varphi$.

A *modulated structure* $\mathtt{A}^{\mathcal{K}}$ for S-sorted signature ρ is an expansion of an S-sorted (first-order) structure $\mathtt{A}$ for ρ, obtained by assigning to each sort s of signature ρ a complex $\mathcal{C}_s$ over the universe $\mathtt{A}[s]$ of sort s (giving the important subsets of $\mathtt{A}[s]$).

The extension of *satisfaction* becomes relativized to sorts accordingly: $\mathtt{A}^{\mathcal{K}} \models (\nabla v : s)\varphi[s]$ iff the set $\{a \in \mathtt{A}[s] : \mathtt{A}^{\mathcal{K}} \models \varphi[\mathtt{S}(v \mapsto a)]\}$ belongs to the family $\mathcal{C}_s \subseteq \wp(\mathtt{A}[s])$.

The *axiom schemata* in the sets $B^k(\rho)$ become sorted as well:

$[\nabla\alpha]_s(\nabla v : s)\varphi \to (\nabla u : s)\varphi(v/u)$, for a new u: $[\forall\nabla]_s : (\forall v : s)\varphi \to (\nabla v : s)\varphi$;

$[\nabla\exists]_S(\nabla v : s)\varphi \to (\exists v : s)\varphi$;

$[\to\nabla]_S : (\forall v : s)(\psi \to \theta) \to [(\nabla v : s)\psi \to (\nabla v : s)\theta]$;

$[\nabla\wedge]_S : [(\nabla v : s)\psi \wedge (\nabla v : s)\theta] \to (\nabla v : s)(\psi \wedge \theta)$;

Much as in classical first-order logic, the sorted and unsorted versions are very similar. So, soundness and completeness carry over to the sorted version, by relativizing to sorts the previous arguments.[45]

6.3. Comparing Relative Notions

We now take a closer look at the proposal of employing distinct notions of sizable subsets. We shall examine how the need for establishing some connections while blocking others leads to comparing relative notions of sizable sets. We will use variations of the previous examples to introduce the ideas and some of its features, emphasizing the filter logic $\mathcal{L}_{\omega\omega}(\rho)^f$ for 'most'.

The next example shows how some (undesired) conclusions can be blocked.

EXAMPLE [*Birds and penguins with unconnected important sets*]. Given that "All penguins are birds", i.e., $P \subseteq B$, consider the following assertions:

"Most birds fly" (the flying birds form a large set of birds, i.e., $B \cap F \in \mathcal{C}_B$),

and

"Most penguins fly" (the flying penguins form a large set of penguins, i.e., $P \cap F \in \mathcal{C}_P$).

Now, neither the former entails the latter (since we may even have $P \cap F = \varnothing$), nor conversely (since $P \subseteq B$ may very well be a small set of *birds*), if the relative notions of large sets are not connected.[46]

This example illustrates the idea of independent notions of large subsets. If the set of penguins is a not a large set of *birds*, then a set $X \subseteq P$ (e.g., of non-flying penguins) may be a large set of *penguins* without being a large set of *birds*.[47]

The next example shows how some (desired) conclusions can be achieved.

EXAMPLE [*Birds and winged birds with connected important sets*]. Given that "All birds have wings", i.e., $W \subseteq B$, consider the following assertions:

"Most birds fly" (as before $B \cap F \in \mathcal{C}_B$ or $B \cap F \approx B$),

and

"Most winged birds fly" (the flying winged birds form a large set of winged birds, i.e., $W \cap F \in \mathcal{C}_W$ or $W \cap F \approx W$).

Given also that "Most birds have wings" ($W \approx B$), the set $B - W$ of exceptional wingless birds is a negligible set (of *birds*). So, it is intuitively plausible that the large subsets of W are the relativizations $W \cap Y$ of the large subsets Y of B.[48] Thus, we shall also assume the following coherence principle: for any set $Y \subseteq B$, $W \cap Y \approx W$ iff $Y \approx B$. In the presence of this principle, assertions $B \cap F \approx B$ and $W \cap F \approx W$ become equivalent.

These two examples illustrate the following ideas. Given $S \subseteq T$ and a proper filter $\mathcal{C}_s$ over S, consider the *relativizable complex* ${}^T\mathcal{C}_S := \{Y \subseteq T : S \cap Y \in \mathcal{C}_S\}$. If $(T - S) \in \mathcal{C}_T$ or $S \notin \mathcal{C}_T$, then we have an independent notion of important subsets of T;[49] if $(T - S) \notin \mathcal{C}_T$ and $S \in \mathcal{C}_T$, then we may take ${}^T\mathcal{C}_S$ as $\mathcal{C}_T$, if we wish to enforce coherence of important subsets.[50]

6.4. SORTED FRAMEWORK FOR RELATIVE NOTIONS

We shall now consider comparison of universes, with distinct notions of important subsets, in a sorted framework. We shall examine how to formulate some ideas related to sub-universes and coherent transfer in this approach.

In our sorted framework, sorts are unrelated: we have equality only over a sort, rather than between distinct sorts. Nevertheless, we can express some relationships among sorts by means of appropriate injections. The idea is that an injection i from s to t establishes a bijection from its domain s onto its image $i[s]$, the latter being a subset of t.

To express that s is a subsort of t, we resort to a unary function i from s to t together with an axiom asserting its injectivity (Meré and Veloso 1995). This gives transitivity of subsorts.

Now, consider an injection $i : s \to t$, where the image $i[s]$ is an important subset of t (i.e., $i[s] \in \mathcal{C}_t$). Then, the non-image $t - i[s]$ is a negligible subset of t, where the distinction between a set $Z \subseteq t$ and its pre-image $i^{-1}[Z]$ is confined. So, we may consider $Z \subseteq t$ an important subset of t iff its pre-image $i^{-1}[Z]$

is an important subset of s. These considerations lead to the following coherent connection principle for an injection $i : s \to t$, where the image $i[s]$ is an important subset of t (i.e., $i[s] \in \mathcal{C}_t$): for any set $Z \subseteq t$, $Z \in \mathcal{C}_t$ iff $i^{-1}[Z] \in \mathcal{C}_S$.[51]

We now formulate this coherent connection principle in our logics for 'generally'. Given $i : s \to t$ and formula $\varphi(z)$ with variable z over t, we can express: "objects of t generally are in the image" (i.e., $i[s] \in \mathcal{C}_t$), "objects of t generally have property $\varphi(z)$", and "objects of s generally give objects in t with property $\varphi(z)$".[52]

This leads to the *coherent connection schema* $[\nabla i : s \subseteq t]$, with the instances $(\nabla i : s \subseteq t/\varphi)$: $(\nabla z : t)(\exists x : s)z \equiv i(x) \to [(\nabla z : t)\varphi \leftrightarrow (\nabla x : s)\varphi(z/i(x))]$.[53]

Let us now examine our preceding examples in this sorted formulation.

EXAMPLE [*Sorted birds, winged birds and penguins*]. Consider three sorts: b (for birds), w (for winged birds) and p (for penguins) and a unary predicate F (for flies) over sort b.

a. First, consider the formulation, with $j : w \to b$, of "All winged birds are birds" and "Birds generally have wings" as $(\forall x', x'' : w)[j(x') \equiv j(x'') \to x' \equiv x'']$ and $(\nabla z : b)(\exists x : w)z \equiv j(x)$, respectively. Then, from the instance $(\nabla j : w \subseteq b/F(z))$ of the transfer schema, we conclude the equivalence between $(\nabla z : b)F(z)$ ("birds generally fly") and $(\nabla x : w)F(j(x))$ ("winged birds generally fly"). Thus, since the winged birds form an important set of birds, "generally flying" is inherited both downwards and upwards.

b. Now, consider the theory Γ with the two axioms: $(\nabla z : b)F(z)$ ("birds generally fly") and $(\nabla y', y'' : p)[k(y') \equiv k(y'') \to y' \equiv y'']$ ("all penguins are birds"), where $k : p \to b$. We will also assume the coherence transfer schema and examine the relative importances of the sets of birds that are penguins and of flying penguins, in the context of the extended theory $\Gamma^* := \Gamma \cup [\nabla k : p \subseteq b]$.

   On the one hand, we can see that the coherence transfer instance $(\nabla k : p \subseteq b/F(z))$ leads from $k[p] \in \mathcal{C}_b : (\nabla z : b)(\exists y : p)\, z \equiv k(y)$ ("birds generally are penguins") to $k^{-1}[F] \in \mathcal{C}_p : (\nabla y : p)F(k(y))$ ("penguins generally fly").

   Thus, if we had assumed that the penguins form an important set of birds then we would have penguins generally flying; but otherwise this conclusion is not forced upon us.[54]

   Also, the coherence transfer instance $(\nabla k : p \subseteq b/\neg F(z))$ can be seen to lead from $P - k^{-1}[F] \in \mathcal{C}_p : (\nabla x : p)\neg F(k(y))$ ("penguins generally do not fly") to $B - k[p] \in \mathcal{C}_b : (\nabla z : b)\neg(\exists y : p)z \equiv k(y)$ ("birds generally are not penguins").

   Thus, if we know that penguins generally do not fly then we can conclude that birds generally are not penguins.[55]

This example illustrates how the coherence axiom schema for an injection provides uniform control based on the relative importance of the sorts.

In the sorted framework for 'generally', we consider sorted theories consisting of the following sets of axioms: a set Σ expressing (basically syntactical) subsort information, a set Δ expressing coherent transfers between some subsorts,[56] and a set Γ expressing the remaining available knowledge. Since the union $\Gamma \cup \Sigma \cup \Delta$ specifies a many-sorted theory within a logic for 'generally', we have soundness and completeness, as before.

7. Concluding Remarks

In this section we will reassess our concepts and results and consider some prospects for our logics for 'generally', such as possible developments and applications. Our development has been motivated by the goal of rigorous qualitative reasoning about (some) vague notions.

We have examined some logical systems intended to cope with some vague notions. Assertions and arguments involving vague notions occur often, in ordinary language and in some branches of science. Our primary motivation is providing a framework where one can express these notions and reason precisely about them. For, this purpose, we have extended classical first-order logic, with generalized quantifiers. As these generalized quantifiers are intended to capture variants of 'generally', such as 'most' and 'many', they are interpreted as ranging over appropriate families of subsets of the universe of discourse.

These (monotonic) generalized logics, with simple sound and complete deductive calculi, are proper conservative extensions of classical first-order logic, with which they share various properties. Technically, these logics for qualitative reasoning are part of a larger family of the so-called modulated logics (Carnielli and Grácio 2000), which are obtained by restricting, or modulating, the logical behavior of the generalized quantifiers by means of mathematical structures like filters, ultrafilters, topological spaces, and so on.

Vagueness may also be, less obviously, conveyed by objects termed 'generic' (such as 'typical' or 'archetypical'). For such kinds of vagueness, special individuals have been introduced by means of 'generally', and internalized as generic constants, thereby producing conservative extensions where one can reason about generic objects as intended, as shown in Section 5.

Some interesting situations, however, require assertions relative to distinct universes, involving "most birds", "many penguins", and "typical eagle", for instance. We have shown how our concepts can be adapted to such cases by means of notions of 'generally' relative to the universes. As natural frameworks for this approach, we have introduced sorted versions of our logics, with families over the universes, which share various properties, such as supporting generic reasoning, with the original versions.

Thus, our logics for 'generally' can be said to provide a rigorous framework for qualitative reasoning about vague notions, such as 'many' and 'most' (and their duals 'few' and 'very few'), as well as about objects termed 'generic' (such as 'typical' or 'archetypical').

The expressive power of our generalized quantifiers paves the way for other possible applications where it may be helpful. For instance, some fuzzy concepts (Zadeh 1975; Turner 1984) may be explained in this manner.[57] Another possible application could be reporting empirical experiments and tests. This arises from the observation that, whereas laws of pure mathematics may be of the form "*M*'s always are *N*'s", one can argue that empirical laws of (natural) sciences could, and perhaps should, be regarded assertions of the – more cautious – form "*M*'s generally are *N*'s".[58] Also, these ideas suggest that similar weakening of some mathematical concepts may be of interest, e.g., "almost" dense, cover and partition.[59]

Some interesting questions can be posed concerning our logics (and, in more general terms, modulated logics).

(1) Which portions of the proof theory and model theory for classical first-order logic can be extended (or adapted) to such logics for qualitative reasoning?[60]
(2) Are there interesting propositional versions of such logics for qualitative reasoning?[61]
(3) Is it possible (and fruitful) to consider quantitative notions in such qualitative realms?

From the philosophical standpoint, there are interesting questions related to the consequences of our treatment of argumentation. We are trying to give mathematical formulations to forms of qualitative reasoning with (some) vague notions. There have been some critiques against the such ambition to cast logic into mathematical form. For instance, Toulmin (1958) criticizes it, recalling that Aristotle, in his Prior Analytics, says that logic is concerned, not only with apodeixis (how conclusions are to be drawn), but also with episteme, i.e., the science devoted to such establishments.

Toulmin (1958, chapter III), in the hope to unite logic and epistemology, accuses logic (and logicians) as responsible for neglecting some distinctions concerning arguments. In particular, he has in mind distinctions between conclusive (necessary) and tentative (provisional or probable) arguments, between arguments that may be accepted as valid but are not formalizable, as well as the fundamental division between analytic and substantial. He considers what he calls 'quasi-syllogisms' (for instance, going from "Petersen is a Swede" and "Scarcely any Swedes are Roman Catholics" to "Almost certainly, Petersen is not a Roman Catholic"), and concludes that this kind of argument is not conclusive (though it is a legitimate and important form of argument).[62] He considers such arguments to be "unruly" (Toulmin 1958, p. 149) to logical methods. Now, our logics for qualitative reasoning do provide ways of formalizing some versions of such arguments.[63] In this sense, they may be relevant to a reassessment of positions related to Toulmin's views.

Important issues concern establishing assertions on the basis of (confirmatory) evidences. There are some well-known problems about inductive arguments and objections against the possibility of legitimate logics of induction. David Hume[64] presents arguments, generally considered to be solid, against the justification of natural or scientific induction: Hume argues that the only possible way to justify the inductive jump would be to appeal to some principle (of regularity of resemblance) ensuring some kind of uniformity in nature.[65] We have seen (in Section 5.3) that this problem is not alleviated by weakening quantifiers from universal to generalized ones: the inference of generalized assertions requires a similar principle about regularity of resemblance assuring a (somewhat weaker) kind of uniformity.

Another aspect of our development refers to confirmatory evidences and the paradox of confirmation, as in the seminal work of Hempel (1945). It seems natural to accept the so-called "equivalence condition" to the effect that (logically) equivalent assertions have the same confirmatory evidences, but this leads to somewhat counter-intuitive consequences (Hempel 1945).[66] We have faced a similar situation in 6.1 when trying to express relative generally assertions and a change of context seemed to be involved. One way we have proposed to overcome this problem is to consider a sorted approach (with notions of 'generally' relative to sorts), as developed in Section 6. Such sorted approaches may provide ways towards diminishing the impact of the some paradoxes (such as the confirmation paradox)[67] or even dissolving them altogether (as seems to be the case with the sorites paradox).

We may now sum up our conclusions as follows.

Our goal is to provide rigorous bases for qualitative reasoning involving (some) vague notions, with the motivation that such kinds of ideas and arguments occur often also in sciences. In the course of this development we have touched upon some questions related to Philosophy of Science. We think that the approaches we have taken seem close to the actual practice in (natural) sciences, as some of the preceding observations indicate. Apparently, approaches and solutions of the kind we propose were not seriously taken into consideration and developed before. We venture to conjecture that this is probably because it was not clear to philosophers of science that such states of affairs were not necessarily "unruly" and at odds with formal logic.

Our results bring some progress towards extending logic into the realm of vagueness. In terms of Aristotle's view of logic (in his Prior Analytics, mentioned above): apodeixis seems to have been given a good start on a secure ground, and in episteme, our approach (mainly as formalized in Section 5.3) may lead to a better understanding of some basic philosophical issues, even if not achieving complete explanations and solutions for them. We would be glad if our proposals and results in the present paper would modestly contribute to further the development of such a science.

Notes

[1] For instance, consider an experiment, say Millikan's experiment for determining the charge of the electron. In such a careful set-up, one would try to keep under control some important factors, such as electric and magnetic fields, humidity and temperature, etc. But what about others, such as the position of Pluto or the humor of the laboratory's cleaning lady? One would probably not be prepared to swear that these are under control. Yet, one would say that "most intervening factors are being kept under control (or, at least, taken into account)".

[2] Concerning technical aspects, our logics are monotonic and conservative extensions of classical logic, in sharp contrast to non-monotonic approaches. As for intended interpretations, one can phrase the difference in terms of positive and negative views (Sette et al. 1999). Our approach caters to a positive view, in the sense that we favor representing 'generally' explicitly, rather than interpreting it as "in the absence of information to the contrary".

[3] Brazil counts nowadays about 170 million inhabitants.

[4] Indeed, any infinite universe V can be partitioned as the union of two sets X and Y with the set cardinality as V; so V, X and Y cannot have all the same probability, even though they have the same size. For a finite universe, it suffices to consider a non-uniform distribution.

[5] An explanation for not accepting the fourth assertion is as follows. The "Brazilians that have their beards shaved" are generally males, whereas the "Brazilians that shave their legs" are generally females. So, the "Brazilians that have their beards shaved *and* shave their legs" form a rather small fraction of the population. (Acceptance of the third assertion should be clear.)

[6] An explanation for accepting the fifth assertion is as follows. Since "Most American males like beer" and "Most American males like sports", the sets of exceptions – namely the sets of non beer-lovers and of non sports-lovers – have very few elements. Thus, it seems reasonable to say that "Very few American males fail to like beer *or* sports", so "Most American males like beer *and* sports". (Acceptance of the fourth assertion should be clear.)

[7] It is the assertion "Real numbers generally are irrational" – in the sense "Most reals are irrational" – that appears to be more reasonable, as explained above (in Section 2.2).

[8] In a given topology, a *dense* subset of the universe is one having the universe as its closure (or equivalently, one intersecting every non-empty open set).

[9] Further examples illustrating this point are as follows. First, consider two sets with the same size: one consisting of a horse and an ox, and another one consisting of a horse and a dog. These sets may be just as important to a conservationist. But, the former may be more important to a farmer, whereas the latter might be preferred by an English gentleman, keen on fox hunting. Now, consider two sets with distinct sizes: one consisting of thirty birds, and another one consisting of a couple of elephants. The Zoo director is likely to consider them equally important. But, an ornithologist might rank the former as more important, whereas a truck driver in charge of transporting them would probably give more attention to the latter. So, a smaller set may be more important than a larger set, or just as important.

[10] These properties can also be explained in terms of a more basic notion of 'having about the same importance' (Veloso 1999, 2002).

[11] These formulas can also be given fuzzy readings: "x is very short" and "x is very tall".

[12] In general, we can define $\mathcal{C}$-*consequence* for a given class $\mathcal{C}$ of complexes: $\Gamma \models^{\mathcal{C}} \varphi$ iff $\mathtt{A}^{\mathcal{K}} \models \varphi$, for every modulated structure $\mathtt{A}^{\mathcal{K}} = (\mathtt{A}, \mathcal{K})$ with $\mathcal{K}$ in $\mathcal{C}$ such that $\mathtt{A}^{\mathcal{K}} \models \Gamma$. By specializing this notion to classes of complexes, we obtain semantic consequences for our logics for 'generally': *upward closed consequence* $\models^{\mathcal{S}}$, *filter consequence* $\models^{\mathcal{F}}$ and *ultrafilter consequence* $\models^{\mathcal{U}}$. This leads to a chain of extensions of classical first-order logic $\mathcal{L}_{\omega\omega}(\rho)$, namely $\mathcal{L}_{\omega\omega}(\rho) \subseteq \mathcal{L}_{\omega\omega}(\rho)^s \subseteq \mathcal{L}_{\omega\omega}(\rho)^f \subseteq \mathcal{L}_{\omega\omega}(\rho)^u$.

[13] Also, the analogue of universal instantiation also does not hold for the new quantifiers: in general, $\varphi(t)$ does not follow from $\nabla v\varphi(v)$.

[14] This hexagon of oppositions has some interesting interpretations in terms of corroboration and refutation: generalized sentences are harder to corroborate and to refute than universal and existential ones. Thus, generalized sentences fail to present the asymmetry between corroboration and refutation, of importance to some views of Popper (1934, 1975).

[15] Alternatively, one can replace the schemata $[\forall\nabla]$ and $[\nabla\exists]$ by the axioms $\nabla vv\exists v$ and $\neg\nabla v\neg v\exists v$, much as in the case of topological logic (Sgro 1972).

[16] The properties of conservative extensions by the addition of witnesses and Lindenbaum extensions for our deductive systems can be established as in classical first-order logic, by relying on the connection $(\vdash^k)$ in 3.3.

[17] In the case of the ultrafilter logic $\mathcal{L}_{\omega\omega}(\rho)^u$, this family ${}_\nabla\Sigma \subseteq \wp(H)$, having the finite intersection property, can be extended to a (proper) ultrafilter $\mathcal{H}$ over H. In the other cases, we can form the closure ${}_\nabla\Sigma^{\supseteq} := \{X \subseteq H : \varphi^\Sigma \subseteq X$, for some $\varphi^\Sigma \in {}_\nabla\Sigma\}$ of ${}_\nabla\Sigma \subseteq \wp(H)$ under supersets. (This extended family ${}_\nabla\Sigma^{\supseteq} \subseteq \wp(H)$ will be upward closed in the case of the logic $\mathcal{L}_{\omega\omega}(\rho)^s$, for 'several', and a filter in the case of the logic $\mathcal{L}_{\omega\omega}(\rho)^f$, for 'most'.) In these cases, we use $\mathcal{H} := {}_\nabla\Sigma^{\supseteq}$ as the complex.

[18] The property $\varphi^\Sigma \in {}_\nabla\Sigma$ iff $\varphi^\Sigma \in \mathcal{H}$ follows from schema $[\to\nabla]$ in the cases of the logics $\mathcal{L}_{\omega\omega}(\rho)^s$ (for 'several') and $\mathcal{L}_{\omega\omega}(\rho)^f$ (for 'most'), and from $[\neg\nabla]$ in the case of the ultrafilter logic $\mathcal{L}_{\omega\omega}(\rho)^u$. (The inductive steps for the propositional connectives as well as for the classical quantifiers $\forall$ and $\exists$ are as in Henkin's proof.)

[19] The apparent conflict with Lindström's results (Lindström 1966; Barwise and Feferman 1985) is explained because we are using a non-standard notion of model (with complexes). This feature may confer to our logics for 'generally' some extra model-theoretic interest.

[20] The strictness of the chain $\vdash \subset \vdash^s \subset \vdash^f \subset \vdash^u$ of inclusions for derivability can be seen by considering the schemata in their axiomatizations (cf. 3.3 in Section 3).

[21] This extra freedom leads to the general question: can one replace the generalized quantifier ∇ by additional extra-logical symbols without losing expressive power? It will become clear that allowing expanded signatures will have little impact on the question. So, if desired for the sake of simplicity, these expansions may be glossed over in a first reading.

[22] More precisely, given a formula φ of $\mathsf{L}^\nabla(\rho)$ and a set Δ of sentences over signature ρ' expanding signature $\rho(\rho \subseteq \rho')$, a ∇-*elimination* of φ will be a formula θ of $\mathsf{L}(\rho'')$ (with $\rho'' \supseteq \rho'$ having the same free variables as φ) such that $\Delta \models^{\mathcal{U}} (\varphi \leftrightarrow \theta)$.

[23] First, the formula $\nabla v\, v \equiv u$ is not 2-eliminable because $\Delta \not\models (\exists v\, v \equiv u \to \forall v\, v \equiv u)$, for any purely first-order theory Δ having models with more than one element. (Similarly, the sentences $\nabla v\, v \equiv c$ and $\neg\nabla v\, v \equiv c$, where c is a constant symbol, are not 2-eliminable.) Now, the sentence $\exists u\, \nabla v\, v \equiv u$ expresses that the ultrafilter is principal. It is not $\aleph_0$-eliminable because there exist principal and non-principal ultrafilters over each infinite universe.

[24] In each case, such a purely first-order sentence σ would provide an elimination of ∇ from $\exists u\nabla v\, v \equiv u$ within the first-order theory Δ. This example can also be used to illustrate the same features in the logics $\mathcal{L}_{\omega\omega}(\rho)^s$, of 'many', and $\mathcal{L}_{\omega\omega}(\rho)^f$, of 'most'.

[25] The term 'generic' has been used in other contexts, such as (Fine 1985), for similar, but not quite the same, ideas. Here, we employ 'generic' as a catch-all name for notions such as 'typical' and 'archetypical' (as introduced in Section 5.1), as well as 'prototypical'.

[26] In conceiving such a 'typical' (or 'generic') bird, one would be likely to have trouble in assigning certain attributes, such as definite size, weight or color to it, probably because they are less characteristic of the class of birds. This indicates that one might be picturing something like the Platonic form of bird.

27 Similarly, one can introduce a *prototypical* element a of $\mathtt{A}^{\mathcal{K}}$ for $\nabla v\varphi$ as one such that $\mathtt{A}^{\mathcal{K}} \models [a]$ or $\mathtt{A}^{\mathcal{K}} \models \nabla v\varphi$. So, we can regard a typical or prototypical element as providing a local test for a generalized sentence, which turns out to be decisive in case it is archetypical.

28 The set of typical elements in $\mathtt{A}^{\mathcal{K}}$ for a set Ψ of (generalized) sentences is the intersection of the extensions $\mathtt{A}^{\mathcal{K}}[\varphi]$, for those $\nabla v\varphi \in \Psi$ such that $\mathtt{A}^{\mathcal{K}} \models \nabla v\varphi$ (i.e., $\mathtt{A}^{\mathcal{K}}[\varphi] \in \mathcal{K}$).

29 In an ultrafilter structure, archetypical elements are indiscernible among themselves, in that they cannot be separated by formulas. Also, the new operator ∇ reduces to relativized classical quantification, in the case of a definable nonempty set of archetypical elements.

30 Here, one might say that the standard naturals are 'very low', the non-standard ones are 'very high' and numbers in the same copy are 'about as high'.

31 In an ultrafilter structure, an element is archetypical for $\nabla v\varphi$ iff it is typical for both $\nabla v\varphi$ and $\nabla v\neg\varphi$.

32 Thus, a theory Γ over $\mathsf{L}^{\nabla}(\rho)$ has conservative extensions $\Gamma[\leftrightarrow c] := \Gamma \cup \varpi[\leftrightarrow c]$ and $\Gamma[\leftrightarrow c] := \Gamma \cup \varpi[\rightarrow c]$ of Γ by a new archetypical or typical constant c, where $\varphi(v/c)$ is equivalent (respectively, follows from) $\nabla v\varphi$ for every generalized sentence $\nabla v\varphi$ of $\mathsf{L}^{\nabla}(\rho)$.

33 The reason for this may be clarified by considering our axiom schemata (cf. 3.3 in Section 3). By replacing each generalized sentence in them by its version with a new generic constant, we obtain valid sentences.

34 In testing a program, one examines its behavior for a (small) set of data and then argues that the program will exhibit this behavior in general. Here, the rationale is that the set of test data is 'representative' in that it covers reasonably well the possible execution paths.

35 More precisely, $(\exists w : e)\varphi(w)$ and $(\forall w : e)\varphi(w)$, respectively, abbreviate $\exists w[e(w) \wedge \varphi(w)]$ and $\forall w[e(w) \rightarrow \varphi(w)]$.

36 This also explains the idea of transfer under similarity suggested above. Using a predicate S for similarity, we express the preceding conditions (d) and (t) on the sample as

δ: $\nabla v \exists u[e(u) \wedge s(u, v)]$ (*e* generally dense for *S*);
τ: $\nabla v \nabla u[(e(u) \vee S(u, v)) \rightarrow (\theta(u) \rightarrow \theta(v))]$ (*S* on *e* generally transfers θ).

We then have both $\{\delta, \tau\} \vdash^{f} \eta$ and $\{{}_{e}\theta, \eta\} \vdash^{f} \nabla u\theta(u)$.

37 Because of this general reflection, such a prototypical object u may be somewhat elusive: reminiscent, perhaps, of a Platonic form.

38 The non-triviality condition $\exists e$: $\exists u e(u)$ prevents vacuous satisfaction of ${}_{e}\theta : (\forall v : e)\theta(v)$. Also, this condition $\exists e$: $\exists u e(u)$ follows from the above resemblance sentence η.

39 If we replace the generalized quantifier ∇ by the classical universal quantifier $\forall$, we obtain a version of Hume's argument, as follows. Given experimental evidence ${}_{e}\theta$ that formula θ holds on non-trivial sample examined (in the past), the only way to infer or consistently hold the universal assertion $\forall v\theta$ (θ will always hold) is to do so for the uniformity assertion η^{*}: $\forall v \exists u[e(u) \wedge (\theta(u) \rightarrow \theta(v))]$ (all objects will always resemble – with respect to θ – those already examined (in the past)). (In this case $\eta^{*} \wedge {}_{e}\theta$ and $\forall v\theta \wedge \exists e$ are logically equivalent.)

40 These auxiliary hypotheses may involve extra concepts as well as boundary conditions. We wish the experiments to be realizable, at least in principle. So, we do not wish to allow the conceived experimental set-up to constrain our present knowledge (or conflict with it).

41 Clearly, the preceding analysis extends to this case with several variables.

42 The motivation is that auxiliary formula ψ is presumably simpler (to test) than formula θ. But, we do not require any such assumption concerning relative complexities.

43 One can consistently hold that "All penguins are winged birds", "Winged birds generally fly" and "Penguins generally do not fly" (or even "No penguin flies"). Apparently, the set of penguins, being a rather negligible set of winged birds, does not constitute an important set of exceptions to the belief that winged birds generally fly.

[44] This issue also appears to be connected to the so-called "confirmation paradox" in Philosophy of Science (Hempel 1945, 1965). Each flying eagle is considered as evidence in favor of "eagles fly", whereas a non-flying non-eagle is not, even though "eagles are fliers" and "non-fliers are non-eagles" are logically equivalent.

[45] For completeness, the witnesses introduced for the existential quantifiers inherit the corresponding sorts and we thus have sorted families of represented subsets (cf. Section 4.1).

[46] If only, say, at most 5% of the birds are penguins, then the penguins have unimportant impact on the likelihood of birds flying.

[47] For an upward closed family $\mathcal{C}_T$, over T, and $S \subseteq T$, $\wp(S) \cap \mathcal{C}_T \neq \varnothing$ iff $S \in \mathcal{C}_T$. Thus, if $S \notin \mathcal{C}_T$ then, for every $X \subseteq S$, we have $X \notin \mathcal{C}_T$ (even for those X in complex $\mathcal{C}_S$ over S).

[48] For a set $Y \subseteq B$, its set of exceptions $W - Y$ and $B - Y$ with respect to W and B, respectively, are connected by $W - Y \subseteq B - Y \subseteq (W - Y) \cup (B - W)$. Also, if, say, 98% of the birds have wings, then the wingless birds have negligible impact on the likelihood of birds flying.

[49] Given a proper upward closed family $\mathcal{C}_S$ over a sub-universe $S \subseteq T$, consider the relativizable complex ${}^T\mathcal{C}_S := \{Y \subseteq T : S \cap Y \in \mathcal{C}_S\}$. Then, $(T - S) \notin {}^T\mathcal{C}_S$, $S \in {}^T\mathcal{C}_S$ and ${}^S\mathcal{C}_T$ will inherit the properties of complex $\mathcal{C}_S$ over S. Also, $\mathcal{C}_S = \wp(S) \cup \mathcal{C}_T$ whenever $\mathcal{C}_T = {}^T\mathcal{C}_S$.

[50] Notice, however, that such questions fall outside the realm of our logics for 'generally'. Whether or not "birds generally are penguins" or "birds generally have wings" are ornithological, rather than logical, matters.

[51] This coherent connection principle can be explained, much as before, by resorting to the family ${}^I\mathcal{C}_S := \{Z \subseteq t : I^{-1}[Z] \in \mathcal{C}_S\}$.

[52] These assertions can be expressed by: $(\nabla z : t)(\exists x : s)z \equiv i(x)$, $(\nabla z : t)\varphi(z)$, and $\nabla x : s)\varphi(i(x))$. Notice that $\neg(\nabla z : t)\neg(\exists x : s)z \equiv i(x)$ is a consequence of $(\nabla z : t)(\exists x : s)z \equiv i(x)$ in filter logic $\mathcal{L}_{\omega\omega}(\rho)^f$, and is equivalent to $(\nabla z : t)(\exists x : s)z \equiv i(x)$ in ultrafilter logic $\mathcal{L}_{\omega\omega}(\rho)^u$.

[53] A more general formulation has antecedent $[(\nabla z : t)(\exists x : s)z \equiv i(x) \wedge \neg(\nabla z : t)\neg(\exists x : s)z \equiv i(x)]$, yielding $(\nabla z : t)(\exists x : s)z \equiv i(x) \rightarrow [(\nabla z : t)\varphi \leftrightarrow (\nabla x : s)\varphi(z/i(x))]$ in filter logic $\mathcal{L}_{\omega\omega}(\rho)^f$, and equivalent to it in ultrafilter logic $\mathcal{L}_{\omega\omega}(\rho)^u$.

[54] More precisely, $\Gamma^* \cup \{k[p] \in \mathcal{C}_b\} \vdash^f k^1[F] \in \mathcal{C}_p$, but $\Gamma^* \nvdash^f k^{-1}[F] \in \mathcal{V}_p$.

[55] More precisely, $\Gamma^* \cup \{k[p] \in \mathcal{C}_b\} \vdash^f B - k[p] \in \mathcal{C}_b$.

[56] As mentioned, such decisions are outside the realm of our logics for 'generally'.

[57] For instance, as suggested in Section 3.1 ('very' short or tall, cf. note 11) and 5.1 ('very' low or high and 'about as high', cf. note 30).

[58] See Section 1.1 (note 1). Notice that this weakening breaks the asymmetry, of importance to some views of Popper (1934, 1975), between corroboration and refutation (cf. Section 3.3, note 14). The appearance of a non-black raven is not enough to refute the belief that "Most ravens are black" or that "A typical raven is black" (cf. Section 5.2).

[59] The basic idea is weakening some $\forall$ to ∇. For instance, an 'almost' partition would amount to a set of blocks 'almost' covering the universe where intersecting blocks would have 'almost' the same elements. (Note that we are not proposing a program: one can expect that only some such weakenings – with qualitative flavor – will be of interest.)

[60] Some examples for ultrafilter logic are the ultraproduct theorem (Grácio 1999), interpolation and modularity results (Veloso 2001), theorem proving (Veloso and Veloso 2002) and a natural deduction system with normalization (Rentería et al. 2001). We believe that the "purely qualitative models" might have surprising properties. Also, the role and aspect of the finite models of this kind are not entirely clear.

[61] Modal versions of our generalized quantifiers can be introduced in a natural manner.

[62] The distinction between deductive and inductive arguments permeates Toulmin's five-class division, and the label 'inductive' should also be attached to substantial, warrant-establishing, tentative and non-formalizable arguments in this classification (Toulmin 1958).

[63] From "Petersen is a 'typical' Swede" and "Swedes 'rarely' are Roman Catholics", one infers "Petersen is not a Roman Catholic" (cf. 5.2). Also, from "Stockholmers are Swedes" and "Very few Swedes are Roman Catholics", one infers "Very few Swedes are Roman Catholic Stockholmers" (cf. 6.2).

[64] See especially Hume's works 'A Treatise of Human Nature' (Book I, part III, Section 6) and 'An Inquiry Concerning Human Understanding' (Section IV, part II).

[65] There have been some objections against Hume's characterization of induction as generalizing from past experience according to some uniformity principle, e.g., the new riddle of induction illustrated by "grue" example due to Goodman (1955). One may take Goodman's objections as forcing us to face new difficulties with respect to induction (a major problem is that natural induction is not well defined) or put aside Goodman's paradox as extreme. The situation appears to resemble to the concept of truth: even though there are several semantical paradoxes related to the notion of truth (the Liar's paradox and others), logicians have found ways to restrict the application of this notion, instead of rejecting semantics as a whole.

[66] The equivalence condition states that if ψ and θ are logically equivalent, then each instance of ψ is one of θ. Then, since "All ravens are black" is logically equivalent to "All non-black things are non-ravens", we are forced to conclude that every non-raven object serves as a confirmatory instance for "All ravens are black": not only black shoes, but also shoes of any other color (which are non-black things and non-ravens). Several authors, including Hempel (1945) himself, regard this as a psychological illusion and not necessarily a paradox, while others argue that considering shoes of any color as confirmatory instances of "All ravens are black" is really nonsensical.

[67] Evidences may be confined to non-raven objects deemed "relevant", say, birds that are not ravens, by considering ravens as a subsort of the universe of birds. In this case, we would express "All ravens are black" as "All ravens are black birds", whereas to express "All non-black birds are non-ravens" one would need the subsort of non-black birds as well.

References

Antoniou, G.: 1997, *Nonmonotonic Reasoning*, Cambridge, MIT Press.

Black, M.: 1967, 'Induction', in Edwards (1967) **4**, 169–181.

Barwise, J. and R. Cooper: 1981, 'Generalized Quantifiers and Natural Language', *Linguistics and Philosophy* **4**, 159–219.

Barwise, J. and S. Feferman: 1985, *Model-Theoretic Logics*, New York, Springer-Verlag.

Bell, J. L. and A. B. Slomson: 1971, *Models and Ultraproducts: an Introduction*, Amsterdam, North-Holland, (2nd rev. pr.).

Besnard, P.: 1989, *An Introduction to Default Logic*, Berlin, Springer-Verlag.

Brewka, G.: 1991, *Nonmonotonic Reasoning: Logical Foundations of Commonsense*, Cambridge, Cambridge University Press.

Brewka, G., J. Dix and K. Konolige: 1997, *Nonmonotonic Reasoning: An Overview*, Stanford, CSLI.

Carnielli, W. A. and M. C. G. Grácio: 2000, 'Modulated Logics and Uncertain Reasoning', in *Proc. Kurt Gödel Colloquium*, Barcelona (to appear).

Carnielli, W. A. and A. M. Sette: 1994, 'Default Operators', *Abstracts of the Workshop on Logic, Language, Information and Computation*, Recife.

Carnielli, W. A. and P. A. S. Veloso: 1997, 'Ultrafilter Logic and Generic Reasoning', in G. Gottlob, A. Leitsch and D. Mundici (eds.), *Computational Logic and Proof Theory*, Berlin, Springer-Verlag (LNCS 1289), pp. 34–53.

Church, A.: 1956, *Introduction to Mathematical Logic*, vol. I, Princeton, Princeton University Press.

Enderton, H. B.: 1972, *A Mathematical Introduction to Logic*, New York, Academic Press.

Edwards, P. (ed.): 1967, *The Encyclopedia of Philosophy*, London, Collier Macmillan, (repr. Macmillan, New York, 1972).

Fine, K.: 1985, *Reasoning with Arbitrary Objects*, Oxford, Basil Blackwell (Aristotelian Society Series, vol. 3).

Frege, G.: 1879, *Begriffsschrift, eine der arithmetischen nachgebildete Formelsprache des reinen Denkens*, Louis Nebert, Halle (English translation in (van Heijenoort 1967, pp. 1–82)).

Gärdenfors, P.: 1988, *Knowledge in Flux: Modelling the Dynamics of Epistemic States*, Cambridge, MIT Press (Bradford Books).

Goodman, N.: 1955, *Fact, Fiction and Forecast*, Cambrige, Harvard University Press.

Grácio, M. C. G.: 1999, *Lógicas Moduladas e Raciocínio sob Incerteza*, D. Sc. dissertation, UNICAMP, Campinas.

Halmos, P. R.: 1963, *Lectures on Boolean Algebras*, Princeton, D. van Nostrand.

Hempel, C.: 1945, 'Studies in the Logic of Confirmation', *Mind* **54**, 1–26, 97–121, (reprinted in (Hempel 1965, pp. 1–51)).

Hempel, C.: 1965, *Aspects of Scientific Explanation and Other Essays in the Philosophy of Science*, New York, Free Press.

Henkin, L.: 1949, 'The Completeness of the First-order Functional Calculus', *Journal of Symbolic Logic* **14**, 159–166.

Keisler, H. J.: 1970, 'Logic with the Quantifier there Exist Uncountably Many', *Annals of Mathematical Logic* **1**, 1–93.

Kelley, J. L.: 1955, *General Topology*, New York, D. van Nostrand.

Lecourt, D. (ed.): 1999, *Dictionnaire d'Histoire et Philosophie des Sciences*, Paris, PUF (Presse Universitaire de France).

Lindström, P.: 1966, 'On Extensions of Elementary Logic', *Theoria* **35**, 1–11.

Lukaszewicz, W.: 1990, *Non-monotonic Reasoning: Formalization of Commonsense Reasoning*, Chichester, Ellis Horwood.

Makinson, D. and Gärdenfors, P.: 1991, 'Relations between the Logic of Theory Change and Non-monotonic Logic', in A. Fuhrmann and M. Morreau (eds.), *The Logic of Theory Change*, Berlin, Springer-Verlag (LNAI 465).

Marek, V. W. and M. Truszczyński: 1993, *Nonmonotonic Logic: Context-dependent Reasoning*, Berlin, Springer-Verlag.

Meré, M. C. and P. A. A. Veloso: 1995, 'Definition-like Extensions by Sorts', *Bulletin of the IGPL* **3**, 579–595.

Montague, R.: 1974, in R. Thomason (ed.), *Formal Philosophy: Selected Papers*, New Haven, Yale University Press.

Mostowski, A.: 1957, 'On a Generalization of Quantifiers', *Fundamenta Mathematicae* **44**, 1236.

Nagel, E.: 1961, *The Structure of Science: Problems in the Logic of Scientific Explanation*, New York, Harcourt Brace.

Peterson, P. L.: 1979, 'On the Logic of 'Few', 'Many', and 'Most' ', *Notre Dame Journal of Formal Logic* **20**, 155–179.

Popper, K. R.: 1934, *Logik der Forschung*, J. C. B. Molir, Tübingen (5. Aufl. 1973) (English translation *The Logic of Scientific Discovery*, New York, Basic Books, 1959 (6th edn. 1972)).

Popper, K. R.: 1975, *Objective Knowledge: An Evolutionary Approach*, Oxford, Clarendon.

Rentería, C. J., E. H. Haeusler and P. A. S. Veloso: 2001, 'NUL: Natural Deduction for Ultrafilter Logic, *Natural Deduction 2001*, Rio de Janeiro (also 'Dedução natural para lógica de ultrafiltros', Res. Rept. MCC 16/02, PUC-Rio, Rio de Janeiro, 2002).

Reiter, R.: 1980, 'A Logic for Default Reasoning', *Journal of Artificial Intelligence* **13**, 81–132.

Rescher, N.: 1962, 'Plurality Quantification', *Journal of Symbolic Logic* **27**, 373–374.

Rosenkrantz, R. D.: 1982, 'Does the Philosophy of Induction Rest on a Mistake?', *Journal of Logic of Philosophy* **79**, 78–97.

Schilpp, P. A. (ed.): 1963, *The Philosophy of Rudolf Carnap*, La Salle, Open Court.

Schilpp, P. A. (ed.): 1973, *The Philosophy of K. R. Popper*, La Salle, Open Court.

Schlechta, K.: 1995, 'Defaults as Generalized Quantifiers', *Journal of Logic and Computation* **5**, 473–494.

Sette, A. M., W. A. Carnielli and P. A. S. Veloso: 1999, 'An Alternative View of Default Reasoning and its Logic', in E. H. Haeusler and L. C. Pereira (eds.), *Pratica: Proofs, Types and Categories*, Rio de Janeiro, PUC-Rio, pp. 127–158.

Sgro, J.: 1972, 'ompleteness Teorems for Tpological Mdels' *Notices of the American Mathematical Society* **19**, A-765.

Shoenfield, J. R.: 1967, *Mathematical Logic*, Reading, Addison-Wesley.

Slanley, J.: 1988, 'A Note on 'Most' ', *Analysis* **48**, 134–135.

Slomson, A. B.: 1967, *Some Problems in Mathematical Logic*, D. Sc. dissertation, Oxford, Oxford University.

Tarski, A.: 1936, 'Der Wahrheitsbegriff in den formalisierten Sprachen', *Studia Philosophica*, pp. 261–405 (English translation in (Tarski 1956, pp. 152–278)).

Tarski, A.: 1956, *Logic, Semantics and Metamathematics: Papers from 1923 to 1938 by Alfred Tarski* (Woodger, J. H. (trans.)). Oxford, Clarendon Press.

Toulmin, S. E.: 1958, *The Uses of Argument*, Cambridge, Cambridge University Press.

Turner, W.: 1984, *Logics for Artificial Intelligence*, Chichester, Ellis Horwood.

van Heijenoort, J. (ed.): 1967, *From Frege to Gödel: A Source Book in Mathematical Logic*. Cambridge, Harvard University Press, (3rd prt).

Veloso, P. A. S.: 1998, 'On Ultrafilter Logic as a Logic for 'Almost all' and 'Generic' Reasoning', Res. Rept. ES-488/98, COPPE-UFRJ, Rio de Janeiro.

Veloso, P. A. S.: 1999, 'On 'Almost all' and Some Presuppositions', in L. C. P. D. Pereira and M. B. Wrigley (eds.), *Logic, Language and Knowledge: Essays in Honour of Oswaldo Chateaubriand Filho*, Manuscrito XXII, pp. 469–505.

Veloso, P. A. S.: 2000, 'On the Power of Ultrafilter Logic', *Bulletin of the Section of Logic* **29**, 89–97.

Veloso, P. A. S.: 2001, 'On Interpolation and Modularity for Ultrafilter Logic', in J. M. Abe and J. I. Silva Filho (eds.), *Logic, Artificial Intelligence and Robotics: LAPTEC'2001*, Amsterdam, IOS Press, pp. 270–278.

IOS Press Veloso, P. A. S.: 2002, 'Issues in Reasoning with 'Generally' and 'Rarely' ', in A. O. Cupani and C. A. Mortari (eds.), *Linguagem e Filosofia: Anais do 2° Simpósio Internacional Principia*, UFSC, Florianópolis, pp. 51–72.

Veloso, S. R. M. and P. A. S. Veloso: 2001, 'On a Logical Framework for 'Generally' ', in J. M. Abe and J. I. Silva Filho (eds.), *Logic, Artificial Intelligence and Robotics: LAPTEC'2001*, Amsterdam, IOS Press, pp. 279–286.

Veloso, S. R. M. and P. A. S. Veloso: 2002, 'On Special Functions and Theorem Proving in Logics for 'Generally' ', in G. Bittencourt and G. L. Ramalho (eds.), *Advances in Artificial Intelligence: 16th Brazilian Symposium in Artificial Intelligence*, (Lecture Notes in Artificial Intelligence 2507), Berlin, Springer-Verlag, pp. 1–10.

Zadeh, L. A.: 1975, 'Fuzzy Logic and Approximate Reasoning', *Synthèse* **30**, 407–428.

LOGIC OF DYNAMICS AND DYNAMICS OF LOGIC: SOME PARADIGM EXAMPLES

BOB COECKE[1], DAVID J. MOORE[2] and SONJA SMETS[3]
[1]*University of Oxford, Oxford University Computing Laboratory, Wolfson Building, Parks Road, Oxford, OX1 3QD, UK, E-mail:* coecke@comlab.ox.ac.uk
[2]*University of Canterbury, Department of Physics and Astronomy, Private Bag 4800, Christchurch, New Zealand, E-mail:* closcat@hotmail.com
[3]*Free University of Brussels (VUB), Department of Philosophy, Pleinlaan 2, B-1050 Brussels, Belgium, E-mail:* sonsmets@vub.ac.be

Abstract. The development of "operational quantum logic" points out that classical Boolean structures are too rigid to describe the actual and potential properties of quantum systems. Operational quantum logic bears upon basic axioms which are motivated by empirical facts and as such supports the dynamic shift from classical to non-classical logic resulting into a dynamics of logic.

On the other hand, an intuitionistic perspective on operational quantum logic, guides us in the direction of incorporating dynamics logically by reconsidering the primitive propositions required to describe the behavior of a quantum system, in particular in view of the emergent disjunctivity due to the non-determinism of quantum measurements.

A further elaboration on "intuitionistic quantum logic" emerges into a "dynamic operational quantum logic", which allows us to express dynamic reasoning in the sense that we can capture how actual properties propagate, including their temporal causal structure. It is in this sense that passing from static operational quantum logic to dynamic operational quantum logic results in a true logic of dynamics that provides a unified logical description of systems which evolve or which are submitted to measurements. This setting reveals that even static operational quantum logic bears a hidden dynamic ingredient in terms of what is called "the orthomodularity" of the lattice-structure.

Focusing on the quantale semantics for dynamic operational quantum logic, we delineate some points of difference with the existing quantale semantics for (non)-commutative linear logic. Linear logic is here to be conceived of as a resource-sensitive logic capable of dealing with actions or in other words, it is a logic of dynamics.

We take this opportunity to dedicate this paper to Constantin Piron at the occasion of his retirement.

1. Introduction

As a starting point for our discussion on the dynamics of logic we quote G. Birkhoff and J. von Neumann, confronting the then ongoing tendencies towards intuitionistic logic with their observation of the "logical" structure encoded in the lattice of closed subspaces of a Hilbert space, the "semantics" of quantum theory (Birkhoff and von Neumann 1936):

> "The models for propositional calculi [of physically significant statements in quantum mechanics] are also interesting from the standpoint of pure logic. Their nature is de-

S. Rahman et al. (eds.), Logic, Epistemology, and the Unity of Science, pp. 527–555.

> termined by quasi-physical and technical reasoning, different from the introspective and philosophical considerations which have to guide logicians hitherto [...] whereas logicians have usually assumed that [the orthocomplementation] properties L71–L73 of negation were the ones least able to withstand a critical analysis, the study of mechanics points to the distributive identities L6 as the weakest link in the algebra of logic." (p. 839)

They point at a fundamental difference between Heyting algebras (the semantics of intuitionistic propositional logic) and orthomodular lattices (the "usual" semantics of quantum logic) when viewed as generalizations of Boolean algebras (the semantics of classical propositional logic). A new intuitionistic perspective on operational quantum logic (see below) provides a way of blending these seemingly contradicting directions in which logic propagated during the previous century (Coecke 2002). In this paper we focus on two new logical structures, namely (intuitionistic) linear logic (Girard 1987; Abrusci 1990) which emerged from the traditional branch of logic, and dynamic operational quantum logic (Coecke nd; Coecke and Smets 2001), emerging from an elaboration on the above mentioned blend. We also briefly consider the "general dynamic logic" proposed in van Benthem (1994). We do mention epistemic action logic e.g., Baltag (1999) and computation and information flow (e.g., Abramsky (1993), Milner (1999)) as other examples of dynamic aspects in logic, which we unfortunately will not be able to consider in this paper. We also won't discuss the "geometry of interaction" paradigm which provides a (promising) different perspective on linear logic (Abramsky and Jagadeesan 1994). Concretely we start in Section 2 with an outline of static operational quantum logic. In Section 3 we survey dynamic operational quantum logic, and investigate the link between the emerging structures and the logical system proposed in van Benthem (1994). In Section 4 we compare dynamic operational quantum logic with linear logic while we focus on formal and methodological differences.

2. Static "Old-Style" Operational Quantum Logic

The Geneva School approach to the logical foundations of physics originated with the work in Jauch and Piron (1963), Piron (1964), Jauch (1968), Jauch and Piron (1969), Piron (1976) and Aerts (1981) as an incarnation of the part of the research domain "foundations of physics" that is nowadays called "Operational quantum logic" (OQL)[1] – see Coecke et al. (2000) for a recent overview of its general aspects. With respect to the intuitionistic perspective and the dynamic aspects which we put forward below, we will further refer in this paper to OQL as "sOQL", emphasizing its static nature. More concretely, sOQL as a theory aims to characterize physical systems, ranging from classical to quantum, by means of their actual and potential properties, in particular by taking an ontological rather than an empirical perspective, but, still providing a truly operational alternative to the standard approaches on the logical status of quantum theory.[2] Since Moore (1999), Coecke et al. ((nd)a,b) and Smets (2001) provide recent and detailed discussions of sOQL, we intend in this paper to give only a concise overview, focusing on its basic concepts and underlying epistemology. We are aware of the strong conceptual restrictions imposed by the rigid foundation of sOQL, necessary in order to obtain a framework with fully well-defined primitive notions. Clearly, in view of still existing conceptual incompatibilities at the foundational level between quantum theory and relativistic space-time and field theoretic considerations, the development of an essentially "towards dynamics directed"-formalism for quantum logicality should range beyond the rigid concepts of sOQL. Therefore, we conceive of the notions inherited from sOQL as a stepping-stone for further development rather than something necessarily "to carry all the way". This is the reason why we use lowercase 'o' in our notation DoQL referring to "dynamic operational quantum logic" and IoQL, when referring to "intuitionistic operational quantum logic". In particular the sOQL assumptions (see below) of "a particular physical system which is considered as distinct from its surroundings", "the specification of states as a (pre-defined) set" and "the a priori specification of a particular physical system itself" need to be reconsidered when crossing the edges of sOQL (which itself was designed to clarify the structural description of quantum systems and justify an ontological perspective for non-relativistic quantum theory).

First we want to clarify that the operationalism which forms the core of sOQL points to a pragmatic attitude and not to any specific doctrine one can encounter in last century's philosophy of science (Coecke et al. (nd)a). Linked to the fact that we defend the position of critical scientific realism in relation of sOQL, operationality points to the underlying assumption that with every property of a physical system we can associate experimental procedures that can be performed on the system, each such experimental procedure including the specification of a well-defined positive result for which certainty is exactly guaranteed by the actual existence of the corresponding property. In particular, while on the epistemological level our knowledge of what exists is based on what we could measure or observe, on the ontological level "physical" properties have an extension in reality and are not reducible to sets of procedures.[3]

Focusing on the stance of critical scientific realism, we first adopt an ontological realistic position and a correspondence theory of truth.[4] Though, contrary to naive realists we do adopt the thesis of fallibilism by which truth in relation to scientific theories has to be pursued but can only be approached. As such we agree with Niiniluoto (1999) that scientific progress can be characterized in terms of increasing truthlikeness. Furthermore, we believe that reality can indeed be captured in conceptual frameworks, though contrary to Niiniluoto we like to link this position to a "weak" form of conceptual idealism. Concentrating for instance on Rescher's conceptual idealism as presented in Rescher (1973, 1987, §11) and revised in Rescher (1995, §8), this position maintains that a description of physical reality involves reference to mental operations, it doesn't deny ontological realism and also adopts the thesis of fallibilism. This position is opposed to an ontological idealism in which the mind produces the "real" objects. As one might expect, we want to stress that we are not inclined to adopt Rescher's strict Kantian distinction between reality out there and reality as we perceive it, though we are attracted by the idea of capturing ontological reality in mind-correlative conceptual frameworks. Once we succeed in giving such a description of reality, it approaches according to us the ontological world close enough to omit a Kantian distinction between the realms of noumena and phenomena. Why it is of importance for us to reconcile critical realism with a weak form of conceptual idealism becomes clear when we focus on sOQL. Firstly, we cannot escape the fact that in our theory we focus on parts of reality considered as well-defined and distinct which we can then characterize as physical systems. Secondly we have to identify the properties of those systems which is a mind-involving activity. To be more explicit, a physicist can believe that a system ontologically has specific actual and potential properties, but to give the right characterization of the physical system he has to consider the definite experimental projects which can test those properties so that, depending on the results he would obtain, he can be reinforced in his beliefs or has to revise them.[5] How sOQL formally is built up, using the notions of actual properties, potential properties and definite experimental projects, will be clarified below. To recapitulate we finish this paragraph by stressing that we focus in our scientific activity on parts of the external world, reality is mind-independent, though the process of its description "involves" some mind-dependent characterizations.

Let us introduce the primitive concrete notions on which sOQL relies, explicitly following Coecke et al. ((nd)a):

- We take a particular physical system to be a part of the ostensively external world which is considered as distinct from its surroundings – see Moore (1999) for a discussion on this matter;
- A singular realization of the given particular physical system is a conceivable manner of being of that system within a circumscribed experiental context;
- States $\mathcal{E} \in \Sigma$ of a given particular physical system are construed as abstract names encoding its possible singular realizations;
- A definite experimental project $\alpha \in \mathcal{Q}$ on the given particular physical system is a real experimental procedure which may be effectuated on that system where we have defined in advance what would be the positive response should it be performed.
- Properties $a \in \mathcal{L}$ of a given particular physical system are construed as *candidate* elements of reality corresponding to the definite experimental projects defined for that system. We as such obtain a mapping of definite experimental projects $\mathcal{Q}$ on properties $\mathcal{L}$.

The notion of an element of reality was first introduced in Einstein et al. (1935) as follows:

> "If, without in any way disturbing a system, we can predict with certainty [. . .] the value of a physical quantity, then there exists an element of physical reality corresponding to this physical quantity" (p. 777).

While this formulation is explicitly the starting point for Piron's early work, we insist that Piron's operational concept of element of reality is both more precise and allows theoretical deduction, being based on an empirically accessible notion of counterfactual performance rather than a metaphorical notion of non-perturbation. The basic ingredient that we inherit from this setting is the agreement that actual properties exist. Before we give a more rigid characterization of how an actual property is conceived within sOQL, we want to remark that Piron adopted the Aristotelian concepts of actuality and potentiality and placed them in a new framework (see e.g., Piron (1983)). For an analysis of how these notions, used within sOQL, are still related to the old Aristotelian ones, we refer to Smets (2001). Let us just briefly mention here that with regard to actuality, an actual property is within sOQL conceived as an attribute which "exists"; it is some realization in reality or in other words; an element of reality. A potential property on the other hand, does not exist in the same way as an actual one, it is conceived merely as a capability with respect to an actualization since there is always a chance – i.e., except for the absurd property – that it could be realized after the system has been changed without destroying it. It will become clear that a property can be actual or potential depending on the state in which we consider the particular physical system. Similarly we can say that certainty of obtaining a positive answer when performing a definite experimental project depends on the state of the system. In order to construct our theory further we need to introduce the following relationship between definite experimental projects, states and properties:

- A definite experimental project α is called certain for a given singular realization of the given particular physical system if it is sure that the positive response would be obtained should α be effectuated;
- A property a is called actual for a given state if any, and so all, of the definite experimental projects corresponding to the property a are certain for any, and so all, of the singular realizations encoded by that state. A property is called potential when it is not actual.

In particular is one of the essential achievements of sOQL that it gives a consistent and coherent ontological account of physical properties contra certain 'overextrapolations' claimed to be inherent in quantum theory. The quantum mechanical formalism itself indeed allows a characterization of the properties of a quantum system as being in correspondence with the closed subspaces of the Hilbert space describing the system in the above sense: Definite experimental projects α expressible in quantum theory are of the form, "the value of an observable is in region $E \subset \sigma(H)$", where $\sigma(H)$ is the spectrum of the self-adjoint operator H describing the observable, thus we can write $\alpha(H, E)$; more explicitly, the definite experimental project $\alpha(H, E)$ consists of measuring the observable H and obtaining an outcome in E; the corresponding property a is then represented by the closed subspace of fixpoints of the projector P_E that arises via decomposition of H according to von Neumann's spectral decomposition theorem since only the states that imply the actuality of that property, that is, the states represented by a ray included in that subspace, will yield a positive outcome with certainty when "we would perform $\alpha(H, E)$".

Given the above notions, it becomes possible to introduce an operation on the collection $\mathcal{Q}$ of definite experimental projects. What we have in mind is the product of a family of definite experimental projects ΠA which obtains its operational meaning in the following way:

- The product ΠA of a family A of definite experimental projects is the definite experimental project "choose arbitrarily one α in A and effectuate it and attribute the obtained answer to ΠA".

More explicitly, given a particular realization of the system, ΠA is a certain definite experimental project if and only if each member of A is a certain definite experimental project. Formally it becomes possible to pre-order definite experimental projects by means of their certainty:

- $\alpha \prec \beta := \beta$ is certain whenever α is certain.

The notions of a trivial definite experimental project, which is always certain and an absurd definite experimental project, which is never certain, can be introduced and play the role of respectively maximal and minimal element of the collection $\mathcal{Q}$. The trivial and absurd definite experimental projects give rise on the level of properties to a trivial and absurd property, the first is always actual while the latter is always potential. Through the correspondence between definite experimental projects and properties, the mentioned pre-order relation induces a partial order relation on $\mathcal{L}$:

- $a \leq b := b$ is actual whenever a is actual.

The product of a collection of definite experimental projects with corresponding collection of properties A provides a greatest lower bound or meet for any $A \subseteq \mathcal{L}$, and as such the set of properties $\mathcal{L}$ forms a complete lattice. Indeed, once we have the "meet" of any collection of properties, we can construct the operation of "join" or least upper bound via Birkhoff's theorem stating that for a given $A \subseteq \mathcal{L}$:

$$\bigvee A = \bigwedge \{x \in \mathcal{L} \mid (\forall a \in A) a \leq x\}.$$

Note however that contrary to the meet, the join admits of no direct operational meaning in sOQL. Even more, we see that the join of a collection of properties need not, and generically does not, correspond to a classical disjunction since the following implication is only secured in one direction,

- ($\exists a \in A$: a is actual) $\Rightarrow \bigvee A$ is actual.

With respect to the meet we do have

- ($\forall a \in A$: a is actual) $\Leftrightarrow \bigwedge A$ is actual,

following directly from the identification of the meet with the product of definite experimental projects.

The property lattice description of a physical system in sOQL allows a dual description by means of the system's states (Moore 1995, 1999) in terms of maximal state sets $\mu(a) \subseteq \Sigma$ for which a common property $a \in \mathcal{L}$ is actual whenever the system is in a state in $\mathcal{E} \in \mu(a)$, these sets being ordered by inclusion. More explicitly, the state-property duality may be straightforwardly characterized once we introduce the forcing relation $\triangleright$ defined by $\mathcal{E} \triangleright a$ if and only if the property a is actual in the state $\mathcal{E}$, by the fact that we can associate to each property a the set $\mu(a) = \{\mathcal{E} \in \Sigma \mid \mathcal{E} \triangleright a\}$ of states in which it is actual, and to each state $\mathcal{E}$ the set $S(\mathcal{E}) = \{a \in \mathcal{L} \mid \mathcal{E} \triangleright a\}$ of its actual properties. Formally $\mu : \mathcal{L} \to P(\Sigma) : a \mapsto \mu(a)$ is injective, satisfies the condition $\mu(\bigwedge A) = \bigcap \mu[A]$ and is called the Cartan map. In particular we now see that actuality can be studied at the level of either states or properties since we have

$$\mathcal{E} \in \mu(a) \Leftrightarrow \mathcal{E} \triangleright a \Leftrightarrow a \in S(\mathcal{E}).$$

We will be a bit more explicit about the dual description of a physical system by means of its states and operationally motivate the introduction of a symmetric and antireflexive orthogonality relation on the set Σ:

- Two states $\mathcal{E}$ and $\mathcal{E}'$ are called orthogonal, written $\mathcal{E} \perp \mathcal{E}'$, if there exists a definite experimental project which is certain for the first and impossible for the second.

If we equip Σ with $\perp$, we can consider $\mathcal{L}_\Sigma$ as the set of biorthogonal subsets, i.e., those $A \subseteq \Sigma$ with $A^{\perp\perp} = A$ for $A^\perp = \{\mathcal{E}' \mid (\forall \mathcal{E} \in A)\mathcal{E} \perp \mathcal{E}'\}$. Whenever $\perp$ is *separating* (Coecke et al. (nd)a), the codomain restriction of the Cartan map μ to the set of biorthogonals gives us an isomorphism of complete atomistic orthocomplemented lattices (see below) extending the dual description of a physical system by its state space Σ and its property lattice $\mathcal{L}$ – for details we refer to Moore (1999) and Coecke et al. ((nd)a). It thus follows that $\mathcal{L}_\Sigma$ is an atomistic lattice where the singletons $\{\mathcal{E}\} = \{\mathcal{E}\}^{\perp\perp}$ are the atoms. Let us first introduce the notion of an atomistic lattice more explicitly:

- A complete lattice $\mathcal{L}$ is called atomistic if each element $a \in \mathcal{L}$ is generated by its subordinate atoms,

  $$a = \bigvee \{p \in \Sigma_\mathcal{L} \mid p \leq a\},$$

  where the $p \in \Sigma_\mathcal{L}$ are by definition the minimal nonzero elements of $\mathcal{L}$.

Under the assumption that states are in bijective correspondence with atoms, $\Sigma = \Sigma_\mathcal{L} \subseteq \mathcal{L}$, each property lattice is atomistic in the sense that

$$a = \bigvee \{\bigwedge S(\mathcal{E}) \mid \mathcal{E} \in \mu(a)\}$$

for each $a \in \mathcal{L}$. Note that $\bigwedge S(\mathcal{E})$ is the strongest actual property of the collection $S(\mathcal{E})$ which as such "represents" the state $\mathcal{E}$.

While we will only work with complete atomistic lattices in the following, we want to finish off this section with an example explaining that a property lattice description for a quantum system will not lead to a Boolean algebra Under the assumption that properties have opposite properties and even more that each property $a \in \mathcal{L}$ is the opposite of another one, we can formally introduce an orthocomplementation, i.e.:

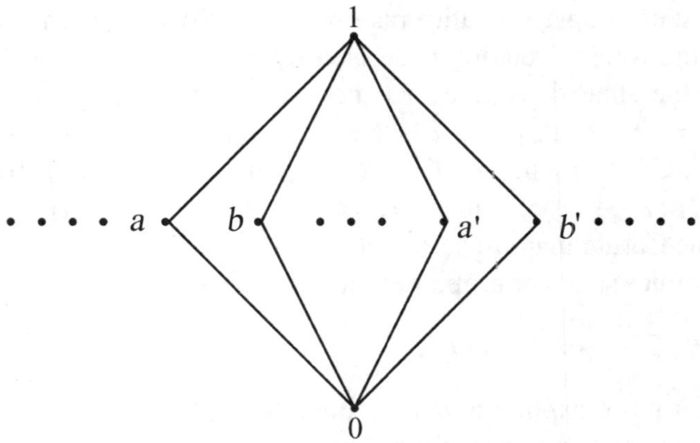

Figure 1. The lattice for a photon.

- A surjective antitone involution $' : \mathcal{L} \to \mathcal{L}$ satisfying $a \wedge a' = 0$, $a \leq b \Rightarrow b' \leq a'$ and $a'' = a$. A lattice equipped with an orthocomplementation is usually abbreviated as an ortholattice.

Consider now the property lattice description of a photon as presented in Piron (1978). Take as a physical system a propagating photon which is linearly polarized. A definite experimental project α_ϕ is then defined by:

- i. The apparatus: A polarizer oriented with angle ϕ and a counter placed behind it; ii. The manual: Place the polarizer and counter within the passage of the photon; iii. If one registers the passage of the photon through the polarizer, the result is "yes" and otherwise "no".

Clearly, a property a corresponding to α_ϕ is called actual if it is certain that α_ϕ would lead to the response "yes", should we perform the experiment. In the diagram (Figure 1) of this photon, we consider some of its properties explicitly: a corresponds to α_ϕ, b to α'_ϕ, a' to $\alpha_\phi + \pi/2$ and b' to $\alpha'_\phi + \pi/2$. One immediately sees that this lattice is not Boolean since distributivity is violated in the sense that $a \wedge (b \vee a') = a$ while $(a \wedge b) \vee (a \wedge a') = 0$.

3. Dynamic Operational Quantum Logic

Contrary to the static approach outlined above we will now analyze how an actual property before alteration will induce a property to be actual afterwards and, conversely, we characterize causes for actuality. The obtained result will then give rise to DoQL, when passing via IoQL. We stress that both these approaches are still under full development. These developments were preceded by a representation theorem for deterministic evolutions of quantum systems as given in Faure et al. (1995) and for which DoQL provides an extension to non-deterministic cases. The new primitive concept (as compared to sOQL) is the notion of induction, defined in Amira et al. (1998) – see also Coecke et al. (2001) and Smets (2001):

- An induction $e \in \epsilon_s$ is a physical procedure that can be effectuated on a particular physical system s. This procedure, when carried out, might change s, modify the collection of its actual properties and thus its state, or even destroy s.

On the collection ϵ_s of all inductions performable on a physical system s we can consider two operations, one corresponding to the arbitrary choice of inductions and one corresponding to a finite concatenation of inductions. Following Amira et al. (1998) and Coecke et al. (2001), the finite concatenation of inductions $e_1, e_2, \ldots, e_n \in \epsilon_s$ is the induction $e_1 \& e_2 \& \ldots \& e_n$ which consists of first performing e_1 then e_2, then ... until e_n. The arbitrary choice of inductions in $\{e_i \mid i \in I\} \subseteq \epsilon_s$ is the induction $\bigvee_i e_i$ consisting of performing one of the e_i, chosen in any possible way.[6] Defined as such, an act of induction can for example be the assurance of a free evolution (Faure et al. 1995), indeterministic evolution (Coecke and Stubbe 1999), e.g., a measurement (Coecke and Smets 2000, 2001), or the action of one subsystem in a compound system on another one (Coecke 2000). For reasons of formal simplicity we will only focus on inductions which cannot lead to the destruction of the physical system under consideration.[7] As such we presume that an induction cannot alter the nature of a physical system, whereby we mean that what can happen is that a system's state is shifted within the given initial state space, i.e., that the actuality and potentiality of the properties in the initial property lattice is changed. This implies that the description of a physical system by means of its state space or property lattice encompasses those states in which the system may be after performing an induction. More explicitly, we can point to the particular type of properties which are actual or potential before the system is altered and which contain information about the actuality or potentiality of particular properties after the system is altered. We introduce this particular type of properties formally (Coecke et al. 2001):[8]

$$(e, a) : \epsilon \times \mathcal{L}^{op} \to \mathcal{L}^{op} : (e, a) \mapsto e.a \tag{1}$$

where the reason for reversal of the lattice order (this is what "op" in $\mathcal{L}^{op}$ stands for) will be discussed below. The property $e.a$ stands for "guaranteeing the actuality of a". Existence of a property $e.a$ is indeed operationally assured in the following way: given that a corresponds to α, $e.a$ corresponds to a definite experimental project "$e.\alpha$" of the form "first execute the induction e and then perform the definite experimental project α, and, attribute the outcome of α to $e.\alpha$" (Faure et al. 1995, Coecke 2000, Coecke et al. 2001). In terms of actuality, following Smets (2001):

> "$e.a$ is an actual property for a system in a certain realization if it is sure that a would be an actual property of the system should we perform induction e."

This explains that if $e.a$ is actual it indeed "guarantees" the actuality of a with respect to e, while if $e.a$ is potential it does not. This expression crystallizes into the idea of introducing a *causal relation*:

$$\overset{e}{\rightsquigarrow} \subseteq \mathcal{L}_1 \times \mathcal{L}_2,$$

where subscript 1 points to the lattice before e and 2 to the lattice after e, as follows (Coecke et al. 2001):

- $a \overset{e}{\rightsquigarrow} b$:= the actuality of a before e induces (or, guarantees) the actuality of b after e.

Against the background of our characterization of $e.a$ we now see that $e.a \overset{e}{\rightsquigarrow} a$ will always be valid and that

$$a \overset{e}{\rightsquigarrow} b \iff a \leq e.b,$$

so $\overset{e}{\rightsquigarrow}$ fully characterizes the action of $e.- : \mathcal{L}_2 \to \mathcal{L}_1$. In case e stands for the induction "freeze" (with obvious significance, given a referential), conceived as timeless, then $\overset{e}{\rightsquigarrow}$ reduces to the partial ordering $\leq$ of $\mathcal{L}_1 = \mathcal{L}_2$.

To link the physical-operational level to a mathematical level, we associate with every induction $e \in \epsilon_s$ a map called property propagation and a map called property causation (Coecke et al. 2001):

(1) Property Causation:

$$\bar{e}_* : \mathcal{L}_2 \to \mathcal{L}_1 : a_2 \mapsto e.a_2 = \bigvee\{a_1 \in \mathcal{L}_1 \mid a_1 \overset{e}{\rightsquigarrow} a_2\};$$

(2) Property Propagation:

$$\bar{e}^* : \mathcal{L}_1 \to \mathcal{L}_2 : a_1 \mapsto \bigwedge\{a_2 \in \mathcal{L}_2 \mid a_1 \overset{e}{\rightsquigarrow} a_2\}.$$

Given those mappings, clearly $\bar{e}_*(a_2)$ is the weakest property whose actuality guarantees the actuality of a_2 and $\bar{e}^*(a_1)$ is the strongest property whose actuality is induced by that of a_1. Further we immediately obtain the Galois adjunction[9] $\bar{e}^*(\bar{e}_*(a_2)) \leq a_2$ and $a_1 \leq \bar{e}_*(\bar{e}^*(a_1))$, denoted as $\bar{e}^* \dashv \bar{e}_*$, since

$$a \leq \bar{e}_*(b) \iff a \overset{e}{\rightsquigarrow} b \iff \bar{e}^*(a) \leq b.$$

In Coecke et al. (2001) this adjunction is referred to as "causal duality", since it expresses the dual expressibility of dynamic behavior for physical systems respectively in terms of propagation of properties and causal assignment.[10] We also recall here that this argument towards causal duality suffices to establish evolution for quantum systems, i.e., systems with the lattice of closed subspaces of a Hilbert space as property lattice, in terms of linear or anti-linear maps (Faure et al. 1995) and compoundness in terms of the tensor product of the corresponding Hilbert spaces (Coecke 2000). It also follows from the above that the action defined in Equation (1) defines a quantale module action (Coecke et al. 2001) – quantales will be discussed below. Crucial here is the fact that $(\bigvee_i e_i) \cdot \alpha$ and $\Pi_i(e_i \cdot \alpha)$ clearly define the same property $(\bigvee_i e_i) \cdot a = \bigwedge_i(e_i \cdot a)$, since both Π for definite experimental projects and $\bigvee$ for inductions express choice. Accordingly, the opposite ordering in Equation (1) then matches $\mathcal{L}$-meets with ϵ-joins.

In the above we discussed the strongest property $\bar{e}^*(a_1)$ of which actuality is induced by an induction due to actuality of a_1, but only for maximally deterministic evolutions this fully describes the system's behavior. In other cases it makes sense to consider how (true logical) disjunctions of properties propagate, as such allowing accurate representation of for example the emergence of disjunction in a perfect quantum measurement (Coecke 2002; Coecke and Smets 2000; Smets 2001) due to the uncertainty on the measurement outcome whenever the system is not in an eigenstate of this measurement.[11] First we introduce the notion of an actuality set as a set of properties of which at least one element is actual, clearly encoding logical disjunction in terms of actuality. As before we want to express propagation and causal assignment of these actuality sets. While we shift from the level of properties to sets of properties, we will have to take care that we don't loose particular information on the structure of $\mathcal{L}$, in particular its operationally derived order. The solution to this problem consists in considering a certain kind of ideal. More specifically we work with the set $DI(\mathcal{L}) \subseteq P(\mathcal{L})$, $P(\mathcal{L})$ being the powerset of $\mathcal{L}$, of which the elements are called property sets and which formally are the so-called distributive ideals of $\mathcal{L}$, introduced in a purely mathematical setting in Bruns and Lakser (1970):

- A distributive ideal is an order ideal, i.e., if $a \leq b \in I$ then $a \in I$ and $I \neq \emptyset$, and is closed under distributive joins, i.e., if $A \subseteq I \in DI(\mathcal{L})$ then $\bigvee A \in I$ whenever we have

$$\forall b \in \mathcal{L} : b \wedge \bigvee A = \bigvee \{b \wedge a \mid a \in A\}.$$

Intuitively, this choice can be motivated as follows (a much more rigid argumentation does exists):[12] i. a first choice for encoding disjunctions would be the powerset itself, however, if $a \leq b$ we don't have $\{a\} \subseteq \{b\}$ so we do not preserve order; otherwise stated, if $a < b$ then the "propositions" $\{a\}$ and $\{a, b\}$ (read: either a or b is actual) mean the same thing, since actuality of b is implied by that of a; ii. we can clearly overcome this problem by considering order ideals

$$I(\mathcal{L}) := \{\downarrow[A] | A \subseteq \mathcal{L}\} \subset P(\mathcal{L});$$

however, in case the property lattice would be a complete Heyting algebra in which all joins encode disjunctions, then A and $\{\bigvee A\}$ again mean the same thing; this redundancy is then exactly eliminated by considering distributive ideals (Coecke 2002; Coecke and Smets 2001). For $\mathcal{L}$ atomistic and $\Sigma \subseteq \mathcal{L}$, $DI(\mathcal{L}) \cong P(\Sigma)$ which implies that $DI(\mathcal{L})$ is a complete atomistic Boolean algebra (Coecke 2002).

Similarly as for properties, for property sets we can operationally motivate the introduction of a causal relation $\overset{e}{\leadsto} \subseteq DI(\mathcal{L})_1 \times DI(\mathcal{L})_2$:

- $A \overset{e}{\leadsto} B :=$ if property set A is an actuality set before e, A induces that property set B is an actuality set after e.

To every induction we associate a map called property set causation and a map called property set propagation:

(1) Property set causation

$$\hat{e}_* : DI(\mathcal{L})_2 \to DI(\mathcal{L})_1 : A_2 \mapsto \mathcal{C}(\bigcup\{A_1 \in DI(\mathcal{L})_1 \mid A_1 \stackrel{e}{\leadsto} A_2\})$$

(2) Property set propagation

$$\hat{e}^* : DI(\mathcal{L})_1 \to DI(\mathcal{L})_2 : A_1 \mapsto \bigcap\{A_2 \in DI(\mathcal{L})_2 \mid A_1 \stackrel{e}{\leadsto} A_2\}$$

where

$$\mathcal{C} : P(\mathcal{L}) \to P(\mathcal{L}) : A \mapsto \bigcap\{B \in DI(\mathcal{L}) \mid A \subseteq B\}.$$

Similar as above we obtain an adjunction $\hat{e}^* \dashv \hat{e}_*$ following from:

$$\hat{e}^*(A)_1 \subseteq A_2 \iff A_1 \stackrel{e}{\leadsto} A_2 \iff A_1 \subseteq \hat{e}_*(A_2)$$

In case the induction e stands for "freeze" we obtain $A \stackrel{e}{\leadsto} B = A \subseteq B$. Note that the join preservation that follows from the adjunction $\hat{e}^* \dashv \hat{e}_*$ expresses the physically obvious preservation of disjunction for temporal processes.

It is however also important to stress here that not all maps $\hat{e}^* : DI(\mathcal{L})_1 \to DI(\mathcal{L})_2$ are physically meaningful. Indeed, any physical induction admits mutually adjoint maps $\bar{e}^*$ and $\bar{e}_*$ with the significance discussed above, the existence of such a map $\bar{e}^* : \mathcal{L}_1 \to \mathcal{L}_2$ forcing $\hat{e}^*$ to satisfy a join continuity condition (Coecke and Stubbe 1999; Coecke et al. 2001; Coecke 2002), namely:

$$\bigvee A = \bigvee B \implies \bigvee \hat{e}^*(A) = \bigvee \hat{e}^*(B), \tag{2}$$

which indeed expresses well-definedness of a corresponding $\bar{e}^*$ since given an actuality set A, the strongest property that is actual with certainty is $\bigvee A$. It is exactly in the existence of a non-trivial condition as in Equation (2) that the non-classicality of quantum theory comes in. As such, both causal dualities, the one on the level of properties and the other on the level of property sets, provide a *physical law* on transitions, respectively condition Equation (2) and preservation of $DI(\mathcal{L})$-joins. We also want to stress a difference here with the setting in van Benthem (1994):

> "The most general model of dynamics is simply this: some system moves through a space of possibilities. Thus there is to be some set [Σ] of relevant *states* (cognitive, physical, etc.) and a family [$\{R_e | e \in \epsilon\}$] of binary *transition relations* among them, corresponding to actions that could be performed to change from one state to another. [...] Let us briefly consider a number of dynamic 'genres', [...] • Real action in the world changes actual physical states. [...] What are most general operations on actions? Ubiquitous examples are *sequential composition* and *choice*." (p. 109, 110, 112)

Thus it seems to us that the author aims to cover also the dynamic behavior of physical systems. He moreover states (van Benthem 1994):

> "The main claim of this paper is that the above systems of relational algebra and dynamic logic provide a convenient architecture for bringing out essential logical features of action and cognition." (p. 130)

In this paper we do not favor the use of relational structures for modelling physical dynamics, even classicaly! Let us motivate this perspective. It follows from the above that transitions of properties of physical systems are internally structured by the causal duality, which in the particular case of non-classical systems restricts possible transitions. Recalling that for atomistic property lattices we have $P(\Sigma) \cong DI(\mathcal{L})$ it is definitely true that any join preserving map $\hat{e}^* : DI(\mathcal{L})_1 \to DI(\mathcal{L})_2$ defines a unique relation $R_e \subseteq \Sigma_1 \times \Sigma_2$, and conversely, any relation $R \subseteq \Sigma_1 \times \Sigma_2$ defines a unique (join preserving) map $f_R : DI(\mathcal{L})_1 \to DI(\mathcal{L})_2$. Next, any relation $R \subseteq \Sigma_1 \times \Sigma_2$ has an inverse $R^{-1} \subseteq \Sigma_2 \times \Sigma_1$ and this inverse plays a major role in van Benthem (1991, 1994) as *converse action*. However, nothing assures that when $\hat{e}^*$ satisfies Equation (2) that the map $f_{R_e^{-1}} : DI(\mathcal{L})_2 \to DI(\mathcal{L})_1$ encoding the relational inverse satisfies Equation (2), and as such, has any physical significance at all.[13] Obviously, this argument applies only to non-classical systems. More generally, however, since it is the duality between causation and propagation at the $DI(\mathcal{L})$-level that guarantees preservation of disjunction we feel that it should be present in any modelization, and although relations and union preserving maps between powersets are in bijective correspondence, they have fundamentally different dual realizations: relations have inverses, and union preserving maps between powersets have adjoints, and these two do not correspond at all, respectively being encoded as (in terms of maps between powersets):

$$f_{R_e^{-1}}(A) = \{p \in \Sigma | \exists q \in A : q \in \hat{e}^*(\{p\})\}$$
$$\hat{e}_*(A) = \{p \in \Sigma | \forall q \in A : q \in \hat{e}^*(\{p\})\}$$

if it was even only by the fact that one preserves unions and the other one intersections. As such, the seemingly innocent choice of representation in terms of relations or union preserving maps between powersets does have some consequence in terms of the implementation of causal duality.

In the remaining part of this section we concentrate on the logic of actuality sets as initiated in Coecke (2002). Introduce the following primitive connectives:

$$\bigwedge_{DI(\mathcal{L})} : P(DI(\mathcal{L})) \to DI(\mathcal{L}) : \mathcal{A} \mapsto \bigcap \mathcal{A};$$

$$\bigvee_{DI(\mathcal{L})} : P(DI(\mathcal{L})) \to DI(\mathcal{L}) : \mathcal{A} \mapsto \mathcal{C}(\bigcup \mathcal{A});$$

$$\to_{DI(\mathcal{L})} : DI(\mathcal{L}) \times DI(\mathcal{L}) \to DI(\mathcal{L}) : (B, C) \mapsto \bigvee_{DI(\mathcal{L})} \{A \in DI(\mathcal{L}) \mid A \cap B \subseteq C\}$$
$$= \{a \in \mathcal{L} | \forall b \in B : a \wedge b \in C\};$$

$$\mathcal{R}_{DI(\mathcal{L})} : DI(\mathcal{L}) \to DI(\mathcal{L}) : A \mapsto \downarrow (\bigvee_{\mathcal{L}} A).$$

While the first three connectives are standard in intuitionistic logic, $\mathcal{R}_{DI(\mathcal{L})}$ should be conceived as a "resolution-connective" allowing us to recuperate the logical structure of properties on the level of property sets (Coecke 2002). In particular, for classical systems we have $\mathcal{R}_{DI(\mathcal{L})} = id_{DI(\mathcal{L})}$. Note that the condition in Equation (2) now becomes:

$$\mathcal{R}_{DI(\mathcal{L})}(A) = \mathcal{R}_{DI(\mathcal{L})}(B) \implies \mathcal{R}_{DI(\mathcal{L})}(\hat{e}^*(A)) = \mathcal{R}_{DI(\mathcal{L})}(\hat{e}^*(B)), \tag{3}$$

restricting the physically admissible transitions. Clearly, this condition is trivially satisfied for classical systems.

When concentrating on the material implication, we want to stress that on the level of $\mathcal{L}$ only for the properties in a distributive sublattice we can say the following:

$$\begin{aligned} p \models (a \to_{\mathcal{L}} b) &\iff \{p\} \cap \mu(a) \subseteq \mu(b) \\ &\iff p \in \mu(a) \Rightarrow p \in \mu(b) \\ &\iff p \models a \Rightarrow p \models b. \end{aligned}$$

In that case, this implication satisfies the strengthened law of entailment:

$$\mu(a \to_{\mathcal{L}} b) = \Sigma \iff \mu(a) \subseteq \mu(b) \iff a \leq b.$$

Note that in the non-distributive case for $a \to_{\mathcal{L}} b = a' \vee b \in \mathcal{L}$ we can only say that $(p \models a \to_{\mathcal{L}} b) \Leftarrow (p \models a \Rightarrow p \models b)$, which goes together with the fact that there are examples of orthomodular lattices for which $a' \vee b = 1$ while $a \not\leq b$. Hence in general, for $T = \{p \in \Sigma \mid p \models a \Rightarrow p \models b\}$ there will not be an element $x \in \mathcal{L}$ for which $\mu(x) = T$. There are of course examples of other implications which do satisfy the strengthened law of entailment in the orthomodular case – see for instance Kalmbach (1983). It is now our aim to focus on the implication for elements in $DI(\mathcal{L})$. First we lift the Cartan map to the level of property sets $\mu(A) := \bigcup \mu[A] \subseteq \Sigma_{\mathcal{L}}$, then we obtain the following semantical interpretation:

$$\begin{aligned} p \models (A \to_{DI(\mathcal{L})} B) &\iff \{p\} \cap \mu(A) \subseteq \mu(B) \\ &\iff p \in \mu(A) \Rightarrow p \in \mu(B) \\ &\iff p \models A \Rightarrow p \models B. \end{aligned}$$

From this it follows that

$$\mu(A \to_{DI(\mathcal{L})} B) = \{p \in \Sigma_{\mathcal{L}} \mid (p \models A) \Rightarrow (p \models B)\},$$

which allows us to reformulate the given static implication $(- \to_{DI(\mathcal{L})} -)$ as follows (Coecke nd; Coecke and Smets 2001):

$$\begin{aligned}(A \to_{DI(L)} B) &= \bigvee_{DI(L)} \{C \in DI(L) \mid \forall D \vdash C : (D \vdash A \Rightarrow D \vdash B)\} \\ &= \{c \in L \mid \forall d \leq c : (d \in A \Rightarrow d \in B)\},\end{aligned}$$

where $D \vdash A \Leftrightarrow \forall p \in \mu(D) : p \in \mu(A)$. Again this implication satisfies the strengthened law of entailment in the sense that

$$(A \to_{DI(\mathcal{L})} B) = \mathcal{L} \iff A \subseteq B$$

and is as operation the right adjoint with respect to the conjunction $\bigwedge_{DI(\mathcal{L})}$, as it is always the case for the implication connective of a complete Heyting algebra – see for example Borceux (1994) or Johnstone (1982):

$$(A \wedge_{DI(\mathcal{L})} -) \dashv (A \to_{DI(\mathcal{L})} -).$$

As for the static material implication defined above, we now want to look for a dynamic propagation-implication satisfying the following:

$$(A \stackrel{e}{\to} B) = \mathcal{L} \iff A \stackrel{e}{\rightsquigarrow} B.$$

The candidate which naturally arises is (Coecke (nd), Coecke and Smets 2001):

$$(A \stackrel{e}{\to} B) := \{c \in L \mid \forall d \leq c : (d \in A \Rightarrow \hat{e}^*(\downarrow d) \subseteq B)\}.$$

indicated by a semantical interpretation as

$$\mu(A \to_{DI(\mathcal{L})} B) = \{p \in \Sigma_{\mathcal{L}} \mid (p \models A) \Rightarrow (\hat{e}^*(\{p\}) \models B)\}.$$

In case e stands for the induction "freeze" we see that $\stackrel{e}{\to}$ reduces to $\to_{DI(\mathcal{L})}$. Similar as in the static case, we can find an induction-labeled operation as left adjoint to the dynamic propagation-implication, i.e.,

$$(A \otimes_e -) \dashv (A \stackrel{e}{\to} -) \quad \text{with} \quad A \otimes_e B := \hat{e}^*(A \wedge_{DI(\mathcal{L})} B).$$

It is important here to notice that this dynamic conjunction is a commutative operation. Since $\hat{e}^*$ preserves joins and since in $DI(\mathcal{L})$ binary meets distribute over arbitrary joins (being a complete Heyting algebra) we moreover have (Coecke (nd)):

$$A \otimes_e (\bigvee_{DI(\mathcal{L})} \mathcal{B}) = \bigvee_{DI(\mathcal{L})} \{A \otimes_e B \mid B \in \mathcal{B}\},$$

so $DI(\mathcal{L})$ is equipped with operations $\stackrel{e}{\to}$ and $\otimes_e$, for every $e \in \epsilon_s$ the latter yielding a multiplicative lattice $(DI(\mathcal{L}), \bigvee_{DI(\mathcal{L})}, \otimes_e)$. Let us give the definition of a quantale (Rosenthal 1990, 1996, Paseka and Rosicky 2000):

- A *quantale* is a complete lattice Q together with an associative binary operation $\circ$ that satisfies $a \circ (\bigvee_i b_i) = \bigvee_i (a \circ b_i)$ and $(\bigvee_i b_i) \circ a = \bigvee_i (b_i \circ a)$ for all $a, b_i \in Q$.

For each induction e we obtain that $(DI(\mathcal{L}), \bigvee_{DI(\mathcal{L})}, \otimes_e)$ has a commutative multiplication since $\otimes_e$ is a commutative operation. In case e stands for "freeze", the mentioned structure becomes a locale $(DI(\mathcal{L}), \bigvee_{DI(\mathcal{L})}, \wedge_{DI(\mathcal{L})})$, i.e., a complete Heyting algebra. Recall here that a locale is a quantale with as quantale product the meet-operation of the complete lattice, and one verifies that this definition exactly coincides with that of a complete Heyting algebra – for details we refer again to Borceux (1994) or Johnstone (1982).

As for the propagation-implications which, when valid, express a forward causal relation between property sets, we can introduce causation-implications. The relation to which these causation-implications match will be a backward relation introduced as follows (Coecke and Smets 2001):

- $A \overset{e}{\hookleftarrow} B$:= If property set B is necessarily an actuality set after e then property set A was an actuality set before e.

Formally we see

$$A \overset{e}{\hookleftarrow} B \iff \hat{e}_*(B) \subseteq A.$$

The causation-implications we want to work with now have to satisfy:

$$A \overset{e}{\leftarrow} B = \mathcal{L} \iff A \overset{e}{\hookleftarrow} B.$$

The candidate which satisfies this condition is:

$$(A \overset{e}{\leftarrow} B) := \{c \in L \mid \forall d \leq c : (d \in A \Leftarrow \hat{e}^*(\downarrow d) \subseteq B)\}.$$

indicated by a semantical interpretation as

$$\mu(A \leftarrow_{DI(\mathcal{L})} B) = \{p \in \Sigma_{\mathcal{L}} \mid (p \models A) \Leftarrow (\hat{e}^*(\{p\}) \models B)\}.$$

As such we see that when $A \overset{e}{\leftarrow} B$ is valid (i.e., equal to $\mathcal{L}$) then it expresses that if property set B is an actuality set after e then property set A was an actuality set before e. Again we have a left adjoint for the causation-implication:

$$(- \,{}_e\!\otimes B) \dashv (- \overset{e}{\leftarrow} B) \quad \text{with} \quad A \,{}_e\!\otimes B := A \wedge_{DI(\mathcal{L})} \hat{e}_*(B).$$

Thus we can additionally equip $DI(\mathcal{L})$ with $\overset{e}{\leftarrow}$ and ${}_e\otimes$, for each $e \in \epsilon_s$ the latter yielding non-commutative multiplicative lattices $(DI(\mathcal{L}), \bigwedge_{DI(\mathcal{L})}, {}_e\otimes)$, i.e., with a distributive property with respect to meets. Note that the preservation of joins for propagation versus that of meets for causation reflects here in a two-sided distributivity respectively with respect to joins and meets. Indeed, since we have:

$$(\mathcal{L} \otimes_e -) = \hat{e}^*(-) \qquad\qquad (\mathcal{L} \,{}_e\!\otimes -) = \hat{e}_*(-)$$

this distributivity truly encodes the respective join and meet preservation and consequently, also the causal duality.

It is important to remark that the semantics we obtain is the complete Heyting algebra of actuality sets $DI(\mathcal{L})$ equipped with additional dynamic connectives to express causation and propagation:

$$\Big(DI(\mathcal{L}), \mathcal{R}_{DI(\mathcal{L})}, \bigvee_{DI(\mathcal{L})}, \bigwedge_{DI(\mathcal{L})}, \neg, \{\otimes_e, {}_e\otimes, \stackrel{e}{\rightarrow}, \stackrel{e}{\leftarrow}, \stackrel{e}{\neg} \mid e \in \epsilon\}\Big).$$

It turns out that we have a forward negation $\neg$ which does not depend on e, and thus coincides with that of "freeze", i.e., the static one, and a backward negation $\stackrel{e}{\neg}$ which by contrast does depend on e – for a discussion of these negations we refer to Coecke (nd) and Smets (2001). Note here that $DI(\mathcal{L})$ is also a left quantale module for $\hat{\epsilon} = \{\hat{e} \mid e \in \epsilon_s\}$ when considering the pointwise action of $(e.-)$. This then intertwines the two multiplicative structures that emerge in our setting. We end this paragraph by mentioning that it is possible to implement other kinds of implications on $DI(\mathcal{L})$ that extend the causal relation. As an example, it is possible to extend $DI(\mathcal{L})$ with bi-labeled families of non-commutative multiplications rendering bi-labeled implications, some of them extending the ones presented in this paper.

In Coecke and Smets (2001) it is argued that the Sasaki adjunction is an incarnation of causal duality for the particular case of a quantum measurement φ_a with a projector (on a) as corresponding self-adjoint operator. This has as a striking consequence, since validity of the Sasaki adjunction is equivalent to "orthomodularity", sOQL embodies a hidden dynamical ingredient which is algebraically identifiable as orthomodularity. One could as such argue that the necessity of the passage from sOQL to DoQL was already announced within sOQL itself, it was just waiting to be revealed. As a more radical statement one could say that due to this hidden dynamical ingredient, it is impossible to give a full sense to quantum theory in logical terms within an essentially static setting. Following Coecke and Smets (2001), this fact can be deduced from DoQL by eliminating the emergent disjunctivity when introducing modalities with respect to actuality and conditioning. We can in that case derive that the labeled dynamic hooks that encode quantum measurements act on properties as

$$(a_1 \stackrel{\varphi_a}{\rightarrow} a_2) := (a_1 \rightarrow_{\mathcal{L}} (a \stackrel{S}{\rightarrow} a_2)) \quad \text{and} \quad (a_1 \stackrel{\varphi_a}{\leftarrow} a_2) := ((a \stackrel{S}{\rightarrow} a_2) \rightarrow_{\mathcal{L}} a_1)$$

where $(- \stackrel{S}{\rightarrow} -)$ is the well-known Sasaki hook and we identify a and $\{a\}$. One could say that the transition from either classical or intuitionistic logicality to "true"quantum logicality entails besides the introduction of an additional unary connective "operational resolution" the shift from a binary implication connective to a ternary connective where two of the arguments have an ontological connotation and the third, the new one, an empirical.

4. Comparison with Linear Logic

We will analyze another logic of dynamics, namely linear logic, while focusing on the differences with DoQL especially with respect to the multiplicative structures mentioned above. We intend to give a brief overview of the basic ideas behind "Linear logic" as introduced in Girard (1987, 1989). It's categorical semantics in terms of a *-autonomous category appeared in Barr (1979) and it is fair to say that already in Lambek (1958) a non-commutative fragment of linear logic was present. We follow the discussion of Smets (2001).

The main advantage linear logic has with respect to classical/intuitionistic logic is that it allows us to deal with actions versus situations in the sense of stable truths (Girard 1989). This should be understood in the sense that linear logic is often called *a resource sensitive logic*. The linear logical formulas can be conceived as expressing finite resources, the classical formulas then being interpretable as corresponding to unlimited or eternal resources. Allowing ourselves to be a bit more formal on this matter, resource sensitivity is linked to the explicit control of the weakening and contraction rules. As structural rules, weakening and contraction will be discarded in the general linear logical framework. Note that in non-commutative linear logic, to which we will come back later, the exchange-rule will also be dropped. We use $\mathfrak{A}$, $\mathfrak{B}$ for sequences of well formed formulas and $\mathfrak{a}$, $\mathfrak{b}$ for well formed formulas. Sequents are conceived as usual:

$$(weakening\,L)\quad \frac{\mathfrak{A} \longrightarrow \mathfrak{B}}{\mathfrak{A}, \mathfrak{a} \longrightarrow \mathfrak{B}} \qquad (weakening\,R)\quad \frac{\mathfrak{A} \longrightarrow \mathfrak{B}}{\mathfrak{A} \longrightarrow \mathfrak{B}, \mathfrak{a}}$$

$$(contraction\,L)\quad \frac{\mathfrak{A}, \mathfrak{a}, \mathfrak{a} \longrightarrow \mathfrak{B}}{\mathfrak{A}, \mathfrak{a} \longrightarrow \mathfrak{B}} \qquad (contraction\,R)\quad \frac{\mathfrak{A} \longrightarrow \mathfrak{a}, \mathfrak{a}, \mathfrak{B}}{\mathfrak{A} \longrightarrow \mathfrak{a}, \mathfrak{B}}$$

$$(exchange\,L)\quad \frac{\mathfrak{A}_1, \mathfrak{a}, \mathfrak{b}, \mathfrak{A}_2 \longrightarrow \mathfrak{B}}{\mathfrak{A}_1, \mathfrak{b}, \mathfrak{a}, \mathfrak{A}_2 \longrightarrow \mathfrak{B}} \qquad (exchange\,R)\quad \frac{\mathfrak{A} \longrightarrow \mathfrak{B}_1, \mathfrak{a}, \mathfrak{b}, \mathfrak{B}_2}{\mathfrak{A} \longrightarrow \mathfrak{B}_1, \mathfrak{b}, \mathfrak{a}, \mathfrak{B}_2}$$

Dropping weakening and contraction implies that linear formulas cannot be duplicated or contracted at random, in other words, our resources are restricted. An important consequence of dropping these two structural rules is the existence of two kinds of ' 'disjunctions" and "conjunctions".[14] We will obtain a so-called additive disjunction $\oplus$ and additive conjunction $\sqcap$ and a so-called multiplicative disjunction $\wp$ and multiplicative conjunction $\otimes$. The following left and right rules will make their differences clear.

$$(\sqcap R)\ \frac{\mathfrak{A} \rightarrow \mathfrak{a}, \mathfrak{B} \quad \mathfrak{A} \longrightarrow \mathfrak{b}, \mathfrak{B}}{\mathfrak{A} \longrightarrow \mathfrak{a} \sqcap \mathfrak{b}, \mathfrak{B}} \qquad (\oplus L)\ \frac{\mathfrak{A}, \mathfrak{a} \longrightarrow \mathfrak{B} \quad \mathfrak{A}, \mathfrak{b} \longrightarrow \mathfrak{B}}{\mathfrak{A}, \mathfrak{a} \oplus \mathfrak{b} \longrightarrow \mathfrak{B}}$$

$$(\sqcap L)\ \frac{\mathfrak{A}, \mathfrak{a} \longrightarrow \mathfrak{B}}{\mathfrak{A}, \mathfrak{a} \sqcap \mathfrak{b} \longrightarrow \mathfrak{B}} \quad \frac{\mathfrak{A}, \mathfrak{b} \longrightarrow \mathfrak{B}}{\mathfrak{A}, \mathfrak{a} \sqcap \mathfrak{b} \longrightarrow \mathfrak{B}} \qquad (\oplus R)\ \frac{\mathfrak{A} \longrightarrow \mathfrak{a}, \mathfrak{B}}{\mathfrak{A} \longrightarrow \mathfrak{a} \oplus \mathfrak{b}, \mathfrak{B}} \quad \frac{\mathfrak{A} \longrightarrow \mathfrak{b}, \mathfrak{B}}{\mathfrak{A} \longrightarrow \mathfrak{a} \oplus \mathfrak{b}, \mathfrak{B}}$$

$$(\otimes R)\ \frac{\mathfrak{A}_1 \longrightarrow \mathfrak{a}, \mathfrak{B}_1 \quad \mathfrak{A}_2 \longrightarrow \mathfrak{b}, \mathfrak{B}_2}{\mathfrak{A}_1, \mathfrak{A}_2 \longrightarrow \mathfrak{a} \otimes \mathfrak{b}, \mathfrak{B}_1, \mathfrak{B}_2} \qquad (\wp L)\ \frac{\mathfrak{A}_1, \mathfrak{a} \longrightarrow \mathfrak{B}_1 \quad \mathfrak{A}_2, \mathfrak{b} \longrightarrow \mathfrak{B}_2}{\mathfrak{A}_1, \mathfrak{A}_2, \mathfrak{a} \wp \mathfrak{b} \longrightarrow \mathfrak{B}_1, \mathfrak{B}_2}$$

$$(\otimes L)\ \frac{\mathfrak{A}, \mathfrak{a}, \mathfrak{b} \longrightarrow \mathfrak{B}}{\mathfrak{A}, \mathfrak{a} \otimes \mathfrak{b} \longrightarrow \mathfrak{B}} \qquad (\wp R)\ \frac{\mathfrak{A} \longrightarrow \mathfrak{a}, \mathfrak{b}, \mathfrak{B}}{\mathfrak{A} \longrightarrow \mathfrak{a} \wp \mathfrak{b}, \mathfrak{B}}$$

By allowing the structural rules and by using ($\otimes$) we can express the ($\sqcap$)-rules and vice versa. A similar result can be obtained for the ($\oplus$) and ($\wp$)-rules.

To understand linear logic, it is necessary to take a look at the intuitive meaning of the above additives and multiplicatives derived from their use by the above rules. First it is important to note that to obtain $\otimes$ in a conclusion, no sharing of resources is allowed, while the contrary is the case for $\sqcap$. Similarly, there is a difference between $\wp$ and $\oplus$. In the line of thought exposed in Girard (1989), the meanings to be attached to the connectives are the following:

- $\mathfrak{a} \otimes \mathfrak{b}$ means that both resources, $\mathfrak{a}$ and $\mathfrak{b}$ are given simultaneously;
- $\mathfrak{a} \sqcap \mathfrak{b}$ means that one may choose between $\mathfrak{a}$ and $\mathfrak{b}$;
- $\mathfrak{a} \oplus \mathfrak{b}$ means that one of both resources, $\mathfrak{a}$ or $\mathfrak{b}$, is given though we have lack of knowledge concerning the exact one;
- $\mathfrak{a} \wp \mathfrak{b}$ expresses a constructive disjunction.

The meaning of $\wp$ becomes clearer *when* we follow J. Y. Girard in his construction that every atomic formula of his linear logical language has by definition a negation $(-)^{\perp}$. Running a bit ahead of our story, the meaning of $\mathfrak{a} \wp \mathfrak{b}$ will now come down to the situation where "if" not $\mathfrak{a}$ is given "then" $\mathfrak{b}$ is given and "if" not $\mathfrak{b}$ is given "then" $\mathfrak{a}$ is given. Of course this explanation is linked to the commutative case where a linear logical implication is *defined* as $\mathfrak{a}^{\perp} \wp \mathfrak{b} := \mathfrak{a} \multimap \mathfrak{b}$ which by transposition equals $\mathfrak{b}^{\perp} \multimap \mathfrak{a}^{\perp}$. In the same sense, only in the commutative case where $\mathfrak{a} \otimes \mathfrak{b}$ equals $\mathfrak{b} \otimes \mathfrak{a}$, does it make sense to say that $\mathfrak{a} \otimes \mathfrak{b}$ comes down to *simultaneous* given resources.

We will be more specific on the linear logical implication and analyze the underlying philosophical ideas as presented in Girard (1989) where of course $\multimap$ is defined by means of $\wp$. What is important is that the linear implication should mimic exactly what happens when a non-iteratable action is being performed, where we conceive of a non-iteratable action to be such that after its performance the initial resources are not available any more *as* initial resources. The linear implication should as such express the *consumption of initial resources* and simultaneously the *production of final resources*. Indeed, as stated in Girard (1995), the linear implication $\multimap$ expresses a form of causality: $\mathfrak{a} \multimap \mathfrak{b}$ is to be conceived as "from $\mathfrak{a}$ get $\mathfrak{b}$". More explicitly (Girard 1989):

> "A causal implication cannot be iterated since the conditions are modified after its use; this process of modification of the premises (conditions) is known in physics as *reaction*." (p. 72)

or in Girard (2000):

> "C'est donc une vision *causale* de la déduction logique, qui s'oppose à la pérennité de la vérité traditionelle en philosophie et en mathé-matiques. On a ici des vérités fugaces, contingentes, dominées par l'idée de ressource et d'action." (p. 532)

If we understand this correctly, the act of consumption and production is called a non-iteratable action while the process of modification of initial conditions, the deprivation of resources, is called reaction. The idea of relating action and reaction is in a sense metaphorically based on Newton's action-reaction principle in physics. Girard uses this metaphor also when he explains why every formula has by definition a linear negation which expresses a duality or change of standpoint (Girard 1989):

> "action of type A = reaction of type $A^{\perp}$." (p.77)

or in Girard (2000):

> "Concrètement la négation correspond à la dualité "action/réaction" et pas du tout à l'idée de ne pas effectuer une action: typiquement lire/écrire, envoyer/recevoir, sont justicibles de la négation linéaire." (p. 532)

In terms of functional programming or categorical semantics, the negation represents an input-output duality. In game terms, it is an opponent-proponent duality. In Girard (1989) the notion of a reaction of type $A^{\perp}$, as dual to an action of type A, is quite mysterious and not further elaborated. The only thing Girard mentions is that it should come down to an "inversion of causality, i.e., of the sense of time" (Girard 1989). It is also not clear to us what this duality would mean in a commutative linear logic context when we consider the case of an action of type $\mathfrak{a} \multimap \mathfrak{b}$ (or $\mathfrak{b}^{\perp} \multimap \mathfrak{a}^{\perp}$) and a reaction of type $\mathfrak{a} \otimes \mathfrak{b}^{\perp}$ (or $\mathfrak{b}^{\perp} \otimes \mathfrak{a}$) – where $\mathfrak{a} \multimap \mathfrak{b} = \mathfrak{a}^{\perp} \wp \mathfrak{b}$ and $(\mathfrak{a}^{\perp} \wp \mathfrak{b})^{\perp} = \mathfrak{a} \otimes \mathfrak{b}^{\perp}$. Thus we tend to agree with Girard (1989) where he says that this discussion involves not standard but non-commutative linear logic. Indeed, against the background of the causation-propagation duality elaborated upon above, if Girard has something similar to causation in mind, we know that what is necessary is an implication and "non-commutative" conjunction which allow us to express causation. As we will show further on, switching from standard to non-commutative linear logic effects the meanings of the linear implication and linear negation.

Leaving this action-reaction debate aside, we can now explain why our given interpretation of non-iterability is quite subtle. First note that in Girard's standard linear logic, $\vdash \mathfrak{a} \multimap \mathfrak{a}$ is provable from an empty set of premises. As such $\multimap$ represents in $\mathfrak{a} \multimap \mathfrak{a}$ the identity-action which does not really change resources, but only translates initial ones into final ones. And although we could perform the action twice in the following sense: $\mathfrak{a}_1 \multimap \mathfrak{a}_2 \multimap \mathfrak{a}_3$ – for convenience we labeled the resources – this still does not count as an iteratable action since $\mathfrak{a}_1$ is to be conceived as an initial resource, different from $\mathfrak{a}_2$, the final resource of the first action. But we can go further in this discussion and follow Girard in stipulating the fact that we may still encounter situations in which the picture of "consuming all initial resources" does not hold. Linear logic henceforth allows also the expression of those actions which deal with stable situations and which are iteratable. In the latter case the use of exponentials is necessary where for instance the exponential ! gives $!\mathfrak{a}$ the meaning that $\mathfrak{a}$'s use as a resource is unlimited.

In Girard (1989), Girard discusses the link between states, transitions and the linear implication. In particular, for us it is the following statement which places the linear implication in an interesting context, thinking of course about the above discussed DoQL, (Girard 1989):

> "In fact, we would like to represent states by formulas, and transitions by means of implications of states, in such a way that the state S' is accessible from S exactly when $S \multimap S'$ is provable from the transitions, taken as axioms." (p. 74)

In Girard (1989) this statement applies to for instance systems such as Petri nets, Turing machines, chessboard games, etc. Focusing "in this sense" on physical systems, and using $\multimap$ for transitions of states, it becomes interesting to investigate how $\multimap$ can be conceived in the context of our logic of actuality sets. While a formal comparison on the semantical level will be given in the next paragraph we have to stress here that there is on the methodological level the following point of difference between DoQL and linear logic: contrary to DoQL, it should be well understood that linear logic is not a temporal logic, no preconceptions of time or processes is built into it. More explicitly (Girard 1989):

> "Linear logic is eventually about time, space and communication, but is not a temporal logic, or a kind of parallel language: such approaches try to develop preexisting conceptions about time, processes, etc. In those matters, the general understanding is so low that one has good chances to produce systems whose aim is to avoid the study of their objects [...] The main methodological commitment is to refuse any *a priori* intuition about these objects of study, and to assume that (at least part of) the temporal, the parallel features of computation are already in Gentzen's approach, but are simply hidden by taxonomy." (p. 104)

As such DoQL started out with a different methodology. The objects of study are well-known "scientific objects" such as physical systems and their properties and the inductions performable on physical systems. In a sense this information has been encrypted in the formulas we used. All dynamic propagation- and causation-implications have been labeled by inductions, and this is different from the linear logical implications which are used to express any (non-specified) transition. In view of quantum theory it is indeed the case that one cannot speak about observed quantities without specification of the particular measurement one performs, and as such, the corresponding induction that encodes von Neumann's projection postulate, or in more fashionable terms, state-update. Exactly this could form an argument against applying the linear logical implications in a context of physical processes. Thus, we are not tempted to agree with the proposal in Pratt (1993) of adding linear logical connectives as an extension to quantum logic, but rather focus on the development of a new logical syntax which will however have some definite similarities with linear logic, in particular with its quantale semantical fragment.

In order to get a grasp on the quantale semantics of linear logic we have to say something about its non-commutative variants. In the literature on non-commutative linear logic, two main directions emerge. In a first direction one introduces non-commutativity of the multiplicatives by restricting the exchange-rule to circular permutations while in a second direction one completely drops all structural rules. Concentrating on the first direction, here linear logic with a cyclic exchange rule is called *cyclic linear logic* and has mainly been developed by D. N. Yetter in Yetter (1990), though we have to note that Girard already makes some remarks on cyclic exchanges in Girard (1989). More explicitly we see that the restriction to cyclic permutations means that we consider the sequents as written on a circle (Girard 1989). This then should come down to the fact that $\mathfrak{a}_1 \otimes \ldots \otimes \mathfrak{a}_{n-1} \otimes \mathfrak{a}_n \vdash \mathfrak{a}_n \otimes \mathfrak{a}_1 \otimes \ldots \otimes \mathfrak{a}_{n-1}$ is provable in cyclic linear logic. Of course the meaning of $\otimes$ with respect to the standard case changes in the sense that it expresses now "and then" (Yetter 1990) or when following Girard (1989) it means that "in the product $\mathfrak{b} \otimes \mathfrak{a}$, the second component is done before the first one". As we will explain in the next paragraph on Girard quantales, it is exactly the difference between $- \otimes \mathfrak{a}$ and $\mathfrak{a} \otimes -$ which leads to the introduction of two different implication-connectives in cyclic linear logic: $\multimap$ and $\multimapinv$.

Given a unital quantale $(Q, \bigvee, \otimes)$, with 1 as the multiplicative neutral element with respect to $\otimes$, it then follows that the endomorphisms $a \otimes -, - \otimes a : Q \to Q$ have right adjoints, $a \multimap -$ and $- \multimapinv a$ respectively:

$$a \otimes c \leq b \Leftrightarrow c \leq a \multimap b \qquad c \otimes a \leq b \Leftrightarrow c \leq b \multimapinv a$$

$$a \multimap b = \bigvee\{c \in Q : a \otimes c \leq b\} \qquad b \multimapinv a = \bigvee\{c \in Q : c > \otimes a \leq b\}.$$

We know that in the standard linear logic as presented by Girard, the following holds $\mathfrak{a} \multimap \mathfrak{b} = \mathfrak{a}^{\perp} \wp \mathfrak{b} = (\mathfrak{a} \otimes \mathfrak{b}^{\perp})^{\perp} = (\mathfrak{b}^{\perp} \otimes \mathfrak{a})^{\perp} = \mathfrak{b} \wp \mathfrak{a}^{\perp} = \mathfrak{b}^{\perp} \multimap \mathfrak{a}^{\perp}$. In cyclic linear logic where we have now two implications, things change in the sense that we obtain:

$$(\mathfrak{a} \otimes \mathfrak{b})^{\perp} = \mathfrak{b}^{\perp} \wp \mathfrak{a}^{\perp} \qquad (\mathfrak{a} \wp \mathfrak{b})^{\perp} = \mathfrak{b}^{\perp} \otimes \mathfrak{a}^{\perp}$$

$$\mathfrak{a} \multimap \mathfrak{b} = \mathfrak{a}^{\perp} \wp \mathfrak{b} \qquad \mathfrak{b} \multimapinv \mathfrak{a} = \mathfrak{b} \wp \mathfrak{a}^{\perp}$$

To be more explicit, the linear negation can be interpreted in the unital quantale $(Q, \bigvee, \otimes, 1)$ by means of a *cyclic dualizing element*, which can be defined as follows (Rosenthal 1990, Yetter 1990):

- An element $\perp \in Q$ is dualizing iff $\perp \multimapinv (a \multimap \perp) = a = (\perp \multimapinv a) \multimap \perp$ for all $a \in Q$. It is cyclic iff $a \multimap \perp = \perp \multimapinv a$, for every $a \in Q$.

Here the operation $- \multimap \bot$ or equivalently $\bot \circ\!\!- -$ is called the linear negation and can be written as $(-)^\bot$. Note that a unital quantale with a cyclic dualizing element $\bot$ is called a *Girard quantale*, a notion having been introduced in Yetter (1990). These Girard quantales can be equipped with modal operators to interpret the linear logical exponentials and form as such a straightforward semantics for the cyclic as well as standard linear logical syntax. In the latter case $a \multimap b = b \circ\!\!- a$. The disadvantage suffered by cyclic linear logic is that it is still not "non-commutative enough to properly express time's arrow" (Yetter 1990). Indeed, if $\multimap$ is to be conceived as a causal implication then it would have been nice to conceive $\circ\!\!-$ as expressing past causality, though this interpretation is too misleading according to (Girard 1989). In a way we agree with him since in cyclic linear logic $\mathfrak{a} \multimap \mathfrak{b}$ and $\mathfrak{a}^\bot \circ\!\!- \mathfrak{b}^\bot$ are the same – in the sense that they are both equal to $\mathfrak{a}^\bot \wp \mathfrak{b}$.

Focusing on the second direction in the non-commutative linear logical literature, we first have to mention the work of J. Lambek. Lambek's *syntactic calculus* originated in Lambek (1958) and as Girard admits, is the non-commutative ancestor of linear logic. However it has to be mentioned that Lambek's syntactic calculus, as originally developed against a linguistic background, is essentially multiplicative and intuitionistic. Later on Lambek extended his syntactic calculus with additives and recently renamed his formal calculus *bilinear logic*. In the same direction we can place the work of V. M. Abrusci who developed a non-commutative version of the intuitionistic linear propositional logic in Abrusci (1990) and of the classical linear propositional logic in Abrusci (1991). Specific to Abrusci's work, however, is the fact that a full removal of the exchange rule requires the introduction of two different negations and two different implications. To explain this in detail we switch to the semantical level of quantales; where we want to note that Abrusci works in Abrusci (1990, 1991) with the more specific structure of phase spaces which as proved in Rosenthal (1990) are examples of quantales. This then leads to the fact that for $a \in Q$: $\bot \circ\!\!-(a \multimap \bot) \neq (\bot \circ\!\!- a) \multimap a$. Here we can follow Abrusci and define $\bot \circ\!\!-(a \multimap \bot) = {}^\bot(a^\bot)$ and $(\bot \circ\!\!- a) \multimap \bot = ({}^\bot a)^\bot$, where on the syntactical level $\mathfrak{a}^\bot$ is called the linear postnegation, ${}^\bot\mathfrak{a}$ the linear retronegation, $\mathfrak{a} \multimap \mathfrak{b}$ the linear postimplication and $\mathfrak{b} \circ\!\!- \mathfrak{a}$ the linear retroimplication. Further on the syntactical level we obviously have $\mathfrak{a}^\bot \wp \mathfrak{b} = \mathfrak{a} \multimap \mathfrak{b}$ while $\mathfrak{b} \wp {}^\bot\mathfrak{a} = \mathfrak{b} \circ\!\!- \mathfrak{a}$. Let us finish of this paragraph with a note on the fact that there is of course much more to say about (non)-commutative linear logic, indeed contemporary research is in full development and heads in the direction of combining cyclic linear logic with commutative linear logic, we however limit ourselves for the time being to the overview given.

Given the above expositions, it follows that $(DI(\mathcal{L}), \bigvee_{DI(\mathcal{L})}, \otimes_e)$, the multiplicative fragment emerging for propagation for a specific induction e, provides an example of the quantale obtained for commutative linear logic. Note that the difference of course lies in the fact that in commutative linear logic no retro-implication different from $\multimap$ is present, not even as an additional structure. In this respect, to obtain a retro-implication in linear logic it is necessary to move to a non-commutative linear logic providing a single quantale in which to interpret $\multimap$ and $\circ\!-$, in sharp contrast to our constructions allowing the interpretation of $\stackrel{e}{\rightarrow}$ and $\stackrel{e}{\leftarrow}$ for each specific e. Finally, let us note that while the implication $\multimap$, when focusing on it as an implication expressing simultaneous consumption of initial resources and production of final resources, is much stronger than $\stackrel{e}{\rightarrow}$, it can nevertheless be reconstructed within the framework of DoQL, for which we refer to Smets (2001).

5. Conclusion

It seems to us that an actual attitude towards the logic of dynamics ought to be pluralistic, as it follows from our two main paradigmatic examples, dynamic operational quantum logic and Girard's linear logic.

The mentioned attempts that aim to integrate the logic of dynamics as it emerges from for example physical and proof theoretic considerations fail either on formal grounds or due to conceptual inconsistency. We indeed indicated a structural difference between dynamic operational quantum logic and van Benthem's general dynamic logic, and focused on the different methodology of linear logic.

We end by mentioning two promising recent alternative approaches in relating quantum features and linear logic, in Blute et al. (2001) in terms of polycategories and deduction systems and in Abramsky and Coecke (2002) in terms of geometry of interaction in categorical format, that is, in terms of traced monoidal categories. It is however to soon to obtain conclusions from these lines of thought.

Acknowledgements

We thank David Foulis, Jim Lambek, Constantin Piron and Isar Stubbe for discussions and comments that have led to the present content and form of the presentation in this paper. Part of the research reflected in this paper was performed by Bob Coecke at *McGill University, Department of Mathematics and Statistics, Montreal* and *Imperial College of Science, Technology & Medicine, Theoretical Physics Group, London*; and part as a Postdoctoral Researcher at the European TMR Network "Linear Logic in Computer Science". Sonja Smets is Postdoctoral Researcher at Flanders' Fund for Scientific Research.

Notes

1 From now on OQL will only refer to Geneva School operational quantum logic.

2 It is exactly the particular operational foundation of ontological concepts that has caused a lot of confusion with respect to this approach, including some attacks on it due to misunderstandings

stemming from identification of "what is", "what is observed". "what will be observed", "what would be observed", "what could be observed", etc. We don't refer to these papers but cite one that refutes them in a more than convincing way, namely Foulis and Randall (1984). We recall here that it were D. J. Foulis and C. H. Randall who developed an empirical counterpart to C. Piron's ontological approach (Foulis and Randall 1972; Foulis et al. 1983; Randall and Foulis 1983). We also quote the review of R. Piziak in Mathematical Reviews of one of these attacks of Piron's approach exposing their somewhat doubtful aims (MR86i:81012):

> "... in fact, they confused the very essence of Piron's system of questions and propositions, the sharp distinction between properties of a physical system and operationally testable propositions about the system. [...] In their reply to Foulis and Randall, HT ignore the list of mathematical errors, confusions and blunders in their papers except for one (a minor one at that). HT simply reissue their challenge and dismiss the work of Foulis and Randall as well as Piron as being "useless from the physicist's point of view". However, when one finds an argument in HTM (the main theorem of their 1981 paper) to the effect that the failure to prove the negation of a theorem constitutes a proof of the theorem, one might form a different opinion as to whose work is 'useless'. It is right and proper for any scientific work to be scrutinized and criticized according to its merits. Indeed, this is a main impetus to progress. But if mathematics is to be used as a tool of criticism, let it be used properly."

See also Smets (2001) for an overview of most of the criticism and its refutal on sOQL. We refer to Coecke (nd) for a formulation of sOQL where a conceptually somewhat less rigid, but more general perspective is proposed, avoiding the notion of test or definite experimental project in the definition of a physical property as an ontological quality of a system. One of the motivations for this reformulation is exactly the confusion that the current formulation seems to cause – although there is definitely nothing wrong with it as the truly careful reader knows, on the contrary in fact.

[3] We are not implying that our scientific theories are to be based on obtained measurement-results. The oft-drawn conclusion from this stating that "ontological existence is independent from any measurement or observation" also holds in our view. It is however specific for our position, which may not be share by every scientific realist, that our *knowledge* of what exists is linked to what we could measure, stated counterfactually. As such we adopt an "endo perspective", measurements are not a priori part of our universe of discourse but incorporated in a conditional way, it is in this sense that e.g., two not simultaneously observable properties which ontologically exist, can both unproblematically be incorporated in our description. We refer to Coecke and Smets (2001) for more details on the "endo versus exo perspective". A somewhat related view we want to draw to your attention is put forward in Ghins (2000), where he argues that from the affirmation of some "existence", under a criterion of existence based on the conditions of "presence" and "invariance", certain counterfactuals should reasonably also be affirmed.

[4] We are well aware of the fact that several correspondence theories of truth have been put forward and have also been criticised. Adhering to some form of scientific realism does not necessarily imply that one accepts a correspondence theory, even more it has been suggested that the debate on the notion of truth can be cut loose from the debate on realism versus anti-realism – see for instance Horwich (1997), Tarski (1944). Still, against the background of sOQL, we are sympathetic towards a contemporary account of a Tarskian-style semantic correspondence theory – see for example Niiniluoto (1999, §3.4).

[5] In the line of C. Piron (1981) we note that in general, the actual performance of a definite experimental project can at most serve to prove the falsity of a physicists assumptions, it cannot hand out a prove for them to be true.

[6] Note here the similarity with the generation of a quantale structure within the context of process semantics for computational systems sensu Abramsky and Vickers (1993) and Resende (2000).

[7] Different approaches however do exist for considering potentially destructive measurements, see for example Faure et al. (1995), Amira et al. (1998), Coecke and Stubbe (1999), Coecke et al. (2001) and in particular Sourbron (2000).

[8] From this point on we will identify those inductions which have the same action on properties, i.e., we abstract from the physical procedure to its transitional effect.

[9] A pair of maps $f^* : L \to M$ and $f_* : M \to L$ between posets L and M are *Galois adjoint*, denoted by $f^* \dashv f_*$, if and only if $f^*(a) \le b \Leftrightarrow a \le f_*(b)$ if and only if $\forall a \in M : f^*(f_*(a)) \le a$ and $\forall a \in L : a \le f_*(f^*(a))$. One could somewhat abusively say that Galois adjointness generalizes the notion of inverse maps to non-isomorphic objects: in the case that f^* and f_* are inverse, and thus L and M isomorphic, the above inequalities saturate in equalities. Whenever $f^* \dashv f_*$, f^* preserves all existing joins and f_* all existing meets. This means that for a Galois adjoint pair between complete lattices, one of these maps preserves all meets and the other preserves all joins. Conversely, for L and M complete lattices, any meet preserving map $f_* : M \to L$ has a unique join preserving left Galois adjoint $f^* : a \mapsto \bigwedge\{b \in M | a \le f_*(b)\}$ and any join preserving map $f^* : L \to M$ a unique meet preserving right adjoint $f_* : b \mapsto \bigvee\{a \in L | f^*(a) \le b\}$.

[10] This dual representation is included in the duality between the categories Inf of complete lattices and meet-preserving maps and Sup of complete lattices and join preserving maps, since this duality is exactly established in terms of Galois adjunction at the morphism level – see Coecke et al. (2001) for details.

[11] See also Amira et al. (1998) and Coecke and Stubbe (1999) for a similar development in terms of a so-called operational resolution on the state set.

[12] Let us briefly describe the more rigid argumentation. Consider the following definitions for $A \subseteq \mathcal{L}$: i. $\bigvee A$ is called disjunctive iff ($\bigvee A$ actual $\Leftrightarrow \exists a \in A : a$ actual); ii. Superposition states for $\bigvee A$ are states for which $\bigvee A$ is actual while no $a \in A$ is actual; iii. Superposition properties for $\bigvee A$ are properties $c < \bigvee A$ whose actuality doesn't imply that at least one $a \in A$ is actual. We then have that ($\bigvee A$ disjunctive $\Leftrightarrow \bigvee A$ distributive) provided that existence of superposition states implies existence of superposition properties (Coecke 2002). Moreover, any complete lattice L has the complete Heyting algebra $DI(L)$ of distributive ideals as its distributive hull (Bruns and Lakser 1970), providing it with a universal property. The inclusion preserves all meets and existing distributive joins. Thus, $DI(\mathcal{L})$ encodes all possible disjunctions of properties, and moreover, it turns out that all $DI(\mathcal{L})$-meets are conjunctive and all $DI(\mathcal{L})$-joins are disjunctive – note that this is definitely not the case in the powerset $P(\mathcal{L})$ of a property lattice, nor in the order ideals $I(\mathcal{L})$ ordered by inclusion. It follows from this that the object equivalence between: i. complete lattices, and, ii. complete Heyting algebras equipped with a distributive closure (i.e., it preserves distributive sets), encodes an intuitionistic representation for operational quantum logic – see Coecke (2002) for details.

[13] The following map $\hat{e}^* : P(\Sigma)_1 \to P(\Sigma)_2$ provides a counterexample: Let $\Sigma_1 = \Sigma_2 := \{p, q, r, s\}$ with as closed subsets $\left\{\emptyset, \{p\}, \{q\}, \{r\}, \{s\}, \{q, r, s\}, \Sigma\right\} \subset P(\Sigma)_1 = P(\Sigma)_2$, and set $\hat{e}^*(\{p\}) \mapsto \{p, q\}$, $\hat{e}^*(\{q\}) \mapsto \{q\}$, $\hat{e}^*(\{r\}) \mapsto \{r\}$, $\hat{e}^*(\{s\}) \mapsto \{s\}$; one verifies that although $\bigvee\{q, r\} = \bigvee\{r, s\}$ we have $\bigvee f_{R_e^{-1}}(\{q, r\}) = \bigvee\{p, q, r\} \ne \bigvee\{r, s\} = \bigvee f_{R_e^{-1}}(\{r, s\})$ since $\bigvee\{p, q, r\}$ yields the top element of the property lattice and $\bigvee\{r, s\}$ doesn't.

[14] Note here that in accordance to the literature on linear logic, contra section 2 and section 3 of this paper, we do use the terms conjunction and disjunction beyond their strict intuitionistic significance. This however should not cause any confusion.

References

Abramsky, S.: 1993, 'Computational Interpretations of Linear Logic', *Theoretical Computer Science* **111**, 3–57.

Abramsky, S. and B. Coecke: 2002, 'Physical Traces: Quantum vs. Classical Information Processing', CTCS'02 Submission.

Abramsky, S. and R. Jagadeesan: 1994, 'New Foundations for the Geometry of Interaction', *Information and Computation* **111**, 53–119.

Abramsky, S. and S. Vickers: 1993, 'Quantales, Observational Logic and Process Semantics', *Mathematical Structures in Computer Science* **3**, 161–227.

Abrusci, V. M.: 1990, 'Non-Commutative Intuitionistic Linear Logic', *Zeit-schrift für Mathematische Logik & Grundlagen der Mathematik* **36**, 297–318.

Abrusci, V. M.: 1991, 'Phase Semantics and Sequent Calculus for Pure Noncommutative Classical Linear Propositional Logic', *The Journal of Symbolic Logic* **56**, 1403–1451.

Aerts, D.: 1981, *The One and The Many, Towards a Unification of the Quantum and the Classical Description of One and Many Physical Entities*, PhD-thesis, Free University of Brussels.

Amira, H., B. Coecke and I. Stubbe: 1998, 'How Quantales Emerge by Introducing Induction within the Operational Approach', *Helvetica Physica Acta* **71**, 554–572.

Baltag, A.: 1999 'A Logic of Epistemic Actions', in W. van der Hoek, J. J. Meyer and C. Witteveen, (eds.), *Proceedings of the Workshop on 'Foundations and Applications of Collective Agent Based Systems' (ESLLI'99)*, Utrecht University.

Barr, M.: 1979, **-Autonomous Categories*, Lecture Notes in Mathematics **752**, Springer-Verlag.

van Benthem, J.: 1991, in S. Abramsky et al, (eds.), *Language in Action: Categories, Lambdas and Dynamic Logic, Studies in Logic and Foundations of Mathematics* **130**, North-Holland, Amsterdam.

van Benthem, J.: 1994, 'General Dynamic Logic', in: D. M. Gabbay (ed.), *What is a Logical System?*, pp. 107–139, *Studies in Logic and Computation* **4**, Oxford Science Publications.

Birkhoff, G. and J. von Neumann: 1936, 'The Logic of Quantum Mechanics', *Annals of Mathematics* **37**, 823–843.

Blute, R. F., I. T. Ivanov and P. Panangaden: 2001 'Discrete Quantum Causal Dynamics', Preprint; arXiv: gr-qc/0109053.

Borceux, F.: 1994, *Handbook of Categorical Algebra 3, Categories of Sheaves*, Cambridge, Cambridge University Press.

Bruns, G. and H. Lakser: 1970, 'Injective Hulls of Semilattices', *Canadian Mathematical Bulletin* **13**, 115–118.

Coecke, B.: 2000, 'Structural Characterization of Compoundness', *International Journal of Theoretical Physics* **39**, 585–594; arXiv: quant-ph/0008054.

Coecke, B.: 2002, 'Quantum Logic in Intuitionistic Perspective' and 'Disjunctive Quantum Logic in Dynamic Perspective', *Studia Logica* **70**, 411–440 and **71**, 1–10; arXiv: math.L0/0011208 and math.L0/0011209.

Coecke, B.: (nd), 'Do we have to Retain Cartesian Closedness in the Topos-Approaches to Quantum Theory, and, Quantum Gravity?', preprint.

Coecke, B., D. J. Moore and S. Smets: (nd,a), 'From Operationality to Logicality: Philosophical and Formal Preliminaries', submitted.

Coecke, B., D. J. Moore and S. Smets: (nd,b), 'From Operationality to Logicality: Syntax and Semantics', submitted.

Coecke, B., D. J Moore and I. Stubbe: 2001, 'Quantaloids Describing Causation and Propagation for Physical Properties', *Foundations of Physics Letters* **14**, 133–145; arXiv:quant-ph/0009100.

Coecke, B., D. J. Moore and A. Wilce: 2000, 'Operational Quantum Logic: An Overview', in B. Coecke, D. J. Moore and A. Wilce (eds.), *Current Research in Operational Quantum Lo-*

gic: Algebras, Categories and Languages, Dordrecht, Kluwer Academic Publishers, pp. 1–36; arXiv:quant-ph/0008019.

Coecke, B. and S. Smets: 2000, 'A Logical Description for Perfect Measurements', *International Journal of Theoretical Physics* **39**, 595–603; arXiv:quant-ph/0008017.

Coecke, B. and S. Smets: 2001, 'The Sasaki-Hook is not a [Static] Implicative Connective but Induces a Backward [in Time] Dynamic One that Assigns Causes', Paper submitted to *International Journal of Theoretical Physics* for the proceedings of IQSA V, Cesena, Italy, April 2001; arXiv:quant-ph/0111076.

Coecke, B. and I. Stubbe: 1999, 'Operational Resolutions and State Transitions in a Categorical Setting', *Foundations of Physics Letters* **12**, 29–49; arXiv: quant-ph/0008020.

Einstein, A., B. Podolsky and N. Rosen: 1935, 'Can Quantum-Mechanical Description of Physical Reality be Considered Complete?', *Physical Reviews* **47**, 777–780.

Foulis, D. J., C. Piron, and C. H. Randall: 1983, 'Realism, Operationalism, and Quantum Mechanics', *Foundations of Physics* **13**, 813–841.

Foulis, D. J. and C. H. Randall: 1972, 'Operational Statistics. I. Basic Concepts', *Journal of Mathematical Physics* **13**, 1667–1675.

Foulis, D. J. and C. H. Randall: 1984, 'A Note on Misunderstandings of Piron's Axioms for Quantum Mechanics', *Foundations of Physics* **14**, 65–88.

Faure, CL.-A., D. J. Moore and C. Piron: 1995, 'Deterministic Evolutions and Schr-dinger Flows', *Helvetica Physica Acta* **68**, 150–157.

Ghins, M.: 2000, 'Empirical Versus Theoretical Existence and Truth', *Foundations of Physics*, **30**, 1643–1654.

Girard, J.-Y.: 1987, 'Linear Logic', *Theoretical Computer Science* **50**, 1–102.

Girard, J.-Y.: 1989, 'Towards a Geometry of Interaction', *Contemporary Mathematics* **92**, 69–108.

Girard, J.-Y.: 1995, 'Geometry of Interaction III: Accommodating the Additives', in J.-Y. Girard, Y. Lafont and L. Regnier, (eds.), *Advances in Linear Logic*, Cambridge University Press, pp. 329–389.

Girard, J.-Y.: 2000, 'Du pourquoi au comment: la théorie de la démonstration de 1950 à nos jours', in J.-P. Pier, (ed.), *Development of Mathematics 1950–2000*, Basel, Birkhäuser Verlag, pp. 515–546.

Horwich, P.: 1997, 'Realism and Truth', in E. Agazzi (ed.), *Realism and Quantum Physics; Poznan Studies in the Philosophy of the Sciences and the Humanities* **55**, 29–39.

Jauch, J. M.: 1968, *Foundations of Quantum Mechanics*, Reading, MA, Addison-Wesley.

Jauch, J. M. and C. Piron: 1963, 'Can Hidden Variables be Excluded in Quantum Mechanics?', *Helvetica Physica Acta* **36**, 827–837.

Jauch, J. M. and C. Piron: 1969, 'On the Structure of Quantal Proposition Systems', *Helvetica Physica Acta* **42**, 842–848.

Johnstone, P. T.: 1982, *Stone Spaces*, Cambridge University Press.

Lambek, J.: 1958, 'The Mathematics of Sentence Structure', *American Mathematical Monthly* **65**, 154–170, reprinted in: W. Buszkowski, W. Marciszewski and J. van Benthem, (eds.): 1988, *Categorial Grammar*, Amsterdam, John Benjamins Publishing Co.

Kalmbach, G.: 1983, *Orthomodular Lattices*, London, Academic Press.

Milner, R.: 1999 *Communicating and Mobile Systems: π-Calculus*, Cambridge University Press.

Moore, D. J.: 1995, 'Categories of Representations of Physical Systems', *Helvetica Physica Acta* **68**, 658–678.

Moore, D. J.: 1999, 'On State Spaces and Property Lattices', *Studies in History and Philosophy of Modern Physics* **30**, 61–83.

von Neumann, J.: 1932, *Grundlagen der Quantenmechanik*, Berlin, Springer Verlag, English Translation: 1996, *Mathematical Foundations of Quantum Mechanics*, New Jersey, Princeton University Press.

Niiniluoto, I.: 1999, *Critical Scientific Realism*, Oxford University Press.

Paseka, J. and J. Rosicky: 2000, 'Quantales', in B. Coecke, D. J. Moore and A. Wilce (eds.), *Current Research in Operational Quantum Logic: Algebras, Categories and Languages*, Dordrecht, Kluwer Academic Publishers, pp. 245–262.
Piron, C.: 1964, 'Axiomatique quantique (PhD-Thesis)', *Helvetica Physica Acta* **37**, 439–468, English Translation by M. Cole: 'Quantum Axiomatics', RB4 Technical memo 107/106/104, GPO Engineering Department (London).
Piron, C.: 1976, *Foundations of Quantum Physics*, Massachusetts, W. A. Benjamin Inc. .
Piron, C.: 1978, 'La description d'un système physique et le présupposé de la théorie classique', *Annales de la Foundation Louis de Broglie* **3**, 131–152.
Piron, C.: 1981, 'Ideal Measurement and Probability in Quantum Mechanics', *Erkenntnis*, **16**, 397–401.
Piron, C.: 1983, 'Le Realisme en Physique Quantique: Une Approche Selon Aristote', in E. Bitsakis (ed.), *The Concept of Reality*, Athens, I. Zacharopoulos, pp. 169–173.
Pratt, V. R.: 1993, 'Linear Logic for Generalized Quantum Mechanics', in *Proc. Workshop on Physics and Computation (PhysComp'92)*, Dallas, IEE, pp. 166–180.
Randall, C. H. and D. J. Foulis: 1973, 'Operational Statistics. II. Manuals of Operations and their Logics', *Journal of Mathematical Physics* **14**, 1472–1480.
Rescher, N.: 1973, *Conceptual Idealism*, Oxford, Basil Blackwell.
Rescher, N.: 1987, *Scientific Realism, A Critical Reappraisal*, Dordrecht, D. Reidel Publishing Company.
Rescher, N.: 1995, *Satisfying Reason, Studies in the Theory of Knowledge*, Dordrecht, Kluwer Academic Publishers.
Resende, P.: 2000, 'Quantales and Observational Semantics', in B. Coecke, D. J. Moore and A. Wilce (eds.), *Current Research in Operational Quantum Logic: Algebras, Categories and Languages*, Dordrecht, Kluwer Academic Publishers, pp. 263–288.
Rosenthal, K. I.: 1990, *Quantales and their Applications*, USA, Addison Wesley Longman Inc.
Rosenthal, K. I.: 1996, *The Theory of Quantaloids*, USA, Addison Wesley Longman Inc.
Smets. S.: 2001, *The Logic of Physical Properties, in Static and Dynamic Perspective*, PhD-thesis, Free University of Brussels.
Sourbron, S.: 2000, A Note on Causal Duality, *Foundations of Physics Letters* **13**, 357–367.
Tarski, A.: 1944, 'The Semantic Conception of Truth', *Philosophy and Phenomenological Research* **4**; Reprinted in S. Blackburn and K. Simmons (eds.), *Truth*, pp. 115–143, UK, Oxford University Press.
Yetter, D. N.: 1990, 'Quantales and (Noncommutative) Linear Logic', *The Journal of Symbolic Logic* **55**, 41–64.

COMPLEMENTARITY AND PARACONSISTENCY

NEWTON C. A. DA COSTA and DÉCIO KRAUSE
Department of Philosophy, Federal University of Santa Catarina, E-mails:
ncacosta@terra.com.br; dkrause@cfh.ufsc.br

Abstract. Bohr's Principle of Complementarity is controversial and there has been much dispute over its precise meaning. Here, without trying to provide a detailed exegesis of Bohr's ideas, we take a very plausible interpretation of what may be understood by a theory which encompasses complementarity in a definite sense, which we term $\mathfrak{C}$-theories. The underlying logic of such theories is a kind of logic which has been termed 'paraclassical', obtained from classical logic by a suitable modification of the notion of deduction. Roughly speaking, $\mathfrak{C}$-theories are non-trivial theories which may have 'physically' incompatible theorems (and, in particular, contradictory theorems). So, their underlying logic is a kind of paraconsistent logic.

1. Introduction

"Ceci met en évidence l'apparence irrationnelle de la complémentarité qui ne se rationalise que par des schèmes logiques nouveaux." P. Fèvrier (1951)

The concept of 'complementarity' was introduced in quantum mechanics by Niels Bohr in his famous 'Como Lecture', in 1927 (Bohr 1927). The consequences of his ideas were fundamental for the development of the Copenhagen interpretation of quantum mechanics and constitutes, as is largely recognized in the literature, as one of the most fundamental contributions to the development of quantum theory (see Beller 1992; Jammer 1966, 1974).

Notwithstanding their importance, Bohr's ideas on complementarity are controversial. In reality, it seems that there is no general agreement on the precise meaning of his *Principle of Complementarity* (see for instance Beller 1992, 148); Bohr's own words, by posing that "I think that it would be reasonable to say that no man who is called a philosopher really understands what is meant by complementary descriptions" (quoted from Cushing 1994, 32), might suggest the difficulties involved in any attempt to search for a 'rationale' for his Principle. Anyhow, this remark invites us to look also at the logico-mathematical grounds, mainly in connection with the paraconsistent program (see da Costa and Marconi 1987; da Costa and Bueno 2001).

So, although it has also been claimed that Bohr apparently understood the Principle of Complementarity from an epistemological point of view only (cf. Jammer 1974, 70 and 89), we think that it is pertinent to ask for the logical structure of a theory which encompasses such a principle in its bases. Then, taking into account that

S. Rahman et al. (eds.), Logic, Epistemology, and the Unity of Science, pp. 557–568.

the intuitive idea of complementarity resembles that of contradiction (see below), the underlying logical structure of such a theory should be made explicit.

As a historical remark, we recall that some authors like C. von Weizsäcker, M. Strauss and P. Fèvrier already tried to elucidate Bohr's principle from a logical point of view (cf. Fèvrier 1951; Jammer 1974, 376ff; Strauss 1973, 1975); Jammer mentions Bohr's negative answer to von Weizsäcker's attempt of interpreting Bohr's principle and observes that this should be taken as a warning for analyzing the subject (ibid., 90). He also mentions that Strauss' intention was to develop a logic in which two propositions, say α and β (which should stand for complementary propositions) may be both accepted as true, but not their conjunction $\alpha \wedge \beta$ (ibid., 335); R. Carnap suggests that Strauss' logic were 'inadvisable' (Carnap 1995, 289).

The introduction of some non-classical logical systems developed more recently may enrich the discussion, and this is what we are doing now. But let us first recall that apparently 'complementary descriptions' are more concerned with 'exclusive descriptions' than with the impossibility of 'simultaneous measurement', as implicitly suggested in some standard books when they 'define' complementarity (see for instance Omnès 1995).

We shall proceed as follows. Without discussing von Weizsäcker's or Strauss' works (only Février's ideas will be mentioned in brief below in order to motivate the paper), we introduce the concept of a theory which admits a *Complementarity Interpretation* (to use Jammer' s words – see below). Then we suggest that under a plausible interpretation of what is to be understood by complementarity, the underlying logic of such a theory is a *paraclassical* logic (first proposed in da Costa and Vernengo 1999). Below we shall sketch the main features of this logic as applied to our purposes.

En passant, let us mention that one thing is to provide an exegesis of Bohr's ideas; another is to pay attention to the underlying logical structure of a theory which encompasses complementarity in some sense. In this paper, although we regard the first topic as very important, we are fundamentally concerned with the second, even if we do not provide all the technical details, which will be postponed to future technical works. So, this paper can be regarded as an adjunct to the speculations on this second point. Concerning the first point, see Beller (1992) for a detailed attempt to 'decipher' Bohr's principle "by uncovering and describing the underlying network of implicit dialogues in the Como lecture".

Finally, let us say that our paper might be also viewed as an attempt to investigate a line of research which was envisaged, but not developed, by P. Février; in short, she attributed a third value (*impossible*) to the conjunction of complementary propositions (*propositions incomposables*) so that her logic resembles Łukasiewicz' three valued logic (Jammer's book provides a general view on these logics; see Jammer 1974, 341ff). Notwithstanding, Février recognized that we could also consider that the conjunction of complementary propositions cannot be performed: "la conjonction 'et' ne peut leur être appliquée" (Fèvrier 1951, 33), but she did not consider such a possibility due to "raisons de technique mathématique" (ibid.). In this paper we articulate a possible way to supersede these 'difficulties', motivated by the paraconsistent program, which at that time had not yet been developed. Our approach runs in the direction of not avoiding that the conjunction of complementary sentences can be performed but, roughly speaking, that such a conjunction cannot be derived as a theorem of the theory.

In our opinion, Bohr's view provides the grounds for defining a very general class of theories, which may be regarded as theories which incorporate axioms that may entail propositions like γ and $\neg\gamma$ (the negation of γ), but such that the theory is not trivial in the sense that this fact does not imply that all the formulas of its language are theorems, as we shall see below. In other words, the theories we shall characterize below are such that from γ and $\neg\gamma$ we cannot deduce $\gamma \wedge \neg\gamma$, that is, a contradiction.

We should still remark that this kind of investigation has not only historical reasons, as one should infer from the fact that nowadays the concept of complementarity seems to be no more popular among physicists. Really, the investigation of the logical foundations of science has a value by itself, and the resulting systems (when they arise, as in the present case), built as sometimes motivated by not so clear intuitions, not only may provide them a sense according to acceptable paterns of rigour, but they also may be useful in other situations as well, which may provide other insights and further developments. Furthermore, our work shows that by taking the concept of complementarity as we have considered it (see the next section), there is a sense in saying that the founders of quantum theory, in particular Bohr, may be referred to as 'inconsistent', as probably are all those who are developing very creative efforts, but for sure their feelings were not trivial in the sense defined below. Maybe we could say, taking the due care: they are paraconsistent.

2. A Way of Understanding Complementarity

In order to explain the sense according to which we shall consider the term 'complementarity' in this paper, let us look at how this concept was analyzed by some authors.

Of course, a few isolated quotations cannot provide evidence for the understanding of concepts, especially regarding the present case, but perhaps we could reinforce our point by showing that complementarity stands more for 'incompatibility' in some sense (the 'sense' being explained in the next sections) than for impossibility of 'simultaneously measuring', an expression which could resemble the use of some kind of temporal logic.

Anyway, it should be remarked that we may also find Bohr speaking about complementary concepts which cannot be used *at the same time* (as we can see in several papers in Bohr 1985), but these situations according to him demand isolated analyses, and perhaps it is not possible to provide a general description which allows us to deal with all of these cases: according to Bohr, "One must be very careful, therefore, in analyzing which concepts actually underly limitations" (ibid., 369).

Pauli, for instance, has claimed that, "[If] the use of a classical concept excludes of *another*, we call both concepts (...) *complementary* (to each other), following Bohr" (Pauli 1980, 7, quoted in Cushing 1994, 33). Cushing has also stressed that, "[W]hatever historical route, Bohr did arrive at a doctrine of mutually exclusive, incompatible, but necessary classical pictures in which any given application emphasizing one class of concepts *must* exclude the other" (ibid., 34–35).

This idea that complementary propositions 'exclude' each other (incompatibility) is reinforced by Bohr himself in several passages:

> The existence of different aspects of the description of a physical system, seemingly incompatible but both needed for a complete description of the system. In particular, the wave-particle duality. (quoted from French and Kennedy 1985, 370)
>
> The phenomenon by which, in the atomic domain, objects exhibit the properties of both particle and waves, which in classical, macroscopic physics are mutually exclusive categories. (ibid., 371–372)
>
> The very nature of the quantum theory thus forces us to regard the space-time co-ordination and the claim of causality, the union of which characterizes the classical theories, as complementary but exclusive features of the description, symbolizing the idealization of observation and definition respectively. (Bohr 1927, 566)

Several other passages from Bohr could be quoted from Scheibe's book (1973), for instance, the following:

> The apparently incompatible sorts of information about the behavior of the object under examination which we get by different experimental arrangements can clearly not be brought into connection with each other in the usual way, but may, as equally essential for an exhaustive account of all experience, be regarded as 'complementary' to each other. (Bohr 1937, 291; Scheibe 1973, 31)

Scheibe also says that

> ... which is here said to be 'complementary', is also said to be 'apparently incompatible', the reference can scarcely be to those classical concepts, quantities or aspects whose *combination* was previously asserted to be characteristic of the classical theories. For 'apparently incompatible' surely means incompatible on classical considerations alone. (Scheibe 1973, 31)

The following quotation is also relevant for the point we are trying to stress here: the characteristic of 'exclusion' of complementarity. Bohr says:

> Information regarding the behaviour of an atomic object obtained under definite experimental conditions may, however, according to a terminology often used in atomic physics, be adequately characterized as *complementary* to any information about the same object obtained by some other experimental arrangement excluding the fulfillment of the first conditions. Although such kinds of information *cannot be combined into a single picture* by means of ordinary concepts, they represent indeed equally essential aspects of any knowledge of the object in question which can be obtained in this domain. (Bohr 1938, 26, quoted from Scheibe 1973, 31, second italic ours).

In other words, it seems perfectly reasonable to regard complementary aspects as *incompatible*, in the sense that their *combination* into a single description may lead to difficulties. In this sense, the quantum world is rather distinct from the 'classical' world.

It should be remarked that in the 'classical world', which at first glance can be described by using standard logic and mathematics, if α and β are both theses or theorems of a theory (founded on classical logic), then $\alpha \wedge \beta$ is also a thesis of that theory. This is what we intuitively mean when we say that on the grounds of classical logic, a true proposition cannot 'exclude' another true proposition.

In classical logic, if from some group Δ_1 of axioms of a theory T we deduce γ, and if from another group Δ_2 we deduce $\neg\gamma$, then $\gamma \wedge \neg\gamma$ is also deductible in T.

Normally, our group Δ of axioms of T is *finite*, so that we may talk of the conjunction of its sentences instead of Δ itself. Then, if α and β are respectively the conjunctions associated to Δ_1 and Δ_2, as above, we are looking for a theory T such that in T we may have $\alpha \vdash \gamma$ and $\beta \vdash \neg\gamma$, but in which $\gamma \wedge \neg\gamma$ is not a theorem of T.

Therefore, our goal is to describe a way to formally avoid that $\Delta_1 \cup \Delta_2$ (or $\alpha \wedge \beta$) entails a contradiction, since we do not intend to rule out 'complementary situations'. Notwithstanding, we emphasize that Bohr's ideas are not completely clear, as the following quotation shows:

> The term 'complementarity', which is already coming into use, may perhaps be more suited also to remind us of the fact that it is the combination of features which are united in the classical mode of description but appear separated in the quantum theory that ultimately allows us to consider the latter as a natural generalization of the classical physical theories. (Bohr 1929, 19)

Anyhow, the treatment of complementarity given below can cope with this more general view of this concept.

3. $\mathcal{C}$-theories

In order to provide a more adequate idea about the manner we consider complementary propositions, let us quote Max Jammer:

> Although it is not easy, as we see, to define Bohr's notion of *complementarity*, the notion of *complementarity interpretation* seems to raise fewer definitory difficulties. The following definition of this notion suggests itself. A given theory T admits a complementarity interpretation if the following conditions are satisfied: (1) T contains (at least) two descriptions D_1 and D_2 of its substance-matter; (2) D_1 and D_2 refer to the same universe of discourse U (in Bohr's case, microphysics); (3) neither D_1 nor D_2, if taken alone, accounts exhaustively for all phenomena of U; (4) D_1 and D_2 are mutually exclusive in the sense that their combination into a single description would lead to logical contradictions.
>
> That these conditions characterize a complementarity interpretation as understood by the Copenhagen school can easily be documented. According to Léon Rosenfeld, (. . .)

> one of the principal spokesmen of this school, complementarity is the answer to the following question: What are we to do when we are confronted with such situation, in which we have to use two concepts that are mutually exclusive, and yet both of them necessary for a complete description of the phenomena? "Complementarity denotes the logical relation, of quite a new type, between concepts which are mutually exclusive, and which therefore cannot be considered at the same time – that would lead to logical mistakes – but which nevertheless must both be used in order to give a complete description of the situation." Or to quote Bohr himself concerning condition (4): "In quantum physics evidence about atomic objects by different experimental arrangements (...) appears contradictory when combination into a single picture is attempted." (...) In fact, Bohr's Como lecture with its emphasis on the mutual exclusive but simultaneous necessity of the causal (D_1) and the space-time description (D_2), that is, Bohr's first pronouncement of his complementarity interpretation, forms an example which fully conforms with the preceding definition. Borh's discovery of complementarity, it is often said, constitutes his greatest contribution to the philosophy of modern science. (Jammer 1974, 104–105)

Jammer's quotation will be interpreted as follows. Firstly, we shall take for granted that both D_1 and D_2 are sentences formulated in the language of a theory T and that they refer to the same universe of discourse, so that D_1 and D_2 can be formulated in its language. So, items (1) and (2) will be considered only implicitly. Item (3) will be understood as entailing that *both* D_1 and D_2 are, from the point of view of T, *necessary* for the full comprehension of the relevant aspects of the objects of the domain; so, we shall take both D_1 and D_2 as 'true' sentences (in an adequate 'model' of T). Item (4) deserves further attention. Jammer says that 'mutually exclusive' means that the "combination of D_1 and D_2 into a single description would lead to logical contradictions", and this is reinforced by Rosenfeld's words that the concepts "cannot be considered at the same time", since this would entail a "logical mistake". Then, we will informally say that 'mutually exclusive', or complementary, are incompatible sentences or propositions whose conjunction lead to a contradiction (in a theory T based on classical logic).

So, following Jammer and Rosenfeld (according to the above quotation), we shall say that a theory T admits complementarity interpretation, or that T is a $\mathfrak{C}$-theory, if T encompasses non equivalent true formulas α and β (which may stand for Jammer's D_1 and D_2 respectively) about its particular universe of discourse such that they are 'mutually exclusive' in the sense that their conjunction yields to a contradiction in T, according to classical logic.

The problem with the above characterization of complementary sentences is that if the underlying logic of T is classical logic or, say, intuicionistic logic, then T is contradictory or inconsistent. Apparently, it is precisely this what Rosenfeld claimed in the above quotation. Obviously, if we intend to maintain the idea of complementary propositions in the sense described above, we must change the underlying logic of T, in particular, the way we 'deduce' things. So, we shall modify the classical concept of deduction, obtaining a new kind of logic, called *paraclassical* logic (cf. da Costa and Vernengo 1999).

4. The Underlying Logic of $\mathcal{C}$-theories

As we have remarked, if a theory T that admits complementarity is based on classical logic or even on the most usual systems of logic, then the existence of mutually exclusive theorems as described in the previous section implies that T is trivial, that is, all formulas of the language of T are theorems of T. But there is the possibility of using a convenient type of logical system to found such a theory T; by this way, we shall be able to treat situations in which γ and $\neg\gamma$ are both theorems of T but $\gamma \wedge \neg\gamma$ is not. So, if conveniently introduced, such logic will allow us to deal, in T, with the desirable 'complementary propositions' without contradiction and triviality or, in Rosenfeld's words quoted above, without danger of a "logical mistake". In what follows we shall delineate the basic ideas of such a logic.

In da Costa and Vernengo (1999), a new way of dealing with nontrivial systems was proposed. The logic presented in that paper can also be useful in situations that encompass complementarity. This kind of logic is a paraconsistent logic (according to the characterization of such logics described in da Costa and Marconi 1987; da Costa and Bueno 2001) and it is very well suited for our purposes. Since this logic is still not well known, we shall recall here its main features and emphasize those aspects that are relevant for our purposes. After this we show how such logic can be used as the underlying logic of $\mathcal{C}$-theories, and in the last section we sketch a way to generalize the ideas presented. As in da Costa and Vernengo (1999), we shall be restricted to the propositional level of the new logic P, but of course it is easy to extent P to a first-order or even to higher-order systems.

Let C be an axiomatized system of the classical propositional calculus. The concept of deduction of C is the standard one; we use the symbol $\vdash$ to represent deductions in C. Furthermore, the formulas of C are denoted by Greek lowercase letters, while Greek uppercase letters stand for sets of formulas. The symbols $\neg$, $\rightarrow$, $\wedge$, $\vee$ and $\leftrightarrow$ have their usual meanings, and standard conventions in the writing of formulas will be also assumed without further comments. All the syntactical concepts and details may be found in Mendelson 1987. In particular, we are interested in the following definitions: a theory T (a set of formulas closed under deduction) is **inconsistent** if it contains a theorem α whose negation $\neg\alpha$ is also a theorem of T; otherwise, T is **consistent**. If $\mathcal{F}$ denotes the set of all formulas of the language of C, then T is **trivial** if the set of its theorems coincides with $\mathcal{F}$; otherwise, T is **nontrivial**.

All syntactical concepts of P are similar to the corresponding concepts of C. The notion of P-deduction is introduced as follows:

DEFINITION 4.1. Let Γ be a set of formulas of P and let α be a formula (of the language of P). Then we say that α is a (syntactical) P-consequence of Γ, and write

$$\Gamma \vdash_{\mathsf{P}} \alpha$$

if and only if

(P1) $\alpha \in \Gamma$, or

(P2) There exists a consistent (according to classical logic) subset $\Delta \subseteq \Gamma$ such that $\Delta \vdash \alpha$ (in classical logic).

We call $\vdash_{\mathsf{P}}$ the relation of P-**consequence**. It is immediate that, among others, the following results can be proved:

THEOREM 4.1.

1. If α is a theorem of the classical propositional calculus C and if Γ is a set of formulas, then $\Gamma \vdash_{\mathsf{P}} \alpha$. In particular, $\vdash_{\mathsf{P}} \alpha$.
2. If Γ is consistent (according to C), then $\Gamma \vdash \alpha$ (in C) iff $\Gamma \vdash_{\mathsf{P}} \alpha$ (in P).
3. If $\Gamma \vdash_{\mathsf{P}} \alpha$ and if $\Gamma \subseteq \Delta$, then $\Delta \vdash_{\mathsf{P}} \alpha$ (The defined notion of P-consequence is monotonic.)
4. The notion of P-consequence ($\vdash_{\mathsf{P}}$) is recursive.
5. Since the theses of P are the theses of C, P is decidable.

DEFINITION 4.2. A set of formulas Γ is P-**trivial** iff $\Gamma \vdash_{\mathsf{P}} \alpha$ for every formula α. Otherwise, Γ is P-**nontrivial**.

DEFINITION 4.3. A set of formulas Γ is P-**inconsistent** if there exists a formula α such that $\Gamma \vdash_{\mathsf{P}} \alpha$ and $\Gamma \vdash_{\mathsf{P}} \neg\alpha$. Otherwise, Γ is P-**consistent**.

DEFINITION 4.2.

1. If α is an atomic formula, then $\Gamma = \{\alpha, \neg\alpha\}$ is P-inconsistent, but P-nontrivial.
2. If the set of formulas Γ is P-trivial, then it is trivial (according to classical logic). If Γ is nontrivial, then it is P-nontrivial.
3. If Γ is P-inconsistent, then it is inconsistent according to classical logic. If Γ is consistent according to classical logic, then Γ is P-consistent.

A semantical analysis of P, for instance a completeness theorem, can be obtained without difficulty, as indicated in da Costa and Vernengo 1999.

We remark that $\{\alpha \wedge \neg\alpha\}$ is trivial in classical logic, but not P-trivial. Notwithstanding, we are not suggesting that complementary propositions should be understood as pairs of contradictory sentences.

DEFINITION 4.4. A $\mathfrak{C}$**-theory** is a set of formulas T closed under the relation of P-consequence $\vdash_{\mathsf{P}}$, that is, $\alpha \in T$ for whatever α such that $T \vdash_{\mathsf{P}} \alpha$. In other words, T is a theory whose underlying logic is P.

THEOREM 4.3. There exist $\mathfrak{C}$-theories that are inconsistent from the point of view of classical logic, though P-nontrivial.

Proof. Immediate consequence of Theorem 4.2. ⊣

In the common applications, the existence of consistent sets of formulas are usually assumed only in an informal way, as an implicit postulate. Intuitively speaking, it makes reference to the fact that some 'classical' (that is, based on usual mathematics) theories and hypotheses scientists accept are thought of as not contradictory (as consistent) in principle.

THEOREM 4.4. Every consistent classical theory, that is, every consistent theory founded in classical logic (and set theory) is a particular case of $\mathfrak{C}$-theories.

Finally, we state a result (the theorem below), whose proof is an immediate consequence of the above definition of P-consequence, that links our logic with the characterization of 'complementary propositions' presented above. Before this, we make a definition:

DEFINITION 4.5. Let T be a $\mathfrak{C}$-theory and let α and β be formulas of the language of T. We say that α and β are T-**complementary** (or simply **complementary**) if there exists a formula γ of the language of T such that:

1. $T \vdash_{\mathsf{P}} \alpha$ and $T \vdash_{\mathsf{P}} \beta$
2. $T, \alpha \vdash_{\mathsf{P}} \gamma$ and $T, \beta \vdash_{\mathsf{P}} \neg\gamma$

It is immediate that contradictory propositions like α and $\neg\alpha$ are complementary in the above sense, but once more we remark that we are not arguing that this particular logical situation constitute a condensed account of all Bohr's ideas, as those involved in the quotation shown in the end of the section 2. The interesting case results from the following theorem.

THEOREM 4.5. If α and β are complementary theorems of a $\mathfrak{C}$-theory T and $\alpha \vdash_{\mathsf{P}} \gamma$ and $\beta \vdash_{\mathsf{P}} \neg\gamma$, then in general $\gamma \wedge \neg\gamma$ is not a theorem of T.

Proof. Immediate, as a consequence of Theorem 4.2. ⊣

This result is in fact interesting, since we may admit propositions (complementary propositions) so that one of them entails a proposition while the another one entails the negation of such a proposition, but we cannot deduce that their conjunction entails a contradiction. As an example of a situation involving $\mathfrak{C}$-theories, suppose that our theory T is classical mechanics, which can be axiomatized by means of a set-theoretical predicate (see Suppes 2002, Chap. 7), and that to the axioms of T we add the following ones:

(Ax1) p is a particle
(Ax2) p is a wave

Since (Ax2) implies the negation of (Ax1) (and reciprocally), T may be viewed as an example of a $\mathfrak{C}$-theory, for in T we can derive both, 'p is a particle' and 'p is not a particle', but we cannot infer 'p is a particle *and* p is not a particle', which of course has no sense in physics. So, it seems reasonable to assume that the underlying logic of T is the paraclassical logic P.

The basic characteristic of T as a $\mathcal{C}$-theory is that in making inferences, we suppose that some hypotheses we handle are consistent. In other words, $\mathcal{C}$-theories are closer to those theories scientists *actually* use in their day-to-day activity than theories encompassing the classical concept of deduction.

5. The Paralogic Associated to a Logic $\mathcal{L}$

The technique used in this paper to define the paraclassical logic associated with classical logic can be generalized to any logic $\mathcal{L}$ (including logics having no negation symbol, but we will not deal with this case here). More precisely, starting from a logic $\mathcal{L}$, we can define the $\mathsf{P}_{\mathcal{L}}$-logic associated to $\mathcal{L}$ (the 'paralogic' associated to $\mathcal{L}$) as follows.

Let $\mathcal{L}$ be a logic, which may be classical logic, intuicionistic logic, some paraconsistent logic or, in principle, any other logical system. The deduction symbol of $\mathcal{L}$ is $\vdash_{\mathcal{L}}$, and it is defined according to the standards of the particular logic being considered. We still suppose that the language of $\mathcal{L}$ has a symbol for negation, $\neg$.

DEFINITION 5.1. A **theory** based on $\mathcal{L}$ (an $\mathcal{L}$-**theory**) is a set of formulas Γ of the language of $\mathcal{L}$ which is closed under $\vdash_{\mathcal{L}}$. In other words, $\alpha \in \Gamma$ for every formula α such that $\Gamma \vdash_{\mathcal{L}} \alpha$.

DEFINITION 5.2. An $\mathcal{L}$-theory Γ is $\mathcal{L}$-**inconsistent** if there exists a formula α of the language of $\mathcal{L}$ such that $\Gamma \vdash_{\mathcal{L}} \alpha$ and $\Gamma \vdash_{\mathcal{L}} \neg\alpha$, where $\neg\alpha$ is the negation of α. Otherwise, Γ is $\mathcal{L}$-**consistent**.

DEFINITION 5.3. A $\mathcal{L}$-theory Γ is $\mathcal{L}$-**trivial** if $\Gamma \vdash_{\mathcal{L}} \alpha$ for any formula α of the language of $\mathcal{L}$. Otherwise, Γ is $\mathcal{L}$-**nontrivial**.

Then, we define the $\mathsf{P}_{\mathcal{L}}$-logic associated with $\mathcal{L}$ whose language and syntactical concepts are those of $\mathcal{L}$ but by modifying the concept of deduction as follows: we say that α is a $\mathsf{P}_{\mathcal{L}}$-**syntactical consequence** of a set Γ of formulas, and write $\Gamma \vdash_{\mathsf{P}_{\mathcal{L}}} \alpha$ iff:

1. $\alpha \in \Gamma$, or
2. There exists $\Delta \subseteq \Gamma$ such that Δ is $\mathcal{L}$-nontrivial, and $\Delta \vdash_{\mathcal{L}} \alpha$.

For instance, we may consider the paraconsistent calculus $\mathcal{C}_1$ (da Costa and Marconi 1987) as our logic $\mathcal{L}$. Then the paralogic associated with $\mathcal{C}_1$ is a kind of 'para-paraconsistent' logic.

It seems worthwhile to note the following in connection with the paraclassical treatment of theories. Sometimes, when one has a paraclassical theory T such that $T \vdash_{\mathsf{P}} \alpha$ and $T \vdash_{\mathsf{P}} \neg\alpha$, there exist *appropriate* propositions β and γ such that T

can be replaced by a classical consistent theory T' in which $\beta \rightarrow \alpha$ and $\gamma \rightarrow \neg\alpha$ are theorems. If this happens, the logical difficulty is in principle eliminable and classical logic maintained.

6. More General Complementary Situations

As it is well known, Bohr tried to apply his principle of complementarity to other fields of knowledge (cf. Jammer 1974). More recently, Englert et al. (1994) have suggested that complementarity is not simply a consequence of the uncertainty relations, as advocated by those who believe that "two complementary variables, such as position and momentum, cannot simultaneously be measured to less than a fundamental limit of accuracy" (op. cit.), but that

> (...) uncertainty is not the only enforce of complementarity. We devised and analysed both real and thought experiments that bypass the uncertainty relation, in effect to 'trick' the quantum objects under study. Nevertheless, the results always reveal that nature safeguards itself against such intrusions – complementarity remains intact even when the uncertainty relation plays no role. We conclude that complementarity is deeper than has been appreciated: it is more general and more fundamental to quantum mechanics than is the uncertainty rule. (ibid.)

If Englert et al. (1994) are right, then it seems that paraclassical logic can be useful also to treat those theories which encompass complementarity in their sense.

Anyway, this kind of logic can be also modified to cope with more general kinds of incompatibility, say 'physical incompatibility', incorporating physical incompatible postulates, so as characteristics of the behaviour of human beings, etc., but we shall leave this topic for another work.

Acknowledgments

The authors would like to thank Prof. Osvaldo Pessoa Jr. and the anonymous referees for useful remarks and comments.

References

Beller, M.: 1992, 'The Birth of Bohr's Complementarity: The Context and the Dialogues', *Stud. Hist. Phil. Sci.* **23**(1), 147–180.

Bohr, N.: 1927, 'The Quantum Postulate and the Recent Development of Atomic Theory', Atti del Congresso Internazionale dei Fisici, 11–20 September 1927, Como-Pavia-Roma, Vol. II, Zanichelli, Bologna, 1928, pp. 565–588, reprinted in (Bohr 1985, 109–136).

Bohr, N.: 1928, 'The Quantum Postulate and the Recent Developments of Atomic Theory', *Nature* **121**(Suppl.), 580–590, reprinted in (Bohr 1985, 147–158).

Bohr, N.: 1934, 'Introductory Survey', reprinted in (Bohr 1985, 279–302).

Bohr, N.: 1934, *Atomic Theory and the Description of Nature*, Cambridge, Cambridge University Press, reprinted in (Bohr 1985, 279–302).
Bohr, N.: 1937, 'Causality and Complementarity', *Phil. Sci.* **4**(3), 289–298.
Bohr, N.: 1938, 'Natural Philosophy of Human Cultures', in *Atomic physics and Human Knowledge*, New York, Wiley, 1958, pp. 23–31 (also in *Nature* **143**, 1939, 268–272).
Bohr, N.: 1958, 'Quantum Physics and Philosophy: Causality and Complementarity', in R. Klibanski (ed.), *Philosophy in the Mid-Century, I*, Firenze, La Nuova Italia, pp. 308–314.
Bohr, N.: 1985, *Collected Works*, E. Rüdinger (general ed.), J. Kolckar (ed.), *Foundations of Quantum Physics I*, Vol. 6, Amsterdam, North-Holland.
Carnap, R.: 1995, *An Introduction to the Philosophy of Science*, New York, Dover Publishing.
da Costa, N. C. A. and D. Marconi: 1987, 'An Overview of Paraconsistent Logics in the 80's', *Monogr. Soc. Paran. Mat.* **5**.
da Costa, N. C. A. and R. J. y Vernengo: 1999, 'Sobre algunas lógicas paraclássicas y el análisis del razonamiento jurídico', *Doxa* **19**, 183–200.
da Costa, N. C. A. and O. Bueno: 2001, 'Paraconsistency: Towards a Tentative Interpretation', *Theoria-Segunda Época* **16**(1), 119–145.
Cushing, J. T.: 1994, *Quantum Mechanics: Historical Contingency and the Copenhagen Hegemony*, Chicago & London, The University of Chicago Press.
Englert, B.-G., M. O. Scully and H. and Walther: 1994, 'The Duality in Matter and Light', *Scientific American* **271**(6), 56–61.
French, A. P. and P. J. Kennedy (eds.): 1985, *Niels Bohr, a Centenary Colume*, Cambridge MA and London, Harward Unversity Pres.
Février, P. D.: 1951, *La structure des théories physiques*, Paris, Presses Université de France.
Hughes, G. E. and M. J. Cresswell: 1996, *A New Introduction to Modal Logic*, London, Routledge.
Jammer, M.: 1966, *The Conceptual Developlemt of Quantum Mechanics*, McGraw-Hill.
Jammer, M.: 1974, *Philosophy of Quantum Mechanics*, New York, John Wiley.
Mendelson, E.: 1987, *Introduction to Mathematical Logic*, 3rd Edn, Monterrey, Wadsworth & Brooks/Cole.
Omnès, R.: 1994, *The Interpretation of Quantum Mechanics*, Princeton, Princeton University Press.
Pauli, W.: 1980, *General Principles of Quantum Mechanics*, Berlin, Springer-Verlag.
Scheibe, E.: 1973, *The logical Analisys of Quantum Mechanics*, Oxford, Pergamon Press.
Strauss, M.: 1975, 'Foundations of Quantum Mechanics', in C. A. Hooker (ed.), *The Logico-Algebraic Approach to Quantum Mechanics*, Vol. I, Dordrecht, D. Reidel, pp. 351–364.
Strauss, M.: 1973, 'Mathematics as Logical Syntax – A Method to Formalize the Language of a Physical Theory', in C. A. Hooker (ed.), *Contemporary Research in the Foundations and Philosophy of Quantum Theory*, Dordrecht, D. Reidel, pp. 27–44.
Suppes, P.: 2002, *Representation and Invariance of Scientific Theories*, Stanford, CSLI Publishing.

LAW, LOGIC, RHETORIC: A PROCEDURAL MODEL OF LEGAL ARGUMENTATION

ARNO R. LODDER
Computer/Law Institute, Vrije Universiteit Amsterdam, The Netherlands,
E-mail: lodder@rechten.vu.nl, www.rechten.vu.nl/~lodder

Abstract. Legal argumentation can be modeled using logic, but in this chapter it is claimed that logic alone does not suffice. A model should also take the rhetoric nature of legal argumentation into account. DiaLaw is such a model: a formal, procedural model in which the logical and rhetorical aspects of argumentation are combined.

The core of this chapter consists of a description of the basic concepts of DiaLaw and an extensive account of why rhetorical, non-logical elements of legal argumentation are essential.

1. Introduction

Rescher (1977, 43) quite boldly claimed that "legal trial is not concerned with the 'real truth of the matter'[1] – else why have categories of 'inadmissible' evidence? – but for making out of a legally proper case". I feel sympathy for the point Rescher wants to make, because he stresses the importance of procedures in law. It is, though, not true that a lawyer is not interested in the real truth of the matter, but he realizes that a legally proper case is all he can get. In order to obtain what could be called procedural truth (as opposite to material or substantial truth) legal procedures (e.g., trials) have to be fair. (cf. the fair trial principle of Article 6 of the European Convention on Human Rights). Pursuing fairness is the reason why certain evidence is inadmissible; evidence may not be obtained by all means (e.g., torture, manipulation).

Not only is a trial a process, but in my opinion any statute, any legal decision, even any legal statement is (the outcome of) a process. Following Ronald P. Loui, who once stated that "everything is a process", we could say: law is a process. The central claim in this chapter is related to this general statement, namely that legal statements are justified if they are accepted in a procedure. The theory I describe and have implemented in DiaLaw (Lodder and Herczog 1995; Lodder 1999) is a general theory of legal justification. The theory is not targeted at a specific procedure such as a trial, although examples might be based on actual cases.

In DiaLaw logic and rhetoric are combined. Partly, my approach is comparable to Perelman and Olbrechts-Tyteca (1971). The main difference is that I do not reject the use of logic. Under circumstances, a participant in DiaLaw can force the other to accept statements based on the statements he is already committed to. This

S. Rahman et al. (eds.), Logic, Epistemology, and the Unity of Science, pp. 569–588.

so-called forced commitment is comparable to derivation in logic. In this chapter I do not concentrate on the logic underlying DiaLaw, but elaborate upon rhetoric, because I consider this to be the most important part of DiaLaw.

The remainder of this chapter is structured as follows. My research is part of Artificial Intelligence & Law. First I briefly introduce the field of AI & Law, in particular the research on legal reasoning and argumentation is relevant (Section 2). Subsequently I will introduce in main lines my theory on legal justification (Section 3). Section 4 provides an informal discussion of the basic concepts of the dialogical model DiaLaw (Lodder and Herczog 1995; Lodder 1999). Subsequently section 5 is dedicated to the limitations of logic, and Section 6 discusses rhetorical arguments. After replying to criticism in Section 7, closing remarks conclude this chapter.

2. Artificial Intelligence & Law Models

Due to lack of space in this section most AI & Law models are merely mentioned. A discussion of a wide variety of AI & Law models can be found in Prakken (1997) and Lodder (1999).

Alan Turing, one of the founding fathers of Artificial Intelligence, claimed back in 1950: "I believe that at the end of the century (...) one will be able to speak of machines thinking without expecting to be contradicted". Neither in general AI, nor in AI & Law, progress was made as quickly as initially thought. In AI & Law, the initial idea of 'machines' with general knowledge of the law was soon replaced by the more modest goal of building so-called expert systems with knowledge of only a small part of the law. One early project was the implementation in ProLog of the British Nationality Act by computer scientists (amongst others Kowalski and Sergot) in 1986. Although the set of rules was fairly limited, the results were not as good as expected. An important reason for this failure is that often rules need to be interpreted before they can be applied, as already Leith (1986) noted. So, legal reasoning could not be simply represented by programming logic. In order to answer the question "how should legal reasoning and argumentation be represented?" a fundamental line of research started in the nineties. This AI & Law research (1991–1999) resulted in several models of non-monotonic logic and theories of argumentation, amongst others by Sartor (1994), Gordon (1995) and Prakken (1997). Gordon was one of the first to present a dialogical model of legal argumentation, a branch in AI & Law that has been very popular in the nineties (e.g., Hage et al. (1992), Loui et al. (1993), Nitta et al. (1993), Lodder and Herczog (1995), Freeman and Farley (1996), Prakken and Sartor (1996), Bench-Capon (1998), Jakobovits and Vermeir (1999)). All these models are rule-based, examples of case-based models of legal reasoning and argumentation are those by Ashley (1990), Branting (1991) and Skalak and Rissland (1992).

Recently, Verheij (1999) has given an interesting account on the topic of legal logic.[2] The question he asks is whether any ordinary logic or argumentation theory suffices to model legal reasoning or whether a special logic or argumentation theory is needed. Verheij distinguishes two lines of research in formalizing legal reasoning. One line is of a general nature and is characterized as the "solution of technical difficulties as they are encountered during the formalization of legal reasoning (...) Topics such as the defeasibility and the dynamics of reasoning, and the dialogical characterization of valid inference (...)". If this line is followed no special legal logic is needed, just as Soeteman (2000) has claimed at several occasions. Soeteman has the opinion that the legal premises (sources of law, in particular codes and precedents) make legal argumentation special, and that these premises can be modeled in any ordinary logic. The second line of research, the legal logic, aims to "search for notions of inference that correspond to actual legal reasoning (...) Topics such as reasoning with rules and principles, rule applicability, and the purpose of rules (...)". In general, Verheij et al. (1997) identify the following notions that were worked out in several AI & Law models: undercutters, rebutters, weighing information, reasoning about weighing information, reasoning about rules, lines of argumentation and dialogues, procedural rules, commitment rules, and burden of proof. As one example of a legal logic model, I mention the work of Hage (1997) and Verheij (1996) who have developed a semantic theory of rules and reasons called Reason-Based Logic (RBL). RBL is a theory of defeasible reasoning that is built on top of (monotonic) FOPL. Special predicates are used to express that rules are valid, that rules apply, etc. Derivation rules define, e.g., that a rule that is applied gives rise to a reason, that a conclusion holds if the reasons pro outweigh the reasons con, etc. The logic has an enormous expressive power. The drawback of the expressiveness is its computability. Not surprisingly, except for some earlier versions of their theory there has been no implementation.

3. Basic Concept of DiaLaw[3]

DiaLaw is a two-person dialog game, in which both players make moves alternately. The goal of the game is to justify statements in a dialog. The statements put forward by one player become justified whenever they are accepted by the other player. Consider the following discussion between Bert and Ernie.

Bert:	My intelligent agent is capable of acting
Ernie:	Is he?
Bert:	Yes, he just surfed the internet and bought a book
Ernie:	I think you're right, Bert

THE BOX DISCUSSION

The example shows a simple, short dialog. In the remainder of this section it will be referred to as the BOX DISCUSSION. The following concepts of the dialog game are informally introduced:

- the participants;
- the moves of the game;
- the burden of proof;
- the role of commitment;
- the dialog rules;
- levels in the dialog.

For a formal definition of DiaLaw the reader is referred to Lodder (1999, 41–80); the implementation of Prolog is described in Lodder (1999, 171–184).

3.1. THE PARTICIPANTS

DiaLaw regulates the discussion between two players. DiaLaw can be played by two (groups of) people, or even by a single person.

If the players cannot agree on a statement, there are two options. First, an independent third party may be asked to decide. In the law the role of this third party is performed by judges, arbiters, etc. The second option is to leave the disagreement, so agree to disagree. In DiaLaw the role of an arbiter is not modeled. If it were, it would imply that there indeed exists an independent criterion to settle conflicts, namely the criterion the judge uses to decide.[4] This would be in contradiction with my observation that such a criterion does not exist in law (Lodder 1999, Ch. 2). An unpleasant consequence of not having an arbiter is that the dialog is not guaranteed to end. However, I prefer an unfinished dialog to one that ends because of a decision based on prefixed criteria. Moreover, in case a dialog does not end, apparently the statement is not justified in the eyes of the opponent. Since justification is defined as acceptance of a statement by the opponent, a decision by an independent third party that forces the opponent to accept is not desirable.

3.2. THE MOVES OF THE GAME

The players make moves alternately. These moves contain two elements: (1) an illocutionary act with (2) a propositional content.[5] The illocutionary act is one of the following four:

a. claim;
b. question;
c. accept;
d. withdraw.

The propositional content of these illocutionary acts is a statement. In the formal definition of moves the illocutionary act and the propositional content are separately modeled. For example, the first move of the BOX DISCUSSION between Bert and Ernie would formally be:

```
... claim, capable_of_acting(agent)...
```

In the examples below, these two elements of the move are not separated, but combined in one informal statement.

Claim

If a player claims a statement, he expresses that he believes that the statement is justified. In principle, a player may claim any statement. Only in some cases the claim of a particular statement is forbidden by the dialog rules. For instance, a player cannot claim a statement if he just claimed the opposite. So, the third move in the following dialog is not allowed.

Bert:	My intelligent agent is capable of acting
Ernie:	Is he?
Bert:	My intelligent agent is not capable of acting

If a player denies a statement claimed by his opponent, this is also modeled as a claim. The propositional content of this claim is the negation of the statement claimed by the opponent. The following dialog is allowed.

Bert:	Intelligent agents are capable of acting
Ernie:	Intelligent agents are not capable of acting

Question

If a player questions a statement, he asks a justification of the statement claimed by his opponent. Question is neither an acceptance, nor a denial, but lies just in between these two acts. A player usually questions a statement if he is not yet convinced. In the BOX DISCUSSION Ernie questioned the claim that an agent is capable of acting, and later became convinced. On another occasion a player may question if he is already convinced, because he wants to hear arguments for the statement. For instance, if Bert is a well-known lawyer who claims that agents are capable of acting, and Ernie is already convinced, he might yet want to question because he then will here the arguments of the expert.

Accept

If a player accepts a statement, he agrees with the statement claimed by the other.

Accepting a statement is comparable to claiming a statement. Both the player who accepts and the player who claims, believe that the particular statement is justified. The difference is that a claim initiates a dialog, and an acceptance ends it.

Acceptance is a reaction to a claim of the other player. In the BOX DISCUSSION Ernie accepted that agents are capable of acting, when he said that Bert was right. An even shorter example is:

Bert:	Agents are capable of acting
Ernie:	I agree

Withdraw

If a player withdraws a statement, he retracts a statement claimed by himself. The player will withdraw a statement if he is no longer willing to defend it. Withdraw is the opposite of a claim: withdrawal of a statement undoes a previous claim of that statement. Withdraw is also similar to accept: just like after an acceptance the discussion ends. A player can withdraw a statement immediately after it is questioned, but almost always there will be several moves between a claim and a withdrawal. Moves during which the player became convinced by counterarguments, made him realize the weakness of his own position, etc.

Bert:	Agents are capable of acting
Ernie:	Are they?
	... *several moves later*
Bert:	I no longer defend that agents are capable of acting

3.3. BURDEN OF PROOF

The burden of proof plays an important role in regulating legal procedures. In DiaLaw the burden of proof is simple. The player who claims has the burden to prove that the claimed statement is justified. This means that if a player has claimed a statement and the statement is questioned, he must adduce other statements that support his claim.

The role of the player who has the burden of proof is usually called the proponent. For each claimed statement the player who claimed it is the proponent of that statement, and the other player is the opponent.

The roles of proponent and opponent can shift during the game. This means that the player who initiated the dialog is not necessarily the proponent of all statements. An example of a situation in which the roles change, is when a claimed statement is denied (see above, the second claim example). In this dialog the burden of proof shifted from Bert to Ernie. After the first move Bert had the burden of proof, after the second move Ernie has. Only if Ernie withdraws his statement, the burden of Bert revives.

3.4. COMMITMENT

Commitment is a central notion in the dialog. Commitment originates when a statement is claimed or accepted. For instance, in the BOX DISCUSSION both Bert, who claimed, and Ernie, who accepted, are committed to the statement that the intelligent agent of Bert is capable of acting.

During the dialog the commitment of the players is recorded in what is called a commitment store.[6] In the commitment store it is exactly indicated which player is committed to what statements.

Commitment starts when a statement is claimed or accepted. Commitment terminates when a statement is withdrawn. The consequence of withdrawing a statement is that the related element of the commitment store is deleted.

Commitment of a player limits him in subsequent moves. An example of such a limitation is that a player may neither claim, nor accept a statement, when he is committed to the negation of that statement.

To avoid that the dialog remains an informal talk, a player has means to force his opponent to accept a statement. This is what is called forced commitment. Forced commitment is comparable to derivation in logic, and occurs when a player is forced to accept a statement, due to the statements he is already committed to. Assume a player is committed to a reason that supports a statement, and there are no reasons against this statement. In case this player is not able to put forward a reason against this statement, he is forced to accept it.

3.5. THE DIALOG RULES

Since DiaLaw is a game, there are also rules telling how to play the game. The dialog rules define:

- which player's turn it is;
- whether a move is allowed;
- the consequences of valid moves in terms of commitment.

The first move of a dialog is a claim by one of the players, so each dialog starts with the illocutionary act of the type claim. For example, the moment Bert in the BOX DISCUSSION claimed that his agent is capable of acting, a dialog started. The dialog rules define for each illocutionary act what moves can follow. This means that for each possible stage of the game it is defined what moves can follow.

The dialog rules also define whose turn it is. Only in a few exceptional cases the same player moves twice consecutively. Normally, the players make moves in turn.

A dialog ends with an acceptance or a withdrawal. For instance, a dialog about whether agents are capable of acting continues until either Bert withdraws this statement or Ernie accepts it. As long as neither Bert withdraws nor Ernie accepts, the dialog remains unfinished.

Finally, for each illocutionary act it is defined what the consequences are for commitment. Recall that basically commitment originates after a claim and after an acceptance, and ends after withdrawal.

3.6. Levels in the Dialog

To structure the argumentation the dialog has levels. The initial level is 0. The dialog turns to a deeper level only after questioning. So, after Ernie questioned 'my intelligent agent is capable of acting', the level becomes 1. On this new level, statements are adduced that are arguments for or against the statement on the previous level. If a statement is accepted or withdrawn, the dialog returns to the level on which this statement was claimed.

Bert:	My intelligent agent is capable of acting (level 0)
Ernie:	Is he? (level 0)
Bert:	Yes, he just surfed the internet and bought a book for me (level 1)
Ernie:	Did he? (level 1)
Bert:	Take a look as this receipt (level 2)
Ernie:	Must have bought the book (level 2)
Bert:	Someone who makes contracts has to be capable of acting (level 1)
	and so on . . .

As an argument for the 0-level statement that intelligent agents are capable of acting, Bert adduces on level 1 the claim that after surfing the internet, his agent has bought a book for him. Ernie questions this statement. The argument that a receipt shows what happened supports the statement on level 1. For Ernie this support is sufficient, so he now accepts that the agent bought a book. He is, however, not yet convinced that agents are capable of acting, so Bert continues the dialog by adducing a second argument on level 1, namely that someone who makes contracts has to be capable of acting.

This dialog will continue until Bert withdraws the statement that his intelligent agent is capable of acting, or Ernie accepts it. An example of a finished dialog is the level-1 dialog about the claim that the agent surfed the internet and bought the book. The moment Ernie accepted this ("must have bought the book"), the level-1 dialog stopped.

4. Logic Alone Does Not Suffice

There is more to legal argumentation than logic. As Govier (1987, 203) putted it quite clearly for argumentation in general: "Either logic includes much that is nonformal or it tells us only a small amount of what we need to know to understand and evaluate arguments". What makes it that logic alone is not fit for the task of modeling legal argumentation? I have struggled with this question for quite some years now. At first sight logic seems suited to model legal arguments. For instance, the Modus Ponens argument represents the two essential characteristics of legal argumentation: universality of rules (if A then B) and the individuality of the case (A). The related syllogism is sometimes even referred to as the legal syllogism. Nonetheless, not all legal cases can be solved by simply applying legal rules. Sometimes rules need to be interpreted. It might be that 'B' can be justified given 'C' and 'if A then B'. One has to argue then why given 'C' the rule 'if A then B' applies. Still, one can use logic to bridge the gap between A and C. Many legal philosophers (e.g., Aarnio, Alexy, MacCormick, Peczenik) make the distinction between an internal and external justification.[7] In main lines the approaches by the various authors are similar. The internal justification is a logical, deductive justification. The external justification aims at the content, not (primarily) at the formal validity of a justification, and is meant as a justification of the premises used in the internal justification. In the above example the deductive justification of 'B' could consist of the following propositions: 'C', 'if C then A', and 'if A then B'. The propositions 'if A then B' (probably with reference to a code), and in particular 'if C then A' should be justified externally. These justifications heavily rely on rules. Rules are important in law. Conclusions can be justified either by applying existing rules ('if A then B') and/or by creating new rules ('if C then A'). This distinction is common, e.g., Toulmin (1958, 120) makes a distinction between 'warrant-establishing' arguments and 'warrant using' arguments. In my opinion there is also a third category: arguments based on no rules at all. This observation was my first step towards a theory of legal argumentation that is not based on logic alone. From a logical perspective these type of arguments, sometimes referred to as enthymemes, are not compelling. These arguments are rhetorical.

5. Legal, Rhetorical Arguments

Classic argument structures are deduction and induction. A type of argument, supplementing these two has been introduced by Wellman (1971). The so-called conductive arguments are characterized as reasoning in which:

(1) a conclusion about some individual case;
(2) is drawn nonconclusively;
(3) from one or more premises about the same case;
(4) without any appeal to other cases.

Obviously, this is neither deduction (2), nor induction (4). There are just premises (3) and a conclusion (1). These kinds of arguments are quite typical for legal reasoning. Peczenik (1996) characterizes it as follows: "A weighing and balancing of *prima facie* reasons is a jump:[8] one has premises, the conclusion and a gap between. The inference is not conclusive." The question is what to do with such a gap, what to do with nonconclusive arguments. Are these arguments actually conclusive, and should they be made conclusive in reconstruction? Govier (1987, 97) says: "Anyone arguing from some premises (CON) to a conclusion (C), must believe that if CON then C. This belief is indicated by his reasoning in the way he does reason, and it is 'assumed' by his argument. However, to say this is not to say that it is a missing premise in the argument." I agree, except that in my opinion the belief 'if CON then C' expresses the relation between premises and conclusion in a way that to explicitly hints at missing premises. In my opinion, arguers do not have to believe 'if CON then C', the mere adduction of premises supporting a conclusion might be all there is to an argument. In the next sections I will further elaborate my point by making a distinction between logical and rhetorical arguments (6.1, 6.2) introducing diarationality (6.3), and giving an actual example (6.4).

5.1. Compelling Arguments, the Logical Perspective[9]

In the logical perspective on argumentation it must be guaranteed that if the premises are accepted, it is warranted to accept the conclusion. The obvious example that meets the requirements is a deductive argument. Not only deductive arguments, however, also contemporary theories of defeasible argumentation use a logical notion of argument. Examples from AI & Law are the theories of Gordon (1995) and Prakken and Sartor (1996). Put simply, in these models an argument is based on the application of a rule. The difference with deductive arguments is that conclusions are derived only provisionally. New information can change the status of the conclusion, while in a deductive argument the conclusion stands once and for all.

Although a procedure can be conducted in order to select premises or arguments, basically procedures play a minor role in the logical perspective. Whether arguments are valid or not, whether arguments do justify or not, is established independent from the procedure. In case of a logically valid argument there is a criterion that can be used to determine whether the conclusion follows from the premises. This criterion does not refer to a procedure. This does not preclude the possibility that arguments are generated in a procedure, as said, but this procedure is irrelevant to the justifying force of the arguments.

Although in the models of Gordon and others arguments are defined to be *used* in a procedure, the definition of the arguments is still independent of the procedure. The definition concerns the structural products of argument. Just like the Modus Ponens argument is defined independent of a procedure, but can nevertheless be used in a procedure (see Lorenz 1961; Alexy 1989). Parts of the arguments can even be established in the procedure, e.g., the premises the argument is based on, or the rule that is applied in the argument. However, still the structure of the argument is prescribed independent from the procedure. For instance, the argument has a condition part, a conclusion part and an underlying rule, or the argument is based on rule application.

5.2. Rhetorical Arguments, the Psychological Perspective

The psychological perspective is one that was, to my knowledge, introduced by Stevenson (1944, 113):

> "The reasons which support or attack an ethical judgement (...) are related to the judgement psychologically rather than logically. They do not strictly imply the judgement in the way that axioms imply theorems (...)"

In the psychological perspective conclusions are justified by effective, convincing arguments. The structure of the argument is not important, but rather is its effect. An argument is effective if the audience accepts the statement that had to be justified.

Argumentation in the psychological perspective is in some respects the opposite of argumentation in the logical perspective. Basically, if statements are justified in the psychological perspective the structure of arguments is irrelevant. In order to establish whether a statement is justified, argumentation has to be conducted in a procedure; in the logical perspective the structure of the arguments determine whether a statement is justified, and a procedure is optional, not necessary.

5.3. Diarational: Logic and Rhetoric

Normally, argumentation is considered rational if the premises are sufficient to justify the conclusion. Logic can be used to determine whether premises are sufficient. Whether the conclusion of the argument actually is accepted is irrelevant. Rules, both inferential and legal, are essential to rational argumentation. The arguments that justify conclusions are based on rules.[10] Currently, most researchers on formal argumentation model arguments that are based on rules (e.g., Vreeswijk (1993), Dung (1995), Gordon (1995), Prakken and Sartor (1996)). In all these approaches an underlying logic is used to define valid inference: given some premises, it can be determined whether a conclusion is acceptable. If conclusions are actually accepted by the audience towards which the argumentation is directed, without the premises being *sufficient* to accept the conclusion, this reasoning is sometimes called a-rational. I would call this rational too, but to avoid confusion I introduce another term that refers to procedural, dialogical models: diarational. The actual acceptation of the conclusion is essential, because the premises are not logically sufficient to accept the conclusion. Rules can play a role in diarational argumentation, but not necessarily have to.

Diarational is comparable to rhetorical, but it is more than that. Actually, rhetoric and logic are combined in this approach to argumentation. Logically compelling (rational) arguments are allowed, as well as convincing, psychological arguments (rhetoric). Although this combination of argument types makes the formal model of legal argumentation DiaLaw special, the inclusion of the logical types of arguments does not deserve further discussion here. I will give an informal example of the rhetorical arguments that can be modeled in DiaLaw. My aim is to convincingly show that these type of arguments are important in law and that changing these arguments into logically compelling ones is wrong.

5.4. An Example of Rhetorical Argument: No Premise Missing

Not only the argumentation of academics or rhetorically well-equipped attorneys, but also the argumentation of courts is regularly not logically compelling. I will argue that it is not meaningful to add premises and even that adding such a premise is erroneous.[11] An example of a not compelling argument by Courts is the adduction of a couple of statements followed by the phrase "in view of the above" (in Dutch: gezien het bovenstaande).

Statement$_1$

Statement$_2$

...

Statement$_n$

"in view of the above".

Conclusion

One or more statements are claimed and subsequently it is assumed that the conclusion follows from these statement(s). Interestingly enough, the argumentation works, the argumentation is effective; mostly the argumentation is convincing, although it is not logically compelling.

By adding a premise the argumentation could be transformed into what is broadly accepted as rational argumentation. Often it is argued that such a premise is already implicitly present. For instance, if someone argues that the conclusion 'B' follows given 'A' the premise '$A \rightarrow B$' can be added. However, this reconstruction of the argument not necessarily models the actual argument. It is not always intended that the conclusion follows logically from the claimed statements. Instead, statements are claimed to convincingly show that the conclusion is justified. The statements are rhetorical means, they are meant to convince. The connection between the statements and the conclusion that is obtained by adding a premise is satisfactory in some cases, and in all cases to those that only accept logically valid arguments. However, such a reconstruction does not represent the actual argumentation, because the rhetoric is not dealt with. From a psychological or rhetorical perspective, the 'missing' premise does not add something to the argument, except an often unnatural connection between a set of statements and a conclusion. Moreover, as the next example shows, in some cases a missing premise should not be added.

In the Dutch Road Traffic Act it is said that in case of an accident between a motorist and other non-motoring road users, e.g., pedestrians or bikers, the motorist has to pay for the damages, except if he acted in force majeure. In the early nineties the Dutch Supreme Court decided that in case a child not older than thirteen is involved in the accident, the motorist always has to pay the damages, except when the child acted recklessly or intentionally.[12] This rule is known as the 100% rule. In a later case the Dutch Supreme Court had to decide whether the 100% rule should not only be applied to young children but also to elderly people.[13] The Supreme Court decided that the 100% rule was not applicable to elderly people, based on the following (abbreviated) statements:

1. the arguments used to define the rule for young people are less striking for elderly people;
2. law knows no fixed age criterion for elderly people (for young people it does);
3. elderly people are less recognizable.

These three premises supported the conclusion that the 100% rule should not be applied to elderly people. These premises, however, are in no way sufficient to accept the conclusion. It could be argued that the missing premise is a rule that can be constructed in order to make the premises sufficient to accept the conclusion. However, instead of creating a rather artificial rule,[14] the argumentation should be modeled procedurally in a dialog. In this representation the conclusion is accepted after the three supporting statements are adduced. So, the argumentation of the Dutch Supreme Court should be modeled in a diarational way, where the statements caused the acceptation of the conclusion.

6. Reply to Criticism

Some critics claim that it might be the case that in construction the argument is not logically compelling, but that in reconstruction the argument should be made conclusive. I already argued that a reconstruction should reflect the original argumentation, not the argumentation that is desired according to some logical theory.

Another objection is that a theory of argumentation should allow to distinguish between good and bad argumentation. For instance, in propositional logic the modus ponens argument is 'good' and an abductive argument (if a then b, b, so a) is not. However, in my procedural model of argumentation there is no criterion outside the procedure, so for any statement that is accepted apparently the argument (in the minimal case only the statement itself) was 'good'. Obviously it is possible to criticize procedurally justified statements. But since there are no criteria outside a procedure then just a new procedure of argumentation starts. The outcome of this new procedure can be criticized too, again in a procedure. An appropriate name for this evaluation of argumentation in still new procedures would be infinite progress (as opposed to infinite regress: the premises that justify should be justified themselves, etc.). I do not intend to say that the outcome of every subsequent procedure is better, but progress should reflect he continuing character.

Yet another objection is that if an argument from a previous case is used in a later case, the argument must have been generalized in order to apply the argument in a new case. However, the example of the previous section could be used in two ways for which no generalization is needed. First, if there is another case with an elderly person involved, one could just use the conclusion of the argument (the 100% rule is not applicable to elderly people) under reference to the precedent. Nor the statements or premises supporting the statement that the 100% rule is not applicable to elderly people, neither some generalization (rule) connecting the premises and the conclusion have to be mentioned. The conclusion of the argument and the source (the previous ruling) suffices. Second, there might be another group of people, e.g. blind people, to which in a new case one does want to apply the 100% rule. The statements that were adduced in support for the 100% rule being not applicable to elderly people could be modified in order to support the current claim. No generalization (rule) is needed.

However, maybe not in our example, but there might be other cases in which a generalization (rule) is appropriate. Moreover, it is widely accepted that arguments (or reasons) adduced in a particular case can be universalized (cf. Hare (1963)). I do not deny that. My point is rather that if the original arguer did not mention this universalization (rule), you can never know whether he meant to adduce it as such. You are free to use this universalization (rule) in a new case, but that is not the same as saying that the argumentation of the previous case was wrong or that it should be reconstructed in a way the universalization (rule) is made explicit. In particular because there are cases (the above example about the 100% rule) in which such a premise is wrong.

One of the referees asked how the work of Perelman and others on formulating criteria for persuasiveness of argument patterns matches with my purely procedural view. First, as I already noted, I do not claim that legal justification is purely procedural. Second, I do believe that the schemes of Perelman could be a really valuable addition to DiaLaw as long as the use of the argument schemes is voluntarily. Just as is the case for the logical part of DiaLaw. The participants can use the logic of DiaLaw to force each other to accept statements, but they can also ignore the logic and only use rhetorical arguments. Under the same constraints schemes developed by Walton for presumptive reasoning could be added, including the additions by Blair (2001).

A final remark concerns a question once asked by someone from the Pragma-Dialectic school about the status of the premises adduced in the above Supreme Court case. There is difference between linked and convergent arguments (a good survey on this topic is Walton 1996, 109–182). In case of a linked argument all premises together are necessary to support the conclusion. In case of a convergent argument each premise alone provides sufficient support for the conclusion. In a procedural approach this distinction is not that relevant. A party may explicitly indicate that premises are conjunctive (by using 'and') or disjunctive (by using 'or'). If he does not use such indicators it depends on the audience what type the argument is. For instance, an assumed linked argument with three premises may be accepted by the opponent after only the first premise has been adduced. In that case this first premise appeared to be sufficient to accept the conclusion, although maybe the proponent did not think it would be. Various other examples can be thought of, for example that the conclusion of a convergent argument with three premises (so each premise suffices to accept the conclusion) is not accepted even after the third premise has been adduced. In a procedural model of argumentation it all depends on acceptation by the opponent. In reconstructing an argument as the one by the Supreme Court, it may be safest to consider it a linked argument if it is not explicitly indicated that it is a convergent argument. But one can never be sure. Moreover, since it all depends on the audience, it might well be that a reader of the argument accepts the conclusion after seeing only one or two premises. So, actually, if the arguer does not indicate what type of argument he uses (linked or convergent), it is best not to label it as either one. You should see it as just a set of premises that support a conclusion. There is nothing more to it (in any case not a missing premise!).

7. My Procedural View on Legal Justification[15]

In this section the main characteristics of a procedural model are reiterated. The problem with a product model of legal argumentation (as opposed to a procedural model) is that an independent criterion by which it can be determined whether an argument justifies does not exist. Another reason why a product model is not satisfactory is that if a conclusion is drawn, it is important to know which exceptions or counter-arguments have been considered. The product of justification does not show this. Namely, if an exception is considered but regarded not relevant enough to apply, or a counter-argument not strong enough to rebut, this cannot be recognized in the product of justification. In a procedural model, all steps that lead to the final conclusion are included, so, even a weak counter-argument. This is an important characteristic of a procedural model, because not only the support of a conclusion is relevant, but also which arguments were defeated while justifying a conclusion.

A procedural model of justification does not concentrate on the specific structure of reasoning schemes (the product of justification). Instead, the focus is moved to the procedure in which statements are justified. In the field of law this is not an uncommon method. For instance, legislation originates after a particular legislative procedure. The content of a statute plays an essential role during the legislative process, but for the validity of a legal norm the content is of marginal interest. If a majority of the Parliament votes for a certain statute, this statute becomes part of the law.

After the norm is enacted objections about the content of a norm can be raised, but only in other procedures, e.g., law suits or elections. The fact that these objections can be substantial, so about the content of the norm rather than about the originating procedure, does not defeat my claim that justification is a process or that no independent criterion exists. Both procedural and material arguments have to be adduced in a procedure. Without the testing of either procedural or material arguments in a procedure, one cannot be sure whether these arguments justify.

Protagoras realized long ago that each case has two sides. He recognized the power of rhetoric and claimed that if someone told his side, he would win the case for him. Because each side of a dispute has to be represented, a procedural model must contain at least two parties. Two parties do not necessarily mean that two persons are involved. Even a single person's attempt to justify a statement can be modeled as a two-person dialog game[16] in which the person alternately plays the role of someone who attacks a statement and the other one who defends it. Or, according to Barth and Krabbe (1982, 12): "Reasoning as carried out by one person should be studied as an (important) special case, viz. the case where the two parties coincide in one person: the self-critical case". Of course, opinions of others can play a role in such a one-person dialog. A procedural, dialogical model has a set of rules that defines when parties are allowed to adduce statements, arguments, etc. The rules guide the procedure, comparable to the rules according to which any other game is played.

Only within the dialog game statements can become justified. The justification is not related to some independent criterion outside the procedure. Instead, justification is defined relative to the parties of the dialog game. Only if the parties want to hear reasons for a statement these reasons have to be adduced, and only if a statement is accepted by the parties, it is justified. The dialog game is a rhetorical procedure (Tindale 1999). Characteristic for such a procedure is that there is no predetermined outcome, the procedure is non-deterministic. By presenting reasons, each party tries to draw the outcome in his direction, but the final result cannot be determined in advance. In AI & Law the non-deterministic character of law has been used as an argument against logical models of legal reasoning (Berman and Hafner 1987). Moles (1992) stated: "The latter's [judge] role involves what may be called a 'performative utterance' which is much more a matter of rule creation than it is of rule following. This (...) is a factor which logical modelling cannot account for". In a dialog model arguments can be based on both existing and newly created rules (or, as in DiaLaw, even on no rule at all).

The outcome of a dialog only holds for the participants at the moment they finish their dialog. The justification is relative to a particular audience and relative in time. In a new game the outcome of a previous game can be discussed. Such a discussion is comparable to, for example, the evaluation of a court decision by legal scholars.

I do not defend the claim that legal justification is purely procedural. In a pure procedure (Rawls 1972) an independent criterion to evaluate the outcome of the procedure does not exist (I agree) and the procedure is guaranteed to lead to desired results (I disagree). In my opinion, statements can be justified only in a procedure, but no procedure can really guarantee that the outcome of the procedure is just.

8. Closing Remarks

Law is a process, and legal argumentation is procedural by nature. In this contribution I explained what role logic can play in modeling legal argumentation, and, especially why rhetoric is so important for a model of legal argumentation. For the modeling of legal argumentation logic and rhetoric should go hand in hand. The formal, implemented model DiaLaw combines these two approaches.

One point I have not particularly paid attention to is that in a procedural model non-logical arguments do have structure. Not a structure by themselves, but the procedure provides the structure of the argumentation. That is the reason the discussion in DiaLaw is layered: the different levels provide the necessary structure. A new line of research is to employ only this element of a procedural model. In Lodder and Huygen (2001) a tool has been developed to support parties in online dispute resolution (ODR). The tool helps the participants to bring forward the statements of the dispute in a structured and concise way. After each added statement the online tool shows the structured layout of the statements. What remains is just rhetoric, logic is left out. Such a pure rhetorical model is very useful in practice. The field of ODR is indeed an interesting area for practical application of theoretical AI & Law models (Lodder and Vreeswijk 2004; and Lodder and Zeleznikow 2005). The editors of this volume stress that the results from one academic discipline could influence other disciplines. My background is interdisciplinary by nature: Artificial Intelligence & Law. In AI & Law there is a mix of people with a legal background (e.g., Allen, Ashley, Gordon, Hage, Lodder, Oskamp, Prakken, Sartor), and a non-legal background (e.g., Aleven, Bench-Capon, Leenes, Loui, Quaresma, Moens, Verheij, Winkels, Zeleznikow). AI & Law builds on research from the fields of (legal) philosophy, argumentation, computer science/AI, and logic, and our research is used in related fields, but both happens too little. I hope that my contribution stimulates the exchange of ideas between different disciplines.

Notes

1 I sometimes put it even more boldly by claiming that *law* has nothing to do with the truth and that it is therefore ironic that probably the best-known legal phrase is "... do you promise to tell the truth, the whole truth, and nothing but the truth ...".

[2] Note that the term legal logic has been used in different meanings. For instance, in the 1970s Rödig tried to axiomatize the legal system. His work and that of similar German scholars is referred to as legal logic, because they used logic to model law. A group of Belgium scholars (amongst others Perelman) also worked on what was called legal logic, but in stead of using existing logic they held the opinion that special legal elements should be added to logic.

[3] This section uses material from Lodder (1999, 35–41).

[4] An alternative is to let the judge decide randomly. Whenever called to decide, he tosses a coin. However, that is something the players could equally do themselves, no judge would be needed for that. Moreover, since justification is based on acceptance, a toss seems not appropriate. An additional problem is how to determine at what moment a player may call the judge, e.g., immediately after the claim of a statement?

[5] The terms illocutionary act and propositional content are taken from Searle (1969, 30). The illocutionary act types are inspired by the work of amongst others Van Eemeren and Grootendorst (1982) and MacKenzie (1979).

[6] The idea of using such commitment stores is Hamblin's (1970), the term commitment store derives from MacKenzie (1979).

[7] Not all authors uses the same words for the two justifications. MacCormick, for instance, uses the terms first and second order justification.

[8] On the so-called 'jumps' see also Peczenik (1989, 115f.).

[9] Parts of 6.1-6.4 are based on Lodder (1999, 148–155).

[10] An example of a logical argument explicitly mentioning rules is (Walton and Krabbe, 1995, 180).

[11] Nutting (2002) claims that legal reasoning makes sense only "against a background of unarticulated (*and, perhaps, unarticuble*) assumptionps". Italics are mine.

[12] Decisions on June 1, 1990 (NJ 1991, 720), and May 31, 1991 (NJ 1991, 721).

[13] Decision on February 28, 1992 (NJ 1993, 566).

[14] I do not claim that it is never possible to model the ratio decendi of a case as a rule, see Loui and Norman (1995).

[15] This section uses some material from Lodder (1999, 24–25 and 163–164).

[16] In the context of dialogs I use the terms model and game interchangeably.

References

Alexy, R.: 1989, *A Theory of Legal Argumentation*, Oxford, Clarendon press.

Ashley, K. D.: 1990, *Modelling Legal Argument: Reasoning with Cases and Hypotheticals*, MIT Press.

Barth, E. M. and E. C. W. Krabbe: 1982, *From Axiom to Dialogue*, Berlin, New York, Walter de Gruyter.

Bench-Capon, T. J. M.: 1998, 'Specification and Implementation of Toulmin Dialogue Game', in: J. C. Hage et al. (eds.), *Legal Knowledge Based Systems: JURIX 1998*, Nijmegen, GNI, pp. 5–19.

Berman, D. H. and C. D. Hafner: 1987, 'Indeterminacy: A Challenge to Logic-based Models of Legal Reasoning', *Yearbook of Law, Computers and Technology* **3**, 1–35.

Blair, J. A.: 2001, 'Walton's Argumentation Schemes for Presumptive Reasoning: A Critique and Development', *Argumentation* **15**, 365–379.

Branting, L. K.: 1991, *Integrating Rules and Precedents for Classification: Automating Legal Analysis*, Doctoral dissertation, University of Texas at Austin.

Dung, P. M.: 1995, 'On the Acceptability of Arguments and its Fundamental Role in Nonmonotonic Reasoning, Llogic Programming and *n*-person Games', *Artificial Intelligence* **77**, 321–357.

Freeman, K. and A. M. Farley: 1996, 'A Model of Argumentation and Its Application to Legal Reasoning', *Artificial Intelligence and Law* **4**, 163–197.

Gordon, T. F.: 1995, *The Pleadings Game – An Artificial Intelligence Model of Procedural Justice*, Dordrecht, Kluwer Academic Publishers.

Govier, T.: 1987, *Problems in Argument Analysis and Evaluation*, Dordrecht, Foris Publications.

Hage, J. C.: 1997, *Reasoning with Rules, An Essay on Legal Reasoning and its Underlying Logic*, Dordrecht, Kluwer Academic Publishers.

Hage, J. C., G. P. J. Span and A. R. Lodder: 1992, 'A Dialogical Model of Legal Reasoning', in C. A. F. M. Grütters et al. (eds.), *Legal Knowledge Based Systems: Information Technology and Law, JURIX '92*, Lelystad, Koninklijke Vermande.

Hamblin, C. L.: 1970, in Richard Clay (ed.), *Fallacies*, Bungay, Suffolk , The Chaucer Press Ltd.

Hare, R. M.: 1963, *Freedom and Reason*, Oxford University Press.

Jakobovits, H. and D. Vermeir: 1999, 'Dialectic Semantics for Argumentation Frameworks', *Proceedings of the Seventh International Conference on Artificial Intelligence and Law*, New York, ACM, pp. 53–62.

Leith: 1986, 'Fundamental Errors in Legal Logic Programming', *The Computer Journal* **29**(6)

Lodder, A. R and A. Herczog: 1995, 'DiaLaw – A Dialogical Framework for Modeling Legal Reasoning', *Proceedings of the Fifth International Conference on Artificial Intelligence and Law*, New York, ACM, pp. 146–155.

Lodder, A. R. and P. E. M. Huygen: 2001, '*Eadr* A Simple Tool to Structure the Information Exchange between Parties in Online Alternative Dispute Resolution', in Bart Verheij, Arno R. Lodder, Ronald P. Loui and Antoinette J. Muntjewerff (eds.), *Legal Knowledge and Information Systems. Jurix 2001: The Fourteenth Annual Conference*, Amsterdam, IOS Press, pp. 117–129.

Lodder, A. R. and G. A. W. Vreeswijk: 2004, 'Gearbi: Proposal for an Online Arbitration Service under the ICC Rules of Arbitration, and a Preliminary Implementation', *ICC International Court of Arbitration Bulletin Special Supplement.*

Lodder, A. R. and J. Zeleznikow: 2005, 'Proposal for an Online Dispute Resolution Environment: Dialogue Tools and Negotiation Systems in a Three Step Model', *Harvard Negotiation Law Review* Spring 2005, to appear.

Lodder, A. R.: 1999, *DiaLaw – On Legal Justification and Dialogical Models of Argumentation*, Dordrecht, Kluwer Academic Publishers.

Lorenz, K.: 1961, *Arithmetik und Logik als Spiele*, Dissertation, Kiel.

Loui, R. P. and J. Norman: 1995, 'Rationales and Argument Moves', *Artificial Intelligence and Law* **3**, 159–189.

Loui, R. P., J. Norman, J. Olson and A. Merill: 1993, 'A Design for Reasoning with Policies, Precedents and Rationales', *Proceedings of the Fourth International Conference on Artificial Intelligence and Law*, New York, ACM, pp. 202–211.

MacKenzie, J. D.: 1979, 'Question-Begging in Non-cumulative Systems', *Journal of Philosophical Logic* **8**, 117–133.

Moles, R. N.: 1992, 'Expert Systems – The Need for Theory', in C. A. F. M. Grütters et al. (eds.), *Legal Knowledge Based Systems: Information Technology and Law, JURIX '92*, Leystad, Koninklijke Vermande.

Nitta, K., S. Wong and Y. Ohtake: 1993, 'A Computational Model for Trial Reasoning', *Proceedings of the Fourth International Conference on Artificial Intelligence and Law*, New York, ACM, pp. 20–29.

Nutting, K.: 2002, 'Legal Practices and the Reason of the Law', *Argumentation* **16**, 109–131.

Peczenik, A.: 1989, *On Law and Reason*, Dorcrecht, Kluwer Academic Publishers.

Peczenik, A.: 1996, 'Jumps and Logic in the Law', *Artificial Intelligence and Law* **4**, 297–329.

Perelman, Ch. and L. Olbrechts-Tyteca: 1971, *The New Rhetoric, A Treatise on Argumention*, London, University of Notre Dame Press.

Prakken, H. and G. Sartor: 1996, 'A Dialectical Model of Assessing in Conflicting Arguments in Legal Reasoning', *Artificial Intelligence and Law* **4**, 331–368.

Prakken, H.: 1997, *Logical Tools for Modelling Legal Argument, A Study of Defeasible Reasoning in Law*, Dordrecht, Kluwer Academic Publishers.

Rawls, J.: 1972, *A Theory of Justice*, Oxford, Oxford University Press.

Rescher, N.: 1977, *Dialectics, A Controversy-Oriented Approach to the Theory of Knowledge*, Albany, State University of New York Press.

Sartor, G.: 1994, 'A Formal Model of Legal Argumentation', *Ratio Iuris* **7**(2), 177–211.

Searle, J. R.: 1969, *Speech Acts: An Essay in the Philosophy of Language*, Cambridge University Press.

Skalak, D. B. and E. L. Rissland: 1992, 'Arguments and Cases: An Inevitable Intertwining', *Artificial Intelligence and Law* **1**, 3–45.

Soeteman, A.: 2000, 'Over de moraal van de juridische argumentatie', in: E. T. Feteris et al. (eds.), *Met recht en redden*, Nijmegen: Ars Aequi Libri, pp. 15–21.

Stevenson, C. L.: 1944, *Ethics and Language*, New Haven and London, Yale University Press, The 1979 reprint of the 1944 edn.

Tindale, C. W.: 1999, *Acts of Arguing, A Rhetorical Model of Argument*, Albany, New York, State University of New York Press.

Toulmin, S. E.: 1958, *The Uses of Argument*, Cambridge University Press.

Van Eemeren, F. H. and R. Grootendorst: 1982, *Regels voor redelijke discussies, Een bijdrage tot de theoretische analyse van argumentatie tot oplossing van geschillen*, Dissertation, Foris, Dordrecht.

Verheij, B.: 1996, *Rules, Reasons, Arguments: Formal Studies of Argumentation and Defeat*, Dissertation, Universiteit Maastricht.

Verheij, B.: 1999, 'Logic, Context and Valid Inference, Or: Can there be a Logic of Law?', in H. J. van den Herik et al. (eds.), *Legal Knowledge Based Systems, JURIX 1999, The Twelfth Conference*, Nijmegen, GNI, pp. 109–121.

Verheij, B., J. Hage and A. R. Lodder: 1997, 'Logical Tools for Legal Argument: a Practical Assessment in the Domain of Tort', *Proceedings of the Sixth International Conference on Artificial Intelligence and Law*, New York, ACM, pp. 243–249.

Vreeswijk, G. A. W.: 1993, *Studies in Defeasible Argumentation*, Dissertation, Vrije Universiteit, Amsterdam.

Walton D. N. and E. C. W. Krabbe: 1995, *Commitment in Dialogue*, Albany, State University of New York Press.

Walton, D.: 1996, *Argument Structure A Pragmatic Theory*, University of Toronto Press.

Wellman, C.: 1971, *Challenge and Response: Justification in Ethics*, Southern Illinois University Press.

ESSENTIALIST METAPHYSICS IN A SCIENTIFIC FRAMEWORK

ULRICH NORTMANN
Universität des Saarlandes, Philosophisches Institut, D-66041 Saarbrücken, Germany,
E-mail: u.nortmann@mx.uni-saarland.de

Abstract. In Section 1, the subject of the article is presented: the prospect of integrating an essentialist metaphysics into the scientific enterprise. Section 2 collects together a number of claims which are characteristic of essentialism. A species of inference rules, called (PFE)-rules, is introduced, referring to an idea of E. Hirsch's. Supplementing classical logic by the conditional schemes corresponding to a choice of such rules yields a first order theory of which it is claimed that it can be used as the core of an essentialist metaphysical theory. Section 3 presents a definition of a property's essentially belonging to an individual. Chiefly in Sections 4 and 5, it is shown in detail how essentialism as described in Section 2 can be based on this definition and a system of (PFE)-rules. In order to completely achieve this end, Section 5 additionally presents a recursive refinement of the original concept of essential belonging. The article concludes by sketching, in Section 6, the role of empirical research work in the formation of systems of (PFE)-rules.

1.

Recently, a philosopher asked what would have to be changed in our picture of human kind if it turned out that human beings are neurobiologically fully determined.[1] He argued that nothing would have to be changed. Whether he is right or wrong about this, a discussion of the thesis which is his concern forms a typical part of what I take to be one of the more important tasks of philosophy today: to try to integrate the best theories available in the various research fields, together with as much as possible of our common sense picture of the world, including phenomena like our pretheoretic practices of attributing responsibility, into a maximally coherent world view – if necessary by making more or less drastic changes in one or more of the different theoretical or pretheoretical strands. Another area where this task is on the agenda is the alleged gap between metaphysics and the scientific enterprise. The particular problem which interests me in this paper is the prospect of uniting an essentialist metaphysics with a world view oriented to scientific and mathematical rigour.

It is well known that essentialism as attributed to Aristotle, e.g., was dismissed on logico-philosophical grounds by scientifically minded 20th century thinkers like W. V. O. Quine. On the other hand, essentialist convictions have continued to form a rather deeply entrenched element in a common-sense picture of the world. In philosophy, neo-essentialists like S. A. Kripke, though very profitably theorizing

S. Rahman et al. (eds.), Logic, Epistemology, and the Unity of Science, pp. 589–600.

within an essentialist framework, expended little effort on trying to impart a solid and scientifically acceptable foundation to essentialism. In fact, Krikpe simply says in his *Naming and Necessity*, in the context of defending claims like "That's the guy who might have lost" (referring to Nixon and the 1968 election), that is, claims which in effect deny the essentiality of electoral victory:

> "Here I am just dealing with an intuitive notion and will keep on the level of an intuitive notion. That is, we think that some things, though they are in fact the case, might have been otherwise" (Kripke (^{2}1980, 39 fn. 11)).

So this gap is still gaping.

2.

I wish to show that the gap can be closed by drawing on an appropriate logical construction. Apart from the general claim that a distinction can be made between accidental and essential properties of an individual, the following more specific claims are common ingredients of an essentialist metaphysics:

(1) For certain properties F and individuals a, no matter how these are described, counterfactuals of the type "if a were not (or: were no longer) an F at time t, a would not exist (would have ceased to exist) at t" can be recognized as true.

(2) For certain properties F, it is a fact that any individual which is an F is so necessarily/essentially (for instance: "for all humans: they are humans with necessity"; "for all portions of H_2O molecules: they (mainly) consist of H_2O with necessity"); and the corresponding generalizations express non-contingent facts.

(3) There can be properties which essentially belong to some individuals and which accidentally belong to other ones.
Being located at a certain place seems to be an example, according to usual essentialist intuitions. My staying in a certain region, by staying in a certain town situated there, is an accident, yet the town's being situated there does not seem to be accidental – or could the town be taken down and completely transplanted to a different place where it would persist as the very same town?

All of these claims can be justified within the framework I am going to develop.

As a starting point, I wish to take an idea advanced by E. Hirsch in an essay on essence and identity of 1971.[2] Thinking of "E terms" as predicates which signify essential properties, Hirsch writes:

> "A general term F is an E term if and only if the statement 'This F (thing) was (or will be) at place p at time t' logically entails 'There was (or will be) an F (thing) at place p at time t' " (Hirsch 1971, 33).

Here it is understood that the use of an expression like "this F thing" for picking out a certain individual entails by itself, regarding the predicate F and its instantiation, not more than the claim that the individual picked out is an F thing just at the time of utterance. As distinguished from this claim, the mentioned conclusion entails, for a certain time t which is earlier or later than the time of utterance, that there exists an F thing at that time.

Hirsch's definition of the concept of E term reflects the essentialist idea that there are properties F which, belonging here and now to an individual (so that the individual can be referred to as "this F (thing)"), have belonged and will continue to belong to it at every past and future time of the individual's existence. For this is what apparently underlies the mentioned transition to an existential statement of the future tensed variety, say, "There will be an F at place p at time t": If the premise is true, the individual in question will be located at place p at time t (and thus still exist then), and since it will still be an F thing *then*, there will be an F thing at that place at that time.

The idea of tying essential belonging to lifelong belonging is good; and it is also good to try to give this linkage a logical basis, since there must be more to essential belonging than mere lifelong belonging, given that lifelong belonging might be a matter of chance. Unfortunately, however, the familiar apparatus of predicate logic (supplemented with definite descriptions and the possibility of expressing tensed predications like "a is F at t" as relational propositions $F(t, a)$) does not yield the type of entailment relation appealed to by the quoted definition. As a way out, one has to think of this logical apparatus as being supplemented by inference rules which, for a certain choice of predicates F, will precisely do the job of licensing the transitions in question (or equivalently, as being supplemented by axioms to the same effect).

I propose to give such rules, in an informal reading, the slightly generalized form (for an F which belongs to the relevant stock of predicates):

(PFE) For every (simple or complex) predicate G:
From
"this F thing will be G at t"
you may move to
"at t, there will be an F thing which is a G thing"
(and correspondingly for the past).

Let us for the sake of simplicity work with this formulation. Actually, the range of predicates G must be somehow delimited so as to exclude predicates which, for instance, express or imply non-existence. (We sometimes truly say things like: some day, this F thing will no longer exist.)[3] The basic idea will not be affected by any such modification. In a rigorous formulation, the augmentation could be achieved by adding axiom schemes of the form

$$(F(t_0, x) \land \exists t_1 > t_0(E(t_1, x) \land G(t_1, x))) \supset \exists x \exists t_1 > t_0(E(t_1, x) \land F(t_1, x) \land G(t_1, x)).$$

Here, $E(t_1, x)$ is taken to mean that x exists at time t_1. Assuming this kind of formulation, the problem mentioned above will no longer arise, since the antecedent will be false for the critical predicates.

"PFE" is short for the "past and future existentials" which (PFE)-rules permit to deduce.

Inference rules like that, or the corresponding axioms, can be used to form the core of an essentialist metaphysics wherever they have been joined to a logical axiom system of the familiar type to make up a more comprehensive theory. This core has been couched in purely extensional terms. In fact, one of our main objectives is to avoid admitting into the *primitives* modal operators and their kinship which scientifically minded theorists could feel bound to reject as illegitimate. Why should it in contrast be illegitimate to have inference rules of the (PFE) variety (and to make them explicit in a regimented reformulation of a natural language, say)? After all, they seem to be rather on a par with semantical conventions like

"From 'this person is a bachelor' you may move to 'this person is unmarried' ".

To be sure, there is a difference: In rules of the (PFE)-variety, elements of an ontology show up. To decide upon certain properties F as belonging to their bearers for a lifetime is only one side of a coin whose other side consists in framing certain sorts of individuals as being endowed with the corresponding persistence properties (as ceasing to exist, e.g., when F is lost).

The questions of interest, regarding the metaphysics-science-gap, are: where do such rules come from (are they or can they in principle be interwoven with science), and can particular claims like (1), (2), or (3) be based upon them?

3.

Let's first turn to the second question. In a series of articles,[4] I have defended the position that a positive answer is possible, provided that a missing link is supplied: an appropriate definition of essential belonging. Here it is (for any condition $H(t, y)$):

(EB) $NH(t, y) \Leftrightarrow_{\text{def.}} H(t, y)$, and $\vdash H(t_0, x) \supset \forall t_1(E(t_1, x) \supset H(t_1, x))$ ("property H essentially/necessarily belongs to individual y at time t iff H actually belongs to y at t, and it is provable, for variable t_0 and x: if H belongs to x at t_0, then H belongs to x at t_1 for all times t_1 which x exists at").

It is assumed that the object language level which expressions like $H(t, y)$ will belong to in a rigorous formulation is contained in the meta language, so that $NH(t, y)$ is a metalanguage formula, enclosing a sentential operator N which has been defined by involving a minimum of intensionality: the only "modal" concept needed is the concept of provability in an extensional theory. For special purposes we may alternatively assume that the provability property finds expression on the objectlanguage level, via arithmetization of the syntax.

Note that the provability part of the definiens requires the derivability of a certain relation independently from any particular description of an individual (which reflects the *de re*-character of the modality being defined). The provability symbol $\vdash$ is meant to relate, in the first place, to a basis which comprises classical predicate logic, a supply of (PFE)-rules for particular predicates, and moreover a rule specifying that, for any time t, the times later than t, the earlier times, and t itself exhaust all time. In a second step, we can supplement (EB) by its counterpart which refers, via another provability sign, to an axiomatic basis including as an additional axiom the object language variant of the initial (EB) biconditional itself. (This sort of step can be iterated, enabling reference to more and more enriched systems.)

Given the mentioned prerequisites, it is easily seen that at least for each predicate F for which a (PFE)-rule has been adopted, the conditional

$$F(t_0, x) \supset \forall t_1(E(t_1, x) \supset F(t_1, x))$$

is in fact provable. For assume the negation. Then x is an F at time t_0, and there is a future time, say, at which x will exist and not be an F. So "this F" (i.e., this thing which now, at t_0, is an F) will be a non-F at future time t_1, whence it follows according to the relevant (PFE)-convention, by choosing $G =$ non-F, that at t_1 (or some other future time) there will be an F thing which is a non-F thing, a contradiction. If there is moreover an actual F thing (at t), it will be an essential F thing (at t) according to (EB).

So there turns out to exist essential properties. There will of course be more of them than just those figuring in (PFE)-rules. It turns out, e.g., that any existing individual essentially exists when it exists. This is a consequence of the trivial fact that $\forall t_1(E(t_1, x) \supset E(t_1, x))$ is derivable already from the logical part of the underlying theory. Essential existence in this sense must not be confused with necessary existence in a stronger sense, as for instance in the sense of $E(\tau, a)$, or even $\forall t E(t, a)$, being provable in an adequately supplemented system, for an individual a and a specific time τ. On the other hand, there will be, as a rule, lots of predicates which don't satisfy the provability condition. So the general claim that there are both accidental and essential properties can be justified within the proposed framework.

4.

Regarding claim (2), it is in place to remark that, at first sight, an attribution of essential belonging is, according to (EB), bound to be an assertion in a language relating as a metalanguage to the underlying object language, due to the metalinguistic character of the symbol $\vdash$ occurring in the definiens. On further reflection, however, a move to the object-language level turns out to be possible. The theoretical apparatus which the object language has been equipped with so far permits an arithmetization of the proof relation, so that the provability condition contained in (EB) can be expressed in arithmetical terms within the object language itself – assuming that the latter has been additionally equipped with Peano arithmetic (= PA). Let us make this assumption now! Then a vindication of claim (2) is within reach.

Take $H(t, y)$ to be any condition for which $NH(\tau, a)$ is true, with an individual a and a time τ. Thence according to (EB), the conditional $H(t_0, x) \supset \forall t_1(E(t_1, x) \supset H(t_1, x))$ as well as its universal closure will be a theorem. Applying the first Hilbert-Bernays derivability condition, the arithmetical version of the proposition saying just this, i.e., the formula

$$Bew(\ulcorner \forall t_0 \forall x(H(t_0, x) \supset \forall t_1(E(t_1, x) \supset H(t_1, x)))\urcorner),$$[5]

is a theorem itself. Moreover, (EB) in its object-language version yields the theorem

$$Bew(\ulcorner \forall t_0 \forall x(H(t_0, x) \supset \forall t_1(E(t_1, x) \supset H(t_1, x)))\urcorner) \supset (H(t, y) \supset NH(t, y)).$$

So

$$(^*)\ \forall t \forall y(H(t, y) \supset NH(t, y))$$

turns out to be a theorem of the correspondingly enriched system. If it is granted that theoremhood implies truth, the hereby established truth of the generalization (*) can be taken to correspond to the first half of claim (2).

The theoremhood of (*) yields the second half of claim (2), and this can be couched in essentialist terms. In order to achieve the latter, we can't simply proceed by applying something like a rule of necessitation so as to yield

$$N \forall t \forall y(H(t, y) \supset NH(t, y)).$$

A combination of an alethic modal logic of the standard type with an axiomatic basis like the one being employed here, including as it does PA, would be threatened by Montague's paradox, to mention one reason. Instead, we have to manage with the only thing we know about N so far, which is the (definitionally established) validity and the theoremhood of the biconditionals introduced by (EB), in its objectlanguage version, and by (EB)'s counterparts relating to the pertinent enriched systems. Abbreviating the formula $H(t, y) \supset NH(t, y)$ by $H^*(t, y)$, we have so far the provability of $H^*(t, y)$, and therewith, of course, the provability of

$$H^*(t_0, x) \supset \forall t_1(E(t_1, x) \supset H^*(t_1, x)).$$

Assuming again the truth of our axiomatic basis, the generalization $\forall t \forall y H^*(t, y)$ is true (because of its provability), so that in sum the generalization

$$\forall t \forall y(H^*(t, y)\ \&\ Bew(\ulcorner H^*(t_0, x) \supset \forall t_1(E(t_1, x) \supset H^*(t_1, x))\urcorner)$$

turns out to be true. Then (EB) entails the truth of $\forall t \forall y N H^*(t, y)$, i.e., the truth of

$$\forall t \forall y N(H(t, y) \supset NH(t, y)),$$

as desired. The argument shows that, given the general setting developed above and the particular assumption that H essentially belongs in at least one case, the more complex property of 'being an essential H if being an H at all' is an essential (and in this precise sense not a contingent) property of all individuals (at all times).[6]

Note by the way that an individual's possessing a (complex) property at a time must not be taken as entailing the individual's existence at that time. $\forall t \forall y \neg(F(t, y)\ \&\ \neg F(t, y))$, e.g., is provable as well, but this does not license us to conclude for each individual that it exists at any time because the property of not being an F and a non-F belongs to it at that time. That some caution is required here is already clear from the very fact that I introduced an existence predicate. If existence can be predicated as a first order predicate, a denial of singular existence must also be meaningful. But such a denial would be disastrous if an individual's possessing the property of not being existent (at a time) entailed its existence (at that time).

5.

The essentialist's desideratum (1) seems at first glance to be even more easily got. Assume once more that H essentially belongs to some individual a (at some time τ). Abbreviating

$$H(t_0, x) \supset \forall t_1(E(t_1, x) \supset H(t_1, x))$$

by $H^\circ(t_0, x)$, our assumption entails the provability of $H^\circ(t_0, x)$, and therewith, by an argument similar to the preceding one, the provability of

$$H^\circ(t_0, x) \supset \forall t_1(E(t_1, x) \supset H^\circ(t_1, x)).^7$$

As a consequence (given the truth of the axioms),

$$H^\circ(\tau, a)\ \&\ Bew(\ulcorner H^\circ(t_0, x) \supset \forall t_1(E(t_1, x) \supset H^\circ(t_1, x))\urcorner)$$

is true, which by (EB) entails the truth of $NH^\circ(\tau, a)$, i.e., the truth of

$$N(H(\tau, a) \supset \forall t_1(E(t_1, a) \supset H(t_1, a))).$$

If the N-operator could be distributed here, the truth of $NH(\tau, a)$ (our assumption) would in turn entail the truth of

$$N\forall t_1(E(t_1, a) \supset H(t_1, a)).$$

The last formula, in virtue of being governed by a modality, could in fact be paraphrased as a counterfactual: "If a existed at t, a would be an H at t." But may the operator be so distributed? As an ingredient in a special logic of essentialism, a corresponding principle would sound intuitively plausible.[8] If the aim is, however, to found essentialism on a scientifically acceptable, distinctly extensional basis, an appeal to a genuine logic of essentialism, separated from and in no way explicable in terms of the received extensional logic, will not do. The best result I could find within such extensionalist limits is a theorem (to be given below) which, however, draws on a certain modification of (EB). This modification is of some interest also with regard to claim (3).

(EB) in its original form admits only those properties as essentially belonging to some individual which moreover uniformly essentially belong to *any* individual to which they belong at all. This is obvious from the formulation of (EB). The relevant provability condition only depends on H. Thus, the definiens as a whole will be true for any individual which is an H thing, given that it is true for at least one individual. The fact is underlined by our preceding argument in defence of claim (2). In connection with this claim, however, we could not be content with the simple truth of $\forall t \forall y(H(t, y) \supset NH(t, y))$, given that H essentially applies to some individual, but needed the theoremhood.

To get rid of this limitation to uniformly essentially belonging properties, a recursive refinement of (EB), permitting degrees of essential belonging, is required:

(EB+) H essentially belongs to y at t with degree 0
$\Leftrightarrow_{\text{def.}}$ $H(t, y)$, and $\vdash H(t_0, x) \supset \forall t_1(E(t_1, x) \supset H(t_1, x))$;
H essentially belongs to y at t with degree $n + 1$
$\Leftrightarrow_{\text{def.}}$ there are a property K and and a natural number $m \leq n$ such that: $H(t, y)$, and K belongs to y at t with degree m, and
$\vdash H(t_0, \iota z(z = x \,\&\, K(t_0, z))) \supset \forall t_1(E(t_1, \iota z(z = x \,\&\, K(t_0, z))) \supset H(t_1, \iota z(z = x \,\&\, K(t_0, z))))$.
H essentially belongs to y at t (*simpliciter*)
$\Leftrightarrow_{\text{def.}}$ there is a natural number n such that H essentially belongs to y at t with degree n.

The idea is obvious: While in the degree 0 case (which precisely corresponds to (EB)) the provability of a lifelong possession of property H by an H thing x is required with x remaining completely undescribed (matching the idea of modality *de re*), the higher degree cases admit x at least as being characterized by predicates K of which it has already been settled that they essentially apply to x; the expression $\iota z(z = x \,\&\, K(t_0, z))$ is the symbolic counterpart of something like "the K thing x", or "the thing which is K at t_0".

A simultaneous supplementation of the stock of (PFE)-conventions with rules of the slightly more complicated type

From
"this (F & K) thing will be G at t"
you may move to
"at t, there will be an F (& K) thing which is a G thing"
(and similarly for the past)

will make it possible to derive the essential belonging of F to those among the actual F things which are also essential K things – while there is room for the existence of non-K things which are actual F things, but no essential ones. This is what claim (3) demands.

Regarding the distribution of the essentiality operator N over the terms of a conditional, the following can be proven about essential applying as defined by (EB+):

PROPOSITION. Let J and K be open formulas each containing a free individual variable and a free time variable such that J and $J \supset K$ essentially apply to y at t. Then K will also essentially apply to y at t.

Proof in Nortmann (2001).

This completes the vindication of claim (1).

6.

Now that it has been shown how much can be achieved by exploiting (PFE)-rules (presupposing an appropriate proof theoretical infrastructure), what remains to be answered is the first question: where do (PFE)-rules come from, and (how) do they relate to the scientific enterprise?

It seems to me to be clear that to a certain extent at least, human beings feel free and are free to decide at will, on the basis of their interests and ends, which kinds of impact, with respect to certain sorts of individuals, to regard as effecting mere changes of these individuals or as effecting destructions, say. This is particularly plausible for artefacts, and works of (applied) art are an especially telling example. If effecting a loss of a property F counts as a mere change of the underlying particular, this amounts to denying F the status of an essential property (of the particular in question). In accordance with this linkage, the amount of freedom claimed above entails a corresponding amount of arbitrariness in delimiting a stock of (PFE)-rules. In other words, ontology (here: matters of essential and accidental belonging) will, as a rule, not be completely determined by nature, in particular not with regard to sorts of things produced by man himself.[9] This may have begun to dawn even on Aristotle when, at the end of book Z of his *Metaphysics*, he seems to suggest that maybe only natural beings should be taken to have an essence: "… not all objects are substances, but only those that are formed naturally and in accordance with their nature" (1041b 28–30).[10]

Nevertheless, it is plausible to assume that, apart from a not negligible conventionalistic share in the formation of inferential behaviour which can be reconstructed as being guided by (PFE)-rules, nature does get involved, via the scientific enterprise, to a considerable extent.

For an explanation remember the following. Human beings aim at being able to cope with the demands of life on an experiential basis, and they seek to read a structure into the world which supports this aim. Hence, they tend to hypothetically conceive the distinctness and the sameness of entities in such a way as to promote the adoption of a maximally comprehensive system of propositions assuring, among other things, that roughly speaking employing the same means will always yield the same result (given suitable circumstances, where it is necessary to be more specific) – as for instance that quenching one's thirst can always be achieved with the same sort of liquid which could be used to this end in the past (after filtering out contaminations etc. if necessary). Empirical research work will clearly take a substantial share in determining such a system of propositions. We learn, for instance, that we can quench our thirst precisely with portions of liquid consisting, apart from harmless or even favorable admixtures, of H_2O molecules and called 'water' in English. Hence there is a strong inclination to unite those portions of liquid which mainly consist of H_2O into one (natural) kind of liquid (a kind, water, whose essence some feel inclined, in a second step, to conceive as consisting of the property of being composed of H_2O).

The above example is for one thing about the sameness of a *kind* of individuals, the individuals being particular portions of liquid. On the other hand, the uniting move in question has its consequences for our views about the sameness of particulars, and thereby for our views about essential and accidental belonging. Given a portion of liquid of the water kind, we will not be willing to conceive it as persisting as one and the same particular, having undergone a mere change, whenever it has been transformed into a portion of liquid no longer belonging to the H_2O kind and, correspondingly, no longer being useful for certain of our ends. (Think for instance of sulphur dioxide as having been dissolved in our portion of liquid; this is not the adding of an admixture, but a manipulation which effects a chemical reaction, producing sulphureous acid, H_2SO_3). Instead, our scientific insight will strongly promote the belief that any portion of liquid consisting mainly of H_2O molecules will during its whole existence consist of (mainly the same particular) H_2O molecules. For this reason, the predicate "... is composed of H_2O" will be a very good candidate for entering a list of (PFE)-rules.

Adding our outline of the step which further leads from (PFE)-rules to claims like (1), (2), and (3), we have designed in sum a picture which, on my view, basically captures the way empirical science and logic flow together to have their share in shaping m etaphysics. According to this picture, a considerable amount of connection can be found between realms which formerly would have been considered as hopelessly disparate.[11]

Notes

[1] A. Beckermann (2001).

[2] This essay is contained in M. Munitz (1971).

[3] Thanks to Helge Rückert for a pertinent hint.

[4] Central among them is Nortmann (2002a). The historical side of the matter is stressed in Nortmann (2002b).

[5] The symbols ⌜ and ⌝, the 'Gödel hooks', are to be understood as making of any expression enclosed by them a name of the Gödel number of that expression, some Gödel numbering having been fixed.

[6] The justifiability of generalizations of the type $\forall y N(H(y) \supset NH(y))$ which admit an essentialist reading is of some interest also on historical grounds. It has been shown that supposing the universal apodeictic *protaseis* in Aristotle's modal syllogistic to have a parallel structure affords the validation of certain central claims of Aristotle's which had been under suspicion for a long time. Cf. Nortmann (1996).

[7] Remember that t_0 is a *variable* (ranging over times). Constants denoting a time are represented by the Greek letter τ.

[8] Remember for instance the familiar possible-worlds explanation of essential belonging (which we cannot make use of here, apart from intuitive considerations, because its unreduced modal character clashes with our foundational aim): H essentially belongs to a if and only if in every possible world in which a exists H belongs to a. Given that both "if J, then K" and J belong to a in this sense, then K will also belong to a in every world containing a.

[9] M. Heller has convincingly argued for the recognition of a large share of conventionalism as being involved in the adoption of persistence conditions for various types of individuals, in his Heller (1990).

[10] Combine this with Aristotle's presumable view that substance (*ousia*) is "what being is for a thing" (that is, essence), and that any thing which has an essence is, in a sense, identical with that essence. It can then be inferred from the quoted assertion that any thing which has an essence, being an essence and therewith being a substance, is naturally formed.

[11] I am grateful to Richard Gaskin (Liverpool) and to Paul Thom (Lismore) for critical comments on an earlier version of the article.

References

Beckermann, A.: 2001, 'Was würde sich an unserem Menschenbild ändern, wenn sich herausstellte, daß wir neurobiologisch determinierte Wesen sind?', in print.

Heller, M.: 1990, *The Ontology of Physical Objects: Four-dimensional Hunks of Matter*, Cambridge, Cambridge University Press.

Hirsch, E.: 1971, 'Essence and Identity', in M. Munitz (ed.), *Identity and Individuation*, New York, New York University Press, pp. 31–49.

Kripke, S. A.: ^{2}1980, *Naming and Necessity*, Oxford, Blackwell.

Meixner, U. (ed.): 2001, *Metaphysik im postmetaphysischen Zeitalter*, Wien, hpt.

Munitz, M. (ed.): 1971, *Identity and Individuation*, New York, New York University Press.

Nortmann, U.: 1996, *Modale Syllogismen, mögliche Welten, Essentialismus. Eine Analyse der aristotelischen Modallogik*, Berlin, de Gruyter.

Nortmann, U.: 2001, 'Essentialistische Konditionale für Extensionalisten', in U. Meixner (ed.), *Metaphysik im postmetaphysischen Zeitalter*, Wien, hpt, pp. 149–160.

Nortmann, U.: 2002a, 'Warum man Essentialist sein kann – eine logische Konstruktion im Schnittfeld von Sprache, Ontologie und Naturwissenschaft', *Erkenntnis* **57**, 1–39.

Nortmann, U.: 2002b, 'The Logic of Necessity in Aristotle – an Outline of Approaches to the Modal Syllogistic, together with a General Account of *de dicto*- and *de re*-Necessity', *History and Philosophy of Logic* **23**, 253–265.

INDEX